GAINING GROUND:

IN PURSUIT OF ECOLOGICAL SUSTAINABILITY

GAINING GROUND:

IN PURSUIT OF ECOLOGICAL SUSTAINABILITY

Edited by David M. Lavigne

Associate Editor, Sheryl Fink

Published by the International Fund for Animal Welfare, Guelph, Canada, and the University of Limerick, Limerick, Ireland.

www.ifaw.org/forum

Library and Archives Canada Cataloguing in Publication

Gaining ground : in pursuit of ecological sustainability / edited
by David M. Lavigne ; associate editor, Sheryl Fink.

Includes bibliographical references and index.
ISBN 0-9698171-7-7

1. Wildlife conservation. 2. Sustainable development. 3. Ecology.
I. Lavigne, D. M. (David Martin), 1946- II. International Fund for
Animal Welfare

QL82.G35 2006 333.95'16 C2006-902295-X

Available in Canada and the United States from:
IFAW
555 Admiral Drive
London, ON, Canada
N5V 4L6
gainingground@ifaw.org

Global Distributor:
NHBS Environment Bookstore
2-3 Wills Road, Totnes, Devon
TQ9 5XN, United Kingdom
nhbs@nhbs.co.uk
www.nhbs.com

Cover photo: The Earth "rising into the night" in Meru National Park, Kenya.
© IFAW / D. Willetts
The original caption on this photo was "Sunset in Meru National Park", confirming – as Ian McCallum writes in his recent book, *Ecological Intelligence: Rediscovering Ourselves in Nature* (Africa Geographic, Cape Town, South Africa, 2005) – that "Copernicus, at a subtle level, has not been fully acknowledged" in our language. McCallum writes, "In our self-centred world of the human animal, we have great difficulty in speaking about the Earth rising into the night ... or of our planet dipping sharply into the morning ... To see the horizon tilting upwards and away from the sun", McCallum continues, "is an entirely different experience to watching the sun going down".

Cover design by Marc Meadows.

Printed by Ampersand Printing, Guelph, Canada.
The paper used in this book is 100% post-consumer recycled, Environmental Choice Certified, processed chlorine free.

A Norwegian PM name of Gro
Took a concept which as we all know
Does not hold water
Though some think that it ought to
But the Planet just won't grow and grow.

Roger G.H. Downer 2004

Table of Contents

Foreword

The book you are holding in your hands is the result of a Forum convened in June 2004 by the International Fund for Animal Welfare (IFAW) in association with the University of Limerick, Ireland. Notable experts from around the world – scientists, social scientists, professors, policy advocates and others – came together to share their perspectives on wildlife conservation and the pursuit of ecological sustainability.

The Limerick Forum marked a first attempt by IFAW to summon leading thinkers and advocates from around the planet – and across the philosophical spectrum – and to invite their best thinking on one of the great challenges of our time: how to reconcile the twin imperatives of improving the human condition, and protecting wildlife and the natural systems upon which all life depends.

Each of the expert presentations that follow shares important insights. Taken together, they provide a rare contemporary study in wildlife conservation and the search for ecological sustainability. While the perspectives differ, the attentive reader consistently is drawn to several important conclusions:

- That the fates and well-being of people, other species, and the planet we share are inextricably linked;
- That our shared world, its inhabitants and natural systems, face new and growing threats; and
- That fresh, far-sighted approaches to wildlife conservation and ecological sustainability are urgently needed if the richness of life as we know and value it is to be passed on to our children and our grandchildren.

Over the course of the past century, attempts to improve the human condition have focused almost exclusively on sustaining economic development, often at the expense of the very natural systems on which human welfare and prosperity fundamentally depend. Now, in the first decade of the 21st century, the triumph of economy over ecology is virtually complete. The result? Burgeoning human population growth, unprecedented atmospheric and marine pollution, persistent poverty within and among nations, inadequate environmental protection in most regions, and a chronic lack of compliance and enforcement where legal protections for wildlife and the environment do exist.

The changes we have witnessed in our lifetime may not seem all that dramatic. They have been slow, subtle, and steady. From one generation to the next, what was a pristine environment becomes quasi-developed. The newly compromised ecosystem in turn becomes the baseline for the next level of development. Bit by bit, piece by piece, step by step, we become inured to lowered baselines of expectation, surrendering to the tyranny of small, incremental decisions that ultimately undermine not only our very own well-being but also the well-being of other species with which we share the planet.

Neither IFAW nor the distinguished Forum participants whose thoughts appear here are content to stay the course. Clearly, there are no mythical "magic bullets" for making the world a better place for all. There are many necessary elements of a new approach contained in these contributions, but none are sufficient to address the enormity of the problem. Yet we must start somewhere, and we need to start now. In fact, the experts have been telling us we need to start "now" since the middle of the last century.

In the essays that follow, each of the invited speakers, and a number of other participants in the Limerick Forum, present their thoughts and ideas about where we have been, and where we need to go, if we are to gain ground in our pursuit of ecological sustainability. They give us much to think about, and even more to do, if we really want to change the course of human history.

Frederick M. O'Regan
President, IFAW

Preface

We are in the midst of an ecological crisis. The human population continues to grow; the demand on nature and natural resources increases daily; more species are being lost more rapidly than ever before; climate change is upon us; sea levels are rising; almost one billion people go to bed hungry every night; the gap between the rich and poor continues to widen; and the threat of a global pandemic seems more likely with each passing day.

To say that it is a "life and death" crisis, not only for many species of plants and animals with whom we share the planet, but also for individual people and various human cultures, may sound like hyperbole, but it isn't. We've known about most of the problems for decades, but we've done little to ameliorate them. Although it's now getting late in the day, remarkably we may still have a bit of time to deal with the issues, reduce the losses, and slow or reverse at least some of the current trends. But this will only happen if we start now. And that, very briefly, is what this book is about.

The book itself actually goes far beyond what was originally envisioned. In 2002, the International Fund for Animal Welfare (IFAW) decided to organize an "IFAW Forum" to focus attention on, and to advance the discussion of, issues of particular relevance to its programs and campaigns. After considerable internal deliberation, it was decided that IFAW's first Forum would deal with ongoing concerns about the loss of biodiversity resulting primarily from human activities. Of special interest were those terrestrial and marine fauna that are currently threatened by commercial exploitation and international trade.

At the time, questions of particular interest to IFAW included: How can a continually increasing human population protect wild animals in their ever-shrinking habitats? In cases where humans exploit wild animals and their habitats, how can we take steps to insure that our use of them is ecologically sustainable in the sense that wild populations are maintained at sufficiently high levels that their ability to recover is not jeopardised, and that exploited ecosystems continue to exist and function?

While these sorts of questions have concerned conservationists for decades, there are precious few examples where ecological sustainability has actually been achieved.[1] In the face of evidence that species are being lost at ever increasing rates, some individuals and organizations now promote a "use it or lose it" philosophy, contending that wildlife must "pay its way" in order to be conserved. Yet, history suggests that the commodification and commercial consumptive use of nature and natural resources usually contribute to species and habitat depletion, not to their conservation or sustainable use. The historical evidence also suggests that there are other approaches that hold more hope for achieving biological and ecological sustainability. Curiously, some of the more promising ones are usually overlooked when promoting conservation today, especially in the developing world.

The first IFAW Forum was designed, therefore, to offer alternatives to the "use it or lose it" philosophy for achieving "sustainable use" of fish and wildlife populations and for pursuing the larger goal of achieving ecological sustainability. These alternatives not only build on the lessons of history, but also upon the most recent developments in the new and growing field of conservation biology. Collectively, conservationists have a good understanding about what needs to be done in order to move towards achieving ecological sustainability. The challenge is to get that message back on the conservation agenda, and then to get conservation back onto the world agenda, in order to make a difference.

The Forum – entitled *Wildlife Conservation: In Pursuit of Ecological Sustainability* – took place from 17-19 June 2004 in Limerick, Ireland, co-hosted by IFAW and the University of Limerick. It consisted of twenty-three invited presentations, with additional papers contributed as posters. The speakers came from a variety of fields and backgrounds including: conservation biology, engineering, ethics, earth sciences, fisheries, geography, law, the social sciences, and zoology. As individuals, they had different opinions on many subjects, but, if they were – and remain – united in anything, it is that they all agree that the pursuit of ecological sustainability is a topic worthy of discussion, debate, and action.

Each of the invited speakers was asked to prepare an essay on their assigned topic. These essays were re-written, reviewed, and revised in the months following the Forum, and edited into this book. Authors of three of the contributed posters were also invited to submit chapters that are included here.

The organization of this volume follows the original Forum program. After a brief introduction to the topic, Part I provides the Global Context for much of the discussions that follow. Martin Willison addresses the ongoing loss of biodiversity and why the maintenance of biodiversity is important, and he offers some suggestions for achieving it. Ward Chesworth reminds us that unsustainable practices have been with us since the dawn of agriculture some 10,000 years ago and notes that an entirely new ethic will be required if we are to change our habits in the future. Sidney Holt examines the very notion of sustainability, drawing heavily on his long career as a scientist dealing with the issue, particularly as it relates to commercial fisheries and whaling. The last chapter in this section, authored by Sharon Beder, recognizes that individuals and organizations attempting to promote ecological sustainability in the 21st century must first understand and appreciate that the conservation movement has changed since the publication of the first World Conservation Strategy in 1980 and its introduction of the now controversial idea of "sustainable development". They must also understand the nature of the "wise-use" movement that has reshaped conservation over the past 20 years. Professor Beder concludes with the suggestion that ethical, political and social changes are required in order to preserve the environment, a suggestion that is reiterated throughout the book.

Part II moves on to examine four specific examples where the pursuit of ecological sustainability has remained an elusive goal: commercial fisheries (Jeffrey Hutchings), commercial whaling (Vassili Papastavrou and Justin Cooke), the trade in elephant ivory (Ashok Kumar and Vivek Menon), and the current bushmeat crisis (Heather Eves). These examples all involve commercial consumptive use. The final two chapters in this section discuss examples of so-called non-consumptive (sometimes more accurately called low-consumptive) use. Arthur Mugisha and Lilly Ajarova examine the benefits and challenges of ecotourism in an African context, while Peter Corkeron tackles the issue of whale watching. All the authors in this section offer suggestions for dealing with the problems identified, including the need to separate science from politics and to recognize explicitly the role of ethics in conservation. They reiterate that "maximum sustainable use" is not, in fact, biologically sustainable, and call for the application of the precautionary approach, and the development of responsible ecotourism. Two additional messages to emerge from this discussion are that commercial consumptive use is not a synonym for biologically sustainable use, and that non-consumptive uses such as ecotourism are not "magic bullet" solutions to all of our present conservation problems, nor do they necessarily achieve objectives such as ecological sustainability.

Part III discusses some of the factors at play in wildlife conservation and the pursuit of ecological sustainability. Such factors include: human attitudes, values and objectives, and other ethical considerations that provide the real basis for conservation initiatives and policies (Vivek Menon and myself); what science can and cannot contribute to conservation in an uncertain world, and the important role of ethics in science and conservation (William Lynn); the limitations of purely economic approaches (William Rees), including subsidies (Michael Earle), to conservation; and the role that our human roots as Darwinian animals play in the process (Ronald Brooks).

The final section, Part IV, presents some ideas on how to put theory into practice and, ultimately, how to facilitate social change in both the developed and developing worlds. It includes a number of initiatives that, if widely implemented, could move us towards achieving the goal of ecological sustainability. Sir Robert Worcester provides insights into how and why societal attitudes change, based on his decades of experience as a pollster. Such information is essential if John Oates is correct in arguing, in the subsequent chapter, that a change in attitudes will be required if we are ever to be successful in our pursuit of ecological sustainability. In particular, Professor Oates highlights the need to divorce wildlife conservation from economic development, a controversial recommendation in this age of "sustainable development". Valerius Geist reviews the lessons learned from the North American approach to wildlife management, an approach that arguably – at least for a time – was well on the way to achieving the goal of ecological sustainability for many wildlife populations on that continent. E.J. Milner-Gulland discusses the development of a new generation of precautionary, risk averse modeling approaches designed to achieve ecologically sustainable results, based on her work on the bushmeat crisis in Africa. Bill de la Mare then draws on his engineering experiences to ask what is wrong with our approaches to fisheries and wildlife management and what can be learned from the engineering profession to improve the current situation. Lawyer Michelle Campbell and biologist Vernon Thomas review some of the history associated with the precautionary approach and discuss how it might be implemented, using marine mammal management in Canada as a convenient example. Brian Czech provides some concrete suggestions of how society might move toward a steady state economy, a prerequisite, in his view, for achieving ecological sustainability. Atherton Martin then proposes what he terms a new "architecture" for wildlife conservation in the developing world, drawing on his experiences internationally and as a resident of the small island nation of Dominica. On a related theme, activist Stephen Best focuses on western democracies in the developed world. He suggests that major changes in the conservation and environmental movements are required if we are to have any hope of achieving ecological sustainability in the future. In the end, he maintains, conservation boils down to politics, and that change will only happen through the acquisition and use of political power.

If there were only one take-home message from these essays it would be that there is an urgent need to reinvigorate, if not reinvent, the conservation movement, if we are to make any progress toward the goal of ecological sustainability. The final chapter – Reinventing Wildlife Conservation for the 21st Century – coauthored by myself, Rosamund Kidman Cox, Vivek Menon, and Michael Wamithi, draws on the earlier chapters to provide some concrete recommendations on what conservationists might consider doing in order to make progress in the coming years.

A few words about style. In editing the various essays, I made little attempt to impose a single style on such a diverse and talented collection of authors. It also became obvious at an early stage that such an eclectic group could never be convinced to prepare endnotes in an entirely consistent manner. Differences from one chapter to the next thus reflect the different traditions of various fields in which the contributors work.

Among the extra-curricular activities at the IFAW Forum was a limerick contest. Presentation and judging occurred at the closing banquet and prizes were awarded based on audience response. When the applause meter failed, we reverted to an informal panel of judges, which included – appropriately enough – one or two Irishmen. Although some participants are still complaining, the judges' decisions were (and remain) final! The winning submissions, and a number of other entries of particular relevance to subject matter under discussion are reproduced throughout the book. Apologies in advance for the sometimes loose interpretation of what constitutes a "limerick".

One of IFAW's goals in holding the Forum and publishing this book was to provide a provocative, wide-ranging, and stimulating discussion of the conservation movement – past and present – and to glean some ideas about how best to proceed from here on in. The contributing authors have done their part magnificently. They have provided a wealth of ideas and suggestions that will undoubtedly inform future discussions, stimulate additional thinking and debate and, hopefully, move people into action. Of course, their interpretations and opinions remain their own and do not necessarily represent those of the publishers: the International Fund for Animal Welfare and the University of Limerick.

[1] e.g. Ludwig, D., R. Hilborn and C. Walters. 1993. Uncertainty, resource exploitation, and conservation: lessons from history. Science 60: 17, 36.

ACKNOWLEDGEMENTS

The organization of a successful international meeting and the production of a book require the involvement of many people. Azzedine Downes, IFAW's Vice President and Chief Operating Officer, first mentioned the idea of a Forum at IFAW's Millennium Meeting on Cape Cod in January 2000. Fred O'Regan, IFAW's President and Chief Executive Officer, soon joined that discussion and, over the next two years, a variety of potential topics were bounced around, principally among Fred, Azzedine, Cindy Milburn, Chris Tuite, and myself. Once the topic was agreed upon, a tentative program was assembled and discussed, and several colleagues – including Kelvin Alie, Vivek Menon, Steve Njumbi, Fred O'Regan, Vassili Papastavrou, Patrick Ramage, and Michael Wamithi – subsequently suggested a number of the eventual speakers.

I thank all the speakers who accepted the invitation to participate in the Forum. They not only all showed up to present their papers and take part in the discussions, but also delivered an essay for inclusion in this book. The Forum would never have happened without their enthusiastic participation. I also thank Jane Goodall, who presented a timely public lecture during the Forum in Limerick, offering reasons for hope during part of the meeting when the participants might have been excused for feeling somewhat gloomy after sitting through a number of papers discussing the current state of the planet. I also thank those who expanded the scope of the meeting by submitting and presenting posters. Four of those individuals subsequently accepted invitations to produce three additional essays for inclusion in this book.

I thank Rosamund Kidman Cox, the former editor of BBC Wildlife magazine, and IFAW's Peter Pueschel, Beatriz Bugeda, Kelvin Alie, and Patrick Ramage, for chairing sessions during the Forum.

I am particularly indebted to my old friend and colleague, Professor Roger G.H. Downer, President, University of Limerick, for inviting us to hold the IFAW Forum at the university, and for co-hosting the meeting and co-publishing this book with IFAW. The university's facilities on the River Shannon – from the residences in Drumroe Village and the conference hall, to Plassey House and the Stables – provided the ideal venue for our meeting. I thank Seán Donlon, Chancellor, for joining us at the opening ceremonies and warmly welcoming the participants to Limerick; and Éamonn Cregan, Director of Corporate Affairs, for his generous hospitality, and for his various contributions before, during, and after the Forum.

IFAW colleagues Sheryl Fink, Carol Cassello, and Christine Jones, looked after many of the organizational and administrative details associated with the Forum. I also thank Nick Jenkins of IFAW for his work on publicity and media relations prior to and during the meeting. The Forum itself was executed by Deborah Tudge and Linda Stevens from Plassey Campus Centre Ltd., University of Limerick, and Louise Mulcahy, and Mary Dunleavy from Limerick Travel, Limerick, who together formed the university's amazing conference team. This team arranged accommodations and meals on campus, handled registration, and generally looked after the logistics from beginning to end. You would be hard pressed to find a more friendly, enthusiastic, and efficient group of professionals to run your next international meeting.

Kate Clere, Second Nature Films, Australia, and Mick McIntyre, IFAW's Asia-Pacific Director, came up with the idea of interviewing many of the speakers and producing a DVD that nicely captures the messages emanating out of Limerick and complements the essays in this book.

My office-mates, Jan Hannah and Sue Wallace reviewed and proof-read various chapters in this book. Ingrid Nielsen, in our Ottawa office, also proof-read numerous chapters. I also thank those who generously gave of their time and provided critical reviews of the submitted essays. Robin Clarke at IFAW Headquarters looked after the cover design from beginning to end.

I must single out my colleague in the next office, Sheryl Fink, for special mention. From the time I started to assemble the draft program to the delivery of the book manuscript to the printer, Sheryl was involved at every step. In addition to things already mentioned, she built and maintained the Forum web page (http://www.ifaw.org/forum), and designed and produced the book of abstracts for the meeting. As the associate editor of this book, she kept track of manuscripts, communicated with authors, prepared figures and tables, designed and formatted the entire manuscript, editing, copy-editing and proof-reading along the way. And, while all that was going on, she continued to dig out obscure references for me as I wrestled with my own chapters. Merely saying, "Thank you, Sheryl", is hardly sufficient. This project would never have been completed without her.

Finally, I thank Fred O'Regan and Azzedine Downes for giving me the opportunity to undertake this project and for providing the funding, both for the Forum and for the production of this book.

David M. Lavigne
Guelph, Canada
February 2006

Chapter 1

Wildlife Conservation and the Pursuit of Ecological Sustainability: a brief introduction[1]

David Lavigne

About the time I was immersed in the planning stages for the IFAW Forum that gave rise to this volume, I happened across a book review in the journal *Nature*, written by Brian Child, Chairman of IUCN's[2] Southern Africa Sustainable Use Specialist Group. "Conservation," Child begins,

> ...is like the old Soviet Union: A few self-appointed wise men – scientists mostly, in this case – decide what is best for everyone, justified by a strong ideological creed of nature preservation, animal welfare and wilderness.[3]

I was not particularly surprised by Child's comments, having first become aware of his views in a 1990 paper, co-authored with his father and presented at the IUCN triennial meeting in Perth, Australia.[4] And I was very familiar with the views of the Southern Africa Sustainable Use Specialist Group,[5] which – among other things – once recommended that:

> The precautionary approach should be applied in this sense, it is risky not to use resources, therefore we should use them.

It was sobering to realize, however, that Child's own ideological "creed of nature preservation" – once associated with those who simply wanted to justify their exploitation of nature and wild living natural resources for personal gain – had become so absorbed by the mainstream that by 2004 it would be printed without question, comment or debate in an invited book review in one of the most prominent scientific periodicals on the planet.

I was reminded of an article published a year or so earlier in *The Economist*, which described "a clash between different flavours of conservationist". The article reads in part,

> Wolf Krug, a researcher on social and economic issues...believes there is an ideological clash between "welfare" and "sustainable use" conservationists. The *sustainable-use lobby* takes the view that the long-term future of a species is best ensured from making money from it [emphasis added].[6]

The "clash" was further described as a conflict between "the poor-world agenda of sustainable use", and the "rich-world agenda of [animal] welfare".

While I would agree that there currently is a "clash" within the conservation movement (if it can any longer really be called a movement, as such), the clash is very different from the one alluded to in *Nature* and depicted in *The Economist*. The clash I see is actually between:

- the "sustainable use lobby" (also known as the "wise-use movement", the commercial consumptive use movement, or – to use its own words – the "anti-environmental movement"),[7] which advocates a "use it or lose it philosophy"; and

- the traditional, progressive conservation movement that views commercial consumptive use as a threat to wildlife, and advocates prudence and precaution when we use nature and so-called natural resources.

On one side of the conflict are those who promote short-term economic gain and often, in reality, biologically *unsustainable* uses of wildlife, and nature, generally – "wise users" such as Child and Krug, quoted above. On the other side are those who argue that *if* or *when* we use resources, we should do so prudently and in a biologically sustainable manner – the traditional conservationists, who dominated the mainstream conservation movement throughout most of the 20th century. The latter continue to build on the legacies of Gifford Pinchot, Aldo Leopold, and their successors in the ongoing pursuit of ecological sustainability.[8]

In order to develop my thesis and to lay the groundwork for the chapters that follow, I begin by making some introductory comments on the nature of "conservation" and the concept of "sustainability". This is followed by a brief history of the conservation movement and its evolution, in order to provide a basis for understanding where we are today in relation to where we have been before. Such understanding is essential if we are ever again going to gain ground in the pursuit of ecological sustainability.

THE MEANING OF WORDS

Words evolve, and in our post-*1984* world, they often evolve in strange ways.[9] This is certainly the case with the lexicon of modern conservation. Words such as "conservation" and "sustainability" (among many others, as we shall see) have become "vacuum words" sucking up any meaning that someone wants to give to them.[10] For this reason, I will begin with definitions of what I generally mean, and what I think most of the contributors to this book usually mean, when these terms are used.

Conservation

While the word "conservation" is still sometimes used as a synonym for "preservation",[11] it has come to mean quite different things to different people and organizations. Today, conservation, like beauty, clearly is in the eye of the beholder.[12]

In the context of traditional, 20th century conservation, the term usually embodies two concepts: *preservation*, and *use*. Definitions of wildlife conservation suitable as a starting point for the discussions that follow might therefore read:[13] the protection (preservation) and careful use of wild animals and plants (usually termed "wildlife" or "natural resources") and the environments in which they live; or, more encompassingly: the restoration, protection and prudent use of natural resources to provide the greatest aesthetic, social, ecosystem and economic benefits for present and future generations.

Sustainability[14]

The notion of sustainability has always been an integral part of conservation.[15] "To sustain" means simply "to keep in existence or maintain"; "sustainable" means "capable of being sustained or maintained", as in sustainable yield or sustainable populations;[16] and "sustainability" refers to "the property of being sustainable".

Hewitt provided an early definition of what we might now call "biological sustainability" as early as 1921, with the simple notion that, "A species must not be destroyed at a greater rate than it can increase".[17] A more modern definition may be found in the *World Conservation Strategy* of 1980, under the heading "sustainable utilization":

> Species and ecosystems should not be so heavily exploited that they decline to levels or conditions from which they cannot easily recover.[18]

Different types of sustainability encountered in the conservation literature are briefly distinguished below.

Biological sustainability versus ecological sustainability

Biological and ecological sustainability are two terms that are often used interchangeably. Some, however, make a distinction between the two, restricting "biological sustainability" to refer to a single species or population.[19] "Ecological sustainability" then becomes the multi-species, or ecosystem equivalent, which can be defined as the maintenance of the structure and function of ecosystems over time and space.

The "structure" and "function" of ecosystems referred to above includes the abundance (population sizes) of individual species, the diversity of species comprising the biotic (living) community (often called biodiversity); the abiotic (non-living) components, such as soil productivity, water quality and quantity, and air quality; ecological processes, including nutrient cycling and energy transfer; and natural evolutionary processes.[20]

Economic sustainability

Of course, not all discussions of "sustainability" in the conservation field refer to biological or ecological entities and, as a consequence, it can be quite misleading to use the word in the absence of a qualifier that denotes the type of sustainability being discussed. A case in point involves the term "economic sustainability", which not only is something completely different from biological or ecological sustainability, but also something that can have very different consequences for individual populations, species or ecosystems. In fact, even a locally extirpated population or an extinct species can provide sustainable economic benefits in perpetuity.

Such a situation appears to exist on the island of Sardinia, where an apparently thriving ecotourism operation takes unsuspecting visitors to view magnificent caves once inhabited by Mediterranean monk seals, *Monachus monachus*.[21] The tourists buy monk seal mementos and t-shirts; they purchase a boat ticket displaying a monk seal and are taken to the *Grotto del Bue Marino* – the grotto of the sea ox. It is only when they reach the cave that they learn that it is unlikely that they will see a live monk seal. Why? Because monk seals are locally extinct in Sardinia. Legend has it that the last known resident seal was shot years ago by a fisherman, angry that one of his entrepreneurial colleagues was making money taking people out to see this elusive and rare animal. Here we have a failure to achieve biological sustainability resulting in a loss of biodiversity through the extirpation of a highly endangered species, leading to a "sustainable economic use" over some, as yet unknown, time period.[22]

A related example involves the enduring popularity of dinosaurs. In this instance, sustained economic growth is associated not just with an extinct population but rather with an entire group of animals that disappeared some 65 million years ago.

A more recent example where the achievement of economic sustainability may be at odds with the goal of biological or ecological sustainability relates to those so-called "natural resources" where the population growth rate in the wild is less than the prevailing bank interest rate. Such "resources" include the large whales,[23] old-growth forests[24] and, likely, elephants (well, at least their tusks).[25] Their economic value actually increases faster in banks than it does in the wild. Viewed in this light, there is no economic incentive for the whaling industry, the ivory traders or the forest industry to conserve stocks in the wild. It makes far more economic sense to over-exploit such resources as quickly as possible and invest the proceeds than it does to "harvest" (a conservation euphemism) the "resource" in a biologically sustainable manner. All of which goes a long way to explain the overexploitation of the great whales and elephants in the 20th century and the continued clear-cutting of old growth and tropical rainforests today.[26]

When people talk about sustainability, it is crucial to ask, therefore, whether they are talking about biological or ecological sustainability or about economic sustainability. Neither biological nor ecological sustainability are necessarily prerequisites for economic sustainability.

CONSERVATION: A BRIEF HISTORY

In order to appreciate where the conservation field is today and where it is likely headed tomorrow, it is necessary to know about its origins and its history (Figure 1–1).[27] From the beginning, there were essentially two traditions:[28] the arcadian (or romantic) tradition of the Reverend Gilbert White[29] that eventually gave rise in the United States to the Protectionist School of Conservation, most often identified with John Muir (Figure 1–1, upper line); and the "utilitarian" tradition of Frances Bacon. Bacon's anthropocentric and imperialistic view of nature – itself a descendent of the Christian tradition in which God gave man "dominion over…"[30] – spawned, in the 18th century, Progressive Scientific Agriculture and, later, the utilitarian School of Progressive Conservation (Figure 1–1, lower line).

Gifford Pinchot, America's first forester and a hunting companion of President Theodore Roosevelt, spearheaded Progressive Conservation in the United States.[31] He has been widely credited – by himself, among others[32] – as having coined the term "conservation" in 1907.[33] "Conservation," Pinchot wrote, "is the wise use of resources."

Although Pinchot was an unabashed utilitarian who "liked to refer to forest conservation as tree farming",[34] he nonetheless laid the groundwork for the brand of conservation that dominated throughout most of the 20th century.

As early as 1910, Pinchot outlined three principles of conservation:[35]

- The first principle was *development* – the use of natural resources, for the people who live here now.
- The second was the *prevention of waste* – something that disturbed Pinchot about American forestry practices at the time.
- The third principle embodied the idea that natural resources must be developed and preserved [i.e. conserved] *for the benefit of many and not merely for the profit of a few.*

Pinchot also talked about sustaining natural resources for future generations. Conservation, he wrote,

> recognizes fully the right of the present generation to use what it needs and all it needs of the natural resources now available, but it recognizes *equally* our obligation so to use what we need that our descendants shall not be deprived of what they need [emphasis added].[36]

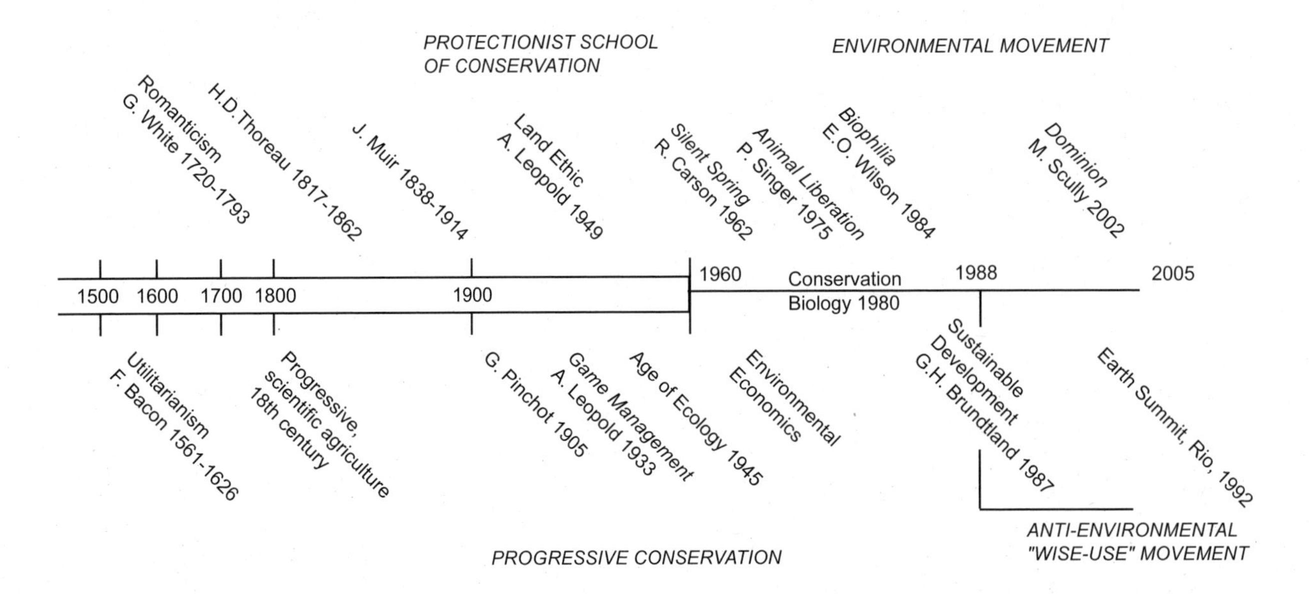

Figure 1–1. The roots of wildlife conservation.

At this early stage in the game, Pinchot's use of words like "development", "conservation", "preservation", and "wise use" was clear; the words meant then what a reasonable person might expect them to mean, even today. But, as implied earlier, these words would take on other meanings as the field – and the language – evolved.

The Dawn of Wildlife Conservation and Game Management

"Wildlife conservation" was born in the early years of the 20th century[37] out of a desire not to protect nature for its own sake but rather to protect certain favoured species – then called "game animals" – from the ravages of 19th century over-exploitation. From the outset, wildlife conservation was an elitist, top-down movement championed mainly by gentlemen hunters in the old British Empire[38] and by influential bureaucrats and politicians in North America. The North American brand of conservation, however, was always somewhat more egalitarian than the British.[39]

With the publication of Aldo Leopold's *Game Management* in 1933 – arguably the first real textbook on wildlife conservation and management – Pinchot's "tree farming" was formally extended to wildlife. Initially, game management was described as "the art of making land produce sustained crops of wild game for recreational use".[40] Of course, at that time, anything that seemed to obstruct that goal – a predator, for example – was targeted for extermination.

Attempts to "develop" natural resources while at the same time preserving them required that the yields removed from forests, fisheries, and wildlife should be sustainable (implicitly meaning biologically and ecologically sustainable). This meant removing so-called surplus production while leaving some natural capital protected. Scientists, particularly in fisheries, but in other fields as well, developed surplus-yield models to predict how much of a particular resource could be removed annually while maintaining a viable and somewhat constant population in the wild in perpetuity. Thus evolved the concept of "sustainable yields", including the mathematically simplistic and appealing idea of the "maximum sustainable yield" (MSY), which became for a time the objective of many fisheries and wildlife-management programs.[41]

Over the decades, the human animal increasingly became an integral part of the conservation agenda. Prior to the early 1960s, progressive conservationists generally regarded nature and natural resources simply as commodities. That view, which in some quarters persists today, is championed particularly by the "wise-users" mentioned earlier. Aldo Leopold summarized the situation in the 1940s when he wrote, "There is no ethic dealing with man's relation to the land and to the animals and plants which grow upon it…Land," he continued, "is still property. The land relation is still strictly economic, entailing privileges but not obligations".[42]

To address this problem, Leopold proposed a "land ethic" to "enlarge the boundaries of the community to include soils, waters, plants, and animals, or collectively: the land". Importantly, Leopold's land ethic, "changes the role of *Homo sapiens* from conqueror of the land community to plain member and citizen of it. It implies respect for his fellow-members, and also respect for the community as such".[43] It would take almost 20 years, however, for Leopold's view that humans were an integral part of nature to become part of the wider public discourse.

The Emergence of an Environmental Movement

Beginning in the early 1960s, both progressive and protectionist schools of conservation[44] were subsumed by a larger, more encompassing environmental movement that began to realize that humans were very much part of the equation. Concern was no longer limited mainly to exploited species but was extended to other species as well. The focus broadened to include animal-welfare concerns [45] and the maintenance of intact and functioning ecosystems. From the mid-1960s to mid-1980s there was an increase in the number of non-governmental organizations, both in the United States and globally, concerned with animal welfare, pro-animal conservation, and the environment. Programs to reintroduce predators were initiated, and attempts were made to restore native ecosystems. And the burden of proof began to shift from those who wished to protect species and ecosystems to those who wished to exploit or develop them.[46]

Reflecting growing public support for such initiatives, there was a proliferation of national legislation and international conventions aimed at reducing our impacts on other species and ecosystems.[47] The publication of the first IUCN *Red Data Book* of endangered mammals in 1966[48] was followed, for example, by the appearance of national endangered species legislation in a number of countries. The United States played a leading role, introducing its first *Endangered Species Act* in 1966; its current Act dates to 1973. Among developed countries, Canada was an anomaly in this regard. It only passed its *Species at Risk Act* (SARA) in 2002.[49]

On the international stage, the world celebrated the first "Earth Day" in 1970, and in 1972, the United Nations Conference on the Human Environment was held in Stockholm.[50] A number of conventions were soon signed and ratified, with the intent of providing increased protection for nature and natural resources. Examples include the *Convention on International Trade in Endangered Species of Wild Fauna and Flora* (1973), the *Convention on the Conservation of Migratory Species of Wild Animals* (1979), the *Convention on the Conservation of Antarctic Marine Living Resources* (1980), the *United Nations Convention on the Law of the Sea* (1982), and its offspring, the *UN Fish Stocks Agreement* (1995).

In retrospect, it was the publication of Rachel Carson's *Silent Spring* in 1962 that not only marked the beginning of the modern environmental movement, but began to involve humans directly in the conservation/environmental equation in a novel way.[51] Carson's book was a warning about the dangers of environmental contaminants, not only to non-human animals, but also to humans themselves. Our concern was no longer limited to conserving nature and natural resources simply for our use.[52]

Paul Ehrlich's 1968 book *Population Bomb*[53] drew further attention to the connection between humans – specifically, our growing numbers – and the exploitation of resources and the environment, a theme developed by Garrett Hardin in his classic paper "The Tragedy of the Commons", published in the same year.[54] Also in 1968, The Club of Rome was established "to come to a deeper understanding of the world problematique".[55] In 1972, it published *Limits to Growth*, predicting dire consequences resulting from the impacts that humans were having on the planet, and generating widespread controversy in the process.[56] The idea began to sink in that our actions not only affected other species and the environment, but also had profound implications for the quality of human life, now and into the future.

In other words, it became clear that wildlife conservation does not and cannot occur in a vacuum (Figure 1–2[57]). Wildlife species are dependent upon plants, soils and water for their habitat. These, in turn, are influenced by climate and climate change. We humans are animals too, and we influence other animals and plants, and soils and water, in a variety of ways. We also use tremendous amounts of fossil sunlight (i.e. oil and natural gas) and, in the process, make our own contribution to climate change. That "everything is connected to everything else" – Barry Commoner's "first law of ecology", first published in 1971[58] – is perhaps even more pertinent today than it was then.

By the early 1970s, it had also become apparent that achieving biological and ecological sustainability was a difficult undertaking. In some quarters at least, it was realized that we really didn't know how to manage nature and natural resources *per se* and that what we really should be trying to do is to manage human activities in a way that minimizes our impacts on wild populations and the environment.[59] Biological models and natural resource policies were modified accordingly in an attempt to reduce human impacts on wild populations and their habitats.[60]

Among other things, limits were placed on how far we were willing to risk reducing exploited wildlife populations. In the United States, for example, marine mammal populations reduced to about 50 per cent of their unexploited stock size were classified as "depleted" and offered protection until such time that they recovered to "optimum sustainable population levels".[61] Similarly, the International Whaling Commission's (IWC) New Management Procedure of the 1970s provided protection for whale stocks when their numbers fell 10 per cent below the level required to produce (in theory, at least) the MSY.[62] Consistently, the *United Nations Convention on the Law of the Sea* states:

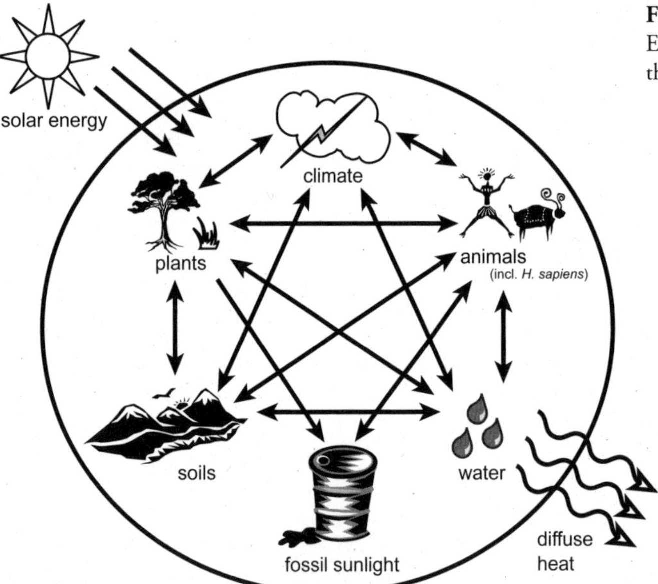

Figure 1–2. Commoner's First Law of Ecology: "Everything is connected to everything else".

> In determining the allowable catch and establishing other conservation measures for the living resources in the high seas, States shall: take measures which are designed, on the best scientific evidence available...*to maintain or restore* populations of harvested species at levels which can produce the maximum sustainable yield... [emphasis added].[63]

The IWC's current Revised Management Procedure[64] not only retains the idea of limiting exploitation when populations are reduced to a predetermined level, but it is also far more precautionary in setting catch limits, should internationally-managed commercial whaling[65] resume at some point in the future.

Many of these developments implicitly or explicitly acknowledged the need for centralized policy formation and management of human activities related to nature and natural resources. Such need was recognized early in the 20th century; it later became fundamental to the success of wildlife-management programs, for example, in North America.[66] Indeed, the need for centralized management was still being acknowledged in some quarters as recently as 1992. At the Earth Summit in Rio, for example,

> States recognized the responsibility of the International Whaling Commission for the conservation and management of whale stocks and the regulation of whaling...[67]

Over the decades, our approach to dealing with nature and natural resources generally became increasingly precautionary: scientific uncertainty was no longer an acceptable reason for postponing measures to prevent serious or irreversible damage to wildlife populations, their habitats or entire ecosystems.[68] In the latter years of the 20th century, national laws (e.g. Canada's *Ocean Act*, 1996) and international agreements (e.g. *United Nations Convention on the Law of the Sea* and Agenda 21, arising from the Earth Summit in Rio) routinely included reference to the precautionary principle or the precautionary approach. The Rio Declaration, for example, states in Principle 15:

> In order to protect the environment, the precautionary approach shall be widely applied by states according to their capabilities. Where there are threats of serious or irreversible damage, lack of full scientific certainty shall not be used as a reason for postponing cost-effective measures to prevent environmental degradation.[69]

Throughout the first 80 years of the 20th century, then, the conservation movement evolved, albeit slowly, and largely through trial and error, into a global movement. Among its success stories, it is credited with halting the decline and promoting the recovery of numerous North American terrestrial wildlife populations devastated by market hunting in the late 1800s.[70] Key to this success was the elimination of markets for meat and other products derived from vulnerable wildlife, including large mammals and waterfowl. Any material benefits derived from wildlife, such as licenses to hunt for personal use, were allocated by law – following consultation with the public, who owned the resource – and not by the marketplace, birthright, land ownership, or social position.[71]

That's where the mainstream conservation movement was headed throughout most of the 20th century. From the early 1960s to 1980, the environmental movement appeared to gain ground in the pursuit of biological and ecological sustainability. But, as we will soon see, that was about to change.

1980 and the Birth of "Sustainable Development"

Two profoundly significant events happened in 1980. The first was the publication by the World Wildlife Fund (WWF), IUCN, and the United Nations Environment Programme (UNEP) of the first *World Conservation Strategy* (WCS).[73] At the time, I don't think many people recognized just how much parts of that strategy deviated from the traditional conservation approaches that had evolved over the previous decades of the 20th century.[73] Most significant was its promotion of an idea it called "sustainable development". The WCS defined development as:

> the modification of the biosphere and the application of human, financial, living and non-living resources to satisfy human needs and improve the quality of human life.

For such development to be biologically or ecologically sustainable, the WCS continued:

> it must take account of social and ecological factors, as well as economic ones; of the living and non-living resource base; and of the long term as well as the short term advantages and disadvantages of alternative actions.

In this context, "conservation" was defined as:

> the management of human use of the biosphere so that it may yield the greatest sustainable benefit to present generations while maintaining its potential to meet the needs and aspirations of future generations.

The concept of sustainable development was soon being described as the marriage between economics and the environment, a magic bullet that would save the environment and simultaneously solve many other world ills arising from economic disparity and social inequity, especially poverty and hunger.

But the idea would only enter the public consciousness and discourse in 1987, after the publication of the Report of the *World Commission on the Environment and Development* (WCED, also known as the Brundtland Report, after its chair, Gro Harlem Brundtland, the former Prime Minister of Norway. Tellingly, in that role, she argued that Norway should resume commercial whaling unfettered by any international regulations or limits). By then, the definition of "sustainable development" had been reduced to:

> ...development that meets the needs of the present without compromising the ability of future generations to meet their own needs.[74]

The circularity of the definition – development was defined in terms of itself – and the lack of clarity about what should be sustainable, other than meeting human needs today and in the future, didn't seem to dampen enthusiasm for the idea.

As soon as the Brundtland Report hit the bookstores, the rush to jump on the bandwagon of sustainable development was immediate – from governments and industry to the academic community and even some prominent members of the international conservation community, including IUCN and the World Wide Fund for Nature (WWF).[75]

The second major development of 1980 was the election of Ronald Reagan to the White House and the appointment of James Watt as his first Secretary of the Interior. While Reagan once apparently claimed that "Trees cause more pollution than automobiles",[76] Watt purportedly described the environmental movement as "A left-wing cult dedicated to bringing down the type of government I believe in",[77] and held the view that God put natural resources on the earth for our use and that we would be abrogating our responsibility if we did not use them.[78]

These two events foreshadowed what was to come for the conservation movement over the next 25 years. One only has to look at what has happened since the appearance of the first *World Conservation Strategy*, the election of Ronald Reagan, and the subsequent publication of the Brundtland Report to appreciate some of the ramifications.

The Rise of "Wise Use"

Since the early 1980s there has been a backlash against the environmental and conservation movements and, particularly, against the gains they extracted between 1960 and the early 1980s. This backlash – variously described as *The War Against the Greens,*[79] *Green Backlash,*[80] or "brownlash"[81] – has been orchestrated by those who wish to exploit nature and natural resources solely for short-term economic gain.

Thus was born the "wise-use" movement. This movement may be seen as an integral part of the shift in western societies towards the conservative right, which has also been promoting with some considerable success, privatization, free trade, economic liberalization, and globalization.[82] This "anti-environmental movement" describes itself as "an informal amalgamation of individuals and groups" that "fights for private property, individual liberties, and free enterprise against environmental oppression".[83]

To accomplish its goals, the movement – in true Orwellian fashion – successfully hijacked and corrupted the language of the traditional conservation movement. Terms such as "wise use", "sustainable use", and the "precautionary principle" no longer mean what they used to mean (Table 1-1).[84] The term "wise use", for example, is obviously borrowed from Pinchot and the original progressive conservation movement. It used to mean, more or less, wise (thoughtful or prudent) use. Now, in reality, it often means unwise or ecologically unsustainable use.

What "wise-users" actually advocate is the very antithesis of the evolving conservation measures noted earlier (Table 1-2). They say that nature "must pay its own way", ignoring all the priceless services provided by the biosphere that make life, including human life, possible in the first place. They argue for the privatization of wildlife and promote its commercial consumptive use and free trade, endangered species included.[85] They lobby for reduced legal protection for wildlife and its habitat and less regulation of human activities generally. They pervert the precautionary principle, arguing that in the face of uncertainty we should err on the side of continued economic growth. Their ultimate goal, says American Ron Arnold – the movement's self-proclaimed founder – is:

> to destroy, to eradicate the environmental movement. We want to be able to exploit the environment for private gain, absolutely. And we want people to understand that this is a noble goal.[86]

And let there be no doubt, the "wise users" are making headway. Agenda 21 abandoned older definitions of sustainable use in order *to promote* the commercial consumptive use of wildlife, an activity that is rarely ecologically sustainable. The World Trade Organization is now positioned to overturn any national environmental legislation that it views as an obstacle to free trade. The *Convention on International Trade in Endangered Species of Wild Fauna and Flora* (CITES), signed in 1973 to protect endangered species from the threats imposed by international trade has, since the mid-1980s, been used increasingly to facilitate trade in endangered species.[87] The most obvious manifestation of this is a recent move, promoted by the CITES Secretariat, to integrate economic incentives into national wildlife trade policies. It recently held a Technical Workshop on Economic Incentives and Trade Policy. At that meeting, the United Nations Conference on Trade and Development's (UNCTAD) Biotrade Initiative "promote[d] trade and investment in products and services derived from biodiversity in support of sustainable development".[88] Most Parties to CITES, however,

Table 1–1. A glossary of modern "wise use" terms. The traditional words of conservation are given in bold, along with modern definitions frequently employed by the "wise-use" or anti-environmental movement.

Conservation: Commercial Consumptive Use.

Ecosystem management: 1. A euphemism for the introduction of predator control programs. 2. Anti-environmentalists sometimes argue that the "ecosystem" should be the unit of concern in conservation. In other words, if we lose a few species along the way due to things such as over-exploitation for commercial purposes, including international trade, such losses are not particularly problematic because the ecosystem (or more correctly, some ecosystem) will still persist. Or, as WWF puts it: "There is a redundancy in the ecological roles of species in most ecosystems…" (see text for details).

Endangered species: Animals that must be killed for commercial purposes in order to be saved (conserved). Use only becomes biologically unsustainable if the species becomes extinct.

Extremists: Anyone who, at any time, questions the ideology of commercial consumptive use. (Such individuals are sometimes also called "environmental terrorists" or "eco-terrorists".) The term "extremists" also has been applied to countries, e.g. non-whaling nations, in meetings of the International Whaling Commission; it can also be applied to any non-governmental organization or individual – such as the author of this chapter – who questions the conservation benefits of commercial consumptive use.

Precautionary approach: In the face of uncertainty, err on the side of continued or accelerated economic growth. As noted in the text, IUCN's Southern Africa Sustainable Use Specialist Group advises that "The precautionary approach should be applied in this sense, it is risky not to use resources, therefore we should use them".

Preservation: Commercial Consumptive Use.

Sustainable Development: Accelerated economic growth, i.e. ecologically unsustainable development.

Sustainable Use of Wildlife: Commercial consumptive use; usually biologically unsustainable use.

"Wise Use": Unwise, imprudent and, usually, biologically or ecologically unsustainable use.

Table 1–2. A comparison of conservation philosophies of traditional progressive conservation and those of the "wise use" movement.

Progressive Conservation	*"Wise Use"*
If it is to be used, it must be conserved.	If it is to be conserved, it must be used.
Devalues wildlife in marketplace	Values wildlife in marketplace
Opposes commercial consumptive use	Promotes commercial consumptive use
Promotes centralized management	Promotes decentralized management
Views wildlife as a public resource	Promotes privatization of wildlife
Favours laws, enforcement, compliance	Promotes less regulation
Places onus on exploiter	Does not place onus on exploiter
Emphasizes precaution	Emphasizes use

have thus far proved reluctant to integrate economic incentives into national wildlife trade policies.

In 1992, "sustainable use" was enshrined among the objectives of the *Convention on Biological Diversity* (CBD). "Wise users" have subsequently attempted to use the CBD to subjugate or "harmonize" older biodiversity-related conventions concluded in the 1970s and 1980s. More recently (late 2003), the US Fish and Wildlife Service proposed a rule change to its *Endangered Species Act* to allow wealthy Americans to import endangered species from the developing world, using the "wise-use" argument that the monies paid could be used to conserve the few specimens remaining in the wild.

The "wise-use" movement's greatest success, however, may be the extent to which it has successfully deceived so many traditional conservation organizations and much of the worldwide media into parroting its dogma unchallenged. As early as 1990, the International Union for the Conservation of Nature and Natural Resources (IUCN), was advocating the exploitation of endangered species as a means of furthering their "conservation",[89] a marked departure from their earlier incarnation as the International Union for the Protection of Nature (IUPN).[90] Today, the organization is more concerned about the "sustainability of uses of biodiversity components" than with the sustainability of biodiversity, *per se*.[91] Its Southern Africa Sustainable Use Specialist Group goes even further, arguing (in addition to perverting the precautionary principle) that, contrary to established conservation principles, "There is no arbitrary population size below which use should be prohibited...", and that "...use can be regarded as sustainable...provided the species population is not reduced to the level that extinction is a threat."[92]

In other words, "wise-users" argue entirely arbitrarily that the level at which "extinction is a threat" – a level that is difficult if not impossible to know in advance – should guide the extent of exploitation, replacing the more precautionary approach (based on the lessons of history) of not reducing populations "to levels or conditions from which they cannot easily recover".[93]

In a discussion paper entitled "The commercial consumptive use of wild species: Managing it for the benefit of biodiversity", even the WWF – in its "guidelines for the commercial consumptive use of wild species" – goes so far as to suggest that, "there is redundancy in the ecological roles of species in most ecosystems" and accepts that "some biodiversity may need to be sacrificed...to make wildlands competitive with alternative land uses".[94]

Sustainable Development in Retrospect

Over the past 25 years, despite much talk about conservation and sustainable development, the global environment has been further degraded. We have continued to lose biodiversity, and there is now even greater disparity between rich and poor nations than there was when the Brundtland Report was published in 1987.[95] As the media reported following the release of the United Nations annual report for 2001:

> Humans are plundering the planet at an unprecedented and unsustainable rate that needs to be curbed quickly to avoid worldwide disaster. More people are using more resources with more intensity than at any point in human history.[96]

The only things we have managed to sustain, it seems, are the economies of developed nations and – despite all evidence to the contrary – the idea that "sustainable development" *is* the solution to the world's ills. In short, sustainable development – as envisioned by the 1987 World Commission on Environment and Development – has failed to achieve its objectives in ways that only a few skeptics predicted it would from the very start.

So why, in the face of all the evidence that things have continued to deteriorate, do academics, politicians, governments, big business, and even some prominent members of the conservation and environmental communities, continue to advocate sustainable development as a viable solution to the problems confronting the human condition? Indeed, why did so many support such an ill-defined and oxymoronic[97] concept in the first place?

A number of critiques of sustainable development have been published over the years, and some of the contributors to this volume have additional thoughts on the topic.[98] Consequently, I will simply make a few introductory comments here.

Sustainable development is not an operationally defined scientific or technical concept, but rather an intentionally vague idea that conveniently can be used by interested parties with diametrically opposed views and objectives. In short, it is a political concept.[99] While some will argue that the lack of a definition of "sustainable development" allows for "constructive ambiguity",[100] I have argued that it is more about "deceptive ambiguity", raising false hopes and obscuring the real agenda of those who promote it.[101]

If it is accepted that "sustainable development" is a political term, things begin to make a bit more sense. Politics involves conflict among individuals, groups and nations about alternative values or competing visions of what is good.[102] Language is one of the weapons used in such conflicts and, as Orwell recognized, those who control the language control the debate.[103] The participants in any political debate or conflict will shade the truth, first for their audiences; then, in many cases for themselves.[104] In other words, politics involves deception and self-deception, and there are elements of both in the sustainable-development debate. As Bill Willers wrote as early as 1994, sustainable development is "the new world deception".[105]

Sustainable development, as defined by Brundtland, cannot possibly raise the standards of living in the developing world, allow the developed world to increase its standard of living, *and* sustain the environment, all at the same time. Standards of living in the developed world are already unsustainable, and it would take several planet Earths to achieve these objectives.[106]

The Brundtland Commission, and particularly the "wise use" movement that followed it, have co-opted and corrupted the language of conservation, deceptively redefining everything from sustainability to precaution. When they "promote" the commercial consumptive use of nature and natural resources, free trade, and globalization, they do so under the guise of providing benefits to poor nations of the South. The fact of the matter is that the North (the developed world) requires the continued exploitation of the South (the developing world), and free trade, to maintain and increase the size of its ecological footprint and to sustain its continued economic growth or, in other words, to continue its unsustainable lifestyles under the mantra of "sustainable development". Viewed in this light, the concept of sustainable development is entirely consistent with a strategy designed, in the words of my colleague Sidney Holt, "to maintain capitalism as the only and permanent economic system", and to ensure the continued domination of the North over the South.[107]

CONSERVATION TODAY

In 1995, journalist Mark Dowie wrote a book about the decline of the American environmental movement in the latter years of the 20th century. The book was titled *Losing Ground* and, in many ways, Dowie's observations apply equally to the worldwide decline of the traditional conservation movement, over much of the same period. [108]

Today (2005), the human population continues to increase, and it will continue to do so for decades, simply as a result of population momentum alone.[109] At the turn of the century we numbered over 6 billion.[110] During the early 1980s, the human ecological footprint[111] surpassed the planet's capacity to maintain our current lifestyles and, by the end of the 20th century, it was estimated that it had exceeded the bio-capacity of the planet by some 20 per cent, i.e. we now require more than 1.2 planet Earths to support our present condition.[112] By 2050, the United Nations predicts that the human population will have increased to about 9 billion.[113]

The problems facing the planet (or, more precisely, the human species and the many other species whose fate is in our hands) are well documented. As the global human population grows, resources continue to be depleted and the environment becomes increasingly degraded.[114] Our unsustainable practices include the clearing of forests,[115] the loss of productive soils,[116] and the over-exploitation of fisheries,[117] all of which are contributing to the ongoing loss of biodiversity that some have characterized as the "sixth extinction".[118] In addition, we are interfering with evolutionary processes through the exploitation of natural resources (e.g. selective hunting, including trophy hunting,[119] fishing, and forestry), the introduction of exotic, alien, or non-native species and, most recently, through the production and release of genetically modified organisms into the environment. We are also depleting oil and natural gas reserves,[120] increasing greenhouse gas emissions and contributing to global climate change.[121] Superimposed on all these realities is the growing social inequity and economic disparity between the North and South, and between rich and poor within many nations.[122] Of particular concern to many is the fact that millions (some say billions) of people, most of whom live in the developing world, are going hungry and suffering from malnutrition.[123]

Despite what you may read in *Nature* or *The Economist*, there are today actually three competing "schools of conservation," each driven by a different set of values and objectives and each using the same terms to convey very different messages.

At one end of the spectrum is "protectionist conservation" – characterized largely by moralistic and humanistic attitudes[124] towards animals and nature. Because protectionist conservationists – especially the animal-rights movement – are basically opposed to the consumptive use of animals, they have been largely marginalized in the sustainability debate.[125] At the other end is the self-styled "wise-use" movement. Motivated by utilitarian and dominionistic attitudes, it has become a major player, if not the major player, on the world stage. *The New York Times* recently touted it as "... an increasingly popular but controversial conservation movement... The philosophy is that saving a species may require commercially exploiting it".[126] Yet, for decades, it has been recognized that those species which people use as commodities are inherently at risk of population depletion or elimination, and that "wildlife conservation is incompatible with global markets and private ownership" (see Table 1–3).[127] Today, these are the very ideals that are being promoted by the "wise-use" movement to achieve its version of sustainable (read unsustainable) use.

Caught between these two extremes is "traditional" progressive conservation. In reality, it is the only "school" of modern conservation that is truly concerned with biologically and ecologically sustainable use.

CONCLUSIONS

James Shaw, among others, has depicted the history of wildlife conservation in North America as a series of distinct eras, spanning the time frame 1600-1984 (Table 1–4).[128] Although the dates may vary from place to place, the progression, particularly since 1850, seems generally applicable to the global situation. Shaw's chronology ended in 1984, just before the publication of his book. In 1999, I argued, along with Victor B. Scheffer and Stephen Kellert, that by the mid-1980s conservation had entered a new era.[129] What was not clear to us then was exactly how that era would be remembered. At the time, we wrote:

Table 1–3. Twentieth century views of traditional conservationists on the topic of commercial consumptive use of wildlife.

"Here there is an inexorable law of Nature, to which there are no exceptions: *No wild species of bird, mammal, reptile, or fish can withstand exploitation for commercial purposes*".
William T. Hornaday 1913

"It is almost a truism that the very best way to exterminate any species of wild life is to put a price on its head. As long as there are dealers in game you will find men who will kill it in spite of anything you may do to the contrary".
Frederick K. Vreeland 1915

"It seems almost axiomatic, however, that a return to wide-open markets would spell the certain doom of wild game, and even edible non-game...With a wide-open market, public game could not exist, and game laws would become useless and unnecessary".
Aldo Leopold 1919

"It is universally recognized now...that the free marketing of wild game is one of the greatest factors tending rapidly to exterminate our native game resources...".
C. Gordon Hewitt 1921

"It is now generally recognized that the commercialization of game means its extinction".
George Grinnell 1925

"The rather strangely assorted leaders of the early conservation movement were unanimous on one point – that the market hunter had to go".
James Trefethen 1966

"The reintroduction of markets in wildlife meat and parts jeopardizes North America's system of wildlife conservation".
Valerius Geist 1988

"Species that people use as commodities are inherently at risk of population reduction or elimination".
Elliot A. Norse 1993

"Virtually all species and stocks of wild living resources... which are being harvested commercially have been or are being depleted".
Lee M. Talbot 1996

"The most severe overexploitation tends to occur when hunting is done to supply markets rather than just to feed hunters' families...when strong market demand exists for a mammal's meat, hide, horns, tusks, or bones, species can decline on a catastrophic scale".
John Tuxill 1998

> If the trends established between 1966 and the early 1980s are anything to go by, it may well end up being remembered as the era of conservation biology, after the emerging field of environmental research of the same name, with its several journals – e.g., *Conservation Biology*, *Biological Conservation*, and text books bearing similar titles.[130]
>
> The other possibility is that the current period will be remembered as the era of "wise use", a time when the conservation gains of the 20th century began to erode, ostensibly to facilitate economic growth and development through the consumptive use of wildlife and the international trade in wildlife products. If this were to occur, the 20th century would be viewed, in retrospect, as an anomaly, a time when people tried, for a while at least, to reduce their impacts on wild populations and their habitats, and, to their own detriment, ultimately failed.

It was too early to say, in 1999, whether the "wise-use movement" would leave a lasting mark on the history of conservation or whether any apparent gains it had made since 1980 would be seen in retrospect only as short-term setbacks to the conservation advances made throughout the 20th century. It seemed clear, nonetheless, that as we approached the end of the millennium, we were at a "crossroads".[131]

Table 1–4. Eras of Wildlife Conservation

1600-1849	Era of Abundance
1850-1899	Era of Over-exploitation
1900-1929	Era of Protection
1930-1965	Era of Game Management
1966-1984	Era of Environmental Management
1985- ????	Era of Conservation Biology or
	Era of Commercial Consumptive Use or
	Era of Sustainable Development or
	Era of Transition...

Writing about "*The Collapse of Globalism*" in 2005, John Ralston Saul put it, perhaps, a better way:

> ...we are transiting one of those moments that separate more driven or coherent eras. It is like being in a vacuum, except that this is a chaotic vacuum, one filled with dense disorder and contradictory tendencies... one of those moments...when...there is furious, disordered activity until one side finds the pattern and the energy to give it control.[132]

Saul goes on to say:

> ...a period of uncertainty is also one of choice, and therefore of opportunity... [and] that we have the power to choose in the hope of altering society for the greater good.

The idea embodied in that last sentence is really what the IFAW Forum – *Wildlife Conservation: In Pursuit of Ecological Sustainability* – was all about: offering an alternative for achieving biological and ecological sustainability to the "wise use" – "use it or lose it" – philosophy, emphasizing that we really do have "the power to choose in the hope of altering society for the greater good".[133]

Today, global threats to wildlife are greater than ever. If modern society really wants to pursue biological or ecological sustainability and preserve as much of the remaining biodiversity as possible, the only viable option is to learn from the lessons of history[104] and to adapt traditional conservation principles to deal with the realities of the 21st century.

Despite losing ground in recent decades, traditional conservation still recognizes that the important issue is not whether the exploitation of wildlife is commercial or non-commercial, consumptive or non-consumptive, but rather whether a particular use is biologically and ecologically sustainable.

Just before the IFAW Forum in Limerick, I wrote the following:

> ...in order to achieve its goals, conservation must become more widely embraced than in the past. [We] must accept that humans are an integral part of organic evolution and that the well being, quality of life, and the survival of all species, including *Homo sapiens*, depends on the maintenance of evolutionary processes and functioning ecosystems. Thus, when humans use nature (as they always will), we must strive with greater resolve to reduce

the risks of causing irreversible damage to
the biosphere and its component parts.[135]

Today, we have a choice. We can rise to the challenge, or not.[136] If we choose the latter, we, but more so our children, our grandchildren, and their children, will soon have no choice but to accept the inevitable consequences.

ACKNOWLEDGEMENTS

Many of the ideas presented in this chapter were developed over the past two decades, primarily as a result of my involvement in teaching undergraduate and graduate courses at the University of Guelph and my various writings in *BBC Wildlife* magazine dating back to the late 1980s – thanks to the opportunities and guidance provided by its long-time editor, Rosamund Kidman Cox. My views have evolved in recent years, largely as a result of my involvement in the 2001 & 2002 Kenneth Hammond Lectures on Environment, Energy and Resources at the University of Guelph and through my experiences as science advisor to the International Fund for Animal Welfare. Over the years, my understanding of the subject matter has benefited from discussions with – among many others – Steve Best, Ron Brooks, Carolyn Callaghan, Ward Chesworth, Sheryl Fink, the late Stuart Innes, Val Geist, Sidney Holt, Vassili Papastavrou, Rick Smith, Vernon Thomas, and Bill Rees. I thank Steve Best, Rosamund Kidman Cox, Ward Chesworth, Michael Earle, Sheryl Fink, Jan Hannah, Sidney Holt, Barry Kent MacKay, Vassili Papastavrou, Rosalind Reeve, and Sue Wallace for their constructive comments and suggestions on earlier drafts of this chapter. It should not be assumed that any of the aforementioned individuals agree with my analysis and, of course, any errors of interpretation or fact are mine.

NOTES AND SOURCES

[1] This chapter traces its roots to a seminar I first delivered as a "Distinguished Visitor" in the Environmental Research and Studies Centre, University of Alberta, Edmonton, Canada in January 2003. At that time I was trying to gather my thoughts on the topic and get them down on paper, prior to finalizing the Forum program and completing the roster of invited speakers. Some of the content of the present chapter draws heavily on earlier writings (e.g. Lavigne 2002a,b; Lavigne *et al.* 1996; 1999; all referenced elsewhere in these endnotes). Parts of an essay I wrote for *BBC Wildlife* magazine in May 2004, just prior to the IFAW Forum in Limerick, are also incorporated here with permission of the then editor, Rosamund Kidman Cox.

[2] IUCN – The World Conservation Union, is otherwise known by its earlier name, the International Union for the Conservation of Nature and Natural Resources (IUCN).

[3] Child, B. 2003. When biodiversity meets humanity. Book Review. Nature 421:113-114.

[4] Child, G. and B. Child. 1990. An historical perspective of sustainable wildlife utilization. Paper prepared for Workshop 7, 18th IUCN General Assembly, Perth, Australia. Also see Lavigne, D.M. 1991a. Your money or your genotype. Special Report. BBC Wildlife, 9(3):204-205.

[5] Southern Africa Sustainable Use Specialist Group. 1996. Sustainable Use Issues and Principles. Southern Africa Sustainable Use Specialist Group, IUCN Species Survival Commission, Gland, Switzerland; also see Lavigne, D.M., V.B. Scheffer, and S. R. Kellert. 1999. The evolution of North American attitudes toward marine mammals. pp. 10-47. In J.R. Twiss and R.R. Reeves (eds.). *Conservation and Management of Marine Mammals*. Smithsonian Institution Press, Washington, DC.

[6] Anon. 2002. Out of the blue. The Economist. 31 October 2002.

[7] For an introduction to the anti-environmental movement, see Dowie, M. 1995. *Losing Ground. American Environmentalism at the Close of the Twentieth Century.* Chapter 4. The MIT Press, Cambridge, MA and London, England.

[8] Pinchot, G. 1947. *Breaking New Ground.* Island Press, Washington, DC and Covelo, CA; Miller, C. 2001. *Gifford Pinchot and the Making of Modern Environmentalism.* Island Press/Shearwater Books, Washington, Covelo, London; Meine, C. 1988. *Aldo Leopold: His Life and Works.* The University of Wisconsin Press, Madison, Wisconsin.

[9] See Orwell, G. 1949. *1984.* Signet Classic Edition. 1961. The New American Library of World Literature, Inc. New York, NY.

[10] I first came across the use of the term "vacuum word" in reference to the concept of globalization, in a discussion of "Globalization and Cinema" (see Peña, R. 2001. The Roots of Globalization in the Cinema. pp. 3-4. In The Cultural Politics of Film. Correspondence. An International Review of Culture and Society. Issue 8. Summer/Fall 2001. Available at www.cfr.org/pdf/correspondence/CORRSum01.pdf.)

[11] See Johnson, W.M. and D.M. Lavigne. 1998. *The Mediterranean Monk Seal: Conservation Guidelines.* Multilingual Edition: English, French, Greek, Spanish, Turkish. International Marine Mammal Association Inc., Guelph, Ontario, Canada. Following the Oxford English Dictionary, we defined "conservation" as preservation from destructive influences...protection from undesirable changes" (OED. 1989. Conservation. pp. 764-765. In *The Oxford English Dictionary*. Second Edition, Vol. III. Prepared by J.A. Simpson and E.S.C. Weiner. Clarendon Press, Oxford).

[12] Here, I have slightly re-arranged an original quotation from Olver, C.H., B.J. Shuter, and C.K. Minns. 1995. Toward a definition of conservation principles for fisheries management. Canadian Journal of Fisheries and Aquatic Sciences, 52:1584-1594.

[13] Both definitions given here are modified from the many definitions of "conservation" found in dictionaries and on the Internet.

[14] Sidney Holt provides a lengthy discussion of the "notion of sustainability" in Chapter 4.

[15] For thorough discussions of sustainability, see the various contributions in Levin, S.A. 1993. Forum. Science and sustainability. Ecological Applications 3(4):547-589; also see Holt, Chapter 4.

[16] The US Marine Mammal Protection Act (1972) switched the emphasis from "sustainable yields" to "optimum sustainable populations." In other words, the focus of management concern was shifted from removing "sustainable yields" from populations in perpetuity to maintaining populations in the wild. See, for e.g. http://www.nmfs.noaa.gov/pr/laws/mmpa.htm.

[17] Hewitt, G.C. 1921. *The Conservation of the Wild Life of Canada*. Charles Scribner's Sons, New York.

[18] IUCN/UNEP/WWF. 1980. *World Conservation Strategy. Living Resource Conservation for Sustainable Development.* Gland, Switzerland. See 7.1 Priority requirements: sustainable utilization. One weakness of this definition – as my colleague Sidney Holt recently reminded me – is that the inclusion of the word "easily" in the definition is both meaningless and misleading.

[19] Such usage is found, for example, in Papastavrou and Cooke, Chapter 7.

[20] The definitions of biological and ecological sustainability that I provided in my opening talk to the IFAW Forum overlooked (explicitly, at least) one important consideration. As Sidney Holt observed during the opening session, the idea of sustainability also should allow for the continuation of evolutionary processes and I have, therefore, added that to the definition given here. To put it another way, human activities (including exploitation) should avoid interference with evolutionary processes. The degree to which the physical, chemical, and biological components (including composition, structure, and processes) of an ecosystem and their relationships are present, functioning, and capable of self-renewal is sometimes referred to as "ecological integrity". Ecological integrity implies the presence of appropriate species, populations and communities and the occurrence of ecological processes at appropriate rates and scales as well as the environmental conditions that support these taxa and processes. Source: http://science.nature.nps.gov/im/monitor/glossary.htm.

[21] Johnson, W.M. 1998. Monk seal myths in Sardinia. Monachus Guardian 1(1):11-15. Available at http://www.monachus.org/mguard01/01mguard.htm. I have recounted this story previously in: Lavigne, D.M. 2002a. Ecological footprints, doublespeak, and the evolution of the Machiavellian mind. pp. 61-91. In W. Chesworth, M.R. Moss, and V.G. Thomas (eds.). *Sustainable Development: Mandate or Mantra.* The Kenneth Hammond Lectures on Environment, Energy and Resources 2001 Series. Faculty of Environmental Sciences, University of Guelph, Guelph, Canada.

[22] For more on ecotourism, see Mugisha and Ajarova, Chapter 10, and Corkeron, Chapter 11.

[23] Clark, C.W. 1973a. The economics of overexploitation. Science 181:630-634; Clark, C.W. 1973b. Profit maximization and the extinction of animal species. Journal of Political Economy 81: 950-961.

[24] Clark, C.W. 1989. Clear-cut economies. Should we harvest everything now? The Sciences 29: 16-19.

[25] Caughley, G. 1993. Elephants and economics. Conservation Biology 7:943-945.

[26] Clark 1973a; Caughley 1993.

[27] See Worster, D. 1994. *Nature's Economy. A history of ecological ideas.* Second Edition. Cambridge University Press, Cambridge, New York and Melbourne.

[28] Ibid.

[29] White, G. 1778. *The Natural History of Selbourne.* Reprinted 1906. J.M. Dent & Sons Ltd., London.

[30] See Genesis 1:26; and, for a different interpretation to the one implied here, see Scully, M. 2002. *Dominion: The Power of Man, the Suffering of Animals, and the Call to Mercy.* St. Martin's Press, New York, NY. (especially pp. 90-99).

[31] Pinchot 1947; Miller 2001. Several colleagues have indicated that my discussion of progressive conservation has a distinct American bias. Readers interested in the history of conservation from the perspective of the old British Empire should consult: Fitter, R. and Sir P. Scott. 1978. *The Penitent Butchers: 75 years of wildlife conservation.* Fauna Preservation Society. London, UK; Threlfall, W. 1995. Conservation and Wildlife Management in Britain. pp. 27-74. In Geist, V. and I. McTaggart Cowan (eds.). *Wildlife Conservation Policy. A Reader.* Detselig Enterprises Ltd., Calgary, Alberta, Canada; Adams, W.M. 2004. *Against Extinction: The Story of Conservation.* Earthscan, London UK. Menon and Lavigne, Chapter 12, also include earlier references to the history of conservation in Asia and elsewhere.

[32] Pinchot 1947; Shaw, J.H. 1985. *Introduction to Wildlife Management.* McGraw-Hill Book Company, New York, NY.

[33] Pinchot G. 1907. The Use of the National Forests. U.S. Department of Agriculture. 42 pp. Of course, the word conservation (as a synonym for preservation) had been around for years before Pinchot and his colleagues gave it new meaning (see for e.g. Pinchot 1947, p. 326).

[34] Worster 1994, p. 267.

[35] Pinchot, G. 1910. *The Fight for Conservation.* Doubleday, Page & Company. New York, NY.

[36] Pinchot 1910, p. 80; emphasis added.

[37] e.g. Hewitt 1921.

[38] Adams 2004.

[39] See, for example, Geist, Chapter 19.

[40] Leopold, A. 1933. *Game Management.* Charles Scribner's Sons, New York. Reprinted 1986. The University of Wisconsin Press, Madison, Wisconsin.

[41] Although the late Peter Larkin wrote "an epitaph to the concept of maximum sustainable yield" in 1977 (Larkin, P. 1977. An epitaph for the concept of MSY. Transactions of the American Fisheries Society, 106(1):1-11), this archaic concept remains embedded in some international conventions (e.g. *United Nations Convention on the Law of the Sea*) and continues to be the stated objective of many regional fisheries organizations. It should be added, however, that reference to MSY is still a more conservative approach than the "use it or lose it" approach adopted by "wise-users". The failures associated with the use of MSY as a management objective are primarily related to the failure to recognize quickly enough when populations are being depleted more than expected or desired (S. Holt, pers. comm.).

[42] Leopold, A. 1970. *Sand County Almanac with Essays on Conservation from Round River.* Sierra Club/Ballantine Books Inc., San Francisco, CA and New York, NY.

[43] Leopold's philosophy toward the end of his career was outlined in his now famous book, *Sand County Almanac*, originally published shortly after his untimely death in 1948. (Oxford University Press, Oxford, UK, 1949). Its republication in the 1970s (see endnote 42) led to Leopold being labelled in some quarters as the father of the environmental movement. For additional details on the remarkable career of Aldo Leopold, see Meine, 1988.

[44] Lavigne 2002a.

[45] Examples include the *U.S. Laboratory Animal Act* (1966) and parts of the *U.S. Marine Mammal Protection Act* (1973). In Canada, the *Seal Protection Regulations*, promulgated under the *Fisheries Act* in the mid-1960s included provisions for regulating the way seals were killed in Canada's annual seal hunt.

[46] e.g. Geiser, K. 1999. Establishing a general duty of precaution in environmental protection policies in the United States. A proposal. pp. xxi-xxvi. In C. Raffensperger and J. Tichner (eds.). *Protecting Public Health & the Environment. Implementing the Precautionary Principle.* Island Press, Washington, DC and Covelo, CA.

[47] See Lavigne *et al.* 1999.

[48] Simon, N. 1966. *Red Data Book*, Vol. 1. International Union for the Conservation of Nature and Natural Resources. Morges, Switzerland.

[49] For more on *SARA*, see Hutchings, Chapter 6.

[50] It is noteworthy – in the context of this brief history – to keep in mind that the 1992 UN Conference in Rio was not just about the environment, but rather about environment and development.

[51] Carson, R. 1962. *Silent Spring.* Houghton Mifflin, Boston MA. Many people argue that the appearance of *Silent Spring* marked the beginning of the environmental movement.

[52] Now, we were directly involved, and there is something about being directly involved that focuses the human mind. This point was made explicitly by Robert Worcester during the IFAW Forum; see Chapter 17.

[53] Ehrlich, P. 1968. *The Population Bomb.* Sierra Club-Ballantine Books, New York, NY.

[54] Hardin, G. 1968. The Tragedy of the Commons. Science, 162:1243-1248.

[55] See http://www.clubofrome.org/organisation/index.php.

[56] Meadows, D.H., D.L. Meadows, J. Randers, and W.W. Behrens III. 1972. *Limits to Growth.* Universe Books, New York, NY.

[57] Reprinted from Lavigne, D.M. 2005. Reducing the Agricultural Eco-Footprint: Reflections of a Neo-Darwinian Ecologist. pp. 119-166. In A. Eaglesham, A. Wildeman, and R.W.F. Hardy (eds.). Agricultural Biotechnology: Finding Common International Goals. National Agricultural Biotechnology Council Report 16. NABC, Ithaca, NY. Available at http://nabc.cals.cornell.edu/pubs/nabc_16/talks/lavigne_corrected.pdf.

[58] Commoner, B. 1971. The closing circle: nature, man, and technology. pp. 161-166. In M.A. Cahn and R. O'Brien (eds.). *Thinking About The Environment.* M. E. Sharpe, Inc., Armonk, NY.

[59] e.g. Holt, S.J. 1978. Opening Plenary Meeting. *Mammals in the Seas.* Report of the FAO Advisory Committee on Marine Resources Research. Working Party on Marine Mammals. FAO Fisheries Series, No. 5, Vol. 1: 262-264. This idea was reiterated on several occasions during the IFAW Forum in Limerick; see, for example, Holt, Chapter 4; de la Mare, Chapter 21.

[60] Lavigne, D.M. C.J. Callaghan, and R.J. Smith. 1996. Sustainable utilization: the lessons of history. pp. 250-265. In V.J. Taylor and N. Dunstone (eds.). *The Exploitation of Mammal Populations.* Chapman and Hall, London.

[61] Baur, D.C., M.J. Bean, and M.L. Gosliner. 1999. The laws governing marine mammal conservation in the United States. pp. 48-86. In J.R. Twiss Jr. and R.R. Reeves (eds.). *Conservation and Management of Marine Mammals.* Smithsonian Institution Press, Washington and London. See p. 56.

[62] Holt, S.J. and N.M. Young. 1991. *Guide to Review of the Management of Whaling.* Second Edition. Center for Marine Conservation, Washington, DC; also see Papastavrou and Cooke, Chapter 7.

[63] *United Nations Convention on the Law of the Sea.* Article 119 (a); also see United Nations Conference on Environment and Development. 1992. Agenda 21, Programme of Action for Sustainable Development, para 17.46(b). United Nations Publications, New York, NY.

[64] Holt and Young 1991; Cooke, J.G. 1995. The International Whaling Commission's Revised Management Procedure as an example of a new approach to fishery management. pp. 647-670. In A.S. Blix, L. Walløe, and Ø. Ulltang (eds.). *Whales, Seals, Fish, and Man.* Developments in Marine Biology 4. Proceedings of the International Symposium on the Biology of Marine Mammals in the North East Atlantic, Tromsø, Norway, 29 November – 1 December 1994. Elsevier Press, New York, NY.

[65] Although there is currently a moratorium on commercial whaling, whales continue to be killed primarily for commercial purposes by Japan, Norway and Iceland. This commercial whaling is not regulated by the International Whaling Commission; catches are determined by the whaling nations themselves (see Papastavrou and Cooke, Chapter 7).

[66] e.g. Geist, V. 1988. How markets in wildlife meat and parts, and the sale of hunting privileges, jeopardize wildlife conservation. Conservation Biology, 2:1-12.; Geist, V. 1989. Legal trafficking and paid hunting threaten conservation. Transactions of the North American Wildlife and Natural Resources Conference, 54:172-178; also see Geist, Chapter 19.

[67] United Nations Conference on Environment and Development. 1992. *Agenda 21, Programme of Action for Sustainable Development.* Para 17.62(a). United Nations Publications, New York, NY.

[68] e.g. Johnston, D.J., P. Meisenheimer, and D.M. Lavigne. 1999. An evaluation of management objectives for Canada's commercial seal hunt, 1996-1998. Conservation Biology, 14: 729-737; Ralls, K. and B.L. Taylor. 2000. Special Section: Better policy and management decisions through explicit analysis of uncertainty: New approaches from Marine

Conservation. Conservation Biology, 14:1240-1242; Taylor, B.L., P.R. Wade, D.P. DeMaster, and J. Barlow. 2000. Incorporating uncertainty into management models for marine mammals. Conservation Biology, 14:1243-1252.

[69] United Nations Convention on Environment and Development. 1992. For additional discussion of the precautionary approach, see Campbell and Thomas, Chapter 22.

[70] This success story is captured by Geist, Chapter 19.

[71] Ibid.

[72] IUCN/UNEP/WWF. 1980.

[73] The late Ian McPhail, one of the founders of WWF, was one of the first people I knew who warned of the dangers implicit in the new *World Conservation Strategy*. I remember him saying something to the effect that 50 years from now we would look back and realize what a mistake it was; especially its promotion of an idea it called "sustainable development". Ian recently died but I think he lived long enough to know that it did not take 50 years before the false promises of sustainable development would be obvious to critical observers in the field. For further discussion, see Beder, S. 1996. *The Nature of Sustainable Development*. Second Edition. Scribe Publications, Newham, Australia; Chesworth, W., M.R. Moss, and V.G. Thomas (eds.). 2002. *Sustainable Development: Mandate or Mantra*. The Kenneth Hammond Lectures on Environment, Energy and Resources 2001 Series. Faculty of Environmental Sciences, University of Guelph, Guelph, Ontario, Canada. Also see Holt, Chapter 4; Beder, Chapter 5; Rees, Chapter 14; Brooks, Chapter 16; Oates, Chapter 18; Czech, Chapter 23.

[74] World Commission on Environment and Development. 1987. *Our Common Future*. Oxford University Press, Oxford and New York. p. 43.

[75] For other views on sustainable development, see Beder, 1996; Lavigne 2002a; and Willers, B. 1994. Sustainable development: A new world deception. Conservation Biology 8:1146-1148. Also see Beder, Chapter 5; Brooks, Chapter 16.

[76] Olive, D. 1992. *Political Babble. The 1,000 Dumbest Things Ever Said by Politicians*. John Wiley & Sons, Inc. New York, NY. p 90.

[77] Ibid.

[78] Here I am paraphrasing an old (and apparently long forgotten) quote from James Watt that was buried in my old lecture notes from the early 1980s, unfortunately without the original source. Quotes expressing similar sentiments do, however, survive, e.g. "The Earth was put here by the Lord for His people to subdue and to use for profitable purposes on their way to the hereafter" (see Deem, R. Is Christianity anti-Environmental?, http://www.godandscience.org/apologetics/environment.html, citing Wolf, R. 1981. God, James Watt, and the Public Land. Audubon, 83(3):65). It is also apparent, however, that many quotations attributed to Watt, are apocryphal.

[79] Helvarg, D. 1994. *The War Against the Greens*. Sierra Club Books, San Francisco. (A second revised edition was published by Johnson Books, Boulder, Colorado, in 2004.)

[80] Rowell, A. 1996. *Green Backlash: Global Subversion of the Environment Movement*. Routledge, London and New York, NY.

[81] Ehrlich, P.R. and A.H. Ehrlich. 1996. *Betrayal of Science and Reason: How Anti-Environmental Rhetoric Threatens our Future*. Island Press/Shearwater Books, Washington, DC/Covelo, CA.

[82] Rowell, A. 1996; also see Lavigne *et al.* 1999.

[83] Center for the Defense of Free Enterprise. 1994. *The Wise Use Address Book. One Thousand Names, Addresses of Activists for Property Rights, Jobs, Communities, and Access to Federal Lands*. A Report by the Center for the Defense of Free Enterprise, Bellevue, WA.

[84] Modified from Lavigne 2002a; Lavigne, D.M. 2004. The Return of Big Brother. BBC Wildlife, May 2004. pp. 66-68.

[85] See Lavigne, D.M. 1991a.

[86] See PEER (Public Employees for Environmental Responsibility). 2001. Wise Use or Abuse? Available at http://www.peer.org/watch/wiseuse/index.php (Accessed 21 September 2005).

[87] I first heard this view expressed at the 1985-CITES Conference of the Parties in Buenos Aires by a delegate from India. He said something to the effect that CITES began as an international convention designed to protect endangered species from the threats imposed by international trade; it is now turning into a convention that facilitates the trade in endangered species.

[88] Lojenga, R.K. and R. Sanchez. 2003. United Nations Conference on Trade and Development. UNCTAD Biotrade Initiative. CITES: Workshop on Economic Incentives and Trade Policy. 1-3 December 2003, Geneva Switzerland. Available at http://www.cites.org/eng/prog/economics/ppt/UNCTAD.pdf.

[89] Now known as "IUCN – the World Conservation Union". See Lavigne 1991a.

[90] Lavigne, D.M. 1991b. Slipping into the marketplace. Special Report. BBC Wildlife, February, pp. 128-129.

[91] IUCN. 2004. IUCN welcomes adoption of sustainable use principles by CBD. Available at http://www.iucn.org/info_and_news/press/cop7sustainableuse.pdf.

[92] Southern Africa Sustainable Use Specialist Group 1996.

[93] IUCN/UNEP/WWF. 1980.

[94] Freese, C. 1996. The commercial, consumptive use of wild species: Managing it for the benefit of biodiversity. A cooperative publication of WWF-US and WWF-International.

[95] Lavigne, D.M. 2002b. In my view. BBC Wildlife, September, pp. 65-66.

[96] This quotation comes from the Globe and Mail (Canada), 7 November 2001, reporting on the publication of the United Nations Population Division. 1991. World Population Prospects: The 2000 revision. Highlights. DRAFT ESA/PWP.165, 28 February 2001. United Nations Population Division, Department of Economic and Social Affairs, United Nations, New York.

[97] e.g. Chesworth, W., M.R. Moss, and V.G. Thomas. 2001. Malthus and sustainability: A codicil. pp. 163-167. In W. Chesworth, M.R. Moss, and V.G. Thomas (eds.). *Malthus and the Third Millennium*. The Kenneth Hammond Lectures on Environment, Energy and Resources. 2000 Series. Faculty of Environmental Sciences, University of Guelph, Guelph, Ontario, Canada.

[98] Beder 1996; Chesworth *et al.* (eds.) 2002, including Lavigne

2002a; also see Beder, Chapter 5; Rees, Chapter 14; Brooks, Chapter 16.

[99] Caccia, C. 2002. The politics of sustainable development. pp. 35-49. In W. Chesworth, M.R. Moss, and V.G. Thomas (eds.). *Sustainable Development: Mandate or Mantra.* The Kenneth Hammond Lectures on Environment, Energy and Resources. 2001 Series. Faculty of Environmental Sciences, University of Guelph, Guelph, Ontario, Canada.

[100] e.g. Robinson, J. 2002. Squaring the circle? On the very idea of sustainable development. pp. 1-34. In W. Chesworth, M.R. Moss, and V.G. Thomas (eds.). *Sustainable Development: Mandate or Mantra.* The Kenneth Hammond Lectures on Environment, Energy and Resources. 2001 Series. Faculty of Environmental Sciences, University of Guelph, Guelph, Ontario, Canada.

[101] Lavigne 2002a.

[102] Donovan, J.C., R.E. Morgan, and C.P. Potholm, 1981. *People, Power & Politics. An Introduction to Political Science.* Addison-Wesley Publishing Company, Reading, UK.

[103] Orwell 1949.

[104] Donovan *et al.* 1981.

[105] Willers 1994.

[106] WWF. 2002. Living Planet Report 2002. Gland, Switzerland: WWF International, Gland. Available at http://www.panda.org.

[107] From Lavigne 2002a.

[108] Dowie 1995.

[109] "Population momentum" refers to the tendency for population growth to continue beyond the time that replacement-level fertility has been achieved because of a relatively high concentration of people in the childbearing years (http://www.uwmc.uwc.edu/geography/Demotrans/demodef.htm). In other words, the human population is expected to increase in the coming decades simply because of the number of people already born who will enter their child-bearing years.

[110] United Nations. 1999. The World at Six Billion. Population Division Department of Economic and Social Affairs United Nations Secretariat. ESA/P/WP.154. 12 October 1999. Available at http://www.un.org/esa/population/publications/sixbillion/sixbilcover.pdf.

[111] Wackernagel, M. and W. Rees. 1996. *Our Ecological Footprint. Reducing Human Impact on the Earth.* Gabriola Island, BC, Canada: New Society Publishers.

[112] Rees, W.E. 2002. Footprint: Our impact on Earth is getting heavier. Nature 420:267-268; Wackernagel, M., L. Onisto, P. Bello, A. Callejah Linares, I.S. López Falfán, J. Méndez Garcia, A.I. Suárez Guerrero, and Ma. G. Suárez Guerrero. 1999. National natural capital accounting with the ecological footprint concept. Ecological Economics 29:375-390; Wackernagel, M., N.B. Schulz, D. Deumling, A.C. Linares, M. Jenkins, V. Kapos, C. Monfreda, J. Loh, N. Myers, R. Norgaard, and J. Randers. 2002a. Tracking the ecological overshoot of the human economy. PNAS Vol. 99, No. 14 9266-9271. Available at http://www.pnas.org/cgi/doi/10.1073/pnas.142033699; Wackernagel, M., C. Monfreda, and D. Deumling. 2002b. Ecological footprints of Nations. November 2002 Update. How much nature do they use? How much nature do they have? Redefining Progress for People, Nature, and the Economy. Sustainability Issue Brief. November 2002. 14 pp. Available at http://www.RedefiningProgress.org; WWF. 2002. Living Planet Report 2002. Gland, Switzerland: WWF International, Gland. Available at http://www.panda.org.

[113] United Nations. 2003. World population totals for 1980 – 2050, according to the United Nations revisions of world population estimates and projections. (Medium-fertility variant projections.) United Nations Department of Economic and Social Affairs, Population Division. Available at http://www.un.org/esa/population/publications/longrange2/worldpoptotals.doc.

[114] Meadows, D., J. Randers, and D. Meadows. 2004. *Limits to Growth. The 30-year Update.* Chelsea Green Publishing Company, White River Junction, VT.

[115] Pimm, S.L. 2001. *The World According to Pimm.* A Scientist Audits the Earth. McGraw-Hill, New York, NY.

[116] Jackson, W. 2004. Agriculture: The Primary Environmental Challenge of the 21st Century. pp. 85-99. In W. Chesworth, M.R. Moss, and V.G. Thomas (eds.). *The Human Ecological Footprint.* The Kenneth Hammond Lectures on Environment, Energy and Resources. 2002 Series. Faculty of Environmental Sciences, University of Guelph, Guelph, Ontario, Canada.

[117] FAO. 2002. The State of World Fisheries and Agriculture. Food and Agricultural Organization of the United Nations. Rome, Italy. Available at http://www.fao.org/sof/sofia/index_en.htm; Pauly, D., V. Christensen, S.Guenette, T.J. Pitcher, U.R. Sumaila, C.J. Walters, R. Watson, and D. Zeller. 2002. Toward sustainability in world fisheries. Nature 418:689-695.

[118] Leakey, R. and R. Lewin. 1996. *The Sixth Extinction: Patterns of Life and the Future of Humankind.* Anchor Books, New York; Eldredge, N. 2001. *The Sixth Extinction.* American Institute of Biological Sciences. Available at http://www.actionbioscience.org/newfrontiers/eldredge2.html; Ward, P. 2004. The father of all mass extinctions. Conservation In Practice 5(3):12-19.

[119] But see Geist, Chapter 19.

[120] Roberts, P. 2004. *The end of oil. On the Edge of a Perilous New World.* Houghton Mifflin Company, Boston, MA.

[121] IPCC. 2001. *Climate Change 2001: The Scientific Basis.* Contribution of Working Group I to the Third Assessment Report of the Intergovernmental Panel on Climate Change. J.T. Houghton, Y. Ding, D.J. Griggs, M. Noguer, P.J. van der Linden, X. Dai, K. Maskell, and C.A. Johnson (eds.). Cambridge University Press, Cambridge U.K. and New York, N.Y.

[122] Elliot, J. 2001. The West knows now there is no wall to hide behind. The Guardian. 13 November 2001. Available at http://www.guardian.co.uk/waronterror/story/0,1361,592449,00.html.

[123] Mittal, A. 2000. Enough food for the whole world. Washington Post, Friday, 15 September, p. A26; Pimentel, D. 2004. Food Production and Modern Agriculture. In From the Ground Up. The Importance of Soil in Sustaining Civilization. Symposium. AAAS Annual Meeting. Seattle, WA. 12-16 February. (Abstract).

[124] Here and elsewhere in this book, I am using Stephen Kellert's typology of human attitudes and values (see Lavigne *et al.* 1999); also see Menon and Lavigne, Chapter 12.

[125] A good example is Hoyt, J.A. 1994. *Animals in Peril. How "sustainable use" is wiping out the world's wildlife.* Avery Publishing Group, Garden City Park, NY. For additional discussion, see Lavigne *et al.* 1996.

[126] Ellsworth, B. 2004. Saving a Species: Can Profit Make the Caged Bird Sing? New York Times, Section F; Column 2; Science Desk; p. 2, 28 December 2004.

[127] Geist, V. 1995. North American Policies of Wildlife Conservation. pp. 77-129. In V. Geist and I. McTaggart-Cowan (eds.). *Wildlife Conservation Policy. A Reader.* Detselig Enterprises Ltd., Calgary, Alberta, Canada; Lavigne *et al.* 1996. Table 1–3 is from research conducted by Sheryl Fink. Sources for the quotations are, in chronological order: Hornaday, W.T. 1913. *Our Vanishing Wildlife.* New York Zoological Society, New York, NY; Vreeland, F.K. 1916. Prohibition of the sale of game. *Conservation of Fish, Birds and Game.* Committee of Fisheries, Game, and Fur-bearing Animals. Commission of Conservation Canada. Proceedings of a meeting of the Committee, November 1 and 2, 1915. The Methodist Book and Publishing House, Toronto; Leopold, A. 1991. Wild Lifers vs. Game Farmers: A Plea for Democracy in Sport [1919]. pp. 62-67. In S.L. Flader and J. B. Callicott. *The River of the Mother of God and Other Essays by Aldo Leopold.* The University of Wisconsin Press, Madison, WI; Hewitt, C.G. 1921; Grinnell, G.B. 1925. American Game Protection. pp. 201-257. In G.B. Grinnell, G. Bird, and C. Sheldon (eds.). *Hunting and Conservation: The Book of the Boone and Crockett Club.* Yale University Press, New Haven, CT; Trefethen, J.B. 1966. Wildlife regulation and restoration. pp. 22-28. In H. Clepper (ed.). *Origins of American Conservation.* The Ronald Book Company, New York, NY; Geist, V. 1988.; Norse, E.A. (ed.) 1993. *Global Marine Biological Diversity.* Island Press, Washington, DC; Talbot, L.M. 1996. Living Resource Conservation: An International Overview. Marine Mammal Commission, Washington, DC; Tuxill, J. 1998. Living strands in the web of life: Vertebrate declines and the conservation of biological diversity. Worldwatch Paper 141. Worldwatch Institute, Washington, DC.

[128] Shaw, J.H. 1985. Table 4 is modified from Lavigne *et al.* 1999, after Shaw, 1985.

[129] Lavigne *et al.* 1999.

[130] Meffe, G.K. and C.R. Carroll. 1994. *Principles of Conservation Biology.* Sinauer Associates, Sunderland, MA.

[131] This idea remains extant. As this chapter was about to go to press, a special issue of *Scientific American* entitled "*Crossroads* for Planet Earth" arrived on the newsstands. The cover carried the message that "The human race is at a unique *turning point.*" (emphasis added; see Scientific American, Special Issue. September 2005).

[132] Saul, J.R. 2005. *The Collapse of Globalism and The Reinvention of the World.* Viking Canada, Toronto.

[133] Ibid.

[134] e.g. Lavigne *et al.* 1996.

[135] Lavigne 2004.

[136] The September 2005 issue of Scientific American referred to in endnote 131 also asked: "Will we *choose* to create the best of all possible worlds?" (emphasis added).

There once was a theory of wise use
For whose logic there is no excuse
In its perversity
It'll kill biodiversity
And turn the planet into refuse.

Chris Tuite 2004

PART I:

THE GLOBAL CONTEXT

Chapter 2

Conserving Biodiversity: Why We Should and How We Can

Martin Willison

Almost everyone has a sense that nature is precious and beautiful. On a summer's day, children sit on the ground and wonder at the ants and beetles that wander about, while their parents stare at the sky and marvel at a hawk in pursuit of prey. Whether we see this diversity of life as a creation of natural evolution, or as God's special creation, we are bound to be struck by its wonderful richness. Furthermore, we know that in order for it to hang together as a whole it must be rich and diverse. An ecosystem consisting only of wheat, cows and humans could not exist.

For all that we may not understand the functional role of a strange-looking beetle, or even know that a slippery protozoan exists, we are intuitively aware that these are necessary for ecosystems to function. While some parts of the life-system may be disposable in theory, others are not, and we have no idea which parts are technically redundant. Furthermore, we have no moral ground for regarding any part as redundant. After all, if there is such a thing as an ecosystem part that is not functionally necessary, then why should that unnecessary part not be the human? We do not consider humans to be disposable, and so we should also not consider any other species or sub-species to be disposable.

While it is easy to conclude that we should care about biological diversity and strive to maintain it, it is less easy to know how to achieve its conservation. We are inextricably a part of nature. We cannot stand apart from nature and "preserve" it like a pickle in a jar – because we are also in the jar. This is the meaning of conservation – living along with something, and using it but not destroying it. The very fact that humans exist as a part of Earth's living system means that we use it all in one way or another – for its integrity keeps us in existence. Indeed, the wisest way to use Earth's life system is to use it as sparingly as possible, for its integrity is currently in jeopardy as shown by the rapid and fundamental changes that are currently occurring as a result of human actions in the atmosphere, biosphere, hydrosphere, and even the superficial layers of the geosphere.

BIOLOGICAL DIVERSITY

We describe the enormous variety of life on the Earth by the term "biological diversity", which is commonly rendered as "biodiversity". Many books have been written on this subject, of which Edward O. Wilson's "The Diversity of Life" is perhaps the most readable.[1] At the introductory academic level, an excellent general reference has been provided by Gaston and Spicer.[2]

Biodiversity is a familiar concept for biologists. It is both a description and a measure of the richness of life and is considered to exist at three main levels of the attributes of life: genes, taxa, and ecosystems. Genetic diversity consists essentially of variations in the DNA molecules that define biological inheritance. Taxonomic diversity refers to variations among individual life forms, notably among species, but it also includes variations that we recognize at higher levels of taxonomic organization. Ecosystem diversity refers to variations among communities of living organisms, including their non-living habitats.

The foundation of genetic diversity is the "gene", which is a unique molecular combination that acts as a blueprint for a protein, one of life's molecular building blocks. Most genes exist in several different forms, and some exist in many different forms. Each individual member of a complex species like the human, or the red oak tree, has thousands of different genes, and each individual's set of genes is unique. Even among organismal clones, such as varieties of strawberries, potatoes, and other plants that reproduce vegetatively, there is a small amount of variation from one individual to another due to the phenomenon of somatic mutation. Sexual reproduction always produces a unique genetic combination, and thus contributes to the enormous pool of life's genetic diversity. The differences among related organisms are small, while those among races within species are greater; those among species are greater still, and so on as we consider higher and higher taxonomic levels.

Humans have an orderly view of nature, and so we have produced taxonomies of all the types of organisms that exist on Earth. A taxonomy is a hierarchical description of the varieties of living organisms. The most generally accepted scientific taxonomy is that due to the Swedish scientist, Carl Linnaeus (born 1707), and his followers. This taxonomy is strictly hierarchical and is described as the "Linnaean system". The foundation of this system is the "species", being usually a set of individuals who freely interbreed and who have features that are recognizably similar. The genetic basis for this similarity is called the "genotype", and the outward appearance is called the "phenotype". Similar species are grouped as a Genus (plural: genera), similar genera are grouped as a Family; similar families are grouped as an Order, and so on until we get to the ultimate level consisting of the "Kingdoms" (usually: plants, animals, fungi, protists, and bacteria). Each level in the taxonomy is called a "taxon" (plural "taxa"). Millions of different species are currently in existence.

Ecosystem, or ecological, diversity is very complex and its units are not readily defined, despite it being the most familiar of the forms of biological diversity. Oceans clearly differ from land. The boundary between a pond and its neighbouring forest is similarly clear, although defining the precise boundary would create anxiety in a graduate student. Sand dunes covered with marram grasses grade delicately into damp slacks, lying between the dunes, with a quite different set of plants. The community of organisms living in a wet forest in Brazil is almost entirely different from the community living in a similarly wet forest in Canada. In each of these cases, the different ecosystems are founded upon physico-chemical distinctions, and these create the conditions for distinctive communities of plants, animals, and micro-organisms.

Taxonomic diversity is often used as a proxy for biological diversity as a whole because it captures to some extent both genetic diversity and ecological diversity.

EVOLUTION AND EXTINCTION

The great richness and complexity of life provides the foundation of Earth's life system and the foundation for all human enterprises. Without it, human existence is impossible. This biological richness did not spring into existence in an instant but has evolved in fits and starts since the dawn of microbial life over three billion years ago. Animals, as we currently recognize them, began their evolution about 550 million years ago at the beginning of the Paleozoic Era.[3]

Ever since the dawn of life, there has been a general tendency for biological diversity to increase with the passage of evolutionary time. This tendency for biological diversity to enrich itself arose largely because each living system builds on the complexity that pre-existing living systems had created. Thus, as new organisms evolved to fill new niches, they created yet more niches to be filled. For example, the evolution of the tree created living niches for epiphytes and for arboreal animals, and these created yet more niches for new organisms.

There has never been a time when all the life forms survived for very long. The general pattern has been that evolution and extinction have gone hand in hand, with a slight general tendency for diversity to increase with time, albeit punctuated by periods of relatively rapid loss or gain. This is revealed by the synthesis of findings of painstaking research on the fossilized shells and other hard parts of marine animals conducted by palaeontologists, notably John Sepkoski and David Raup,[4] as shown in Figure 2–1.

It is estimated that currently existing species of plants and animals represent between 2% and 4% of all the species of plants and animals that have ever existed, thus the vast majority of all such species that have existed in the entire Phanerozoic period have become extinct. Figure 2–1 shows several periods of mass extinctions, of which there are five great ones at roughly 65, 205, 240, 370 and 435 million years ago. It is thought that about 5-10% of all extinctions occurred in these massive spasms of extinction.[5]

The explanation for these past extinction events is that Earth was probably assaulted by large meteorites or comets. It is thought that the impacts forced huge amounts of dust and gas into the atmosphere, and the resultant environmental changes led to the extinction spasms. If we think of Earth as a living being, as in the Gaian view of the world,[6] this is the equivalent of being

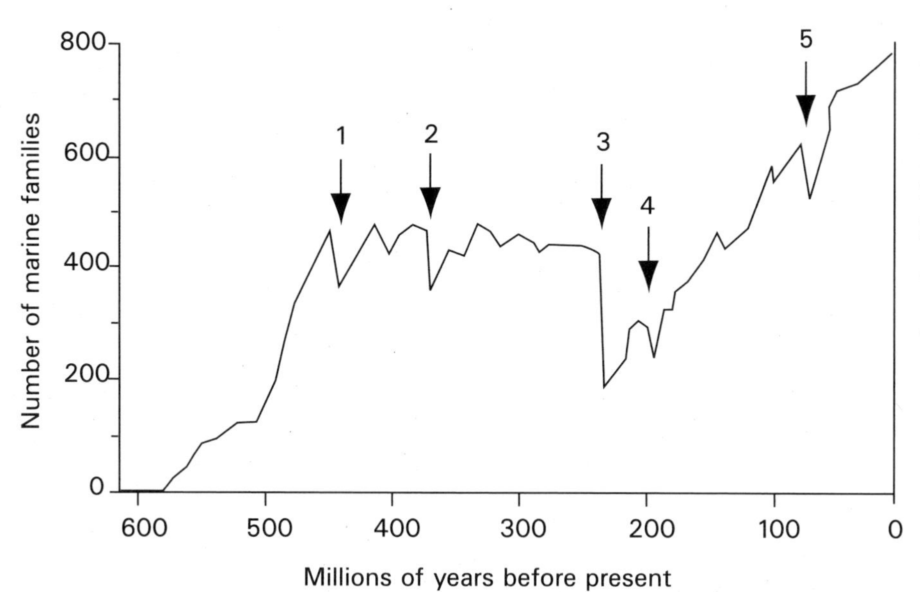

Figure 2–1.

Variation in the diversity of marine animal families with time since the origin of large complex marine animals about 550 million years ago (modified from May *et al.*, 1995; after Sepkoski 1992). The entire period illustrated is called the Phanerozoic. Numbers 1 to 5 identify five major mass extinction events.

injured by a bullet. After each such event, the life system of Earth recovered from the injury, but it took millions of years to do so. Healing created numerous new niches, and as a result a new arrangement of life arose each time there was recovery from a mass extinction event. That which occurred 65 million years ago was perhaps sufficient to permit mammals to escape from their domination by reptiles, leading to the emergence of larger mammalian forms, including primates such as humans.

Many conservation scientists think that the sixth great extinction event is currently underway.[7] By contrast with extinction events caused by an agent external to the Earth, in the Gaian view of the world the current extinction event is an attack from within. Today's problem is the equivalent of illness rather than injury, and the enterprising human is the biological agent of that illness.

Some scientists have suggested that the belief that the Earth exists for human taking underpins the enterprises we conduct and that threaten the life system of Earth. This drive to prosper is probably not different, however, from the biological forces that drive any other organism to prosper. If there is a difference it is that the resource-use technologies modern humans have developed are sufficient for us to take with ease whatever natural resources we want. The recent evolution of human behaviour has been extremely rapid, particularly during the Industrial Revolution of the last 200 years. As is widely recognized, the situation is unlikely to be sustainable.

To counter this rapacious tendency, conservation biologists and others concerned about the future of life on Earth are developing improved methods of conserving both natural resources and biological diversity. In addition, the conservation community is finding appropriate social systems for the adoption of these new conservation measures. Such a new conservation approach may be critical not only for much of Earth's biological diversity, but also for the long-term survival of the human species.

EARTH'S PULSE

As noted above, the rate of extinction is a basic vital sign for planet Earth. Recovery of biodiversity after each major meteorite strike appears to have taken millions of years. There is no reason to expect that the current human-caused extinction spasm will be significantly different, except perhaps that we are beginning to adopt strategies for the maintenance of "endangered species" at minimal population levels, a last-ditch strategy that could eke out existence for a greater range of biological diversity than would be the case in the absence of this approach.[8] The sheer number of humans, our consumptiveness, and our carelessness are in combination a powerful brew of threats to biodiversity.[9]

Recent research has shown that about 30% of global net primary production (NPP) now supports human enterprises on Earth, leaving only about 70% for other species. This estimate does not include the roughly 20% of NPP that has been lost due to degradation of productivity by conversion of lands to fields,[10] highways, towns and other domestication projects. In densely populated regions of the world, such as south Asia and Europe, 70% of NPP is appropriated for human use leaving very little for other species. Since biological diversity has strong regional differences, this has led to very high levels of threat to biological richness that is endemic to regions with very dense human populations.[11] This finding is consistent with ecological footprint analyses that indicate that Earth's carrying capacity for modern humans has already been exceeded.[12]

No single reliable estimate of the current extinction rate can be reasonably made. About 1.8 million species have been given scientific names to date, and there are more species to name than have been named to date. Estimates of the total number of living species vary from 3.6 million to 112 million.[13] Only 0.1% of the scientifically named species have definitely gone extinct since 1600, when naming began, but this number is seriously misleading as an indication of the actual extinction rate for many reasons. For example, rare species are less likely to be named than common species, but are more vulnerable to extinction. Similarly, ecosystems that have been destroyed are not available for biological surveying, and so their endemic biodiversity was probably not described either. In addition, conservation efforts are focused on what we know about, not on what we have never seen, and so a large portion of biological diversity is vulnerable to ignorance.

More realistic estimates of extinction rates can be obtained by considering those groups of organisms that we know particularly well. Birds, for example, fascinate humans. They are remarkably visible when in flight, and flight itself is a skill we envy. Virtually every living bird species has been named and studied, and much effort has gone into studying the skeletons of birds found in caves, middens, and other places where bones are preserved naturally. Richard Holdaway, an expert on extinct birds of the southern hemisphere, estimates that about a quarter of the bird species of 4000 years ago are now non-existent, mostly as a result of depredation on remote Pacific islands where evolution of small founding populations led to local endemism, and then to extinction when the isolation barrier broke down.[14]

Another way to illustrate the quickening pace of extinction is to estimate the average span of existence of a species, as discussed in detail by May and colleagues.[15] Fossil records indicate that an average species has lasted for about 1 to 10 million years before it has succumbed to extinction. For those birds and mammals in the recent historical period for which clear records have been kept, the span of existence is some 100 to 1000 times shorter, at about 10,000 years. If, however, we calculate probable extinctions based on the biogeographical principle of species-area relationships, average contemporary species-existence spans of 200-500 years have been estimated. Based on risk estimates for particularly vulnerable groups at the present time, such as palm trees, the estimated average future existence span of a species in the vulnerable group falls as low as 50 years.

Extinction rates of this magnitude may seem a little hard to believe, and so it is worth examining a specific well-understood case more carefully. Native Hawaiian land birds are such a case, albeit relatively extreme. The following data are derived from records kept by the Bishop Museum in Hawaii.

The first humans arrived in Hawaii about 1300 years ago. At that time, at least 110 land bird species were breeding on the Hawaiian islands, almost all of which were to be found nowhere else in the world. Shortly after the arrival of Polynesian people, easily hunted birds began to go extinct. To date about 35 of these lost species have been identified, but it is thought that there were several others.

Captain Cook was the first European sailor to visit Hawaii, arriving in 1778. Since Cook's arrival, 23 or 24 additional native bird species have gone extinct. Of the 51 native species that remain, 30 are listed as threatened or endangered. If these also become extinct, over 80% of Hawaii's native bird species will have been lost since humans arrived there.

In order to be able to develop conservation technologies, we need to know why extinction occurs. In the case of birds, the major causes are well known thanks to painstaking research by William King who analyzed the world's most recent 175 bird extinctions.[16] Roughly 93% of these occurred on oceanic islands. The three most significant causes of extinction are: depredation by exotic species, habitat disturbance, and hunting by humans. The importance of each of these differs considerably between continental regions and oceanic islands. Invasion of exotic species is particularly serious in islands. Only one of the 175 extinctions was not caused by humans, but by a hurricane.

We can't now do anything about species that have already gone extinct, but data on species at risk of extinction are kept by many nations so that protective actions can be taken. National data are brought together and assessed collectively by the World Conservation Union (IUCN) and kept in their Red List of Threatened Species (Table 2–1).[17]

For the well-evaluated groups, birds and mammals, 12 to 24% of species are currently at risk of extinction, not counting those that have already been lost. For both plants and invertebrates there has probably been a bias towards evaluating the conservation status of those species that may be vulnerable, which accounts for the apparently high proportion that are judged to be at risk of extinction (58% for invertebrates and 70% for plants). Only an extremely small proportion of the world's invertebrate species have been evaluated at all, which can be fairly attributed to lack of caring. Vertebrate animals are often described as being "charismatic", and so they have received more attention than invertebrate animals such as insects, worms, and snails.

Ocean ecosystems are particularly rich in invertebrate animals, and most of these are completely unknown and

Table 2–1. Species at risk as of January 2004.

Taxon	Described species in the world	Number evaluated by IUCN to 2003	Number judged at risk in 2003	% at risk among those evaluated
Mammals	4,842	4,789	1,130	24%
Birds	9,932	9,932	1,194	12%
All vertebrates	56,586	17,127	3,524	21%
All invertebrates	1,190,200	3,382	1,959	58%
All plants and relatives	297,655	9,708	6,776	70%

mysterious. A good example of the impact on biological diversity of this level of ignorance has been the treatment of corals. While most people romantically imagine that coral reefs are found only in the tropics around palm-fringed islands, the truth is quite different. Most corals, in terms of both abundance and diversity, are found in deep cold oceans throughout the world.

Reefs of the stony coral, *Lophelia pertusa,* serve as a good illustration of cold water corals. Although the species was described by Linnaeus in 1758, almost no attention was paid to its huge reefs until the 1980s. This is despite the fact that *Lophelia* is common around the European margin at depths of around 300 metres beneath the ocean surface. In the case of Canada, although a few specimens of this coral had been recovered from the Atlantic Ocean in the early 1900s, its presence was largely ignored until the fragmentary remnants of a living reef was discovered using a submersible camera in 2002. This reef is the only one of its type known in Canadian waters. It had obviously been quite extensive at one time, but by 2002 it had been almost completely destroyed by aggressive fishing using mobile trawls. An estimated 95% of the living reef had been converted to dead rubble.[18]

Living *Lophelia* reefs provide rich habitat for fish, and it is impossible that the fishermen who damaged this reef did not know that it was present and was being damaged by their fishing gears. Bits of the coral and many other strange invertebrate organisms would always have come up in their nets among the fish. Within months of the announcement that this reef had been found, and following a public outcry, fishing on the reef was prohibited. Had some care been taken earlier by marine scientists and fishery regulators, this unfortunate loss could have been largely averted.

CONSERVATION ETHICS

We can only properly address the tragic situation of declining biological diversity with fundamental change. While it is both natural and desirable to use living resources, almost no one considers it sensible to use them unsustainably, even while doing so. Rather than seeing life as primarily an inexhaustible resource, unlikely to be threatened by our actions, we therefore need to regard it as precious regardless of whether the life force is vested in a human or some other living being. Many conservation scientists have, through simple logic like this, concluded that the answer to the moral conundrum of using while protecting living resources lies in developing an ecocentric conservation ethic. Ethics of this kind have been espoused by several philosophers and biologists, such as Stan Rowe and Ted Mosquin in "ecospherics" and the Manifesto for Earth[19] (Table 2–2), and also by deep ecologists like Arne Naess.[20]

Table 2–2. Principles of the "Manifesto for Earth".

CORE PRINCIPLES

1. The ecosphere is the centre of value for humanity.
2. The creativity and productivity of Earth's ecosystems depend on their integrity.
3. The Earth-centered worldview is supported by natural history.
4. Ecocentric ethics are grounded in awareness of our place in nature.
5. An ecocentric worldview values diversity of ecosystems and cultures.
6. Ecocentric ethics support social justice.

ACTION PRINCIPLES

7. Defend and preserve Earth's creative potential.
8. Reduce human population size.
9. Reduce human consumption of Earth parts.
10. Promote ecocentric governance.
11. Spread the message.

A rational way to explain the critical importance of environmental ethics becomes apparent if we consider the simple idea that the impact of humans on Earth can be quantified in the form ***I* = *P·A***, in which ***I*** is Impact, ***P*** is Population and ***A*** is Affluence, or consumption of materials.[21] Some environmentalists argue that if this simple relationship was properly understood then people might curb their desires to reproduce and consume materials. Others have since drawn attention to the simplistic nature of this formula, and the critique can be summarized as follows.

If there are more people [***P***], there will be more impact. If the people are more affluent [***A***], they consume more, and they will have more impact.

So, ***I*** increases as ***P·A*** increases. But the manner in which we spend our affluence can change the impact. We can be wasteful, or cautious (think of recycling, and environmental impact assessments, and "sustainable technology"). Let's call this technology, **T**.

So, ***I*** increases with ***P·A***, but is moderated by ***T***. If we assume that ***P*** and ***A*** are always greater than one, then ***T*** can be smaller than, or greater than, one. Thus ***I*** is a function of ***P·A·T***.

But let's imagine a world full of wonderful technology, but careless people. Then think of a world full of wonderful technology and careful people. Clearly the outcomes are quite different. Caring is an ethical matter; let's call it ***E***. If we add ***E*** to the equation it becomes: ***I*** is a function of ***P·A·T·E*** (or in simple form, ***I* = *PATE***).

Technology is the knowledge of how to do something, but ethics affects whether we will do the thing that we know how to do. Take contraception for example. This is a technology (***T***) that reduces fertility (***P***), but it is used under circumstances that are regulated by wealth (***A***) and ethics (***E***). Similarly, buying wood products that will have a lower impact on biodiversity by being Forest Stewardship Council certified requires knowledge of the Council (***T***), and a desire to do it (***E***). The ecocentric lifestyle choice of "voluntary simplicity", in which people choose for ethical reasons to live simple lives with minimal consumption of materials is the most direct illustration of the influence of ***E*** on ***A***.

Population is simple to quantify: 5 billion, 6 billion, 7 billion. Money measures affluence adequately, but technology and ethics are relatively difficult to quantify, even though we know they exist as independent terms. What might we use as measures?

For technology we can try to count those changes that are having an effect. For example, we can measure things like the rate of emission of defined pollutants per person, and rate of deforestation per person. By defining several measurable parameters, and combining these, a crude measure could be obtained.

For ethics, we might use laws and policies as a guide. For example, Canada's recent passage of the *Species At Risk Act* shows an ethical shift. This came decades after a similar shift in the United States. On the other hand, Canada has ratified the Kyoto Accord on addressing carbon dioxide emissions, while the government of the United States has stated that it will not do so. Similarly, both Canada and the United States have, in the last few years, developed laws and policies related to the creation of marine protected areas. Indicators like these are quantifiable in principle, although it would be difficult to find standards that would be agreeable to everyone.

Is the formula useful even if we can't do a good job of quantifying the terms in it? Perhaps someone will argue that "There's nothing we can do about environmental problems; it's too complex and people aren't going to change anyway." The ***I* = *PATE*** formula provides four answers: we can change population growth rate; we can change our rates of material consumption; we can change our technology; and we can change our ethics (laws, behaviour, politics, and so on). Any change in one of these terms changes the value of ***I***, and combinations of small changes in each term have compound effects.

The common sense arguments I have made here are similar in many ways to those made independently by Schulze[22] who has proposed modifying the formula to ***I* = *PBAT***, where ***B*** is "behaviour".

APPLIED ETHICS: CONSERVATION BIOLOGY AND ECOLOGICAL ECONOMICS

Conservation biology is a science that provides concepts and techniques for protecting biological diversity by considering all three of its levels: genes, species, and ecosystems. In applying their knowledge, conservation biologists seek ways to avoid extinctions. They recognize that a rational approach to protecting biological diversity requires knowledge and actions that go well beyond the generally accepted boundaries of "biology" itself. Michael Soulé is the founder of the discipline, and his leadership is well illustrated by his strategic framework for choosing the right conservation tactic.[23] He argues that in order to minimize the loss of biodiversity, we need to be able to choose among an armoury of conservation tactics and to select the most appropriate tactic for the local situation.

Soulé reduces the conservation choices to "eight paths to biotic survival": *in situ* reserves (wilderness areas, national parks and nature reserves); *inter situ* projects (where native species and their natural systems are present, but outside reserves); extractive reserves (where biotic resources are harvested in a sustainable manner in a wild setting); ecological restoration projects, of which there are a large variety; zooparks (where local and exotic species are mixed and managed in semi-natural environments); agroecosystems and agroforestry projects (in which resource use is modified so as to take account of ecological principles, including biodiversity conservation); living *ex situ* projects (such as zoos and botanic gardens); and suspended *ex situ* programs (such as seed banks and other "germ banks").[24] He also provides strategic advice for making rational choices among these by taking account of socio-economic and political factors.

Protected Natural Areas, the Precautionary Approach

The most simple and direct response to declining biological diversity is to protect all three of its levels within protected natural areas, such as national parks, nature reserves and protected wilderness areas. The term "protected natural area" has been variously defined in legislation around the world. Some of the official definitions are clumsy, but this one from the New Zealand National Parks Act captures the idea well:

> A legally protected area, characterised by indigenous species or ecosystems or landscape features, in which the principal purpose of management is retention of the natural state.

The values people place in protected areas have changed over time. When the first national parks were established in Canada, the focus was on the protection of direct economic values. Banff National Park was developed in 1887 as a tourist destination, with protection put in place so that the scenic values would be retained, and tourists would continue to use the railroad to visit a beautiful place. The importance of such places grew as visitors appreciated the value of passive recreation and spiritual renewal offered by protected places.

The list of uses for protected areas has grown since these early beginnings. They have been seen as economic engines, recreational space, conservation centres, and part of the global environmental research infrastructure. In order to achieve all of these objectives it has become increasingly important to expand the amount of protected space and also to rationalize the planning of protected areas systems.[25]

Protected areas are classified by the World Conservation Union in six main categories,[26] the first of which is subdivided. Only the first two classes are very significant for conserving biological diversity, although all have at least some functional importance. The first two classes are: strictly protected nature reserves (Class 1a), wilderness areas (Class 1b), and national parks and their equivalents that are retained in a natural state (Class 2).

The pace of human development, including the growth in human numbers, has been such that the protection of natural values has generally been left as an afterthought in land-use planning. Today, we establish protected areas as islands (pockets) of wilderness in a sea of developed land. Even in a good situation, conservation planners are often severely constrained in the choices they can make as they seek to protect representative fragments of the wild environment, despite the fact that this is probably the most important thing that could possibly be done by a modern municipal or regional land-use planner.

It is reasonable to argue that if humans fully co-existed with other species, there would be islands of exploited land, connected by narrow transportation corridors. This ideal option is, however, now available only in remote parts of the world.

How Much Land Should we Protect?

In 1987, the World Commission on Environment and Development, or Brundtland Commission, set a target of 12% of the area of any jurisdiction to be set aside as protected areas. This target has been adopted by international nature conservation organizations, such as World Wide Fund for Nature (WWF), which advocate the protection of nature and biodiversity. The target is arbitrary, however, and is regarded as inadequate by conservation biologists. It was set as a matter of political expediency only. Soulé and Sanjayan state, for example, "In the few detailed studies available, the typical estimate of the land area needed to represent and protect most elements of biodiversity, including wide-ranging animal species, is about 50%".[27]

In many parts of the world where human populations are dense the option to protect half the land as parks and wilderness is not available. Wilderness values have already been too severely compromised. As a result, protected area planners battle to find compromises, such as in plans for core protected areas, buffer zones and connecting corridors[28] that are arranged so to maximize the representation of a region's ecological diversity in the protected areas system. Finding compromises of this sort reaches its highest level of sophistication in Biosphere Reserves, a concept promoted in the Man and the Biosphere Programme (UNESCO-MAB) in which community-based sustainable economic development is combined in a fundamental way with the conservation of nature.

Ecological Economics

Recently, some economists have tried to bring economics and ecology together, and have founded a branch of economics called ecological economics.[29] In principle this is not a new field. Thomas Malthus (1766-1834), for example, was a political economist who can be considered an ecological economist. Nevertheless, the new wave of ecological economists is different in that they are building their theories beyond the foundations of modern (neoclassical) economic theory by using modern ecological theory. Ecological economics should not be confused with so-called "sustainable development", which is an attempt to marry neo-classical economics with environmental theory for international development purposes.[30]

Ecological economists consider that standard economic measures of prosperity are misleading, as discussed later in this book by Brian Czech.[31] In order to address the need for measures of progress that are robust, ecological economists have created alternatives, such as the Genuine Progress Index. This has been applied in Atlantic Canada with some measure of political success by a research group called GPI Atlantic, which is a leading group in Canada for innovation in economics.

If we try to count the monetary value of biological diversity, in neo-classical economics it has no value. In contrast, ecological economists have placed an annual value of about $300 billion on the economic and environmental benefits of biodiversity in the United States alone.[32] This contrast provides a good illustration of the difference between neo-classical economic accounting and the "full cost accounting" preferred by ecological economists.

In neo-classical economics the extinction of a species is regarded as an "externality" (i.e. lying outside the economic system). Even if the species has utility value, it is not counted. This is because in conventional economic accounting it is assumed that the lost good will be replaced by another one, of equivalent value, in the process of market substitution. This makes good sense when applied to a manufactured gadget that is replaced by another one having greater utility, but is effective for living organisms in the long term only if the rate of creation of new life forms by evolution is equal to, or greater than, the rate of extinction. In practice extinction greatly exceeds evolution and our biological capital is rapidly depreciating. The fact that conventional economic accounting takes no account of this is a serious flaw that brings one of the foundations of conventional economics into serious question.

Monetary valuation is so effective for commodities and human services that it has become normal to use it as a simple method for valuing everything. It would appear worthwhile therefore to attempt to place a monetary value on an average species in order to assess the lost value when extinction occurs, despite the ethical absurdity of the proposition.

Estimating the value of a species at the time of loss is made difficult by the economic process of substitution. In this process, the lost commodity value of one species is substituted by another species. Thus the commodity value of initial losses appears relatively low, and that of later losses relatively high. This method of accounting therefore distorts the true value of each species and hides the lost value. Thus, a theoretical average should be calculated by using some other method, such as my attempt that follows.

The current human population is about 6 billion, and the average person generates about $6,000 towards gross world product annually. For convenience, let us assume that under optimal circumstances this might continue at the same rate for as long into the future as humans might reasonably expect to exist. I will take this to be 500,000 years, which is a modest expectation for the existence of an ordinary mammalian species.[33] This gives us a future cumulative value of the human economy of $\$18 \times 10^{15}$. It would be much greater than this if we allow for expansion of the economy in the future and for humans living an average species lifetime of about 10 million years.

Biologists estimate that there are between 2 and 100 million living species on Earth, as discussed previously. Let's take this number to be 10 million, which is a commonly used number for practical purposes. The rate of evolution of new species is generally low; let's take it to be 2 per year. The rate of extinction is currently high; at least 10,000 per year according to several estimates. For practical purposes, let us take this rate to be 10,000 per year. At that rate, there will be no species in one thousand years. Let us assume that humans will become extinct at the 50th percentile, or 500 years from now. Under this circumstance, there will be no human economy after 500 years from now, representing a loss in future economic value of about $\$18 \times 10^{15}$.

Each species therefore has an average existence value of about $\$36 \times 10^{8}$, or one tenth the value of the estimated annual gross world product. At the current rate of extinction, their lost value is 10,000 times higher than this, or $\$36 \times 10^{12}$. This is about one thousand times greater than the annual gross world product, and indicates that the current world economy has an enormous annual deficit. In effect, by this measure the world economy should be regarded as running backwards if the rate of extinction exceeds the rate of evolution by a factor greater than ten.

Visioning the Future

Robert Costanza is an influential ecological economist who has examined "the importance of visioning" in our personal philosophy. Regardless of what we do, we will change the world by what we do, so it is better done with a vision in mind.

In a fascinating research report, Costanza argued that people today tend to have a worldview of either technological optimism (technological progress will deal with all future challenges) or technological skepticism (technology is limited and ecological carrying capacity can be exceeded).[34] Having recognized this, he asked: what will happen if the technological optimists are correct, or incorrect, in their prediction; and what will happen if the skeptics predict correctly, or incorrectly?

Costanza approached several selected groups of people with these choices by asking them to rate four scenarios of the future that are based on this two-by-two matrix (optimism and skepticism, correct and incorrect). He then analyzed the outcome using basic game theory in order to recommend a rational path forward. His conclusion was clear – it is safer to be a technological skeptic. In his wise words: "a cooperative, precautionary policy set that assumes limited resources is shown to be the most rational and resilient course in the face of fundamental uncertainty about the limits of technology."

NOTES AND SOURCES

[1] Wilson, E.O. 1992. *The Diversity of Life.* W.W. Norton, New York.

[2] Gaston, K.J. and J.I. Spicer. 1998. *Biodiversity, an introduction.* Blackwell Science, Oxford. 113 pp.

[3] For a recent review, see Jackson, J.B.C. and K.G. Johnson. 2001. Measuring past biodiversity. Science 293: 2401-2404.

[4] See Sepkoski, J.J. Jr. 1992. Phylogenetic and ecologic patterns in the Phanerozoic history of marine biodiversity. pp. 77-100 in: N. Eldredge (ed.). *Systematics, Ecology, and the Biodiversity Crisis.* Columbia University Press, New York.

[5] For detailed reviews, see May, R.M., J.H. Lawton and N.E. Stork. 1995. Assessing extinction rates. pp. 1-24. In J.H. Lawton and R.M. May (eds.). *Extinction Rates.* Oxford University Press, Oxford; Jackson, J.B.C. and K.G. Johnson. 2001. Measuring past biodiversity. Science 293: 2401-2404.

[6] Lovelock, J.E. 1995. *The Ages of Gaia: a biography of our living Earth.* 2nd Edition. Oxford University Press, Oxford.

[7] See for e.g. Leakey, R. and R. Lewin. 1996. *The Sixth Extinction: Patterns of Life and the Future of Humankind.* Anchor Books, New York; Ward, P. 2004. The father of all mass extinctions. Conservation In Practice 5(3): 12-19.

[8] For a review, see Beazley, K. and R. Boardman, (eds.) 2001. *Politics of the Wild, Canada and Endangered Species.* Oxford University Press.

[9] For a review, see Soulé, M.E. 1991. Conservation: tactics for a constant crisis. Science 253: 744-750.

[10] See Chesworth, Chapter 3.

[11] For details, see: Imhoff, M.L., L. Bounoua, T. Ricketts, C. Loucks, R. Harriss, and W.T. Lawrence. 2004. Global patterns in human consumption of net primary production. Nature 429: 870-873; Rojstaczer, S., S.M. Sterling, and N J. Moore. 2001. Human appropriation of photosynthesis products. Science 294: 2549-2552; Vitousek, P. M., H.A. Mooney, J. Lubchenco, and J.M. Melillo. 1997. Human Domination of Earth's Ecosystems. Science 277: 494-499.

[12] Wackernagel, M. and W. Rees. 1996. *Our Ecological Footprint: Reducing Human Impact on the Earth.* New Society Publishers, Gabriola Island, British Columbia, and Philadelphia.

[13] For details, see Lawton. J.H. and R.M. May. 1995. *Extinction Rates.* Oxford University Press, Oxford; Gaston, K.J. and J.I. Spicer. 1998. *Biodiversity, an introduction.* Blackwell Science, Oxford.

[14] Holdaway, R.N. 1989. New Zealand's pre-human avifauna and its vulnerability. New Zealand Journal of Ecology 12 (supplement): 11-25.

[15] May, R.M., J.H. Lawton and N.E. Stork. 1995. Assessing extinction rates. pp. 1-24. In J.H. Lawton and R.M. May (eds.). *Extinction Rates.* Oxford University Press, Oxford.

[16] King, W. B. 1980. Ecological basis of extinction in birds. *Acta Congressus Internationalis Ornithologici* 2: 905-911.

[17] Available at www.redlist.org.

[18] P. Mortensen, personal communication.

[19] Available at www.ecospherics.org.

[20] Sessions, G. (ed.). 1995. *Deep Ecology for the Twenty-First Century.* Shambhala Publications Inc., Boston; also see Foundation for Deep Ecology, www.deepecology.org.

[21] See Holdren, J.P. and P.R. Ehrlich. 1974. Human population and the global environment. American Scientist 62: 282-292.

[22] Schulze, P.C. 2002. I=PAT. Ecological Economics 40: 149-150.

[23] Soulé, M.E. 1991. Conservation: tactics for a constant crisis. Science 253: 744-750.

[24] Soulé, M.E. 1991.

[25] Beazley, K. and R. Boardman (eds.). 2001. *Politics of the Wild, Canada and Endangered Species*, Oxford University Press, Oxford.

[26] See www.iucn.org/themes/wcpa/wcpa/protectedareas.htm.

[27] Soulé, M.E. and M.A. Sanjayan. 1998. Conservation targets: do they help? Science 279: 2060-2061.

[28] For a review, see Noss, R.F. and A.Y. Cooperrider. 1994. *Saving Nature's Legacy, Protecting and Restoring Biodiversity.* Island Press, Washington, D.C.

[29] Costanza, R. 1989. What is ecological economics? Ecological Economics 1: 1-7; also see The International Society for Ecological Economics, www.ecoeco.org.

[30] See for e.g. Beder, S. 1996. *The Nature of Sustainable Development.* Second Edition. Scribe Publications, Newham, Australia; Chesworth, W., M.R. Moss, and V.G. Thomas. (eds.). 2002. *Sustainable Development: Mandate or Mantra.* The Kenneth Hammond Lectures on Environment, Energy and Resources. 2001 Series. Faculty of Environmental Sciences, University of Guelph, Guelph, Canada.

[31] See Czech, Chapter 23.

[32] See Pimentel, D., C. Wilson, C. McCullum, R. Huang, P. Dwen, J. Flack, Q. Tran, T. Saltman, and B. Cliff. 1997. Economic and environmental benefits of biodiversity. Bioscience 47: 747-757.

[33] See May, R.M., J.H. Lawton and N.E. Stork. 1995. Assessing extinction rates. pp. 1-24. in J.H. Lawton and R.M. May (eds.). *Extinction Rates.* Oxford University Press, Oxford.

[34] Costanza, R. 2000. Visions of alternative (unpredictable) futures and their use in policy analysis. Conservation Ecology 4(1): 5. http://www.consecol.org/vol4/iss1/art5.

Chapter 3

The Enemy Within

Ward Chesworth

> *"How many millennia of deforestation, dust storms and soil erosion has it taken for us to realize that our agricultural methodology has had serious flaws in it from the start."*
>
> Angus Martin 1975[1]

Felipe Fernandez-Armesto defines civilization as "a relationship to the natural environment, recrafted by the civilizing impulse, to meet human demands".[2] On the face of it, we have been astonishingly successful in meeting these demands and optimists believe we can continue to do so. Bjorn Lomborg[3] is only the latest to tell us that in 1798 Malthus[4] was wrong to predict a miserable future for humanity as the exponential growth of population outstripped the arithmetic growth of the food supply.

Over the past 10,000 years, keeping the food supply growing has depended mostly on being able to convert an increasing area of soil to agriculture. Now that we have commandeered most of the best soils for this purpose, we are frantically developing chemical and genetic strategies to keep a Malthusian fate at bay for a little while longer. Civilization is completely dependent on soil and the surplus of food the farmer takes from it, and one of the great geological determinants of human settlement and history has been soil quality. Given a reasonable climate, the factor of crucial importance has been the content of plant nutrients in the soil, that is, the inherent fertility of the soil. Inherent fertility is a function of the nature of the parent materials and the length of time that the soil has been developing on them.

As an introductory example, consider the loessial soils of the world (Figure 3–1). Loess is a silt, carried by wind from the heterogeneous scrapings of the recent continental ice sheets, and dumped and redistributed by water in many places over the land surface. The world's great grasslands are largely on the soils that have developed on loess over the last 15,000 years or so. It isn't surprising then, that farmers have almost completely taken over these young, nutrient-rich soils for our "grassy" crops. As far as the good soils of the planet are concerned we have essentially reached the limit in grabbing what is available. We need to take stock now and see (to use Fernandez-Armesto's words) whether our "civilizing impulse" has resulted in a "recrafting" of the biosphere to such a degree that we are a clear and present danger to the web of life. Indeed, some have suggested that we are implicated in a "sixth great extinction" on the planet.[5]

Here, my specific purpose is to show that the growing threat that our use of soil in agriculture presents within the biosphere raises doubts about our ability to sustain the civilization we currently enjoy. In this essay, I will examine the issue from the perspective of soil science, and will introduce the topic by way of a brief account of the Tanzania-Canada Agrogeology Project.[6]

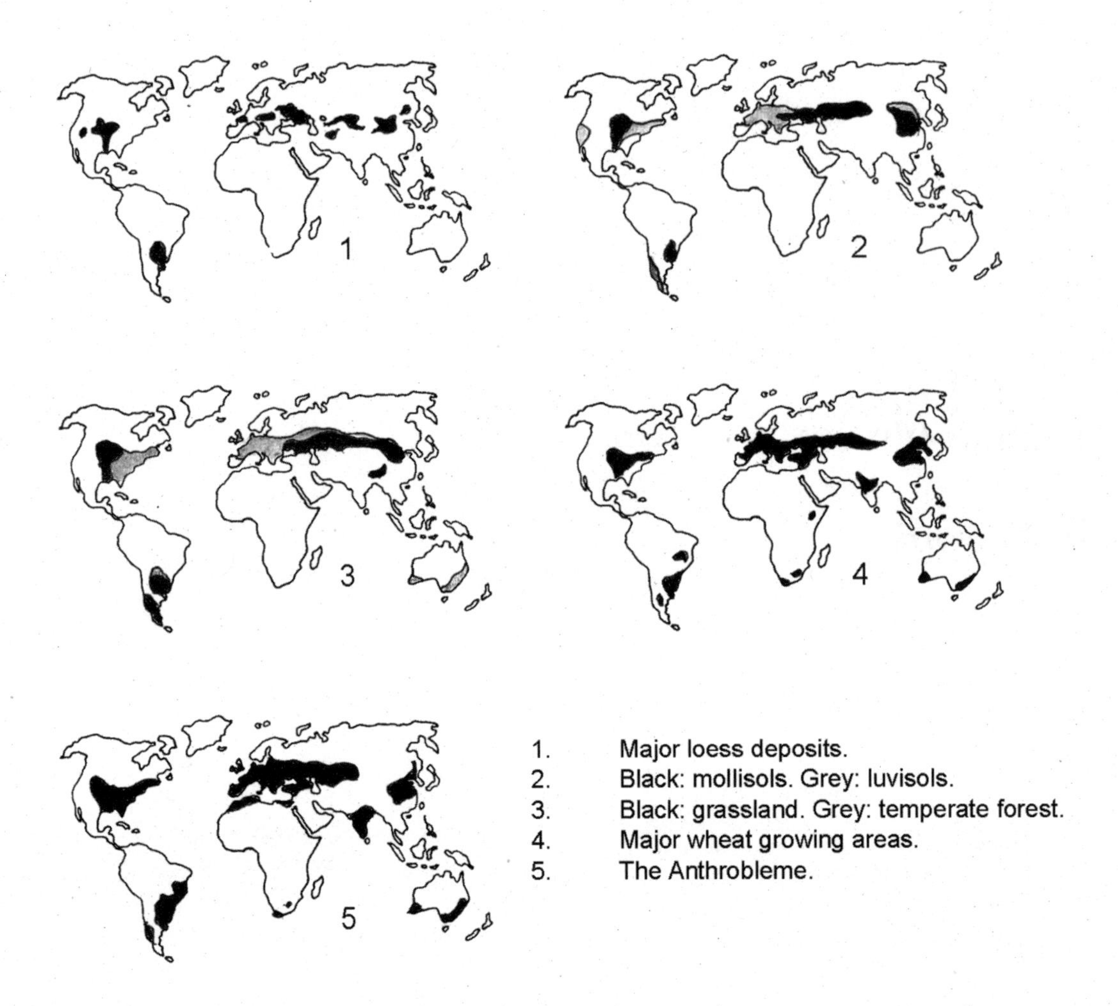

Figure 3–1. From Loess to Anthrobleme.

The parts of the earth's surface that are the most highly modified by human activity are referred to here as the anthrobleme or "human scar", by analogy to astrobleme ("star-scar") the term coined by Bob Dietz to describe the marks left by asteroid impacts. Wes Jackson provided the impulse to modify Dietz's term in this fashion when he introduced the term agrobleme ("agricultural scar"), which he attributes to his colleague Jerry Glover at the Land Institute, Kansas.[43] The extent of the anthrobleme is based on the distribution of land described as "almost fully transformed" in the AAAS Atlas of Population and the Environment.[44] The sequence of diagrams is analogous to a palimpsest – a parchment which has been written on several times with later writings obscuring earlier ones. The earliest inscription is geological, the loess. The next overlay comprises the soils developed on loess: mostly mollisols, but including some alfisols. Next comes a layer of vegetation with mollisols supporting grassland or forest depending mostly on rainfall, and alfisols commonly supporting temperate forest. Map 4 shows the extent to which the soils of Map 3 have been taken over as our major wheat producing areas. Finally the anthrobleme (Map 5), the area modified not only by agriculture, but also by all the human activities that developed from agriculture – urbanization and industry in particular – largely continues the pattern set by the fundamental geology and pedology that predates all human civilization.

THE TANZANIA-CANADA AGROGEOLOGY PROJECT

In the 1980s, Johnson Semoka of Sokoine University and I were coordinators of the Tanzania-Canada Agrogeology Project. It involved a team of geologists, soil scientists and agronomists from both our universities, and from the Geological Survey of Tanzania, as well as the Uyole Field Station in Mbeya District, Tanzania. The International Development Research Centre in Ottawa provided the funds. Our objective was to find local, low-tech ways of maintaining the fertility of soil, and therefore of sustaining the food supply. Josephat Kamasho, in Mbeya District, performed many of the field trials, and I will use the natural features of his area to introduce some fundamental soil science as a prelude to the subsequent discussion of the impact of agriculture on the biosphere.

Mbeya District, lies between Lake Nyasa and the Rukwa Trough. It is the site of a "triple point," a tectonic feature where three geological rifts, or graben, come together. The convergence zone of the rifts has been the site of copious volcanic activity, resulting in the construction of a vast pile of ash, basalt and a few minor rock types, on the much older granitic and metamorphic basement. The extrusive rocks are known collectively as the Rungwe Volcanic Complex, or simply Rungwe Volcano, topographically, the most prominent feature (Figure 3–2). The natural development of the soils of the area follows three lines of evolution. In areas of high precipitation, on well drained materials, a progressive acidification has taken place leading to so-called ferralitic soils. They are best developed on older, granitoid landscapes beyond the edges of the volcano's ash plume. Oxidation of iron to the ferric state gives them a reddish colour in superficial horizons. A second evolutionary path occurs under more arid conditions, where evapotranspiration is greater than rainfall. In this case, alkalinisation produces sodium-rich soils and even surficial salt deposits as an end point. These are especially characteristic of the Rukwa Trough, northwest of Rungwe. The third line of development is found where the soil has become waterlogged and reduced. Under these hydromorphic conditions, iron is changed from the immobile ferric, to the mobile ferrous state. As a consequence, it may be flushed from the soil, which thereby acquires a subdued grey to light grey colour. Hydromorphic soils are best developed in a small area south of Rungwe, bordering the northern tip of Lake Nyasa.

The Mbeya District illustrates one of the ways that geological forces maintain the fertility of soil, and hence of the terrestrial biosphere. Weathering over the long term tends to cause the cumulative loss of plant nutrients by leaching – old soils generally have a low inherent fertility. The ferralitic (iron and aluminum rich) soils, which have been weathering on the southern Tanzanian landscape for several million years, have reached this state. If there were no countervailing processes, the surface of the earth as a whole might be expected to be equally low in plant nutrients by now. Clearly, there must be ways by

Figure 3–2.

A. The Rungwe Volcanic Complex, Mbeya District, Tanzania. This is technically part of a plate tectonic triple point where the Eastern and Western Rifts of East Africa come together then proceed southwards as a single rift.
B. Outlying cinder cone, badly eroded subsequent to deforestation and adaptation to agriculture. Terracing would minimize the loss of soil, and though the technique was introduced in colonial times, it is not common in Tanzania.
C. Detail from B: the light volcanic ash, unprotected by vegetation, is easily eroded, especially by running water in the rainy season.

which the planet maintains the fertility of soil and one way is by volcanic activity.

Over the last two million years volcanic activity, centred on Rungwe, has continually added nutrient-rich material to the soil. The hard, compact basalt has not yet weathered enough to produce a well developed soil, but the easily weathered ash is continuously releasing nutrients into the biosphere and has sustained the original rainforest, now seen only in remnants on the upper slopes of the complex. Most of the rainforest has been removed and is now farmland, the most productive of which is closest to Rungwe itself, where the ash tends to be deepest.

The elevated topography of the volcanic complex has further consequences for the farmer. First, it has the positive effect of producing a zone of high rainfall in contrast to semi-arid regions at lower elevations. Consequently, local agriculture is rainfed rather than irrigated. However, topography is also the cause of a major physical problem. Farming on the flanks of the volcanic complex invariably leads to erosion and the progressive loss of fertile topsoil to expose the less fertile subsoil. The problem is accentuated by the easily transported nature of the ashy soils.

The local crops are corn, potato, banana, coffee and tea, and the soils are managed in traditional ways. This means that the inherent fertility of the soil is supplemented by manures to replace nutrients removed by cropping. In the past, and particularly in colonial times, application of artificial fertilizers was practiced only for the cash crops, coffee and tea. The thin margins of the ash, however, show high enough nutrient deficiencies now that fertilizers are used more generally when available.

Around the immediate boundary of the volcanic ash, rainfall is still sufficient to support farming, though the inherent fertility of the prevailing ferralitic soils is low. Where fertilizers rather than manures have been used for any length of time, a notable problem is acidification. In other words, the practice of agriculture has emphasized the natural direction of evolution in these soils and pushed them towards a pathological state. In the coffee growing area of Mbozi for example, the acidity of topsoil has been increased tenfold (i.e. one pH unit) by comparison with unfertilized soils.

This brief safari into East Africa suggests five generalizations:

1. Soils evolve chemically in one of three directions: towards acid, alkaline or reduced endpoints.
2. Long-term weathering leads to nutrient-poor soils, e.g. the ferralitic soils.
3. The most productive farmland depends on recent additions of weatherable, nutrient-rich material – in this case, volcanic ash.
4. In taking over the soil for agriculture we inevitably eliminate other species – most obviously in this case the species of the native forest.
5. Farming has caused degradation of soil, both physically (by erosion) and chemically (by acidification).

All of these generalizations are extended to the global scale below, but first, a few general remarks about soil and its role in the biosphere.

THE ROLE OF SOIL IN THE BIOSPHERE

No generally accepted definition of soil exists – a somewhat desperate situation that led one eminent soil scientist to conclude "soil is anything so-called by a competent authority".[7] It is a material with all the ingredients of a Hollywood blockbuster: sex, violence and a cast of billions. The billions, mostly microbial, flourish in a sort of natural midden of organic wastes, destroying, recycling, and rebuilding endlessly. All this activity takes place in a frenzy of Darwinian competition that could serve as a paradigm for capitalism in its purest form. As a working definition I will take soil to be:

> a combination of mineral and organic constituents, containing water and gases, and organized into a loose, porous, horizonated, plant-bearing material, which constitutes the natural covering of the land surface of the earth, and which is formed by weathering – a complex set of interactions between lithosphere, hydrosphere, atmosphere and biosphere.

This set of interactions produces three evolutionary trends as we have seen in Tanzania. Technically the trends are:

1. an overtitration to produce acid soils,
2. an undertitration to produce alkaline soils, and
3. reduction to produce hydromorphic soils.

E. O. Wilson has used the bottleneck as a metaphor for a period when life was almost squeezed out of existence.[8] A similar metaphor is useful in illustrating the role of soil in the biosphere. On the land surface, virtually all of the fluxes of energy and matter that keep the biosphere going are "squeezed" through the soil. In other words, soil can be considered conceptually as sitting in the neck of an hourglass (Figure 3–3).

In directing and controlling energy and matter in soil, three constituents are of particular importance – organic matter, the clay fraction, and water. The organic matter that accumulates in soil and which is loosely referred to as

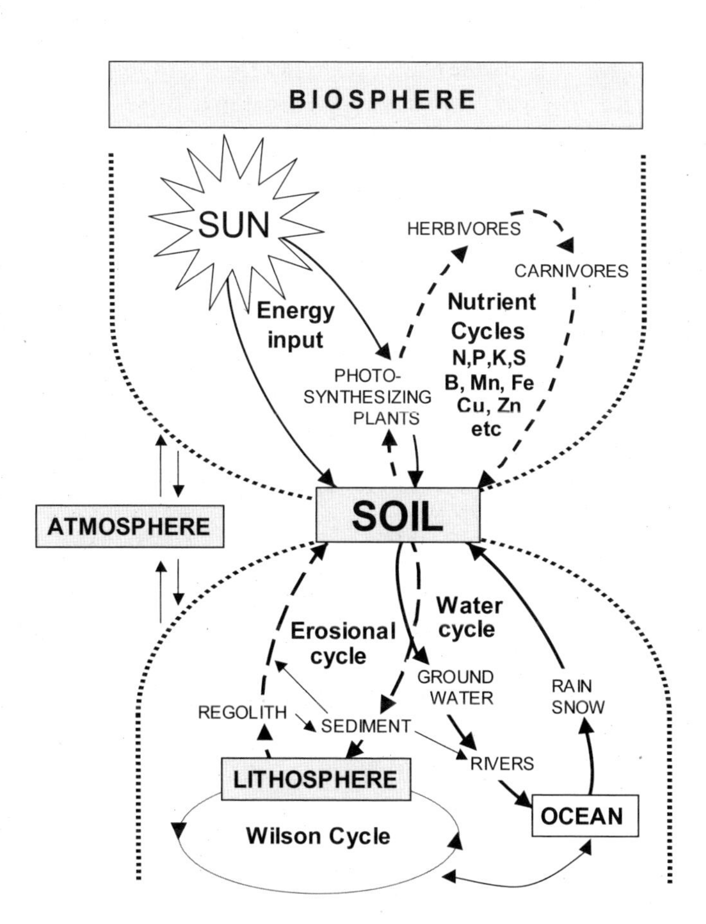

Figure 3–3. Biospheric Cycles and the Hourglass Paradigm.

A schematic view of the biosphere that emphasizes the soil as a vulnerable zone of constriction through which energy and materials pass between different compartments of the biosphere. Not shown are the individual compartments in the soil, where energy and materials are also exchanged.

humus, is mostly the decaying remains of photosynthesizing plants. As such, it constitutes a carbon source for other organisms. Above all it is the fuel "burnt" by bacteria and other decomposers, which themselves become the fuel for other organisms in the food web. In this way, energy originally from the sun is distributed to organisms elsewhere in the biosphere that do not themselves photosynthesize.

The clay fraction is the most chemically active product of weathering. At an early stage of weathering an amorphous material dominates this fraction. At a more mature stage, sheet silicates ("clay" in the restricted meaning of the word) develop and provide a large surface area where nutrient elements may be stored in readily available forms. With increased weathering the type of clay changes, and its ability to serve this function is diminished.

The importance of water to life in and on the soil goes without saying. Provided that the planetary water cycle does not fail, the water in soil is a renewable resource. In numerical terms it appears negligible. Only about 2.5% of the hydrosphere is considered to be freshwater, and only 0.2% of this is contained in the soil at any one time. In soil it is both an important reaction medium for chemical and physical change, as well as a reactant in its own right. Leaving aside metabolic functions within an organism, water is important to life in the soil because without it, chemical weathering to release nutrients cannot occur. Furthermore, as a solvent, water is the medium by which nutrients are released. The ability of soil to retain water is largely a function of the other two important constituents. Soils poor in organic matter and/or clay minerals, tend to be droughty. Even if rainfall is adequate, the soil lacks the physical structure to retain it.

THE AGRICULTURAL REVOLUTION

What I am calling the Agricultural Revolution normally appears in the literature as the Neolithic Revolution, a label that philosophers of history have carped and caviled over almost since it was first coined. I carp and cavil too, though from a different perspective. The philosophers' hang-up relates to the word revolution. Revolutions are supposed to be sudden upheavals and the phenomenon under discussion took a few thousand years (around 10,000 years ago) to happen in diverse places over the globe (Figure 3–4). I have no problem with this, and not simply because as a geologist I consider a few thousand years a blink of the eye. More importantly, with the invention of farming *Homo sapiens* underwent a radical change in life-style, from an unsettled, foraging existence, to the sedentary one of cultivating the soil. This brought about a true, radical revolution in the way we live. By settling down we were able to take the leap into civilized life. You cannot invent civilization on the go; you need to settle down as Jakob Bronowski shows in his *Ascent of Man.*[9]

My hang-up concerns the term Neolithic. Technically it refers to the "new stone age" and yet the chief material resource that came into use at that time was soil not stone. With agriculture we began a 10,000-year experiment with soil, and through it, the biosphere. Using this new resource, we have latterly become the only "big fierce animal"[10] currently increasing in number, and in those ten millennia we have grown from a few million people to more than 6 billion.

So on the whole I am more comfortable calling the changes that took place in this time-period the Agricultural, rather than the Neolithic Revolution. Bronowski goes further and calls it the Biological Revolution. Alfred North Whitehead once wrote that the European philosophical tradition "consists of a series of footnotes to Plato".[11] In a similar spirit of hyperbole, I believe that our history over the last 10,000 years, is little more than a footnote to the Agricultural Revolution. All our magnificent cultural artifacts from cathedrals to efficient plumbing systems are a direct result. The dark side is that the material civilization we have built up over ten millennia now dominates the biosphere in a way that gives rise to real concern, and it all comes down to the exploitation of soil.

EXPLOITING THE SOIL

Clausewitz, the great Prussian philosopher of war, teaches that to prevail in battle a general needs to command the strategic heights of the battlefield.[12] In pursuing the activity of farming and by "commanding" the requisite soils, the human race is in a fair way to accomplishing just that.[13] So far, we have taken over at least a third of the earth's soils more or less completely, while considerably modifying another third. The commanding heights of the Darwinian battlefield we call the terrestrial biosphere, are thereby ours.

The beginnings of large-scale agriculture required soil with a high inherent fertility, and a means of maintaining, storing, and delivering water and soil nutrients to plants. We have already seen that the presence in the soil of organic matter and the right type of clay minerals are important in this regard. On the scale of the landscape, weathering and the erosional cycle are the natural maintenance systems of soil fertility that the early farmers depended upon.

A characteristic locality, which brings all the prerequisites together, is a large river system, draining an area where freshly exposed and lightly weathered geological materials provide nutrient-rich sediment for distribution lower down the valley (Figure 3–4). Early civilizations in the valleys of the Tigris and Euphrates, the Nile, Indus and Huang Ho, had this kind of physical basis. Because of the association with rivers, they are often referred to as "hydraulic civilizations". Each appears to have been preceded by a period of experimentation in nearby highland regions where rainfed agriculture was practiced.[14] Some of the oldest known rainfed systems are found in southern Turkey, Catal Huyuk, for example, through northern Syria and Iraq to the Zagros Mountains in the highland rim of the "Fertile Crescent". After considerable trial and error, farming was introduced into the valleys of the Tigris and Euphrates, where the earliest cities were built soon after 5000 B.C. Their support came from the surplus of food that the farmers produced. All later civilizations, including our own, have a similar foundation: the farmers' surplus is crucial. The necessity to manage river

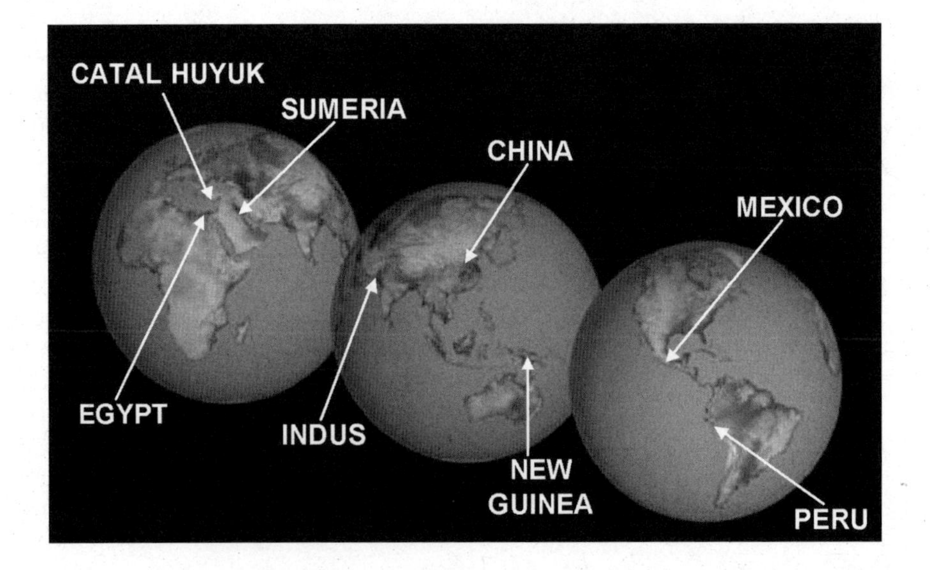

Figure 3–4.

Early centres of settled agriculture: agriculture started in all the areas shown about 10,000 years ago, plus or minus a thousand years or so. The extensive continental glaciations were over; climate had ameliorated; and the food supply available to human hunters was in decline. The latter may have been a kind of Malthusian drive to the invention of farming. (Representation of Earth, courtesy of NASA.)

water for irrigation of crops was probably a major factor in developing the kinds of cooperation and division of labour necessary for a civilized existence.

When civilization spread out from the river valleys it diffused into areas where atmospheric precipitation was reliable and irrigation unnecessary. The materials and processes of surficial geology were of paramount importance. The diffusion of agriculture into Europe for example, appears to have been brought about by the migration of farmers from the Middle East moving at the rate of about 1 km per year. As Jared Diamond points out, it was helped by the fact that latitudinal diffusion is made easy by the absence of north-south barriers.[15] Moving domesticated crops in an essentially east-west direction minimizes the amount of climatic variation the farmers must deal with in terms of the crop varieties they grow.

They moved across to the Atlantic coast by two routes – one via the Mediterranean, the other via the easily worked soils of the alluvial plains of the Danube valley and the loessial deposits of northern Europe. Before the introduction of the iron plough, only light soils were worked. The iron plough was introduced between 2000 and 2500 years ago, and since then virtually any soil, no matter how heavy, has become amenable to cultivation given appropriate climatic conditions, and help from the agricultural engineer. Since the nineteenth century, the determining factor in agricultural expansion has been the availability of a means of transport for getting the harvest to distant markets in timely fashion.

If we look at the state of agriculture now, we find a global pattern similar to the one described for the Mbeya district of Tanzania. Productive farmland correlates well with regions where geological processes of one kind or another have added fresh nutrient-rich minerals to the weathering regime.[16] Egyptian agriculture, for example, has been dependant on the erosion of young soils from recent volcanics in the Ethiopian Highlands. Similarly, the early centre of Chinese civilization did not arise directly on loess, but on alluvial sediment eroded from a loessial source in the upper Huang Ho valley. In the case of the Tigris-Euphrates, Indus and Ganges valleys, agriculture depended on the sediments derived from young fold-mountains. In all but very few instances, the natural fertility exploited by early farmers was derived from geological source materials fresh enough to contain a reasonable store of nutrients that could be released by weathering.

The very productive agriculture of Australia appears to disprove this – Australian soils generally, have soil fertilities that are amongst the lowest of any in the world. This is because Australia, of any of the settled continents, is the one which has not recently undergone a natural refertilisation by volcanism, glaciation or orogeny. However, Australians are an affluent people and can afford to employ expensive agrichemicals and irrigation systems to produce good harvests – though not without major problems developing, as we shall see later.

PROBLEMS ARISING IN AGRICULTURAL SYSTEMS

The earliest agricultural disaster story pertains to the earliest seat of sedentary agriculture, the Tigris-Euphrates valley. A net deficit in precipitation with respect to evapotranspiration means that the natural weathering system in this region follows the alkalizing trend. Irrigation, which maintained a sequence of societies in Mesopotamia (the earliest being Sumeria, with the Akkadian dynasty, and the Babylonian and Assyrian empires as later examples), emphasizes this trend. After about 2000 years of increasing salinisation of the soil, coupled with climatic cooling, there was a change from wheat to the more salt and cold tolerant barley, followed by the abandonment of extensive farming altogether in the lower reaches of the two rivers. The irrigators had also to contend with excessive sedimentary loads, which caused changes to the river channels and blocked irrigation canals. The excessive loads were themselves the result in part of forest clearances by the farming populations of the upper reaches of the river system.

Salinisation brought about, or at least accentuated, by irrigation played a role in the demise of the agricultural basis of the Harrapan civilization, which lasted from about 2500 to 1500 B.C. in the Indus valley. But just as climatic change added impetus to agricultural problems in Mesopotamia, an impersonal force, totally outside human control added its effect in the Indus valley. Tectonic uplift along the coast appears to have caused changes in the course of the Indus, leading to ponding and flooding in and around major city centres such as Mohenjodaro.[17]

Induced salinisation remains a problem in modern agriculture, in farming and social systems as varied as California, Australia and Kazahkstan. In the latter region, irrigation of the vast cotton fields not only produced salt affected soils, it also destroyed the livelihood of fishermen in the slowly shrinking, and increasingly polluted, Aral Sea. As Kennedy says, this particular agricultural disaster was "the result of decisions by Soviet planners whose need for cheap fibre outran their good sense".[18]

Rainfed agriculture is practiced where the acid trend dominates the system of soil formation. Again, farming emphasizes the natural tendency, in this case by removal of cations from the soil by harvesting a crop. It is equivalent to increasing the rate of chemical weathering by between one and two orders of magnitude. Forest clearance by fire, practiced by all Neolithic farming communities, as well as by earlier, pre-agricultural peoples colonizing new territory, such as the original population of Australia,[19] also boosts the trend towards acidification in humid climatic regions. The acid heathland of western Europe is testament to this.[20] In the present century, acidification has also been brought about by the use of acidifying fertilizers, as we saw with the example of Mbozi in Tanzania.

An example of agriculture causing extreme acidification in the biosphere is found where mangroves have been cleared for farming. Pyrite, which commonly forms in the water-saturated root zones of the mangrove, weathers by reaction with oxygen in the aqueous environment to produce sulphuric acid and a pH of 3 or less. Nowadays, the cash crop is shrimp for the tables of the affluent, as aquaculture takes over the mangrove swamps of Southeast Asia. Neo-classical economists make much of aquaculture as a reasonable approach to global hunger,[21] ignoring the environmental costs and, more fundamentally, the thermodynamic costs. For example, aquaculture can take as much as 50 calories input for 1 calorie of food.[22]

Increased rates of erosion appear to attend all attempts to cultivate soil, especially in situations where forest is first cleared. Britain, as a typical European example, was largely wooded until 5000 years ago.[23] As the Neolithic population grew, the chalklands in the south were cleared, occasioning the first bout of anthropogenic erosion. In the north at about the same time, the acid heaths mentioned in the last paragraph were in process of formation. Under Roman occupation, population was about 4 million and manuring and marling were practiced to revitalise the soil. Winter wheat was grown, a procedure that further encourages erosion. When the Romans withdrew, population dropped to about half and erosion was greatly diminished. Population grew again until the Black Death in the mid-1300s. By this time there was no extensive forest left in southern Britain. Fallowing had now been introduced to deal with problems of soil fertility. This practice leaves fields vulnerable to erosion until a cover of vegetation has been established. After the plague, population was down to about 2 to 2.5 million. From 1500, population began to build again and, after 1750, with the prosperity brought about by the agricultural and industrial revolutions, it soared. Enclosures and grazing, rather than arable farming, gave protection against erosion.

In the twentieth century increased erosion has been encouraged by socio-political forces. In Britain in the 1920s, the Ministry of Agriculture encouraged the farmers of East Anglia to grow sugar beet, a crop that provides poor protection against erosion. So-called market forces created a greater problem of erosion after about 1960, when a more intensive, monocropping form of industrial farming became common, in which hedgerows were removed to facilitate the use of heavy machinery. It is more than a little ironic that the removal of hedgerows should cause problems, in view of the fact that their introduction during the enclosure movements of the eighteenth and early nineteenth centuries was itself attended by considerable social upheaval. Today, water erosion, resulting from the planting of autumn cereals, has replaced wind erosion as the principal problem in the United Kingdom.

The New World has been as disaster prone as the Old, though in techniques of harvesting water – chinampas, ridging, runoff farming, and so on – much ingenuity has been employed in devising systems in equilibrium with local environments.[24] However, in spite of the fact that Spanish colonists have been largely blamed for the destruction of what some claim was an environment tended with loving care by the pre-Columbian population,[25] accumulating evidence from limnological studies in central Mexico, show that there – as usual – the introduction of farming brought about massive erosion a good 2000 years before the arrival of Europeans.[26] In fact soil loss may have been one of the factors that led to the demise of Mayan civilization. Three-quarters of the soils that the Mayans depended upon are now classified as especially prone to erosion.[27]

Water management was an important feature of the agriculture of the Andean civilizations. Elaborate terracing helped manage water in the semi-arid Peruvian highlands. Irrigation of terraces also provided fertilising silt – Knapp notes that silt in the northern Andes has up to five times the available phosphate as nearby cultivated soils.[28] A major problem in managing water in the Andes, however, is caused by neotectonic uplift. The Kingdom of Chimor, for example, which flourished in northern Peru about 1000 A.D., brought water by canal from up to 70 km away. The water flow was interrupted and finally terminated by uplift of the land.[29] In addition, the inhabitants of Chimor had to contend with advancing dunes and *El Niño* floods – factors beyond human control.

So far, the problems that have been emphasized have all been terrestrial. Modern agriculture has essentially taken over the world's temperate grasslands and much of the temperate forest. The business is now so gigantic that impacts are not only felt on the land, but also at sea. The most impressive extensive impact is associated with the agricultural takeover of the Mississippi watershed, about 70% of which is used for agricultural purposes. Fertilizers (especially nitrogen-containing) and pesticides have been increasingly used since the end of WWII, with impressive results when judged only in terms of agricultural productivity.

Furthermore, the need to support an increasing population worldwide has brought an increasing acreage of erosion-prone marginal land into production west of the hundredth meridian. Irrigation using non-renewable fossil water from underground sources such as the Ogalalla aquifer has made this possible. Sediment and water carried down the river enters the Gulf loaded with agrochemicals, as well as with additional materials picked up from the sewage systems of a number of major conurbations. As a result, a zone of hypoxia is spreading at the mouth of the Mississippi, and is currently 500 km in length.[30] Similar "dead zones" are now growing off the mouths of major rivers on the east coast of North America, in the North Sea, especially at the mouth of the Rhine, and off Japan.

THE QUESTION OF SUSTAINABILITY

There is no doubt that agriculture is laying a growing stress on the biosphere and, while population continues to rise, there is every reason to expect the stress to increase and the environmental horror stories to proliferate. Even a collapse of civilized society is not beyond the bounds of possibility.[31]

What happens when a population outstrips the ability of its environment to support it, the process that Malthus saw as inevitable, can be seen over and again in human society, if only at a localized level. The collapse of a pre-Neolithic village in northern Syria,[32] the breakdown of the society that produced the monuments on Easter Island,[33] and the increase in the population of Java from 4 to almost 100 million in 100 years, with all the attendant pressure on land, food and other resources,[34] suggest lessons that don't hold out much hope for the future. Kennedy for one believes that Malthus' prediction of 200 years ago is now close to a reality on a global scale.[35]

In this context, Easter Island is worth looking at more closely. As well as Ponting, Jared Diamond and Bill Rees have used the story of Easter Island's demise as a warning for the future of the human race.[36] The original inhabitants probably got to the island by chance as part of the diaspora of Melanesian navigators that populated the south Pacific in the first few centuries of the Common Era. Because of the isolated nature of the place, they were essentially marooned once they got there. That was okay to begin with. They found themselves in a well-forested land, where the soils that had formed on recent volcanic ash were especially well developed and fertile – as with Rungwe. The new colonists cleared land for agriculture, built sturdy boats to fish the immediate seas, prospered for a couple of hundred years and built up a complex society. Finally the population outgrew the resources of the island itself, and having chopped down the last palm tree, no longer had the wherewithal to fish the surrounding ocean. As a consequence the society collapsed into barbarism and cannibalism.[37]

Is the well-established Malthusian collapse of Easter Island a reasonable model for the collapse of society on a global scale? Messers Ponting, Diamond and Rees think so, but not Bjorn Lomborg. That "skeptical environmentalist" states that out of 10,000 Pacific islands, only 12 underwent such a decline so that the odds are vastly in favour of a rosy future.[38] What he's missing here is that Easter Island, unlike most Pacific islands, was essentially an isolated system. There were no nearby territories to provide help when indigenous resources grew scarce, nowhere for the islanders to extend their ecological footprint. And it's precisely this point that makes the fate of Easter Island germane to the fate of the Earth: we are isolated in the same sense. There are no nearby planet Earths to overrun or trade with when we've degraded all the low entropy resources on this one. And, we have taken over and degraded about a third of the soils on the planet, the best ones in fact – the low entropy ones – in the sense that they were the ones best fitted for agricultural purposes.

THE END OF HISTORY?

In 1992 Francis Fukuyama announced that we had reached "The End of History",[39] and that it basically consists of an American style, egalitarian democracy, where we can expect to lead a prosperous existence and enjoy the inalienable rights of "life, liberty and the pursuit of happiness". He was dead wrong of course, as the "War on Terror" is currently emphasizing.

In any case, the only conceivable end is extinction. As John Maynard Keynes once said, in the long run we're all dead.[40] That goes for species as well as individuals. In the long run, human occupation of planet Earth is no more sustainable than a trilobite's. Natural Selection acts as the Grim Reaper, helped by the kind of major disaster favoured by Hollywood – catastrophic impact of asteroid-sized meteorites, massive floods, volcanic eruptions and a climate swinging between greenhouse and icehouse, events that punctuate the history of the Earth like bursts from a cosmic Gatling gun.

Yet, short of this inevitable end, what "end" might we foresee in the Fukuyama sense? Although farming systems have been known to fail within two or three generations (the Midwest dustbowl, the Aral Sea), two have lasted for 5,000 years or more (Egypt, northern China). So, can we take this as a best case scenario and assume that by employing our undoubted human ingenuity we can sustain our civilized existence for at least several thousand years? The quick answer is no, and the reason is that the persistence of agriculture in Egypt and northern China is not so much a result of human ingenuity as of a fortunate geological situation that has reliably provided water and natural fertilizer to the soils of those regions throughout their history. But, as the recent history of Egypt demonstrates, such geologically reliable delivery systems are not completely proof against human folly. The Nile transports some 100 million tonnes of silt per year, much of it due to soil erosion in the Ethiopian highlands. The Delta (14,000 square km) is Africa's most productive farmland, and was built up by this annual load of silt. For over 7,000 years it was a reliable annual supply of fertilizer for the farmer. Since 1964, the Aswan Dam has trapped about 98% of this free fertilizer, so that Egypt has had to become a major importer of the manufactured variety. In addition, evaporation losses in the dam have diminished the flow of water in the lower Nile to the extent that farmers in the Delta have to contend with sea-water incursion into groundwater.

How long a given system may survive is determined by a variety of factors, the resilience of the biosphere to the stresses and strains imposed upon it being the most important. Our poor understanding of the variables involved, and especially our limited ability to juggle scientific and technical factors with social, political and economic ones, largely accounts for the fact that we have not yet invented an agricultural system that is truly sustainable, i.e. one that, in principle, leaves the ecosystem that contains it continuously viable and recharged after each harvest. We seem to operate mostly by a form of crisis management in keeping a farming system going – the longevity of the system being a measure of a kind of band-aid persistence on the part of the farmer, rather than real sustainability. Eventually, one or other of the determining factors that we are maladroitly juggling, is pushed to a breaking point – or reaches such a point through natural planetary change. Human societies have a hard time remaining stable for as much as a few hundred years, and more often than not it is a failure of the food supply that promotes the breakdown.

No doubt in the industrialised world, we can and will enjoy our high material standard of living for a while longer, by continuing to draw down the global capital of fossil fuel reserves, conceivably ensuring the supply by means of wars of the Gulf variety. Currently, we are perched precariously on the top of "Hubbert's Mountain" ready to say goodbye to another low entropy resource as we slide downhill into the post-oil and gas future.[41] And no matter how much we dig and delve and drill, we will never find on this planet the resources we would need to achieve Gro Harlem Brundtland's ambition of raising less fortunate nations to our standard of living.[42] Four or five planet Earths would be needed for that, according to Bill Rees.

In the short term we are looking at a future dominated by a scramble for resources. It will not be a Hobbesian-Darwinian "war of all against all" as much as a war of the United States against all. In the long term, when all of the low entropy resources including the best soils, have been used up, our descendants will probably be left with only two options: a highly regimented and closely controlled population reminiscent of Mao's China, or (much more likely) an Easter Island future, where, in spite of any environmental smarts we may have (or develop), somebody will exercise their "right" to chop down the last palm tree.

And after that? The biosphere will continue without us. All the soils we have degraded will be renewed and recycled over and over by the impersonal forces of geology. It will continue for billions of years as if we had never lived.

SUMMARY AND CONCLUSION

Agriculture, more obviously than any other part of the human enterprise, is strategically situated to consume the biosphere from within. The prerequisites – soil, water, and genetic resources – are all extracted from the biosphere, and although crops rely on contemporary sunlight for most of their energy requirements, oil and gas, biospheric energy from the geological past, have become the necessary subsidy for modern agriculture.

In effect the Neolithic invention of agriculture started a 10,000 year experiment with the biosphere that makes us the only "big fierce animal" increasing in number. During this experiment we have drawn down the Earth's capital stock of habitat, thereby diminishing the biodiversity of the planet. In maintaining the surplus of food necessary to support our dominance, agriculture has spread from places like the mountain fringe of the Fertile Crescent, into the great river valleys of the early civilizations, gradually to take over the temperate grasslands and much of the forest and wetland biomes. An agricultural scar now covers 35% of the land surface as cropland and pasture.

Most of the expansion has happened since the steam locomotive opened up the grasslands of the New World to the markets and bellies of the Old. Cheap energy and the exponential growth of the human population has also meant that we are now a potent geological force in our own right, unique to the Holocene. The natural tendencies for the soil component of the biosphere to erode or to become acidified, saline, or hydromorphic, depending on factors such as climate, texture and drainage, have been magnified and exaggerated into pathological states by the practice of farming. Ploughing currently moves some 10^{15} kg of soil per year worldwide, an intervention into the erosional cycle which modifies river systems and ultimately impacts the oceans in the form of fertilizer-loaded sediment that causes hypoxia along the coasts of North America, Europe and East Asia.

Before farming was invented our numbers were controlled by competition for limited resources. By commandeering more and more of the biosphere, and by squandering a finite resource of cheap fuel, we have fallen prey to the "no limits" illusion of neo-classical economics. This supports a belief that agriculture is currently sustainable, whereas the fact is that a sustainable agriculture has yet to be invented. A further irony is that the clearly unsustainable high input/high output agriculture of the developed world, now has an even less sustainable offspring in the form of aquaculture. If a sustainable food supply ever is devised it will be based on a much smaller human population than the present one, and it will utilise renewable resources – human intelligence, energy obtained from contemporary sunlight, water, and carefully managed soil fertility.[43]

Above all it will require a long term conservation ethic, a bottleneck that our Darwinian hard-wiring may preclude; in which case, the fate of our civilization may be the same as Easter Island's, whereas our own fate will inevitably be that of the Dodo.

NOTES AND SOURCES

[1] Martin, A. 1975. *The Last Generation: the End of Survival?* Fontana, Glasgow. p. 180.

[2] Fernandez-Armesto, F. 2000. *Civilizations.* Key Porter Books. Toronto. 636 pp.

[3] Lomborg, B. 2001. *The Skeptical Environmentalist.* Cambridge University Press. 515 pp.

[4] Malthus, T. 1798. An Essay on Population. 1976 reprint of first edition, edited by Philip Appleman. Norton, N.Y. 260 pp.

[5] e.g., Lewin, R. and R. Leakey. 1996. *The Sixth Extinction: Patterns of Life and the Future of Humankind.* Anchor Press. 288 pp.

[6] Chesworth, W., P. Van Straaten, J.M.R. Semoka and E.P. Mchihiyo. 1985. Agrogeology in Tanzania. Episodes 8: 257-258.

[7] Fitzpatrick, E.A. 1986. *An Introduction to Soil Science.* 2nd edition. Longmans, London. 255 pp.

[8] Wilson, E.O. 2002. *The Future of Life.* Knopf, New York, 254 pp.

[9] Bronowski, J. 1973. *The Ascent of Man.* Little, Brown and Co., Boston. 448 pp.

[10] Colinveaux, P.A. 1978. *Why Big Fierce Animals are Rare.* Princeton University Press. 256 pp.

[11] Whitehead, A.N. *Process and Reality.* Macmillan, N.Y. p. 63.

[12] Clausewitz, Carl von. 1832. *On War.* 1968 reprint. Aeonian Press, Mattituck, N.Y. 460 pp.

[13] Chesworth, W. 1996. Agriculture as a Holocene process and its bearing on sustainability. pp. 31-36 in A. Perez Alberti, P. Martini, W. Chesworth and A. Martinez Cortizas (eds.). *Dinamica y Evolucion de Medios Cuarternarios.*

[14] See for example Jarriage, J.F., and R.H. Meadown. 1980. The antecedents of civilization in the Indus valley. Scientific American 243: 122-133; Moore, A.M.T. 1979. Pre-Neolithic farmers' village on the Euphrates. Scientific American 241: 62-70.

[15] Diamond, J. 1997. *Guns, Germs and Steel.* W. W. Norton, N.Y. 480 pp.

[16] Chesworth, W. 1982. Late Cenozoic geology and the second oldest profession. Geoscience Canada 9: 54-61.

[17] Dales, G.F. 1965. New investigations at Mohenjodaro. Archeology 18: 145-150.

[18] Kennedy, P. 1993. *Preparing for the Twenty-First Century.* Random House, N.Y. 428 pp.

[19] Goudsblom, J. 1992. *Fire and Civilisation.* Allen Lane, London. 247 pp.

[20] Dimbleby, G. 1967. *Plants and Archaeology.* Paladin. London. 190 pp.

[21] Lomborg 2001. p. 63.

[22] Lavigne, D. 2002. Ecological footprints, doublespeak, and the evolution of the Machievellian mind: pp. 63-91 in W. Chesworth, M.R. Moss and V.G. Thomas (eds.). *Sustainable Development: Mandate or Mantra.* Faculty of Environmental Sciences, University of Guelph, Guelph, Canada. 140 pp.

[23] Evans, R. 1992. Erosion in England and Wales – the present is the key to the past. pp. 53-56 in M. Bell and J. Boardman (eds.). *Past and Present Soil Erosion.* Oxbow Books, Oxford. 250 pp.

[24] Pearce, F. 1991. Ancient lessons from arid lands. New Scientist 132: 42-43, 46-48.

[25] Sale, K. 1990. *The Conquest of Paradise.* Knopf, NY. 453 pp.

[26] O'Hara, S.L., F.A. Street-Perrot, and T.P. Burt. 1993. Accelerated soil erosion around a Mexican highland lake caused by prehispanic agriculture. Nature 362: 48-51.

[27] Ponting, Clive. 1991. *A Green History of the World.* Penguin, London. 430 pp.

[28] In a comment on page 420 of Guillet, D. 1987. Terracing and irrigation in the Peruvian highlands. Current Anthropology 28: 409-422.

[29] Ortloff, C.R. 1988. Canal builders of pre-Inca Peru. Scientific American 259: 100-107.

[30] Downing, J.A., J.L. Baker, R.J. Diaz, T. Prato, N.N. Ra-bal-ais, and R.J. Zimmerman. 1999. Gulf of Mexico hypoxia: land and sea interactions. Council for Agricultural Science and Technology, Ames, Iowa, USA. 40 pp.

[31] Catton, W. R. 1980. *Overshoot: the Ecological Basis of Revolutionary Change.* University of Illinois Press. 298 pp.

[32] Moore, A.M.T. 1979. Pre-Neolithic farmers' village on the Euphrates. Scientific American 241: 62-70.

[33] Ponting 1991.

[34] Forsyth, A. 1992. No new worlds. Equinox. V11: 84-95.

[35] Kennedy 1993.

[36] Ponting 1991; Diamond, J.A. 2005. *Collapse: How Societies Choose to Fail or Succeed.* Viking Press, N.Y. 591 pp.; Rees, W.E. 1999. Consuming the earth: the biophysics of sustainability. Ecological Economics 29: 23-27; Rees, W. E., 2003. A blot on the land. Nature 421: 898. Recently, Rolett and Diamond (2004) in Environmental predictors of pre-European deforestation on Pacific Islands. Nature 431: 443-446, and Diamond (2005) in *Collapse* (p. 118), have nuanced Diamond's original view of the demise of Easter

Island society from an interpretation weighted heavily in the direction of the lack of foresight of the Islanders in the use of their resources, to "Easter's collapse was not because its people were especially improvident but because they faced one of the Pacific's most fragile environments" (Rolett and Diamond p. 445).

[37] Diamond 1997.

[38] Lomborg 2001, p. 29.

[39] Fukuyama, F. 1992. *The End of History and the Last Man.* Free Press, Toronto. 418 pp.

[40] Keynes, J.M. 1924. *Monetary Reform.* Harcourt, Brace, N.Y. p. 88.

[41] Campbell, C. 2001. Newsletter # 2001/2-1, M. King Hubbert Center For Petroleum Supply Studies, Colorado School of Mines, 3 pp.

[42] World Commission on Environment and Development. 1989. *Our Common Future.* Oxford University Press. 400 pp.

[43] Jackson, W. 2004. Agriculture: the primary agricultural challenge of the century. pp. 85-99 in W. Chesworth, M.R. Moss and V.G. Thomas (eds.). *The Human Ecological Footprint.* Faculty of Environmental Sciences, University of Guelph, Guelph, Canada. 202 pp.

[44] The *AAAS Atlas of Population and Environment* is freely available online at http://atlas.aaas.org.

Chapter 4

The Notion of Sustainability

Sidney Holt

> *It is sometimes said that we should never believe a scientific theory until it is verified by experience. But...also...we should never believe an observation until it is confirmed by a theory.*
>
> João Magueijo 2003[1]

> *la tenacière des quattre saisons*
> *l'avare ficelée dans ses longitudes*
> [Proprietress of the four seasons
> the miser tied up in her longitudes]
>
> *Jules Supervielle*, inveighing, it is said, against the meanness of Earth, a limited and unexpandable Heritage[2]

In this chapter, I look mainly at the last one hundred and fifty or so years during which the idea of sustainable use of living resources has emerged, crystallised and been intellectualised and provided with a theoretical cloak and a political mission. Its roots reach down to an earlier period in which hunters, fishers and elites expressed their concerns that adequate breeding stock should be left for future benefit, and fishes, birds, mammals and other animals be given the chance to survive to sexual maturity and/or grow bigger and hence more valuable. I begin with Thomas Malthus, the much misunderstood parson/mathematician who, writing in mid-nineteenth century, is said to have presumed – although in fact he didn't – that biological populations (he focused mainly on humans) tend to increase geometrically until they outrun the resources on which they depend – principally food but also space – and then may crash. Thus, it was thought, the Malthusian hypothesis would lead systems – particularly human societies – towards catastrophe and revolution.

An almost immediate counter to this was the observation that populations seemed to increase along an S-shaped curve and, to "explain" that, the idea of "density dependence" was put forward. Several decades later this led to the presumption that wild animal populations could be "harvested sustainably" at rates commensurate with the slope of the S-curve at any point along it and, furthermore, that a "maximum sustainable yield" could be obtained by cropping at such a rate as to maintain the population at some intermediate level between its natural, unexploited abundance, and its extinction, at the size corresponding with the inflection of the curve. This density-dependence paradigm was, and still is, the core of efforts to manage exploitation for sustainable yields, modulated by considerations of periodic natural fluctuations and cycles, and other transient processes. It has been behind almost all efforts to manage sea fisheries (one of the first applications was to the blue whales, *Balaenoptera musculus,* of the southern hemisphere).

In recent decades the idea of sustainability has been incorporated in national and international policies for economic and social "development", but in several respects such policies – or, rather, the activities they promote – are the antithesis of *biological* sustainability. To justify them a new mythology of control of natural and human processes is being written.

Here I explore some of the mysterious corners of the biological core of the sustainable use notion – with little more than glances at the economic, social and political superstructures that have been erected on it – and conclude that a critical reappraisal of the notion and its implications is now due. The efforts by scientists, politicians and administrators to invent a way of bringing about a sustainable whaling industry (even though few people now want that to happen!) are taken to be the flagship bearing such a reappraisal.

Some implications of transformations from one population or ecosystem state to another, or changes in patterns of variability, are explored, in addition to the concept of precaution, and recent challenges to the root assumptions by the originators of the notion of sustainable use of living resources.

ORIGINS

In choosing a title my first thought was to treat sustainability and sustainable use as a myth. Indeed, among the earliest creation myths of the Eastern Mediterranean, the Levant and the Fertile Crescent Paradise is an apple orchard yielding fruit year after year.[3] However, as will be seen, it seemed more appropriate for this essay to begin, not at the beginning, but with the European Enlightenment and its children. So, is sustainability a modern myth? Not quite, I think, if we abide by the definition of myth in the Oxford English Dictionary (OED):

> A traditional narrative *involving supernatural or fancied persons*, and embodying popular ideas on natural or social phenomena (my italics).

But the next section of the OED gives us, *notion, n*:

> View, opinion, theory, vaguely held or insecurely based.[4]

That's more like it! Glancing over our shoulder, the links between the largely Protestant Enlightenment to the decidedly Roman Renaissance and from there back to the Moslem civilizations and the transfer of Greek ideas both through that religion and through Christianity, and the various versions of the Creation myths, is perhaps worthy of historical treatment, but not here.

We shall see that the idea of paradise gained and lost, possibly repeatedly, persists; history repeating itself – but as tragedy or farce? After the Enlightenment the early horrors of the burgeoning Industrial Revolution, seemingly bringing another loss of rural Paradise, was to lead – the early Socialists thought – to its restoration through the victory of the proletariat and poor peasants in a class war that would bring about a classless society which might change over time but would in principle persist for ever.

Things did not turn out quite like that, especially as the twentieth century spawned and endured global wars that were more wars between empires than among classes. But after the second big one the vague new-old idea of the regaining of paradise began to emerge from its chrysalis of environmentalism, and was ultimately baptized as *Sustainable Development*, its Law being given to us not in tablets of stone but as "Agenda 21". Arguments, continuing, about the interpretation of The Law belong more in the realm of religion than of science. Here we look only at the scientific – or is it pseudo-scientific? – kernel: sustainable use of the Earth's physical and biological resources.

BIRTH, GROWTH AND DEATH

Sustainability is the hybrid (bastard?) child of Steady-State Systems and of Change – "generally used to mean the fact of becoming different". But "natural environmental properties" vary on many time-space scales, so that further definition is necessary. With regard to climate-ocean-biology change we mean a shift from one pattern of variability to another, or a transformation from one state to another. In order to detect the direction, magnitude and frequency of such deviations, we must have a norm or a baseline from which to compare the departures. Since very large spatial changes happen only rarely and slowly very long baseline sets of rather frequent measurements are necessary. There are few such time-series in marine biology, chemistry or physics. An objective for the future is to establish such monitoring systems, because serial measurements of these properties are essential to our understanding of the word "change".[5]

In this essay I shall be writing, on the face of it, almost exclusively about aquatic – especially marine – populations; that is the bulk of my personal experience. But the idea of sustainability, as formally, and largely mathematically, expressed, began with problems of the sea fisheries, and it is in that arena that I think much of its conceptual development has been performed. I am writing, too, as a scientist. I offer no excuse for that since sustainability is a concept invented by, and mostly advocated by, scientists. But it has, of course, a moral and even a religious content, in seeking to take proper account of the desires and likely needs of future human generations, and I – like, I think, most scientists – am sensitive to that. Recently the word has begun to acquire, in some parts of the world, in some societies and in certain contexts, the implication not only of a "light ecological human footprint" but additionally of a light touch with respect to animals that may be "used"

– for which I think the phrases "animal welfare" and "animal rights", though increasingly pronounced, are not fully adequate. I am sensitive to that, too, and it has scientific aspects – relating to the nature and evolutionary "purpose" (sorry, Darwinians!), of pain and discomfort, animal consciousness and culture,[6] the exploration of which has scarcely begun, but I shall leave that out of this essay.

The original of my title – as given me by the organizers of this Forum – was the rather heavy-duty *The History of Managing Population Exploitation: Theory vs Practice; Myth vs Facts.* And that remains my main theme. Notice reference to managing *exploitation*, **not** to managing *populations*, resources, whales, fishes etc. I've spent much of my life opposing the hubris of "managing" the ocean and its contents; a losing battle I'm afraid.

Anyway, that observation opens my way to more pedantic self-indulgence. "Sustainability" is a word given many – somewhat contradictory – meanings, even in its ecological/biological context; in its "development" context it has virtually no meaning as far as I can see; but I'm still open to correction.[7] Here I cannot hold back from quoting a very recent article by Alex Gillespie – a New Zealander and specialist in international law – who describes "sustainable development" as having "a mantra-like quality": "Although this claim [*he is here referring to the notion that human-induced climate change hinders sustainable development – sjh*] has an intuitive appeal, it too is doomed to failure. In an ideal world, the phrase 'sustainable development' could be aired and all would agree and know what was meant by it. However, we do not live in an ideal world and the term has become increasingly lost in a labyrinth of political and philosophical considerations".[8] But at least there have been, and continue to be, serious attempts at *objective* definition. That is not true of many other weasel words and phrases so commonly used in the context of conservation, such as: a "healthy" ecosystem/environment or whatever; "fragile" and "stressed" ditto; "abundant" or "rare"; "balance of nature";[9] "robust";[10] "protection"; "precautionary"; "diversity"; "depleted"; "endangered"; "threatened"; and many more,[11] including, of course, "conservation" itself.[12] I found all these terms in a few pieces of paper produced by various arms of the World Conservation Union (IUCN) – such as its Commission on Ecosystem Management, by the UN Commission on Sustainable Development and under the Convention on Biological Diversity, and even in technical papers by well-intentioned and competent scientists.[13] I think I use the qualifier "objective" advisedly; some of the terms I have listed serve principally a propaganda purpose of those who wish to exploit living resources – "healthy" (about an exploited population of wild animals; what a Cockney merchant might casually describe as "in good nick", just before he sells you a flawed used item!), used a lot by whalers, is a case in point – but others also have a political background, coming from the sections of society that wish to bring exploitation under a modicum of control: witness the IUCN *Red Lists* and the CITES criteria.

"Sustainability " is a term that has now, loosely and mostly ambiguously, been stretched to apply to time-frames of "development" in the politico-socio-economic domain. It is not the only important term stolen – and, I think, misused – from its genesis in relation to the use and states of renewable natural resources. Another, more recently appropriated, is "the precautionary principle", used to justify pre-emptive war. This deviant has recently been admirably deconstructed by David Runciman.[14]

This Forum is really about all the above-mentioned notions, and more besides, but I want to start with a closer look at sustainability itself, with a focus on its biological meaning(s). Our troubles begin with Thomas Malthus – child of the Enlightenment – or, rather, with succeeding interpretations and misinterpretations of what he published in the period 1798 to 1830.[15]

Those works are long, complex and thoughtful. The essence of what concerns us here is captured in words early in *The Second Essay*:

> Population, when unchecked, increases in a geometrical ratio. Subsistence increases only in an arithmetical ratio.

Let us not stop there, however:

> A slight acquaintance with numbers will show the immensity of the first power in comparison with the second. By that law of our nature which makes food necessary for the life of man, the effects of these two unequal powers must be kept equal. This implies a strong and constantly operating check on population from the difficulty of subsistence. The difficulty must fall somewhere and must necessarily be severely felt by a large proportion of mankind.

That is, of course, all about humans. But then:

> Through the animal[16] and vegetable kingdoms, nature has scattered the seeds of life abroad with the most profuse and liberal hand. *She has been comparatively sparing in the room and the nourishment necessary to rear them* (my emphasis).

First, let me dispose of the main misinterpretations.[17] One is that Malthus said it was "wrong to bring in measures of social amelioration, for preventing the death of infants and for keeping the people healthy, for if that were done the problem would become worse". So said Lord (then Sir

John) Boyd Orr in 1948 at a conference on Population and World Resources in Relation to the Family. This "tendentious distortion", as Anthony Flew called it, was made by a physiologist and nutritionist, a Nobel Prize winner, and the first Director-General of the Food and Agriculture Organisation (FAO) of the United Nations. Malthus said no such thing.

A second misinterpretation was that Malthus considered that the destructive interaction between birth rate and nutrition arose from there being an upper limit to the latter, that is the existence of *finite* nutritional resources. Malthus' studies were almost entirely related to political economy, and he saw the conflict as being between two forms of *rate* of increase – geometric (what we often now call *exponential*) and arithmetic (that we would more usually now call *linear*). Even in an infrequent reference to animals and plants Malthus only hinted at the existence of upper limits, and then – as in the above quotation – only to the extent of believing that nature was sparing in the provision of living space and nutritional resources.[18]

Malthus was firmly embedded in the newly emerging tribe of British and French philosophers, historians and political economists, thus influenced by the writings of Adam Smith (1723-1790), Edward Gibbon (1737-1794; Jean-Jacques Rousseau (1712-1778,) and David Hume (1711-1776), these last two being close friends of his father. He had taken Holy Orders but he also graduated as a *Wrangler* at his Cambridge college, Jesus, which was about as good as one could get in mathematics at that time, and he had been elected a Fellow of the Royal Society of London.

The First Essay was essentially a polemic against a then famous French author, Marquis de Condorcet, and an English one, William Godwin, father-in-law of Percy Shelley. Both were enthusiasts for the French Revolution, 1789+, and their writings had been pressed upon Thomas Malthus by his enthusiastic father.[19] Godwin and Condorcet both foresaw and hoped for societies of social and economic equality. While Thomas rejected what might be considered to be programmes for a perfect and egalitarian society, he did so not on ideological grounds but, as he would see it, mathematical and scientific ones. Boyd Orr said Malthus warned against the imminent reaching of a historical stage "where there was not sufficient food to feed the people of the world". But, as Flew noted: "Malthus himself saw his principle of population differently", and observed that "Malthus did never at any stage, see his discoveries as providing a warrant for abandoning piecemeal and realistic efforts for improvement"; thus: "this constantly subsisting cause of periodical misery has existed ever since we have had any histories of mankind, does exist at present, and will for ever continue to exist, unless some decided change takes place in the physical constitution of our nature".[20] Note "our nature", not "Nature". But note also the very significant appearance of "periodical".

Malthus next considers the ways in which "the constant operation of the strong law of necessity (the need for food) acts as a check upon the greater power (reproduction). He wrote, regarding the nature of the check among animals and plants, "Wherever there is liberty, the power of increase is exerted, and the super-abundant effects are repressed afterwards *by want of room and nourishment*, which is common to animals and plants, *and among animals by becoming the prey of others*"(my emphasis). And then: "The effects of this check on man are more complicated". After some discourse on morality, monogamy, social constraints on early marriage and the like, Malthus describes a *cyclical* process in human societies of contentment and misery. And this, he clearly sees, results from the class-structure of societies with which he is familiar: the coexistence of a numerous poor and the far less numerous rich, combined with inter- and intra-generational time-delays in the system.[21]

"This sort of oscillation (*the repetition of the same retrograde and progressive movements with respect to happiness*) will not be remarked by superficial observers, and it may be difficult even for the most penetrating mind to calculate its periods".[22] Malthus then asserts that "...in all old states some such vibration does exist, though from various transverse causes, in a much less marked, and in a much more irregular manner than I have described it...". He then offers reasons why such oscillations have not been remarked upon, one of them being that "the histories of mankind that we possess are histories only of the higher classes". In his "class" theory of population Malthus, the mathematical priest, is a couple of decades ahead of Karl Marx and Freidrich Engels,[23] and, as we shall see, pretty much in line with modern ideas about the "vibrations" of populations.[24]

Although he was writing from an English/European perspective Malthus' data came mainly from censuses of the population of the British colonies in North America. This was of course increasing by immigration as well as by the operation of "the passion between the sexes" of the colonists. Some numbers are given in *A Summary View*: the white US population increased by 36.3% from 1790 to 1800 (average population 3,738,500); by 35.9% from 1800 to 1810 (average 5,087,500) , and by 34.1% from 1810 to 1820 (average 6,861,900). Malthus took this as evidence of geometric increase, with no "vibration" in that period. I would happily regard it as evidence of a slowing of the population increase rate as the population increases, although of course the census data could not have been very accurate (as modern censuses also are not).[25] Such a regression is commonly regarded as evidence of *density dependence of the rate of population increase.*[26]

Before I leave Thomas Malthus in peace I should draw attention to his efforts to promote a rational approach to deep political and philosophical questions. He rails equally against "advocates of the present order of things" and "advocates of the perfectability of man, and of society". Both, he considers, "offend against the cause of truth".[27] And he makes a step towards the dialectic expressed in the sub-titular quotation I attribute to João Maguiero: "It is an acknowledged truth in philosophy that a just theory will always be confirmed by experiment. Yet so much friction, and so many minute circumstances occur in practice, which it is next to impossible for the most enlarged and penetrating mind to foresee, that on few subjects can any theory be pronounced just, till all the arguments against it have been maturely weighed and clearly and consistently refuted".

MATHEMATICAL POPULATION MODELING

Although Malthus was a mathematician he did not express his theories mathematically, at least not in his published works. But less than a decade after publication of "*A Summary View*" a Belgian mathematician, P.-F. Verhulst, had published what he called his *logistique* equation (He later changed the term to "logistic" because of possible confusion with the French meaning: "the art of calculation") describing the growth of a biological population in which the *rate* of increase (strictly, the *proportional* increase rate, that is the numerical increase in a period of time expressed as a fraction – percentage – of the population number during that time) is continuously and progressively reduced as the population grows.[28] This equation predicts a smooth population growth up an S-shaped ("sigmoid") curve to a stable upper limit, usually designated as K and called, variously, the *environmental carrying capacity* or *saturation level.*[29]

Verhulst's equation, as originally expressed, assumed the decrease in the growth rate was a constant value for a given increase in population, that is that the increase rate is a linear function of the population size, with negative slope. This can be regarded as a process of feedback, called *compensation.* The simple logistic can be generalized to take into account any growth function such that the proportional growth rate is a (continuously) decreasing function of the population size, not necessarily linear. Such are called pure compensation models, and we shall encounter some of them later in connection with the regulation of whaling and some kinds of fishing.[30,31]

Verhulst's formulation must have come sweetly to the ears of the European establishments and businessmen of the mid-nineteenth century – if they ever heard of it – since it described *stability*, not Malthus's and Godwin's nasty "vibrations" and "oscillations". This was Good News at a time – mid-century – of serious labour disturbances in nearly every country, while politicians and the new capitalists revered the new-fangled idea of "progress" and needed social continuity for profit and capital accumulation.

Eighty years after Verhulst's seminal paper an American demographer, Raymond Pearl, was busily applying a logistic equation to data for the population of the United States of America as well as more generally.[32] Pearl was a re-inventor of the logistic curve; he was at the time unaware of Verhulst's work. While writing the first draft of this essay my attention was drawn to a masterly historical review by Sharon Kingsland.[33] Kingsland unravelled the 1930s and 1940s, but focused on the role of Pearl in the development of ideas about population growth and the strong scientific controversies engendered by his ideas. Pearl actually drew his inspiration from the work of T. B. Robertson,[34] who applied a sigmoidal curve which he called an *autocatalytic* or *self-accelerating* curve but which was identical with the logistic, to data for the growth of individual plants and animals, including humans. His name for it came from an analogy with the dynamics of certain chemical reactions.[31]

At about this time an Italian mathematician, V. Volterra,[35] and an American, A. J. Lotka,[36] were generalizing the logistic model to apply to the interaction between two animal species, such as a predator and its prey, and demonstrating that such a system would generate oscillations.[37] In the same period, however, three Norwegian scientists – Johan Hjort, Per Ottestad and G. Jarn – were applying the logistic theory to the problem of overfishing in the Northeast Atlantic and also particularly to Antarctic whaling, with reference to the rapid diminution of blue whales on which that vast industry had been founded.[38] The idea was pursued by Michael Graham at the English Fisheries Laboratory in Lowestoft with respect to the cod, *Gadus morhua*, and other groundfish species in the North Sea,[39] and later especially by M. B. ("Benny") Schaefer and Oscar Sette, in California, with respect to tunas in the eastern tropical Pacific, and the pilchard (sardine), along the US West Coast.[40] Then, throughout the 1970s the Scientific Committee of the International Whaling Commission (IWC) tried, with limited success, to use a modified form of the logistic to provide advice on the regulation of all commercial whaling.[41] It is remarkable that Kingsland seems to have been completely unaware of this sort of application of the logistic equation; her review concentrates, in the later years, on its use in the analysis of laboratory experiments.[42]

The idea behind these applications is simple in principle. It rests on the assumption that the trajectory of population growth is, in effect, *reversible.* What does that mean? The 18th and 19th century students of human population were close to catastrophic events such as plague, famine and war that brought about sharp declines in population number. After them populations increased again. Technically, these were *perturbations*, the study of which is perhaps the most important way in which we learn about the *dynamics* of processes, whether in physics, population biology or economics. We watch small perturbations to see if the system returns to "normal", and large ones to see whether the system is disrupted. The ease with which a population returns to its earlier state when the force causing a perturbation is removed is called its *resilience*, and the trajectory of that return is a *transient.*[43] In biology a very large perturbation may lead to the extinction of the population, and simple logistic theory is not able to deal with that. But a large one that does not lead to extinction is treated as if nothing very important has happened – a sharply reduced population will, given time and peace, simply bounce back.

Malthus and some of his successors realized that a population that has been reduced by an external event or special internal process does not have the same characteristics as a growing population of the same size, because it is *structurally* different: the sex ratios may be different, the frequencies of children, adolescents, adults will be different, and – in human populations – the class composition will be different, and these differences must be taken into account in predicting the future trajectory.

The biologists studying fish stocks in the 1920s-1970s faced a situation in which populations were being reduced steadily by application of a force – fishing effort – which was prolonged and sometimes sustained, to the point of industrial disaster. Monitoring the population decline itself, together with the collection of statistics on catches and some kind of point estimate of the numbers of fish or whales (or their total biomass) could be used to assess the states of the stocks and provide a basis for formulation of management advice.[44] That can only succeed, however, if some way is found to take structural differences between the growing and the declining population into account. This is partially achieved in the IWC application by devising a population model BALEEN II which was developed from the simple modified logistic originally used.[45] This modification has important implications for management but does not alter the basic properties of the logistic model.[46]

Fisheries scientists are usually much more interested in the weight of catches than the numbers of fish in them, and so more in the biomass of the exploited population than in the numbers of fish in the sea. Expression of this interest calls for knowledge of the age structure of the population and composition of the catches, and about the growth rate of the animals as well as their reproduction and mortality. Throughout the early history of the International Council for the Exploration of the Sea (ICES) the attention to be given to estimating ages of fishes (and of whales, too) was a matter of some controversy. Johan Hjort was a strong advocate of routine and comprehensive age-determination; others, including especially the great Scottish scientist, W. D'Arcy Thompson, were – to say the least, unenthusiastic.[47] It is, I think, worth quoting D'Arcy Thompson's 1899 definition of the task of ICES, just about to be established.[48] Scientists studying fisheries for the smaller pelagic species, such as herring, *Clupea harengus,* were preoccupied by the large annual fluctuations in catches, attributed to variations in mortality of eggs and larvae and reflected in variations in recruitment of young, but catchable and marketable, animals to the exploited population. Great effort was for decades put to efforts to predict these fluctuations and to finding relations between recruit numbers and the numbers of female parents – spawners. Those studying bottom fishes, such as cod, flatfishes, and haddock, saw less year-to-year variability and were thus more focused on the survival of still-growing fish in the exploited part of the population.[49] The development of mathematical models with this in mind began with a Russian scientist, F. I. Baranov.[50] His fundamental work was virtually lost until "rediscovered", in English translation, in the "western" world in the late 1930s, by Michael Graham and others.[51]

The differences that nature presented to different groups of scientists led to the parallel development of two general types of mathematical models, commonly labeled as *production models* (derived from the logistic equation approach) and *age-structured models* (derived from Baranov's approach, taken up by, among others, W. E. Ricker, in Canada, and by R. J. H. Beverton and me, both in the 1950s).[52] The two have rarely come together; I think for the first time in a 1957 book by Beverton and me, and called therein "a self-regenerating model". Limited computing facilities at that time precluded more than a cursory exploration of its properties[53] over a small range of parameter space. We do know, however, that this model reveals that a fish population can be driven to extinction if the fishing effort/mortality rate exceeds a critical level.[54]

For the purpose of considering sustainability in the framework of single-species population models it is convenient to confine discussion to production models (including those with limited structural content such as IWC's BALEEN II.) We must keep in mind, however, that some age-structured models of the self-regenerating type turn out to have some rather different properties.

Through most of the twentieth century biologists studying the conservation of terrestrial (and freshwater) systems were concerned about the possibility of population – or even species – extinctions resulting from intensive "use", while marine biologists had no such concerns. Their aim in "conservation" was primarily prevention of the *economic* extinction of fish stocks, and the maintenance – and, where necessary, *restoration* of depleted populations to ensure future economic value. There have been two rather different approaches to this self-imposed task. One, associated especially with the European side of the North Atlantic, sought to find ways, through regulation, to ensure *continuity* and *stability* in the face of growing fishery pressures generated by changing markets.[55] The other, associated mainly with the North American side, sought *optimization* of fishing, especially through setting the target of Maximum Sustained (or Sustainable) Yield (MSY); this is now embodied in international law of the sea, particularly at a global level. Let us see what that meant in practice. But, first, another look at the logistic function.

If the formula for population growth is in fact reversible then we can say that for any population size a catch could be taken in perpetuity, periodically (usually we think in terms of years) of a size equal to the *slope* of the population growth curve. That is one definition of a *sustainable yield.* According to the logistic, modified Pella-Tomlinson style, *a sustainable yield can be taken when the surviving population is at **any** level of abundance* other than its pristine state (at carrying capacity). This is easily seen if one plots the slope of the growth curve against the population size, to give one of the two graphs commonly used for illustrative purpose in fisheries research and management. (The other is a plot of sustainable yield against the intensity of fishing – fishing effort or something closely related to that, such as the mortality rate of fish generated by fishing). This yield-population curve is characteristically dome-shaped, with the sustainable yield going to zero at zero population (of course!) and also at maximum population. With the simple logistic the peak of this curve, which is the MSY, is located at a population level (MSYL) equal to half the carrying capacity. In modified versions the peak may be at higher or lower levels; in the whaling assessments of the 1970s it was assumed (rather arbitrarily, as a political compromise, with little scientific basis) to be at 60% of the carrying capacity. In many fishery assessments the implied MSYL is well below 50%, often nearer 30%.

The MSYL is the population size at which the sigmoid curve of population growth begins to turn, its *inflexion* that is, where its slope, which has been increasing, starts to decrease, and where the actual growth, in number – not the relative rate of growth – is highest. Theoretically, if a catch taken at a certain population size is less than the sustainable catch at that size, the population will increase. If that size of catch is continued the population will increase until it reaches a level for which the sustainable yield is the same as the catch being taken; that level must be above the MSY level. When MSY is being identified as the target for management then it is sometimes said that a stock in this state is being *underutilized.* If, however, a catch is taken that is higher than the sustainable yield at that population level then the population will diminish. What happens next, if catches continue at the same size, depends on the starting point. If the population was initially above the MSYL, and if the catch was no higher than MSY the population will stabilize at MSYL or above it. If, however, the population was initially below the MSYL then continued catches of the same size will result in further reduction of the population and eventually its extinction.

So much for simple theory. In practice, however, there are great uncertainties about the yield-population curve, its shape, vertical scale and the values of other parameters defining it. In practice, too, these uncertainties cannot be fully resolved, no matter how much research is carried out. And, further, external factors – such as environmental changes that might affect the carrying capacity, and the impact of other species on the target species (competitors, prey, predators, parasites) are usually unknown or cannot be predicted. The logistic model is essentially *deterministic*, too, while in the real world even well-estimated parameters exhibit statistical variability, to take account of which corresponding *stochastic* models are required.

These considerations have led fisheries scientists to pay close attention to the notion of *precaution.* How this is done can be illustrated by the approach taken by IWC scientists in the 1970s when told to implement what the IWC called the New Management Procedure (NMP) in the wake of the UN Conference of 1972, in Stockholm, on the Human Environment, which had been very critical of the IWC's conservation record. Until then – but only since 1960 – the IWC had a policy of trying to set annual catches at no more than sustainable levels, a policy that had failed dismally for a number of reasons, one being a refusal, despite 20 years of pleas from the scientists, to set catch limits for each species and population separately instead of as a multi-species bloc – the Blue Whale Unit (BWU). In 1974, policy shifted to targeting MSY, but with a precautionary flavour in consideration of the dramatic consequences of over-estimating sustainable yields in the previous era.[56]

Precautions took three forms: (1) classify populations with respect to their level of depletion below the carrying capacity, and allow whaling only on those above or close to the MSY level, approximately (with MSYL assumed to be at 60% no whaling was to be permitted on any population judged to be below 54% of carrying capacity); (2) allow maximum catches to be only a certain fraction of the estimated MSY; this was in practice taken to be 90% (in retrospect not very precautionary!); and (3) as a later amendment, permit no exploitation of hitherto unexploited populations and species until there had at least been satisfactory estimates of the sizes of such populations.[57]

These precautions were very far from adequate. The final blow to the NMP was given by William de la Mare who showed by computer simulation (the first time such a method was used in fisheries studies) that the Procedure would not be conservative of whale populations even if the model was structurally correct and its parameters perfectly estimated.[58] This revelation was instrumental in encouraging the IWC scientists to look at completely different possible ways of managing future whaling for sustainability and with due precaution and safeguards. The declaration by the IWC of an indefinite moratorium on all commercial whaling in 1982, becoming effective in 1986, gave the scientists the opportunity to devise and test (by computer simulation) a Revised Management Procedure (RMP) which has been accepted by the IWC at the political level but not yet implemented, pending agreement on other elements of a Revised Management Scheme (RMS), including such features as an International Observer Scheme and a database of DNA profiles of caught whales.[59, 60]

We need not worry here with the details of the RMP, but the objectives it is designed to meet, as set by the IWC itself, based on the specification of feasible options by the Scientific Committee, are worth noting:

1. There must be only a very low probability of any population being reduced, by accident, over a period of one hundred years, to below a certain minimum size, that size to be much greater than a theoretical or observed level at which biological extinction of that population might happen.
2. The population must, at the end of the hundred-year period, be fairly close – as specified by "tuning" – to the carrying capacity at that time.
3. Providing conditions (1) and (2) are met the *cumulative* catch over a period of 100 years should be permitted to be as high as possible.
4. The catch limits should not vary greatly from year-to-year except to meet unexpected emergencies, this to satisfy industrial operational needs.

The period of one hundred years was originally set by the limitations of computer capacity in the simulation trials. It is, however, also related to the life-spans of the baleen whales, which are of the same order as the human life-span.

Conditions (1) and (2) essentially provide a new definition of *sustainability*; and condition (3) defines *sustainable use*. Condition (4) simply eases some of the organizational and economic difficulties that could be faced by the industry in being subject to such a rigorous management scheme. In addition the Scheme would provide for all annual catch limits to be zero unless otherwise decided, and where non-zero limits are set they would be a small fraction of what would have been permitted under the NMP. The "small probabilities" define the degree of *precaution* adopted. Finally, while it has been tested exhaustively by computer simulation using "data" generated from various population models (the "base model" being the BALEEN II model previously developed for implementation of the NMP), the catch limit algorithm (CLA) does not itself incorporate or depend upon any particular population model.[61]

Successful application of the RMP approach in the future would depend critically on rigorous management and a long-term commitment. All non-zero catch limits must be those specified by the CLA; there would be little if any room for negotiations. Population surveys must be conducted periodically by specified means; failure to do so, or to provide other required data, would lead to an automatic reduction in catch limits and eventual shut-down of the whaling operation. It remains to be seen whether the international community can abide by such constraints, plus some effective enforcement and control measures, for at least several decades.

ALLEE AND HIS EFFECT

Per Ottestad, in commenting on his application of the logistic for assessment of the sustainable yield as a function of population size or fishing intensity, remarked that Raymond Pearl's analysis of the human population series did not prove the validity of the hypothesis that generated it. Many other mathematical functions would fit the data just as well; goodness of fit does not prove that the data have been generated by the processes assumed in the derivation of the equation. Feller elaborated this view.[62] Subsequent modifications of the simple logistic that have been employed in fisheries studies have usually affected the middle and upper bounds of population abundance or density. Apart from the modifications of the Pella-Tomlinson type, which affect the location of the MSYL, the main difficulty found at high population levels is that if for some reason the population over-shoots the notional carrying capacity the logistic model predicts a negative growth rate. As the growth rate is the *difference* between natural mortality and reproduction this is not impossible, since the (natural) mortality rate could perhaps increase in an unexploited population at high density, so as to exceed reduced reproduction. But that can become "uncomfortable" because a rather small over-shoot leads to rapidly accelerating and unrealistic increase in mortality. It is for that reason that Witting (see below) replaced, in his new model, the unruly Pella-Tomlinson function by two others – a function proposed by William Ricker[63] and a power function, neither of which force negative growth at high densities.

In the context of this Forum, and the concept of sustainability, what happens when a population is reduced to a very small size by exploitation (i.e. is *depleted*) is more important. Both the original and modified logistic models force a population to extinction only if the intensity of exploitation is continually increased; this seems unrealistic. In using the Pella-Tomlinson modification in their BALEEN II model the IWC scientists simply selected an arbitrary level below which the model might not apply, and devised management procedures aimed at avoiding the reduction of the population to such a level, whether deliberately (as for the NMP) or by accident (as for the RMP).

Age-structured models such as the Beverton-Holt self-regenerating model can imply thresholds for "safe" population levels; Tony Pitcher has devoted a short paper to this[64] and the definition of a quantity F_{ext}, the exploitation rate that will lead to extinction. But it is not easy to see how this could be estimated from the usual data, and its experimental determination is unlikely to be proposed!

The usual logistic-style production models can easily be modified to generate what has been called the Allee effect,[65] most easily by adding another parameter to the stock-recruitment function; John Shepherd has shown the way,[66] but, again, it is not obvious how that parameter can be estimated. However, theoretical approaches may help us understand the potential dangers of ignoring the effect.[67]

In fisheries studies the Allee effect may be applied either to the relation of recruitment to parent population size or more generally to the curve of yield against population size. In the latter case the simple dome-shaped curve given by the logistic is called a *pure compensation* model; the proportional (or relative) growth rate continuously increases with declining population. But if the proportional growth rate then begins to decrease as the population declines further the model is said to exhibit *depensation.*[68] This introduces an inflexion into the yield-population curve near its lower end. If the depensation process is strong enough, a *critical depensation* curve is generated. The important feature of this is that there then exists a *minimum viable population level.* The Beverton-Holt self-regenerating model exhibits critical depensation.

There have been numerous efforts to suggest what might be the possible biological mechanisms of depensation; they include cannibalism, for example.[69] Ricker put forward hypotheses in connection with the dynamics of Pacific salmon. But how can sustainable use be assured if we do not know whether to expect depensation to occur or, if we do know that, know at what population level it may manifest itself – that is at the level of an inflexion in the yield curve or in stock-recruitment curves? One approach is that taken by the IWC scientists, and also in effect by the scientific advisers to IUCN in the preparation of criteria for designating species as threatened, endangered, etc. – merely to assume it cannot be higher than some *guessed* level. One criterion which may set an upper limit to this threshold is provided by consideration of the natural variability of the unexploited population, if that is known – which is rarely. Clearly, if the threshold were very high the population would be unstable and at some time go spontaneously extinct. Another approach would be simply to observe, by close monitoring, if and when an exploited population shows signs of behaving differently from what is predicted by simpler models, and take rapid corrective action; but a difficulty there is that wild populations can rarely be monitored with sufficient precision in a timely manner.

These considerations lead, I think, to the practical conclusion that the only practicable way to avoid risk of inadvertent extinction or even of undesirable depletion, is to adopt extremely precautionary management measures (and enforce them, of course). That is what has been done in the development of the IWC's RMP; the CLA provides that this lower threshold is as high as the MSYL of the simple logistic – a depletion of 50% from the carrying capacity. Some experience would justify such a choice: whale populations have many times been depleted to substantially less than that level, and have not shown signs of imminent extinction. And even those that have in the past been reduced to, say, less than 10 or even 5% of their unexploited level are still with us, thus not exhibiting *critical depensation.*[70] However, in some cases they do seem to recover very slowly, for several decades, before beginning to show real signs of recovery; possibly they had been in a phase where depensation is manifest, but not critically so. It must be said, however, that this threshold for the RMP was not set only – or even mainly – to avoid possible extinction, but rather to meet the criterion of high cumulative yield provided that was "safe", and that can only come from an abundant stock.

I end this section with the observation that even the simplest assessment models have on occasions been seriously misused by scientific advisers to management authorities. Such misuse has arisen from attempts to simplify assessment procedures, to avoid recognition of the broad lack of relevant information and, sometimes, from carelessness. Collapses of sea fisheries have often been due to refusal of management authorities to heed otherwise good scientific advice. The most common scenario is a sort of compromise: the scientists say sustainability could be attained by setting catch or fishing effort limits at such and such a level; fishers say they need limits substantially higher than that in order to profit from their activity; authorities agree on something in between. But one cannot compromise with Nature; one cannot take more than Nature is ready to yield, at least, not in the long term.

However, the scientific advice has sometimes not been good, quite apart from the consequences of uncertainty and ignorance. The most common such failure has been to ignore the density-dependent reproductive processes and how they behave when a population is being depleted. A commonly used procedure called Virtual Population Assessment or Analysis (VPA) does just that; it is akin to the error, discussed later, of managing by setting catch limits at the level of replacement yield; its temporary effects when used as an interim measure are not usually damaging, but repeated over time, unthinkingly, they can be catastrophic. Another error has been to regulate catching in such a way as to ignore the fact that a population which has appeared robust when it is abundant – because recruitment shows little relationship with parent density – becomes vulnerable when it is depleted and the inevitable stock-recruitment relationship emerges[71] A third pitfall is the use of simplified assessment models that, in certain conditions, can drive a population to depletion if not extinction.[72]

THE MAGIC OF SIGMAS, BELLS AND DOMES

Jugglers with the logistic in the 1930s and the first decades after World War II – at least in the fisheries arena – were struck by the similarity of the first derivative of the simple, original version of that function to the "Gaussian" or "Normal" curve of statistical theory (when they are scaled appropriately) and, conversely, with the near identity of the logistic itself with the integral of the normal curve – accumulated probability – despite their being algebraically quite different. Both are symmetrical and long-tailed, and the logistic derivative is only slightly more "peaked" (more "compact") than the normal curve. Both were applied, for example, to the data from trials to determine the selectivity of fishing nets with respect to size (length) of fish caught.

Although scientists studying whales and dolphins, and also those looking at terrestrial wildlife, usually concern themselves with the numbers of animals, fisheries scientists, and most ecologists are, as noted earlier, interested primarily in the weights of catches and the biomasses of populations. It is perhaps worth recalling here that the applications of the logistic – which was derived from consideration of numbers – was commonly applied, in the fisheries context, as if it equally defined changes in biomass/catch by weight. While that is perhaps harmless in the context of general popular exposition, it has occasionally been confusing when unthinkingly applied in management/conservation calculations. The first application, to whales and whaling, was exceptional in that practical interest was in the *numbers* of particular species of whale caught, the slide into applying the logistic, without modification, to – for example – the catches of North Atlantic cod or to yellowfin tuna, expressed in thousands of tonnes, is questionable. So we should look at the relation between the two.

Many, perhaps most, fishes seem to be able to grow continuously through life, albeit more slowly as they increase in size. The curve of weight of an individual against age is typically S-shaped, with an inflexion commonly at between about 30 and 40% of the asymptotic weight. Naturally, there are wiggles in the curve caused by seasonal changes in growth rate within years, but I shall ignore those here. Various functions have been used over the years to fit annual growth data, the most commonly applied being a function due to von Bertalanffy or some modification of that.[73, 74]

Now consider the total weight (biomass) of a cohort of animals, what fishery biologists usually call a year-class. If it is not exploited the number of animals in the year-class will diminish by natural causes, and if those causes act uniformly through life the decline will be geometric/exponential. The number surviving to a certain age multiplied by the average weight of the individual at that age gives the biomass of the cohort; this is typically an asymmetrical bell-curve with a longish right-hand tail. We could envisage that if we could arrange to capture an entire cohort when it had reached the age shown at the peak of that curve we could maximize our catches. But of course we could not do that in practice. If we were fishing more or less continuously then we would catch some fishes at all ages. It is not too difficult to calculate how intensely we should fish to maximize the catch from the cohort throughout its natural life, though that would be somewhat less than we could get theoretically by catching all at the same "optimal" age.

A little conjuring trick allows us to say, in certain specified conditions, that what could be taken from one cohort throughout its life-span is the same as what could be taken in a single year from all existing cohorts combined. This gives us the Sustainable Annual Yield per Recruit curves (***Y/R***) familiar to fisheries scientists, where the number of "recruits" is the average number of fish attaining "fishable" size each year and so becoming liable to capture. ***Y/R*** values are usually plotted against a measure of *fishing effort* or *intensity*, or the mortality rate due to fishing (***F***). Typically such curves are bell-shaped but skewed to the left; the location of the peak of the curve gives an indication of what has been referred to as an "optimum" fishing intensity – though it is optimal only in a special restricted sense.

Y/R calculations have been widely used to provide scientific advice for managing fisheries for both sustainability and high production. Their usefulness depends on the validity of an assumption that the number of recruits is on average essentially unrelated to the number of sexually mature animals that produced them. For many fishes this is not an unreasonable assumption provided that the population is rather lightly exploited, or if it is increasing; in this latter situation the benefits of management for sustainability might be greater than as calculated. But, of course, it would be a dangerous procedure to apply to a declining population and/or an intensely exploited and hence probably depleted one.

Although ***Y/R*** curves are typically peaked, that is not always so; the existence or otherwise of a peak depends on the relation between the natural mortality rate and one of the parameters of the growth function, and also on the size or age at which fish become liable to capture, relative to their final (asymptotic) size. If the natural mortality rate is relatively high the ***Y/R*** curve against fishing intensity becomes asymptotic. This is dangerous! It leads one to expect that the sustainable catch can be continuously increased – though perhaps only slightly by an unlimited increase in fishing intensity. But, of course, such increase leads more and more rapidly to a decline in the population, and especially of mature adults, so that eventually the numbers of recruits will diminish and the population will crash.[75]

This difficulty can in theory be resolved by combining yield-per-recruit calculations with an appropriate function relating recruit number to the size (number, biomass or other appropriate measure taking into account variations of fecundity with age or size of fish) of the preceding parent "stock". (There is a voluminous literature on so-called "stock-recruitment relationships" in connection with fisheries management.) As I have already mentioned this combination provides what have been called *self-regenerating yield curves*, the general properties of which are not unlike those provided by generalized logistic models.[76] However, they do differ in two specific ways: they take explicit account of the age-structure of the population, and they are used to calculate weight of fish in catches and populations rather than numbers. They can also have an additional property of predicting population extinction under certain circumstances of excessive fishing intensity, without the stock-recruitment relationship itself having *depensation* specifically written into it.[77]

MORE MODELING

We have seen that the logistic models would "allow" some catches to be taken repeatedly and sustainably at any level of population, although approaches such the IWC's NMP and RMP would in principle set limits to that by prescribing minimum population sizes and/or levels to below which there must be minimal or no risk of the population being driven towards extinction by exploitation. There are two quite different reasons for trying to prevent use – driven by search for profit – from depleting populations of whales, fishes or other wildlife to "dangerously" low levels. One is, of course, the fear of accidental extinction. The other is to provide for exploited populations to remain abundant or to recover to abundant levels. There can be several reasons for this latter wish. One, obviously, is that abundant populations can in principle provide higher absolute yields and also better returns for effort than depleted ones. For certain types of so-called non- or low-consumptive use, such as whale watching and other forms of ecotourism, high abundance may have an intrinsic value – though this is not necessarily so: much effort is put into spotting rare species. But increasingly it is being appreciated that controlling exploitation so that abundances of primary target populations are high may be a safeguard against unwanted "damage" (however that might be defined) to the ecosystems of which those populations are integral elements. To evaluate this we need very different kinds of models, and those we have now are, I think, far from adequate for that purpose.

Nowadays, these considerations of possible biological "damage" fall under the rubric of "preserving biological diversity" around which an enormous volume of documentation has been produced in recent years, and an astronomical number of words spoken. Several dimensions of diversity have been identified: genetic diversity within the species, or population; diversity of species populations; diversity of species; diversity of ecosystems and, I suppose, global diversity.[78] To these I would add potential for continued evolution, for which genetic diversity is a necessary but not sufficient condition, and also "cultural diversity" with respect to the animal kingdom. Sustainability should therefore encompass protection of all these forms, and sustainable use must in future provide for the possibility of *substantial* use while maintaining diversity in all its forms as far as practicable.[79] This calls for balancing a variety of objectives, with priorities set among them, just as the IWC's RMP aims to do in a much more restricted field. A brief look at some of the emerging ways of dealing with this situation seems to be in order. It is also timely in view of growing pressures to "cull" some species in order to "save" others.[80] In particular, considerable efforts are being made to convince us that the present global crisis in sea fisheries is, at least in part, due to consumption of fishes and squids by whales, dolphins, seals and perhaps by other wild predators. All such efforts are directed at sub-optimal use of the species to be culled (in single-species terms, and if it has a market value) and, generally, *unsustainable* use. These pressures are therefore a potential threat to the notions of sustainability and precaution as they are commonly now understood, even if not always in precise terms.[81]

The orthodox approach to this issue is the construction of multi-species models comprised of two or more – sometimes very many more – single-species elements.[82]

The elements, or primary modules, are usually quite simple ones, sometimes without the internal density-dependences that would provide for their individual stability or resilience. Such models involving interactions between only two or three species may be constructed from logistic elements, and the differential-integral equations defining them may be formally soluble.[83] When there are more elements to be taken into consideration solutions must be found for specific situations by computation, usually involving iteration – that is, a process of repeated calculation and successive approximation until the answer converges, hopefully, to a correct value, number or pattern. Modern computers permit this, but a price paid can be that one may be "unable to see the wood for the trees", and to know whether the particular solution obtained has any more general validity, and especially whether it is a unique one.

In most current many-species models the variables are not population numbers and corresponding reproductive and mortality rates, but rather biomasses, sometimes of species populations but also of groups of species assumed to occupy the same or similar *niches* in the ecosystem in question, or even certain *trophic levels*, such as "primary producers" (usually by photosynthesis), "herbivores", "primary predators", "top predators", "scavengers", "reducers" (anaerobic bacteria for example) and the like.[84] Such models deal with the transfers of organic matter (biomass, essential and critical elements such as molecules containing nitrogen, carbon, silicon, phosphorus, iron) and of metabolic energy among the elements.[85]

The complex structure of such models is well illustrated in an ecosystem diagram provided by Peter Yodzis.[86] Even if numbers and formal relationships among elements and modules (groups of elements) were to be specified in such a diagram, and iterative solutions were to be found by very long computation, the results would always be extremely sensitive to any errors in estimates of parameter values or in definitions of structural relationships. This matter of sensitivity has often been over-looked, even though testing is in principle simple, if laborious – merely by running the model over and over again with a suite of different values. In reviewing this problem Anne Hollowed and her co-authors had this to say after their examination of four types of models, which they classed as descriptive multi-species, dynamic multi-species, aggregate system, and dynamic system:

> Populations are regulated by competition for food (food limitation), predation and environmental variability. Each factor may influence different life-history stages, locally or regionally.[87] However, most multi-species models address only a sub-set of these factors, often aggregated over functionally different species or age groups. Models that incorporate the important interactions at specific stages and scales will be necessary if they are to continue to supplement the information provided by single-species models.[88]

I think these authors are over-optimistic. Even the dynamic stability of such models is practically impossible to determine, although examination of their structure might indicate whether they will certainly or possibly exhibit *chaotic*, that is unpredictable, behaviors in response to perturbations.[89]

In recognizing such problems the IWC scientists took a different route to taking into consideration multi-species issues for the practical purpose of providing scientific advice for a possible sustainable and precautionary exploitation of baleen whales. This involved, during the development of the RMP, testing the robustness of various proposed single-species population CLAs in situations of hypothesized drastic environmental change. The scenarios tested included sudden changes not only in the general biophysical environment but also changes – reductions or increases – in "carrying capacity" and/or population growth rate, and gradual changes of similar magnitude over many years. In this case "the environment" includes significantly interacting competitors, predators and the like, be they other types of whale or other kinds of animals. The validity of this approach depends upon a more or less correct determination of the time-scales of change – which includes especially the life spans of the animals involved – and periodic monitoring of events, also on a commensurate time-scale. This approach is described in numerous Annexes to several reports of the IWC Scientific Committee, and some of its implications are described by W. K. de la Mare.[90]

GEOGRAPHY

So far we have concentrated on one dimension only of the problem of sustainability – potentially unlimited and continuous time. Another dimension – space – has been given much less attention, particularly in theoretical studies. Recently, however, the idea of declaring *marine protected areas* (MPAs), particularly as places where exploitation will not be permitted (*closed areas* in other terminologies) as a conservation measure has been explored by a number of authors. Ray Beverton and I were, I think, the first to look at the idea mathematically (and quite crudely!) by hypothesizing that bottom fishes such as plaice, diffused randomly as well as moving periodically between spawning and feeding grounds in adjacent areas), and we applied the well-known techniques of physical diffusion to do this. The idea was eventually picked up again by Sylvie Guenette *et al.*[91] and reviewed by Sumaila *et al.*[92]

Although potentially important for fisheries management this literature scarcely mentions sustainability as such. In general the idea of the reserve is to set limits on fishing effort and to redistribute it, and the issues of sustainability are, I think, essentially the same as consideration in the time dimension. Examples of this limited modeling approach are provided by Carl Walters and others.[93]

More directly relevant to our task here is, I think, the work of Alec McCall[94], based on a large body of theoretical literature about population dynamic models in heterogeneous environments, reviewed by S. A. Levin.[95] In this context the central concept of Density Dependent Habitat Selection (DDHS) refers to a special category of habitat selection by animals (much of the basic work is on birds) in which population size and local density are important factors influencing choice of habitat and hence the relative distribution of the population among habitats. Most models portray arrays of discrete habitats, but in application to aquatic systems a subset of the theory models, spatially, continuously varying habitat; the difference is fundamentally no more than envisaging on the one hand geometric changes with time and on the other exponential changes with time. The basic approach is to imagine a matrix of small "cells" within which changes in local population density/number are described by the logistic or a variant of it. In each place the values of one or the other or both of the two parameters that specify the logistic – the carrying capacity, $\boldsymbol{K}$, and the intrinsic rate of population growth, $\boldsymbol{r}$ – define the habitat suitability there.[96] Another parameter or function describes the dynamics of movement between one cell and another.

A "commonly observed phenomenon associated with DDHS", writes McCall, "is *expansion and contraction of population range* or differential utilization of marginal habitat with changes in population abundance".[97] This process can be envisaged as follows (again I quote McCall):

> If habitat suitability (or per capita growth rate) is depicted geographically as increasing downward, habitats can be described as a continuous geographic suitability topography having the appearance of an irregular basin, whose shape may also vary over time. According to the ideal free distribution, the population will fill this basin as if it were a (viscous) liquid under the influence of gravity.[98]

This "basin model" can be used to examine the relationship between range and population abundance and so has important consequences for the consideration of sustainability.[99]

If sustainable use in the time dimension is considered, for example, to be a level of use that does not drive populations close to an inflexion of the density-dependence curve where depensation may begin (see below), then the equivalent in the spatial dimensions would be a threshold beyond which the target's range should not be forced to reduce further for fear of subsequent spontaneous contraction. Similarly, if sustained use involves taking no more than will cause the population to decline no further than it may already have done, then the geographical equivalent would be the maintenance of the current range. *Restoration* of a depleted population similarly has both temporal and spatial dimensions. What emerges clearly from DDHS theory is that population abundance and distribution are inseparable features of dynamics.

This is an appropriate place for a warning about the notions of the population size, or *abundance*, and *density*. These terms frequently appear in the literature of population dynamics and related management theory as virtual synonyms. This may not be a problem if range and distribution do not change over time. But, if they do, then the two quantities must be carefully distinguished. Ultimately, that is probably nearly always, since concurrent changes in both total abundance and range are surely the norm, not the exception.

Here we have been considering the expansion and/or contraction of ranges of populations. As with the transition of the concept of natural population growth to that of reduction by exploitation – regarded, by default, as reversible – so it is tempting to see the reduction of the range of a population as simply the reverse of the natural expansion of a population over a new area. This might be a reasonable starting hypothesis in looking at sustainability, but has pitfalls. Where, in the time dimension, a population that has been reduced and then "protected" may encounter a situation where the resources available to it are very different from those pertaining before – for example another, competing, species has expanded into the "niche" previously occupied – then, in the spatial dimension, other species may have moved in to occupy part of the original range.

Theoretical and field studies of the biology of the two dimensions – time and space – have developed largely independently, and in different contexts. (They might be coming together with increasing interest in "protected areas" and the like as conservation and management tools.) Indeed McCall[100] observed that, even within the studies of the spatial dimension,

> Cross references among the various treatments of DDHS are exceedingly rare, even in review articles. This lack is especially

> surprising, given that the Fretwell-Lucas and Verner-Orians theories first appeared and are well known in ornithological contexts. These reviewers, and hence many ecologists, may not have recognized fully the unifying principles of DDHS underlying the various models.

Nor, I would add, have the unifying principles underlying the separate studies of the time and the spatial dimension of population dynamics yet been sufficiently appreciated. In this connection, and with reference to the historical starting point of this essay – with Malthus in the eighteenth century – it is interesting to note (again I rely on McCall's words):

> the independent development of concepts akin to DDHS in the field of economics, extending back to the work of David Ricardo, who pioneered the concepts of diminishing returns and economic rents in the early nineteenth century.[101]

McCall further notes that F. H. Knight explored these ideas as early as 1924, in relation to social welfare, and discussed the "marginal value theorem", that was independently discovered in the biological context of optimal foraging some forty years later by E. L. Charnov.[102]

A distinguishing feature of the spatially focused population theories, in contrast with the main thrusts of the temporally focused studies, is the importance they give to understanding the "mechanics" of biological evolution. So it may be that it is by the confluence of the two intellectual streams that we shall find the way to use populations and ecosystems sustainably while preserving the evolutionary capacities of those systems.

GOOD VIBRATIONS

Oscillations, sometimes sliding into unpredictable chaos, are an almost universal feature of even moderately complex systems, as Robert May demonstrated more than thirty years ago.[103] Their existence and ubiquity deny any universal interpretation of sustainability in terms of the generalized logistic model, where a sustainable catch can be taken from a population whatever its size (other than at carrying capacity), and a maximum sustainable catch taken at some intermediate size. The possible existence of processes generating depensation – particularly critical depensation – and overcompensation, also preclude such a simplistic definition.

We have seen that fisheries scientists, especially those studying the smaller pelagic species, were long preoccupied with annual fluctuations in catches, caused mainly by fluctuations in the numbers of young recruits entering the fishery. Ray Beverton and I, with others of the time, showed – in the 1950s – that if such fluctuations were more or less random then this would not, in itself, necessarily invalidate a management process based on consideration of steady states of populations.[104] The existence of oscillations, especially perhaps those with harmonics, whether caused by processes intrinsic to the population itself, or by interactions with other populations, or by periodic environmental changes more generally, force us – if we want to retain the notion – to consider more subtle definitions of what was once called "wise use" – conservation based on maintenance of viable exploited populations at relatively high, essentially unchanging, levels.[105] To begin to do that we must appraise the scales and frequencies, and causes, of a wide variety of types of oscillation in marine systems. We begin with the ocean itself, and as whole, and for examples of this I am especially indebted to the remarkably comprehensive and accessible summary of modern findings in oceanology by Bruno Voituriez.[106]

The vast world ocean circulation, generated by the interactions of changing salinity and temperature, and hence density (the thermohaline circulation, known more popularly as "the conveyor belt"), involves the movement of most of the water in the ocean through a cycle, in part near the surface, in part at the ocean bottom, from and between the North Atlantic, North Pacific and tropical Indian Ocean, and the great Southern Ocean.[107] This circulation has a period of about 1500 years. It is generated by the interactions of the ocean with the atmosphere in terms of water and heat transfers, and is particularly driven by the Antarctic Circumpolar Current which is generated by prevailing westerly winds at high southern latitudes, themselves the result of the great torrent, at up to 200km/hr, of cold, heavy air falling from the high Antarctic ice-cap. The details of this circulation have changed many times during and between a succession of ice-ages; the last was about 18,000 years ago, but these ages have been traced back through the preceding interglacial period of 120,000 years and eventually back to over 500,000 years ago.

Within this general circulation there are local and regional oscillations, the first of which to be identified was the so-called El Niño – La Niña, which appeared to affect alternately the two sides of the North and Equatorial Pacific and was linked with large changes in fish catches, especially of anchovy, off Peru and northern Chile. Subsequently it was found that this was a partial regional manifestation of a large regional process called the El Niño-Southern Oscillation (ENSO); El Niño is the warm phase of ENSO; La Niña its cool phase.[108] While it is repeated every few years ENSO has no clear periodicity. But it can be identified and characterized by the Southern Oscillation Index (SOI), which is based on the difference of atmospheric pressure at sea level between Tahiti and Darwin, and from its oceanic component using sea-surface temperature anomalies in the eastern equatorial Pacific, or the sea-level anomalies along the equator.[109] Similarly a North Atlantic Oscillation (NAO) is defined as the difference in atmospheric pressure between the Azores anti-cyclone and the low sub-polar pressure near Iceland. Like the SOI the NAO also varies – when the pressure off Iceland falls that at the Azores rises, and *vice versa*. Reconstruction of the NAO back to about 1700 AD, using data from the growth rings of trees, reveals oscillations of periods of 2, 8, 24 and 70 years. The climates of Europe, northwestern Asia and the northwestern coast of North America are, especially in winter, closely linked with the NAO index. Thus, in this sub-system we are looking at periodic changes on decadal scale.

From these analyses, backed by theory and observation, it emerges that the regional oscillations are in reality vibrations in the great conveyor, with, necessarily, differences in timing of the various consequences. This recognition gives us now a basis for interpretation of vast periodic changes in fisheries over scales of decades and centuries. Before that we should note that the details of biological production processes, providing the basis for the fluctuations of fisheries, are largely determined by the existence of *divergences* between water masses (zones within a current or, more often, at the interface between two currents), towards which subsurface water flows, causing the thermocline – the layer of sharp temperature change[110] – to be near the surface, and convergences (corresponding zones towards which surface water flows, causing a deepening of the thermocline). The divergences and other upwellings (mainly caused by westerly winds impacting the west coasts of the continents) are usually, but not always, associated with high primary biological production.[111] *Eddies*, or *gyres*, on the other hand, are usually, especially in their centers, markers of oceanic "deserts"; they are whirlpool-structures characteristic of mid-scale turbulence, major loops of the oceanic circulation associated with the main subtropical anticyclonic circulations in the Atlantic and Pacific.[112]

With all that in mind let us look first at the dramatic collapse of the cod stocks in the North Atlantic in recent years.[113] There is no doubt that this is primarily the result of excessive fishing. After the Basque whalers discovered the great Newfoundland bank stock, after they had pursued the North Atlantic right whale, *Eubalaena glacialis*, to near extinction (in about the middle of the second millennium AD), that fishery was sustained for several centuries. Even in the 1970s it seemed that the recruitment to this stock or stocks was essentially independent of the parent stock size, as was commonly believed in relation to groundfishes (demersal species, traditionally caught by bottom trawls). However, as fishing pressure increased further (in this case mainly as a result of the opening up of larger markets for frozen fish that had previously been served by fresh and then iced fish) the parents were reduced to levels at which an effect on recruitment was induced (but not noticed, and not immediate because it takes cod a few years to reach sexual maturity) and this was followed by the beginning of serious decline. But, even though the normal annual variation in cod recruitment is not so great as in, for example, the herrings, sardines and anchovies, one or more "unlucky" years of low recruitment were sufficient to trigger a catastrophic decline. But there was no evidence of large-scale natural oscillations in this species.

In 1970 the fishery off Peru for anchovy, the world's largest fishery by far – by weight though not by value – collapsed, at the peak of its production. This was attributed at the time to the onset of an El Niño that year. But the previous El Niño events in 1965 and 1969 had not affected catches. Nevertheless the fishermen and authorities were content to blame the El Niño scapegoat rather than look to a consequence of overfishing. This theory led to a great deal of scientific research on a topic of obvious high economic importance. But, still, the catches increased again from about 1973 and were not much if at all diminished by an El Niño event in 1997-98. Why not?

The anchovy does not live alone in these waters; it shares them with the sardine. When the anchovy is scarce the sardines are fished more intensively. But the sardine appeared to be immune to the El Niño phenomenon. The expansion of the Peruvian anchovy fishery was triggered by the collapse of the Californian sardine fishery between 1945 and 1950. Until then the anchovy had been "protected" in the interest of the guano producers, but guano gradually lost its value with the advent of synthetic phosphate fertilizers. So the anchovy fishery could expand dramatically. BUT, it was later noticed that large swings in sardine production in Japan preceded by a certain time lag the vicissitudes of the Californian industry, and subsequently the swings of sardine catches in Peru followed the same pattern. The fishery data alone do not permit a conclusion as to whether there is a 50-60 year cycle in alternating anchovy and sardine production in the Pacific, but analysis of scales of both species in the ocean sediments off California, provide a history from 270 to 1970 AD showing a fairly regular cycle with a 60-year period. Analysis of catches in the long history of Japanese sardine fishing has revealed a corresponding, but apparently less regular cycle since the early 17th century, with peaks in 1650, 1710, 1820, 1935, and 1985.

So, sustainable use of the anchovy and sardine resources of the Pacific would apparently involve understanding such natural oscillations and predicting them. However, the fishery for salmon species in Alaska also shows somewhat dramatic changes, coincident with changes in sardines and anchovies. It turns out that the warm phase of the Pacific Decadal Oscillation (PDO)[114] favours the salmon of Alaska and the sardines of California, Peru and Japan, while the cool phase favours only the anchovies.

Let us now return to the North Atlantic. The once great herring fishery collapsed in the 1960s and 1970s. Thereafter there was much controversy over whether that was due to fishing or to natural causes. There is now no doubt that it was triggered by the onset of fishing for young herrings from which to make fishmeal and oil. But the European herring fishery is very ancient, and it was undoubtedly "sustainable", in the common sense, for many centuries. Still, as Voituriez writes:

> The long history was not without incident and, in certain cases, notably that of the Norwegian Sea stock, boom periods for the fishery alternated with others in which the herring almost completely disappeared.

Overall records are not of long enough duration to allow unquestioned appraisal of these variations but there exist one thousand years of good records for the gatherings of herring in the Swedish coastal region of Bohuslän, in the Skagerrak between the Baltic and the North Sea. There have been nine periods of great abundance; in more recent years these were: 1556-90 1660-80, 1747-1809, 1877-1906. These so-called Bohuslän episodes are indicators of perturbations which, beyond the Skagerrak, concern the whole of the North Atlantic and have an impact on the fishery, not only of the herring *but also of the North Atlantic sardine*. Sardines inhabit different waters from the herring; they do not thrive in cold water. Catches of the two species taken near the moving boundary of warmer and cooler water, in the Southern North Sea and The Channel, will tend to alternate, and these alternations have now been shown to be related to the NAO. The NAO fluctuates on various time-scales, and the positive and negative anomalies may extend over several years or even decades. We do not have direct measurements of the NAO prior to the mid-nineteenth century, but the history has been reconstituted indirectly from the length of the season in which the coasts of Iceland are ice-bound; this is possible from the analysis of tree-growth rings and from the cores of the Greenland icecap. Voituriez comments, significantly:

> The last recorded Bohusiän episode dates back to 1877-1906. The NAO did not cease to oscillate then and another should have occurred in the 1960s when the negative anomaly of the NAO was at its maximum. But it did not. Why? Probably because the considerable increase in the fishing effort which led to the collapse of all the stocks in the 1960s-1970s made the climatic signal quite secondary and insignificant.

The last fisheries example given by Voituriez is for the bluefin tuna fished in the Mediterranean, for which we have reliable catch data since the sixteenth century.[115] These show oscillations of very wide amplitude (but no long-term trend), and period of about 100-120 years that cannot be attributed to changing social or economic conditions, nor to repeated overfishing since the fishing effort has been nearly constant. The causes remain unclear; hypotheses range from climatic variations modifying the migratory path, periodic changes in conditions governing larval survival, and the intrinsic dynamics of the species amplifying by resonance the random variability of recruitment. In a sense this fishery has been sustained, even though it is highly and regularly variable. But a change in the effort – for example by introduction of another kind of fishing gear and an expansion of markets – could change that situation rapidly. What is very clear however is that too short-term a view could lead to completely wrong conclusions about management, depending on circumstances; in this case long, almost continuous, declines in catches for half a century (repeatedly) can look like serious overfishing demanding restrictive regulation. On the other hand, focus on a long period of steady increase can, in different circumstances, encourage more intensive fishing. Such considerations may also remind us of a common attitude of fishers in the cases of fisheries with highly variable annual recruitments, which is that when there is a good year-class it should be exploited very intensively, before the fish all die from natural causes; this has sometimes been expressed in religious terms: God gave us such good luck, it is sinful not to take advantage of that.

In all the examples given there appears to have been something of a climatic break around 1975, a period when we passed from the "classical scenario" of the El Niño to a more peculiar regime, with a predominance of negative anomalies in the SOI. It was also a period when the PDO switched from a cold to a warm phase (with serious consequences for salmon, sardines and anchovies), and when the NAO inverted, passing from a negative anomaly (cold) to a positive one (warm). The climate dynamics are of planetary scale, so it is not surprising that these oscillations are not independent of each other. Indeed, air-temperature data analysed by the UN Intergovernmental Panel on Climate Change reveal an oscillation for the entire northern hemisphere with a period of 65-70 years. A Russian meteorologist, L. B. Klashytorin, has looked at these data through definition of an Atmospheric Circulation Index (ACI), calculated from the atmospheric pressure field over the Atlantic and Eurasia. With that tool Klashytorin analysed global fish catches over the past 100 years and showed that the 60-year variations in the ACI closely reflected the yields of salmon, herring, sardines and anchovies. This global synchrony, linking the oscillations of fisheries to atmospheric conditions, is not surprising, considering that they are mediated through oceanic processes.[116]

So, in our efforts to attain sustainability we have to cope with natural "vibrations" of various periods and types and amplitudes, generated by external events, that are to be expected but are not entirely regular and therefore predictable. And also oscillations derived from the internal structure of the exploited populations as well as their interactions with other populations. Furthermore we always encounter extreme difficulty in disentangling these two general types and causes of change. But before we leave these problems arising from various kinds of variability we need to confront yet another challenge to our ability to use living resources sustainably.

CHALLENGING THE ROOTS

In recent years Lars Witting, a Danish scientist working in Greenland, has cogently argued that the assumption that the rate of population growth when the population is small is geometric/exponential is not valid. His argument is derived from considerations of population genetics. If there is some genetically inherited element of variation in the reproductive powers or mortality probability among individuals, then this will have survival value and the natural population as a whole will evolve in the expected direction. It turns out that the intrinsic rate of population increase would not be exponential but, rather hyper-geometric in form.[117] When this function is used to replace the exponential assumption in otherwise traditional population models, it is found that they exhibit oscillations.[118] This might look like a purely academic finding, but it has great practical implications.

Under long protection, the gray whale population of the Northeast Pacific and the Arctic Ocean recovered from near extinction by the American commercial whaling of the late nineteenth century. That recovery has been monitored by sightings from shore stations on its migration route to and from the breeding areas of Southern California and Mexico; such monitoring is facilitated by the fact that this route passes close to the western coasts of North America for several thousand miles. The trouble is that, despite continued substantial catches by indigenous people in Siberia (authorized by the IWC), and despite the severe reductions over the past century in its available breeding grounds, the gray whale has recovered too far. It is now two or three times as numerous as would be expected by the usual kind of back-extrapolation of population size taking account of the recorded catches over a long period. Furthermore, the difference cannot be explained by hypothetical large unrecorded catches, historically, by the indigenous people of North America or the Arctic. The "standard" population models simply do not work. However, Wittings modified model *does* account for the present high numbers, and it predicts that the population is now close to a peak and will soon begin to decline, regardless of any exploitation pressure.[119] Based on the standard BALEEN II model the IWC Scientific Committee has advised that "a take of up to 482 (gray) whales per year is sustainable." Witting's model, on the other hand, predicts, according to its author, that constant future annual catches of 50 or 170 whales (the present authorized annual "aboriginal subsistence" catch is about 180 whales) may cause the extinction of the population during the current management regime, regardless of any possible adverse environmental change. The calculated population trajectories fit the data well for the thirty-year period of sightings surveys. *The period of oscillation is of the order of two centuries.*

So it seems that the gray whale history lends considerable support to Witting's hypothesis. If indeed it is valid it has profound implications for our concepts of sustainability, particularly in terms of precautionary assessments of acceptable future catches. As it happens, the IWC Scientific Committee's view has no immediate practical consequence, mainly because three of the four main range States – Mexico, USA and Canada – within whose Exclusive Economic Zones this species mostly lives do not wish to engage in commercial whaling. Nevertheless, continuation of the large "aboriginal subsistence" catches by Russian nationals in the Arctic could eventually have a devastating effect.

The gray whale is the only species for which we have a long time-series of population estimates made by what would now be regarded as a consistent and reliable method. This species does have a life-style distinctly different from that of other baleen whales (it is, for instance, not a pelagic but a benthic feeder), but that does not justify an assumption that the population dynamics just described are unique to it. At this point it would be reasonable to assume that, until we have contradictory evidence, such dynamic behaviour would probably be demonstrated by other species.

Evidently, the existence of cycles, in general, whether derived from Witting inertial dynamics, internal population structure (Malthus-Clark vibrations), interactions with other species (Volterra-Lotka-May type), regional or global swings in ocean climate (as summarized by Voituriez) or sustained ocean changes (such as the retreat of the Antarctic sea-ice edge which also might have a long-term cyclic component) suggest that it is time to discard simple ideas about steady states, equilibrium, and stable sustainable yields that might be altered only by casual environmental changes.

ECONOMIC SUSTAINABILITY

This book is focused on ecological sustainability, not its economic – possibly illegitimate – cousin. Nevertheless the two are related and some words may be appropriate here.

Economists, when creating mathematical models intended for application in managing the use of renewable natural resources, especially living resources, naturally include in their constructions modules pertaining to the dynamics of the resources themselves. Unfortunately, the crudest and simplest models available have almost always been used for this purpose. Sometimes they do not even have density dependence or other regulatory feedback processes incorporated. More usually the simple logistic is the chosen formulation. In certain circumstances this may not matter, but sometimes the choice is crucial. We have seen some of the failings of the simple logistic, and also of some of its modified forms – for example that it does not confront the possibility of the resource being driven to extinction, except under infinite exploitation pressure, which is hardly realistic in any circumstances.

A use that is biologically sustainable (regardless of the precise meaning given to the phrase, in context) will not necessarily be economically sustainable. The history of industrial whaling is a case in point. If such whaling had been limited to biologically sustainable levels in the days when the great whales were enormously abundant – say the 1920s – then, given the available hunting and processing technologies and the types of markets, an industry would probably have been economically sustainable because it was profitable and because it permitted both operating costs to be covered and capital to be accumulated for replacement of ships and equipment and for research, both on the natural resources and on the improvement of the technologies. If, however, a precautionary regime aimed at biologically sustainable use, such as the RMP/RMS, is put into effect with respect to whale stocks that are still very far below their carrying capacities there is no guarantee that such use will be economically sustainable. Even hunting for the still rather abundant minke whales in the Southern Ocean, that is now carried out by a single "pelagic expedition" (the operating time of which is now split between the Antarctic and the North Pacific, for economic reasons), has to be heavily subsidized by the Government of Japan and is surely not economically sustainable.

Continuation of an industry after the resource has been depleted, through seeking biological sustainability by severe restraints on exploitation (catch limits and the like), will usually be extremely painful, socially and economically. The vast profits that may have been acquired during the biologically unsustainable preceding phase have usually been spirited away into other enterprises; this is what happened with respect to Norwegian and British Antarctic whaling up to their end, in the 1960s-early 1970s. Those profits helped fuel, respectively, the growth of the Norwegian shipbuilding industry and the road haulage industry of the United Kingdom. So a sort of "sustainable development" was ensured. Before the end "exit strategies" were prepared (secretly, of course) by companies in both countries and, while lip service was paid to the IWC's efforts to stabilize the industry through sharply reduced catch limits, the reality was that it was economically more desirable to continue for a limited period at a biologically unsustainable level.

With a global crisis in the fishing industry now, mainly caused by overfishing, it must be very tempting for corporations (usually with sympathetic governments behind them), to opt for continued unsustainable levels of catch while exit strategies are prepared. It is not so easy, politically, to do this since the broad adoption of the sustainability philosophy by the international community in recent decades, mainly through the series of UN special conferences since 1972. Unfortunately, the new focus on multi-species and ecosystem management policies offers a loophole for those desiring to pursue the unsustainable-exit course, as they would appear to legitimize the over-exploitation of some resources in order to maximize the use of others and "optimize" the total use.

Naturally, those resources that are economically most valuable would be the ones chosen for selective over-exploitation; that, in the ocean, is usually the bigger predators. A provision in the *UN Convention on the Law of the Sea* to impede this has received scant attention since it was adopted in the 1980s; it provided, clearly though rather cryptically, for the limitation of the exploitation of prey species so that the biologically dependent predators would not suffer the loss of their food to the degree that it might reduce their reproduction and growth. At the same time a shift has stealthily been brought about with respect to the politics of sustainability. The World Conservation Strategy promulgated by the United Nations Environment Programme (UNEP) and the World Conservation Union (IUCN), with the support of such bodies as the World Wildlife Fund (now the World Wide Fund for Nature, WWF) originally proposed that if a wild living resource is used, such use should be (biologically) sustainable. But this requirement is now commonly interpreted – even by its proponents – as meaning that such resources *must* be used, albeit sustainably.

Rather late in the day economists began to correct the errors they had made in the first decades after World War II by failing to consider the effect of the *discount rate* in their models. They were brought up sharply by the work of a Canadian mathematician, Colin W. Clark. Clark's publications are voluminous, and many are applications to whaling and fishing; the most accessible one on this particular topic is perhaps his chapter in May (ed.).[120] Clark writes:

> It is a standard hypothesis of dynamic economic theory that firms behave in such a manner as to maximize the present value of their profits, discounting future revenues at a rate equal to the 'opportunity cost' of capital – that is, the rate of interest that the firm's owners or stockholders would expect to earn on alternative investments. From the social point of view, discounting implies a preferential treatment of present over future (human) generations.

Thus it is a powerful force counter to the pursuit of sustainability. In their recent analysis of the fisheries of the North Atlantic, Daniel Pauly and Jay Maclean have expressed the effect of discounting in a slightly different, but equally valid, way.[121]

VULNERABILITY

This is a commonly mentioned notion in discussion of conservation and sustainable use of wild living resources, and may relate to their responses to natural environmental changes or to human depredation or other intrusive behaviour. It may be applied to individuals but, in the present context, usually to a *type* of animal, species and population. It seems obvious that types of animals that are easily visible to, or otherwise detected by, humans are more vulnerable than those that are not. Whales and dolphins are more vulnerable than deep-sea squids, for example. The vulnerability of species is most usually now expressed by biologists in terms of whether they have evolved to follow a ***K***-life history strategy or an ***r*** one (***K*** is the so-called carrying capacity [asymptote] in the logistic and related models; ***r*** is the intrinsic rate of natural increase, commonly presumed to be a constant exponent, possibly genetically determined). The energy available to an organism capable of reproduction may be directed towards the survival and growth of that organism, or towards the production of offspring, or towards some partitioning between the two. Any change in the allocation of energy will influence the species' bionomic strategy, and hence the parameters of the population model applied to it.[122] T.R.E. Southwood (1977)[123] considered that "Each organism will have a bionomic strategy, expressed through its size, longevity, range and migration habit – which are not, of course, all independent variables – that is summarized by the parameters of appropriate population models. This strategy", he wrote, "will evolve to maximize the *fitness* of the organism to its environment. Hence the organism's *habitat* may be viewed as a template against which evolutionary forces fashion its bionomic or ecological strategy". He then defined three qualities of habitat: *duration stability* (the length of time a particular habitat remains in a particular geographical location, and favorable to the organism); *temporal variability* (the extent to which the carrying capacity, *K*, of the habitat varies during the time that a site is tenable by the organism); and *spatial heterogeneity* (continuity versus patchiness).

First, consider the matter of size. The size of a species is positively correlated with generation time; generally the bigger the animal the longer between generations.[124] As a species evolves a change in size – commonly to be larger – will move it one way or another along the ***r*** - ***K*** spectrum, and ***r*** is more sensitive to changes in generation time than is ***K***. Then, larger species tend to have an advantage over smaller ones in inter-specific competition, and the allometric (out of proportion) growth of offensive or defensive appendages (legs and arms, tentacles, claws) will enhance this ability and that, combined with longevity, allows the possibility of a high level of parental care and protection. And the size of an individual will influence the size of its habitat, including the possibility of moving within it.

K-strategists have a stable habitat (the ratio ***r/H*** being relatively small – ***H*** is the length of time a habitat remains favorable) and so evolve towards maintaining their populations at equilibrium levels. So, they will be selected for large size, long generation times and lower ***r***. So high levels of fecundity are not needed for survival; if low birth rates can be matched by higher survival; the young will be few but relatively large. However, populations of ***K***-strategists suffering perturbations need to return quickly to equilibrium levels in the face of inter- and intra-specific competition for habitat/resources. Because their mortality rates are low this is accomplished by changes in the birth rate. Thus the birth rate is sensitive to population density and will rise rapidly if density falls. ***K***-strategists are unlikely to be well adapted to recover from population densities driven far below their carrying capacity, and are thus vulnerable even to extinction.[125]

In contrast, ***r***-strategists are continually colonizing habitats of a temporary nature (***r/H*** is not small) and they are exposed to selection at all population densities. They are opportunists. High ***r*** is achieved by high fecundity and short generation time. They are typically small. Competitive ability is not important; other than high fecundity their defence against predators comes from synchrony (shoaling, flocking) and 'hide and seek' facilitated by short range mobility. ***r***-strategists tend to be resilient rather than vulnerable in the sense we are using here.

K-strategists tend to have stable populations at high densities or abundances, but to exhibit depensation at low densities. ***r***-strategists would not be expected to exhibit depensation, but would fluctuate about high population numbers and exhibit over-compensation. Other things being equal (which, of course, they rarely are in practice!) ***K***-strategists will be more vulnerable to human depredation than ***r***-strategists.

There is, however, another dimension to the matter of vulnerability. That is the particular nature of species' interactions with humans. Curiosity may place them in a difficult position in confrontation with a predator who is cleverer or much more powerful – but unobtrusively – than they are. If the individuals, or parts of them, have a high market value and are readily transportable to distant markets they become especially vulnerable. Changes in those conditions can be disastrous for them or, sometimes, their successors.[126]

Consider the herring of the Northeast Atlantic. A fishery for that ***r***-selected species was continuous for many centuries; to all intents and purposes it was biologically sustainable, even though it fluctuated – sometimes quite dramatically – and probably oscillated. At one time in the late nineteenth century, and even well into the twentieth century, there were several thousand sailing vessels hunting for the herring, as food for humans, and marketed fresh or salted. Many scientists thought the species was virtually inexhaustible. Even the introduction of steam- and diesel-powered vessels did not result in population collapse. However, in the second half of the twentieth century the processing technology changed drastically and over just a few years. Herring were turned into fish-meal and oil, especially to supply protein and fats, mainly for livestock feed supplements. The market price was low, compared with that for human food; this called for greatly increased catching to maintain profits, the introduction of new types of fishing gear and, most importantly, the catching of smaller, younger animals. In a few years (relative to the previous long period of sustainability) the herring population crashed, and it has only partially recovered – probably, in part, because the economic and social pressure has been so great that fishing has been permitted to resume before recovery had proceeded far enough.

The lesson? Evolving towards an ***r***-strategy does not confer invulnerability to modern human predation. Regaining economic sustainability, long-term, even if rigorous catch limits had been promulgated and fully enforced, would be difficult, painful and perhaps even impossible. At very low population levels even small catches, misjudged with respect to uncertainties of many kinds, would be likely to lead to further depletion, at some time. Recovery to high population levels, and a severely precautionary limit to catches, would not supply the fish-meal and oil market competitively due to low prices per tonne and the existence of competitive products of another kind. I think only a return to restrained catching of adult fish for human consumption from a virtually fully recovered stock could possibly ensure renewed sustainability.

THE TIMEFRAME AND QUALITY OF SUSTAINABILITY

Our subject is the sharing of use of a resource amongst users. The users might be different nation states or a variety of social groups differing in the manner in which they use the resource and/or the purposes to which the fruits of use are put. The users might be the present human generation or population on the one hand, and generations to come on the other. The fact that "generations" are overlapping in time is a significant consideration in this matter. In recent times a matter of concern to some is our sharing of this planet, specifically the biosphere, with other species.

Time I

The problems of sharing across generations and between current users are often closely linked. The FAO Department of Fisheries recently published a series of case studies of fisheries in which the allocation of fishing rights was the focus.[127] I provided an analysis of the process of "sharing" the Antarctic baleen whale resources, especially in the 1960-1970s during which success in obtaining an agreement to reduce catches to a currently sustainable level was completely dependent on a simultaneous agreement on the sharing of such reduced catches among five states and several groups of companies engaged in industrial-scale pelagic whaling in the Antarctic.[128] Gerald Elliot, who worked for most of his life in the world's largest whaling company – Christian Salvesen, of Leith, Scotland – and was eventually its Chairman, has published a fascinating first-hand account of how this worked, with reference to decisions made by the then Chairman, Harold Salvesen, which included preparing an exit strategy from whaling on the assumption that it would never be made sustainable[129] through international agreement and enforcement. This is worth quoting here:

> Harold had no illusions about the decline of the whale stocks. As early as 1945 he had warned Salvesen shareholders that the whaling industry had a limited life. Without taking the initiative in proposing a reduction of catch, he accepted the general policy of the British Government of getting catch quotas down to sustainable levels. But Harold, true to himself, did not swallow intellectually the generally accepted thesis that a self-renewing natural resource should only be harvested at a level that allowed it to be sustained for the future. He would point out to me that in economic terms it might be right to use the exist-

> ing capital equipment to take as many whales as possible up to the point where there were so few left that they were not worth catching. That would give immediate benefit to the whaling companies and the world's supply of wealth. Whale stocks would be brought to a new level, but not so far as to be in danger of disappearing, and they would eventually recover. Against this had to be set, in world terms, the loss of the far greater wealth that the whales could bring in future if harvested as a sustainable resource. But, he argued, even if we rate the interests of future generations as high as our own, it is by no means certain that whales will have any economic value in the future. Their products may well be replaced by others that can be produced at less cost. Harold never pursued this line of thought in public. He was as conscious as anyone of long-term world interest, but he felt that sustainable harvesting, like any other policy, should be argued out and not accepted as self-evident.[130]

Salvesen ceased whaling after making sure that the company would be able to sell its vessels and equipment to others who decided to carry on for longer, *and that those sales would carry with them the agreed shares of quota* (these had, in fact, a much higher market value than the hardware). The proceeds were used to start a large, and until now fairly successful, road-haulage company.[131] Similarly, proceeds from Norwegian Antarctic whaling created a "sustainable" city – Sandefjord – and a renowned shipbuilding industry. Whales and whaling provide a spectacular example, but they are not unique; uses of other marine resources have followed similar lines. Indeed, it was argued by some scientists and authorities until as late as the 1950s that restraints on fishing were unnecessary because the activity would cease as soon as stocks had been sufficiently reduced.

There were two errors in this line of thinking. One was failure to recognize that in industries based on a mixture of species, caught more or less simultaneously, exploitation of the more vulnerable (and usually more valuable) species would continue as virtually a "by-catch" of an industry now targeting mainly other, less valuable but still abundant, species, thus really threatening the continued existence of the former. In Antarctic whaling this was the mix of blue, fin, sei and humpback whales included in the notorious Blue Whale Unit (BWU) by which catch limits were set. In the North Sea trawl fisheries it was the skate (a species of ray that, like other elasmobranches, has a low reproductive rate but, unlike them, fetched high prices in major markets) that drew the booby prize for being the first exterminated marine commercial fish, as an unregulated by-catch of the industry focused on flatfishes and gadoids. The other error was failure to see that greater profit could be derived from a still abundant resource, or from a depleted one allowed to recover, provided the discount rate was not too high. ("Too high" is measured by the ratio of the discount rate to the net reproductive rate of the resource; whales, like hardwood trees, have slow net rate growth/reproductive rates and are hence especially vulnerable to high discount rates.)

Time II

Time is virtually the lost dimension of both the theory and the practice of sustainable use of renewable resources. The logistic equation and its derivatives predict that a population will attain its final abundance only after an infinite time has passed. They also generally predict that the population will stabilize, following a perturbation, only after an infinite time. It is obviously not possible to base management practice formally on such a paradigm.[132] One practical way around this dilemma may be to define a finite time by which a "recovering" population would be expected to reach some fraction of an "equilibrium" or "steady state" size (the so-called carrying capacity) – say 99 or 95%.

More generally one might arbitrarily terminate the period of transition from one steady state to another at the point where say 99 or 95% of the difference between the two states has been achieved. Another way of evading the issue is, occasionally, to deploy the terms "indefinite" or "indeterminate". These have been used in relation to administrative decisions (such as those defining moratoria on "use" or the creation of protected areas or designation of protected species and the like) meaning "until the authority concerned may decide otherwise". I think it is obvious, however, that whatever theorists do with integro-differential equations solved with respect to an infinite time of expression, a finite time limit must be specified and set in any attempted implementation of the notion of sustainability.

In terms of the human users of renewable resources the reality is that present users may have a primary interest in continued use *by themselves* later in life, and maybe by their families. Next priority may be use by the next generation in their society, but particularly by their own descendents. Concern and advance care might apply to the generation after that (grandchildren) in societies where family structures persist and where the descendents are likely to continue to base their survival on continuation of the family occupation. Sometimes, in discussion of this matter, it is assumed that the next generation can be left to make its own decisions about leaving resources in good order for the following one. And, of course, we cannot know what might be the preferences and choices of future generations – such as whether they will want to eat whale meat or a butter substitute from whale oil or to use them as a resource for tourism or basic scientific research, or even not to "use" whales at all.

The interests of the actual users – fisher-folk, whalers – are usually not the same as those of the society whose markets they supply, although both are linked by the behaviour of those markets. This was clearly understood by Michael Graham, one of the first students of the logistic, who argued that the interests of the fishermen and their families came first, and that if their continuity could be assured by restraints on fishing effort then the nation(s) would benefit from a good steady supply of fish.[133] But he could not have anticipated in the 1930s that in the not so distant future a main use of fish would be for conversion to fish-meal to provide food supplements for livestock raising and aquaculture (which, being largely based on the catching of smaller, younger fish, could threaten the existence of the resource in a way that old-style fishing never did – well, hardly ever), and which in turn would be largely replaceable by similar products from soy beans and sunflower seeds. Nor did many people, in and before the 1930s, expect that few sons of fishermen would become fishermen themselves, at least in Western Europe.

So, our view ahead is quite limited and perhaps can only practically be thought of in terms of a few overlapping human generations, at most. Also, realistically, it must be expected that no management regime can last for ever; in fact modern societies are sufficiently unstable and their practices rather arbitrarily variable that any such regime will probably endure, at most and in rather exceptional circumstances, for just a few decades. But in considering the sustainable use of a living resource, account equally has to be taken of the generation duration of the resource organisms themselves. In some cases, such as whales and perhaps some fishes, and terrestrial mammals and birds, and trees, this may be of the same order as that of humans. In other cases, such as herrings and anchovies, the generation time is much shorter, relatively, and modeling and simulation testing for a regime of sustainable use (and possibly restoration of an already depleted resource) will take a somewhat different form. The regime will also depend critically on the degree of natural variability – usually annual – of the resource, and this as we have seen, is correlated (negatively) with the life-spans and sizes of the animals in question.

A complicating factor is the possible natural oscillatory behaviour of the exploited population. In cases where the cycles may be of short duration it could be reasonable to fix a duration lasting a few cycles, and focus on the mean population levels through those cycles. If, however, the expected oscillations are of long duration, as, for example, in the case of whales assessed by Witting's models then the practical management horizon might be no more than one cycle, or even less. In this case the decision about what is "sustainable" and what is not is distinctly more complicated.

This linking between the generation times of humans and of exploited wild animals is understood by most developers of (hopefully) sustainable management regimes. However, what is still not usually taken into consideration is the possibility of long natural cycles of growth and abundance – long, that is, in relation to the generation interval. Witting's study of the gray whale provides a good example. Would we regard a whaling management regime that provided a high probability that the wave-length (perhaps a couple of centuries) and amplitude (3-or-more-fold swings), and the very-long-term mean of abundance were not substantially altered, as ensuring sustainable use? Similarly if the vast long cycles of ocean features, apparently reflected in oscillations of fish abundances and hence catches, are substantiated and general, fisheries managers and scientists really do have to go back to their drawing boards.

There are also limits to our calculating ability, as well as to our imagination and foresight. In developing the RMP the IWC scientists decided they could only reasonably conduct simulations through one hundred years. Now, a decade later, they could certainly work on a longer time-scale, but would it be worth it? Could anyone seriously think that some agreed management regime would endure for more than a century? A consequence of this is that management objectives would not be defined in terms of permanent sustainability or of more or less steady year-to-year catches, but rather in terms of the cumulative catch over the full period of simulation, and how high it could be without risking inadvertent resource depletion. There can be no reference to such theoretical but hazy notions as that of MSY and similar "sustainable" targets, even if the simulation models have that concealed within them.

In summary I would suggest that, among the many temporal features demanding that we not consider sustainability in terms of infinite time, and considering the new approaches to management illustrated by the IWC's RMS/RMP, the most important is the likely duration of the instruments of human societies that would operate any management scheme for sustainable use of wild living resources.[134]

Quality

In, I think, 1948, at a Plenary Session of the Permanent Commission for Northeast Atlantic Fisheries, in London, the Commissioner for France – a biologist, M. Furnestin – waved two frying pans, a larger one and a smaller one. "This" he said (if my memory serves me correctly), shaking the smaller pan, "is a French pan; French housewives like to sauté small soles. And this – waving the bigger one – is an English (he didn't even say British) pan in which your housewives like to fry plaice, and bigger soles if they can get them".

Plaice and soles were caught in the same trawling operations in the North Sea (although astute fishermen knew where to go to take advantage of some species segregation on the fishing grounds). The UK wanted larger-meshed nets to be legislated (to benefit the plaice fishery but which would have allowed most of the slimmer soles to escape), and the fishing effort at least stabilized. Any idea that there could be agreed an optimal catch that would satisfy both parties flew out of the window that day. Following it, Michael Graham asked Ray Beverton and me to try to work out "equivalent" regulations to resolve this dilemma, by looking at what change in fishing intensity would be equivalent in its results to a certain change in mesh size. Some rough judgments could be made, but an adjustment to give equal expectation of total yield would not provide the same catch composition, and vice-versa. We developed the notion of "eumetric (well-measured) fishing", but its application to multi-species fisheries was difficult.

It is usual in considering marine resources to discuss some optimization of a sustainable catch by weight. In a few cases – whaling is one – discussion turns on numbers rather than biomass. (In the whaling case, this is not unreasonable – at least as far as baleen whales are concerned – because last year's calves have generally grown to an economically useful size by the time they have gone to the Antarctic to feed and, being concentrated, become vulnerable to whalers, and because individuals grow rather quickly to near maximum size.) But whalers have always been interested in quality. Southern hemisphere whaling locations and seasons were determined and legalized for many decades in such a way as to try to optimize individual weight and hence oil production. This is one reason why whaling – and fishing – regulations can become complicated: catch limits, protected areas, closed seasons, size limits for the market, preference for certain gears and operational methods, and so on. One-dimensional regulation by catch limits will not lead to the desired end. And the market alone will not lead to some sub-optimisation when different users have differing quality requirements. Such considerations apply also when the various uses are qualitatively different. For example, if whaling and touristic whale watching are carried on together, the tourists' aim to see whales that do not flee from ships may be frustrated.[135]

In an exploitation system such as trawling it might be assumed that the catches are close to being a random sample of the available fish, or at least of those big enough to be retained by the net. Even such unselective extraction nevertheless changes, over time, the size- and age-distribution of the population, and hence its "quality", and hence the quality of future catches if, for example, the market prefers bigger fish to smaller fish. Such changes can be calculated in advance, at least to a first approximation and taken into consideration in the search for both biological sustainability and continued profit. However, all fishing and hunting operations are selective, and many extremely so, but rarely in known ways. (Even bottom-trawling is not really unselective; fishermen go to where they know some segregation of smaller and larger fish will occur on the grounds; some individuals will be more able than others to escape the net, and so on.) The maintenance of a resource population that not only will not diminish under exploitation but will retain and persist with desired qualities, such as age-composition or "condition" (fatness; ratio of weight to an appropriate power of a linear dimension) presents serious obstacles.

Hunters of terrestrial animals – especially the larger mammals – may place quality over abundance. For instance, like whalers, they do not want emaciated individuals that have lost their body fat nutritional reserves for one reason or another. However, their definitions of quality go further; for example they do not wish to see numbers of sick, lame, emaciated, or just plain elderly individuals in the population they are exploiting. This desire commonly leads to highly selective "culling" by resource managers; a sort of "tidying-up" of the population in question, as well as selectivity rules to be followed by the licensed hunters. It is interesting to note that such operations have been necessitated in many situations because of the elimination, by humans, of the species such as scavengers and predators that would perform that function in more natural conditions, especially by selectively taking the more vulnerable individuals, such as the young and the old and infirm. I think, from slight experience working in the Scottish Highlands in the 1950s, of red deer and the extermination of the wolf, and grouse and the extermination of the wild cat, as well as the near-eliminations of predatory and scavenging birds; and also the attempts to "manage" vicuña populations in Perú, for wool or meat, in a situation of diminishing populations of feline predators. It is little wonder that the theoretical studies I have outlined here have been more applied to marine than to terrestrial situations, although in the latter there have been some rather naïve applications of logistic theory.

Selective elimination or depletion of preferred, larger species, can lead to apparent large changes in the trophic structure of marine ecosystems. A well-known example is the sequential depletion of the southern great whales in their order of size: first the blue whale, then the fin and humpback, then the sei and Bryde's, and the process halted, perhaps only temporarily, at the smallest, the minke. This obvious fact has led some scientists to hypothesize that the smaller species have increased at the expense of the depleted larger ones, but it is difficult – even misleading – to deduce this from the sequence of catch statistics. A more important example, and one that is more difficult to interpret, is the fishing down, and almost out, of larger fish species in a region, with corresponding and sequential rises in the catches of smaller species. This can look as if the smaller ones are "taking over" the ecosystem, and that might be so, but is extremely difficult to either demonstrate or dismiss. Such hypotheses have been explored by Daniel Pauly and his collaborators, and will not be pursued further here; they merely, in our context, emphasise that selective removals in a search for sustainability, will occur at the ecosystem level as well as the population level, with considerable practical consequences.[136]

RECAPITULATION AND CONCLUSIONS

The word "sustainable" has been used with several meanings, not all of them well defined, and not all mutually compatible. Let us first consider the taking of a "small" yield repeatedly from a large previously unexploited population. Standard logistic type theory would tell us that this will reduce the average population size, even if only slightly; we probably wouldn't notice it. If the population was fluctuating before, or oscillating, it will continue to do that, and after a long period we might begin to notice the latter behaviour. The North Sea herring fishery went on like that for centuries, and by most standards the catches were very large, on average. Many other such examples could be cited. The idea that some people have, from repeatedly bad experiences, that sustainable use is never feasible, does not withstand the scrutiny of history.

Standard theory also tells us that even if we intensify our exploitation, to the degree that we begin to notice the population is getting smaller, we can still have a sustainable yield if we are careful. This conclusion depends on the assumed reversibility of the population growth curve implied by our theory. My own skepticism about reversibility has been well expressed by a physicist, Arnold Sommerfeld:

> Reversible processes are not, in fact, processes at all; they are sequences of states of equilibrium. The processes which we encounter in real life are always irreversible processes.[137]

However, as soon as we look at a population with an age-structure (or, for that matter, any other sort of meaningful structure) things are not quite like that. A growing population, when it has reached a given number, will have a particular age-composition, and that will not be a stable composition. Stability of age composition, if it comes at all, arrives at the same time (actually, rather later) as the population closely approaches carrying capacity. A population driven down by exploitation to the same given number will also not have a stable age-composition, and neither will it have the same age-composition as that of the growing population of the same numerical size. To achieve either of those states the process of extraction would have to be very selective, and carefully so. Random selection, even if attainable, will not do the trick, and nor will the kind of selection normally encountered in, say, a fishery, which depends on the nature of the fishing gear, and the location and timing of its deployment, among other things. For these reasons such an exploitation pattern will not be formally sustainable.

But, to continue, we might think that if we simply observe the rate at which the population tends to increase after we have reduced it, and then relaxed our exploitation "pressure", and then continue to take just that much each year or whatever exploitation period we chose, the population will not change and we shall have achieved sustainability. That has been called a strategy of estimating and taking the *replacement yield*, which has been on occasion adopted by the IWC (always said to be a temporary measure, pending better assessments, but commonly continued by default) and in some fisheries. But, taking no account of the characteristics of the out-of-equilibrium age-structure this can be a disastrous policy if pursued blindly; in commonly met circumstances the population is likely to continue to decline. A naïve replacement yield management strategy is to be avoided; it does not in itself lead to sustainability.[138]

If our daring and greed pushes the population down further, we begin to enter the realm of instability, whether or not depensation is then the name of the game. Then not only is sustainability elusive, but the eventual climb back will be painful – for us. And if Nature is playing tricks with "vibrations", as it appears is her wont, her habit, then our calculations become more difficult, more uncertain in their outcomes. The *logistique* – the art of calculation – has its own limits.

In our era, sustainability is more easily advocated than achieved, easier said than done. Existing mathematical models, as well as apparently rational arguments, that do not truly express "laws of nature", are not even framing testable hypotheses, are not even plausible when examined closely, and do not match our admittedly limited historical experience, are not what we need in a quest for sustainability. If that *is* the Holy Grail it will be found by humility, not secured by brute force, stealth or confidence tricks.[139]

A quotation from Witting's aforementioned reconstruction of the population dynamics of the gray whale is, I think, appropriate here; it follows his discussion of evidence that the vibrating population of this "over-recovered" species has already begun to decline:

> With the abundance estimates and catch histories being consistent with cyclic population dynamics it would seem to be a wise strategy if candidate Strike Limit Algorithms *[SLAs, the aboriginal subsistence equivalents of the commercial CLAs – sjh]* were tested against the hypothesis of cyclic dynamics, particularly as the abundance data indicate a decline that is faster than the decline of the fastest declining inertia model (so far examined). ...the accepted SLA must be able to cope even with a crashing population.
>
> Both the management objectives of the IWC and the framework used to evaluate candidate SLAs *[and CLAs – sjh]* have been developed only for traditional density-regulated dynamics. Traditional management objectives are often defined relative to the MSY and the MSY level, which are concepts that apply only to cases where the intrinsic growth rate is constant. For inertial dynamics, the intrinsic growth rate is an initial condition so that most population abundances are associated with a suite of both positive and negative realized... growth rates. Among other things this implies that there is no single curve of sustainable yields and, thus no well-defined harvest optimum as assumed by the IWC.

And, of course, not only by the IWC, but as now – unfortunately – embedded in the United Nations Convention on the Law of the Sea. As a management objective MSY had neither sound theoretical foundations nor empirical justification. It reflects an urge for the institutionalization of greed as a governing principle, and was rightly mocked by Peter Larkin in a well-known polemic.[140]

The vigour with which MSY has been "marketed" internationally, especially by successive Governments of the United States, has long puzzled me. Recent declassification of hitherto restricted documents from the first decade after the end of World War II throws some light on this, and perhaps warrants an extended endnote. But, first, let me quote from a letter, dated 29 May 1956, from Michael Graham (then Director of Fisheries Research, England and Wales) to Milner Schaefer (then with the I-ATTC) who had sent Graham the manuscript of a paper on the application of logistic curve theory to Pacific tuna fisheries. Graham, in his usual charming way, first remarks that asking for a frank and critical appraisal of a manuscript "is a sure opening gambit to killing quite a promising friendship"! While appreciating Schaefer's clarity, Graham observes that "I am not sympathetic to it because the problem of fish conservation is not so clear that we can write about it in an authoritarian way without running the risk of emphasizing what may in the future turn out to be less important points. For example, I think that yield curves may sometimes be very flat-topped ... and I wonder whether we have hold of sufficient theory to take account of the interaction of fish species, which might be of superlative importance.[141] I am still teaching this", writes Graham; "Find which direction to go in and take a small step that way".[142]

I see only two possible ways forward, out of this dilemma. Perhaps they can both be taken, in differing circumstances. One is the traditional one, really: just take what we truly *need*, and define "need" frugally, or even parsimoniously. The other is the route offered by the IWC scientists and – if they can only get around to agreement – by the collective governments of IWC Member states. That is a truly precautionary regime for managing resource use, adequately monitored and universally enforced[143] and complied with, based – in our everlasting ignorance – on extensive computer simulation of as many scenarios as we can imagine.[144]

And forget the "balance of nature". *Gaia* does not "balance"; she vibrates and swells, sometimes explodes, but always with ripples, and her sometimes smiling calm can be deceptive. We do not, and cannot "manage" her; we use her and in doing so stress her, though we cannot destroy her because we are part of her and we shall not endure so long. If we want to use her sustainably, to ensure a little more of our own future, then we have to tread very, very carefully or she will rumble and grumble, and might even spit at us.[145]

In this essay I have sought to trace one strand of descent, as it were, from Malthus and some of his 19th century contemporaries, followers and critics, through some of the highways and byways of population dynamics. There is another strong strand, perhaps two: the evolution of Evolution – through Charles Darwin, Robert Chambers, Alfred Russel Wallace, Herbert Spencer *et al.*, and the discovery of scientific genetics, through Gregor Mendel, Francis Galton, Karl Pearson *et al.* The redoubtable T. H. Huxley was caught up in both, and Karl Marx and Freidrich Engels mixed in for good measure. Those along both the strands were much concerned, from time to time, about applications of theory to both humans and to other animals, as were the participants in the development of ideas about population growth and sustainability. Is it going too far, perhaps, to think that the three strands might at last become plaited together, creating a new synthesis through the approach I have identified with the work of Lars Witting? [146]

Damasio wrote, a few years ago, in *Descartes' Error: Emotion, Reason and the Human Brain*:

> Although Biology and culture often determine our reasoning, directly or indirectly... we must recognize that humans do have room for freedom, for willing and performing actions that may go against the apparent grain of biology and culture.[147]

Now, as we have seen, Malthus was a child of the Enlightenment, a core paradigm of which was the "abyssal separation between body and mind", characterized by the notion of "rationality". In his thoughtful study, "Waking the Sleepwalkers: Globalisation and Sustainability," William Rees has made a plea for reuniting these human faculties under the rubric *enlightened rationality* which, he says,

> recognizes, among other things, that modern humans remain myth-making creatures and that *to achieve sustainability we must create a new cultural myth* (my emphasis)... Our present, increasingly global, myth actually reinforces our biological propensity to expand and fill (to overflowing) all available ecological space. Our once successful evolutionary strategies are now propelling (us) along a destructive course on a finite planet.[148]

What might be the constituents of such a myth, which could redefine the "Notion" which is the subject of my essay? In "Waking the Sleepwalkers" Rees expresses succinctly an idea I have expressed during many discussions about the relationship between the objectives of "conservation" and those of "animal welfare", and elaborated in an unpublished paper to the Fourth International Congress on Bioethics, held in London a few years ago. "The bottom line", he writes, is that we must come to value and defend the Earth for more than the economic gains of exploitation." Fair enough! But, Rees reminds his readers, "Scientists in particular, their feelings numbed by Cartesian objectification, often feel queasy when talking about love and respect, empathy and compassion, particularly when applied to their objects of study...but, following Damasio, the equally human capacities for reason and emotion...cannot be ripped asunder". Rees then asserts that:

> Most rational people will agree that we still need nature to survive and... that **we generally do not protect or save that which we do not love** (again, my emphasis).[149]

In this light it may be both rational and necessary for modern society to acquire a sense of genuine love for the natural world, a world, I will reiterate, of which we are an intrinsic part.

Such love obviously has to embrace both the human and the non-human. Our love for the non-human might range from care for individual sentient animals to care for, for example, biodiversity.[150] But here my focus is on the extent of love for humanity. This has many dimensions, not least in space and time. In recent decades we have heard much public debate about mankind (I prefer *humankind*) and its "common heritage", especially in the context of the deep ocean, the Antarctic continent and the moon, regions of relative inaccessibility and mystery. Much of that debate is about the *sharing* of the heritage, among presently existing humans. But humankind is a concept that embraces both the past – our past – and all conceivable futures.

Why is this latter so important when we talk about sustainability? It is, I suggest, because it determines the real timeframe of our definition of sustainability once we have discarded the idea of it stretching to the World's End or even – perish the thought – to infinity. In discussing this I have referred to the practical limitations of computer simulation, the generation times of that which we wish to use sustainably, and to the expected duration of any administrative/regulatory/political system – or even ethical system – that is supposed to ensure such use. Perhaps it would be fruitful to talk about how far into the future can we really extend our love for humanity, as judged by the degree to which we are prepared to modify our present behaviour so that it may be fulfilled – beyond ourselves, now; our own old age; our friends and younger contemporaries; our children and their children? Grandchildren? How far indeed?

Now, I end by returning to some impassioned words from Bruno Voituriez (whom I have already quoted in several places in this essay), with a little of my editing in consideration of its context:

> The growing 'humanisation' of our planet is not, contrary to what is often said, a mortal menace for Earth and the life on it, which has seen worse and is certainly capable of resisting humanity's efforts tending to change it, even to destroy it. This 'humanisation' is, above all, a menace to mankind itself, thus exposed to grave difficulties that are generators of deadly conflicts. We can no longer say – as the scientists of the nineteenth century might have said: ...if it does not master its own evolution. The fact is we do not master anything, and we know that in our stochastic world, our capacity for prediction will always have limits whatever may be our efforts to reduce the uncertainties, as has been attempted for fisheries, for exam-

ple, by the adoption of the 'precautionary principle'. Far from providing a methodology for decision-making this approach is only a principle of action based on admission of ignorance, albeit a recognized and no longer camouflaged ignorance. The objective of scientific research is to reduce that ignorance... The problem is now planetary and scientists are being called upon to simulate the functioning of the planet Earth, by going back through their archives so as to know Earth's history better, by multiplying observation networks, and by modeling Earth so as to propose future scenarios. Given the results of research, the scenarios the research proposes, and the uncertainties they hide, two extreme extrapolations are possible. First of all, to deny a phenomenon on the pretext that it has not been formally proved... A second plausible attitude is the precautionary approach, in the strongest use of the term, or more precisely the negation of this approach as a principle of action, thus turning it into a principle of reaction opposing any management and development plans in the name of an obscure ecology that makes humanity the enemy of the planet and a kind of outlaw on the earth. Even if 'all is for the best in the best of all possible worlds' let us guard against thinking that all goes wrong in the worst of all possible worlds. We are the future of the human race and we have only the resources offered by the planet on which we live and our intelligence to make the best of it. If there is anything that can be labeled 'Common World heritage' it is precisely the Earth, which we must bequeath to future generations and which we are (and they will be) obliged to manage together.

LAST WORDS

Just as I thought I had finished the final revision of this review I happened across John Bellamy Foster's scholarly "Marx's Ecology: Materialism and Nature" (Monthly Review Press, New York, 2000), containing a chapter ("Parson Naturalists") on 19th century discussions about population among the likes of Malthus, David Ricardo, Percy Shelley, Pierre Proudhon, Karl Marx, Freidrich Engels and Charles Darwin. Foster points out that although Malthus, in his first *Essay* on population was attacking the ideas of Godwin and Condorcet, the controversy had its roots in an earlier publication: a work by Scottish Minister, Robert Wallace, in 1761, affirming exponential growth and worrying about the eventual exhaustion of the finite resources on which humans depend. What we see here is the beginning of doubts about the notion of eternal "progress" coming from Enlightenment optimism.

What Marx didn't like about the Malthusian package was that as far as humans are concerned he believed – I think rightly – that the increase in *available* resources was not linear but rather itself exponential, as a consequence of the rise of science and of industrialisation. In the first half of the nineteenth century no one seems to have realised that the rate of population increase might itself one day slow down, smoothly. One difficulty in this debate was that Parson Malthus substantially changed his mind on several matters between the first and second *Essays*, and again in the third, *Summary View*.

In these series of corrections and amendments, Malthus made it clear that he was not talking about absolute limits to the resources themselves but about limits of access to them: as he put it "easy means of subsistence". Both Marx and Malthus understood that this was a social phenomenon, deriving from the class structure of society.

In his chapter on "The Metabolism of Nature and Society" (Marx's theme), Foster reviews Marx's analysis of "sustainability", defined in much the same way as in the Report of the Brundtland Commission. Marx's formulation was "An entire society, a nation, or all existing societies taken together, are not owners of the earth. They are simply its beneficiaries, and have to bequeath it in an improved state to succeeding generations as *beni patres familias* [good heads of the household]". The passing on of resources in good order Marx repeatedly referred to as "the chain of human generations", while being reminded by Engels, the chemist, that non-renewable resources were being squandered irretrievably at an increasing rate. Marx thought that what we would now call *sustainable development* was attractive but "of very little practical relevance to capitalist society" which depended, for stability and eventually its survival, on continuous economic growth. In this he reminded me of Mahatma Ghandi's purported response to the question:

"What do you think of Western civilisation?"

"That would be a good idea!"

ACKNOWLEDGEMENTS

I have benefited greatly, during the years in which this essay gestated, from conversations – and sometimes arguments – with several friendly colleagues, and occasionally with less friendly opponents. Among the former I mention especially the late Ray Beverton, from the 1940s, the late Geoffrey Kesteven from the 1950s, the late Douglas Chapman from the 1960s, Bill de la Mare, Kees Lankester and Justin Cooke from the 1970s, and assorted members of the IWC Scientific Committee and participants in some other international scientific assemblies. Late night rapping with the late David McTaggart, neighbour and founder of Greenpeace International, fuelled by the good wines of Tuscany and Umbria, also played its part, as did hours spent with my assistant over three years, Olga Nève, walking in the Umbrian and Welsh countrysides.

I am grateful, too, to the International Fund for Animal Welfare (IFAW), and especially its luminaries through the period I have been associated with it: Brian Davies, its Founder; Fred O'Regan, now its President; Chris Tuite and Vassili Papastavrou, for their prolonged support. David Lavigne, now IFAW's Science Advisor and Executive Director of the International Marine Mammal Association (IMMA) has been the most persistent of the arguifiers for a couple of decades. Leslie Busby, now CEO of the Third Millennium Foundation created by McTaggart, frequently listened, apparently attentively, when I have tried to put into words the slight and elementary mathematics of which I am still capable. Mary Carmel Finley, a sharp-eyed and enthusiastic historian of fisheries matters, has introduced me to some aspects of the MSY-debates/scams of which I had only been half aware. Then I am in debt to the people of Blackburn Press, New Jersey, who, by inviting me to write a Foreword to a new edition of a book I wrote half a century ago with Raymond Beverton,[151] unwittingly caused me to think again about the idea of sustainability in the fisheries context. Likewise, conversations with Mario Ruivo, especially in the context of our work together in the Soares Commission (Independent World Commission on the Ocean – IWCO) stimulated reflections on the role of the theories of political economy in this business. Sheryl Fink, also of IMMA and IFAW, has been oh, so patient, in finding and sending me copies of scientific papers that my self-imposed isolation in the hills of Italy and Wales made difficult of access. And lastly, I take this opportunity to express my appreciation to the late Sir Peter Scott, as well as to my friend, the late Professor Peter Jewell, for introducing me to the arguments within the IUCN/SSC about whether elephants and other bulky animals such as whales should be "culled for their own sakes" as well as for the being of the ecosphere.

NOTES AND SOURCES

[1] Quoting an unnamed "famous astronomer", in *Faster than the Speed of Light: the Story of a Scientific Speculation.* Heinemann, London, 2003. Thanks to David Lavigne, the astronomer can be identified as Sir Arthur Eddington. David also has drawn my attention to another version of this important idea: "Scientists are perennially aware that it is best not to trust theory until it is confirmed by experience. It is equally true, as Eddington pointed out, that it is best not to put too much faith in facts until they have been confirmed by theory". *[Unfortunately many scientists seem not to be "perennially aware" of this. – sjh]* The writer – Robert A. McArthur ["Coexistence of Species". pp. 253-259 in *Challenging Biological Problems.* Edited by J.A. Behnke. AIBS, Univ. Oxford Press, NY, 1972] – continues "Ecology is now in the position where the facts are confirmed by theory and the theories at least roughly conformable by facts. But both the facts and the theories have serious inadequacies providing stumbling blocks to present progress." I think the first of these two propositions is debatable.

[2] Quoted by Bruno Voituriez, 2003; see below.

[3] Perhaps the most accessible reference is to the highly individualistic *The Greek Myths*, by Robert Graves, in two volumes, Penguin, 1955.

[4] This is actually the second of four meanings; the others are not applicable.

[5] John McGowan and John Field, in their contribution to a recent book: *Oceans 2020: Science, Trends, and the Challenge of Sustainability.* Eds J. G. Field, G. Hempel and C. P. Summerhayes, Island Press, 2002, 365 pp. The quotation is from p. 41 in the Field and McGowan chapter pp. 9-48, "Ocean Studies").

[6] See for example H. Whitehead *Sperm Whales: Social Evolution in the Ocean.* Univ. Chicago Press, 431pp, 2003. See also Michael Bond, "A way with whales". *New Scientist,* 15 May 2004:43-5. [Interview with H. Whitehead.]

[7] But possibly only by someone who has read and absorbed all of *Sustainable Development: Mandate or Mantra?* edited by W. Chesworth, M. R. Moss and V. G. Thomas, Faculty of Environmental Sciences, Univ. of Guelph, 2002, 140 pp.

[8] "Small island states in the face of climatic change: the end of the line in international environmental responsibility". *UCLA Journal of Environmental Law and Policy* 6/22/2004.

[9] The ugly head of this phantom is still being raised by those connected with hunting, shooting and fishing – especially shooting whales – who say something like "We have spoiled the balance of nature by unsustainable killing; now let us try to restore it by killing more of something else". Charles Elton poured justified scorn on this 'balance' notion nearly eighty years ago in his *Animal Ecology*, 1927 (Sidgwick and Jackson, London).

[10] Here I mean the term as used to identify the character or state of a resource that someone wants to be allowed to exploit, for gain, not its technical use in relation to the testing of mathematical models and algorithms.

[11] Some of these terms have been assigned reasonable definitions in the context of classifying the status of species and populations for CITES and the Red Data Books, now Red Lists, produced by IUCN (*threatened, endangered* are among these). Depletion is commonly used in an economic context also with respect to stocks of non-renewable resources, and in relation to management of use of marine living resources,

it has taken specific meanings in particular cases. Such emotive words as "healthy" and phrases such as "in good shape" as applied to populations or ecosystems are purely propagandistic and essentially meaningless, the sort of language recently castigated by Francis Wheen in his *How Mumbo-Jumbo Conquered the World: A Short History of Modern Delusions*, Fourth Estate, 2004, 342 pp.

[12] See David Lavigne "The return of Big Brother". *BBC Wildlife Magazine*, May 2004: 70-72. A relevant quotation from Alexis de Tocqueville (1846) was given a couple of years ago by the Editors of a group of lectures on Environment, Energy and Resources: "An abstract term is like a box with a false bottom: you may put in it what ideas you please, and take them out again without being observed" ["Introduction: Clearing Away the Undergrowth", by W. Chesworth, M. R. Moss and V. G. Thomas, to *Sustainable Development: Mandate or Mantra*, University of Guelph, 2002, 140 pp.] D.M. Lavigne has a valuable paper in that book (pp. 63-91), with a useful bibliography: "Ecological Footprints, Doublespeak, and the Evolution of the Machiavellian Mind".

[13] The nearest to hand is the otherwise excellent "Design of operational management strategies for achieving fishery ecosystem objectives". by K. J. Sainsbury, A. E. Punt and A. D. M. Smith, *ICES Journal of Marine Science* 57:731-41, 2000. This is one of many in a special ICES volume on "Ecosystem effects of fishing".

[14] "The Precautionary Principle: Tony Blair and the language of risk". *London Review of Books.* 1 April 2004: 12-5. This idea was originally expressed as a desirable prohibition of release of chemicals into the planetary environment before sufficient testing had been carried out to ensure that no irreversible harm would be done to it. It has recently, frequently been confused with *mere caution in the face of uncertainties and the unknown.* One of the very few uses of the Principle in the realm of marine conservation and protection was made by the International Whaling Commission (IWC) in the late 1970s – i.e. before the term came into popular and political use – when it amended its so-called New Management Procedure (NMP) for setting annual "sustainable" catch limits for whales to provide that there should be no commercial exploitation of a previously unexploited whale population until certain prior conditions had been met, namely, in this case, provision of an acceptable estimate of the number of whales in the putative "stock".

[15] "An Essay on the Principle of Population as it affects the future Improvement of Society, with Remarks on the Speculations of Mr Godwin, M. Condorcet, and Other Writers" (published anonymously in 1798, called here *The First Essay*); "An Essay on the Principle of Population; or a View of its Past and Present Effects on Human Happiness; with an Inquiry into our Prospects respecting the Future Removal or Mitigation of the Evils which it occasions" (published under Malthus' own name in 1803, called here *The Second Essay*); "A Summary View of the Principle of Population" (1830, called here *A Summary View*). In between were several substantively different versions of *The Second Essay* as well as numerous articles on political economy, the general features of which were mostly incorporated in *A Summary View.*

[16] In *A Summary View,* Malthus states his position that "Elevated as man is above all other animals by his intellectual facilities, it is not to be supposed that the physical laws to which he is subjected should be essentially different from those which are observed to prevail in other parts of animate nature" but "The main peculiarity which distinguishes man from other animals, in the means of his support, is the power which he possesses of very greatly increasing these means." In the collection of essays Malthus acknowledges his debts particularly to David Hume, Adam Smith and Henry Wallace: "The most important argument I shall adduce (concerning 'the perfectability of man') is certainly not new, and it may probably have been stated by many writers that I never met with. I should certainly not think of advancing it again, though I mean to place it in a point of view in some degree different from any that I have hitherto seen..."

[17] In this task I must acknowledge debt to the illuminating *Introduction* by Anthony Flew to the edited versions of *The Second Essay* and *A Summary View* published in 1970 and 1982 by Penguin, under his sub-title "A household word, but misunderstood". Flew quotes an unnamed cynic: "The classics are like the aristocracy; we learn their titles and thereafter claim acquaintance with them".

[18] Malthus was explicit about this: "No limits whatever are placed to the production of the earth; they may increase for ever and be greater than any assignable quantity; yet still the power of population being a power of a superior order, the increase of the human species can only be kept commensurate to the increase of the means of subsistence by the constant operation of the strong law of necessity acting as a check upon the greater power" (Chapter II of *The First Essay*).

[19] Condorcet was active and influential in the Revolution's first phases but fell out with the revolutionaries and told the Convention that Robespierre had "...neither ideas in his head nor feelings in his heart". Godwin, as Flew writes, was "similarly inspired, albeit from a safer distance".

[20] Actually Malthus was more forceful and eloquent in his commentary on ideological matters: "The advocate for the present order of things is apt to treat the sect of speculative philosophers either as a set of artful and designing knaves who preach up ardent benevolence and draw captivating pictures of a happier state of society only the better to enable them to destroy the present establishment and to forward their own deep-laid schemes of ambition, or as wild and mad-headed enthusiasts whose silly speculations and absurd paradoxes are not worthy of the attentions of any reasonable man". Malthus was particularly ironic concerning Godwin's "solution" to the population-nutrition problem, which was, apparently, to conjecture that "the passion between the sexes may in time be extinguished"!

In this passage Malthus argues that "a satisfactory history, of one people, and of one period, would require the constant and minute attention of an observing mind during a long life" and he continues to list the particular population features that should be monitored.

[21] Here I am using modern terminology of population dynam-

ics in preference to quoting long passages from the *Essays*.

[22] The originator of the notion of oscillation was in fact Condorcet. "When the increase of the number of men surpassing their means of subsistence the necessary result must be either a continual diminution of happiness and population, a movement truly retrograde, or, at least, a kind of oscillation between good and evil? In societies arrived at this term will not this oscillation be a constantly subsisting cause of periodical misery? Will it not mark the limit when all further amelioration will become impossible ...?" [In modern terminology a *(stable) limit cycle*. See for example *Theoretical Ecology: Principles and Applications*. Edited by R. May, Blackwell, Oxford. Second edition 1981. And, in a broader context *Mathematical Bioeconomics* by C. W. Clark. John Wiley & Sons, 1976].

[23] Their classic joint effort, *The Communist Manifesto*, was published in 1848. It is not widely known that Karl, also, was a considerable mathematician: *Mathematical Manuscripts of Karl Marx*, New Park Publications, London, 1983. 280 pp. This English translation contains a revealing technical Preface by S. A. Yanovskaya, written for the 1968 Russian edition: especially with regard to Marx's efforts to explain obscure features of calculus to his slightly less numerate friend, Engels.

[24] Oscillations, in single- and few-species population theory, arise from interactions between two or more components of the system (competitors, predators and prey, different age-classes of the population). They can also arise – and this is new – from intrinsic rates of natural increase not being, as Malthus – and many others after him – thought, geometric/exponential. We shall come to that revolutionary idea later. Here, we note that in Malthus' design, the upper, land-owning and smaller class (the predator) and the larger, newly mostly landless and more numerous class (the prey), must be distinct from each other, i.e. there must be no or negligible "diffusion" between the classes, to produce the predator-prey system oscillations that later became familiar to population biologists.

[25] Remarkably, a linear plot of rate of increase against average population size in each decade, implying logistic growth, "predicts" an upper limit to the US population being reached at about 480 million, about double the present population Still plenty of room, then!

[26] *Density dependence* may be regarded as a *constitutive metaphor* – a term familiar to historians of science – of the discipline of population dynamics. This is "a metaphor that constitutes, at least for a time, an irreplaceable part of the linguistic machinery of scientific theory. Where no literal paraphrase is known, usually at the inception of a new theory, scientists depend upon this metaphor to develop their thoughts. It is, in turn, articulated, probed and extended by other scientists, to the extent that it ceases to be simply a metaphor and enters the process of thought itself, performing the vital cognitive function of framing the discursive patterns of the theory. Eventually, as the new science becomes established, and it is increasingly possible to assign more literal names to processes, so the importance of the metaphor declines. A similar pattern can be seen in social thought. For example nineteenth century physics provided the constitutive metaphor for neo-classical economics" (David Stack, 2003. For full reference, see endnote 146). Similarly, classical economics, from Adam Smith on, provided the powerful metaphor of "competition" to ecology.

[27] "The speculative philosopher", he writes, "... with eyes fixed on a happier state of society, the blessings of which he paints in the most captivating colours, he allows himself to indulge in the most bitter invectives against every present establishment, without applying his talents to consider the best and safest means of removing abuses and without seeming to be aware of the tremendous obstacles that threaten, even in theory, to oppose the progress of man towards perfection." A true Social Democrat, it seems, our Thomas!

[28] "*Notice sur la loi que la population suit dans son accroissement*". *Correspondance Mathématique et Physique* 10:113-21, 1838, followed by "*Recherches mathématiques sur la loi d'accroissement de la population*". *Mem Acad Roy. Belg.* 18:1-38, 1845, and "*Deuxieme mémoire sur la loi d'accroissement de la population*". *Mem. Acad. Roy. Belg.* 20:1-32, 1847. The young Verhulst began his work on population growth at the instigation of his mentor, Adolphe Quetelet, who proposed, in 1835, that the resistance to growth was proportional to the square of the speed at which the population increased; this notion appealed to him because he saw in it a direct analogy to the resistance that a medium opposes to a body traveling through it. Verhulst reasoned that in the early stages of growth a population would increase exponentially until such time as crucial resources became limited. He called that "break" level the *normal* population, and numbers exceeding that a *superabundant* population. His mathematical formulation did not, however, follow his verbal reasoning, since the logistic does not follow from the idea that growth shifts abruptly from an exponential rate to a slower "logistic". Verhulst did realize that his logistic was only one of several possibilities; in his 1847 memoire he suggested that the obstacles to growth might be proportional to the ratio of the superabundant to the total population.

[29] In published papers of the 1930-50s the S-curve is commonly referred to either as an *ogive*, though this usage seems to have ceased. An ogive is a term used in Gothic architecture, referring either to the intersection of vault ribs where the direction of curvature changes, or a window that comes to a point at the top. In English this also became an *ogee*, apparently from the difficulty that medaeval English masons had in pronouncing French words. In the New World (Webster's dictionary) the derived second meaning became more explicit: "A graph each of whose coordinates represents the sum of all the frequencies up to and including a corresponding frequency in a frequency distribution", i.e. the integral of such a distribution, though only if the distribution is essentially bell-shaped. In fisheries science the term was mainly used to describe the curve of probability of a fish being retained in a trawl cod-end as a function of its length, approximated to in population equations by so-called "knife-edge selection" concentrated at the inflection (turning point) of the ogive.

[30] Many functions that have been used to describe either (or both) growth of individuals (increase in size – length or weight) and populations (increase in number or biomass) have an asymmetrical sigmoid form; examples are the growth

functions due to B. Gompertz (*Phil. Trans. Roy. Soc. London* 115: 513-85, 1825) and L. von Bertalanffy ("Untersuchungen ueber die Gestzlichkeiten des Wachstums I." *Roux' Arch. Entwicklungsmech.*, 131: 613-52, 1934). The integral of the familiar "bell curve" (*Gaussian* or *normal* distribution) is a symmetrical sigmoid that looks like the logistic in some applications and over a fairly wide range of time and population size, but has distinctly different properties. The properties of many of these functions are usefully tabulated by R. Christensen *Data Distributions. A Statistical handbook*. Entropy, Lincoln MA, 300 pp, 1984. Pearl and Reed were not entirely satisfied with only a symmetrical growth pattern, and in the process of freeing the logistic from its restrictive symmetry generalized their original equation by adding arbitrary terms to it; this allowed better fits to more data series but, of course, destroyed any pretence that the simple logistic expressed a universal law.

[31] An important but I think little-known discussion of the properties of various sigmoid curves is given by W.E. Ricker "Growth rates and models", pp. 677-743 in *Fish Physiology, Volume VIII, Bioenergetics and Growth*, Edited by W.S.Hoar, D.J.Randall and J.R.Brett, Academic Press, 1979. Ricker is concerned primarily with equations for describing the growth in size of individual animals, but notes and comments upon the parallel use of similar – even identical – equations to graduate population data.

[32] "On the rate of growth of the population of the United States since 1750 and its mathematical representation". *Proceedings of the National Academy of Science USA*, 6: 275-88, 1920; and *The Biology of Population Growth*, Knopf, New York, 260 pp, 1925; see also Pearl, R. and Reed, L. J. "On the mathematical theory of population growth" *Metron* 3(1): 6-19, 1923. It is significant that the US census data that Pearl fitted with a simple logistic did not extend beyond the inflexion – about 1915 – of the fitted curve. And even as a purely empirical process it did not have serious predictive power, even though it happened to predict that the US population would reach about 200 million by the twenty-second century. (The present population is about 250 million.)

[33] "The Refractory Model: the logistic curve and the history of population ecology". *Quarterly rev. Biol.* 57: 29-52, 1982. Pearl had some reservations about this because he noticed that Robertson's data did not in fact follow the *symmetrical* curves given by his equation, and because the similarity between chemical and growth curves implied nothing about the underlying mechanisms of growth. Both Robertson, and Pearl and Reed, realized that populations and individuals occasionally or periodically underwent bursts of growth that were not *fitted* or "explained" by the logistic or its variants. To deal with this they postulated epochs or episodes (they called them, erroneously, "cycles") of successive logistic increase when a major change, such as an industrial revolution (in the case of population growth) or in available resources (in the case of chemical reactions) created the opportunity for growth beyond the limiting value dictated under the existing system.

[34] "On the normal rate of growth of an individual and its biological significance". *Arch. Entwicklungsmech. Org.* 25: 581-613, 1908; and, in the same year, "Further remarks on the normal rate of growth of an individual and its biological significance". *Arch Entwicklungsmech. Org.* 26: 108-18.

[35] "*Variazioni e fluttuazioni del numero d'individui in specie animali conviventi*" [Variations and fluctuations in the number of individuals in animal species living together]. *Memorie Academia Lincei* 2:31-113, 1926; and *Leçons sur la Théorie Mathématique de la Lutte pour laVie.* Gauthier –Villars, Paris, 1931. See also "Population growth , equilibria and extinction under specified breeding conditions. A development and extension of the logistic curve". *Human Biology* 10(1): 1-11, 1938.

[36] *Elements of Physical Biology*. Williams and Wilkins, Baltimore, 1925.

[37] I have given a short summary of these and subsequent developments in "The experience of Antarctic whaling: – Appendix: Reflections on sustainability and precaution" in *Management of Shared Fish Stocks*, pp. 131-50. Edited by A. I. L. Payne, C. M. O'Brien and S. I. Rogers. Blackwell, Oxford, 2004. This book constitutes the Proceedings of a symposium organized by CEFAS in July 2002 to celebrate the centenary of fisheries research at its Lowestoft laboratory, in Suffolk, England.

[38] "The Optimum Catch: Essays on Population". *Hvålradets Skrifter* 7: 92-127. Hjort was the Grand Old Man of fisheries science in Norway, and a major luminary in North and Western Europe; his name on a paper gave it great weight at the time, but this application was probably devised by Ottestad: "A mathematical model for the study of growth". Essays on Population, *Hvålradets Skrifter* 7: 30-54, 1933. Hjort and his colleagues referred to the difference between reproductive and natural mortality rates as *regeneration*, a difference that IWC scientists have commonly referred to as *net reproductive rate*.

[39] "Modern theory of exploiting a fishery, and application to North Sea trawling". *Journal du Conseil International pour l'Exploration de la Mer* 10: 264-74, 1935; "The sigmoid curve and the overfishing problem". *Rapports et Procès-Verbaux des Réunions du Conseil International pour l'Exploration de la Mer*, 110: 15-20, 1939. This and several other papers mentioned in this essay, published between 1935 and 1983 are conveniently reproduced in *Key Papers on Fish Populations*, Edited by D.H.Cushing. IRI Press, 1983, 403 pp. This volume has useful linking commentaries by the editor., whose own important 1983 contribution on "Stock and Recruitment" closes the anthology.

[40] "Some aspects of the dynamics of populations important to the management of the commercial marine fisheries". *Inter-American Tropical Tuna Commission Bulletin*, 1(2): 26-56, 1954 [by Schaefer]; "Growth of the Pacific coast pilchard fishery to 1942". *US Fish and Wildlife Service Research Report*, 29, 31 pp, 1951 [by Schaefer, Oscar Sette and John Marr].

[41] S. J. Holt "Whales and Whaling". Chapter 112 in *Seas at the Millennium: an Environmental Evaluation*, Vol 3: 73-88. [Edited by C. R. C. Sheppard].

[42] An interesting feature of Kingsland's treatment of the earlier history is her account of the controversy about whether the logistic was a scientific *law*, as Pearl maintained, or was sim-

ply an equation that could be used to graduate data? Could it validly be used to make predictions, or not? Kingsland quotes F. E. Smith who, in 1952, expressed ambivalence towards these existing models (Verhulst-Pearl, Lotka-Volterra) – "Experimental methods in population dynamics: A critique". *Ecology,* 33: 441-450. Smith recognized the value of armchair thinking along deterministic lines as a way of generating concepts, but was strongly critical of the lack of correspondence between the logistic theory and the experiments which purported to verify it. In writing that "The degree of acceptance of such concepts...is astonishing". Smith was responding justifiably to a prolific scientific literature on the logistic curve in which the notions of what constituted a law, theory, model, hypothesis or proof were frequently confused. And Kingsland writes that the logistic "was put forward not as a convenient description, but as a law of growth, and was vigorously criticized by statisticians, economists and biologists" over the next fifteen years, before being for the most part discarded. Yet it survived and finally emerged in a different context as one of the central models of experimental population biology in the late 1930s and 1940s. I would add that a similar 'emergence' occurred in the context of fisheries and wildlife conservation and management, and has remained a powerful but hidden force to this day.

[43] These terms are sometimes given slightly different formal meanings.

[44] The most sophisticated way of doing this was devised by an Australian scientist with engineering training, William K. de la Mare, in the context of the management of whaling: "Fitting population models to time series of abundance data". *Rep. Int. Whal. Commn* 36: 399-418,1986. The computer program is given in: "The model used in the HITTER and FITTER programs". *Rep. Int. Whal. Commn* 39: 150-151, 1989 [Report of the Scientific Committee, Annex I]. de la Mare applies his engineering background to the field of fish and wildlife management in Chapter 21.

[45] The main modification adopted was to introduce a non-linear dependence of the population growth rate on density, using an equation put forward in a fisheries context by J. J. Pella and P. K. Tomlinson: "A generalized stock production model". *Bull. Inter-Am. Trop Tuna Commn* 14: 421-496, 1969. This particular modification of the simple logistic was originally devised by F. J. Richards in a different, botanical context: "A flexible growth function for empirical use". *J. Exp. Bot.* 10: 290-300, 1959. Regarding the BALEEN II model, see Punt, A. E. "A full description of the standard BALEEN II model and some variants thereof". *J. Cetacean Research and Management* (Suppl), 1: 267-76, 1999.

[46] For several biologists in the first half of the twentieth century the logistic and related sigmoid curves, with their upper and lower asymptotes and inflections, had an almost mystical significance, beyond their role in fitting equations to empirical data series. Hints of this are to be seen in Michael Graham's classic of fisheries science and management, *The Fish Gate* (Faber & Faber, 1943). Graham sought to explain these ideas in simple language, and to blend or reconcile them with the then current alternative treatment of the growth and deaths of animals in a cohort, in lectures given to third year undergraduate students at the new University of Salford (ex-Royal College of Advanced Technology). Edited versions of those lectures were post-humously published by Manchester University Press, in 1971 as *A Natural Ecology,* an inspiring but sadly neglected work. A different approach was taken by Graham's associate, J.Z.Young, in his magisterial *The Life of Vertebrates*, published by Oxford U.P. in 1950, reprinted 1952. In his exposition of Graham's *Great Law of Fishing* – that unlimited fisheries become unprofitable – Young comments "The aim of regulation is nowadays not so much to save the stock from undue reduction or extinction, but rather to crop it in such a way as shall make fishing profitable". However, he bases his argument essentially on presumed density dependences of growth and perhaps also of natural mortality in the animals of exploitable size, both sexually mature and immature, rather than simply on differences in the age composition of lightly fished and more intensively fish stocks, as did Graham and, before him, E.S.Russell, in *The Overfishing Problem* (Cambridge U.P., 1942).

[47] *ICES Science Symposia* Vol. 215, August 2002 "100 Years of Science under ICES" contains a wealth of documentation relating to this controversy as well as to the evolution of stock assessment procedures more generally. Of particular note are "Overfishing, science and politics: the background in the 1890s to the foundation of the International Council for the Exploration of the Sea" (J. Smed and J. Ramster, pp 13-21); "Realising the basis for overfishing and quantifying fish population dynamics" (Ø. Ulltang, pp 443-452); "ICES involvement in whaling and whale conservation, and implications of IWC actions" (S. J. Holt, pp 464-73); "From fisheries research to fisheries science, 1900-1940. The Bergen and the ICES scenes: tracing the footsteps of Johan Hjort" (G. Saetersdal, pp. 515-22).

[48] Arthur Went, in his insider's history of ICES, quotes D'Arcy Thompson's text that became the definition of what we might now refer to as 'the ICES mission statement': "That...it be recognized as a primary object to estimate the quantity of fish available for the use of man, to record the variations in its amount from place to place and from time to time, to ascribe natural variations to their natural causes, and to determine whether or how far variations in the available stock are caused by the operations of man, and, if so, whether, when and how measures of restriction and protection should be applied". (A. E. J. Went "Seventy Years Agrowing. A History of the International Council for the Exploration of the Sea 1902-1972". *Rapp. Proc.-Verb. Cons. Int. Explor. Mer*, 165, 252 pp., 1972).

[49] Unlike mammals and birds most fishes continue to increase steadily in weight throughout their lives, and also to remain fecund throughout adulthood.

[50] "*Kvoprsy o biologicheskikh osnovanlyahk rybnogo khozyaistva*" *(On the question of the biological basis of fisheries). Nauchnyi Issledovatelskii Ikhtiologicheskii InstitutIsvestya (Report of the Division of Fish Management and Scientific Study of the Fishing Industry)*1: 81-128, 1918.

[51] See R. J. H. Beverton (posthumous) and E. D. Anderson "Reflections on 100 years of fisheries research". *ICES Marine Science Symposia* 215: 453-63, 2002; and Ulltang, Ø.

"Realizing the basis for overfishing and quantifying fish population dynamics" *ICES Marine Science Symposia* 215: 443-452, 2002.

[52] D. W. Skagen and K. H. Hauge "Recent development of methods for analytical fish stock assessment within ICES". *ICES Marine Science Symposia* 215: 523-31, 2002.

[53] This is explained in my paper "Fifty years on" in a special volume of *Reviews in Fish Biology and Fisheries*, 8:357-66, 1998 and in my Introduction to the third printing (Blackburn, New Jersey, 2004) of the 1957 monograph *On the Dynamics of Exploited Fish Populations* written jointly with Beverton.

[54] See T. J. Pitcher "A cover story: fisheries may drive stocks to extinction". *Rev. Fish Biol. and Fisheries* 8: 367-370, 1998.

[55] A contributory factor to this difference between North American and European fisheries management strategies was undoubtedly the greater diversity of the fishing operations, preferences and markets in western Europe.

[56] Apart from the fact that the governments procrastinated over the imposition of much lowered catch limits, and then compromised between what the scientists advised and the industry wanted, the scientific assessments of the 1960s were later found to be in error (over-optimistic) because of a twofold mistake in the interpretation of data for ages of baleen whales in catches, as well as in the unreliability of estimates of the numbers of whales.

[57] For more on the topic of commercial whaling, see Papastavrou and Cooke, Chapter 7.

[58] "Simulation studies on management procedures". *Rep. Int. Whal. Commn* 36: 5-60, 1986.

[59] For an account of the RMP Catch Limit Algorithm (CLA) see papers by its author, J. G. Cooke, especially "The International Whaling Commission's Revised Management Procedure as an example of a new approach to fishery management". In *Developments in Marine Biology 4, Whales, Seals, Fish and Man.* Proceedings of the International Symposium on the Biology of Marine Mammals in the North East Atlantic, Tromsø, Norway, November/December, 1994. [Edited by A. S. Blix, L. Walløe and Ø. Ulltang] pp. 647-57, 1995. Elsevier, Amsterdam, 720 pp. Also: "Improvement of fishery management advice through simulation testing of harvest algorithms". *ICES J. Mar. Sci.* 56: 797-810, 1989.

[60] An crucial feature of the RMS/RMP as currently being negotiated is the provision that all catch limits are set by default to zero, and can only be made non-zero under a suite of quite stringent conditions. One way of looking at this is the application of the Precautionary Principle to all stocks, rather than only to hitherto unexploited stocks as in the amended NMP. However, additional data are required other than an estimate of population number, principally a series of data for historical catches.

[61] It is in the development of the Revised Management Procedure (RMP) that the term *algorithm* came into common use in the context of the management of whaling and the quest for sustainability. In "The Advent of the Algorithm" (Harcourt, New York, San Diego, London, 2000). David Berlinski provides a logician's definition: "A finite procedure; written in a fixed symbolic vocabulary; governed by precise instructions; moving in discrete steps, 1,2,3...whose execution requires no insight, cleverness, intuition, intelligence, or perspicuity; and that sooner or later comes to an end." Berlinski also gives us a looser definition "for the wider world from which mathematics arises and to which the mathematician must like the rest of us return", thus: A set of rules; a recipe; a prescription for action; a guide; a linked and controlled injunction; an adjuration; a code; *an effort made to throw a complex verbal shawl over life's cluttering chaos.*

Justin Cooke argues (pers. lett., 2004) that the so-called New Management Procedure (NMP) – adopted by the IWC in 1975, and used to set catch limits and classify whale stocks until the commercial moratorium came into effect in 1996 – is properly speaking not a "procedure" since it tells the manager and the scientist what is intended to be done but not how to do it. In this it shares properties with virtually all the current "rule systems" of fisheries management.

[62] Feller, W. "On the logistic law of growth and its empirical verification in biology". *Acta Biotheoretica* 5: 51-66, 1940. The practical consequences of not understanding this can be very great. For example, the United Nations regularly publishes predictions of future global and regional human population growth. These are based on a model not unlike the BALEEN II used by whaling biologists, a logistic derivative. The UN corrects the parameter values occasionally, having noted, for example that the birth rate in China has unexpectedly fallen. However, Serge Kapitza (son of the internationally famous Soviet physicist, and himself a physicist), produced a population model having as a major feature *self-similarity*, which, in other contexts may be called a fractal dimension ("World Population Growth". Doc. C4.1.Kap to Pugwash Meeting No. 197, 43rd Pugwash Conference on Science and World Affairs, A World at the Crossroads: New Conflicts; New Solutions. Sweden 1993.) This produces a classic-looking sigmoid growth curve but one which, when fitted to the same data produces very different predictions for the coming decades. These kinds of projections are produced to give flesh to the idea that "we must do something about all this", so a little more attention to the ways they are made would be in order.

A nice example of the ways that two people presented with the same information can reach very different conclusion comes from a quip I first heard from Sir Peter Scott, famous conservationist, painter, sailor and flier, a Renaissance-Man if ever there was one. "An optimist is someone who thinks this is the best of all possible worlds; a pessimist fears that may be so."

[63] Ricker, W. E. "Handbook of computations for biological statistics of fish populations". *Bull Fish. Res. Bd Canada*, 119, 1958.

[64] Pitcher, T. "A cover story: fisheries may drive stocks to extinction". *Rev. Fish Biol. Fisheries* 8: 367-70, 1998.

[65] From Allee, W. C. *Animal Aggregations.* Univ. Chicago Press, 1931.

[66] Shepherd, J. G. "A versatile new stock recruitment relationship for fisheries and the construction of the sustainable

yield curve". *J. Cons. Int. Explor. Mer* 40: 67-75, 1982.

[67] Stephens, P. A. and Sutherland, W. J. "Consequences of the Allee effect for behaviour, ecology and conservation". *Trends in Ecology and Evolution* 14: 401-405, 1999; Courchamp, F., Clutton-Brock, L. and Grenfell, B. "Inverse density dependence and the Allee effect". *Trends in Ecology and Evolution* 14: 405-410, 2001.

[68] I adopt here the terminology used by Clark, C. W. *Mathematical Bioeconomics: The Optimal Management of Natural Resources.* John Wiley and Sons, 1976, 352 pp. The feature *depensation* is called *inverse density dependence* by Courchamp *et al.,* 1999.

[69] Cannibalism in fishes usually involves consumption of eggs, larvae and juveniles by adults. Examples are cod in the North Atlantic, Pacific salmons, and anchovies (Hunter, J. R. and Kimbrell, C. A. "Egg cannibalism in the northern anchovy, *Engraulis mordax*". *US Fishery Bull.* 78: 810-816, 1980).

[70] Perhaps the Northern right whale is an exception, as it appears thus far not to be recovering from extreme depletion, although this might be the result of increases in other threats from human activities, such as entanglement in fishing gear and lethal collisions with ships.

[71] See my Foreword to the fourth printing of Beverton, R. J. B. and Holt, S. J. *On the Dynamics of Exploited Fish Populations,* Blackburn, 2004.

[72] An example is that used to estimate world fishery potential, using a derivation of the simple logistic, by J. A. Gulland in "The Fish Resources of the Ocean", *FAO Fish Tech. Pap.* 97: 1-4, 1970.

[73] See discussion in Beverton, R. J. H. and Holt, S. J. (1957 and 2004) *On the Dynamics of Exploited Fish Populations,* HMSO, UK and Blackburn Press, New Jersey.

[74] For a deeper critical analysis of this matter see "The Biology of Fish Growth", by A.H.Weatherley and H.S.Gill, Academic Press, 1987, and Ricker, 1979, referenced in endnote 31.

[75] Misjudgment of this has led to the demise of valuable fisheries, such as for herring in the Northeast Atlantic, hastened by a switch away from catching mature adults as food for humans to very young fish for conversion to fishmeal and oil as feed-supplements for livestock.

[76] I have reviewed these in my Foreword to the fourth edition (2004) of my monograph written with R. J. H. Beverton (1957).

[77] See Pitcher, T. J. (1998). "A cover story: fisheries may drive stocks to extinction". *Rev Fish Biol.Fisheries* 8: 367-70.

[78] For more on biological diversity, see Willison, Chapter 2.

[79] Unfortunately – as so often – science is not always in accord with "common sense". The latter tells us that diversity – or rather, perhaps, *complexity,* the difference being that complexity involves significant relationship among the parts – will enhance "stability". There is considerable scientific literature, as well as experimental studies and some field experience, suggesting just the opposite. This is thoroughly examined in Robert May's chapter "Patterns in Multi-species Communities", pp. 197-227, in May *et al.,* 1981. The scientific jury is, I think, still out on that question.

[80] See Boyd, I. "Culling predators to protect fisheries: a case of accumulating uncertainties". *Trends in Ecology and Evolution* 16: 281-282, 2001.

[81] I have elaborated this concern in an article in the new newsletter of the ACCOBAMS agreement under the Convention on Migratory Species (CMS): "Sharing our seas with whales and dolphins" *FINS* 1(1): 2-4, 2004 [Monaco, and at http://www.accobams.org/download/newsletter/fins_01.pdf].

[82] An overall review of this matter is given in May, R. M. *Exploitation of Marine Communities.* Report of the Dahlem Workshop on the Exploitation of Marine Communities, 1984, Springer, Berlin.

[83] The classic among such models is, I think, Anderson, E. P. and Ursin, E. "A multi-species extension to the Beverton and Holt theory of fishing, with accounts of phosphorus circulation and primary production". *Meddelelser fra Danmarks Fiskeri- og Havundersgelser* 7:319-415, 1977. This already hints at a transition towards biomass-, critical nutrient- and energy exchange modeling. An important contribution was made a few years later by Ola Flaaten, "The Economics of Multispecies Harvesting: Theory and Application to the Barents Sea Fisheries", *Studies in Contemporary Economics,* 162 pp, Springer-Verlag, Berlin, 1988. Flaaten reminds us that "the recognition of the necessity of harvesting the predator to increase the yield of the prey is not entirely new", and cites the belief of the Italian biologist, Umberto d'Ancona – associate of Volterra – that "the predators of this [Adriatic] sea, the sharks, ought to be decreased by increasing harvest intensity. That would make it possible to increase the yields of the more valuable prey stocks" (*Dell' influenza della stasi peschericchia del periodo 1914-1918 sul patrimonio ittico dell' Alto Adriatico. Memoria CXXVI,* Comitato Talassografico Italiano, 1926.) Flaaten also noticed that Charles Darwin made a similar assertion in *The Origin of Species,* 1859, regarding "game" birds and mammals and the "vermin" that preyed on them.

[84] *Trophodynamics* is the study of ecosystems from that point of view. This is associated originally with R. L. Lindeman, "The tropho-dynamic aspect of ecology". *Ecology* 23(4):399-418, 1942.

[85] Mass and energy transfer models include ECOPATH, ECOSIM and ECOSPACE, associated especially with the names of Daniel Pauly and Carl Walters. A review of the applications of these is given in "Ecopath, Ecosim and Ecospace as tools for evaluating ecosystem impact of fisheries." Pauly, D., Christensen, V. and Walters, K. *ICES J. Mar. Sci* 57:697-706, 2000. Other basic papers are: Pauly, D. and Christensen, V., "Stratified models of large marine ecosystems, a general approach, and an application to the South China Sea", in *Stress, Mitigation and Sustainability of Large Marine Ecosystems,* pp. 148-74, 1993. (Edited by K. Sherman, L.M. Alexander, and B.D. Gold, AAAS, Washington DC); Pauly, D. "Use of Ecopath with Ecosim to evaluate strategies for sustainable exploitation of multi-species resources". *Fisheries Centre Reports* 6 (2) 1998, 49 pp. Vancouver, BC, Canada; Pauly, D., Soriano-Bartz, M.L. and Palomares, M.L.D.. "Improved construction, parameterisation and interpretation of steady-state ecosystems". pp. 1-13, in "Trophic Models of Aquatic Ecosystems", *ICLARM Conference Proceedings* 26, 1993.; Christensen, V. and Pauly, D. "Changes in models of aquatic ecosystems approaching

carrying capacity". *Ecological Applications* 8(1), (Suppl):104-9, 1998.

[86] Yodzis, P. "Must top predators be culled for the sake of fisheries?" *Trends in Ecology and Evolution* 16(2):78-84, 2001. See also Yodzis's "The indeterminacy of ecological interactions as perceived through perturbation experiments". *Ecology*, 69:508-15,1988.

[87] I think it is of interest that Michael Graham's "A Natural Ecology" (referenced above) includes a diagram, somewhat similar to more recent and complicated ones by Yodzis, centred on the herring of the North Sea, and featuring the differences of diet between various age classes of that species. His Diagram 3, illustrating synecology, deals with the herring and the trophic levels "below" it, down to the diatoms and flagellates. Beyond that, however, what Graham writes in the accompanying text is relevant to our present arguments about "culling" predators for human benefit: "As I sketch this diagram (due to Sir Alistair Hardy) on the board in the lecture room I add the cod, which feeds on herring, and feeding on the cod there would be seals, and feeding on the seals there would be polar bears and Eskimo. There would also be the numerous fishermen who in historic times have fished the cod...Competing with the cod for herring we should have to add gannets and toothed whales, and competing with the seals for cod we should add Greenland sharks. Competing with the herring for krill are the whalebone whales and many (species of) pelagic birds. It is evident that the more one knows about any ecosystem, the more difficult it is to show a tidy diagram of its synecology, and that is true of the web of life as it exists in nature". Graham might have added, to complicate things further but not unnecessarily, that cod eat themselves – bigger cod commonly feed on smaller ones.

[88] A. B. Hollowed *et al.* "Are multispecies models an improvement on single-species models for measuring fishing impacts on marine ecosystems?" *ICES J. Mar. Sci* 57:707-19, 2000.

[89] These authors are clearly aware of the intractable difficulties, despite their guarded optimism, and their conclusions are worth recalling in full "What is the implication of these ideas? First, it is clear that multi-species interactions need to be placed within the context of the myriad other factors and processes influencing these systems. Second, multi-species interactions occur within a spatial and ontogenetic structure – models that lack this structure are unlikely to have any predictive capacity because of the complexity of the interactions. Third, this complexity means that models with quite different properties may be developed for any one system. Therefore, predictions need to be addressed within a rigorous and testable modeling framework. Finally, the multi-species interactions of most interest in determining the impacts of fishing on marine ecosystems are those that cause marked departures from the current conditions – models constrained by equilibrium processes are unlikely to capture these departures".

[90] "Marine ecosystem-based management as a hierarchical control system". *Mar. Policy* 2(1):5-68, 2005.

[91] Guenette, S., Lauck, T. and Clark C. "Marine reserves: from Beverton and Holt to the present". *Revs. Fish Biol. And Fisheries* 8: 251-272, 1998.

[92] Sumaila, U. R., Guénette, S., Alder, J. and Chuenpagdee, R. "Addressing ecosystem effects of fishing using marine protected areas". *ICES J. Mar. Sci.* 57: 752-60, 2000.

[93] Walters, C. J., Christensen, V. and Pauly, D. "Structuring dynamic models of exploited ecosystems from trophicmass-balance assessments". *Revs Fish. Biol.* 7:127-42, 1997; Walters, C.J., Pauly, D. and Christensen, V. "Ecospace: a software tool for predicting mesoscale spatial patterns in trophic relationships of exploited ecosystems, with special reference to impacts of marine protected areas". *ICES CM* 1998/:S:4 20 pp., 1998 [a spatially explicit model that includes movement rates to compute exchanges between grid cells and habitat preferences]; Watson, R. and Walters, C. "Ecosim and MPAs: a quasi-spatial use of Ecosim". pp. 15-18 in *Use of Ecopath with Ecosim to Evaluate Strategies for Sustainable Exploitation of Multi-species Resources*. Edited by Pauly, D., Fisheries Centre, UBC, Vancouver. 6(2), 1998. [This is a simple model based on ECOPATH with quasi-spatial relations between biomass and fishing mortality.]

[94] McCall, A. D. *Dynamic Geography of Marine Fish Populations*, 1990, Books on Recruitment Fisheries Oceanography, Univ. Washington, Seattle, 153 pp.

[95] "Population models and community structure in heterogeneous environments", pp. 295-320 in *Mathematical Ecology*, edited by T. G. Hallam and S. A. Levin, Springer-Verlag, Berlin, 1986. See also Rosenzweig, M. I. "A theory of habitat selection". *Ecology* 62: 327-35, 1981.

[96] S. Fretwell proposed this term, in *Populations in a Seasonal Environment*, Princeton Univ. Press, 1972, 217 pp. The general theory was spelled out by Fretwell and H. Lucas in "On the territorial behavior and other factors influencing habitat distribution in birds". *Acta Biotheoretica* 19: 16-36, 1970. Other terms for the notion of *suitability*, such as "environmental density", "fitness", "reproductive value, even "goodness" have been used by various other authors.

[97] A phenomenon examined in depth by Brown, J. H. "On the relationship between abundance and distribution of species". *American Naturalist* 124: 255-279, 1984.

[98] This leads to four statements: 1. The free surface of the liquid will be approximately level, corresponding to the uniform realized suitability resulting from the ideal free distribution; 2. The "shoreline" corresponding to habitats whose basic situation is exactly equal to the uniform realized suitability, establishing the range of the population; 3. The depth of the liquid at any location is proportional to the density-dependent reduction in its realized suitability at that location, and is proportional to local density; 4. The total volume of liquid in the basin is thereby functionally related to total population size.

[99] It was used in the IWC scientists' examination of the *robustness* of proposed CLAs. H. S. Gordon analysed fishermen's choices among alternative fishing grounds and produced habitat utilization diagrams remarkably like those of Freewell, "An economic approach to the optimum utilization of fishery resources". *J. Fish. Res. Bd Canada* 10: 442-57, 1953. Beverton and Holt, 1957, supposed that fishing effort/intensity would distribute itself in such a manner as to lead to fish densities on all grounds affected by the fishing being reduced to about the same value, with allowance being

made for factors arising from the fact that the human effort starts from and returns to certain places on the margins of the fish distribution.

[100] McCall's book has nearly 150 bibliographic references on the subject.

[101] David Ricardo (1772-1823), author of *The Principles of Political Economy* (1817), was a contemporary of Thomas Malthus, and had an interesting interaction with him. He wrote in 1815 that "In all that I have said concerning the origin and progress of rent I have briefly repeated, and endeavored to elucidate , the principles which Malthus has so ably laid down on the same subject…".

[102] "Optimal foraging, the marginal value theorem". *Theoretical Population Biology* 9: 129-36, 1976.

[103] See R. M. May "Stability and complexity in model ecosystems", *Monographs in Population Biology* VI, Princeton, 1973; "Biological populations with non-overlapping generations: stable points, stable cycles, and chaos". *Science* 186: 645-647, 1974.

[104] In fact this is only strictly true for deterministic models; stochastic models can behave very differently.

[105] For more on the evolution of the conservation movement, see Lavigne, Chapter 1.

[106] "The Changing Ocean. Its effects on climate and living resources". *IOC Ocean Forum Series* IV, UNESCO, Paris, 169 pp., 2003.

[107] First described by Broeker *et al.* in *Nature* 315: 21-26, 1985. The Indian and Pacific Oceans' mechanism is described thus: "Warm and salty surface water moving northward though the Atlantic is driven into the North Atlantic and the Norwegian Sea where it is cooled. Its density increases causing it to sink to great depths and return to the South Atlantic, then to the Indian and Pacific Oceans. [This sinking is enhanced by influx of melted sea ice water from the North Polar ocean and freshwater from the Greenland Icecap which, though cold, is less dense]. This deep water diffuses slowly towards the surface where it is taken up by surface currents and carried back to the North Atlantic". The eastward-flowing Antarctic Circumpolar current is deflected to its left (i.e. northward) by the Cariolis force arising from the earth's spin, until it comes into contact with warmer and less dense water of sub-tropical origin.

[108] Further information about the ENSO phenomenon is given in Voituriez, B. and Jacques, G. "El Niño: Fact and Fiction", *IOC Ocean Forum Series*, I, 2000, UNESCO, 128 pp.

[109] An *anomaly* is defined as the difference between the value of a parameter at a given moment and its long-term mean value. It is thus a deviation in time rather than in space, but the concept can be extended to the spacial dimension.

[110] More formally, a layer of water in the ocean in which the temperature decreases rapidly, and commonly limits vertical mixing between the surface mixed layer and the deeper water, and thus impedes the movement of essential nutrients. The *pycnocline*, the layer of water in which the density increases rapidly with depth, may be coincident with the thermocline. The *nutricline*, in turn associated with the pycnocline, is the layer in which there is a rapid change in the concentration of nutrients with increasing depth.The periodic but irregular changes in the location of boundary phenomena – depth of thermocline etc.– and geographic location of divergences and convergences in the open ocean are usually referred to as *undulations*.

[111] The Antarctic Divergence, at about 65ºS., is anomalous, because it brings up water from a very great depth. The high biological productivity is not manifest until the newly-surfaced water reaches about 40ºS., 2000 km to the North, about 200 days later. In that time production has been limited by complex interaction of deficits of a series of nutrient elements, including especially silicon (necessary for the growth of diatoms) and iron. The supreme importance of iron in this region was proven in 1999 by the Southern Ocean Iron Release Experiment (SOIREE) carried out in the Circumpolar Current south of Tasmania. The other purpose of that experiment was to verify or otherwise the role of the Southern Ocean as "Manager of the Atmosphere's Carbon-dioxide".

[112] In general, oceanic *fronts* where different water masses meet are biologically productive (other things being equal), and recent studies have shown that many predatory animals – including whales – are able to detect them and congregate along those lines. Another important boundary system, especially in the Antarctic, is the limit of sea-ice, which, of course, shifts latitudinally through the seasonal cycle. It is also known to have changed the latitude of its average summer location through the past century – the Antarctic sea-ice boundary at a given season is retreating towards the pole and thus is shortening. (Satellite observations over a few decades confirm what has been revealed from historical studies of the geographical distribution of whaling vessels "hugging" the ice edge, carried out by W. de la Mare.) The significance of this is that diatoms are held frozen in the ice-edge during the winter and released as the ice melts in spring, providing nutrition for grazing zooplankters such as krill and so driving their multiplication, which in turn attracts the larger predators.

[113] For more on the collapse of cod stocks, see Hutchings, Chapter 6.

[114] Out-of-phase oscillations in the sea-surface temperature of the Central North Pacific, on the one hand, and of he eastern boundary and the eastern part of the inter-tropical Pacific on the other.

[115] The fish are caught in traps, mostly in Sicily, and the catches have been monitored closely by fiscal authorities, collectors of the ecclesiastical tithe, customs officers and investment bankers!

[116] Klyashtorin went further. He reasoned that the shifts of fluids on the planet modify the speed of the Earth's rotation; indeed the El Niños of 1982-83 and 1997-98 are associated with a tiny reduction in the rate of rotation and hence a lengthening of the day. The ACI is well correlated with the rotation rate, so, Voiteruez asks "Could the speed of rotation become an index of the evolution of marine ecosystems?"

[117] See Witting, L. *A General Theory of Evolution, by Means of Selection by Density Dependent Competitive Interactions.* Peregrine, Århus, Denmark, 1997, 332 pp.

[118] Witting, L. "Population cycles caused by selection by density dependent competitive interactions". *Bull. Math. Biol.*, 62:

1109-1136, 2000.

[119] Witting, L. "Reconstructing the population dynamics of eastern Pacific whales over the past 150-400 years" *J. Cetacean Res. Manage.* 5(1): 45-54, 2003. [Published revised version of a 2001 IWC document.] See also "On inertial dynamics of exploited and unexploited populations selected by density dependent competitive interactions", *IWC Doc SC/D2K/AWMP6* (Rev.), 2001.

[120] Clark, C. 1981."Bioeconomics". pp 387-408 in Theoretical *Ecology: Principles and Applications.* R.M. May, ed. (Blackwell).

[121] *In a Perfect Ocean: The State of Fisheries and Ecosystems in the North Atlantic Ocean.* 175 pp. Island Press, 2003: "A larger reason why we find ourselves with an impoverished ocean is because past generations did not care about us and discounted what appeared to them as future benefits to be gained from exploiting the North Atlantic. Our generation is no better... A factor in policy-making known as the discount rate essentially calculates that the value of an unexploited resource declines over time (by the same percentage for each year), and eventually approaches zero, at least as far as our generation is concerned." Some economists seek to justify the failure to consider future generations by claiming that our descendents will have found substitutes for depleted resources, and there is some validity in this for some kinds of resources. However, the result of policy analysts discounting future fishery benefits, in making cost-benefit analyses, is that they are skewed toward exploitation of the resources by the present generation. "This", say Pauly and Maclean, "puts the burden of proof in policy debates on those who would argue for resource restoration." And, I would add, severely impedes efforts towards sustainable use now.

[122] These are the words of T. R. E. Southwood "Bionomic Strategies and Population Parameters". pp. 30-52 in *Theoretical Ecology: Principles and Applications*, Edited by R. M. May, second edition, 1981.

[123] "Habitat, the temple for ecological strategies". *J. Anim. Ecol.* 46: 337-366, 1977.

[124] Southwood suggested this might be because longevity is inversely related to total metabolic activity per unit body weight.

[125] The fossil record reveals many instances where an evolutionary line has evolved to increased size, until extinction occurs ("Cope's rule"; Cope, E. D. *The Origin of the Fittest*, Appleton and Co., New York, 1887). These lines have become more and more closely adapted to specialized and hitherto stable habitat through progressive ***K***-selection. A profound analysis of these issues has been provided recently, in evolutionary-genetic-mathematical terms, by L. Witting: "Major life-history transitions by deterministic directional natural selection", *J. Theoretical Biol.* 225: 389-406, 2003.

[126] For a general account of this matter see especially Ludwig, D., Hilborn, R. and Walters, C. "Uncertainty, Resource Exploitation, and Conservation: Lessons from History". *Ecological Applications* 3(4): 547-549, 1993.

[127] "Case studies on the allocation of transferable quota rights in fisheries". Edited by Shotton, R. *FAO Fish. Tech. Pap.* 411, 2001, 373 pp.

[128] Holt, S. J. "Sharing the catches of whales in the Southern Hemisphere". *FAO Fish. Tech. Pap.* 411: 322-373, 2001.

[129] Elliot, G. *A Whaling Enterprise: Salvesen in the Antarctic.* Michael Russell, Norwich, UK, 1998, 190 pp. [See especially p. 129.]

[130] It is worth recalling here that several fisheries scientists have advocated consideration of such a strategy, the most well-known and articulate being, I think, Martin Burkenroad. It is, however, associated with Burkenroad's and other's claims that the effects of fishing on fish stocks had not been convincingly demonstrated, except perhaps with respect to the effects of two world wars on demersal species in the North Sea. Some decades later the swarming of large factory trawlers, with European registries, off the coasts of developing countries, especially off western Africa, led to the concept of "pulse fishing" and hence revived questions regarding the link between sustainability and continuity of exploitation.

[131] For more on Salvesen, see Papastavrou and Cooke, Chapter 7.

[132] Readers wishing to explore further the nature of infinity and its multiple forms, and especially of infinite time, might find pleasure in reading *Infinity (A Brief History of): The Quest to Think the Unthinkable*, by Brian Clegg, Robinson, London, 2003, 255 pp.

[133] Graham, M. *The Fish Gate* Faber, London, 1943.

[134] In his review of "The State of Fisheries Science" (pp 25-54 in *The State of the World's Fisheries Resources*, 1994, edited by C. Voigtlander, Lebanon, New Hampshire, International Science Publications), R.J.H. Beverton noted that long-range planning by the fishing industry [*and hence, presumably, a practical time-horizon for both management and management related research - sjh*] "extends over the 20-year or so life-span of its major capital facilities".

[135] For more on whale watching, see Corkeron, Chapter 11.

[136] See Pauly, D. *et al.* "Fishing down marine food webs". *Science* 279: 860-863, 1998; Jackson, J. B. C. *et al.* "Historical overfishing and the recent collapse of coastal ecosystems". *Science* 283: 629-38, 2001; Hempel, G. and Pauly, D. "Fisheries and Fisheries Science in Their Search for Sustainability", pp. 109-35, 2002 in *Oceans 2020.*

[137] *Thermodynamics and Statistical Mechanics. Lectures in Theoretical Physics*, Vol V, trans by J. Kestin. New York Academic Press, 1956, p. 19. Quoted in the biological context by G. Tyler Miller Jr: *Energetics, Kinetics, and Life: an Ecological Approach.* Wadsworth Publishing, 1971, p. 143. In turn quoted and expounded by Wes Jackson "Agriculture: The Primary Environmental Challenge of the 21st Century". pp. 85-99 in *The Human Ecological Footprint*, Kenneth Hammond Lectures on Environment, Energy and Resources, 2002 Series, Faculty of Environmental Sciences, University of Guelph, Ontario, Canada, 2004. Edited by W. Chesworth, M. R. Moss and V. G. Thomas.

[138] This is not the place to discuss the details of why an estimate of so-called replacement yield is not the same as the theoretical sustainable yield. It has been applied especially in situations where there is more than the usual uncertainty as to the dynamics of the population, and especially about its

age-composition and the relation between parent numbers and subsequent recruitment. Most commonly a presumed current rate of increase in number – were exploitation to pause – is calculated from minimal data; which, multiplied by an estimate of the current population size, is taken as the replacement yield. But the rate of increase depends, in part, on the incoming recruitment, which is some function of an earlier population size. While little harm will be done in many cases if this is used to establish an *interim* management measure, say a Total Allowable Catch (TAC) for one or two years, this is usually – whatever might be said at the time – repeated for several years; that is because the original uncertainties are rarely if ever over-come in the iterim. So, *repeated* application of this measure can endanger the population over time. The danger is especially acute for species in which there is a strong correlation between the number in the sexually active part of the parent population and the subsequent number of recruits into the exploited phase of the population, as in marine mammals and, possibly, the relatively slowly reproducing fishes such as elasmobranches (sharks and rays). As this is being written it seems that the Government of Canada has made exactly this mistake in announcing the "rationale" for its recent large increase in its permitted "cull" of harp seals in the Northwest Atlantic.

[139] After reading an early draft of this essay David Lavigne challenged me to redefine our *Holy Grail*, if it is not sustainability. That is perhaps something for the Forum as a whole to tackle. But I would begin with something that has a broad, long-term (but not infinite) idea of sustainability embedded in it but bearing also the notions of precaution, frugality and parsimony. In his unusual little book entitled *Muddling Toward Frugality* (Shambhala, Boulder, 1979), Warren Johnson, one time professor of geography at San Diego State University, California, noted that the word frugality came originally from the Latin *frugalior*, meaning useful or worthy, and *frux*, meaning fruitful or productive, giving, he said "a nice feeling". "However" he observes, "the word has changed over the years, and has come to mean thriftiness, the abstention from luxury and lavishness". In his book he stays with the original meaning, "to suggest economic conditions in which society is obliged by force of circumstances to make full and fruitful use of all its resources". Professor Johnson was a super-optimist; in his Preface he wrote: "If the Earth is to be a true home for us, a place of refuge and nurture, we may as well start to think about how we can make it such a place. The task will not be as difficult as it may sound, and requires no wishful thinking about technological breakthroughs, effective government, or heightened human consciousness. We can move towards a sustainable way of life easily if we accept the logic of frugality."

[140] Larkin, P. A. "An epitaph for the concept of maximum sustained yield". *Trans. Am Fish. Soc.* 106: 1-11, 1977. Peter Larkin was, however, as much concerned with the failure of managers to act on scientific advice as with the conceptual failings of the MSY paradigm and management strategy.

[141] Here Graham confirms that the work by Ray Beverton and myself, under his direction, was not pedagogic in intent, but exploratory.

[142] An alternative to finding an advantageous direction and moving in it is commonly thought to be to define an end target, a "reference point" in current jargon. MSY was pressed as more suitable than, say, some economically or socially determined optimum especially in any international context; something that perhaps every country could agree on regardless of its economic system or state. A very different light is, however, thrown on this by state documents of the immediate post-war period concerning fisheries of the USA and Japan, especially as they concern whales, Pacific salmon and tuna. These have been analysed most thoroughly by a historian at the University of California at Berkeley, Prof. Harry Scheiber. (*Inter-Allied Conflicts and Ocean Law, 1945-53: The Occupation Command's Revival of Japanese Whaling and Fisheries.* Institute of European and American Studies, Academia Sibica, Taipei, Taiwan, 2003, 233 pp.)

Just as it had long been UK and Norwegian policy, while competing with each other, to work together to prevent any other countries engaging in pelagic whaling in the Antarctic, the USA (and Canada) sought ways of keeping Japan out of the then very lucrative salmon fisheries off their west coasts. The occasion was the negotiation of an International North Pacific Fisheries Convention (INPFC), focused on salmon. In that context an abstention principle was invented. The idea was that if existing fishing states – primarily coastal states – were already exploiting a resource at the level of MSY then other states should agree to abstain from entering that industry. Evidently, Japan was unlikely to accept a scarcely concealed discriminatory rule without a *quid pro quo*. That which was granted was an assurance that the Government of the Occupying Power would not impede Japanese fisheries expansion elsewhere – including as it happened Antarctic whaling, but not only that.

The situation changed somewhat when the value of the US Pacific tuna fishery vastly exceeded that of the salmon fisheries, in which – it turned out later – the US would come into vigorous conflict with Latin American competitors and, eventually, Japan, Taiwan and other Asian states. However, for the time being the abstention principle could be turned upside down. The tunas being initially under-exploited, it could be argued that the MSY target justified – even demanded – that the populations be **reduced** to the supposed MSY levels. Much later this same argument was used by Japanese delegations to the IWC to claim that since the minke whales of the southern hemisphere had not been exploited before 1970, good management would ensure that they were reduced by whaling as fast as possible to a putative MSY level; anything other than that – even a more cautious depletion – would be "wasteful".

The last stage in this extraordinary political process concerning exclusion/abstention/non-exclusion/sharing is found in its converse form in Article 62 of UNCLOS, regarding Exclusive Economic Zones: "The coastal State shall determine its capacity to harvest the living resources of the EEZ. **Where the coastal State does not have the capacity to harvest the entire allowable catch it *shall*,** through agreements or other arrangements etc., **give other States access to the surplus of the allowable catch**..." Note that in this case it is the coastal State (and eyes were really on *developing* coastal

States) that has no option but to use "fully" or to share.

[143] As to compliance, in the fisheries context, see for example W. Edeson, D. Freestone and E. Gudmundsdottir "Legislating for sustainable Fisheries. A guide to Implementing the 1993 FAO Compliance Agreement and the 1995 UN Fish Stocks Agreement". *Law, Justice and Development Series*,.The World Bank, 2003, 151 pp.

[144] Although not affecting the principle of the management scheme developed by the IWC some practical problems have arisen, not thought about when this initiative was launched. It seems likely that, even if the specified conditions for setting non-zero catch limits were to be met, the (precautionary) limits would be too small to justify commercial interest. In the time that has elapsed since the RMP/CLA was agreed whaling administrations have surely worked this out for themselves, even though the IWC and its Scientific Committee have not officially performed the calculations. In such circumstances it seems to me unlikely that any of the countries still seriously interested in resuming legitimized commercial whaling would, in the end, accept the proposed RMS. There could therefore be a reversion to the Salvesen's exit strategy: continue depleting the remaining fairly numerous whales (essentially only the minke) and then pull out. This is indeed what the authorities of Norway appear to have decided, as they steadily increase catches of minke whales from the depleted population in the Northeast Atlantic, under their objections to the IWC's closure decisions of the 1980s, and now using the absurd excuse that these whales are threatening the North Atlantic cod and herring stocks. Current Norwegian catches are already several-fold higher than would be granted under the RMP as approved by the IWC. At the present time it seems the only impediment to this policy might be the limitations of the market for minke meat and blubber, rather than restrictions on catch imposed by any international authority

[145] I feel at one with Lynn Margulis, who wrote: "We people are just like our planet-mates. We cannot put an end to nature, **we can only pose a threat to ourselves**". The notion [*here comes that word again! sjh*] that we can destroy all life, including bacteria thriving in the water tanks of nuclear power plants or boiling in hot vents, is ludicrous. I hear our non-human brethren snickering: "Got along without you before I met you, gonna get along without you now." They sing about us in harmony. Most of them, the microbes, the whales, the insects, the seed birds, are still singing. The tropical forest trees are humming to themselves, waiting for us to finish our arrogant logging so they can get back to their business of growth as usual. And they will continue their cacophonies and harmonies long after we are gone." (Closing paragraph of "The Symbiotic Planet: A new look at evolution", Phoenix, 1998.) And, I would say, evolution will continue to plod its relentless way.

[146] For much more on the evolutionary and genetic threads in the history of political economy see *The First Darwinian Left: Socialism and Darwinism 1859-1914*, by David Stack New Clarion Press, Cheltenham, England, 2003, 149 pp.

[147] New York, Avon Books, 1994. Damasio is a neurologist who has spent much of his life studying the function of the human brain and the relationships between reason and emotion, thought and feelings.

[148] Pp. 1-34 in Chesworth, Moss and Thomas (eds.), 2004. See endnote 137, above.

[149] For more on the subject of love, see Lavigne *et al.*, Chapter 26.

[150] Perhaps "biodiversity", a current, all-pervasive buzz-word, examined recently by Ronald Brooks who asked the question "The Paradox of Biodiversity – Is Conservation Ethical, Aesthetic, Utilitarian or an Adaptive Strategy?" ("Earthworms and the Formation of Environmental Ethics and Other Mythologies: a Darwinian Perspective". pp. 59-91 in *Malthus and the Third Millennium*, Kenneth Hammond Lectures on Environment, Energy and Resources, 2000 Series, University of Guelph, Ontario, Canada, 2001. Edited by W. Chesworth, M.R. Moss and V.G. Thomas.) Brooks labels biodiversity as "a virtually holy grail not just for biologists, but for naturalists, environmentalists, politicians and policy makers. But why? Certainly not for any scientific reasons. As I have argued for earthworms, there is little scientific evidence that biodiversity is necessary, beneficial or even natural". Further on, Brooks' comments link clearly again to the issue of sustainability and sustainable use. "One could argue that protection of biodiversity is about as unnatural as true altruism, in other words as unnatural as it gets. Perhaps this conclusion can guide us to a true and fulfilling conservation ethic, one that requires sacrifice rather than self-serving and hypocritical platitudes. But, do we want to give nature the respect we sometimes give people, or is nature a commodity to be managed and placed firmly in the free market?"

Biodiversity and/or other fashionable and related buzz-phrases such as "ecosystem management" and "precautionary principle" could perhaps be themes for a second IFAW Forum?

One participant was Sidney
He says sustainable reaches infidney
He drew lots of curves
That unsettled our nerves
But he looked ever so clever, did'n he?

William de la Mare 2004

CHAPTER 5

THE CHANGING FACE OF CONSERVATION: COMMODIFICATION, PRIVATISATION AND THE FREE MARKET

Sharon Beder

Environmentalists in the late 1960s and 1970s argued that the exponential growth of populations and industrial activity could not be sustained without seriously depleting the planet's resources and overloading the planet's ability to deal with pollution and waste materials. They argued that new technologies and industrial products, such as pesticides and plastics, also threatened the environment. Following the protest mood of the times, they did not hesitate to blame industry, western culture, economic growth and technology for environmental problems. They questioned western paradigms of development and industrialisation, and criticised the inequitable distribution of wealth and resource use.

Although the environment movement was easily characterised as being anti–development, their warnings captured the popular attention, resonating with the experiences of communities facing obvious pollution in their neighbourhoods. As a result, confidence in business declined, contributing to a crisis for the legitimacy of capitalist hegemony.

In any country there will be groups and individuals who do not accept the prevailing ideological hegemony and contest individual governmental and institutional decisions.[1] Frequently they are weak and poorly resourced and can be ignored and marginalised. However, as opposition groups gain popular support, some degree of accommodation is necessary to deal with them. This accommodation may include the appropriation of elements of their discourse and implementation of minor reforms that don't seriously threaten the *status quo*.[2]

In this case, accommodation took the form of the cooption of environmental discourse as well as environmental legislation to which business acquiesced.[3] Although many governments did not recognise the importance of global environmental problems, they were forced by community pressure to respond to local pollution problems. During the 1970s many governments introduced new environmental legislation to cope with the gross sources of pollution. They introduced clean air acts, clean water acts, and legislation establishing regulatory agencies to control pollution and manage waste disposal. These accommodations were partially successful in placating community concerns, and environmental issues remained on the back burner through the 1980s.

New scientific evidence of global problems, such as global warming and ozone depletion, and the perceived failure of environmental legislation to protect local environments, renewed the perceptions of an environmental crisis towards the end of the 1980s once again threatening the hegemony of global capitalism. However, the confrontational, radical potential of the environmental movement had waned by this time and a new strategy of cooperation and negotiation was in place.[4]

SUSTAINABLE DEVELOPMENT DISCOURSE

The sustainable development discourse which emerged during the 1980s accommodated economic growth, business interests and the free market and therefore did not threaten the power structure of modern industrial societies. Many of the ideas associated with sustainable development had been previously articulated in the 1980 World Conservation Strategy produced by the International Union for the Conservation of Nature and Natural Resources (IUCN) in collaboration with the UN Environment Programme (UNEP) and the World Wildlife Fund (WWF, now the World Wide Fund for Nature). This document, which was circulated to all governments, defined conservation as:

> the management of human use of the biosphere [the thin covering of the planet that sustains life] so that it may yield the greatest sustainable benefit to present generations while maintaining its potential to meet the needs and aspirations of future generations.

The World Conservation Strategy argued that while development aimed to achieve human goals through the use of the biosphere, conservation aimed to achieve those same goals by ensuring that use of the biosphere could continue indefinitely.

The World Conservation Strategy and its national equivalents had little impact and few people have even heard of them. However in the mid–1980s the World Commission on Environment and Development rejuvenated the concept in its report Our Common Future (also referred to as the Brundtland Report). In October 1987, the goal of sustainable development was largely accepted by the governments of one hundred nations and approved in the UN General Assembly. The commission defined sustainable development as: "development that meets the needs of the present without compromising the ability of future generations to meet their own needs".[5]

Earlier environmentalists had used the term "sustainability" to refer to systems in equilibrium: they argued that exponential growth was not sustainable, in the sense that it could not be continued forever because the planet was finite and there were limits to growth. "Sustainable development", however, sought to make economic growth limitless by making it "sustainable", mainly through technological change.

The language of sustainable development was clearly aimed at replacing protest and conflict with consensus by asserting that economic and environmental goals are compatible. For more conservative environmentalists and for economists, politicians, business people and others, the concept of sustainable development offered the opportunity to overcome previous differences and conflicts, and to work together towards achieving common goals rather than confronting each other over whether economic growth should be encouraged or discouraged. Instead of being the villains, as they were in the 1970s, technology and industry were now expected to provide the solutions to environmental problems.

Sustainable development, as an attempt at conflict resolution, spawned a number of consensus decision–making processes including Round Tables on Environment and Economy in Canada, the President's Council on Sustainable Development in the USA and the Ecologically Sustainable Development Working Groups in Australia. These processes involved bringing together the various interest groups or stakeholders to reach a consensus about how sustainable development should be achieved and how business interests, economic interests and environmental protection could be reconciled.

> Political conflict over unavoidable trade–offs was perceived as unproductive, and consensus could be achieved only if a proper process was designed to allow the competing stakeholders an opportunity to communicate with each other.[6]

Although sustainable development represents a departure from the previous government view of environmental management that dealt in an *ad hoc* way with the most obvious environmental problems after they occurred,[7] sustainable development nevertheless seeks measures that do not interfere unduly with business activity. It does not include a cultural critique of modern society nor of industrial progress, consumption or limitless economic growth.[8] Rather, it incorporates meanings and assumptions that legitimise rather than challenge the existing capitalist order (see Table 5–1).

Sustainable development discourse takes place within an informal coalition that includes policy makers from national governments and international development and financial organizations; professionals in the area of environmental science, engineering, law, and management; and most mainstream environmental groups. A discourse–coalition does not necessarily share goals, values and interests but it does share concepts and terms, that is, a discourse.[10]

ENVIRONMENTAL ECONOMICS DISCOURSE

The sustainable development discourse was soon colonised by environmental economists. Neo-classical economists have long argued that the "most effective means of dealing with environmental problems is to subject them to the discipline of the market mechanism."[11] They argue that environmental degradation has resulted from the failure of the market system to put any value on the environment, even though the environment does serve economic functions and does provide economic and other benefits. Some environmental resources—such as timber, fish and minerals—are bought and sold in the

Table 5–1. Assumptions of Sustainable Development Discourse[9]

1. Environmental protection and economic growth are compatible. This means that modern capitalism is compatible with environmental protection.
2. Intergenerational equity is about ensuring future generations can meet their needs, which are vaguely defined and can be interpreted as economic needs.
3. A major contributor to environmental decline is poverty in developing countries, to be dealt with through more economic growth and free trade, rather than questioning distribution issues and lifestyles in affluent countries.
4. Technological and scientific progress can solve most environmental problems, including potential biophysical limits of economic and population growth, so there is no need for radical social and political change.
5. The environment needs to be managed for its use/utilitarian value, as opposed to saved for its intrinsic value – it is a system of resources that needs to be looked after.

market but their price usually does not reflect the true cost of obtaining them because the damage to the environment has not been included. Other environmental resources such as clean air are not given a price at all and are therefore viewed by businesses as free.

These economists argue that environmental assets tend to be overused or abused because they are too cheap. Their solution is to create a pricing mechanism so that environmental values are internalised by businesses that harm the environment. The extra costs involved in price–based economic instruments such as charges, taxes and subsidies are supposed to provide an incentive to change environmental damaging behaviour. Such pricing mechanisms are also supposed to prevent depletion of natural resources. John Hood, a visiting fellow at the Heritage Foundation and Vice–President of the John Locke Foundation, maintains:

> For natural resources over which property rights are relatively easy to establish, such as oil, minerals, or timber, prices serve as an early warning signal to companies about scarcity. If the price is rising, that suggests more demand for the resources than can be met by available supply. Companies then have a financial incentive either to find new supplies or to reduce its need by developing alternatives or ferreting out waste. This market process amounts to a sort of ongoing environmental research project seeking an answer to this question: What is the most efficient and least resource-depleting method of producing the goods and services people need?[12]

Some think tank economists also argue that there is little incentive to protect environmental resources that are not privately owned. So economists and free market think tanks came up with creative ways to artificially create property rights to the air and the waters. Rights–based economic instruments such as tradeable pollution rights, for example, "create rights to use environmental resources, or to pollute the environment, up to a pre–determined limit" and allow these rights to be traded.[13] Rights–based measures are also a way of providing a pricing mechanism for environmental resources.

While legislation is aimed at directly changing the behaviour of polluters by outlawing or limiting certain practices, market–based policies let the polluters decide whether to pollute or not. Polluters are not told what to do; rather, they find it expensive to continue in their old practices and they have a choice about how and whether they change those practices.

Governments have traditionally favoured legislative instruments over economic instruments for achieving environmental policy. Economic instruments were thought to be too indirect and uncertain (aimed at altering conditions in which decisions are made rather than directly prescribing decisions). Governments were also concerned that additional charges would fuel inflation and might have the undesirable distributional effect of most severely hitting low–income groups. They have been concerned that the public might see charges as giving companies a "right to pollute" that they had paid for.

Similarly, businesses have preferred direct regulation because of concerns that charges would increase their costs, and also because of perceptions that they would be able to have more influence on legislation through negotiation and delay. However, the threat of a new wave of environmental regulations in the early 1990s caused businesses to rethink this preference. Also, business groups, think tanks and economists were heavily influenced by the resurgence of economic fundamentalism during the 1980s and the trend towards increasing deregulation and privatisation in Western capitalist economies.

The influence of economic fundamentalism on environmental discourse and policy has manifest in different ways in different countries. In Australia, the infiltration and domination of the Canberra bureaucracy by economic rationalists pushing neo-classical economic solutions influenced the framing of sustainable development policy.[15] In Britain, economists such as David Pearce were particularly influential in the sustainable development debate. In the United States, corporate–funded think tanks have been influential. Although economists have long advocated economic instruments for environmental regulation, their popularity today owes much to the work of these think tanks, which have effectively marketed and disseminated these policies.

Conservative think tanks in various nations have consistently opposed government regulation and promoted the virtues of a "free" market unconstrained by a burden of red tape. Think tanks sought to discredit environmental legislation, giving it the pejorative label "command and control", and highlighting its deficiencies and ineffectiveness. The market solutions being advocated by conservative think tanks provided corporations and private firms with an alternative to restrictive legislation and the rhetoric to make the argument against that legislation in terms that were not obviously self–interested.

Think tanks recommended using the market to allocate scarce environmental resources such as wilderness and clean air and replacing legislation with voluntary industry agreements, reinforced or newly created property rights and economic instruments. The idea is to incorporate the commons into the market system through the use of economic instruments and the creation of artificial property rights. The Washington–based Cato Institute, for example, states that one of its main focuses in the area of natural resources is "dismantling the morass of centralized command–and–control environmental regulation and substituting in its place market–oriented regulatory structures..."[16]

According to Heritage Foundation's policy analyst, John Shanahan, the free market is a conservation mechanism. He urged the use of markets and property rights "where possible to distribute environmental 'goods' efficiently and equitably" rather than legislation, arguing that "the longer the list of environmental regulations, the longer the unemployment lines."[17]

Think tanks have popularised and promoted the work of neo-classically–oriented environmental economists, and many of the leading scholars in this area are associated with think tanks, including one of the foremost proponent's of tradeable pollution rights, Robert Hahn, a resident scholar of the American Enterprise Institute; Terry Anderson, a senior associate of the San Francisco–based Pacific Research Institute for Public Policy; Robert Stavins and Bradley Whitehead, authors of a Progressive Policy Institute study; Alan Moran, from the Australasian Tasman Institute; and Walter Block from Canada's Fraser Institute.[18]

The free market environmentalism promoted by these think tank economists emphasises the importance of market processes in determining optimal amounts of resource use. Anderson and Leal argue that the political process is inefficient, that is it doesn't reach the optimal level of pollution where costs are minimised:

> If markets produce 'too little' clean water because dischargers do not have to pay for its use, then political solutions are equally likely to produce "too much" clean water because those who enjoy the benefits do not pay the cost... Just as pollution externalities can generate too much dirty air, political externalities can generate too much water storage, clear–cutting, wilderness, or water quality.[19]

The changing consensus wrought by conservatives has meant that economic instruments, once associated with market economists and conservative bureaucrats, have now been widely accepted. The influence of think tanks and neo-classical economists has been so pervasive that the rhetoric of free–market environmentalism is no longer confined to conservative governments. Bill Clinton said in 1992, prior to becoming US President, that he believed it was "time for a new era in environmental protection, which used the market to help us get our environment on track – to recognize that Adam Smith's invisible hand can have a green thumb..."[20]

Government sustainable development policies today embrace the environmental economics discourse (see Table 5–2). They aim to incorporate environmental assets into the economic system to ensure the sustainability of the economic system. They incorporate the idea that wealth creation can substitute for the loss of environmental amenity; that putting a price on the environment will help us protect it unless degrading it is more profitable; that businesses should base their decisions about polluting behaviour on economic considerations and the quest for profit; that economic growth is necessary for environmental protection and therefore should take priority over it.

Such thinking has spread throughout the world. In 1991 the Organisation for Economic Cooperation and Development (OECD) issued guidelines for applying economic instruments and an Economic Incentives Task force was established by the US Environmental Protection Agency "to identify new areas in which to apply market–based approaches".[21] Similar units have been established in regulatory agencies in other countries. At the Earth Summit in Rio in 1992, business groups pushed for the wider use of economic instruments in conjunction with self–regulation.[22] As a result, the assumptions and language of neo-classical economists in this

Table 5–2. Assumptions of Environmental Economics Discourse.[14]

1. Not only is environmental protection compatible with economic growth but economic growth is also necessary for environmental protection.
2. The environment can assimilate a certain degree of pollution and that assimilative capacity can be predicted reasonably safely.
3. Sustainable development can be achieved by incorporating environmental assets into the economic system; internalising environmental externalities into business accounting.
4. The market is the best and most efficient way of allocating scarce environmental resources. Therefore putting a price on the environment will help protect it.
5. When environmental resources are priced and environmental damage internalised, businesses can continue to base their decisions on economic considerations, that is, markets together with the profit motive can be harnessed for environmental protection.
6. Most losses of environmental amenity can be replaced by wealth creation.

Environmental Economics Discourse is found clearly in Agenda 21, the Action Plan for Sustainable Development, signed by over 100 nations at the Earth Summit.

ENVIRONMENTAL MANAGEMENT DISCOURSE

Just as governments have had to accommodate environmental concerns into their discourse and their governance, so have businesses. In various business meetings during the 1970s corporate executives lamented their decline in influence: "The truth is that we've been clobbered" the chief executive officer of General Motors told chiefs from other corporations. Throughout the 1970s US corporations became politically active, getting together to support a conservative, non–interventionist, economic agenda[23] and financing a vast public relations effort aimed at regaining public trust in corporate responsibility and freedom from government regulation.

> For business, the turbulence of change was a nightmare of new regulations and increasingly vocal interest groups that needed pandering to. The rules of the game had changed, and new ways had to be found to at once get what one needed from government, shout down the opposition, and harness the power of interest groups for one's own benefit through persuasion.[24]

In response to government regulations, brought on by the activities of environmentalists and public interest groups, businesses began to cooperate with each other in a way that was unprecedented, building coalitions and alliances and putting aside competitive rivalries. They joined local, national and international business coalitions and alliances.

Corporations managed to achieve a virtual moratorium on new environmental legislation in many countries throughout the late 1970s and most of the 1980s. However, towards the end of the 1980s when public concern about the environment rose again, business groups had to find ways of accommodating environmental concerns. Corporate funds poured into anti–environmentalist causes such as the Wise Use Movement (see below) and a range of front groups. They contributed to the coffers of neo-conservative think tanks pushing neo-liberalism and market solutions to environmental problems (discussed earlier).[25] And they employed battalions of public relations experts to show how they were now socially and environmentally responsible.

The relatively new field of environmental management bloomed. Like sustainable development and environmental economics discourses, environmental management incorporates assumptions that assume that business goals can be in harmony with environmental goals (see Table 5–3).

Environmental management accommodates the environmental challenge by dealing with the worst instances of environmental degradation and, at the same time, utilises a discourse aimed at "deflecting the demands for more radical change". It is therefore aimed at political sustainability rather than environmental sustainability.[27]

Of course there are other related discourses associated with environmental management including those of triple bottom line accounting, stakeholder engagement, risk management and assessment, and green consumerism, which generally share the assumptions shown in Table 5–3. For example triple bottom line accounting seeks to incorporate environmental and social considerations into company management strategies. In a business context, risk management and assessment refers to the risks to the company rather than risks to the environment and con-

Table 5–3. Assumptions of Environmental Management Discourse[26]

1. The environment should be managed, controlled and dominated by humans for the long–term use of humans.
2. Environmental management should be carried out by corporate managers who are generally well meaning and have the knowledge and resources to provide a stewardship role on behalf of corporate stakeholders.
3. Environmental protection and economic growth are compatible so that environmental management is about finding win–win solutions. This means there is little need for regulation of firms, and markets together with the profit motive can be harnessed for environmental protection.
4. Traditional management tools can be used and extended; for example, total quality management and accounting methods can be extended to include triple bottom line and life–cycle analysis.

siders ways that the environment, can be accommodated and managed to reduce such risks.

As a form of political sustainability, environmental management necessarily requires the "management of environmentalists" and other stakeholders too, which requires stakeholder engagement. As part of this many businesses have attempted to bring environmentalists into its discourse coalition using various means, including donations to environmental groups and causes, joint projects and partnerships, employment and consultancies for individuals.[28]

WISE USE MOVEMENT DISCOURSE

The Wise Use Movement got its name from a Gifford Pinchot saying that "conservation is the wise use of resources". Pinchot, an early 20th century conservationist and head of the US Forest Service, argued for the "wise use" of natural resources and promoted principles of multiple use, scientific management and sustained yield. Pinchot clearly embraced the environmental management discourse.

For him, natural resources, like time and money, were limited and therefore should be used wisely. This was in contrast to his friend, preservationist John Muir, who founded the Sierra Club. Muir didn't view nature as something to be used; rather he held nature as sacred, having inherent value outside of its commercial potential. The two men parted ways over plans to flood the scenic Hetch Hetchy Valley to supply water and electricity to San Francisco. For Pinchot the scheme was a wise use, for Muir it was sacrilege.[29]

The US–based Wise Use Movement is stage–managed by Ron Arnold and Alan Gottlieb from their base at the Center for the Defense of Free Enterprise, a non–profit "educational" foundation "devoted to protecting the freedom of Americans to enter the marketplace of commerce and the marketplace of ideas without undue government restriction". There is no formal structure for the Wise Use movement and cohesion comes from shared enemies (environmentalists) and a few key leaders. Says Arnold, "We provide the Jello mold…The rest of the movement fills it". He sees himself as the Wise Use Movement's thinker and philosopher.[30]

Arnold and Gottlieb formed the Wise Use Movement in 1988 by organising a conference for this purpose. The 250 groups attending the conference included the American Mining Congress, the National Rifle Association, the American Motorcyclists Association and the National Cattlemen's Association, as well as corporations such as DuPont, Macmillan Bloedel, Louisiana–Pacific, Georgia Pacific and Weyerhauser. The conference was co–sponsored by groups such as the National Association of Manufacturers, the United 4–Wheel Drive Association, the Independent Petroleum Association of America, the National Forest Products Association, the American Sheep Industry, Exxon USA and the American Pulpwood Association. Canadian groups attending included the Council of Forest Industries, MacMillan Bloedel, Carriboo Lumber Manufacturers Association, and the Mining Association of British Columbia.[31]

The Wise Use Movement is a broad ranging, loose–knit, coalition of hundreds of groups in the United States that promotes a conservative agenda. Many groups within the movement have received substantial industry funding and support but the movement prefers to portray itself as a mainstream citizens' movement. Indeed its extended membership includes farmers, miners, loggers, hunters and landowners as well as corporate front groups. Local versions of the wise use movement emerged in countries such as Canada and Australia, but they have not had the same impact as the Wise Use Movement has had in the US.

There is a wide diversity within the Wise Use Movement, but members share a dislike of environmental regulations that affect parts of the movement in various ways: by constraining what they could do on private property and how they could use public land and water. Ralph Maughan and Douglas Nilson, academics from Idaho State University, argued that the Wise Use agenda stemmed from an ideology that combined laissez–faire capitalism with "cultural characteristics of an imagined Old West"[32] (see Table 5–4).

Table 5–4. Assumptions of Wise Use Discourse.[33]

1. Human worth should be measured in terms of productivity and wealth. Status and power are a reward for hard work.
2. Nature is there for the use of humans.
3. Real wealth derives from extracting and adding value to primary material resources.
4. Productive lands and waters should be owned (or at least controlled) and tamed by producers. Regulations should be kept to an absolute minimum.
5. Free markets benefit both producers and consumers and constraints on these free markets should be eliminated.
6. Depletion of energy and mineral resources is not the problem that environmentalists make it out to be.
7. Government's role is to protect property and property rights.
8. The quintessential Western person is self reliant, male, and tough.

The term "wise use" covers two main types of groups:

(i) those who advocate opening up of public lands for logging, mining and cattle as well as off–road vehicles and motorcycles, and

(ii) those who lobby against any restriction of use of private lands – property rights advocates.[34]

Wise Use groups in the west of the USA have been dominated by "western ranchers, corporate farmers, and business people whose margin of profit is directly threatened by any fee increases on grazing, water reclamation, and other uses of public lands".[35] Meanwhile, Wise Use groups in the east also took up the theme of private property rights and their protection. They argued that environmental regulations impeded their ability to develop their land in the way they wanted. Arnold and Gottleib argued that private property rights are sacred. They claimed that the environment movement was "actively destroying private property rights on a massive scale" through preventing people from using their land.[36]

Whilst the Wise Use Movement was essentially anti–environmentalist, they liked to portray themselves as the "real" environmentalists and their discourse of multiple–use of land and the sanctity of property rights suited corporate interests and became part of the establishment environmental discourse.

ESTABLISHMENT ENVIRONMENTAL DISCOURSE

Although environmentalists set the agenda and were primary shapers of the discourse of early environmental debates, that is no longer the case. The discourses of conservationism, political ecology and deep ecology have little impact in today's public discourse on the environment.[37] Even more mainstream policy discourses, such as that associated with the precautionary principle, play little role outside of policy documents. Rather, established environmental groups have adopted many elements of discourses outlined above.

Common to all of these environmental discourses that emerged during the 1980s are the more generic, overlapping discourses of mainstream modern environmentalism:

- A neo-liberal discourse where market mechanisms are seen as preferable to government action; property rights are sacrosanct; the environment is an asset that needs to be appropriately valued/priced; decisions are based on an analysis of costs and benefits; and competition, efficiency, individualism, and private enterprise are essential to achieving environmental protection whilst ensuring economic growth.[38]
- A discourse of ecological modernisation, where technological and scientific knowledge and efficient expert management of resources are seen as the key to solving environmental problems and ameliorating the impacts of economic growth and capitalism. The technological solutions proposed are those that can be undertaken without too much disruption, risk and cost. They tend to be incremental rather than radical. In this discourse "conservation" involves the control and domination of nature. It is anthropocentric and instrumental rather than ecocentric and ethical. It assumes that environmental and economic interests are compatible and that major environmental problems can be solved within the current industrial/economic development trajectory without radical social or political change.[39]

In fact, many environmentalists have been persuaded by the rhetoric of free market environmentalism. They have accepted the conservative definition of the problem, that environmental degradation results from a failure of the market to attach a price to environmental goods and services, and the argument that these instruments will work better than outdated "command–and–control" type regulations.[40] The US Environmental Defense Fund has been at the forefront of the push for tradeable pollution rights and the Natural Resources Defense Council has also supported them. The Australian Conservation Foundation (ACF) hired economists to enable them to talk the language of neo-classical economics in their negotiations with government over sustainable development policies. And as we will see below, the Nature Conservancy promotes many elements of the Wise Use, Environmental Management and Environmental Economics discourses.

Many environmentalists have willingly accepted that "all possible instruments at our disposal should be considered on their merits in achieving our policy objectives, without either ideological or neo-classically–inspired theoretical judgement."[41] In fact the ideological and political shaping of these instruments has been hidden behind a mask of neutrality. Stavins and Whitehead have argued that "market–based environmental policies that focus on the means of achieving policy goals are largely neutral with respect to the selected goals and provide cost–effective methods for reaching those goals."[42] Yet the preference for market solutions is obviously an ideologically based one:

> Its first pillar comes squarely out of a philosophical tradition that grew from Adam Smith's notion that individual pursuit of self–interest would, in a regime of competitive markets, maximise the social good. That tradition is so firmly embedded in economics by now that most economists probably do not realize, unless they venture out into the world of noneconomists, that it is a proposition of moral philosophy...[43]

The promotion of market–based instruments is viewed by many of its advocates as a way of resurrecting the role of the market in the face of environmental failure. Given the workings of the market in reality, and the well–elaborated imperfections and problems associated with it, what is surprising is that neo-classical economics has not only dominated environmental economics but has also increasingly dominated the whole public discussion of sustainable development. In the name of free market environmentalism, conservative think tanks have enabled the conservative, corporate agenda of deregulation, privatisation and an unconstrained market to be dressed up as an environmental virtue.

This adoption of mainstream discourse by environmentalists can be explained in various ways:

- ***Institutionalisation*** – As advocacy groups grow they become more professionalised, bureaucratic, centralised and moderate. They hire people on the basis of their professional skills and qualifications (in terms of management, fund raising, public relations etc) rather than their commitment to the environmental cause. Their discourse becomes more aligned with professional environmental bureaucrats in government and business.[44]
- ***Strategy*** – Environmentalists feel that they have to adopt these discourses in order to be taken seriously in the policy arena and to have a voice in the public debate.[45] In particular, participation in consensus processes, such as round tables and working groups require the adoption of common discourse.
- ***Cooption and Marginalisation*** – Some environmentalists are coopted into the dominant discourse coalition and others who cannot be coopted are marginalised.[46]

Although many mainstream environmental groups have joined the dominant discourse coalition, many environmental and anti–globalisation activists, environmental justice groups and resident action groups have not. And those that have joined still retain some of the rhetoric of deep ecology and ecocentrism as they attempt to bridge their various constituencies. The development of the environmental justice movement means that a "populist discourse" which portrays "global capitalism, transnational corporations and colonial powers as villains" also has some currency and is in fact gaining renewed vigour from the growing anti–globalisation movement.[47]

Nevertheless some of the most established and successful environmental organizations in the world have indeed joined the dominant discourse coalition. And some, like The Nature Conservancy (TNC), consciously promote its values.

CASE STUDY – THE NATURE CONSERVANCY

In terms of size and wealth, TNC seems to be the most successful environmental organisation in the world. It has 3,200 employees in 528 offices across the US and in 30 countries.[48] In 2002, the total revenue for TNC was close to a billion dollars (up from $547 million in 2001). This included over $655 million from donations ($225 million from corporations), grants and dues, $105 million from government consulting fees and other payments, and another $182 million from sales of land. Its total assets including nature preserves are now valued at over $3 billion. It is US–based but operates all over the world, particularly in Latin America, the Caribbean and Asia/Pacific.[49] It claims to have "protected more than 14 million acres in the United States, and an additional 80 million acres worldwide".[50]

The TNC protects areas by purchasing them and then exercising its property rights, which may include managing the area or trading it for another area that needs preserving. In recent years, "more stress is being put on techniques that keep land in private ownership, such as conservation easements,[51] leases, and cooperative agreements".[52] An example is an agreement with forestry corporation Georgia–Pacific to jointly manage 21,000 acres of hardwood forest in North Carolina. In return for a Georgia Pacific's promise to conserve 6,000 acres of land along a river, TNC helps manage its timberlands in the area.[53]

The TNC relies on corporate donations and individual subscriptions just as the Wise Use Movement does and they have many corporate donors in common. TNC's 1900 corporate sponsors include ARCO, BHP, BP, Chevron, Chrysler, Coca–Cola, Dow Chemical, DuPont, General Electric, General Mills, General Motors, Georgia–Pacific, McDonalds, Mobil, NBC, Pepsi–Cola, Procter and Gamble, Toyota, and Pfizer. Each state chapter has a Corporate Council for the Environment made up of corporate associates. Some of these companies, including Monsanto, make up TNC's International Leadership Council.[54] And on occasion corporations lend executives to TNC as in the case of Georgia Power which loaned Gordon Van Mol, from its External Affairs Department, to be a member of TNC's development team for a year.[55]

In return for corporate support, TNC promises corporate donors publicity as corporations that care about the environment. For example, the Indiana branch offers members of its Corporate Council for the Environment the opportunity to be listed on stationery, named in advertisements in magazines and newspapers across Indiana, and mentioned by radio stations, as well as listed in the TNC Annual Report.[56] Corporations hope that such donations and the accompanying publicity will improve their reputation with the community and consumers and consequently "translate into greater shareholder value".[57]

Neo-liberal Discourse

TNC's market–based approach to conservation helps to promote property rights and free enterprise as well as provide PR to individual companies. Rather than lobbying the government to implement regulations to ensure the environment is protected, or highlighting the activities of those same corporations in degrading the environment, TNC uses the market to purchase the land it wants to protect. According to former CEO John Sawhill:

> We have made a conscious strategic decision to rely on individual donors and not become too heavily dependent on government, because we want to be clearly identified as a private organization, one that is financed privately and uses free–market techniques. We think of ourselves as Adam Smith with a green thumb.[58]

TNC is conscious of its role in providing environmentalist support for the free market cause. It calls itself "Nature's real estate agent" and indeed it has traditionally employed real estate agents in its top ranks.[59] For example former TNC president Bill Blair had been a deputy assistant secretary in the State Department and chairman of a real estate agency. Sawhill used the language of free–market advocates in talking about TNC's "market–oriented strategy" and "conservation through private action".[60]

TNC helps property–rights advocates not only by demonstrating that property rights can be used to protect the environment, but also by championing an approach that doesn't reduce the rights of the property owner to do what they want with their property. An executive from Consolidation Coal Company, which had donated 8000 city acres to TNC worth tens of millions of dollars, said of TNC: "They acquire land for, I believe, a very good purpose, but do so within the framework of the free–market system. They do not seek to change the law or public opinion so as to deprive individuals or businesses of their just property rights."[61]

TNC provides a useful example to free–market advocates in their arguments for market–based solutions to environmental problems.[62] It represents the free enterprise, corporate autonomy, and small government agenda of a conservative think tank but with the bonus that they have sound environmental credentials. In an article on TNC in the business magazine Forbes, Morgenson and Eisenstodt claim that self interest can be used to protect the environment: "Instead of seeking to curb the profit motive and freedom of individual choice, we would do well to stimulate them both in ways that let the free market reconcile the industrial revolution with the age of environmentalism".[63]

Conservative think tanks have sought to have the conservative, corporate agenda of deregulation, privatisation and an unconstrained market dressed up as an environmental virtue. TNC is a great example for them. For example, Terry Anderson argues that the fight between loggers and environmentalists over government–owned forested areas should be solved by putting them up for auction. Environmentalists could bid against timber companies, and in this way environmentalists would have to face up to the costs of conservation. He gives TNC as a precedent for this type of activity.[64] The Heritage Foundation also cites TNC, along with US Ducks Unlimited and National Wild Turkey Federation (both organisations primarily made up of hunters who protect habitat in order to have enough birds to kill), as a good example of how private cooperative efforts are protecting the environment.[65]

Realty Times, in an article headlined "Nature Conservancy Conserves the Right Way", argued that "the good news is that reasonable environmentalism and the rights of property owners can co–exist... rather than tell other people what to do with their land, the Conservancy has a better idea..." Stroup and Shaw argue that the "beauty of such private efforts is that people who do not care for ducks or egrets need not pay for their upkeep, as taxpayers do when the government is in control."[66]

Organisations such as the Center for Private Conservation (CPC) argue that environmental groups who call for the creation of national parks are asking others to pay the price of what they want and their unwillingness to use their own money for this purpose shows that they shouldn't be taken seriously.[67] TNC, however, uses the market rather than the political process and this is where, according to free–market proponents, environmental choices should be exercised:

> For those who don't like chemicals, more and more stores are offering organic pesticide–free produce, and even Wonder Bread is made with unbleached flour and no preservatives. Vegetarians and meat–eaters shop side–by–side with no rancor....Milk, for example, is sold as whipping cream, half–and–half, whole, 2 percent, 1 percent, nonfat, powdered, and evaporated milk, and is also made into many varieties of yogurt, ice cream, and cheese. Yet you never see anyone chaining themselves to the milk counter demanding more ice cream or suing a dairy to force it to make cheese instead of yogurt.[68]

Ecological Modernist Discourse

The private market strategies engaged by TNC clearly fit within ecological modernist discourse. They deflect attention away from arguments that we are facing a "socio–economic crisis" and suggest that all that is required to protect the environment is good management by private owners; that major environmental problems can be solved within the current industrial/economic development trajectory without radical social or political change.

TNC seeks out win–win solutions that ensure economic growth and environmental protection are compatible so that there is no need for regulation of firms, and markets – together with the profit motive – can be harnessed for environmental protection. It finds this approach is attractive to donors because it means that it does not advocate social change or sue those who don't obey environmental laws or draw media attention to environmental problems. TNC will accept donations from any company no matter what their record on the environment, no questions asked. What is more, TNC is a safe vehicle to invest in because it will not turn around and expose a corporation's dirty record or damaging activities: rather it aims "to forge strong productive partnerships based on mutual benefit and trust".[69]

Whilst TNC seeks to preserve areas of forest, for example, it does not publicly speak out against practices such as clear–cutting. It preserves areas of land for grizzly bears but it does not oppose hunting or developments that endanger those bears and destroy their habitat.[70] Hunting is even allowed on some of TNC's own land and TNC officers acquiring land may go hunting with potential donors as part of the negotiation process.

TNC takes a narrow reductionist/managerialist approach to conservation. It embraces a form of managerialism that views the environment as something to be managed rather than conserved or saved. Management is best undertaken by managers who supposedly have the knowledge and resources to provide a stewardship role.[71] TNC's approach is anthropocentric and instrumental rather than ecocentric and ethical as is clear in Sawhill's statements:

> Some people at the Conservancy think our customers are the plants and animals we're trying to save, but our real customers are the donors who buy our product, and that product is protected landscapes... They like the fact that we use private–sector techniques to achieve our objectives, that we protect the environment the old–fashioned way: we buy it.[72]

(But, in fact, the old–fashioned way to protect the environment was for the commons to be protected collectively, not through purchase!) The solution to environmental problems for TNC is not to halt environmentally damaging economic activity in sensitive ecosystems, but to utilise technology, science and environmental management to enable economic activity to coexist with nature.

TNC works with loggers and ranchers to promote "sustainable" forestry and grazing in conservation areas, thereby endorsing the concept of "multiple use". For example the Wyoming TNC has joined together with the Montana Stock Growers Association and the Wyoming Stock Growers Association as well as the National Cattleman's Beef Association, Children for the West and others to produce a booklet for children entitled "Ranching for Nature" that "showcases grasslands and grazing animals – both domestic and wild."

Two TNC programs promote the idea of multiple use, more usually associated with the Wise Use Movement. One is the Last Great Places program which aims to "show that economic, recreational, and other development can be compatible with preserving nature."[73] It integrates economic development such as farming, forestry and ranching with conservation. The other TNC program is the Center for Compatible Economic Development (CCED), established in 1995, to "provide a vehicle for multiplying these efforts on an international scale".[74] CCED's home page is hosted by the Corporation for Enterprise Development (CFED). CFED's goal is to "create incentives and systems to encourage and assist all American individuals and families to acquire and hold assets".

Timber companies such as Weyerhaeuser and Georgia–Pacific are allowed to log on TNC preserves in several states.[75] "Nearly half of the 7 million acres that the conservancy said it is protecting in the United States is now being grazed, logged, farmed, drilled or put to work in some fashion. The money earned from such activities – about $7 million this year".[76]

However the TNC argues that it is not promoting "working" landscapes for the money and indeed the $7 million it earns from these activities represents less than one percent of the group's annual income. Rather, TNC is concerned with providing examples of private, multiple–use conservation. In some cases it is even paying ranchers and farmers to continue working the land, but in a more environmentally sound way.[77]

Morgenson and Eisenstodt claim that since TNC allows oil drilling and other corporate activities to take place on its land, it proves that "commercial and environmental interests can coexist". They give the example of the Welder Wildlife Foundation in Texas, where 7,800 acres incorporate an active oilfield and a cattle ranch which help to finance the wildlife management.[78]

COMPROMISE AND CRITICISM

TNC's partnerships with forestry and paper companies and ranchers have led to criticisms that TNC is too ready to compromise environmental values. Whilst environmental groups have been lobbying for national parks in New England, TNC has done deals with a number of paper companies involving almost a million acres of forests in Maine and New Hampshire.[79] Most of the land involved in the various deals continues to be logged, although TNC claims that "sustainable" forestry is being practiced.[80] The forestry industry and many local people prefer such a solution because national parks would preclude forestry as well as restricting motorised access. However the easements that these deals incorporate to protect a portion of the land may be stopping development of the land but not protecting their wilderness values.

Jym St. Pierre, director of Restore in Maine, argues that: "These aren't 'forever wild' easements… Some people call them 'forever logging' easements" because the timber companies are replacing native forests with plantation style forests that feature monocultures and therefore biodiversity is not being protected.[81] However TNC likes easements because they are cheaper than buying the land outright, the land does not have to be managed by TNC staff, it can continue "to contribute as an economic enterprise" and the "private owner (rather than the state agency) takes on the responsibility for maintaining improvements."[82]

In one example of multiple use, Mobil Oil Corporation gave TNC an area of land where they had been drilling for oil because it was no longer productive enough for them. The land was significant because it was "the last known breeding ground on Earth for one of North America's most endangered birds" – the Attwater prairie chicken. However, TNC drilled new natural gas wells and grazed cattle on the area so as to make a healthy income from the land – over $5 million between 1995 and 2002.[83] Spokesperson Niki McDaniel argued that TNC couldn't overlook the opportunity to "raise significant sums of money for conservation… maybe it's time we all took a walk in the oilman's shoes".[84]

The TNC argued that these activities would not harm the prairie chickens and called the area a "working" landscape where commerce and conservation could coexist harmoniously. However, environmentalists disagreed, arguing that it was the risk of accidents associated with the gas extraction and pipeline that threatened the prairie chickens and that it was such developments that had put the prairie chickens on the endangered lists in the first place.[85] The number of prairie chickens did indeed decline to less than half the original number, and the TNC's Texas–based science director admitted in an internal report that TNC activities on the land had subjected the birds to a "higher probability of death".[86]

In 1999 its Arizona branch sided with the livestock industry in a coalition called the Arizona Common Ground Roundtable, "dedicated to helping save Arizona's diminishing grasslands". It called for more government subsidies to ranchers using public land and "reform" of the Endangered Species Act (ESA). TNC defended this position by arguing that the laws were forcing ranchers to sell their private land for housing subdivisions and other developments that would fragment the grasslands and open spaces.[87] The Roundtable was held up as a prime example of environmental conflict resolution with TNC arguing: "Americans are tired of hearing about confrontations in the environmental area. They would prefer organizations with environmental goals utilize the strengths of one another and work toward the common goal of environmental protection."[88]

Nevertheless 18 environmental groups put their opposition in writing to the TNC:

> Preserving open space, however, is not the be–all and end–all of environmental protections. It is also absolutely necessary to preserve clean water, recreational opportunities, wildlife habitat, and endangered species.[89]

The groups argued that the ESA was not causing extra costs to ranchers although ranchers often blamed it for their troubles. However, grazing was causing declining water quality, damaging ecosystems and endangering species. They pointed to a study done in partnership with TNC scientists that found that livestock grazing threatened a third of all species listed as endangered by the US Fish & Wildlife Service, the National Marine Fisheries Service, and TNC. It was only marginally behind development as a national threat and a Forest Service report found that in Arizona it was the main threat to species.[90]

A former TNC science director, Jerry Freilich, also recognises that the pounding hooves of cattle degrade fragile environments and claims that in 2000 he was pressured and physically bullied by his boss at TNC to sign documents certifying that specific cattle ranches, which he had never visited, were environmentally sound. (He signed and subsequently resigned and made a complaint to the police. His complaint led to a settlement with TNC a year later.) All but three of the remaining 95 scientific staff at TNC headquarters were subsequently dispersed to branch offices or reassigned to a new organization that services TNC and sells TNC biological data.[91]

TNC even markets "Conservation Beef" with the message to consumers that their "purchase will help save the great Western landscapes for future generations".[92] Ranchers who raise the beef for this program have development restrictions on their grazing land and follow "stewardship plans" that they have put together themselves based on guidelines provided by TNC. The Madison Valley Ranchlands group, made up of Montana ranchers, monitors the program.[93] This self–monitoring by vested interests does not engender confidence in the stewardship.

CONCLUSION

The TNC is not a typical environmental group and some might say that it is more like an environmental business than an environmental advocacy group. However it does represent the leading edge of a trend amongst the more institutionalised mainstream groups to embrace the dominant business–oriented environmental discourses. TNC is more aware and open than most groups about its acceptance of the assumptions and philosophies embedded in those discourses. However, other groups that still purport to be motivated by environmental ideals are also drifting towards uncritically accepting the same assumptions as they increasingly adopt elements of the neo-liberal and ecological modernisation discourses.

The concept of "sustainability", promoted by the environmentalists of the 1960s and 70s, is being replaced by a commodified, privatised, anthropocentric, utilitarian free market version of sustainable development.

Far from being a neutral tool, the promotion of market–based instruments is viewed by many of its advocates as a way of resurrecting the role of the market in the face of environmental failure. They claim that economic instruments provide a way that the power of the market can be harnessed to environmental goals. They serve a political purpose in that they reinforce the role of the "free market" at a time when environmentalism most threatens it.

Market–based measures grant the highest decision-making power over environmental quality to those who currently make production decisions now. A market system gives power to those most able to pay. Corporations and firms, rather than citizens or environmentalists, will have the choice about whether to pollute (and pay the charges or buy credits to do so) or clean up. Tradeable pollution rights mean that permission to pollute is auctioned to the highest bidder.[94] Very polluting or dirty industries can stay in business if they can afford the pollution charges or can buy up credits. In this way, companies can choose whether or not to change production processes, introduce innovations to reduce their emissions, or just pay to continue polluting.

Yet the market, far from being free or operating efficiently to allocate resources in the interests of a globalising society, is dominated by a relatively small group of large multinational corporations that aim to maximise their private profit by exploiting nature and human resources.

Groups like TNC which embrace the neo-liberal and ecological modernisation discourses are not only ignoring the power and equity dimensions of current social institutions and capitalist culture, but are endorsing and facilitating them. They are granting legitimacy to the primacy of the profit motive and property rights thereby compromising environmental protection and the long–term future of the planet. They are denying the essential conflicts and contradictions between economic growth and environmental protection. They are also party to the marginalisation of those environmentalists who seek the ethical, political and social changes necessary to preserve rather than manage the environment.

NOTES AND SOURCES

[1] David L. Levy. 'Environmental Management as Political Sustainability.' *Organization & Environment* 10 (2). June, 1997.

[2] Daniel Egan. 'The Limits of Internationalization: A Neo–Gramscian Analysis of the Multilateral Agreement on Investment.' *Critical Sociology* 27 (3). 1 October, 2001.

[3] Sharon Beder. *Global Spin: The Corporate Assault on Environmentalism*, 2nd edn. Devon: Green Books. 2002; Levy. 'Environmental Management as Political Sustainability.'

[4] Sharon Beder. 'Activism Versus Negotiation: Strategies for the Environment Movement.' Paper presented at the Ecopolitics V. Centre for Liberal and General Studies, UNSW, Sydney. 1992.

[5] World Commission on Environment and Development. *Our Common Future*, Australian edn. Melbourne: Oxford University Press. 1990, p. 81.

[6] George Hoberg. 'Environmental Policy: Alternative Styles'. In *Governing Canada: Institutions and Public Policy*. ed. Michael Atkinson. Toronto: Harcourt Brace. 1993, p. 318.

[7] David Harvey. 'The Environment of Justice'. In *Living with Nature: Environmental Politics as Cultural Discourse*. ed. Frank Fischer and Maarten A. Hajer. Oxford: Oxford University Press. 1999.

[8] Ibid.

[9] Sharon Beder. *The Nature of Sustainable Development*, 2nd edn. Melbourne: Scribe Publications. 1996; David Caruthers. 'From Opposition to Orthodoxy: The Remaking of Sustainable Development.' *Journal of Third World Studies* 18 (2). Fall, 2001; Fernanda de Paiva Duarte. ' 'Save the Earth' or 'Manage the Earth'.' *Current Sociology* 49 (1). January, 2001; Harvey. 'The Environment of Justice'. Paper presented at the Social and Political Theory Environment Workshop. In *Green States and Social Movements: Environmentalism in Four Countries.'* eds. Christian Hunold and John Dryzek. ANU, Canberra, 15 June, 2001; Wolfgang Sachs. *Global Ecology*. London: Zed. 1993; Delyse Springett. 'Business Conceptions of Sustainable Development: A Perspective from Critical Theory.' *Business Strategy and the Environment* 12, 2003.

[10] Maarten A. Hajer. *The Politics of Environmental Discourse: Ecological Modernization and the Policy Process*. Oxford: Oxford University Press. 1995.

[11] E. Savage and A. Hart. 'Environmental Economics: Balancing Equity and Efficiency.' Paper presented at the 1993 Environmental Economics Conference. Canberra, November. 1993, p. 3.

[12] John Hood. 'How Green Was My Balance Sheet: The Environmental Benefits of Capitalism.' *Policy Review* Fall, 1995.

[13] Commonwealth Government of Australia. *Ecologically Sustainable Development: A Commonwealth Discussion Paper*. Canberra: AGPS. 1990, p. 14.

[14] Sharon Beder. 'Charging the Earth: The Promotion of Price–Based Measures for Pollution Control.' *Ecological Economics* 16, 1996.

[15] Michael Pusey. *Economic Rationalism in Canberra*. Cambridge: Cambridge University Press. 1991; Hamilton, 1991.

[16] 'Natural Resource Studies: Energy and the Environment', The Cato Institute, www.cato.org. Accessed 25 November 1995.

[17] John Shanahan. 'How to Help the Environment without Destroying Jobs, Memo to President–Elect Clinton #14.' The Heritage Foundation. January 19, 1993.

[18] Barbara Ruben. 'Getting the Wrong Ideas.' *Environmental Action* 27 (1), 1995; Jeremy D. Rosner. 'Market–Based Environmentalism.' *Los Angeles Business Journal* 14 (40), 1992; Robyn Eckersley, ed. *Markets, the State and the Environment: Towards Integration*. South Melbourne: MacMillan Education. 1995, pp. xi–xii.

[19] T. Anderson and D. Leal. *Free Market Environmentalism*. San Francisco: Pacific Research Institute for Public Policy. 1991.

[20] Quoted in Rosner. 'Market–Based Environmentalism.'

[21] R. Stavins and B. Whitehead. 'Dealing with Pollution: Market–Based Incentives for Environmental Protection.' *Environment* 34 (7), 1992, p. 29.

[22] S. Schmidheiny and The Business Council for Sustainable Development. *Changing Course: A Global Business Perspective on Development and the Environment*. Cambridge, Mass: MIT Press. 1992, Chapter 2.

[23] Jerome L. Himmelstein. *To the Right: The Transformation of American Conservatism*. Berkeley, California: University of California Press. 1990, p. 138.

[24] Jeff Blyskal and Marie Blyskal. *PR: How the Public Relations Industry Writes the News*. New York: William Morrow and Co. 1985, p. 153.

[25] Neo-liberalism, a term favoured in the UK and Europe, is referred to as economic rationalism in Australia and neo-conservatism in the US.

[26] Levy. 'Environmental Management as Political Sustainability.' Springett. 'Business Conceptions of Sustainable Development.'

[27] Levy. 'Environmental Management as Political Sustainability.'

[28] Beder. *Global Spin: The Corporate Assault on Environmentalism*. ; Levy. 'Environmental Management as Political Sustainability.'

[29] William Kevin Burke. 'The Wise Use Movement: Right–Wing Anti–Environmentalism.' *Propaganda Review* (11), 1994; Richard Stapleton. 'Green Vs. Green.' *National Parks* November/December, 1992, pp. 32–3.

[30] Kate O'Callaghan. 'Whose Agenda for America?' *Audubon* September/October, 1992, pp. 84–6; Alan Gottlieb, ed. *The Wise Use Agenda: The Citizen's Policy Guide to Environmental Resource Issues*. Bellevue: The Free Enterprise Press. 1989, pp. 86–7; William Poole. 'Neither Wise nor Well.' *Sierra* Nov/Dec, 1992, p. 88; David Helvarg. *The War against the Greens: The "Wise–Use" Movement, the New Right, and Anti–Environmental Violence*. San Francisco: Sierra Club Books. 1994, p. 126.

[31] Dan Baum. 'Wise Guise.' *Sierra* May/June, 1991; Gottlieb, ed. *The Wise Use Agenda: The Citizen's Policy Guide to Environmental Resource Issues*. 1989, p. 158.

[32] Ralph Maughan and Douglas Nilson. 'What's Old and What's New About the Wise Use Movement.' *GreenDisk* 3 (3), 1994.

[33] Ibid.

[34] Michael Satchell. 'Any Color but Green.' *U.S.News & World Report*. October 21, 1991, p. 74.

[35] Helvarg. *The War against the Greens: The "Wise–Use" Movement, the New Right, and Anti–Environmental Violence*, p. 11.

[36] Ron Arnold and Alan Gottlieb. *Trashing the Economy: How Runaway Environmentalism Is Wrecking America*. Bellevue, Washington: Free Enterprise Press. 1993, pp. 19, 23.

[37] Maarten A. Hajer. 'Ecological Modernisation as Cultural Politics'. In *Risk, Environment and Modernity: Towards a New Ecology*. ed. Scott Lash, Bronislaw Szerszynski and Brian Wynne. London: Sage. 1996.

[38] John Barry. 'The Beginning or the End of Environmentalism? From Green Politics to Green Political Economy in the 21st Century.' Paper presented at the ECPR Joint Sessions Workshop 'The end of environmentalism?' 22–27 March. 2002; Beder. 'Charging the Earth: The Promotion of Price–Based Measures for Pollution Control.'

[39] Lucy Ford. 'Challenging Global Environmental Governance: Social Movement Agency and Global Civil Society.' *Global Environmental Politics* 3 (2). May, 2003, p. 125; Stanley A. Deetz. *Democracy in an Age of Corporate Colonization: Developments in Communication and the Politics of Everyday Life*. Albany, NY: State University of New York Press. 1992, p. 187; Beder. *The Nature of Sustainable Development*; Levy. 'Environmental Management as Political Sustainability.'; Duarte. ' 'Save the Earth' or 'Manage the Earth'.'; Robyn Eckersley. *Environmentalism and Political Theory: Toward an Ecocentric Approach*. London: UCL Press. 1992; Hajer. *The Politics of Environmental Discourse*; Richard Welford. *Hijacking Environmentalism: Corporate Responses to Sustainable Development*. London: Earthscan. 1997; Brian Wynne. 'Risk and Environment as Legitimatory Discourses of Technology: Reflexivity inside Out?' *Current Sociology* 50 (3). 1 May, 2002.

[40] Beder. 'Charging the Earth: The Promotion of Price–Based Measures for Pollution Control.'

[41] Michael Jacobs. 'Economic Instruments: Objectives or Tools?' Paper presented at the 1993 Environmental Economics Conference. Canberra, November. 1993, p. 7.

[42] Stavins and Whitehead. 'Dealing with Pollution: Market–Based Incentives for Environmental Protection.' p.8.

[43] S. Kellman. 'Economic Incentives and Environmental Policy: Politics, Ideology, and Philosophy'. In *Incentives for Environmental Protection*. ed. T. Schelling. Cambridge, Mass: MIT Press. 1983, p. 297.

[44] Klaus Eder. 'The Institutionalisation of Environmentalism: Ecological Discourse and the Second Transformation of the Public Sphere'. In *Risk, Environment and Modernity: Towards a New Ecology*. ed. Scott Lash, Bronislaw Szerszynski and Brian Wynne. London: Sage. 1996; Hein–Anton van der Heijden. 'Political Opportunity Structure and the Institutionalisation of the Environment Movement.' *Environmental Politics* 6 (4), 1997; O. Seippel. 'From Mobilization to Institutionalization?' *Acta Sociologica* 44 (2536). 1 June, 2001.

[45] Barry. 'The Beginning or the End of Environmentalism?'; Beder. 'Activism Versus Negotiation: Strategies for the Environment Movement.'; Duarte. ''Save the Earth' or 'Manage the Earth'.'; Hunold and Dryzek. 'Green States and Social Movements: Environmentalism in Four Countries.'

[46] Beder. *Global Spin: The Corporate Assault on Environmentalism*. ; David L. Levy and Daniel Egan. 'A Neo–Gramscian Approach to Corporate Political Strategy: Conflict and Accommodation in the Climate Change Negotiations.' *Journal of Management Studies* 40 (4). June, 2003; Andrew Rowell. *Green Backlash: Global Subversion of the Environment Movement*. London and New York: Routledge. 1996; John Stauber and Sheldon Rampton. *Toxic Sludge Is Good for You! Lies, Damn Lies and the Public Relations Industry*. Monroe, Maine: Common Courage Press. 1995.

[47] W. Neil Adger, Tor A. Benjaminsen, Katrina Brown and Hanne Svarstad. 'Advancing a Political Ecology of Global Environmental Discourses.' *Development and Change* 32, 2001, p. 704.

[48] David B. Ottaway and Joe Stephens. 'Nonprofit Land Bank Amasses Billions.' *Washington Post*. 4 May, 2003, p. A01.

[49] TNC. 'Annual Report 2002.' Arlington, VA: The Nature Conservancy. 22 March 2003, p. 20; Ottaway and Stephens. 'Nonprofit Land Bank Amasses Billions'. p. A01.

[50] TNC. 'Answers to Frequently Asked Questions.' The Nature Conservancy. Accessed on 20 February 2003. http://nature.org/pressroom/files/faqs.pdf.

[51] In law, an "easement" is the right to use land (property) owned by someone else for some limited or specified purpose.

[52] John C. Sawhill. 'The Nature Conservancy.' *Environment* 38 (5). June, 1996.

[53] Georgia–Pacific. 'The Nature Conservancy and Georgia–Pacific Join Forces.' Georgia–Pacific. Accessed on 17 May 1999. http://www.gp.com/enviro/conservancy.html.

[54] TNC. 'Annual Report Fiscal Year 1998.' Arlington, Virginia: The Nature Conservancy 1998.

[55] TNC. 'Corporate Partnerships.' The Nature Conservancy. Accessed on 17 May 1999. http://www.tnc.org/infield/State/Georgia/corppart.htm.

[56] TNC. 'Corporate Council for the Environment.' The Nature Conservancy of Indiana. Accessed on 17 May 1999. http://www.tnc.org/infield/State/Indiana/corpcncl.htm.

[57] Arjun Patney. 'Saving Lands and Wildlife: Corporations and Conservation Groups in Partnership.' *Corporate Environmental Strategy* 7 (4), 2000, p. 368.

[58] Alice Howard and Joan Magretta. 'Surviving Success.' *The McKinsey Quarterly* Autumn, 1995, p. 156.

[59] Bil Gilbert. 'The Nature Conservancy Game.' *Sports Illustrated* 20 October, 1986.

[60] Sawhill. 'The Nature Conservancy.' , p. 103.

[61] Gilbert. 'The Nature Conservancy Game.'

[62] See for example Jarret B. Wollstein. 'Liberty and the

Environment: Freedom Protects, Government Destroys.' *Freedom Daily* May, 1993, p. 23.

[63] Gretchen Morgenson and Gale Eisenstodt. 'Profits Are for Rape and Pillage.' *Forbes* 5 March, 1990.

[64] Cited in Daniel Seligman. 'The Green Solution.' *Fortune* 3 May, 1993, p. 103.

[65] Brian Jendryka. 'Make Way for Ducklings: The Sky Is Not Falling.' *Policy Review* Spring, 1994.

[66] Richard L. Stroup and Jane S. Shaw. 'How Free Markets Protect the Environment.' PERC. Accessed on 25 March 2003. http://www.perc.org/publications/articles/freemarket.html.

[67] Randy T. Simmons, Fred L. Smith and Paul Georgia. 'The Tragedy of the Commons Revisited: Politics Vs. Private Property.' Center for Private Conservation. October 1996, p.9.

[68] Randy O'Toole quoted in Ibid. , p. 11.

[69] TNC. 'Corporate Council for the Environment.'

[70] Gilbert. 'The Nature Conservancy Game.'

[71] Levy. 'Environmental Management as Political Sustainability'.

[72] Howard and Magretta. 'Surviving Success.' , p. 156.

[73] Sawhill. 'The Nature Conservancy.'

[74] Ibid.

[75] Monte Burke. 'Eco–Pragmatists.' *Forbes* 3 September, 2001, p. 63.

[76] Janet Wilson. 'Wildlife Shares Nest with Profit.' *LA Times*. 20 August, 2002.

[77] Ibid.

[78] Morgenson and Eisenstodt. 'Profits Are for Rape and Pillage.'

[79] Wilson. 'Wildlife Shares Nest with Profit.'

[80] Ibid.

[81] Wayne Curtis. 'Easement Does It.' *Grist Magazine* 8 January, 2001.

[82] Christine Conte. 'A Conservation Victory: Protecting the San Rafael Valley.' Arizona: The Nature Conservancy. 30 March 1999.

[83] Wilson. 'Wildlife Shares Nest with Profit.'

[84] Quoted in Ibid.

[85] Ibid.

[86] Joe Stephens and David B. Ottaway. 'How a Bid to Save a Species Came to Grief.' *Washington Post*. 5 May, 2003. p.A01.

[87] Christine Conte. 'Linking Ranching, Open Space & Sustainable Ecosystems – AZ Common Ground Roundtable.' Arizona: The Nature Conservancy. 28 January 1999.

[88] Roper poll quoted in Ibid.

[89] Southwest Centre for Biological Diversity *et al.* 'The Nature Conservancy Guts ESA to Save Ranchers.' Southwest Centre for Biological Diversity. Accessed on 29 January 1999. http://www.sw–center.org/swcbd/activist/tnc.html.

[90] Ibid.

[91] Joe Stephens and David B. Ottaway. 'Conservancy Scientists Question Their Role.' *Washington Post*. 3 May, 2003.

[92] W. William Weeks. 'Conservation Beef.' Conservation Beef. Accessed on 6 May 2003. http://www.conservationbeef.org/nature.html.

[93] Joe Stephens and David B. Ottaway. 'The Beef About the Brand.' *Washington Post*. 5 May, 2003, p. A10.

[94] Beder. 'Charging the Earth: The Promotion of Price–Based Measures for Pollution Control.'; Robert Goodin. 'The Ethics of Selling Environmental Indulgences.' Paper presented at the Australasian Philosophical Association Annual Conference. University of Queensland, July, 1992.

An economic school from Chicago
Is a trickle down cult of the cargo
Where nature is market
Wild animals cark it
And empathy's held on embargo.

Sharon Beder 2004

Part II:

Modern Examples of "Sustainable Use"

CHAPTER 6

ECOLOGICAL AND FISHERIES SUSTAINABILITY: COMMON GOALS UNCOMMONLY ACHIEVED

Jeffrey A. Hutchings

The words "fisheries sustainability" conjure a variety of associations: laudable objective, government policy, attainable goal, socioeconomic necessity. Ten years ago, the phrase "all of the above" would have appropriately captured most sentiments associated with fisheries sustainability; today, the word "oxymoron" increasingly comes to mind. Ecological sustainability and fisheries sustainability are inextricably linked; the latter is not achievable in the long term without the former. Yet, as the title of this essay suggests, commonality in purpose has not been matched by commonality in achievement.

The failure of developed nations to achieve ecological sustainability in the oceans is most aptly reflected by historically unprecedented declines in the abundance and distribution of various forms of marine life,[1] most notably fishes.[2-8] Persistently unsustainable rates of targeted and incidental exploitation have reduced marine fish populations to such an extent that their ability to recover to previous levels of abundance has been severely compromised;[2-4] simply put, small populations are less able than large populations to persist in face of the vagaries posed by natural environmental variability. Compounding often-negligible recovery rates are fisheries management strategies, political objectives, and mixed messages from those who develop public policy that further erode the probability of achieving ecological or fisheries sustainability.

At its core, I would argue that the collapse of marine fishes can be attributed to a dissociation of public policy from science, an estrangement that seems to recur with increasing frequency in the management and conservation of natural resources. Malleable links between science and public policy have had especially negative socio-economic, financial, and biological consequences for commercial marine fisheries. I would argue that science-based policies and management strategies should not be implemented, or acted upon, if the science required to support such actions does not exist. In this regard, interference with the communication of science to society and to decision makers has had negative ramifications for major fishery resources, as discussed below in the context of fishery collapses in Canada.

With some exceptions (e.g., the Northeast Pacific fishery for Pacific halibut, *Hippoglossus stenolepis*), attempts to achieve fisheries sustainability concomitant with increased subsidies,[9] fishing effort, and technological capability have been abject failures. This essay begins with the empirical justification for this assertion; it will be comparatively brief given the number of recent publications on this topic.[2-8,10] It then focuses on the collapse of Canada's fisheries for Atlantic cod (*Gadus morhua*), followed by an exploration of how the integration of science within government has the potential to limit expression of scientific uncertainty, constrain breadth of scientific expertise, and marginalize scientific input to public policy. The cumulative effects of these deleterious consequences include reduced public confidence in the ability of governments to deal appropriately with science-based issues of import to society, and an inability to address fisheries and ecological sustainability in a scientifically defensible and socio-economically meaningful way. After exploring various means by which science can be more effectively communicated to decision makers and to society, and identifying a Canadian model that might be usefully emulated elsewhere, I describe in some detail how Canada has responded legislatively to the conservation concerns raised by the collapse of directly and incidentally exploited fishes.

MARINE FISH POPULATION COLLAPSES: A BRIEF OVERVIEW

The world's oceans are experiencing biological change at an unprecedented rate, reflected in large part by the fact that 75% of the world's major fish stocks are now fully exploited, over-exploited, or depleted.[11] Potentially permanent influences on species interactions, food web structure, and trophic dynamics are most dramatically reflected by staggering declines in the abundance of marine fishes.

Among species subject especially to incidental catch, large pelagic sharks, including threshers (*Alopias* spp.), great whites (*Carcharodon carcharias*), and hammerheads (*Sphyrna* spp.), have declined more than 75% in the Northwest Atlantic since 1986.[5] The porbeagle (*Lamna nasus*), a shark subjected to directed and incidental bycatch in Canadian fisheries, has declined by an estimated 90% since the early 1960s.[12] Pelagic sharks in the Gulf of Mexico have experienced even greater reductions: silky sharks (*Carcharhinus falciformis*) and oceanic whitetip sharks (*C. longimanus*) are estimated to have declined 90% and 99%, respectively, over the past 40 to 50 years.[7]

Similar reductions over the past half-century have been reported for many other fishes. It has been estimated, for example, that the present biomass of large predator fish, such as tuna (*Thunnus* spp.), may be only 10% of pre-industrial levels.[6] Consistent with this conclusion is a recent analysis of more than 230 marine fish populations that revealed a median 83% reduction in the number of breeding individuals from historic levels.[8] Among 56 populations of clupeid fish (including Atlantic herring, *Clupea harengus*), 73% had experienced historic declines of 80% or more. Within the Gadidae (including cod and haddock, *Melanogrammus aeglefinus*), of the 70 populations for which there are data, more than half had declined 80% or more. And among 30 pleuronectids (flatfishes, including soles, flounders, halibuts), 43% had exhibited declines of 80% or more. These declines, as great as they are, have almost certainly been underestimated in many instances, given that these fishes have been exploited for a considerably longer period of time (typically centuries) than that over which abundance data are available (typically decades).[1]

Among these fishes, the Atlantic cod is not only of great socio-economic and historical importance throughout the North Atlantic, but it is among those species that have experienced the greatest rates of decline. Cod in the North Sea have declined by almost 90% since the early 1970s.[13] More dramatic has been the decline experienced by cod ranging from eastern Labrador south to the northern half of the Grand Bank off the coast of Newfoundland. Numbering as many as 2 billion breeding individuals in the early 1960s, Canada's northern cod stock has declined by as much as 99.9% since that time.[14]

The consequences of such declines are no longer limited to the socio-economic and political spheres of society; increasingly they permeate those aspects of public policy concerned with the protection and recovery of endangered species and the conservation of biodiversity. Inevitably, failure to arrest population declines will result in the loss of species from parts of their current geographical ranges; increasingly, hopes of achieving fisheries sustainability have been replaced by means of preventing commercial and biological extinctions.[8,10] Notwithstanding suggestions otherwise,[15,16] it is becoming clear that marine fish do not possess life history characteristics which render them less vulnerable to extinction than other species.[2-4,8,10,17,18] It also seems clear that lower extinction rates of marine fish in the past two or three centuries can be attributed to the considerably greater financial and technological challenges in harvesting marine fish than those required to hunt terrestrial and aquatic birds and mammals. Notwithstanding these technological limitations, directed or incidental fishing has been responsible for the local extinction, or extirpation, of more than 55 marine fish,[10] including the common skate, *Dipturus batis*, and angel shark, *Squatina squatina*, from the Irish Sea,[10,19-21] and quite possibly populations of Atlantic cod off the northeastern United States.[22]

THE COLLAPSE OF CANADIAN ATLANTIC COD

Historical framework

The Newfoundland fishery for Atlantic cod was once the largest, most productive codfishery in the world.[2,23] The "northern cod" component of this fishery constituted upwards of 70% of all Newfoundland catches since 1954 and probably did so for most of the 500-year history of the fishery.[25] The geographical range of northern cod extends from southern Labrador (55°20'N) southeasterly along the Northeast Newfoundland Shelf to include the northern half of the once biologically rich Grand Bank (46°00'N). Northern cod have probably been fished since the late 15th century,[26,27] although the earliest extant documentation of a Newfoundland fishery dates from 1504.[26] Total harvests appear to have been less than 100,000 tonnes until the late 18th century whereafter catches increased to as much as 300,000 tonnes in the 1880s and

1910s before declining to less than 150,000 tonnes in the mid-1940s.[25] Following the expansion of European-based factory trawlers in the late 1950s and early 1960s, particularly in the virtually unfished offshore waters off southern Labrador (e.g., Hamilton Bank), reported catches increased dramatically to a historical maximum of 810,000 tonnes in 1968 before collapsing in equally dramatic fashion to 1977 when Canada extended its fisheries jurisdiction to 200 miles. Controlled in part by catch quotas established by the Canadian government, catches increased gradually to a post-1977 high of 268,000 tonnes in 1988 prior to the imposition of a moratorium on the northern cod fishery in July 1992.

The collapse of Canadian Atlantic cod has been extraordinary.[14] Among the eight stocks under sole or primary Canadian management, five have declined more than 90% percent since the late 1960s and early 1970s. Northern cod experienced the greatest decline. In the early 1960s, northern cod probably numbered almost 2 billion breeding individuals and comprised an estimated 75-80% of all Canada's cod. Estimates of the decline of northern cod since the late 1960s range between 97% and 99.9%.[14]

In 1994, I co-authored a paper[28] whose title optimistically posed the question: "What can be learned from the collapse of a renewable resource?" The primary purpose of the paper was to explore the relative importance of environment change and fishing to the collapse of Newfoundland's northern Atlantic cod. Surprisingly controversial at the time, the paper concluded that the collapse of a fishery once responsible for almost 2% of the world's capture fisheries production[29] could be attributed to over-exploitation. The socio-economic consequences of the commercial fishing moratorium imposed on northern cod in 1992 (and on most other Canadian cod stocks in 1993 and 1994) were devastating; the ecological and ecosystem consequences equally so.[30] It seemed unfathomable that society would not have learned a very great deal about what needed to be done to prevent such collapses from recurring and to take all reasonable steps to ensure that recovery occurred as rapidly as possible.

Causes and consequences of population collapse

The primary cause of the reduction of Atlantic cod throughout its Canadian range was over-exploitation.[28,31-34] In some areas, reductions in individual growth,[35,36] attributable to the environment or size-selective fishing,[37] may have exacerbated the rate of decline; in some areas, increased natural mortality may also have contributed.[35]

Threats to recovery have included directed fishing (a consequence of the setting of quotas), unreported catch (a consequence of illegal fishing, catch misreporting, and discarding of fish at sea), and bycatch from other fisheries. Additional threats include altered biological ecosystems and concomitant changes to the magnitude and types of species interactions (such as a possible increase in cod mortality attributable to seal predation[38]), all of which appear to have resulted in increased mortality among older cod. Selection against late maturity and rapid growth rate, induced by previously high rates of exploitation, may also be contributing to the higher mortality (caused by the reproductive costs associated with earlier maturity[39,40]) and slower growth observed in some areas today.

In theory, removal of the dominant source of anthropogenic mortality (fishing) should have resulted in population recovery. However, with one exception (St. Pierre Bank cod off Newfoundland's south coast), recovery has not been forthcoming in the decade since the fisheries were initially closed. Empirical analyses of these issues suggest that factors other than fishing may be of greater importance to recovery than fishing alone. Indeed, recent work suggests that lack of recovery is not unusual among marine fish populations that experience 15-year rates of decline greater than 80%,[2,3] even when associated with dramatic reductions in fishing mortality.[4,8]

The tremendous influence that even small catches (in the absolute sense) can have on recovery can be illustrated by considering the lack of recovery for northern cod. Despite having experienced declines in excess of 97%, the Canadian Department of Fisheries and Oceans (DFO) accepted the advice of the quasi-independent Fisheries Resource Conservation Council and allowed directed commercial fishing on this stock from 1999 through 2002, a measure that has had a clear and deleterious impact on the recovery of the world's most depleted cod stock. Although the quotas from 1999 through 2002 were small relative to those set in the mid-1980s (5600-9000 tonnes compared with quotas exceeding 200,000 tonnes), the impact on northern cod was high because the size of the population was correspondingly low.[41] Exploitation rates exacted by these fisheries exceeded, sometimes by more than two-fold, estimates of the maximum rate of population growth for northern cod.[14,41]

If physical structure is critically important to the survival of juvenile cod, notably in the form of plants, bottom heterogeneity, and possibly corals in some regions, then physical alterations in suitable habitat might also be affecting recovery. The reduction in physical heterogeneity on the bottom, and the loss of potentially important deep-sea corals, have been attributed to the bottom-trawling gear used to catch groundfish.[42-44] In Canadian waters, it has also been reported that bottom-trawling may have relatively little impact on invertebrate macrofauna inhabiting sandy bottoms,[45] although similar studies on the potential effects of trawling on fishes and fish habitat have not been undertaken. Regarding predation by marine mammals, an independent expert panel concluded that the recoveries of northern cod in the Newfoundland & Labrador Population and that of northern Gulf cod in the Laurentian North Population may have been negatively affected by seal predation.[38]

The possibility that the intense fishing pressure experienced by cod in the late 1980s and early 1990s resulted in genetic changes to heritable life history traits cannot be discounted.[40, 46-48] There is evidence to suggest that age at maturity and growth rate is lower in several cod stocks at present than it was prior to the stock collapses. The observed changes in age at maturity cannot be explained as phenotypic responses to changes in population density,[40] leaving genetic responses to selection (selecting against late-maturing genotypes) as the most parsimonious explanation for the earlier maturity observed in some areas.[47,48] Similarly, smaller weights-at-age among cod in some areas can also be explained as a result of selection against fast-growing genotypes during periods of intensive fishing.[37]

In summary, the primary cause of the reduction of Atlantic cod throughout its Canadian range in the Northwest Atlantic was over-exploitation; in some areas, the rate of decline may have been exacerbated by reductions in individual growth and increases in natural mortality. Identifiable threats to the recovery of Atlantic cod include directed fishing, unreported catch, and bycatch from other fisheries for bottom-dwelling species. Additional threats to recovery include altered biological ecosystems, and concomitant changes to the magnitude and types of species interactions. These ecosystem-level changes appear to have resulted in increased mortality among older cod. Selection against late maturity and rapid growth rate, induced by previously high rates of exploitation, may also be contributing to the higher mortality and slower growth observed in some areas today.

ESTRANGEMENT OF SCIENCE FROM PUBLIC POLICY: CONSEQUENCES FOR FISHERIES SUSTAINABILITY

> *I recently came across an article written by a Norwegian scientist during the 1970s, when I was Norway's Minister of the Environment. In the article he argued that there was no such problem as acid rain and that 'facts' and 'science' did not belong in the arena of politics and policy. This assertion was counter to my own beliefs and made me react strongly. Politics that disregard science and knowledge will not stand the test of time. Indeed there is no other basis for sound political decisions than the best available scientific evidence. This is especially true in the fields of resource management and environmental protection.*
>
> Gro Harlem Brundtland 1997 [49]

Gro Harlem Brundtland's prescient observation serves as an instructive point of departure for an exploration of the risks to achieving ecological and fisheries sustainability in dissociating public policy from science, an estrangement that has recurred with distressing regularity in the management and conservation of natural resources. Norway's former Prime Minister is well placed to comment on the necessity of underpinning public policy with science. However, strong as her assertion is, it is weakened by the realisation that rarely is it necessary for politics *per se* to "stand the test of time", unless time is measured by the brief intervals that separate successive governments. Rather, it might be more instructive to observe that "*Policies* that disregard science will not stand the test of time".

For the most part, policies fail to contribute effectively to society either because of deficiencies in their formulation or because of deficiencies in their enactment. It is the latter to which I address my remarks, specifically within the context of linking science with public policy pertaining to the exploitation, sustainability, and conservation of fisheries resources. I would also acknowledge the obvious constraint that political decisions founded ostensibly upon science-based policies need not always reflect the science that informs such policies. But I would argue, quite strongly, that the greater the risks to society of a failure in science-based policy, the greater the need for the necessary science and for its appropriate communication to decision makers and to society, and the greater the need to ensure that the policy decisions are consistent with that science.

Brundtland referred to the risks of "disregarding" science. How might a disregard for science be manifested in public policy? Such a disregard might be reflected in one of two ways. The first occurs when a science-based policy decision is made in the absence of the science required to support that decision. The second is more blatant, and is reflected by policy decisions that are, to varying degrees, inconsistent with science. The consequences of both can be equally injurious. Furthermore, they both reflect a failure in communication, the science pertaining to a particular decision having been improperly communicated between scientists and decision makers, between decision makers and the public, or both. Improper communication among scientists, decision makers, and society has had serious consequences for the conservation of natural resources, most recently in the harvesting of commercially exploited marine fishes.

Ineffective communication among scientists, decision makers, and society

The collapse of Canada's Atlantic cod in the early 1990s prompted several examinations of the means by which fisheries science had been interpreted and publicly communicated by the Department of Fisheries and Oceans (DFO), the federal government department responsible for the sustainable exploitation of Canada's fisheries for marine fishes. The title of one of these[50] was framed in the form of a question: "Is scientific inquiry incompatible with government information control?" It addressed the question of whether the conservation of natural resources is best ensured by having science fully integrated within a political body. It was concluded that the greater the input from independent, arm's-length scientists and science advisory bodies, the stronger and more transparent the links between science and policy in natural resource management, and the greater the benefits to society.[51]

In my view, the perceived need for scientific consensus and an "official" position on science matters has often limited the ability of government-based research to contribute effectively towards the sustainable exploitation of fishery resources and to achieve a comprehensive understanding of the factors that affect their persistence. In the context of the demise of Atlantic cod, and based on examples drawn from the mid-1980s to the mid-1990s, Hutchings *et al.*[50] argued that the communication of science to decision makers, and to the public, had been hindered by a number of factors, including government marginalisation of independent work, misrepresentation of alternative hypotheses, interference in scientific conclusions, disciplinary action against government scientists who communicated publicly the results of peer-reviewed research published in the primary scientific literature, and misrepresentation of the scientific basis of public reports, management strategies, and government statements.

If anything, bureaucratic interference had been even more pronounced in DFO's ill-fated attempts to balance the needs of an industry (Aluminum Company of Canada, or ALCAN) and the needs of Pacific salmon (*Oncorhynchus* spp.) for the same water in the Nechako River in British Columbia.[50] Notwithstanding two arguments to the contrary (one by government bureaucrats,[52] the other by a former DFO scientist[52]), Leiss[53] concluded that "none of the replies published to date challenge seriously the evidentiary basis of [Hutchings *et al.*'s[50]] indictment."

Recurring "disregard" for science in the management of Canada's Atlantic cod

One might argue that it would be naive to expect scientific information to flow unimpeded or uninterrupted from scientists to decision makers and then to society. In the field of health and environmental risk, for example, good government practice can require the bureaucracies that serve ministers to manage policy-relevant information flow.[51] Under some circumstances, the benefits of such information management might well outweigh the costs. It is unlikely, however, that such an asymmetric payoff, in which the benefits of information control outweigh the costs, will be realised in the management of natural resources, the conservation of biodiversity, or in the protection of species at risk. This point can be illustrated by examples, drawn from the 1990s, of how DFO communicated scientific information on the conservation status, biology, and recovery of Atlantic cod.

The first concerns an example of a government statement that gave the appearance of being based on science, when in fact it was not. In the late 1980s, the Newfoundland Inshore Fisherman's Association (NIFA) initiated legal action against DFO, arguing that the department was in violation of the federal *Fisheries Act* because it permitted the fishing of cod during spawning, an activity that NIFA asserted was disruptive of cod spawning behaviour and injurious to successful cod reproduction. In February 1990, Fisheries Minister Tom Siddon informed the House of Commons that, "scientists advise me that there is no recorded evidence in the scientific literature or our own research which states that fishing on the spawning grounds does measurable damage to the cod stocks".[55] The Minister had in effect performed the classic "absence of evidence = evidence of absence" sleight of hand. The statement that there was no evidence that fishing during spawning damages cod stocks was truthful, but only because there had been no scientific investigations of the problem, either by DFO or anyone else.

The fishery for Canada's northern cod was closed on 2 July 1992 by Fisheries Minister John Crosbie. The primary reason for the closure was a decline in the size of the breeding part of the population to an historic low, a decline of 99% since the early 1960s.[28] A little more than one year prior to the closure (April 1991), briefing notes prepared for the Minister stated, "Scientific advice is that lower inshore catch rates and smaller fish in the inshore fishery in recent years do not indicate stock decline".[55] The fact that such scientifically misleading information, particularly given its potentially enormous socio-economic and biological consequences, found its way to the Minister's desk reflects an extraordinary degree of information control and even contempt for science.

Inclusion of science within a political body can allow information to be presented as having a scientific basis, and potentially legitimizing government decisions, even when the science on which the decisions are based has not been subjected to independent peer review, a process fundamental to the integrity of science. One example of such a portrayal of "science" as science was presented to the Canadian public in 1992 when the DFO announced that a moratorium would be imposed on the commercial fishery for northern cod. The federal government predicted the moratorium would be in place from 1992 to 1994,[56] a two-year time frame that provided the temporal basis for an income support package for displaced workers (the sum total of all financial assistance resulting from the moratorium has been estimated at C$2 to C$3 billion,[57] roughly £0.8 to £1.2 billion, or €1.2 to €1.8 billion).

The two-year time frame for recovery included projections which indicated that, in the absence of fishing, northern cod spawner biomass would increase more than sixfold between 1992 and 1994 to a level that had not existed since 1972. For the DFO's projections to be realised, the cod stock would have had to grow between 126 and 200% per annum,[50] rates of increase that are biologically unrealistic for northern cod, given that they do not reproduce until they are 6 or 7 years of age. Indeed, based on analyses of data that were available in 1992, a scientifically defendable range of maximum population growth for northern cod was 9 to 19% per annum,[50] one order of magnitude lower than the DFO's predictions, rendering the government's recovery time frame logically suspect and scientifically ill-founded.

The fourth example describes a resource management decision that was scientifically indefensible, a conclusion reached even by the government advisory body that proffered the advice (27 May 1999 letter from the Fisheries Resource Conservation Council to Fisheries Minister David Anderson[58]). Against advice from government and academic scientists, the Fisheries Minister reopened the fishery for northern cod in 1999. In effect, the government was sanctioning the hunt of a population that had declined 99%. The reopening was described as a "limited fishery" because catch quotas were low relative to historic levels. But, of course, it is not the size of the quota that matters, it is the size of the quota *relative to what is available to be caught* that is of importance when evaluating the precautionary nature of any harvesting decision. This critically important distinction, however, was not communicated to the media or to society by government spokespersons. Between 1999 and 2002, the rate at which cod were extracted by this "limited" fishery exceeded the rate at which the cod population was increasing.[14,41] As a consequence, the slow recovery of the stock was halted and the fishery was "re-closed" in 2003; today, northern cod remains at less than 1% of its abundance in the early 1960s.

Strengthening the communication of science and its links with public policy

Malleable links between science and public policy have had highly negative socioeconomic, financial, and biological consequences for commercial marine fisheries in Canada and elsewhere. The examples discussed above underscore the necessity, and utility, of having the science pertaining to a particular issue clearly communicated to decision makers and to the public. There are a number of ways of accomplishing this, most if not all of which require the inclusion of independent scientists unaffiliated with government. I touch upon three here.

One comparatively simple means of better communicating scientific advice to decision makers is to involve individual scientific experts from academic institutions directly in the regular affairs of government. The American government recently took small steps in this regard with the establishment of a programme that allows senior researchers from universities to interact directly, and regularly, with officials in the US State Department.[59] After spending a year in the State Department, the researchers return to their universities where they serve as consultants with government for a further five years. Such an arrangement should increase the level of scientific expertise directly available to decision makers; it may increase the breadth of independent advice considered by government; and it will better inform academic scientists of the specific demands of policy makers and of the constraints that can prevent effective incorporation of science advice in government decisions. However, by involving relatively few scientists in any one discipline, there is a risk that decision makers will receive biased or unduly narrow perspectives.

Scientific advice can be communicated effectively through the findings of expert panels established by independent national scientific academies, such as the Royal Society (United Kingdom), the Royal Society of Canada, and the National Academy of Sciences in the United States. Examples of publications from such academies that address questions of import to fisheries sustainability include Royal Society Responses to

(a) the Royal Commission on Environmental Pollution Consultation on the environmental effects of marine fisheries in May 2003,[60] and

(b) the Prime Minister's Strategy Unit Consultation on UK fisheries,[61] and the US National Academy of Sciences publication on science and its role in the National Marine Fisheries Service.[62]

When such expert panels are properly constituted, decision makers and society benefit from the knowledge that the advice they are receiving is as unfettered as possible by non-science influences. A second benefit lies in the scientific expertise that such panels can offer, an expertise that, for a variety of reasons, might not be widely available from amongst government scientists. Another key advantage to expert panels is that, in general, their work is undertaken in response to clear and (preferably) unambiguous questions or mandates from policy makers. There is a directness in this approach that is often lacking in communications between decision makers and scientists.

Notwithstanding these benefits, the effectiveness of expert panels as a regular means of providing independent advice to government can be problematic. The panels are generally constituted several months, but often several years, after the problem at hand has been identified; this can negatively affect the timeliness of the panel's response. By necessity, the membership of expert panels changes with the subject matter; this can increase the variability in the quality of expert panel reports, and can reduce the likelihood that the advice will be communicated to government in a consistent manner among panels (although an overseeing National Academy of Science can minimize this influence). It is also important to acknowledge the reality that advice proffered by expert panels that *exclude* government employees is less likely to be accepted than that proffered by bodies to which government has explicitly invested expertise. Finally, under most circumstances, expert panel advice, although communicated to the public, can often be ignored by government when there is no legislative requirement for such advice to be part of the decision-making process.

A third means by which the communication of science to decision makers can be better effected is through science advisory bodies comprised of members who act independently of government, have the capacity to respond rapidly to policy "crises" when situations demand it, and are well-informed of the policies to which they are contributing scientific advice. Unfortunately, such bodies are exceedingly rare. However, in the context of ecological sustainability, I would argue that one of the best examples of such an advisory body is the Committee on the Status of Endangered Wildlife in Canada, or COSEWIC. It is the role of COSEWIC, legally recognised in 2003 in the national legislation embodied by the *Species At Risk Act*, to assign status to species at risk in Canada. Although it may be too early to judge the degree to which such an advisory body can effectively provide advice to government and have that advice heeded by government, the model of providing scientific advice it embraces bears examination, given its rarity among countries, particularly in the realm of matters pertaining to ecological sustainability.

PROVISION OF INDEPENDENT SCIENTIFIC ADVICE TO GOVERNMENT: A CASE STUDY

The assessment of species at risk in Canada

The Committee on the Status of Endangered Wildlife in Canada (hereafter, COSEWIC) is the national advisory body responsible for assessing the risk of extinction of native Canadian fauna and flora. COSEWIC was created in 1977 in response to a recommendation made the previous year at a conference of federal, provincial, and territorial Wildlife Directors; it made its first species status designation in April 1978 and has met at least annually ever since. Until the passage of the *Species At Risk Act* (hereafter, *SARA*), COSEWIC designations bore no legal consequences. Despite this, COSEWIC-listed species were usually accorded special consideration by the provinces and territories where they occurred and in environmental impact assessments of projects that may have directly or incidentally harmed designated species.

Initially, the committee was comprised almost entirely of government officials. Between 1978 and 2003 (the year in which the *SARA* was proclaimed by parliament), the number of Species Specialist Subcommittees (SSCs, groups of scientific experts responsible for the preparation and review of the species status reports upon which COSEWIC's assessments are based) had increased to eight: Birds, Terrestrial Mammals, Marine Mammals, Freshwater Fishes, Marine Fishes, Amphibians and Reptiles, Plants and Lichens, and Molluscs and Lepidoptera. A ninth SSC was established in 2004, with lepidopteran species now being assessed by the new Arthropod SSC.

Species status assignments are conducted once or twice annually. A minimum of two-thirds of the electronically cast votes must be achieved before a specific status can be assigned. There are 30 votes on COSEWIC: one for each of four federal organisations (Department of Fisheries & Oceans, Canadian Wildlife Service, Parks Canada, and Canadian Museum of Nature on behalf of the Federal Biodiversity Information Partnership), one for each of the nine Species Specialist Subcommittees, one for each of the ten provinces and three territories, one for each of three non-government members, and one for the Aboriginal Traditional Knowledge Subcommittee. Although governments are represented at COSEWIC, members do not represent their governments when species are being assessed; all members of COSEWIC are expected to act independently and to base their assessments on the best available scientific, community, and aboriginal knowledge, *irrespective of the consequences of that advice.*

As of June 2005, COSEWIC had assigned status to 500 species in Canada (excluding those deemed Not At Risk and Data Deficient) in the following categories: Extinct (n=13 species), Extirpated, i.e., no longer found in the wild in Canada (n=22), Endangered (n=184), Threatened (n=129), and Special Concern (n=152).[63] Extant species or populations assigned a status of Endangered or Threatened are afforded the greatest protection under *SARA*.

Scientific and political realities: uneasy or unwanted bedfellows

A central thesis to this chapter is that the implementation of government policy can be measurably strengthened by the inclusion of an independent, expert science advisory body in the decision-making process. Such a model seems to be particularly appropriate in those circumstances in which political considerations are likely to take precedence over scientific considerations. Although political considerations will always be prominent (and appropriately so, given that elected officials are ultimately accountable to society and that scientific consequences are but one of many considerations in the decision-making process), there is a real risk that the under-valuing of science can reap irreparable harm to ecological sustainability in general and to fisheries sustainability in particular.

At face value, COSEWIC would appear to represent the type of independent advisory body that many would consider an ideal means of infusing government decisions with sound, independent scientific advice. It is a body that includes individuals from academia, several levels of government, non-governmental organisations, and the aboriginal community. Members of COSEWIC act and vote in accordance with their expertise in the science, conservation and management of endangered species, *not* in accordance with the institutions with whom they are employed. By virtue of their membership on COSEWIC, government departments are inextricably linked to the species assessment process, rendering them less able to discount COSEWIC's assessments outright. The species assessment process is open and transparent; status reports are typically subjected to at least one year of review, with input from all individuals and groups who have information bearing on the status of species at risk. The results of COSEWIC's assessments are communicated publicly at the same time that they are communicated to government, thus fulfilling a key requirement of having scientific advice communicated directly to society, unaffected by the various communication filters often used to smooth the rough edges of scientific advice, or to eliminate them completely. As a consequence, society is fully knowledgeable of the status of endangered species in Canada from a scientific perspective, unfettered by political considerations. Finally, COSEWIC is a legally and legislatively recognised advisory body integral to the protection of endangered species in Canada; in other words, it has real influence.

To what degree then has this Canadian model of providing independent scientific advice in the arena of ecological sustainability been successful? It may be too early to tell, given the comparatively brief time that has elapsed since *SARA* was proclaimed in 2003. But the "growing pains" associated with the implementation of *SARA*, particularly as they pertain to the communication of COSEWIC's assessments to government decision makers, bear examination. Regrettably, an early assessment of the process would suggest that aquatic species in general, and marine fishes in particular, might be afforded considerably less protection under *SARA* than might have been initially presumed.[64]

Barriers to timely communication and implementation of independent scientific advice

The prohibitions identified under *SARA* (it is illegal to kill, harm, harass, capture or take legally listed endangered or threatened species, and to damage or destroy their residence) came into force on 1 June 2004. All 233 species assigned Endangered or Threatened status by COSEWIC prior to the passage of *SARA* by both houses of parliament in December 2002 were automatically placed on the *SARA* legal list; those assessed by COSEWIC after the passage of *SARA* are subjected to a lengthy, and potentially indeterminate (see below), period of time before the political decision is made as to whether they will be legally listed or not.

In January 2004, as required by law, COSEWIC communicated its species assessments of the previous year – 91 in total – to the federal Minister of the Environment. On 23 April 2004, Minister David Anderson, presented the Governor in Council (GIC), a subcommittee of the federal Cabinet, with the assessments of 79 of these 91 species. The 12 species whose assessments were not communicated to GIC were all aquatic (note that the *SARA* definition of "species" includes "subspecies" and "geographically or genetically distinct populations"); their exclusion from the GIC submission was made upon request by the federal Fisheries Minister Geoff Regan.

Among the 12 species that the Minister of the Environment subjected to an extended listing process were six marine fish (cusk [*Brosme brosme*], bocaccio [*Sebastes paucispinis*], and four populations of Atlantic cod), one salmonid (one population of coho salmon [*Oncorhynchus kisutch*]), two marine mammals (harbour porpoise [*Phocoena phocoena*], northern bottlenose whale [*Hyperoodon ampullatus*]), two freshwater fish (channel darter [*Percina copelandi*], shortjaw cisco [*Coregonus zenithicus*]), and one freshwater mollusc (Lake Winnipeg physa snail [*Physa* sp.]). These species were eventually submitted to GIC in July 2005, fifteen months later than most had anticipated, based on the time-lines detailed in *SARA* (see discussion below). The Environment Minister has indicated that these 12 species will be submitted to GIC in January 2005. In addition, the Minister declined to make an emergency listing, as assigned by COSEWIC, of two Endangered populations of sockeye salmon (*O. nerka*), populations that had declined as much as 99% in the past decade and whose spawning population sizes in recent years numbered as small as single digits.

SARA identifies specific time-lines for the listing process. Upon receipt of the previous year's assessments by COSEWIC (which the Minister of the Environment will receive annually in July), the Minister has 90 days to indicate how he/she will respond to those assessments. The Minister received the first set of COSEWIC assessments in January 2004. The second and third assessments were communicated to the Minister in July 2004 and August 2005. Upon receipt of COSEWIC's assessments from the Minister, the GIC has nine months to decide whether to

(a) accept the assessment and add the species to the SARA legal list,

(b) not add the species to the list, or

(c) refer the matter back to COSEWIC for further information or consideration.

These 90-day and 9-month time-lines are clearly specified in the Act. What is not specified in the Act is the time period during which the Minister of the Environment must submit the COSEWIC assessments to GIC. Herein lies some highly regrettable flexibility in *SARA*. It is this flexibility that allowed the Minister of the Environment, on request by the Minister of DFO, to postpone the submission of 12 aquatic species (all of which are under DFO's jurisdiction) to GIC in 2004.

The Fisheries Department argued that an "extended consultation process" was required for these 12 species because of the anticipated complexities associated with implementing *SARA* for aquatic organisms that may be directly or incidentally harmed by the fishing industry.

However, the prohibitions under *SARA* were well-known as early as the mid-1990s, and debated endlessly in the House of Commons Standing Committee on Environment and Sustainable Development from the late 1990s through 2002. It was no secret that COSEWIC would be assessing, on a regular basis, high-profile marine species; for example, the status report preparation process for Atlantic cod was initiated in September 2001. Three to four years, a time frame long enough to encompass most terms of national government office, should have been sufficiently long for the DFO to have prepared for the implementation of *SARA* for aquatic species. Rather than reflecting perceived complexities in implementing *SARA*, one could interpret the delay in submitting aquatic species assessments to GIC as either an extraordinary lack of preparedness on the part of the federal fisheries department, or an intent to disrupt the *SARA* listing process, either by minimizing the number of aquatic species assigned to the legal list or by extending indefinitely the listing process by taking advantage of the unintended, but now well-underscored, time "flexibility" in *SARA* discussed above.

EPILOGUE

To be optimistic about the degree to which meaningful fisheries recovery can be achieved for many depleted stocks in the next two or three decades is almost to invite ridicule. One would have reasonably thought that the ecological debacle reflected by the collapse of Newfoundland's northern cod in 1992 would have fundamentally changed the way that governments managed their ocean resources, not simply in Canada but in many parts of the developed world. For the most part, however, real fundamental change has not been forthcoming.

The United States, not normally considered a pantheon for progressive change on the ecological front, might be said to be at the forefront of establishing means by which fisheries sustainability can be achieved. The *Magnuson-Stevens Fisheries Conservation and Management Act* identifies targets, or reference points, against which over-exploitation and recovery for a specific fish stock can be identified. It also makes explicit the time frame over which recovery targets are to be achieved. One can argue about the reliability of the methods used to identify these targets, and many have, but one cannot argue that the establishment of specific targets for over-fishing and recovery, coupled with a legislative commitment to avoid the former and achieve the latter, is not consistent with the goal of attaining fisheries sustainability.

In Canada, the signals have been considerably mixed, at least with respect to the prospects of achieving ecological and fisheries sustainability in the marine environment. Although Canada's national legislation to protect species at risk has the potential to protect and to allow for the recovery of fish and fisheries – where the *Fisheries Act* and *Oceans Act* have been found wanting – the fisheries department's responses to COSEWIC's assessments in the early stages of the implementation of *SARA* have not engendered a great deal of confidence. For example, Canada's sanctioning, from 1999 to 2002, of the harvest of a population (northern cod) that had declined 99.9% cannot be said to be consistent with any conservation or precautionary principle.

The DFO Minister's re-opening of previously closed fisheries for two other populations of Atlantic cod – in the Northern and Southern Gulf of St. Lawrence – in May 2004 seemed to be similarly inconsistent with a precautionary approach to fisheries management. Consider one of these stocks for which the fishery was re-opened: Northern Gulf of St. Lawrence. The stock had been estimated to be 7% of its level in the mid-1970s, is predicted to decline further in the presence of a harvest,[65] was assessed as Threatened by COSEWIC in 2003,[14] and represents one of the 12 aquatic species afforded an "extended consultation period" in the *SARA* legal listing process (discussed above).

Sadly, but not surprisingly, the 2004 re-opening of these cod fisheries is entirely consistent with the time-honoured practice of linking fishery management decisions with politics. The Gulf of St. Lawrence fisheries were re-opened less than one month before the widely anticipated 2004 Canadian federal election was announced. Inconsistencies between political objectives and fisheries science have rarely been viewed as problematic by governments. Rather, purported short-term benefits associated with such political decisions are often peddled with an alarming dismissal of concerns related to their scientific legitimacy or their long-term consequences to fish, fisheries, and fishing communities.

One can argue, as I have done here, that a failure to address existing deficiencies in the communication of science will erode public confidence in the ability of governments to deal effectively with science-based issues of import to society. These deficiencies, both real and perceived, necessitate fundamental change to the means by which science is integrated with public policy.

Ultimately, of course, it is neither government bureaucrats, nor fish harvesters, nor politicians, nor industry to whom the blame for the collapse of marine fish, and to whom the responsibility for their recovery, can be ascribed. Blame and responsibility falls to all of these groups, but it falls to all of society as well. The unprecedented collapse of Atlantic cod in Canadian waters, and to marine fishes worldwide, has happened on our collective watch. The best of legislative intentions and most appropriately constituted science advisory bodies may be for naught if the will to achieve ecological and fisheries sustainability among the populace is lacking.

NOTES AND SOURCES

[1] Jackson, J.B.C., M.X. Kirby, W.H. Berger, K.A. Bjorndal, L.W. Botsford, B.J. Bourque, R. Bradbury, R. Cooke, J.A. Estes, T.P. Hughes, S. Kidwell, C.B. Lange, H.S. Lenihan, J.M. Pandolfi, C.H. Peterson, R.S. Steneck, M.J. Tegner, and R.R. Warner. 2001. Historical overfishing and the collapse of coastal ecosystems. Science 293: 629-638.

[2] Hutchings, J.A. 2000. Collapse and recovery of marine fishes. Nature 406: 882-885.

[3] Hutchings, J.A. 2001. Conservation biology of marine fishes: perceptions and caveats regarding assignment of extinction risk. Canadian Journal of Fisheries and Aquatic Sciences 58: 108-121.

[4] Hutchings, J.A. 2001. Influence of population decline, fishing, and spawner variability on the recovery of marine fishes. Journal of Fish Biology 59 (Supplement A): 306-322.

[5] Baum, J.K., R.A. Myers, D.G. Kehler, B. Worm, S.J. Harley, and P.A. Doherty. 2003. Collapse and conservation of shark populations in the Northwest Atlantic. Science 299: 389-392.

[6] Myers, R.A., and B. Worm. 2003. Rapid worldwide depletion of predatory fish communities. Nature 423: 280-283.

[7] Baum, J.K., and R.A. Myers. 2004. Shifting baselines and the decline of pelagic sharks in the Gulf of Mexico. Ecology Letters 7: 135-145.

[8] Hutchings, J.A., and J.D. Reynolds. 2004. Marine fish population collapses: consequences for recovery and extinction risk. BioScience 54: 297-309.

[9] For more on fishery subsidies, see Earle, Chapter 15.

[10] Dulvy, N.K., Y. Sadovy, and J.D. Reynolds. 2003. Extinction vulnerability in marine populations. Fish and Fisheries 4: 25-64.

[11] FAO. 2004. *The state of world fisheries and aquaculture 2004.* FAO, Rome.

[12] COSEWIC. 2004. COSEWIC assessment and status report on the porbeagle, *Lamna nasus* in Canada. Committee on the Status of Endangered Wildlife in Canada, Ottawa.

[13] ICES. 2003. Cod in Subarea IV (North Sea), Division VIID (Eastern Channel), and Division IIIa (Skaggerak). pp. 53-64. In Report of the ICES Advisory Committee on Fishery Management. International Council For The Exploration of the Sea, Copenhagen.

[14] COSEWIC. 2003. COSEWIC assessment and update status report on the Atlantic cod *Gadus morhua* in Canada. Committee on the Status of Endangered Wildlife in Canada, Ottawa. Available at www.sararegistry.gc.ca.

[15] Musick, J.A. 1999. Criteria to define extinction risk in marine fishes. Fisheries 24: 6-14.

[16] Powles, H., M.J. Bradford, R.G. Bradford, W.G., Doubleday, S. Innes, and C.D. Levings. 2000. Assessing and protecting endangered marine species. ICES Journal of Marine Science 57: 669-676.

[17] Sadovy, Y. 2001. The threat of fishing to highly fecund fishes. Journal of Fish Biology 59 (Supplement A): 90-108.

[18] Reynolds, J.D., N.K. Dulvy, and C.M. Roberts. 2002. Exploitation and other threats to fish conservation. pp. 319-341. In P.J.B. Hart and J.D. Reynolds (eds.). *Handbook of Fish Biology and Fisheries. Volume 2.* Blackwell, Oxford.

[19] Brander, K. 1981. Disappearance of common skate *Raja batis* from Irish Sea. Nature 290: 48-49.

[20] Rogers, S.I., and J.R. Ellis. 2000. Changes in the demersal fish assemblages of British coastal waters during the 20th century. ICES Journal of Marine Science 57: 866-881.

[21] Dulvy, N.K., J.D. Metcalfe, J. Glanville, M.G. Pawson, and J.D. Reynolds. 2000. Fishery stability, local extinctions and shifts in community structure in skates. Conservation Biology 14: 283-293.

[22] Ames, E.P. 2004. Atlantic cod stock structure in the Gulf of Maine. Fisheries 29: 10-28.

[23] McGrath, P.T. 1911. Newfoundland in 1911. Whitehead, Morris, & Co., London. 271 pp.

[24] Thompson, H. 1943. A biological and economic study of cod (*Gadus callarias*, L.). Res. Bull. No. 14, Dept. Nat. Res, Newfoundland Government, St. John's, NF. 160 pp.

[25] Hutchings, J.A, and R.A. Myers. 1995. The biological collapse of Atlantic cod off Newfoundland and Labrador: an exploration of historical changes in exploitation, harvesting technology, and management. pp. 37-93. In R. Arnason and L.F.Felt (eds.). *The North Atlantic Fishery: Strengths, Weaknesses, and Challenges.* Charlottetown, PEI, Canada: Institute for Island Studies.

[26] Quinn, D.B. (ed.). 1979. *New American World. Vol. 1. America From Concept to Discovery. Early Exploration of North America.* Arno Press and Hector Bye, New York.

[27] Cell, G.T. (ed.). 1982. *Newfoundland Discovered: English Attempts at Colonisation, 1610-1630.* The Hakluyt Society, London. 310 pp.

[28] Hutchings, J.A, and R.A. Myers. 1994. What can be learned from the collapse of a renewable resource? Atlantic cod, *Gadus morhua*, of Newfoundland and Labrador. Canadian Journal of Fisheries and Aquatic Sciences 51: 2126-2146.

[29] Based on data provided by Vannuccini, S. 2003. *Overview of fish production, utilization, consumption and trade.* FAO, Rome. 10pp.

[30] Bundy, A. 2001. Fishing on ecosystems: the interplay of fishing and predation in Newfoundland-Labrador. Canadian Journal of Fisheries and Aquatic Sciences 58: 1153-1167.

[31] Hutchings, J.A. 1996. Spatial and temporal variation in the density of northern cod and a review of hypotheses for the stock's collapse. Canadian Journal of Fisheries and Aquatic Sciences 53: 943-962.

[32] Myers, R.A., J.A. Hutchings, and N.J. Barrowman. 1997. Why do fish stocks collapse? The example of cod in Atlantic Canada. Ecological Applications 7: 91-106.

[33] Fu, C., R. Mohn, and L.P. Fanning. 2001. Why the Atlantic cod (*Gadus morhua*) stock off eastern Nova Scotia has not recovered. Canadian Journal of Fisheries and Aquatic Sciences 58: 1613-1623.

[34] Smedbol, R.K., P.A. Shelton, D.P. Swain, A. Fréchet, and G.A. Chouinard. 2002. Review of population structure, distribution and abundance of cod (*Gadus morhua*) in Atlantic Canada in a species-at-risk context. Department of Fisheries and Oceans Canadian Science Advisory Secretariat 2002/082, Ottawa, ON.

[35] Dutil, J.-D., and Y. Lambert. 2000. Natural mortality from poor condition in Atlantic cod (*Gadus morhua*). Canadian Journal of Fisheries and Aquatic Sciences 57: 826-836.

[36] Drinkwater, K.F. 2002. A review of the role of climate variability in the decline of northern cod. American Fisheries Society Symposium 32: 113-130.

[37] Sinclair, A.F., D.P. Swain, and J.M. Hanson. 2002. Measuring changes in the direction and magnitude of size-selective mortality in a commercial fish population. Canadian Journal of Fisheries and Aquatic Sciences 59: 361-371.

[38] McLaren, I., S. Brault, J. Harwood, and D. Vardy. 2001. Report of the Eminent Panel on Seal Management. Department of Fisheries and Oceans, Ottawa, ON.

[39] Beverton, R.J.H., A. Hylen, and O.J. Ostvedt. 1994. Growth, maturation, and longevity of maturation cohorts of Northeast Arctic cod. ICES Journal of Marine Science Symposium 198: 482-501.

[40] Hutchings, J.A. 2005. Life history consequences of overexploitation to population recovery in Northwest Atlantic cod (*Gadus morhua*). Canadian Journal of Fisheries and Aquatic Sciences 62: 824-832.

[41] DFO. 2003. Northern (2J+3KL) cod. Department of Fisheries and Oceans Stock Status Report 2003/018, Ottawa, ON.

[42] Collie, J.S., G.A. Escanero, and P.C. Valentine. 1997. Effects of bottom fishing on the benthic megafauna of Georges Bank. Marine Ecology Progress Series 155: 159-172.

[43] Collie, J.S., S.J. Hall, M.J. Kaiser, I.R. Poiner. 2000. Shelf sea fishing disturbance of benthos: trends and predictions. Journal of Animal Ecology 69: 785-798.

[44] Kaiser, M.J, and S.J. de Groot (eds.). 2000. Effects of fishing on non-target species and habitats: biological, conservation and socio-economic issues. Blackwell, Oxford.

[45] Kenchington, E.L.R., J. Prena, K.D. Gilkinson, D.C. Gordon, K. MacIsaac, C. Bourbonnais, P.J. Schwinghammer, T.W. Rowell, D.L. McKeown, and W.P. Vass. 2001. Effects of experimental otter trawling on the macrofauna of a sandy bottom ecosystem on the Grand Banks of Newfoundland. Canadian Journal of Fisheries and Aquatic Sciences 58: 1043-1957.

[46] Hutchings, J.A. 1999. Influence of growth and survival costs of reproduction on Atlantic cod, *Gadus morhua*, population growth rate. Can. J. Fish. Aquat. Sci. 56: 1612-1623.

[47] Hutchings, J.A. 2004. The cod that got away. Nature 428: 899-900.

[48] Olsen, E.M., M. Heino, G.R. Lilly, J.M. Morgan, J. Brattey, B. Ernande, and U. Dieckmann. 2004. Maturation trends indicative of rapid evolution preceded by the collapse of northern cod. Nature 428: 932-935.

[49] Brundtland, G.H. 1997. The scientific underpinning of policy. Science 277: 457.

[50] Hutchings, J.A., C. Walters, and R.L. Haedrich. 1997. Is scientific inquiry incompatible with government information control? Canadian Journal of Fisheries and Aquatic Sciences 54: 1198-1210.

[51] Also see Leiss, W. 2000. Between expertise and bureaucracy: risk management trapped at the science-policy interface. pp. 49-74. In G.B. Doern and T. Reed (eds.). *Risky Business: Canada's Changing Science-Based Policy and Regulatory*

Regime. University of Toronto Press, Toronto.

[52] Doubleday, W.G., D.B. Atkinson, and J. Baird. 1997. Comment: Scientific inquiry and fish stock assessment in the Canadian Department of Fisheries and Oceans. Canadian Journal of Fisheries and Aquatic Sciences 54: 1422-1426.

[53] Healey, M.C. 1997. Comment: The interplay of policy, politics, and science. Canadian Journal of Fisheries and Aquatic Sciences 54: 1427-1429.

[54] Leiss, W. 2001. *In the Chamber of Risks: Understanding Risk Controversies*. McGill-Queen's University Press, Montreal and Kingston. 388 pp.

[55] Harris, M. 1998. *Lament for an Ocean*. McClelland and Stewart, Toronto. 342 pp.

[56] DFO. 1992. Crosbie announces first steps in northern cod (2J3KL) recovery plan. News Release, 2 July 1992, NR-HQ-92-58E. Department of Fisheries and Oceans, Ottawa.

[57] Commission for Environmental Cooperation. 2001. The North American Mosaic: A State of the Environment Report. Commission For Environmental Cooperation, Montreal, Canada.

[58] FRCC. 1999. 1999 Conservation requirements for 2J3KL cod. Report FRCC.99.R.3 to the Minister of Fisheries & Oceans. Fisheries Resource Conservation Council, Ottawa. 30pp.

[59] Brumfiel, G. 2003. Adviser gears up to bring scientists to matters of state. Nature 425: 439.

[60] Royal Society. 2003. Environmental effects on marine fisheries. Policy Document 18/03, The Royal Society, London. 13pp.

[61] Royal Society. 2003. Strategy Unit Consultation on the UK Fisheries. Policy Document 19/03, The Royal Society, London. 11pp.

[62] National Academy of Sciences. 2002. *Science and Its Role in the National Marine Fisheries Service*. National Academies Press, Washington. 84 pp.

[63] COSEWIC. 2005. Canadian Species at Risk, August 2005. Committee on the Status of Endangered Wildlife in Canada, Ottawa.

[64] VanderZwaag, D.L., and J.A. Hutchings. 2005. Canada's marine species at risk: science and law at the helm, but a sea of uncertainties. Ocean Development and International Law 36: 219-259.

[65] DFO. 2004. The northern Gulf of St. Lawrence (3Pn, 4RS) cod in 2003. Department of Fisheries and Oceans Stock Status Report 2004/19, Ottawa, ON.

There was a Prime Minister named Paul
Whose prospects for election looked small
For his Atlantic wish
He cried "I need more fish!"
And his Minister said "No problem at all."

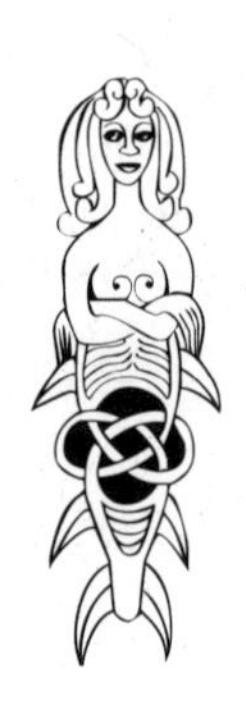

Jeff Hutchings 2004

CHAPTER 7

SUSTAINABLE USE OF OCEANIC WILDLIFE: WHAT LESSONS CAN BE LEARNED FROM COMMERCIAL WHALING?

Vassili Papastavrou and Justin Cooke

The history of commercial whaling is a widely cited example of the failure to manage the use of renewable natural resources sustainably. For centuries, one whale population after another has been brought to the verge of extinction, while technological and market developments have enabled the industry to expand to new whaling grounds and new species. Industrial whaling in the 20th century severely depleted the vast whale populations of the Southern Ocean. *The International Convention for the Regulation of Whaling*, which was concluded in 1946, established an International Whaling Commission (IWC) whose objectives were to manage whaling so as to preserve whale stocks for future generations without ignoring the current interests of those involved in the utilization of whales. For its first quarter-century the IWC merely presided over the decline of whale stocks, but after most of its members had abandoned whaling, the Commission adopted increasingly strict measures to conserve whale stocks, culminating in a moratorium on commercial whaling that took effect from 1986 and the creation of a Southern Ocean Sanctuary in 1994.

Despite being nominally prohibited by the IWC, whaling for essentially commercial purposes has expanded from a low of 326 whales in 1989 to approximately 1,500 whales expected to be caught in 2005 (not counting whales caught by indigenous peoples, nor by-catches in fishing operations). The current IWC membership of 66 countries (on 3 October 2005) is almost evenly divided between countries advocating protection of whales and those advocating their exploitation. The IWC takes decisions either by consensus or on a one-country-one-vote basis, although the adoption of binding regulations on whaling requires a three-quarters majority.

During the moratorium period the IWC has been developing procedures, which are quite advanced by fishery management standards, aimed at ensuring that any whaling that occurs does not pose a risk to whale populations. However, it has been unable to implement these in the face of resistance both from whaling countries, who oppose international control of their operations, and from non-whaling countries, who are concerned that a re-introduction of international control of whaling would be interpreted as an endorsement of the practice, and would put them in a position of joint responsibility for an activity from which they derive no benefits.

In this chapter we summarize the past and recent history of the exploitation of whales, the international attempts to manage it, and the scientific work that has been undertaken to develop more reliable approaches to the management of whaling. We identify some reasons why it has proven difficult to apply the conventional paradigm of sustainable consumptive use[1] to whales. A central conclusion is that the kind of management regime required to ensure the biological sustainability of exploitation appears not to be politically sustainable. This underscores the importance of pursuing other approaches to conservation that depart from the conventional sustainable consumptive use paradigm.

HISTORY OF COMMERCIAL WHALING AND THE ATTEMPTS TO MANAGE IT

There are many accounts of the history of whaling that document the sequential over-exploitation of whale populations, through more than three centuries of whaling.[2,3,4] One species of whale after another was hunted until it became too rare to support continued exploitation or in some cases was actually extirpated (Figure 7–1). As each stock was depleted, whalers moved on to exploit new stocks further afield, or developed new technology to gain access to previously uncatchable species.

The era of unrestricted whaling

Basque whalers had virtually exterminated northern right whales (*Eubalaena glacialis)* in the eastern North Atlantic by the end of 15th century when they ventured west to seek new quarry. In the 16th and 17th centuries Basque, Dutch, British and other whalers all but exterminated the right and bowhead (*Balaena mysticetus*) whales throughout the North Atlantic and eastern Arctic. The Atlantic gray whale (*Eschrichtius robustus*) is thought to have been exterminated by whalers by the early 18th century.[5] Later in the 18th century, whalers spread to the Southern Hemisphere where they severely depleted the southern right whales.

In the 19th century, American whalers learnt to catch sperm whales (*Physeter macrocephalus*). After depleting stocks in the Northwest Atlantic they moved to the Southern Hemisphere and thence to the North Pacific, opening up contact with the previously isolated Japan in the 1860s. Right, gray and bowhead whales in the North Pacific and Bering Sea were hunted to commercial extinction. Although sperm whales had become noticeably scarcer by the mid-19th century, the discovery of petroleum as a luminant reduced the demand for sperm whale oil and this hunt ceased on economic grounds around 1870 before the populations were dangerously reduced. Sperm whaling was revived in the 20th century when industrial uses for sperm whale oil were found, leading to a further wave of depletion.[6]

In the early 20th century the focus of whaling shifted to the Southern Ocean where, following the exploitation of right and humpback whales (*Megaptera novaeangliae*), the other large baleen whales were sequentially exploited in order of size – first blue (*Balaenoptera musculus*), then fin (*Balaenoptera physalus*), then sei (*Balaenoptera borealis*) and finally, minke whales (*Balaenoptera acutorostrata* and *Balaenoptera bonaerensis*).[7] In the 1930 season alone, over 30,000 blue whales were killed. By comparison the total number of blue whales in the Antarctic today is still only 1,000-2,000 despite over 30 years of protection.[8]

First attempts at international regulation of whaling

Attempts to regulate whaling internationally began in the 1930s under the League of Nations but these were abandoned with the advent of World War II, which gave a few years of relative respite to the whales. A new initiative was started after the war, resulting in the *International Convention on the Regulation of Whaling* (ICRW), signed in 1946. This established the International Whaling Commission (IWC) that focused on the regulation of whaling in the Antarctic, which was pursued mainly by British, Norwegian, Dutch, Japanese and Soviet expeditions. Catch levels rapidly returned to levels far above anything the stocks could sustain. By the mid-1950s, blue and humpback whales had become scarce and fin whales were suspected to be declining. But little serious effort was made to rein in catches. No country seemed willing to accept major cutbacks in their catches because

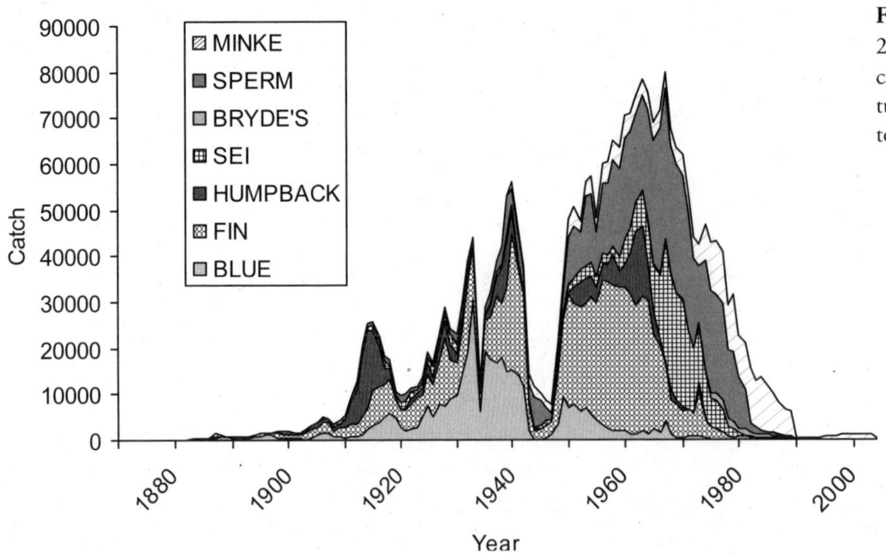

Figure 7–1. Modern Whaling, 1870-2000. The history of modern whaling catches shows that as each species in turn became scarce, attention shifted to the next until it too was depleted.

it would have meant writing-off much of their investment in catching equipment. The Commission set an overall limit on the catch of baleen whales in the Antarctic, but it was far too high and wasn't even divided by species.

A trend was already evident that has been widespread throughout fishery management and indeed environmental management in general. While the failure to take management action was mainly based on the rational self-interest of the participants, it was rare for them to admit this openly. Instead, opponents of catch limit reductions disputed the scientific evidence for the need for reductions. Disputes, which were really about objectives and intentions, were waged as if they were disputes about the facts.

Although the Commission had a Scientific Committee to advise on the state of the whale stocks, even the Commission accepted that its own Scientific Committee was too politicized to provide any objective advice. It often couldn't reach consensus, which was used by the Commission as an excuse for inaction. Although the more honest scientists recognized that catches were greatly exceeding the levels that the stocks could sustain, they could not provide unequivocal advice on how much the catches needed to be reduced.

Eventually, in 1961, the Commission appointed a special Committee of Three Scientists (later four) to produce an independent report.[9] The group confirmed that the blue whale was depleted to such a small fraction of its original abundance that it should be given complete protection for at least 50 years, and that humpbacks should also be completely protected. It estimated that the catches of fin whales would need to be reduced by about two-thirds to be biologically sustainable. Catches of sei whales, which had been expanding as other species became scarce, needed to be held in check until more data were available. At around this time, most countries withdrew from the industry such that by the end of the 1960s only Japan and the USSR were still catching whales in the Antarctic. By that time sperm and sei whales were the main species being caught. From the late 1970s the small minke whale, previously ignored by whalers, became the main target. This is an example of the industry staying one step ahead of efforts to regulate it: the Committee of Three was told to focus on fin and blue whales, but at exactly that time the industry began to exploit sei whales, in lower latitudes.

In short, the whaling industry took care of itself rather than the populations of whales that it needed to continue in the long term. Many authors cite Colin Clark[10] who, in 1973, examined the concept of profit maximization for those exploiting wild living resources. He explained why the most profitable option for those exploiting most ***K***-selected species (those with slow population growth rates such as whales), where the sustainable yield is low, was simply to "mine" the species to commercial extinction. The simple logic was that the money earned would grow faster in the bank than the species could reproduce.

The multi-species nature of Antarctic whaling exacerbated the situation for the more valuable and vulnerable species such as the blue whales. Even after blue whales had become too scarce to be worth hunting in their own right (i.e. were commercially extinct in Clark's sense) they were nevertheless still taken opportunistically in operations directed primarily at other species. True blue whales (not counting the so-called pygmy blue whales *Balaenoptera musculus brevicauda*) are estimated to have numbered around 250,000 in the Southern Ocean at the start of the 20th century, and appear to have been reduced to only a few hundred by the beginning of the 1970s when legal protection was finally enforced, although there is some evidence of limited increases since then.[11]

After the collapse of the Soviet Union, it emerged that the Antarctic whaling situation had been even worse than appreciated at the time. Through the 1950s and 1960s, whaling fleets of the USSR had secretly killed large numbers of protected species, including right whales which had been legally protected since 1935, and systematically falsified the catch data that were provided to the international community.[12] The fraud continued until a scheme for the international exchange of observers on factory ships came into effect in 1972.

The preamble to the 1946 Convention implicitly recognized some basic notions of the theory of population management in that it talked about stocks being able to sustain exploitation and about the optimum level of whale stocks, but the Convention lacked a strategy for putting such notions into practice. The Convention referred to the goal of achieving the optimal level of whale stocks as rapidly as possible, but qualified it with the proviso "without causing widespread economic and nutritional distress". During the 1950s and 1960s, the economic hardship provision was used to reject any measure that would have reduced the return on investments in whaling ships. Much needed conservation measures were delayed until after severe damage had been done to whale populations.

In the event, when Antarctic whaling finally collapsed, no "widespread economic and nutritional distress" came to pass. Food production in the affected countries was already back to normal following the disruptions of World War II, and those employed in the industry had little trouble finding other jobs, because their countries were enjoying a period of rapid economic expansion and full employment.[4]

Much of the whaling industry itself saw the writing on the wall and took the steps they deemed necessary to secure their businesses. However, these did not involve conservation. One such company was Scottish-based Christian Salvesen, which became the largest whaling company in the world. Gerald Elliot, who retired as its Chairman in 1988 wrote of the preparations that his company made to leave whaling: "Our company went on from whaling into many new activities, most of them away from the sea".[13] "In the thirty years that have passed since the end of our whaling, Salvesen has grown steadily and continues to flourish as a public company, though in areas far removed from our original trades". Salvesen is now a large Europe-wide distribution company that would appear to have achieved economic sustainability.

Management based on the science of sustainable use

From the late 1960s the growing global consciousness of the need to safeguard humankind's finite environment directed public attention to the whaling issue, a prime example of the wrong relationship between people and their environment. This led to a resolution at the first UN Conference on the Human Environment in 1972 calling for a 10-year moratorium on commercial whaling, to allow time both for research into the state of whale stocks and for stocks to recover. However, the UN resolution deferred to the IWC to implement this demand. Although some of those IWC members who had recently withdrawn from commercial whaling, such as the UK and USA, supported the call, the Commission did not adopt it. Instead, it adopted in 1974 an Australian amendment to the moratorium proposal, whereby it would only apply to those populations of whales which had been depleted below the levels providing the Maximum Sustainable Yield (MSY). Other populations could continue to be exploited subject to catch limits based on estimates of the MSY. The new policy came to be known as the New Management Procedure (NMP) and came into effect in 1976. With the NMP came at last the agreement to set catch limits separately by species and stocks, and also for all regions, not just the Antarctic.

The NMP was the first systematic attempt to put the management of whaling on a sustainable footing, and was for its time an advanced approach to resource management.

The basic idea behind the NMP, and indeed much of the theory of sustainable use in general, is that a natural population, left to its own devices, will neither expand without limit nor disappear, but hover around a level which the environment can support, where births and natural deaths will balance on average (Figure 7–2).[14] This level is the carrying capacity of the environment for the population, usually denoted by the letter ***K***. A population below this level will, other things being equal, tend to increase back towards that level while a population above that level will on average tend to decrease. The maximum net excess of births over deaths will occur at intermediate population levels. If this net annual increment is taken as offtake then the population should stabilize at this level and the exploitation will be sustainable.

The idea of the NMP is to move whale populations into the range where the net annual increment is greatest, or keep them there. Under the NMP, which is still nominally in force, whale populations were classified into three categories:

- *Protection Stocks* (PS): whale stocks estimated to be more than 10% below MSY level, are accorded complete protection;
- *Sustained Management Stocks* (SMS): whale stocks estimated to be between 10% below and 20% above MSY level – catches are limited to 90% of MSY level;
- *Initial Management Stocks* (IMS): whale stocks subject to little past exploitation estimated to be more than 20% above MSY level. These may be brought down to the MSY level by temporarily unsustainable catches.

By convention the IWC Scientific Committee took the optimal level for baleen whale populations (often referred to as MSYL) to be 60% of ***K***, and assumed that the pre-whaling abundance corresponded to ***K***. A potential complicating factor, ignored in the NMP, is that the carrying capacity of the environment may fluctuate in an unpredictable manner due to natural ecological variability. By amendment to the original NMP it was provided that exploitation of previously unexploited whale stocks should not commence until a satisfactory estimate of the size of the population was available. This was perhaps the first provision of the IWC that could be described as precautionary in the modern sense.

Although the new procedure led, as intended, to the protection of those stocks which were obviously severely depleted, it proved hard in practice to determine levels of sustainable catch for the other populations, due to a lack of data and knowledge on whale populations. The Scientific Committee of the IWC, charged with determining the requisite catch levels, had to rely largely on guesswork or on supposedly scientific methods that did not stand up to critical scrutiny. In most cases, the data did not exist to determine the size of a population relative to its MSY level, nor what the MSY would be. Furthermore, the NMP did not provide much incentive to collect the required data. Indeed, it contained an escape clause which permitted catches to continue at current levels in the absence of positive evidence to the contrary.

The moratorium

As the difficulties in applying the NMP became increasingly apparent, attention turned to complementary approaches to conservation, including regional sanctuaries.

Figure 7–2. Theory of Sustainable Offtake. A natural population above the carrying capacity level ***K*** will tend to decrease towards ***K*** (deaths exceed births), while a population below ***K*** will tend to increase towards ***K*** in the absence of exploitation (births exceed deaths). At population levels intermediate between zero and ***K***, the excess of births over deaths can in theory be taken as sustainable offtake.

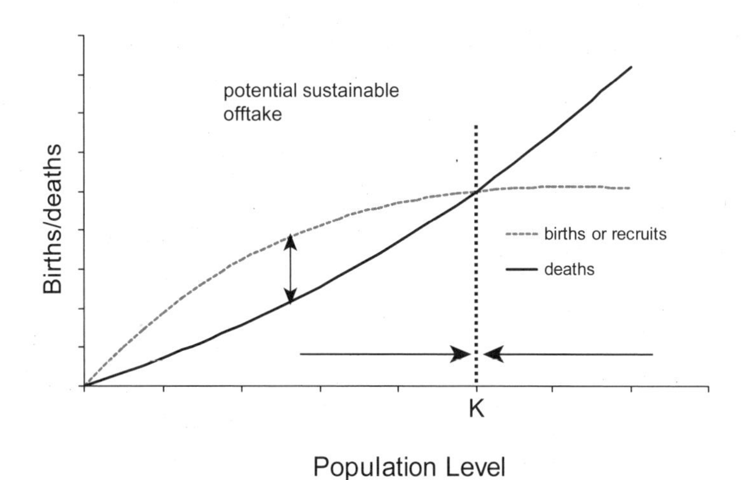

At the initiative of the Seychelles, the IWC in 1979 designated the Indian Ocean north of 55°S as a whale sanctuary in which whaling was prohibited. Also in 1979, the IWC banned the operation of factory ship whaling, except for minke whales.[15] In 1981, a moratorium on the killing of sperm whales was adopted by the IWC, but it excluded the western North Pacific, the most important sperm whaling area in both historical and recent times.

In 1982, the Commission adopted an indefinite moratorium on commercial whaling to take effect from 1986 onwards. The moratorium specified that all catch limits for commercial whaling would be zero until the decision was reviewed, by 1990 at the latest, when alternative catch limits could be considered. When the moratorium came into effect in 1986, the IWC embarked on a "comprehensive assessment" of the state of the world's whale stocks and the development of a Revised Management Procedure (RMP) (discussed further below), that would overcome the deficiencies of the NMP.

The moratorium did not apply to whaling by indigenous peoples for which the IWC has special provisions. Whaling by or for indigenous peoples in Alaska, Russia and Greenland continues under this provision.

The *Convention on International Trade in Endangered Species of Wild Fauna and Flora* (CITES) supported the IWC decision by banning the international commercial trade in whale products (from species that were not already protected) from 1986 onwards, through the inclusion of the affected species on its Appendix I.

The moratorium decision was controversial and several countries lodged formal objections that exempted them from it: Japan, Norway, the then USSR, and Peru. A main argument raised against the moratorium decision was that it ignored the differences in status between whale stocks and was therefore not scientifically justified.

In the event, whaling nations Brazil, Spain and South Korea and Peru (despite its initial objection) accepted the decision and phased out their whaling operations. Brazil adopted a national policy of exclusively non-lethal utilization of whale resources, such as tourism and research. The non-lethal policy has been backed up by the designation of specially protected areas for southern right whales, the most prized species in Brazilian waters, and the Brazilian President has recently appealed for international backing for their policy in the form of support for the South Atlantic Whale Sanctuary proposed by Brazil and Argentina to the IWC. Brazil has taken a range of further measures to protect whales such as banning seismic surveys on the Abrolhos Bank.[16]

The USSR lodged an objection to the moratorium decision but subsequently ceased whaling, although the Russian Federation inherited this objection and hence the right to conduct commercial whaling. Norway objected to the moratorium decision, but suspended commercial whaling in 1987. It resumed whaling again in 1992 and, since 1994, has managed its whaling using successively modified versions of the RMP, to allow steadily increasing catches, as explained in more detail below.

Japan withdrew its objection to the moratorium decision in 1987, but then embarked on a program of "scientific whaling" on minke whales in the Antarctic, making use of a clause in the whaling Convention that exempts catches for "scientific purposes" from all regulations. Since 1994, Japan has also conducted scientific whaling in the North Pacific. In addition to minke whales, this operation also takes, since 2000, sperm whales and Bryde's whales (*Balaenopetera edeni*),[17] and, since 2002, sei whales. The number of whales caught has been increasing year by year and is now on a scale comparable with ordinary commercial whaling. The Commission has on several occasions recommended by majority vote that such use of the

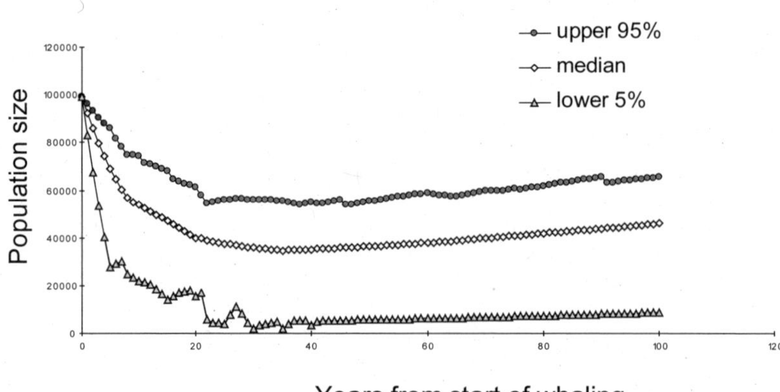

Figure 7–3. Simulation of a whale population, catches set to NMP rules. 100 computer simulations were conducted of the application of the New Management Procedure (adopted in 1975 and nominally still in force) to an hypothetical whale population. The figure shows the median, 5th lowest and 5th highest outcomes from 100 runs. While the population trend is satisfactory on average, the span of possible outcomes is very wide, with the population driven to near-extinction in "unlucky" cases. The NMP is, therefore, not a safe management procedure, even in theory.

scientific whaling exemption be discontinued, but these exhortations have not been heeded.

Iceland did not object to the moratorium decision but initially pursued a scientific whaling program until 1990. It then ceased whaling, and withdrew from the IWC in 1992 in protest at the Commission's failure to end the moratorium. In a legally controversial move, Iceland rejoined the IWC in 2002 with a reservation on the moratorium. Iceland resumed whaling in 2003 under the scientific whaling provision, without invoking its reservation, whose validity is not recognized by many parties.[18]

Because the moratorium was conceived as a temporary measure, the IWC continued to consider other conservation measures and, in 1994, it designated most of the waters south of 40°S as the Southern Ocean Sanctuary which is to remain in effect indefinitely, subject to review at 10-year intervals. This measure effectively establishes a long-term moratorium in a region which had been the major whaling ground of the 20th century. Recent proposals for adjoining sanctuaries in the South Pacific and South Atlantic have failed to attract the three-quarter majorities required for adoption.[19]

THE REVISED MANAGEMENT PROCEDURE

Some pioneering work in the mid-1980s by William de la Mare,[20] showed that to some extent it is possible to simulate the operations of a management procedure such as the NMP on a computer. However, one can only simulate a highly idealized picture of the whale stocks and the management procedure, but if the simulations show that even under idealized circumstances a management procedure is not likely to behave as intended, then clearly there is something wrong with it. The NMP itself was not a fully specified procedure, since it did not specify exactly how catch limits were to be calculated, but the Scientific Committee had developed some *ad hoc* rules of thumb to implement it. Figure 7–3 shows some example simulations of the Committee's method of implementation of the New Management Procedure (NMP).

This figure shows the range of outcomes of 100 replicates in a typical, highly simplified scenario. The reason for running 100 replicates is that the management process is inevitably subject to random, unpredictable factors, such that no two simulations of a given scenario will yield the same result, not least because the data (including surveys of whales in the ocean) on which management decisions are based are subject to random sampling error. The range of trajectories of population size over an 100-year period are shown. The results show that there is a high risk of the whale population being severely depleted. Not shown on this figure are the catch limits under the NMP, which would be liable to fluctuate wildly, such that the procedure would be effectively unworkable in practice. It should be emphasized that this is a highly idealized scenario in which all the assumptions of the NMP are met. In more difficult, and possibly more realistic scenarios, such as where the whale population does unexpected things, performance of the management procedure could only be worse still.

The scientific demonstration of the inadequacy of the NMP led the Commission to approve the development of a Revised Management Procedure (RMP) that should overcome the deficiencies of the NMP.[21] The finding also provided a retrospective justification of the 1982 moratorium decision. Even though not all whale stocks were necessarily overexploited, a blanket moratorium could be considered a rational management measure in a situation where the management procedure itself was deficient, and thus placed all stocks at potential risk. When the moratorium came up for its scheduled review in 1990, this argument was accepted by the Commission, but only just. Whaling countries argued that the moratorium had run its course and proposed non-zero catch limits. Others argued that the moratorium should remain in place until a revised procedure was ready. By a majority of only 2 votes, the latter view prevailed.

The Commission specified in 1989 three objectives for the RMP:

- Stability of catch limits (to permit the orderly development of the whaling industry).
- Acceptably low risk that the stock (= population) is not depleted below a specified level (so that the risk of extinction of any stock is minimal).[22]
- Making possible the highest continuing yield from the stock.

The IWC Scientific Committee considered that the RMP should be a fully-specified procedure, that is amenable to the kind of simulation testing developed by de la Mare. Furthermore it should work directly with data that could actually be obtained in practice, instead of requiring guesses of unobservable quantities such as MSY.

Following an iterative process of testing and development, the Committee was in 1992 able to recommend a procedure that achieved a reasonable balance between the above objectives.[23] The Commission accepted the procedure in principle as the basis for the management of any future commercial whaling that might be authorized.

REVISED MANAGEMENT SCHEME

The RMP is merely a rule for specifying catch limits, and the Commission decided that it should be embedded in a more comprehensive management framework, called the Revised Management Scheme (RMS), which was to include additional elements such as arrangements for inspection and enforcement.

The need for inspection arrangements has been underscored not only by the extensive falsification of catch records by the former USSR (see above) but also by more recent revelations including the falsification of Japanese coastal whaling catch figures in the 1980s. For example, Kondi reports 1157 Bryde's whales killed by Japan in 1986-87 as compared to the official figure of 634.[24]

Negotiations on the RMS have continued since 1995, at varying levels of intensity, within the IWC itself and in various working groups that it has appointed. By October 2005, little progress towards its adoption has been made. A proposal to adopt the RMS that was put to the vote in the Commission in 2002 failed to achieve a majority, receiving votes against, from both supporters and opponents of whaling.[25] Supporters of whaling see the RMS as an overly restrictive scheme, while countries opposing whaling are concerned that its adoption would amount to international endorsement of whaling, for which they would be held jointly responsible. Both sides appear to see the current arrangements, whereby commercial whaling continues in practice without international endorsement, as preferable to having an international management scheme in place.

In the meantime, commercial whaling, under objections to the moratorium and through extensive use of the "scientific purposes" exemption, has continued to expand, outside international regulation, from a low of 328 minke whales in 1991 to a projected catch of around 1,500 minke, sei, sperm and Bryde's whales in 2005, with plans further to increase the number and species composition of "scientific catches" in the near future.

However, the non-adoption of the RMS does not mean that the key players are necessarily content with the *status quo*. Japan has over the last 20 years run an ongoing project to recruit new parties into the IWC, typically small island states and poorer countries, to support its policy. For example, Mauritania, which attended its first meeting of the IWC in 2004, cast 10 votes, all of which were identical with those of Japan. St Kitts and Nevis voted with Japan on 71 out of 72 votes between 1998 and 2003. Japan continues to recruit new members, the latest being Mali in August 2004 and Kiribati in December 2004, and its representatives have expressed optimism that they will soon achieve a majority in the Commission for their position.[26] What Japan calls its "vote consolidation program"[27] has been criticized by non-governmental observers as vote-buying.[28] Though concerns regarding vote-buying have arisen in many fora, this is perhaps the only example of a twenty-year campaign by one country, seeking to buy an entire convention.[29] The Commission addressed the vote-buying issue obliquely at its 2001 meeting, noting the importance of transparency in the Commission's affairs and the right of members to exercise their votes free of coercion,[30] but did not take specific action.

As of the 2005 IWC meeting, the Commission of 66 members is about evenly divided between proponents and opponents of commercial whaling and new members are joining on both sides. It is, therefore, hard to predict what direction the body will take. In the event of proponents of whaling gaining the upper hand, the most likely action of the Commission would be to issue some form of international endorsement or approval of ongoing nationally-managed whaling activities, rather than to reintroduce international management of whaling (which would require a three-quarters majority).

CONCEPTS OF SUSTAINABILITY

The history of whaling, and the attempts to manage it, provide potentially useful lessons relevant to the issue of sustainable exploitation of living resources in general.

Elsewhere in this book, Sidney Holt[31] has indicated that there are many definitions of sustainability, including the definition in the World Conservation Strategy[32] produced jointly by IUCN (now the World Conservation Union), UNEP (United Nations Environment

Programme), and WWF (now the World Wide Fund for Nature) that, "if an activity is sustainable, for all practical purposes it can continue indefinitely". However, this definition does not provide guidance as to how to achieve sustainability in practice, nor how to determine whether or not a particular use is indeed sustainable. It is also unclear whether it refers only to biological sustainability, or also to ecological, economic, technical or political sustainability, or to the combination of all these aspects.[33]

The conventional sustainable use paradigm is that wild living resources, including whales, can be harvested sustainably provided that the exploitation is appropriately managed.

This concept of sustainability was critically scrutinized in a short paper in *Science* in 1993.[34] Like Colin Clark (see above), Ludwig and co-authors believe that exploitation of resources is largely driven by economic factors (the more valuable the resource, the more likely the over-exploitation). The paper attacks "sustainability" as a concept that can rarely be applied in practice, and proposes that in general it is more appropriate to think of resources managing humans than *vice versa.* The larger and more immediate the prospects for gain, the greater the pressure for unlimited exploitation. When the resource has been over-exploited, governments tend to subsidize continued exploitation as substantial investment and many jobs are often at risk. Large levels of variability (in both the distribution of the resource and the pattern of its exploitation) tend to mask problems. Scientific uncertainty is identified as one reason for lack of action to limit exploitation, however, the authors cite instances where, despite scientific certainty, unsustainable practices continued.

The paper concludes with some principles of effective management (the two-page paper itself does not use the history of whaling as an example but mentions instead the exploitation of fish stocks and forests). These principles are as follows:

- Include human motivation and responses as part of the system to be studied.
- Act before scientific consensus is achieved – often calls for more research are simply delaying tactics.
- Rely on scientists to recognize problems but not to remedy them – and be aware that scientists and their judgments are subject to political pressure.
- Distrust claims of sustainability.
- Confront uncertainty. There is an illusion that science can provide all the answers.

However, it is possible to make rational decisions while taking uncertainty into account, by constructing a variety of plausible hypotheses, considering a variety of strategies, favoring actions that are reversible, monitoring results and modifying actions accordingly and erring on the side of precaution.

SCIENTIFIC LESSONS FROM THE DEVELOPMENT OF THE REVISED MANAGEMENT PROCEDURE

We have already seen how, left to itself, the exploitation of whales has been almost universally biologically unsustainable. The example of Salvesen shows that a corporation can achieve economic sustainability based on biologically unsustainable activities, provided that it is sufficiently flexible and adaptive. Capital accumulation obtained from the profitable destruction of a resource then leads to continuity of another activity by the same corporation.[35]

Although the specified objectives of the Revised Management Procedure did not include biological sustainability *per se*, they are quite closely related to concepts of sustainability: low risk of depletion; stable catches; and highest continuing yield.

The experience gained from the IWC's process of developing and trying to implement a Revised Management Procedure echoes some of the problems identified by Ludwig *et al.* with respect to the sustainability concept. There follow some of the key points to emerge from this exercise.[36,37]

Good intentions are not enough

As noted above, the NMP was designed to ensure biologically sustainable exploitation, but when it was subjected to simulation testing, it was found to have a low probability of providing sustainability in practice. Likewise, the early versions of candidates for a Revised Management Procedure were found to provide poor performance in the first round of simulation tests, forcing the developers to modify them substantially.

This experience confirms the wisdom of Ludwig *et al.*'s advice to "distrust claims of sustainability". It is not sufficient that a management approach merely be *intended* or *designed* to be sustainable. Evidence is needed that it would actually perform as expected.

Reaction is not enough: pro-action is needed

Candidate management procedures for whaling that relied on detecting negative trends in populations before limiting catches, did not perform well in tests. Depending how they were tuned, they either tended to act too late, and ran a high risk of severely depleting populations before a decline was detected; or they were too sensitive, responding too readily to "false alarms". In neither case was biological sustainability achieved.

The only procedures that performed reasonably satisfactorily in simple tests were those that required direct estimation of the size of the population to be exploited, and which set the allowed catch limit to a sufficiently small fraction of the abundance estimate, such that rapid depletion of the resource could not occur, even after allowing for some error in estimation.

This experience is consistent with Ludwig *et al.*'s advice to "act before scientific consensus is achieved", which could be phrased more strongly generally as "act before the need is evident".

Other than by setting a trivially low catch limit, it is not possible to provide an advance guarantee of long-term biological sustainability, but it is possible to set a catch limit low enough to ensure that a stock will not be impacted by exploitation too fast to allow time for remedial action. Extensive simulation testing during RMP development showed that "low enough" means typically less than 1% of the estimated population size. Thus a safe catch limit is not much greater than what many would regard as a "trivially" small one.

History matters: the need for a reference point

A further result to emerge from the RMP development process was that it is quite difficult to meet the management objectives without taking past catches into account in some way. This provides an implicit reference point. When a population is small compared with past catches, it can be considered depleted and protective action should be taken. If one is allowed to forget the past, and focus on maintaining populations at their current levels, then it is hard to prevent a slow but sure drift towards extinction. This is an example of the "shifting baseline" problem of fisheries noted by Daniel Pauly.[38] Biological sustainability requires some form of reference point such that populations below this level are restored.

This finding justifies the ongoing effort by the IWC Secretariat to reconstruct from various original sources, the approximate catches actually taken by Soviet era whaling fleets during the 1950s and 1960s, whose reports filed at the time were largely fabricated.

Incentives to collect data

The precision of an estimate of the number of whales in an ocean is directly related to the amount of survey effort expended to obtain it. The more time spent counting the animals, the closer will the estimate be to the true number, on average. The RMP takes this into account in setting the allowable catch limit: the less precise the estimate, the lower the allowed catch, such that the risk of depletion is held to an approximately constant, low level regardless of the amount of data. In the absence of any data, the catch limit is zero. This places a positive value on data and is one way of implementing a precautionary approach. This is in contrast to traditional fishery management approaches where restrictions are only imposed when there are sufficient data to justify them.

This is one aspect of Ludwig *et al.*'s point about the need to include human motivation in the system being studied: relevant data will not be submitted[39] unless there is a pay-off for doing so.

Whale populations cannot be managed

A more subtle but significant result to emerge from the process of developing the RMP is that it is not actually possible to "manage" populations of whales, in the sense of holding them at some desired level. Attempts to do so run a high risk of depleting the population far more severely than intended.

A natural population fluctuates in unpredictable ways. When exploitation is limited to the levels found to be safe, the result is that the dynamics of the population under exploitation are not expected to be appreciably different from those of an unexploited population. Setting the exploitation high enough to substantially impact the population engenders a high risk of serious accidental overexploitation.

Figure 7–4 shows an hypothetical whale population which is either (i) protected; or (ii) subject to a safe level of exploitation; or (iii) subject to an unsafe level of exploitation. The safely exploited population tracks the unexploited population quite closely. A conclusion is that we can at best manage *whaling* so as to limit its impact on whale populations, but we cannot manage the *populations* themselves.

The Maximum Sustainable Yield is not sustainable

Although this finding sounds like a contradiction, it reflects an important truth. While it is possible to identify a safe level of catch, it is hard to identify a *maximum* safe level of catch. An attempt to extract the *maximum* sustainable yield from a stock, as opposed to merely a sustainable yield, runs a high risk of excessive depletion of the stock such that the catch will not be sustained.

How much risk of depletion to accept is partly a value-judgment. The Scientific Committee, somewhat arbitrarily, put forward three different "tunings" of the RMP with different risk levels.[40] These were labeled the 0.60, 0.66, and 0.72 tunings based on the expected depletion of an hypothetical whale stock in a specific reference scenario. The Commission endorsed the 0.72 tuning level which gave the lowest risk level, and hence the lowest allowable catches.

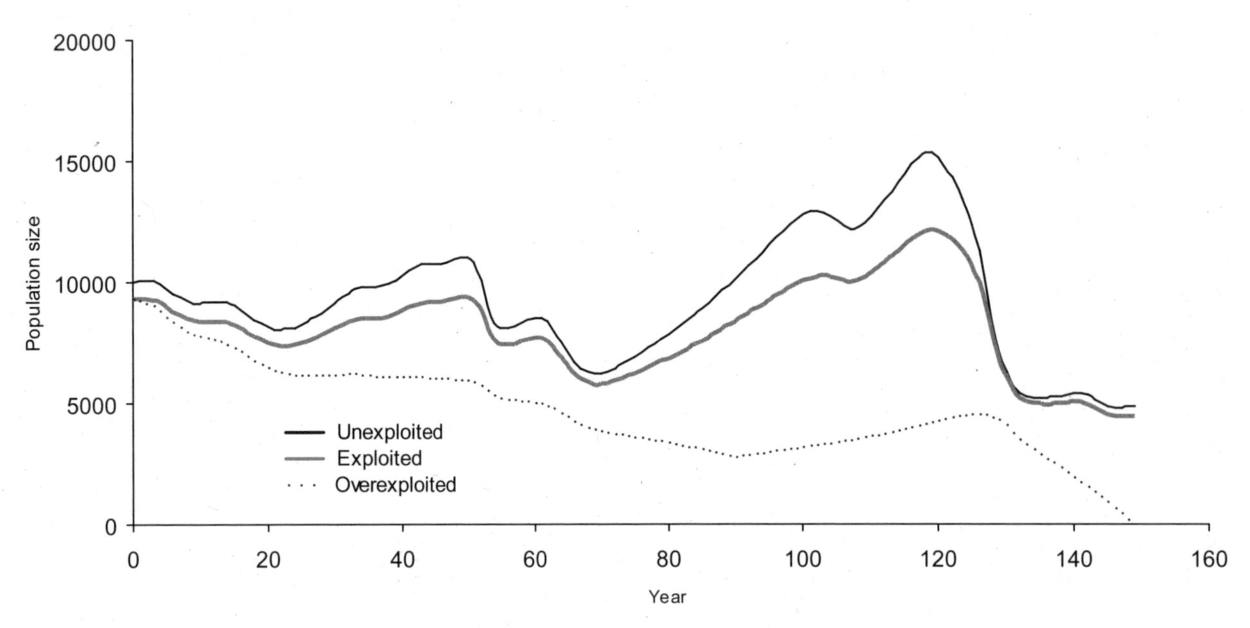

Figure 7–4. Managing populations? A natural population will fluctuate in unpredictable ways (upper curve). A low, safe level of exploitation will reduce the population slightly but leave it subject to much the same level of fluctuation. A level of exploitation high enough to substantially impact the population has a high risk of being unsustainable and driving the population towards extinction (lower curve). It has not proven possible to devise management procedures that hold a population at some desired, stable level.

IMPLEMENTATION OF THE RMP: THE NORWEGIAN EXPERIENCE

The RMP has not been adopted internationally, and Norway is the only country so far to apply a version of it to the management of whaling. Norway filed a formal objection to the moratorium on commercial whaling, so is not bound by that decision.

During 1996-2000 Norwegian fishery authorities set national catch limits for minke whaling calculated from the RMP with the IWC's approved tuning of 0.72. From 2001, the catch limit would have been reduced under RMP rules had the 0.72 tuning been retained. This was because the procedure called for a reduction to compensate for the unduly high proportion of females in the 1996-2000 catch. The Norwegian authorities avoided a catch limit reduction by changing to the 0.66 tuning from the 2001 season. In 2003, a further catch limit reduction was implied by the RMP rules because of a new, lower abundance estimate. This catch limit reduction was avoided by switching the tuning to 0.62 (Figure 7–5).

In 2004, the Norwegian Parliament adopted a policy on marine mammals which proposed a substantial rise in minke whale catches, motivated by the reasoning that reducing the numbers of whales would benefit fisheries for prey species.[41] The government set the catch limit for 2005 at 797 whales, obtained by applying the RMP with the 0.60 tuning. This tuning is the lower end of the range recommended by the IWC Scientific Committee. The Norwegian delegation to the IWC announced that Norwegian scientists would be tabling a radical revision of the RMP in the near future that would allow yet higher catches.

Arguably, each version of the RMP with a given tuning represents a biologically sustainable management procedure, albeit with different risk levels. However, the safety of each tuning level has been tested by simulating it for 100 years assuming that the management procedure would not be changed. The process actually followed by the Norwegian authorities has been to adjust the tuning level to maintain or increase catches whenever the RMP itself called for a decrease. To the extent that the tuning level is only changed in one direction, *this* process cannot be sustainable: there will come a point where the tuning cannot be adjusted any further without abandoning the goal of sustainability.

The Norwegian parliamentary resolution states that sustainable management of minke whale populations may soon be abandoned in favor of a deliberate policy to reduce the populations in order to reduce the amount of commercially valuable fish they consume.

When and whether the management objective is changed from sustainable use to population culling may depend on how much longer the desired catch level can be justified on sustainable use grounds. The desired catch level itself probably depends on market conditions. The

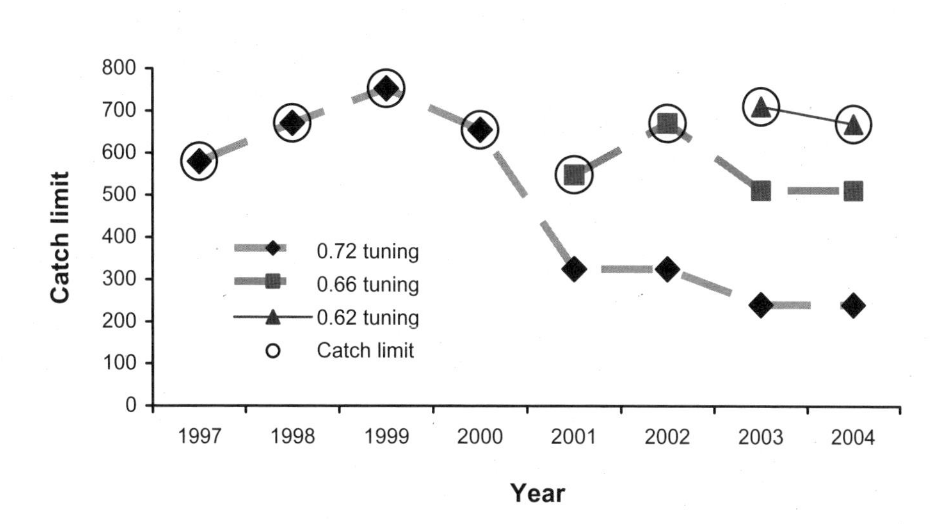

Figure 7–5. Norwegian whaling catch limits. From 1997-2000 Norway set catch limits for whaling using the Revised Management Procedure (RMP) with the IWC's adopted tuning level of 0.72. From 2001 the RMP rules indicated a reduction in the catch limit (lower curve), but Norway changed the tuning to 0.66 to maintain catch levels (middle curve). In 2003 a further cutback was indicated, which Norway avoided by changing the tuning to 0.62.

government's stated policy is to try to boost domestic demand for whale products, and at the same time to try to re-establish export markets, especially in Japan. The government of Japan is not currently permitting the import of Norwegian whale products.

The example of Norway's implementation of the RMP illustrates two points:

- Even the best designed procedure for sustainable management can be implemented in unsustainable ways that the developers of the procedure did not expect; and
- Instead of the management procedure determining the catch level, the reverse process occurs: the desired catch level determines the choice of management procedure.

BIOLOGICAL VERSUS POLITICAL SUSTAINABILITY

The Revised Management Scheme, in which the RMP was to be embedded, has at the time of writing not been adopted into force by the IWC despite over 10 years of negotiations. Until this is done, the IWC is effectively unable to actively manage commercial whaling. Issues of disagreement include, *inter alia*, arrangements for observation and inspection of whaling operations. These would not in principle be a difficult issue to resolve, because in the meantime the placement of observers has become normal practice in many other international fishery management organizations. The failure to reach agreement is, therefore, due primarily to the lack of an incentive to do so.

Whaling countries see the RMS as unduly restrictive and can take more whales under current arrangements. Without an agreed RMS, non-whaling countries can take a position of disapproval of commercial whaling, while leaving responsibility for its management to the minority of countries which conduct it. In view of the general obligation under international law on states to co-operate in the conservation and management of cetaceans, countries are reluctant to be seen explicitly to abandon the notion of international management of whaling, and hence negotiations have continued despite the lack of progress.

The RMS concept fails to satisfy the first key principle enunciated by Ludwig *et al.*, which is to include human motivation in the system to be studied. The standard sustainable use paradigm often assumes the existence of an altruistic management authority, which will take the decisions needed to ensure sustainability. In the case of whaling, this idealized manager does not exist.

The history of whaling management shows that in the time when the IWC was dominated by whaling interests, they acted in their own interests to maximize returns at the expense of biological sustainability. When the IWC became dominated by countries without whaling interests, it adopted increasingly protective measures, but these members have little interest in implementing international regulation for the whaling that does occur. This is rational to the extent that there is no incentive for a government to take joint responsibility for managing an activity from which it derives no benefit itself.

While the current lack of consensus within the IWC may represent rational behavior of those involved, it appears not to be a politically sustainable situation, given the vote-buying phenomenon noted above.

The whaling experience suggests that the kind of management required to ensure biological sustainability may not be politically sustainable. This finding calls the conventional paradigm of managed sustainable use into question. It does not rule out the possibility that specific cases of exploitation can be fortuitously sustainable, simply because the demand for the products happens not to exceed the productive capacity of the resource.

In view of the problems with implementing the sustainable use concept for whaling, players on both sides of the debate have been seeking alternative bases for action.

FROM SUSTAINABLE USE TO CULLING – A NEW JUSTIFICATION FOR INCREASING CATCHES

The Governments of Japan, Norway and Iceland are all vigorously promoting the idea that whales and other marine mammals need to be culled in order to protect fisheries resources.[42] This argument is being pushed partly in order to justify greater catches through culling than would be permitted under a sustainable management regime[43] such as the RMP. If the objective is to deliberately reduce the size of the population in order to decrease the alleged competition, there is an opportunity for higher catches that need not be sustainable.

A number of brochures that have been prepared by the Government of Japan are intended to justify this notion. Press releases are also regularly issued on the subject. In the words of Joji Morishita from the Fisheries Agency of Japan, "This is direct competition with fisheries to feed humans".[44] Then alternate Japanese IWC Commissioner Masayuki Komatsu and his co-author are more strident, "excessive protection of whales is evil, it gives adverse effects to the marine ecosystem as a whole".[45]

The issue of competition between whales and fisheries is frequently cited by the representatives of countries that have been brought into the IWC in the context of Japan's vote consolidation program. It provides an ostensible rationale for their support for whaling in the IWC, despite their countries having no direct interest in whaling. However, some of these countries are located in areas where there is little or no overlap between species targeted by commercial fisheries and the diet of whale species that are potential targets for whaling.[46]

The 2004 Norwegian White Paper explains the need to kill whales in order to benefit fisheries,[47] and the main justification for both Iceland and Japan's "scientific whaling" is to determine the impacts of whales on commercially important fish.

It is not a simple matter to determine what net benefit would actually accrue to fisheries from the reduction of cetacean populations, nor even whether the net effect would be positive or negative in each case.[48] When the IWC examined this issue in a scientific workshop it recognized the complexity of marine ecosystems and stated that, "...there is currently no system for which we have suitable data or modeling approaches to be able to provide reliable quantitative management advice on the impact of cetaceans on fisheries or fisheries on cetaceans".[49] Whether our understanding can improve significantly in the foreseeable future is not yet clear. The use of ecosystem arguments for justifying the culling of whales is likely to generate considerable controversy at both the scientific and the political level.

CONSERVATION-ORIENTED MANAGEMENT

The history of whaling shows that once unsustainable exploitation is occurring, it can be hard to rein it in to biologically sustainable levels. This suggests that it is better to pre-empt the development of dependencies on consumptive exploitation altogether. One way to achieve this is to lock in consensuses for the non-exploitation of cetaceans where these exist.

The sustainable use paradigm is not the only basis for measures to protect whale populations from exploitation, and more protective measures are permitted under international law.[50]

For example, a relatively new agreement, ACCOBAMS (Agreement on the Conservation of Cetaceans in the Black and Mediterranean Seas) came into force in 2001 under the Convention on the Conservation of Migratory Species.[51] It requires parties to co-operate in the conservation of cetaceans in the Black and Mediterranean seas and to prohibit the deliberate exploitation of cetaceans in this area. Until the 1980s, small cetaceans in the Black Sea were subject to heavy exploitation including directed purse-seine catches that depleted them. However, no direct exploitation of cetaceans was occurring when ACCOBAMS was negotiated, and although not all range states have adhered as yet, the Agreement provides a way of entrenching the current consensus for no exploitation. In regions where some countries are currently involved in exploitation, such a consensus is much harder to achieve.

The Southern Ocean Sanctuary, adopted by the IWC in 1994, is an international area, outside exclusive economic zones (EEZs) except those of a few states that are strongly supportive of the sanctuary. When it came up for review in 2004, the IWC Scientific Committee recognized that the sanctuary concept in the Southern Ocean needed to be developed further to provide for long-term protection for whales beyond the simple prohibition of direct exploitation. Nevertheless, this collective decision by the international community not to exploit the whales in these waters appears to have been widely accepted as legitimate, even in the absence of complete consensus, and is one of the few decisions of the IWC that has become widely known outside the organization.[52]

Currently, Japan, the only country to have voted against the establishment of the Sanctuary, is also the only country pursuing whaling there, and no other countries have a concrete current interest in joining in. The Sanctuary may help to entrench the current near-consensus on protection of whales in this region. This could become important in the future, if some whale populations exhibit a substantial recovery and tempt a resumption of the 20th century-style exploitation of whales in the area.

Likewise, the Indian Ocean Sanctuary, established for a limited period in 1979 and made indefinite in 1992, obtained legitimacy through a broad consensus of support amongst Indian Ocean states, and has contributed to ensuring that substantial exploitation of large whales in this region was ended and has not been resumed. Even though the maintenance of an active interest in the Sanctuary by the bordering states has proved harder than anticipated, the Sanctuary has resulted in substantial research activities in this region where the cetacean fauna was previously very poorly known.[53]

Political support for policies of no consumptive use is more likely to be achieved where demonstrable benefits from non-consumptive uses are obtained such that whales have an economic and political value other than as an exploitable resource. Several Southern Hemisphere countries including Brazil, South Africa, Australia and New Zealand have adopted national policies of strictly non-lethal use of cetacean resources, including whale-watching. Whale watching is a rapidly expanding branch of the tourism industry with an estimated global value of US $1 billion per annum growing at approximately 12% per year.[54] Ensuring that non-consumptive uses of whales are managed so as to make a positive net contribution to whale conservation, will be one of the challenges to be met in the coming years.

Although the term whale-watching normally refers to the *in situ* form, *ex situ* enjoyment of whales through films, books and other media may be the larger industry in terms of global turnover. Conducted responsibly, *ex situ* whale watching potentially represents the form of use that involves the least disturbance to whales and to the environment generally. The challenge is to ensure that the benefits from these uses of whales feed back at least in part to support for whale conservation.

It is not necessarily valid to regard consumptive and non-consumptive uses of whales as alternatives, because a situation that involves a direct choice between the two types of use would rarely arise. However, the link exists to the extent that non-consumptive uses help to cement political support for policies that exclude consumptive use.

Traditional fishery management bodies, whose activities are limited largely to the regulation of exploitation, as the IWC was originally conceived, are at a disadvantage when it comes to implementing management approaches that are more conservation-oriented than the conventional sustainable use philosophy, because it can be hard to maintain a functional institutional machinery for the sole purpose of prohibiting an activity. There is, however, considerable precedent for international conventions and institutions to evolve to meet new challenges and responsibilities.[56]

In the case of the IWC, a number of member countries, led by Mexico, launched in 2003 the so-called "Berlin Initiative".[57] The aim was to bring the organization into the 21st century by transforming it from a traditional fishery management body to a modern conservation organization with a comprehensive agenda covering all aspects of the conservation of whales including protection from environmental threats. In addition to helping to address these other threats in their own right, it was hoped that the broadening of the IWC's agenda would reduce the risk that its failure to achieve consensus on the regulation of exploitation would lead to the organization becoming dysfunctional to the detriment of whale conservation. Although it is too early to judge the success of this initiative, there will clearly be a need for a global forum to address the full range of threats to whales and to work to ensure their long-term survival.

CONCLUSIONS

Despite its poor record of conserving whales in the past, the IWC has in recent times made a serious attempt to develop a management procedure that would implement the goal of managed sustainable use to the extent possible. The resulting Revised Management Procedure is widely regarded as one of the most advanced procedures ever developed in the fishery management field.

The process of RMP development has revealed that the conventional goal of extracting the Maximum Sustainable Yield (MSY) from a population is not obtainable. Although very low levels of exploitation can be sustained, there is no clear-cut maximum level. The higher the catches, the greater the risk that they will not be biologically sustainable. Levels of catch sufficient to "manage" populations in the sense of steering them towards a target level, engender a high risk of unintended depletion. One cannot manage whale populations: at best one can manage the exploitation to limit its impact on populations. To do this it is necessary to set catch limits sufficiently low in advance that they do not have a major impact on the exploited populations, without waiting for evidence of negative trends.

The approach of scientifically-managed sustainability is only viable if there is a sufficiently strong constituency with an interest in pursuing this approach and making it work. The political experience from both the history of whaling, and from the attempts to conclude and implement a comprehensive regulatory regime for the management of whaling, reveals that this is not the case. Internationally-managed exploitation of whales would at best be an uneasy compromise between those seeking to exploit whales and those seeking to protect them, with no-one standing to gain significantly from implementing and enforcing the kind of management required. The kind of management that is required to ensure biological sustainability is not politically sustainable.

These conclusions underline the importance of pursuing alternative approaches to whale conservation, based on avoiding the development of dependencies on consumptive exploitation. Such an approach is gradually taking shape in parts of the world, through for example the designation of the Southern Ocean Sanctuary, and regional agreements that exclude consumptive exploitation of cetaceans, such as in the Mediterranean and Black Seas.

Management approaches that are more protective than those based on conventional sustainable use criteria are explicitly permitted for marine mammals under international law, but their acceptance is still far from universal. Where a consensus for the protection of cetaceans exists, the opportunity can be used to formalize their protected status to help make the protection permanent.

Those favoring the exploitation of whales have also begun to shift their public strategy away from the conventional paradigm of sustainable use towards ecosystem-based considerations, with a particular emphasis on arguments that whales need to be culled to protect the resources which they exploit. Considerable controversy in both the scientific and political arenas can be expected in the coming years.

ACKNOWLEDGEMENTS

Several colleagues have contributed to the ideas that have been expressed in this paper. We are grateful to Peter Corkeron, Sidney Holt, David Lavigne and Russell Leaper, for reviewing the manuscript, and Nic Davies for proof reading.

NOTES AND SOURCES

[1] For a discussion of the conventional conservation paradigm, see Lavigne, Chapter 1.

[2] Clapham, P and C. Scott Baker. 2002. Whaling, Modern. In *Encyclopaedia of Marine Mammals*. Edited by W Perrin, B Würsig and J.G. M Thewissen. Academic Press. 1414 pp.

[3] Gambell, R. 1999. The IWC and the Contemporary whaling debate. pp. 179-198. In *Conservation and Management of Marine Mammals*. Edited by J. R. Twiss and R. R. Reeves. Smithsonian Institution Press, Washington.

[4] Tønnessen, J.N. and A. O. Johnsen. 1982. *The History of Modern Whaling*. C. Hurst and Company, London. 798 pp.

[5] Lindquist, O. 2000. The North Atlantic Gray whale (*Eschrichtius robustus*): an historical outline based on Icelandic, Danish-Icelandic, English and Swedish Sources dating from c.a. 1000 A.D. to 1792. Occasional papers, Centre for Environmental History and Policy, University of St Andrews. 1:1-53.

[6] Whitehead, Hal. 2002. Estimates of the current global population size and historical trajectory for sperm whales. Marine Ecology Progress Series 242: 295-304.

[7] It is now agreed that there are at least two species of minke whale.

[8] Report of the Scientific Committee of the International Whaling Commission. 2003. Journal of Cetacean Research and Management 5 (Suppl.):41.

[9] The Commission decided "...to obtain the services of three scientists qualified in population dynamics and drawn from countries not engaged in pelagic whaling in the Antarctic... to report... on the condition of Antarctic whale stocks". See IWC. 1961. Twelfth Report of the IWC. p. 8.

[10] Clark, C. W. 1975. *Mathematical Bioeconomics: The Optimal Management of Renewable Resources*. New York, John Wiley.

[11] Report of the Scientific Committee of the International Whaling Commission. 2004. Journal of Cetacean Research and Management 6(Suppl.): 23.

[12] Yablokov, A.V. 1994. *Validity of whaling data*. Nature 367:108.

[13] Elliot, G. 1998. *A Whaling Enterprise – Salvesen in the Antarctic*. Michael Russell Publishing Ltd, Norwich, Great Britain. For more on Salvesen, see Holt, Chapter 4.

[14] For a detailed discussion of the "notion" of sustainable use, see Holt, Chapter 4.

[15] The factory ship moratorium (Paragraph 10d in the IWC Schedule) is an important conservation provision because no country has filed a formal objection to this paragraph, so therefore all are bound by it. In addition, both the text of this paragraph and the records of the meeting (the Chairman's Report and the Verbatim Record) suggest that this measure was designed to be permanent.

[16] In 2004, the IWC Scientific Committee commended the Brazilian Government for this action. See IWC 2004. Report of the Scientific Committee Section 12.2.5.

[17] Bryde's whales were first distinguished from sei whales from catch statistics in the 1960s but it is still unclear how many different species of Bryde's whale exist. Three species have been proposed (Nature 426: 278-281).

[18] The IWC Schedule indicates in a footnote that, "Iceland's instrument of adherence to the International Convention for the Regulation of Whaling and the Protocol to the Convention deposited on 10 October 2002 states that Iceland 'adheres to the aforesaid Convention and Protocol with a reservation with respect to paragraph 10(e) of the Schedule attached to the Convention'." The manner of Iceland's accession caused a great deal of controversy, with Iceland's conditions for membership being rejected through votes at the 2000 and 2001 IWC meetings. Iceland was eventually admitted as a member at the 5th Special meeting of the IWC that was held in Cambridge in October 2002. For more detail, see the Chair's report of the 5th Special meeting in Annual Report of the International Whaling Commission, 2003, pp. 139-148.

[19] For additional discussion of proposed sanctuaries, see Worcester, Chapter 17, and Martin, Chapter 24.

[20] de la Mare, W.K. 1986. Simulation studies on management procedures. Reports of the International Whaling Commission 36:429-450.

[21] Cooke J.G. 1995. The International Whaling Commission's Revised Management Procedure as an example of a new approach to fishery management. pp. 647-659. In *Whales, Seals, Fish and Man.* Edited by A.S. Blix, L. Walloe and O. Ulltang. Elsevier, Amsterdam.

[22] For historical reasons the reference level for assessing risk was later specified by the Commission to be 54% of pre-exploitation numbers (the old Protection Stock threshold under the NMP, corresponding to 10% below an assumed MSYL of 60%).

[23] Kirkwood, G.P. 1992. Background to the development of revised management procedures. Rep. Int Whal. Commn 42:236-43.

[24] Kondi, I. 2001. *The rise and fall of Japanese coastal whaling.* Sanyo-sha, Tokyo. 449 pp. (In Japanese).

[25] A proposal to adopt the RMS was also tabled by Japan, but it was bundled with additional measures, including the abolition of the sanctuaries in the Southern and Indian Oceans. It was rejected by countries which generally favor the RMS.

[26] IWC audio verbatim record.

[27] Yomiuri Shimbun 13th April 1993. The article described the visit of the IWC Commissioner from Dominica to Japan as a "product of the vote consolidation operation by government and industry circles together".

[28] See for example http://www.ifaw.org/ifaw/dimages/custom/Publications/JapanVoteBuying.pdf, or http://whales.greenpeace.org/reports/IWC_japanese_vote_buying.PDF.

[29] For more information on vote buying at the IWC see Transparency International's annual report (http://www.globalcorruptionreport.org/download/gcr2004/07_Vote_buying.pdf) for a short summary and Third Millennium Foundation (http://www.3rdmf.org/VbuyingRep%20May02.pdf) for a detailed examination of the links between voting patterns and Japan's foreign aid packages.

[30] Resolution on Transparency within the International Whaling Commission. Annual Report of the IWC, 2001, p.54.

[31] See Holt, Chapter 4.

[32] IUCN/UNEP/WWF. 1991. *Caring for the Earth: a strategy for sustainable living.* Gland Switzerland. 228 pp.

[33] The term "biological sustainability" is used in the case of a single species whereas ecological sustainability is used in the case of several species or an ecosystem.

[34] Ludwig, D, Hilborn, R and C Walters. 1993. Uncertainty, resource exploitation and conservation: lessons from history. Science 260: 17 and 36.

[35] In contrast, the imposition of an extremely precautionary regime, such as the RMP/RMS, is intended to ensure biologically sustainable use. However, the catches obtained may be too small to be economically sustainable (i.e. in which income does not cover costs plus amortisation of the capital/hardware). This is likely to be a "normal" consequence of applying a precautionary regime to already greatly depleted resources.

[36] Cooke, J.G. 1994. The management of whaling. Aquatic Mammals 20(3):129-135.

[37] Cooke, J.G. 1999. Improvement of fishery management advice through simulation testing of harvest algorithms. ICES Journal of Marine Science 56:797-810.

[38] Pauly, D. 1995. Anecdotes and the shifting baseline syndrome of fisheries. Trends in Ecology and Evolution 10(10): 430.

[39] The situation prior to the RMP (and for non-RMP related matters such as humane killing) is that whaling countries have indeed been collecting data, but they have generally kept it to themselves, submitting only analyses of it that appear to support their case, not releasing the actual data. Only when the RMP came along, did the whalers in principle stand to gain from submitting the data.

[40] During the development of the RMP large numbers of computer simulation trials were conducted to test the performance of the procedure relative to different management objectives (listed on p.119) in a range of hypothetical scenarios each involving different assumptions about whale population dynamics and other factors. The procedure could be adjusted to place greater emphasis on the different objectives (e.g. higher catches or reduced risk of depletion) by a process known as tuning. The Commission accepted the procedure now known as the RMP, with a tuning of 0.72. This tuning refers to the adjustments made to the procedure to achieve a median final population size after one hundred years of 72% of the assumed natural population level or carrying capacity, in a specific reference scenario that was selected for the purpose of comparing different candidate management procedures.

[41] A summary of the Norwegian White Paper (in English) can be found at http://odin.dep.no/archive/fidvedlegg/01/03/marin044.pdf; the complete version of the White Paper (in Norwegian) at http://odin.dep.no/fid/norsk/publ/stmeld/008001-040014/index-dok000-b-n-a.html.

[42] See Lavigne, D.M. and S. Fink. 2001. *Whales and Fisheries.* International Fund for Animal Welfare, Yarmouth Port, Mass. USA. 12 pp.

[43] See Lavigne, D.M. 2003. Marine Mammals and Fisheries: The Role of Science in the Culling Debate. pp. 31-47. In *Marine Mammals: Fisheries, Tourism and Management Issues.* Edited by N. Gales, M. Hindell and R. Kirkwood. CSIRO Publishing. Collingwood, VIC, Australia.

[44] Fisheries Agency, Ministry of Agriculture, Forestry and Fisheries, Government of Japan. Press Release. Rome February 27, 2003. Available at http://www.icrwhale.org/eng/fisheries.pdf.

[45] Komatsu, M and S. Misaki. 2001. *The Truth Behind the Whaling Dispute.* Published by the Institute of Cetacean Research, 176 pp.

[46] Young, J.W. 2000. Do large whales have an impact on commercial fishing in the South Pacific Ocean? Journal of International Wildlife Law and Policy 3(3):1-32.

[47] Anon. 2004. Norway's policy on Marine mammals. Report 27 to the Storting. Ministry of Fisheries. Available at http://odin.dep.no/filarkiv/202967/marine_mammal_sum mary_final.pdf.

[48] Yodzis, P. 2001. Must top predators be culled for the sake of

fisheries? Trends in Ecology and Evolution 15(2): 78-84.

[49] IWC 2004. Report of the Scientific Committee. Journal of Cetacean Research and Management 6 (supplement). p. 30.

[50] The *UN Convention on the Law of the Sea* extends a special status to cetaceans in two respects. Most cetaceans, apart from the *Phocoenidae*, are listed as "highly migratory species" in Annex A to the Convention: Article 64 obliges states to co-operate internationally for their conservation even when their catches are within their own Exclusive Economic Zone (EEZ). Articles 65 and 120 require that, with respect to all cetaceans both within EEZs and on the high seas, states shall work through the appropriate international organizations for their conservation, management and study. These articles further authorize coastal states and international organizations to prohibit, limit or regulate exploitation of cetaceans more strictly than is required under the Convention's general prescriptions, for living resources, which is that populations be maintained or restored at levels that provide the maximum sustainable yield. These provisions have been cited as providing legal justification for IWC measures such as the moratorium and the Southern Ocean Sanctuary, although they were not explicitly invoked in either proposal.

[51] See http://www.accobams.mc. In particular, Article II indicates that, "...Parties shall prohibit and take all necessary measures to eliminate, where this is not already done, any deliberate taking of cetaceans...".

[52] See, for example, the report by the U.N. Secretary-General (E/CN.17/1996/3) to the Commission for Sustainable Development's 4th Session which cites the creation of the Southern Ocean Sanctuary by the IWC as one of five international agreements considered to be achievements in implementing Chapter 17 of Agenda 21.

[53] De Boer, M.N., Eyre, L., Jenner, K.C.S., Jenner, M-N., Keith, S.G., McCabe, K.A., Parsons, E.C.M., Rosenbaum, H.C., Rudolph, P. and Simmonds M. 2001. Cetaceans in the Indian Ocean Sanctuary: a preliminary review. Doc. IWC SC/53/O6.

[54] Hoyt, E. 2001. *Whale watching 2001: Worldwide tourism numbers, expenditures, and expanding socio-economic benefits.* IFAW, Yarmouth Port, MA, USA. pp. 1-158.

[55] For more on this and whale watching generally, see Corkeron, Chapter 11.

[56] Birnie, P. 1997. Are Twentieth-Century Marine Conservation Conventions Adaptable to Twenty-First Century Goals and Principles? Parts 1 and 2. The International Journal of Marine and Coastal Law 12(3): 307-339.

[57] The Berlin Initiative on strengthening the conservation agenda of the International Whaling Commission. Annual Report of the IWC 2003, pp. 58-77.

There once was a Briton called Holt
Whose studies gave whalers a jolt
He said "MSY
Is pie in the sky
And if you don't believe that you're a dolt".

William de la Mare 2004

Chapter 8

Ivory Tower Sustainability: An Examination of the Ivory Trade

Ashok Kumar & Vivek Menon

INTRODUCTION

The ivory trade is a centuries old business that has created wealth for a limited few, using a slowly renewing, yet finite,[1] natural resource that is obtained most often from living beings. The roots of the ivory trade have very deep cultural and traditional moorings and, both in the 50 nations of Africa and Asia that harbour elephants, and in the dozen or so cultures of Asia that constitute the principal markets, these traditions have played a significant role in determining the context in which such trade has taken place. The sustainability of the trade has been a topic of major debate and the fate of the three extant species of elephants, and of those whose business it is to trade in ivory, continues to hinge around this debate and how it is resolved.

This chapter attempts to demonstrate using available data and a theoretical framework that a sustainable ivory trade is an unattainable abstract: a chimera. There are several arguments that try to legitimize the ivory trade as being good for elephants, good for resolving conflicts between elephants and humans, good for the development of elephant range states, and good for preserving tradition. However, ecological, economic and ethical rationales fail to justify the international ivory trade. To show this, we first provide a short introduction both to the international ivory trade and to the notion of sustainability.[2] Then we discuss the various issues linking the two.

BACKGROUND TO THE IVORY TRADE AND THE POACHING OF ELEPHANTS

Trade in ivory is part of ancient history. Aurignacian Venuses (rotund ivory figurines) were carved 27,000 years ago and ivory diptychs were common in church art from the 4th century AD.[3] Ivory craft in India is at least 4000 years old.[4] Whereas in Asia, elephants were essentially valued to be captured alive and tamed as war machines and later to extract timber from forests, in Africa the lure of elephant ivory (or white gold) resulted in the killing of elephants for the ivory trade. From the ivory carvings and *objets d'art* that were the initial uses of ivory, in the late 18th and 19th centuries it became a raw material for mass-produced items like billiard balls, piano keys and chopsticks. In the last 200 years or so, such uses have declined, giving way to the far eastern markets of Asia. The *hanko* (or *chop*, in Chinese),[5] *jitsuin*,[6] and the *netsuke*,[7] and a few other assorted items that are both mass-produced and crafted individually, are now the major uses of ivory. Trade in ivory figurines continues nationally in many African and Asian countries,[8] adding to the volume of ivory currently in global trade. No estimates are available on total volumes of such trade and there are varying reports on the decline of this industry in Japan, and the concurrent rise of demand in China. What can be said with certainty is that a large volume of illegal ivory continues to be traded around the world. Between 2000 and 2002 alone it was estimated that nearly 50,000 kg of ivory were seized, thereby representing the detected illegal contraband.[9] A minimum of 3,641 elephants was reported poached during the same period.[10]

Milner-Gulland and Beddington provide a comprehensive analysis of the effects of hunting on elephant population dynamics over time.[11] They show that the colonial period in Africa was one of steady decline in elephant numbers (the 19th century showed a rate of decrease of 2% per annum) while the second wave of hunting fuelled by the East Asian ivory trade was far more dramatic.

Today, the international trade of ivory is inextricably linked with the *Convention on International Trade in Endangered Species of Wild Fauna and Flora* (CITES). This relationship began in July 1975 when CITES banned the trade in Asian elephant (*Elephas maximus*) ivory. It took fourteen and a half years for the African elephant (*Loxodonta africana*) to join the Asian elephant on Appendix I of the Convention, which it did in January 1990, following the 1989 Conference of the Parties (CoP) in Lausanne. For the next nine years there was a complete ban on international trade in all ivory. In April 1999, 50 tonnes of ivory were sold by Zimbabwe, Botswana and Namibia to Japan following a decision taken at the 1997 CITES CoP in Harare to have a one-off sale of ivory and the down-listing of the elephant populations in these countries to Appendix II. In April 2000, at the CoP in Nairobi, these countries were joined by South Africa (but with a zero quota), although no additional ivory sales were approved.

At the 2002 CoP in Chile, three African elephant range states, namely Botswana, Namibia and South Africa, were given permission to conduct one-off sales of their specified ivory stockpiles (60 tons in total), but not before May 2004. The delay was intended to provide the CITES Secretariat with time to assess whether: a) national legislation and trade controls in importing countries were adequate and, b) the monitoring system known as MIKE (Monitoring Illegal Killing of Elephants) provided reliable baseline information. The Parties to CITES rejected similar proposals from Zimbabwe and Zambia, and Kenya and India withdrew their joint proposal to put all elephant populations back onto Appendix I.

WHAT CONSTITUTES SUSTAINABILITY?

The word "sustainable" has clear dictionary meanings but a thousand interpretations based on the sort of sustainability desired, and the measures employed in gauging whether it (however defined) has been achieved.[12] Of particular importance to the present discussion is the observation that the achievement of economic sustainability is very different from "ensuring that natural systems are maintained in such a state that they remain functional".[13] Given this, any discussion of sustainability of the ivory trade has to be dealt with under two broad headings: Is it ecologically sustainable for elephants and their habitats? And is it an economically sustainable activity? To these two, we would like to introduce a third element and ask: Is the ivory trade – whether sustainable or not – ethical?

An Ecological Perspective

Ivory is a slowly renewing, finite resource extracted from endangered species.

The ivory trade depends on the availability of raw material. In the recent past, legal trade has only been proposed for the ivory of the African savannah elephant (*Loxodonta africana*). The recently described African forest elephant (*Loxodonta cyclotis*) and the Asian elephant are both more endangered than *L. africana* and both will necessarily be threatened (the degree of threat being debatable) by an increase in trade in look-alike ivory. The ivory from both these species (known as hard ivory) is preferred by the largest global consumer of ivory, Japan. The potential burden of responsibility that international trade would put on more than three quarters of the 50 elephant range countries that harbour these endangered species, countries that can ill afford to put more resources aside for protection, is enormous.

African elephants are currently estimated to number 400,000 – 660,000 animals in the wild,[14] whereas there are only an estimated 35,000 - 45,000[15] Asian elephants remaining. The latter was placed in the Endangered category of the World Conservation Union – IUCN's Red Data Book in 1996[16] and has remained there ever since. A recent study into the ivory trade of Asia lists as one of its key findings:

> Wild elephant populations in Cambodia, Laos and Vietnam from 1988 to 2000 have declined by over 80% in total from an estimated 6250 to 1510, largely due to the trade in ivory and other elephant products. Myanmar, with the largest wild elephant populations left of the countries visited (estimated to be 4280) has suffered a net loss of over a thousand elephants since 1990.[17]

The same report examined habitat loss as a possible reason for population declines and concludes that:

> There is still enough natural habitat left in each of the South and Southeast Asian countries surveyed (and those not surveyed) to support existing, and in many cases, larger elephant populations.[18] Therefore, the most immediate threat to elephant populations in these countries is

> poaching for ivory and other elephant products and not habitat loss.[19]

In India, poaching is threatening elephant populations by skewing sex ratios. To give one example, the former Chair of the Asian elephant specialist group estimates that the number of tuskers of breeding age in an overall population of 25,000–27,000 Asian elephants in India is just 1,200.[20]

The recently described forest elephants of central and West Africa pose even more of a problem. As much less information is available on these populations, their levels of threat are poorly understood except in the general consensus that they are more endangered than either the southern or eastern populations of elephants on the African continent. If, as was believed until a few years ago, all African elephants were of one species, the effect of removing some of them would not have raised as much conservation concern as when this could impact – by heightened poaching – more endangered populations that are now considered as belonging to a separate species.

Furthermore, recent studies have raised questions on whether the West African populations should be considered yet another species.[21] These studies recognize the genetic and ecological distinctions among the three African populations and, although they stop short of proposing a new species, they clearly state that West African elephants require separate conservation management strategies.

Just as the separation of *L. cyclotis* from *L. africana* by genetic studies has raised several management and conservation questions, the theory of a single species of Asian elephant is also being re-examined. Fernando *et al.* have questioned the validity of including the thousand or so elephants in Borneo as part of *E. maximus*[22] and presented the first evidence that may eventually split the species into two. This would mean that the ivory trade would impact the Indonesian elephants even more for, as a species, they have an extremely restricted range and small population size.

To date, IUCN has yet to categorize the threats faced by the newly recognized African forest elephants. The IUCN Red List for 2004[23] classifies both the African savannah elephant and the Asian elephant as Endangered. Following the IUCN listing criteria, the savannah elephant was listed because of a population reduction in the form of an observed, estimated, inferred or suspected reduction of at least 50% over the last 10 years or three generations, whichever is longer, based on an index of abundance appropriate for the taxon. The Asian elephant was classified as Endangered for the same reasons and also because of a decline in area of occupancy, the extent of occurrence and/or quality of habitat and actual or potential levels of exploitation.

Estimates of elephant numbers in southern Africa are the largest (246,000–300,000); eastern Africa follows (118,000–163,000); central Africa has the most uncertain estimate (16,500–196,000); whereas West Africa has the lowest numbers (5,500–13,200). This degree of variability, combined with massive trade involving even the smallest populations, makes any off-take a risky proposition. A recent report from the region with the smallest population, i.e. West Africa, found that there was more illegal ivory in three countries than there were live elephants. The survey, in Nigeria, Ivory Coast and Senegal, unearthed 4000 kg of ivory (or the ivory of more than 760 elephants) on display for sale, considerably more than the 543 elephants that are supposed to inhabit these countries.[24]

Under the current situation, application of the precautionary approach would render illegitimate any international trade that has the potential to further threaten such endangered populations.

Elephant biology does not support a traditional demand-supply model.

"Elephants are not beetles" was part of the title of a seminal paper by Poole and Thomsen[25] that was, in part, responsible for the ivory trade ban at the CITES CoP in Lausanne. And indeed they are not. Elephants are slow growing, slow breeding, long-lived, and socially complex animals that are strongly sexually dimorphic. None of these life-history traits support a traditional demand-supply model that allows the elephant to be a "sustainable" source of ivory for a growing or even steady demand from the Far East. In 1989, Poole and Thomsen argued that given that female reproduction takes place between the ages of 10 and 20,[26] and male reproduction between 25 and 30,[27] and given the particular pattern of off-take of the ivory trade, the exploitation of elephant populations has been biologically unsustainable since 1950. They calculated at the time that even an off-take of up to 4% would drive the species to extinction and that even if the trade were banned, it would take 20-30 years for heavily poached populations to recover. That was in 1989 and, as noted earlier, the respite for some of these elephant populations, courtesy of a complete ban on international ivory trade, lasted only nine and a half years.

Also, in *Elephas*, only the males are tusked, whereas in *Loxodonta*, both sexes have tusks, the weight of the male tusk reaching six times that of the female.[28] In both cases, therefore, older males are the targets of poachers. This selective hunting of large tuskers affects the population in a number of ways. One of these ways is the effect on reproductive rates, as recent studies[27] demonstrate that females prefer males with longer tusks, possibly due to the fact that longer tusks indicate lower parasite levels and therefore healthier mates. An off-take that involves the healthier and fitter males of a population also reverses the "selection of the fittest" theory and therefore can be thought of as unnatural selection.[28] In Asia, hunting for male elephants leads to highly skewed sex ratios, such as the 1:100 (male:female) ratio observed in some parks in Southern India;[29] anything beyond 1:5 is a cause for worry.[30]

These features of megaherbivore biology make the elephant an extremely unsuitable candidate to be a supplier of raw material for commercial trade. If ecological sustainability requires ecological systems to remain functional despite an off-take, it is difficult to achieve this in a species with such biological characteristics.

Carrying Capacity and Culling

The issues of carrying capacity, and elephant management by culling, are neither directly related to the issue of opening international trade in ivory nor firmly grounded in sound science. Most proponents for the resumption of the ivory trade link the issue of re-opening of the trade with that of carrying capacity. "There are too many elephants now", is a common refrain, and "there is a need to control populations by lethal culling" that would generate legal ivory for the international trade. It is very important, however, to examine both parts of this statement. If one begins with the latter, it may be interpreted that the ivory generated by culling operations is directly linked to international trade. Of course, this is far from being true. While control of elephant populations and management techniques, including controversial ones such as lethal culling, are purely national prerogatives and depend on the mandate that the people of a country give their government, the opening of international trade goes beyond national mandates. It is completely possible for countries to cull in order to manage their elephants without putting any ivory into the international trade, thereby dealing with a national problem without endangering other elephant populations and respecting the views of other nation states. The argument that if there are too many elephants they must be culled (i.e. killed), that culling generates a resource (ivory) that must be used, and that this use is only maximized by international trade, is both vexatious and profit oriented. It can just as easily be argued that the number of elephants that constitute "too many" is a value judgment, that culling is not the only means of managing numbers of elephants, and that, even if culled, the ivory thus generated need not be put into a trade that further enhances poaching and illegal trade. Even the Zambian proposal to the CITES CoP 2002 in Santiago, Chile admitted that "the impact on habitats caused by elephants while feeding and moving depends on the observer's value system and the management goals."[31]

Even if the issues of carrying capacity and culling need to be addressed it is clear that they are very contentious. Several reports and papers, however, provide the logic that de-links culling from carrying capacity.[32] Most of the debate revolves around four fundamental arguments:

a) Today, the idea of stable equilibria in ecology has been abandoned by most scientists in favour of interactive and dynamic equilibrium theories[33] that do not dictate human interference in order to bring back an existing state to that of a "stable state". For example, if elephants are destroying a woodland, management practices to halt it because elephants have exceeded their carrying capacity are not essentially prescribed, as this is considered part of a natural change.

b) There is a tendency among traditionalists to view increases in elephant numbers, or range extensions, as exceeding a prescribed carrying capacity when shorter term analyses are done. However, this view can be changed by studying longer data sets and it is often seen that such increases correct themselves naturally. For example, Gillson and Lindsay[34] point out how, in Botswana's proposal to the CITES CoP in Harare, the statement that the geographical range of elephants is expanding to areas such as western Okavango Delta is a misrepresentation of historical fact. The same areas had elephant populations in the late 1890s that were extirpated and the elephants moving back a century later were only re-colonizing a part of their former range. Management authorities tend, however, to view elephant distribution and population dynamics in the short span of human generations that tend to rely on published data rarely exceeding the past few decades.

c) Culling also introduces the dilemma of the elephant being a keystone species. A reduction in elephant numbers would necessarily impact other species as the elephant is "one upon which many other species in an ecosystem depend, and the loss of which could cause a cascade of local extinctions".[35] For example, a recent study shows that *Balanites wilsoniana*, a forest canopy tree in Africa, has no effective seed dispersers besides elephants,[36] making it a potential candidate for local extinction, were the elephant to disappear.

d) It is also now felt that the economic carrying capacity model[37] is based on a principle of maximizing productivity of individual animals for economic gain (borrowed almost completely from the animal husbandry and livestock industry). The moment those goals and values are changed, the very definition of "carrying capacity" can also change.

The arguments are best summarized by Gillson and Lindsay who conclude that "culling elephants, particularly as part of a broad ecosystem-wide population control strategy, is a very blunt, wasteful and almost certainly counter-productive instrument unlikely to maintain or increase biodiversity in the longer term".[38]

An Economic and Social Perspective

The economic value of elephants can be examined in two ways. Firstly, it can be examined by calculating economic values of elephants and examining the kinds of uses that provide the highest values. Secondly, it can be measured by computing social values that have indirect economic ramifications. The first three parts of the following discussion deal with the former situation, the last two with the latter.

Non-consumptive utilization has a better chance of sustainability.

The economics of creating large parks and utilizing the elephant wealth of a nation in a non-consumptive way through tourism, as demonstrated by countries like Kenya, can generate benefits in a far more sustainable manner than following a consumptive utilization model. Consider, for example, the amount of money that can be raised, the employment versus handout possibilities for local communities, and the longevity of each individual elephant, which can contribute to the nation's economy throughout its lifetime.[39] Elephant parks also provide the country with biodiversity catchments, water catchments and climatic moderators.

The total economic value of elephants can be calculated as the sum of direct use, indirect use and non-use values. Geach[40] has studied the economic value of elephants to the Eastern Cape region of southern Africa and determined that in this region direct use values are equal to the non-consumptive use or tourism-related revenues from elephants. Geach lists indirect uses as being related to ecological and ecosystem services provided by elephants, including their contributions to maintaining biological diversity. He also lists among non-use values the donations that come from non-elephant range countries or organizations based in such countries due purely to global concerns for the species. This study, from a part of Africa where consumptive utilization is the nationally accepted model, does not even compute the value of this form of use as a direct value of elephants since "availability of (and demand for) elephant products such as meat, hides and ivory are low". Kiss, for example, pointed out that an elephant as a trophy would yield Zimbabwe $21,900, equivalent to US$4.[41] It is also interesting to note that Geach does not compute non-use values such as spiritual, existence and bequest values that would most definitely play a role in parts of Asia. He concludes that "the use it or lose it" approach to wildlife conservation[42] in favour today is open to question, since this approach is inadequate given the fact that markets are imperfect and many values of a species and ecosystems go uncaptured.

Watts[43] provides a pithy analysis of the tourism versus trade debate. The Zimbabwean government ivory stockpile was valued in the 1997 proposal to the CITES CoP in Harare at US$3.5 million. Watts compares this to the Department of National Parks and Wildlife Management's annual budget, estimated at US$12 million. She calculates that ivory sales would support the Department only for a maximum of 15 weeks. It must be remembered, however, that the ivory in question was stockpiled over 7 years. Zimbabwe's estimate of annual earnings from the ivory trade, were it to be legalized, was at that stage US$500,000, while the government earned more than US$200 million from tourism, albeit not from elephant or wildlife tourism alone. There have been past estimates pegging the value of a live elephant in Kenya at over US$1 million given its estimated life span and the services it could render the wildlife tourism industry.[44] Most of these observations point to non-consumptive utilization models, such as tourism, to provide a more economically (and ecologically) sustainable use of elephants than the ivory trade.

Any money generated from the ivory trade and ploughed back into conservation is not evenly distributed and is too insignificant to have any real impact.

As shown by the Zimbabwean example above, the revenues generated by the ivory trade are miniscule even for developing country economies. This is also borne by evidence that ivory for such trade would essentially come only from natural mortality[45] and the limited culling of certain populations. If this were subtracted from the amount of resources that need to be ploughed into culling,[46] or used in the collection of natural death ivory, the anticipated profit would be whittled down even more. If this profit were further depleted by expensive international monitoring mechanisms such as Monitoring of the Illegal Killing of Elephants (MIKE) and the Elephant Trade Information System (ETIS), which are mandated by the international community as a precursor to legal ivory trade – the costs of which run into millions of U.S. dollars – the profit that can be arguably ploughed back into conservation would be too small to have any real impact at national or elephant population levels. Many writings point to both game viewing tourism and sport hunting as generating larger revenues than ivory sales.[47]

The small amounts of money derived from the trade in ivory will only be of some benefit if directly used at very local, community levels for the conservation of the species. However, there has never been any hard data to support the claim that the money made from legal international ivory trade and ploughed back into conservation has benefited either local communities through an equitable and transparent sharing mechanism, or even the target species, elephants. This, Gillson and Lindsay[48] argue, is largely due to the mechanism of revenue disbursement from ivory sales, where national governments are the principal channel through which money has to be routed. Direct benefits to local communities, such as that which occurs with tourism, is not possible in this case and, as in most examples of revenue disbursement around the world, the communities that live closest to the elephants (read furthest away from the capital) are the ones that are least benefited. The economic sustainability of an activity that generates small amounts of money, with a relatively poor capacity to be evenly distributed, is low.

A shift in demand will render the ivory trade unsustainable in the long run.

What about the traditional consumers of ivory? Do they need the ivory? It is known that nearly three quarters of all ivory that reaches Japan is used for making signature seals or *hankos*.[49] *Hankos* were traditionally made of wood and stone; the use of ivory is a more recent phenomenon, dating back not more than 200 years. Societies in their evolution drop certain traditions regularly and this non-essential use of elephant teeth is already considered old fashioned in the Far East. Many young Japanese and Chinese prefer to sign rather than use the seal and, for those who do not, a number of ivory alternatives are available.

In a detailed study of ivory markets in Japan, Sakamoto[50] reports that the volume of domestic sales by 59 members of the ivory importers association fell from 181.3 tons in 1989 to 82.5 tons in 1990 to 69.9 tons in 1991. Similarly an analysis of the fiscal-year transactions for 1996 and 1997 show that 64% of respondents felt that the volume of trade between manufacturer and wholesaler had decreased for the period, while 80% of wholesalers interviewed reported a decline in volume transacted between wholesalers and retailers.[51] Supporting this fact, the most recent study of the Japanese markets by Martin and Stiles[52] documents that the

> largest decrease in the ivory industry has been in the quantity of tusks used. From 1980 to 1985, Japanese used about 300 tonnes on average per year. In the late 1990s and in 2000 and 2001 the average had dropped to around 10-15 tonnes annually, a decline of at least 95%... By early 2002 the Japanese ivory dealers had come to terms with this low supply of tusks and had accepted that they could survive on a constant supply of 15 tonnes a year...

All this clearly indicates a downward spiral in one of the most important markets for ivory, which may somewhat explain why nearly 178.8 tonnes of ivory remained stockpiled with 200 traders in Japan in November 1996.[53] Despite the fact that 52.6% of the population of Osaka city surveyed said that they preferred ivory for *jitsuin*, or official name seals,[54] there is a general tendency for younger consumers to use other means of signing documents, such as by pen and ink, or using artificial *hankos*. Ivory *hankos* are neither an essential commodity nor do they embody Japanese culture, which used non-ivory *hankos* long before ivory became fashionable. The current persistence by the Japanese government in importing soft ivory from African savannah elephants, which is not preferred at all by the ivory carvers or users,[55] is more a statement of defiance to international pressure than a response to genuine national demand.

The costs of conflicts between humans and elephants

The costs of human-elephant conflicts are one of the largest burdens borne by the human community to maintain elephants on this Earth. And those costs will remain high as long as the symptoms continue to be addressed rather than the causes.

In Southern Africa, Geach[56] states that in Namibia and Botswana, the costs of land purchase, costs of fencing and other elephant proof mechanisms, staff salaries and associated costs related to conflict management, etc. are "one of the most important costs associated with elephants". Project Elephant, for example, run by the Government of India for *in-situ* conservation of elephants, had an annual budget of US$680,000 for habitat management and human-elephant conflict mitigation during 2003-2004.

Thus, it is critical that the root causes of human-elephant conflicts be solved to minimize costs and add to the net economic value that elephants provide. The argument that the international ivory trade is essential to control burgeoning populations of elephants is both illogical (as noted earlier, national elephant population control methods need not necessarily be linked to international trade) and unethical (to reduce conflict a number of land use strategies can be put in place without necessarily resorting to culling, which is the only way ivory trade can be legally sustained). If long-term holistic solutions were attempted, the need to cull would be decreased, and even this supply of legal ivory would necessarily die out. If more holistic solutions are not applied, a small amount of ivory will still enter the market legally from culled sources, but the human-elephant conflict scenario that is undoubtedly one of the key issues in elephant conservation across its range will not be addressed adequately.

Culling merely addresses the fact that elephants are damaging vegetation, human property and lives and, in response to this, a management decision is taken to cull that prescribes a threshold limit for the number of elephants that a landscape can accommodate. Such a decision does not address the threshold limits of an ecosystem for the number of human beings that can be accommodated. Nor does it address the core issues of conflict – why, in the first place does conflict take place? Why are elephants turning aggressors and eating crops or pulling up trees or taking human lives? In most part, be it explained by loss or fragmentation of habitat, loss of traditional movement corridors or preference of palatable agricultural crops to natural vegetation, it has to do with the anthropogenic effects to the landscape. It would thus seem rational that a management plan be drawn up to monitor, regulate and even limit such anthropogenic effects. Culling is a shunning of that rationale and a simplistic, unethical approach to resolving conflict. A large part of the rationale for the ivory trade is thus based on a shortsighted solution to a much more complex problem.

Enforcement issues

As long as the ivory trade continues, it will continue to put pressure on some elephant populations, thereby increasing the threats to their survival and the cost of their protection.

Given the cultural preference of the Japanese for hard ivory or *indo-khiba* and *togata* (traditionally believed to be ivory that is produced by African forest elephants and Asian elephants, respectively) over soft ivory or *shiromono* (ivory that is produced by African savannah elephants) and the difficulty faced by enforcement agencies in telling the two apart, the legalization of soft ivory consignments would result in an enforcement nightmare. It is clearly documented in several of our earlier reports[57] that *shiromono*, *indo-khiba* and *togata* are three different commodities as viewed by the Japanese trader. If vast quantities of soft ivory from savannah or southern African elephants were to flood the market, they would not lower the demand for hard ivory from western African and Asian elephants. This trade perception has also been clearly documented in two Japanese studies. Sakamoto[58] surveyed wholesalers and retailers and found that 100% of wholesalers and 75% of retailers could distinguish (using traditional visual means that are yet untested by modern science) between different forms of ivory. Nishihara records that,

> Recently, in Santiago, Chile, at the 12th CITES CoP, a Japanese ivory dealer said to me, We need the hard ivory that comes from forest elephants, not the soft ivory from southern Africa's savanna elephants, to make our products.[59]

The preference for hard ivory is also demonstrated by the arrest of a board member of the Tokyo Ivory Arts and Crafts Association in 2000 while trying to circulate smuggled hard ivory in Japan. Nishihara also documents that,

> Legal importation of soft ivory does not drive down the price of the preferred hard ivory, also stimulating a market for the contraband forest elephant product.[60]

Menon has shown through a series of investigations in Japan that the demand for hard ivory far exceeds that of soft ivory, even if the latter is legal and the former illegal. In fact illegal ivory has always surpassed legal ivory trade in volume. In the late 1980s it was estimated that as much as 90% of the 1000-odd tons of ivory that entered the global market was illegal. This puts an enormous economic burden on elephant range states to protect elephants from the activities of poachers. India's Project Elephant, for example, had an annual budget of US$320,000 for anti-poaching and anti-depredation in India during 2003-2004.

An Ethical Perspective

To the two concepts of ecological and economic sustainability, it is important to add a third dimension, that of ethics. It has been shown in the above discussion that both ecological and economic reasoning are at best debatable and probably unsustainable. All this should also be woven into an ethical fabric where discerning societies and nations of humankind must examine the ethics of revenue generation from killing "near-persons" such as elephants.

The raw material for the ivory trade comes largely from living, sentient beings.

Whether endangered or not, ivory comes from highly intelligent, social animals that are affected by death and are bound by close familial ties. This is demonstrated by a plethora of elephant studies.[61] In the most comprehensive documentation of elephant sociability and complexity, Poole has written,[62]

> ...relative to other mammals, including humans, elephants are unusually long-lived, exhibit a high degree of social complexity and show a fast rate of change in social setting. Their development includes social learning and behavioral innovation, both of which are manifested in the use and modification of rudimentary tools and in vocal learning. Elephants have high neo-

> cortical development, very good memory and are adept users of Machiavellian intelligence.[63] Mirror self-recognition by elephants indicates self-awareness and numerous observations suggest that elephants may have theory of mind and anticipatory planning capabilities that may include imagining future events, such as pain to themselves and others and possibly their own death. While many other species may rival elephants in one capability or another, there are few that, in totality, equal or surpass elephants in social and behavioral complexity.

While some of the ivory that enters trade comes from natural mortality, the amount is far overshadowed by that which is obtained from unnatural mortality. In Asia this is largely from poaching. Records from India – the country that best documents the poaching of Asian elephants – show that 36.4% of total elephant mortality in a five year span from 1997 to 2001 was natural, while 63.6% represented unnatural deaths.[64] Unnatural deaths include poaching, conflict-related deaths and electrocution. Poaching alone constituted 37.4% of all deaths, marginally more than natural deaths. In Africa, culling is an added means of procuring ivory.

Whether through poaching or culling, ivory sourced from non-natural mortalities originates from the killing of sentient individuals. In a recent paper on elephant personhood and memory, Varner concludes that elephants are "near-persons" based on biographical consciousness, Machiavellian intelligence and encephalization quotients among other traits.[65] He argues that although "person" is normally considered synonymous with human beings, that "among ethicists, the descriptive component usually refers to certain cognitive capacities which may or may not be unique to humans such as rationality, self-consciousness, or moral agency". This scientific yet philosophical rationale goes beyond religious, spiritual and nationalistic callings, which also have their own place in the debate. Douglas Chadwick states the ethical reasoning very simply:

> If a continuum exists between us and such beings in terms of anatomy, physiology, social behaviour and intelligence, it follows that there should be some continuum of moral standards.[66]

Such moral standards would most certainly abhor the conversion of a living elephant into its utilitarian parts. "From a utilization perspective, an elephant is worth the sum of its ivory, its hide, its mountain of meat and few other parts such as feet and tail hairs".[67] This reasoning would thus not support trade in the parts of a species that so closely resembles our own selves.

There is a strong ethical, religious and spiritual basis in many Asian countries that renders the ivory trade illegitimate.

In some Asian countries where elephants are worshipped or revered, killing elephants for trading in their parts is considered unethical. Even in most countries where ivory trade is prevalent there are long standing beliefs that ivory that is traded comes from dead elephant graveyards and does not require the killing of elephants. Menon documents a Japanese ivory carver of more than 40 years – upon realizing that elephants were shot at to gain ivory – worshipping a broken tusk with an accidental bullet embedded in it, by placing it on an altar meant for ancestor worship.[68]

Menon and Sakamoto brought together a collection of Asian philosophers, leaders and conservationists who argue from all corners of the continent that ethical prerogatives are important in determining elephant management.[69] In this collection of essays, the late Prince Sadruddin Agha Khan, a leader of the Shiite Ismaili Muslim community of West Asia pondered,

> What is the human perversity that condones the killing of animals merely to decorate our persons or surroundings with their remnants? We are aghast when so called 'backward' societies indulge in practices like human head-hunting for trophies. By what strange logic can we, then, justify the killing of magnificent animals merely for their tusks, horns or skins?

Maneka Gandhi, an Indian political leader gave an Indian perspective:

> India has most definitely framed its policies keeping ethics, morality and a certain spiritual aspiration of the people in mind, and sheer economics is not the only guiding principle of species conservation in the country. In our country, the species do not always have to pay to survive. True to its beliefs, India has always strongly stood for the precautionary principle at CITES, and seen that flagship species like the elephant are not treated as mere commodities.

Queen Noor of Jordan injected pragmatism into her ethical sensibilities when she wrote:

> ...we must make a distinction between those commodities that are essential for

> providing food, shelter and health and those meant only for luxury and the display of wealth – ivory, tiger skin, shahtoosh shawls, etc. Civilized society must condemn theses senseless killings whose only purpose is to feed the vanity of the few, rather than the genuine hunger of the many.

There is definitely a strong feeling in many quarters, communities and societies of Asia, therefore, that killing elephants is neither ethically nor spiritually acceptable.

CONCLUSION

Collectively, the preceding arguments lead us to the conclusion that a sustainable ivory trade is an unattainable abstract: a chimera. Having used an approach that includes ecological, economic and ethical perspectives, our examination of the sustainability of the trade in elephant ivory is a holistic one. In most cases, there already exist a plethora of writings and academic findings that vindicates our conclusion. In the more nascent field of ethical reasoning, more work needs to be done but sufficient preliminary studies exist to support a non-consumptive approach to elephant conservation. Elephants, by virtue of nature and their biology, simply cannot be a steady supplier of ivory over any great period of time. Harvesting ivory from a single elephant provides only short-term benefit, while the living animal not only generates revenue through non-consumptive uses like tourism and usage in culture, but also helps to maintain ecological relationships in the environs that it inhabits. This ecological benefit could also be translated in terms of the economic benefits that an elephant can yield.

While neither economic nor ecological rationales in our opinion support the consumptive use of elephants, we believe that ethical reasoning alone should be enough to mandate rethinking the consumptive use of elephants. This conclusion stems from our own convictions, and the Asian cultures that we hark from, and points to the need for a more progressive conservation movement that takes into account the welfare of individual animals alongside the existence of their population or race.

NOTES AND SOURCES

[1] While it may be more traditional to refer to elephants as "renewable resources" we consider all individual living beings, and the populations to which they belong, to be "finite" (as opposed to "infinite", which clearly they are not).

[2] See Holt, Chapter 4, for a detailed discussion of the "notion" of sustainability.

[3] Redmond, I. 1993. *Elephant.* Dorling Kindersely Books. London, UK.

[4] Menon, V. 2002. *Tusker: The Story of the Asian Elephant.* Penguin Books India, New Delhi, India.

[5] A *hanko* (or *chop*) is a name seal, customarily used in Japan instead of a handwritten signature when conducting business or other affairs. *Hankos* may be used by both businesses and individuals.

[6] "*Jitsuin,*" or "*true seals*" are registered with the local authorities and used on important legal documents.

[7] A small, ornamental toggle, often elaborately carved, that was used to attach purses and other items to the sash of traditional Japanese male garments.

[8] Menon, V. and A. Kumar. 1999. Signed and Sealed. Asian Elephant Conservation Centre and Wildlife Protection Society of India; Martin, E. and D. Stiles. 2002. The South and South East Asian Ivory Markets. Save the Elephants, Nairobi, Kenya; Martin, E. and D. Stiles. 2003. The Ivory Markets of East Asia. Save the Elephants, Nairobi, Kenya.

[9] Born Free Foundation. 2002. A Global Problem: Elephant poaching and ivory seizure data 2000-2002. Born Free Foundation, West Sussex, U.K.

[10] Ibid.

[11] E.J.Milner-Gulland and J.R. Beddington. 1994. The relative effects of hunting and habitat destruction on elephant population dynamics over time. Pachyderm 17:75-90.

[12] For a discussion, see Lavigne, Chapter 1.

[13] Lavigne, D.M. 2002. Ecological footprints, doublespeak, and the evolution of the Machiavellian mind. In W. Chesworth, M.R. Moss, and V.G. Thomas (Eds.). *Sustainable Development: Mandate or Mantra.* The Kenneth Hammond Lectures on Environment, Energy and Resources 2001 Series. Faculty of Environmental Sciences, University of Guelph, Guelph, Canada; also see Lavigne, Chapter 1.

[14] Blanc, J.J, C.R. Thouless, J.A. Hart, H.T. Dublin, I. Douglas-Hamilton, G.C. Craig and R.F.W. Barnes. *African Elephant Status Report 2002: An Update from the African Elephant Database.* IUCN/SSC African Elephant Specialist Group. IUCN, Gland, Switzerland and Cambridge, UK. Available at http://www.iucn.org/afesg/aed/aesr2002.html.

[15] Sukumar, R. 1998. The Asian Elephant: Priority Populations and Projects for Conservation. Report to WWF-USA by Asian Elephant Conservation Centre.

[16] IUCN. 1996. *1996 Red List of Threatened Animals.* IUCN, Gland, Switzerland.

[17] Martin and Stiles 2002.

[18] Kempf, E. and C. Santiapillai. 2000. Asian elephants in the wild. 2000 – A WWF Species Status Report. WWF International, Gland, Switzerland, 32 pp.

[19] Martin and Stiles 2002.

[20] Menon, V., R. Sukumar, and A. Kumar. 1997. A God in Distress: Threats of Poaching and the Ivory Trade to the Asian Elephant in India. Asian Elephant Conservation Centre and WPSI.

[21] Woodruff, D.S. 2001. Declines of biomes and biotas and the future of evolution. Proc. Nat. Acad. Sci. USA. 98: 5471-5476; Eggert, L.S., A.R. Caylor, and D.S. Woodruff. 2002. The evolution and phylogeny of the African elephant inferred from mitochondrial DNA sequence and nuclear

microsatellite markers. Proc. R. Soc. Lond. Biological Sciences. 269(1504): 1993-2006.

[22] Fernando, P., T.N.C. Vidya, J. Payne, M. Stuewe, G. Davison, R.J. Alfred, P. Andau, E. Bosi, A. Kilbourn, and D.J. Melnick, 2003. DNA Analysis Indicates That Asian Elephants Are Native to Borneo and Are Therefore a High Priority for Conservation. PLoS Biology 1: 1-12.

[23] IUCN 2004. *2004 IUCN Red List of Threatened Species.* Available at http://www.redlist.org.

[24] Courouble, M., F. Hurst, and T. Milliken. 2003. More Ivory than Elephants: domestic ivory markets in three West African countries. TRAFFIC International, Cambridge, United Kingdom.

[25] Poole, J. H. and J.B. Thomsen. 1989. Elephants are Not Beetles: Implications of the Ivory Trade for the Survival of the African Elephant. Oryx 23(4): 188-198.

[26] Moss, C. J. Population Dynamics of the Amboseli Elephant Population, an unpublished manuscript cited in Poole and Thomsen 1989.

[27] Poole, J. H. 1989. The Effects of Poaching on the Age Structure and Social and Reproductive Patterns of Selected East African Elephant Populations. Final Report to the African Wildlife Foundation.

[28] Parker, S.P. 1979. *A Guide to Living Mammals.* McGraw Hill, New York, NY.

[29] Ramakrishnan, U., J.A. Santosh, U. Ramakrishnan, and R. Sukumar. 1998. The population and conservation status of Asian elephants in the Periyar Tiger Reserve. Current Science 74(2): 110-113.

[30] Menon 2002.

[31] Anonymous. 2002. CITES CoP 12 Proposal 9. Transfer of the Zambian population from Appendix I to Appendix II for the purpose of allowing: a) trade in raw ivory under a quota of 17,000 kg of whole tusks owned by Zambia Wildlife Authority (ZAWA) obtained from management operations; and b) live sales under special circumstances.

[32] e.g. Gillson, L. and K. Lindsay. 2003. Ivory and ecology – changing perspectives on elephant management and the international trade in ivory. Environmental Science and Policy 6: 411-419.

[33] McNaughton, S. J., R.W. Ruess, and S.W. Seagle. 1988. Large mammals and process dynamics in African ecosystems. Bioscience 38: 794-800; Pimm, S. L. 1991. *The Balance of Nature? Ecological Issues in the Conservation of Species and Communities.* University of Chicago Press, Chicago and London.

[34] Gillson and Lindsay 2003.

[35] Calow, P. 1999. *Blackwell's Concise Encyclopedia of Ecology.* Blackwell Science Ltd. Oxford, U.K.

[36] Cochrane, E.P. 2003. The need to be eaten: *Balanites wilsonia* with and without elephant seed-dispersal. J. Trop. Ecol. 19: 579-589.

[37] Macnab, J. 1985. Carrying capacity and related slippery shibboleths. Wildlife Society Bulletin 13: 403-410.

[38] Ibid.

[39] For more on ecotourism, see Mugisha and Ajarova, Chapter 10, and Corkeron, Chapter 11.

[40] Geach, B. 2002. The economic value of elephants – with particular reference to the Eastern Cape. In Kerly, G., S. Wilson, and A. Massey (Eds.). Proceeding of workshop on "Elephant Conservation and Management in the Eastern Cape". Zoology Department, University of Port Elizabeth.

[41] Kiss, A. (Ed.). 1990. Living with Wildlife: Wildlife Resource Management with Local Participation in Africa. The World Bank, Washington DC., USA.

[42] For more on the "use it or lose it" philosophy, see Lavigne, Chapter 1.

[43] Watts, S. 1997. Elephants paying their way: Tourism vs Ivory Trade. Proc. African Elephant Conference. EIA. Johannesburg, SA.

[44] Cited in Watts 1997.

[45] WTI unpublished data.

[46] Gillson and Lindsay 2003.

[47] e.g. Bond, I. 1994. The importance of sport-hunted elephants to campfire in Zimbabwe. TRAFFIC Bull. 14: 117 – 119; Child, B., S. Ward, and T. Tavengwa. 1997. Zimbabwe's CAMPFIRE Programme: Natural Resource Management by the People. IUCN-ROSA Environmental Issues Series No. 2, IUCN-ROSA, Harare; Gillson and Lindsay 2003.

[48] Gillson and Lindsay 2003.

[49] Japan Wildlife Conservation Society. 1998. Results of the Ivory Hanko/ Inzai Market Survey. Unpublished Report of Japan Wildlife Conservation Society, Tokyo, Japan. Also see endnote 5.

[50] Sakamoto, M. 1999. Analysis of the amended management system of domestic ivory trade in Japan. Japan Wildlife Conservation Society, Tokyo, Japan.

[51] Ibid.

[52] Martin and Stiles 2003.

[53] Menon 2002.

[54] See endnote 6.

[55] Menon 2002; Sakamoto, M. 1998. Analysis of the Amended Management System of Domestic Ivory Trade in Japan. Unpublished Report. Japan Wildlife Conservation Society, Tokyo, Japan; Sakamoto 1999; Nishihara 2003. What's wrong with selling southern African ivory to Japan? Wildlife Conservation Society Magazine, December.

[56] Geach 2002.

[57] Menon, V., R. Sukumar, and A. Kumar. 1997. A God in Distress: Threats of Poaching and the Ivory Trade to the Asian Elephant in India. Asian Elephant Conservation Centre and Wildlife Protection Society of India; Menon and Kumar. 1999; Menon 2002.

[58] Sakamoto. 1998.

[59] Nishihara 2003.

[60] Ibid.

[61] Varner, G. 2003. Personhood, Memory and Elephant Management. Paper presented in "Never Forgetting: Elephants and Ethics". International Conference, March 19-20, 2003, Smithsonian National Zoological Park, USA.; Moss, C. J. and J.H. Poole. 1983. Relationships and Social Structure in African Elephants. In R.A. Hinde (Ed.). *Primates and Social Relationships, an Integrated Approach.*

Blackwell Scientific Publications, UK.

[62] Poole, J. H. 2003. Elephant Sociability and Complexity: The Scientific Evidence. Paper presented in "Never Forgetting: Elephants and Ethics". International Conference, 19-29 March, Smithsonian National Zoological Park, USA.

[63] For more on the topic of Machiavellian intelligence, see Whiten, A. and Byrne, R.W. (Eds.). Machiavellian Intelligence II. Extensions and Evaluations. Cambridge University Press, Cambridge UK.

[64] WTI unpublished data.

[65] Varner 2003.

[66] Chadwick, D. H. 1994. *The Fate of the Elephant.* Sierra Club Books, San Fransisco, CA.

[67] Price, S. 1997. Valuing elephants: the voice for conservation. Swara 20(3): 29-30.

[68] Menon 2002.

[69] Menon and Sakamoto (Eds.) 2002.

Chapter 9

The Bushmeat Trade in Africa: Conflict, Consensus, and Collaboration

Heather E. Eves

Sustainable development has been a guiding principle for both conservation and development professionals for more than two decades. It remains the focus of manifold projects, organizations, and international efforts.

The notion of sustainable development was formally introduced in the 1980 *World Conservation Strategy.*[1] The World Commission on Environment and Development, established by the UN General Assembly in 1983, subsequently popularized the concept. This commission reviewed the state of the environment and the issue of globalization, including strategies for "sustainable development" to the year 2000 and beyond. This landmark process and its resulting report, entitled *Our Common Future* and frequently referred to as the "Brundtland Report",[2] have framed a movement linking the environment with economic development activities for the past two decades.[3]

The term sustainable development has since become widely used by political leaders,[4] industry personnel, development and conservation professionals alike,[5] who frequently quote directly from *Our Common Future* for a definition of sustainable development:

> Humanity has the ability to make development sustainable to ensure that it meets the needs of the present without compromising the ability of future generations to meet their own needs.[6]

Unfortunately the caveats associated with that definition in the original document have long been forgotten. Without those caveats, such a definition sends inaccurate messages regarding the real limitations of the concept.[7] The remainder of the original definition is as follows:

> The concept of sustainable development does imply limits – not absolute limits but limitations imposed by the present state of technology and social organization on environmental resources and by the ability of the biosphere to absorb the effects of human activities...thus sustainable development can only be pursued if population size and growth are in harmony with the changing productive potential of the ecosystem.[8]

To see the ubiquitous application of this concept in present-day society, one need only to conduct a search on the Internet to find thousands of websites, projects, publications and programs that focus on sustainable development and the sustainable use of natural resources. "Sustainable use" of natural resources, especially wildlife, is another concept where both the definition and the application are elusive; it is nevertheless considered an important component of sustainable development.[9] The real challenge, however, is to find case studies where sustainable use, much less sustainable development, is actually happening, particularly where wildlife is concerned. Even the Convention on Biological Diversity (CBD) website section dealing with the sustainable use of biodiversity contains only 14 documents submitted by parties in the last three years as case studies of sustainable use. And many of these are not actual examples of sustainable use but rather steps that institutions are taking to investigate the sustainability of certain resource uses.[10]

The IUCN-SSC Sustainable Use Specialist Group (SUSG), established in 1991, provides a forum for experts and members to share information regarding sustainable use. The working group for Central Africa has recently reviewed literature on the sustainable use of natural resources and human livelihoods.[11] Its overall conclusion is that,

> With the advent of modern firearms, improved communication and transport, subsistence hunting has given way to anarchic exploitation of wildlife to supply the growing human population.[12]

It takes very little time in the field to realize that, despite considerable efforts by many dedicated individuals and organizations, sustainable use of wildlife is not being achieved across the majority of global landscapes. This is particularly evident in Africa where, in the last decade, the commercial exploitation and trade in wildlife for food may have surpassed habitat loss as the most rapid and pervasive cause of wildlife population declines. Thus, we are now faced with what is widely known as the "bushmeat crisis".

In recent years, concerns regarding the theory and practice of sustainable development and sustainable use of natural resources, including wildlife, have emerged.[13] Addressing the bushmeat crisis requires, therefore, an examination of the assumptions and realities of human development and conservation efforts that seek to achieve biological or ecological sustainability.

Recently I met with an executive from an oil company operating in Central Africa who remarked that, during a visit to Bioko Island in Equatorial Guinea, she was stunned at the silence in the forests she walked through. She stated she had thought that because she was in Africa she was sure to see lots of animals – or at least hear them. I explained to her that she had experienced what has been called "Empty Forest Syndrome".[14] Despite the existence of natural habitat for wildlife – the loss of which was previously identified as the leading threat to wildlife – the habitat is devoid of mammalian and even avian, reptilian, and amphibian life due to the over-harvesting of wildlife. Such empty forests are commonplace in West Africa following rapid logging and accompanying over-hunting decades ago, and now occur with increasing frequency across the Central African landscape.[15] Over-hunting leading to wild areas devoid of wildlife has now spread into East and Southern Africa as well.[16]

Although sustainable use has been a priority focus by both conservation and development organizations, wherever one looks in the literature, or in the field, the scientific evidence points to a trend that wildlife populations are declining and, in many areas, are under threat of local extinction. The evidence is mounting that for a significant portion of wildlife around the globe, despite thousands of conservation projects and millions of dollars over several decades, sustainable use is not being achieved, and sustainable development goals relying on this foundation are rarely being met.

The question is: why?

A QUESTION OF PRIORITIES

Despite years of development efforts seeking to balance the needs of people and nature, it seems that biodiversity conservation and wildlife are seldom adequately accounted for. Robinson's 1993 critique of *Caring for the Earth: A Strategy for Sustainable Living*[17] gives context to this problem and is prescient of what has since come to pass.[18]

Caring for the Earth, the explicit successor to the *World Conservation Strategy* and produced by the same respected conservation organizations, was written with the goal of guiding development and conservation policy, and recommended improving "the condition of the world's people" by developing a sustainable society and integrating conservation and development. Robinson acknowledges the many merits of the document, but also its major flaw:

> [*Caring for the Earth*] does not acknowledge that the goals of development are different from the goals of conservation, and it offers no general principles by which we might resolve conflicts and balance contradictory demands.[19]

The concept of sustainable use is formally linked to sustainable development in this report, but the fact that resource use can be sustainable at several levels of biological degradation is not addressed. Furthermore, the distinctions among ecological, economic, and sociopolitical aspects of sustainability, and their often competing values, are not clear.[20] In hindsight, Robinson's conclusion was a reliable predictor:

> Sustainable use is very appropriate in certain circumstances, but it is not appropriate in all. It will almost always lower biological diversity, whether one considers individual species or entire biological communities, and if sustainable use is our only goal, the world will be the poorer for it.[21]

Countless organizations and efforts are directly linked with sustainable use, though it is often challenging to identify how effective such programs are with specific regard to the sustainable use of wildlife. It is surprising to note that although sustainable use of biodiversity is such a widespread focus of international efforts, wildlife is frequently ignored as an item of concern. Bennett, for example, has identified the lack of focus on wildlife-specific concerns relating to sustainability criteria in timber certification efforts.[22] Despite years of extensive development of timber certification criteria,[23] not one formal certification process has explicitly incorporated the exploitation of wildlife as a key component of certification.

In a recent review I was asked to conduct concerning environmental response criteria for disaster relief protocols, I was surprised to learn that wildlife was not once considered in any of the documentation being reviewed. There were numerous references to fisheries and to soils, water, and timber, but never wildlife. It has been determined that during times of crisis (war, physical disaster, etc.) relief efforts frequently focus on providing supplemental carbohydrates to refugees. Bushmeat, however, is often a primary protein source for these same displaced people. Such omissions are, sadly, commonplace.[24]

At the 7th Conference of the Parties (CoP-7) of the *Convention on Biological Diversity* (CBD) in February 2004, the *Addis Ababa Principles and Guidelines for the Sustainable Use of Biodiversity* were officially adopted. The document includes 14 separate practical principles followed by detailed guidelines for achieving them.[25] Yet, not once in the official text of the entire document is wildlife (or fauna, or animals) mentioned. When reference is made specifically to the sustainable use of natural resources upon which rural communities are dependent for their livelihoods, only timber and fisheries are cited as examples. While wildlife may be implied in the term biodiversity, this example underscores the trend in policy and international development activities to disregard wildlife entirely.

It is unclear how, after many years of focus, including complex political processes aimed at conserving wildlife, the CBD document could still have such glaring omissions. Nonetheless, there are continued calls to promote the use of wildlife (the assumption, of course, being that it is used in a "sustainable" manner, although this is often not explicitly specified). Yet, the overwhelming scientific evidence suggests that we do not currently have sustainable use of wildlife in most areas of the globe nor do we have the capacity to achieve it.[26]

Robinson and Bennett identified three key criteria for sustainability to exist where direct harvesting of wildlife is concerned:

- Harvested populations cannot show a consistent decline in numbers;
- Harvested populations cannot be reduced to densities where they are vulnerable to local extinction; and
- Harvested populations cannot be reduced to densities where the ecological role of the species in the ecosystem is impaired.[27]

There are few, if any, cases today that demonstrate the sustainable harvest of terrestrial vertebrate wildlife in tropical forests in Asia, Africa, and Latin America.[28] Even hunting for subsistence purposes has been found to be unsustainable in many areas. Over-hunting in savannah ecosystems is also emerging as a significant problem and many species are being heavily impacted. Despite years of effort to conserve biodiversity, the goal of balancing demand with production has yet to be achieved.

THE BUSHMEAT TRADE

The bushmeat trade is an example of how far we still need to go in order to achieve the sustainable use of wildlife and wildlife habitat, an implicit component of sustainable development and biodiversity conservation. Bushmeat is most simply defined as meat that is derived from wildlife. The bushmeat trade is less easily defined, but may be characterized by one or more of the following:

- Illegal hunting methods (snares, poison, unregistered firearms),
- Prohibited species (endangered, threatened, or protected),
- Removals of wildlife from unauthorized areas,
- Unsustainable removals of wildlife, and
- Subsistence use and barter by rural and/or indigenous people, or commercial exploitation by professional hunters and traders supplying urban markets, as part of an informal or formal economy.

The bushmeat trade can be linked to a host of ecological, economic, and social issues including: human population growth, poverty, food security, nutrition, health, industry/infrastructure development, indigenous rights, land tenure and access to resources, governance, hunting technology, animal welfare, corporate ethics, land-use planning, and the life history traits of the species involved.[29] It is not a simple issue to address with so many different factors influencing the success or failure of any particular intervention or view.

Looking at the issue from a community perspective, the focus on neoclassical economics in the 1980s[30] with regard to wildlife conservation driven by the notion of sustainable development, particularly in Africa, created a shift toward integrated conservation and development projects (ICDPs). While the value and success of ICDPs has since been discussed in the literature,[31] what remains is a focus of understanding and consideration that local communities adjacent to areas abundant in wildlife – and the target of numerous conservation programs – must be included as part of the management decision-making process.[32] That this is only a recent focus of conservation activities is a misconception. Evidence suggests that although there are reports in the literature of inequitable land-use policies by colonial administrations, there are cases where efforts were made for decades to incorporate local communities in the wildlife management process.[33] The real question is not a matter of *if* communities should be part of decision-making but rather *how* can they most effectively be stakeholders in the decision-making process.

Community-based Conservation

Community-based conservation (CBC) has been strongly encouraged in the quest for sustainable use but unfortunately most communities do not have the technical, financial or social resources to maintain or support complex wildlife management systems. Hackel states,

> ...CBC programs can work to produce a better relationship between wildlife and people, but only a vast improvement in the lives of rural Africans will ultimately produce a more secure future for the continent's wildlife.[34]

This is an important perspective. While many individuals and groups are calling for a focus on livelihoods and the use of bushmeat to support such livelihood needs,[35] this view sacrifices long-term sustainability for the short-term needs of the current generation over the next few years. Furthermore, assuming a "vast improvement in the lives of rural Africans" is measurable in part by increased wealth, research elsewhere indicates we should expect meat consumption also to increase.[36] In the absence of livestock, fish and plant protein alternatives, per-capita bushmeat consumption will rise as incomes rise. This scenario demonstrates how promoting bushmeat use as a development solution is self-defeating, as the foundation of sustainable development is eroded by its own success.

An enormous challenge to wildlife conservationists is the inadequate pool of resources available for conservation activities. The development sector has access to significantly greater financial resources for their activities and with the obvious links between bushmeat and development, some efforts have emerged to gain support for biodiversity conservation by development agencies.[37] This, in theory, creates the resources necessary to do some useful conservation work. The problem is that the purpose of such funding is to meet human needs, as well it should. But when one is dealing with populations of wildlife whose productivity cannot sustainably meet growing human needs (as in the case of many species involved in the bushmeat crisis), a choice eventually has to be made. And that choice is seldom, if ever, favorable to the interests of wildlife.[38]

Unsustainable Use

Hunting wildlife for food can be a biologically sustainable activity, but usually only in the very few places where human population densities are low, perhaps less than 2 people/km^2,[39] and where the hunter only hunts certain species, and only to feed his family. In Africa, population growth rates of 2-3% per year are expected to double the human population within 25 to 35 years.[40] The vast majority of field studies investigating the bushmeat trade in Africa and globally have determined that current activities are unsustainable in the long term. Unless the future generations of new consumers have something else to eat, wildlife will likely be consumed until none is left.

The data backing up these statements are impressive. In Africa, Asia, and Latin America, case after case identifies hunting levels that exceed production levels, with local wildlife populations being driven to extinction.[41] In a review of 18 hunting sites in tropical forests around the globe, high harvest rates were associated with high levels of commercialization. However, at the most commercialized sites, harvest rates were low, reflecting the likely depletion or extirpation of species.[42] Estimates range from 31% to 100% of wildlife populations being hunted unsustainably in tropical forests.[43] Despite this evidence, wildlife use continues to be put forward as a component to addressing human livelihood needs.[44]

Analysis shows that distance from village is positively correlated with wildlife densities, i.e. the closer one gets to people, the lower the densities of wildlife.[45] With growing human populations and centers of activity, there are fewer and fewer places where wildlife can escape human hunting pressure. In the last 200 years the global human population has grown from 800 million to over 6 billion – a greater than 7-fold increase. Worldwide, human population is currently increasing by 81 million people per year, with 95% of new births occurring in developing countries.[46] While urbanization is a growing trend that usually reduces direct reliance on natural resources, urban centers in Africa lack services and infrastructure needed to make this transition.[47] The African urban growth rate of 4.4% outpaces overall population growth, but rather than reducing demand for bushmeat, urbanization is consolidating markets and increasing the profitability of the trade.

Urban growth notwithstanding, the majority of bushmeat is still consumed by people in rural areas. Bushmeat is a staple or emergency source of protein for many rural people. Over 30 million consumers in Central Africa alone need 2.5 million tonnes of meat annually, which is equivalent to 22 billion quarter-pound hamburgers.[48] How will already marginalized consumers satisfy their protein needs when wildlife is gone? Clearly, solutions will need to be developed to address the dramatic loss of wildlife that is seemingly inevitable. That being the case, one wonders why alternatives can't be developed now to avoid the catastrophic losses in the first place?

Understanding and Addressing the Bushmeat Crisis

Wildlife has been a food resource for millennia, but the scale of bushmeat consumption in West and Central Africa only reached crisis proportions in the 1980s and 1990s. Now, the geographic center of the African bushmeat crisis is a moving target. At the dawn of the 21st century, most large mammal populations in West Africa had been extirpated, while the bushmeat trade continued to be a growing concern in East and Southern Africa. Today, Central African countries remain at the center of the crisis, with high rates of wildlife exploitation, pervasive rural poverty, and the least capacity to address either.

Why does this crisis exist? The primary driving forces are uncontrolled access to wildlife, new hunting technologies, rising demand, lack of economic alternatives, absence of substitutes, lack of good governance, and minimal capacity to enforce laws. These factors must be addressed simultaneously and require involvement at all levels, from local to international. However, the resources required to enable such engagement are simply not currently available. That said, research and experimentation in the past five to ten years has substantially developed a collective understanding of each factor, and ideas for addressing each.

The logging industry expanded dramatically in Central Africa in the 1980s and is one of the driving forces behind the expansion of the bushmeat trade.[49] Logging transportation infrastructure opens access to forests, and reduces transport costs for hunters and traders. Industry vehicles and an extensive road network turn what may once have been a 3-4 day hunting trip on foot into an eight-hour excursion, with far greater capacity to carry out carcasses.[50] Furthermore, logging employees can afford to buy bushmeat in large quantities, making the village complex of personnel and their families thriving markets in their own right. Despite the predictions of the ICDP approach and the "wildlife must pay for itself" school of thought, the opposite impact on natural resources, particularly wildlife, occurs in many areas when a slight improvement in livelihoods is created through increased incomes and higher standards of living for workers in the timber industry. Logging companies and others in the business of natural resource exploitation, despite providing employment and social services exceeding host government standards, must go further to provide alternative foods for employees, to monitor wildlife and other non-target resources, and to prevent unsustainable, illegal or commercial hunting within their concessions.

Over the years, hunting weapons have shifted from traditional bows and snares made from natural materials, to guns and wire snares. Today, the majority of bushmeat is acquired using wire snares, which are cheap, durable, easy to use and, for signatories of the 1968 (revised in 2003) *African Convention on the Conservation of Nature and Natural Resources*,[51] illegal. While snares made of natural fibers must be checked frequently as they will decompose over time and break when encountered by non-target large mammals (e.g. lions, elephants, great apes), wire snares are non-selective, long-lasting and, as a result of the latter, likely to maim or kill large, slowly reproducing species, and/or be checked so infrequently that animals die and decompose without being recovered (wastage rates of 25-94% have been documented).[52] In terms of guns, large-game rifles and their ammunition – for example, .458 and 10-75 caliber – should be better regulated or, in the case of rifles only suitable for hunting elephants, banned. Automatic weapons are common and efforts to remove them following civil and regional wars should be implemented and increased for the sake of both people and wildlife.

Demand for bushmeat has increased largely because of growing human populations and urbanization. More people are demanding more resources from an ever smaller land-base without a simultaneous increase in agricultural (including protein production) productivity. When development opportunities arise, in the absence of regulations and enforcement, wildlife exploitation dramatically increases.[53] Protein and income demands by growing human populations must be identified, addressed, managed, and monitored.

There are virtually no alternative sources of income for many rural communities outside of exploiting non-timber forest products (NTFPs), and limited land, infrastructure and technical capacity for the production of cash crops. With few off-farm jobs, low crop prices, deteriorating roads (in many countries), and increased demand, wildlife and fish have become a short-term solution to a long-term economic problem. There are virtually no readily recognizable product outputs aside from agricultural goods, local crafts, or NTFPs that make their way to urban markets; likewise, manufactured products such as aluminum cooking pots, knives, and other basic kitchenware, along with clothing, radios, batteries, soft drinks and alcohol, currently make their way from capital cities to rural communities. Large- and small-scale income-generating activities that are economically and logistically competitive with the current bushmeat trade need to be established for rural communities. That is, for any solution to complete successfully with bushmeat it must provide training, trade routes and marketing networks as well as control of wildlife hunting by those who may try to take advantage of the system (i.e. by participating in new income generating activities while continuing to commercially harvest and trade bushmeat).

Protein substitutes usually do not exist in the quantity or quality to compete successfully with bushmeat. What animal protein is available in rural areas (e.g. livestock) is often expensive compared to bushmeat or used as financial security for emergencies such as medical treatment. Even if alternatives were to be made available, they would not likely be successful without disincentives to illegal wildlife use, including law enforcement. For protein alternatives to compete with bushmeat they would have to satisfy both the nutritional and taste demands of millions of consumers.

Current law enforcement capacity is inadequate in most areas. Even if wildlife officials are present they are frequently ill informed about existing laws, or reluctant or powerless to enforce them. The judicial process is equally flawed. As a consequence, even if offenders are apprehended, most arrests do not end in convictions. The development of legislative capacity and enforcement are essential if the bushmeat crisis is to be adequately addressed. Without such capacity-building, it is likely that the trade will reduce an otherwise rich biodiversity to a few highly-reproductive rodent species in degraded habitats – as is the case in much of West Africa. Many laws on user rights and access are unclear or unenforceable and, at the very least, they should be clarified and explained to the millions of rural hunters, traders and urban consumers who don't know the legal status of the animals they use, or the consequences of their activities.

CHALLENGES TO MAKING PROGRESS

Without simultaneously and effectively addressing these causal factors through long-term commitments of support by governments, industry, development agencies, non-governmental organizations (NGOs), and communities, it is unlikely that any significant impacts can be made to mitigate the bushmeat crisis, and the best hope for biodiversity conservation on a limited scale is support for protected areas. It is essential that policy makers, practitioners, and the public collaborate to assure that the crisis is addressed sufficiently at all levels and with a commitment for the long term.

Critics argue that conservationists have failed in meeting their objectives and have effectively enabled the bushmeat crisis to exist,[54] but the fact remains that the resources available to do conservation work are orders of magnitude less than resources available for development efforts or private industry investments. If development funds are insufficient to meet the income- and protein- and capacity-building demands of rural Africa, then conservation funds are simply delaying the inevitable extinction of many species.

What is needed is a comprehensive paradigm shift enabling realistic resource needs for wildlife and biodiversity conservation to exist – a paradigm that mobilizes the resources for conservation on a scale comparable to religious institutions, medical research, professional sports, or the arts. We must "mainstream" wildlife as a household consideration and responsibility. We are willing to spend hard-earned cash on our places of worship, to support our long-term medical well-being, for our entertainment and cultural enlightenment. But when it comes to spending on wildlife conservation, pockets are empty. It must be understood and agreed that in fact wildlife does NOT have to pay (indeed, cannot pay) for itself, and that the global community must accept the responsibility to heavily subsidize the existence of wildlife for future generations of people around the world.

While communities, governments, conservationists and development professionals agree that bushmeat is currently an important component of rural livelihoods, there is disagreement regarding strategies to accommodate the sometimes conflicting objectives of conserving biodiversity and supporting human livelihoods – both for present and future generations. Recommendations for improved management of the bushmeat trade to achieve sustainable levels of off-take include inputs of expertise, funding, management mechanisms and resources that are either currently non-existent or unrealistic for the long-term.

Whether the goal is regulation or mitigation, the costs of addressing the bushmeat crisis are considerable. Either strategy involves commitment of resources and time by local communities, wildlife and development experts, local, national and international governments, private industry, and institutions. Local communities are asked to divert limited resources toward conservation activities that they are not equipped through traditional training to address. Local and national governments maintain long lists of social, development and security demands that take funding priority. Conservationists have a limited pool of funding from which to draw to enable effective implementation of conservation actions. Development agencies are increasingly being approached to engage on the bushmeat issue, but many view the bushmeat trade as an existing, cost-effective solution to decades of failed effort to alleviate poverty and food insecurity. Wildlife is continuously being asked to meet a demand that biologically cannot be met. How do professionals manage conflicting priorities to create viable strategies for addressing the bushmeat crisis?

First, we need to agree that conservationists, development experts, politicians, communities, and private industry personnel approach the topic of wildlife from completely different perspectives and with often conflicting goals. It is difficult for conservationists to understand how those engaged in development fail to recognize the biological limitations of the wildlife being exploited. Hearing recommendations to use the bushmeat trade to support livelihood needs flies in the face of all the evidence indicating the extreme level of unsustainable exploitation occurring in commercial trade.

As Mabogunje notes:

> One can go on listing the many failures of development efforts in sub-Saharan Africa, failures that have left many of the people poorer today than they were some twenty or thirty years ago. What this will not adequately capture are the multiple pressures these failures have compelled the population to put on the environment.[55]

It is challenging for development practitioners to reconcile arguments for conserving wildlife while being mandated with enabling the improvement of livelihoods, alleviating poverty, building capacity and all the other extremely important societal tasks such development agencies undertake. At the same time, however, using wildlife to shore up failing poverty alleviation strategies is ill advised,[56] as the problems will only become greater once the wildlife has disappeared and the habitats where they were once found are further compromised.

Conservation professionals have long recognized the importance of human livelihoods and collaboration with local communities. Suggesting that policies built on unsustainable wildlife use are inappropriate does not mean that conservationists are unconcerned about human welfare, but rather that they believe that the welfare of both present and future human generations must be included in current policy considerations.

Local communities have their own perspectives regarding wildlife, its use, and their development needs. Most communities identify health care and education as top priority services they would like to see available in addition to more jobs. When asked if they want or expect wildlife to exist for future generations, most individuals agree.[57] Unfortunately, present day needs and lack of information regarding the impacts of the actions they take force most individuals to make decisions and exhibit behaviours that will not allow for such attitudes to be supported. That is, a poor farmer will choose to hunt the last antelope he or she finds while maintaining the desire for his children to be able to hunt wildlife in the future. Wildlife retains deep cultural value for many communities in Africa but such values will cease to be supported in future generations if wildlife continues to disappear.

Bringing these conflicting viewpoints together is essential if the bushmeat crisis is to be effectively addressed. There must be consensus on the unsustainable nature of current activities and a collective commitment to change behavior to achieve biodiversity conservation goals. How to appropriately respond once that consensus is reached will require the full power of a coordinated collaboration of development and conservation agencies, private industry, communities, and governments working together. An effective network can achieve important advances resulting from collaboration in information management, education, public awareness, and in the engagement of key decision-makers.

While the on-the-ground priorities that have been outlined earlier are essential if the bushmeat crisis is to be effectively addressed, these must be backed by consensus building on an enormous scale. Support for conservation activities on the ground can be found in a variety of ways. Priorities are established by a complex academic and practical process that involves not only African experts but also experts from around the globe. There must be a continuous development of understanding about how conservation can best be achieved but with the moving goalposts and the ever-changing "cutting edge" research that grabs scarce research and conservation funds, it is a major challenge to settle on a course of action that can be reasonably funded for the long time required for actual conservation to take place. Nonetheless, by the end of the 20th century, conservation experts recognized that a new and dedicated effort was required in order to facilitate collaboration and to reduce the threats to species involved in the bushmeat crisis throughout much of Africa. Thus emerged the Bushmeat Crisis Task Force.

THE BUSHMEAT CRISIS TASK FORCE

The Bushmeat Crisis Task Force (BCTF) was established in 1999 to facilitate collaboration on the bushmeat issue particularly in the areas of awareness, education, influencing key decision makers, and information management. A set of Operational Guidelines was initially drafted and approved by the full General Membership of the BCTF. It is periodically reviewed and updated by the membership. Supporting and Contributing Members provide financial and technical resources for staff and selected BCTF projects; additional funding is obtained through grants. Close collaboration with government agencies and groups in the United States and in Africa whose mandates include addressing the bushmeat issue were a priority in establishing the BCTF. A Steering Committee composed of senior level experts representing the Supporting Member institutions with an interest and background in the bushmeat issue guide the setting of priorities and the activities of the BCTF and its staff. BCTF is a broad and active network which communicates via listservs, meetings, projects, and the BCTF website.

The BCTF Vision is to eliminate the illegal and/or unsustainable commercial bushmeat trade through the development of a global network that actively supports and informs nations, organizations, scientists and the general public. The BCTF Mission is to build a public, professional and government constituency aimed at identifying and supporting solutions that effectively respond to the bushmeat crisis in Africa and around the world. The primary BCTF Goals are to:

1. Enable information sharing and create an information sharing mechanism on the bushmeat issue;
2. Engage key decision makers in the United States, Europe and Africa;
3. Build awareness and provide education across sectors; and
4. Foster collaboration among member and partner institutions.

The BCTF Phase I Report[58] provides a listing of measurable action items toward achieving its priority goals. Within each primary goal area some of the reported accomplishments of the BCTF collaboration are:

Information Management: BCTF has created databases on the bushmeat issue from peer-reviewed publications, field reports, projects, media, documentaries and other resources. It has partnered with the World Resources Institute to provide a GIS-based information sharing mechanism for its members, key decision makers, media and the public. This project is called the Bushmeat Information Management and Analysis Project (IMAP) and has already proven an important tool in assisting decision making and priority setting on the bushmeat issue. BCTF has created a world-class website on the bushmeat issue that receives well over 10,000 visitors per month and hosts the many documents and publications BCTF produces (Fact Sheets, the Bushmeat Quarterly, the Bushmeat Bulletin, and occasional papers). In addition, BCTF maintains a professional bushmeat listserv started in December 1998 that provides opportunity for information exchange among the world's leading experts in conservation, development, media, government, private industry, and academia.

Public Awareness: BCTF has contributed to the dramatic increase in media reporting on this issue (from an estimated 35 articles before 1999 to over 800 articles during the period 2000-2004) and has been directly responsible for countless articles in peer-reviewed and professional publications, major mainstream print media as well as radio and television. In addition to the media, the BCTF has created *The Bushmeat Promise*, a tool for individuals to make personal commitment toward action to address the bushmeat crisis. BCTF has participated in numerous professional meetings to raise awareness among the conservation and development community as well as key decision makers regarding the bushmeat issue. In Africa, BCTF supports the efforts of its Supporting Members, the CITES Bushmeat Working Group and member institutions of the Pan African Sanctuaries Alliance (PASA) to develop and implement effective public awareness tools.

Education: BCTF has been instrumental in enabling formal education development on the bushmeat issue for those addressing the bushmeat crisis on the ground. The most significant initiative it supports is integrated curriculum development among Africa's three regional wildlife colleges: *École pour la Formation des Spécialistes de la Faune de Garoua* (Garoua Wildlife College), Cameroon; College of African Wildlife Management, Mweka, Tanzania; and the Southern Africa Wildlife College in South Africa. In the US, with a team of over 30 volunteers from 25 institutions, BCTF created the Bushmeat Education Resource Guide (BERG), which provides resources for educators to reach out to the public about the bushmeat issue. The BERG includes resources that can be used to create unique bushmeat education materials, including a training guide, program and activities, middle school curriculum, signage templates, evaluation tools and the Bushmeat Promise.

Key Decision Makers: BCTF was established to be the professional authority on the bushmeat issue. As such, BCTF has responded to direct information requests from the U.S. Secretary of State, the U.S. Secretary of Agriculture, US Fish and Wildlife Service Director, White House Council on Environmental Quality, and numerous other administration officials in Washington. We have provided information and logistical support to Members of Congress hosting briefings, including the International Conservation Caucus. BCTF played significant roles in the development of the IUCN Bushmeat Resolution passed in October 2000, the July 2002 Congressional Hearing on Bushmeat, the official Africa Forest Law Enforcement and Governance Agreement (AFLEG) process in 2003, and the Congressional commissioned report providing recommendations to improve U.S. policy toward Africa – the Africa Policy Advisory Panel (APAP).[59] The final APAP report, reviewed and endorsed by Secretary of State, Colin Powell, included recommendations for directly addressing the bushmeat crisis, including the formation of a US Interagency Bushmeat Task Force. BCTF has worked very closely with the CITES Bushmeat Working Group by supporting its formation at CITES CoP 11 in April 2000 in Kenya and by securing operational funds for three years of the group's activities.

BCTF has also commissioned reports reviewing potential areas of intervention with private industry in Central Africa, and participates in meetings on industrial development and biodiversity conservation, specifically regarding the development of wildlife management plans in resource exploitation areas.

BCTF has brought a wide range of the wildlife conservation community together in a unique collaborative effort. Not only has the BCTF enabled numerous important actions toward addressing the bushmeat crisis to take place, it has made strong efforts to reach out to government, private industry, the media, and development agencies. BCTF members work directly with communities on the ground and guide the BCTF in the implementation of appropriate and priority actions. BCTF staff are now making progress towards working with international development professionals committed to supporting rural communities and helping them achieve a better quality of life.

A CASE STUDY

Among the many projects that are currently taking place to address the bushmeat crisis, one effort in the northern Republic of Congo stands out unlike any other for effectively and concurrently addressing needs for wildlife management, land use planning, national development, partnership, income and protein alternatives, education, awareness, enforcement, and policy development – the Project for Ecosystem Management of the Peripheral Zone of the Nouabalé-Ndoki National Park (PROGEPP).[60] A collaboration of the Government of the Republic of Congo, the *Congolaise Industrielle des Bois* (CIB) logging company, the Wildlife Conservation Society (WCS), and the local communities associated with the concession, PROGEPP provides important lessons concerning multi-stakeholder engagement and the application of several solutions towards long-term viability.

In the early- to mid-1990s, surveys into the Kabo Logging Concession (then *Société Nouvelle des Bois de Sangha*, SNBS) identified a pervasive bushmeat trade.[61] No controls on wildlife hunting, trade, or export existed. Wildlife populations in the region surrounding human habitation were depleted. Communities associated with the logging concession had little organization or control regarding management of wildlife hunting. There was virtually no expertise or capacity within the logging concession for managing wildlife exploitation. The list of factors acting against the sustainable use of wildlife in the concession was extensive.

PROGEPP was established to address the bushmeat trade within the Kabo logging concession. Based on years of relationship-building among concession managers, local and national level government authorities, the Kabo community and WCS personnel, the project was initiated in 1999. Initially, there was a change to internal logging company regulations, including three key management criteria long successful with the communities associated with a national park to the north:

1. Bans on hunting using snares;
2. Prohibition of bushmeat export outside the local community; and
3. No hunting of endangered or threatened species.

These three basic rules were further supported by establishing community hunting zones and education and awareness programs, providing jobs and training (eco-guards), as well as supplying protein alternatives to bushmeat to the people in the local community and logging company workers. All of these activities were further supported by project staff monitoring the concession both for illegal hunting activities as well as for estimating animal populations. The impact of these combined efforts was that wildlife management was effected over an area of more than 500,000 ha, at a cost of about $1 per hectare where before there had been no wildlife management at all. Endangered species continued to survive in hunting zones where other commonly-hunted species were found in relatively stable numbers and there was a dramatic reduction in snare use. This management model now has been adopted into Congolese law and applies to all logging concessions. It has also been shared with (although not yet widely adopted by) neighbouring countries.

This project addresses the main causes of the bushmeat crisis – access, increased demand, new hunting technologies, lack of economic alternatives, lack of protein alternatives, and lack of law enforcement capacity. It provides many of the solutions including protected (no-hunting) zones, policy/regulations, awareness, income and protein alternatives, as well as education and training. It does all of this in a specific location with targeted funds and personnel, and high levels of collaboration among all key stakeholder groups. The project required high initial fiscal and technical inputs from the international NGO community. It also required a large commitment by private industry, linked to consumer demand for "bushmeat-free" products.

While the success of PROGEPP is certainly encouraging, the reasons for its success may not be readily apparent. The main reasons this project has so far succeeded, in addition to the excellent design and multi-tiered effort involving alternatives, awareness, enforcement, and capacity building, is that trusting relationships among the partners were built over many years of working in the same region together. Equally important, there is long-term financial and technical support for the project's operations, and a growing pressure from consumers demanding environmentally and socially responsible exploitation of the forests.

CONCLUSION

Despite many years and billions of dollars of effort, sustainable use of wildlife (not to mention, sustainable development which cannot technically be attained if sustainable use is not first in effect) in Africa, and many other parts of the world, has yet to be achieved. Issues such as the bushmeat crisis that have emerged around the globe are examples of the limitations of policies and current actions with regard to biodiversity conservation. Putting more pressure on already strained wildlife populations will only make the situation worse. What is needed is an immediate global commitment that recognizes the ecological limitations upon which development activities depend – as was cautioned in *Our Common Future* nearly 20 years ago. Since then a whole new generation of development and conservation workers has emerged with the ideal of sustainable development deeply rooted – but with few examples to support it.[62] The bushmeat trade is extremely complex and demonstrates the challenge of linking basic human livelihoods with the objective of biodiversity conservation. It is essential that conservation and development organizations establish systems of communication and collaboration in their efforts with key decision makers, private industry and local communities. If we are to realize the objectives of *Our Common Future*, we must shift our perspective to one that is based on the limited capacity of the planet to support human population growth and development. As the example from Kabo effectively demonstrates, we must also openly identify areas of potential conflict, commit to collaboration, and confirm our common ground.

ACKNOWLEDGEMENTS

The author wishes to extend enormous gratitude to the many reviewers of this document whom greatly improved the final outcome, including David Lavigne, Natalie D. Bailey, Andrew Tobiason, Elizabeth L. Bennett and Richard G. Ruggiero. Many thanks also to Kelvin Alie and IFAW for supporting my participation in this fascinating Forum. Appreciation is also extended to the BCTF Executive, Steering Committee and Network for providing the time and support to engage in researching, writing, and learning further about the bushmeat crisis. It should be noted, however, that while this paper could not have been possible without the support of BCTF and IFAW, the views expressed here and any errors contained in this chapter are the author's own.

NOTES AND SOURCES

[1] IUCN/UNEP/WWF. 1980. *World Conservation Strategy. Living Resource Conservation for Sustainable Development.* Gland, Switzerland. For more on the World Conservation Strategy, see Lavigne, Chapter 1. For an historical account of the notion of sustainability, see Holt, Chapter 4.

[2] World Commission on Environment and Development (WCED). 1987. *Our Common Future.* Oxford University Press, Oxford and New York. The report was accepted "as a guideline to be taken into account in further work of the United Nations Environment Programme" on 19 June 1987 (Decision 14/14, see http://www.unep.org/Documents.Multilingual/Default.Print.asp?DocumentID=100&ArticleID=1643&l=en).

[3] For more on the Report of the World Commission on Environment and Development, see Lavigne, Chapter 1. For more on sustainable development generally see Beder, S. 1996. *The Nature of Sustainable Development.* Second Edition. Scribe Publications, Newham, Australia; Chesworth, W., M.R. Moss, and V.G. Thomas (eds.). 2002. *Sustainable Development: Mandate or Mantra.* The Kenneth Hammond Lectures on Environment, Energy and Resources 2001 Series. Faculty of Environmental Sciences, University of Guelph, Guelph, Ontario, Canada. Also see Beder, Chapter 5; Brooks, Chapter 16; and Oates, Chapter 18.

[4] See for e.g. Caccia, C. 2002. The politics of sustainable development. pp. 35-49. In W. Chesworth, M.R. Moss, and V.G. Thomas (eds.). *Sustainable Development: Mandate or Mantra.* The Kenneth Hammond Lectures on Environment, Energy and Resources. 2001 Series. Faculty of Environmental Sciences, University of Guelph, Guelph, Ontario, Canada.

[5] Dawe, N.K. and K.L. Ryan. 2003. The faulty three-legged-stool model of sustainable development. Conservation Biology 17(5):1458-1460.

[6] WCED 1987, pp. 24-25.

[7] See for e.g. Brooks, Chapter 16.

[8] WCED 1987, pp 24-25.

[9] For further discussion of definitions, see Lavigne, Chapter 1.

[10] Convention on Biological Diversity (CBD). 2004. Sustainable Use of Biodiversity Case Studies. Available at UNEP website http://www.biodiv.org/programmes/socio-eco/use/cs.aspx. Accessed September 2004.

[11] CASUSG. 2003. CASUSG Literature Overview Summaries. March. IUCN Central Africa Sustainable Use Specialist Group. Available at http://www.iucn.org/themes/ssc/susg/susgs/centralafrica.html.

[12] This quotation is very similar to a comment made in Juste, J., J.E Fa, J.P. Del Val, & J. Castroviejo. 1995. Market dynamics of bushmeat species in Equatorial Guinea. Journal of Applied Ecology 32: 454-467. Also see http://bushmeat.net/afprimates98.htm.

[13] e.g. Beder 1996; Lavigne, D.M., C.J. Callaghan and R.J. Smith. 1996. Sustainable utilization: the lessons of history. pp. 250-265. In V. Taylor and N. Dunstone (eds.). *The Exploitation of Mammal Populations.* Chapman & Hall, London; Lavigne, D.M. 2002. Ecological footprints, doublespeak, and the evolution of the Machiavellian mind. pp. 61-91. In W. Chesworth, M.R. Moss, and V.G. Thomas (eds.). *Sustainable Development: Mandate or Mantra.* The Kenneth Hammond Lectures on Environment, Energy and Resources 2001 Series. Faculty of Environmental Sciences,

University of Guelph, Guelph, Canada.

[14] Redford, K.H. 1992. The empty forest. BioScience, 42:412-422.

[15] Robinson, J.G., K.H. Redford, E.L. Bennett. 1999. Wildlife harvests in logged tropical forests. Science 284: 595-596.

[16] Barnett, R. 2000. *Food for Thought: The utilization of wild meat in Eastern and Southern Africa*. Nairobi, Kenya: TRAFFIC East/Southern Africa. 264 pp.

[17] IUCN/UNEP/WWF. 1991. *Caring for the Earth: a strategy for sustainable living*. Gland, Switzerland.

[18] Robinson, J.G. 1993. The limits to caring: sustainable living and the loss of biodiversity. Conservation Biology 7:20-28.

[19] Robinson, J.G. 1993. p. 21.

[20] See Lavigne, Chapter 1.

[21] Robinson, J.G. 1993. p. 26.

[22] Bennett, E.L. 2000. Timber certification: where is the voice of the biologist? Conservation Biology 14: 921-923.

[23] Ibid.

[24] However, see WWF's *Green Reconstruction Policy Guidelines* for Aceh (Sumatra, Indonesia), approved by the governor of Aceh Province, Sumatra and the President of Indonesia, following the devastating tsunami of December 2004. Available at http://www.worldwildlife.org/news/displayPR.cfm?prID=196.

[25] CBD. 2004. *Addis Ababa Principles and Guidelines for the Sustainable Use of Biodiversity*. Convention on Biological Diversity. Available from: http://www.biodiv.org/programmes/socio-eco/use/addis-principles.asp.

[26] See section 2 of this book for additional examples.

[27] Robinson, J.G. and E.L. Bennett. 2000. *Hunting for Sustainability in Tropical Forests*. Columbia University Press, New York. 582 pp.

[28] Ibid.

[29] Wilkie, D.S., E.L. Bennett, H.E. Eves, M. Hutchins, C.M. Wolf. 2002. Roots of the bushmeat crisis: eating the world's wildlife to extinction. *Communiqué*. November 2002, pp. 6-7.

[30] For more on this topic see Rees, Chapter 14; Brooks, Chapter 16; and Czech, Chapter 23.

[31] Barrett, C.B. and P. Arcese. 1995. Are integrated conservation-development projects (ICDPs) sustainable? On the conservation of large mammals in sub-Saharan Africa. World Development 23:1073-84; Berkes, F. 2004. Rethinking community-based conservation. Conservation Biology 18: 621-630.

[32] For more on this topic, see Oates, Chapter 18.

[33] Astle, W.L. 1999. *A History of Wildlife Conservation and Management in the Middle Luangwa Valley, Zambia*. British Empire and Commonwealth Museum. 170 pp.

[34] Hackel, J.D. 1999. Community conservation and the future of Africa's wildlife. Conservation Biology 13: 726-734.

[35] Brown, D. 2003. Bushmeat and poverty alleviation: implications for development policy. ODI Wildlife Policy Briefing Number 2, November.

[36] Eves, H.E. and R.G. Ruggiero. 2000. Socioeconomics and the Sustainability of Hunting in the Forests of Northern Congo (Brazzaville). pp. 427-454. In J.G. Robinson and E.L Bennett (eds.). *Hunting for Sustainability in Tropical Forests*. Columbia University Press, New York.

[37] The Bushmeat Campaign, see http://www.bornfree.org.uk/primate/primnews006.htm; and USAID Central African Regional Program for the Environment (CARPE), see http://carpe.umd.edu.

[38] See Oates, Chapter 18.

[39] Bennett, E.L. and J.G. Robinson. 2000. *Hunting of Wildlife in Tropical Forests: Implications for Biodiversity and Forest Peoples*. Biodiversity Series – Impact Studies Paper No. 76. The World Bank, Washington DC. 56 pp.

[40] Brennan, E. 1999. *Population, Urbanization, Environment, and Security: A Summary of the Issues*. Woodrow Wilson International Center for Scholars. Occasional Paper Number 22. Washington DC.

[41] Robinson and Bennett (2000).

[42] Bennett, E. H. Eves, J. Robinson, and D. Wilkie. 2002. Why is eating bushmeat a biodiversity crisis? Conservation in Practice 3(2):28-29.

[43] Bennett *et al.* 2002.

[44] Brown (2003).

[45] For more on this topic, see Milner-Gulland, Chapter 20.

[46] Brennan 1999.

[47] Ibid.

[48] Wilkie, D.S. and J.F. Carpenter. 1999. Bushmeat hunting in the Congo Basin: an assessment of impacts and options for mitigation. Biodiversity and Conservation 8: 927-955; Fa, J.E., C.A. Peres, and J. Meeuwig. 2002. Bushmeat exploitation in tropical forests: an intercontinental comparison. Conservation Biology 16: 232-237.

[49] Wilkie, D.S., J.G. Sidle, and G.C. Boundzanga. 1992. Mechanized logging, market hunting and a bank loan in Congo. Conservation Biology 6: 570-580.

[50] Bennett *et al.*(2002).

[51] Available at http://www.africa-union.org/Official_documents/Treaties_%20Conventions_%20Protocols/Convention_Nature%20&%20Natural_Resources.pdf.

[52] Muchhaal, P.K. and G. Ngandjui. 1999. Impact of village hunting on wildlife populations in the Western Dja Reserve, Cameroon. Conservation Biology 13: 385-396.

[53] Eves, H.E. and R.G. Ruggiero. 2000. Socioeconomics and the Sustainability of Hunting in the Forests of Northern Congo (Brazzaville). pp. 427-454. In J.G. Robinson and E.L Bennett (eds.). *Hunting for Sustainability in Tropical Forests*. Columbia University Press, New York.

[54] Peterson, D. 2003. *Eating Apes*. University of California Press. 333 pp.

[55] Mabogunje, A. L. 2004. Framing the Fundamental Issues of Sustainable Development in Sub-Saharan Africa. CID Working Paper No. 104. Cambridge, MA: Sustainable Development Program, Center for International Development, Harvard University; also published as TWAS Working Paper 1. Trieste, Italy: Third World Academy of Sciences. p. 2. Available at: http://www.cid.harvard.edu/cidwp/104.htm

[56] For more on this topic, see Oates, Chapter 18.

[57] Mordi, A.R. 1991. *Attitudes Toward Wildlife in Botswana*. Garland Publishing, Inc. New York. 217pp.; Eves, H.E. 1996. Socioeconomic study 1996. Nouabalé Ndoki

National Park Northern Congo. Unpublished Report to The Wildlife Conservation Society and The World Bank. 150pp.

[58] BCTF 2004. *BCTF Phase I Report.*: Bushmeat Crisis Task Force. Washington DC. 155 pp. Accessed 15 April 2005: www.bushmeat.org/cd.

[59] Lapham, N.P. 2004. A natural resource conservation initiative for Africa. pp. 88-103 In: W.H. Kansteiner III and J. S. Morrison (eds.). *Rising US Stakes in Africa: Seven Proposals to Strengthen U.S.-Africa Policy.* A Report of the Africa Policy Advisory Panel (APAP). Center for Strategic and International Studies. Washington DC.

[60] Elkan, P. and S. Elkan. 2002. Engaging the Private Sector: A case study of the WCS-CIB-Republic of Congo project to reduce commercial bushmeat hunting, trading and consumption inside a logging concession. *Communiqué.* November 2002, pp. 40-42.

[61] Auzel, P. and D.S. Wilkie. 2000. Wildlife Use in Northern Congo: Hunting in a Commercial Logging Concession. pp. 413-426. In J.G. Robinson and E.L. Bennett (eds.). *Hunting for Sustainability in Tropical Forests.* Columbia University Press, NY.

[62] See for example, Lavigne 2002.

A CHIMP LYING ON A STORE SHELF
MUSED SADLY ON THE NATURE OF WEALTH
TO USE ME AIN'T WISE
I'M NOT TASTY WITH FRIES
AND AT RISK IS WISDOM ITSELF.

MARTIN WILLISON 2004

Chapter 10

Ecotourism: Benefits and Challenges – Uganda's Experience

Arthur R. Mugisha and Lilly B. Ajarova

Tourism is generally recognized as the fastest growing industry in the world. Globally – according to the World Tourism Organization[1] – the industry generated income of US$476.0 billion in 2000, compared to US$2.1 billion in 1950. International tourist arrivals increased from 25.3 million to 698.3 million people over the same period.

In East Africa (Uganda, Kenya and Tanzania), tourism receipts increased from US$452 million in 1980 to US$1.64 billion in 1995, while tourist arrivals increased from 1.3 million to 4.2 million over the same period, representing an average annual growth rate of 8.25%.

The growth of the tourist industry has generated concerns from environmentalists and others, who are worried especially about the negative impacts that tourists have on the environment and on local communities. In many places, tourism aggravates problems of inadequate sewage disposal. In coastal regions, for example, it is not unusual for hotels to pipe their sewage directly into the sea. In addition, almost all large tourist centers face problems of rubbish disposal and consume large amounts of electricity and water. Furthermore, critics point out the high rate of land use (for golf courses, for example) and the damage to biological diversity. The social and cultural aspects can hardly be overestimated. Tourism in developing countries is said to strengthen the influence of foreign cultures and behaviours and to destroy established social and cultural structures. Critics also claim that tourism also aggravates social imbalances and, in parts of the world, promotes the sex trade, child prostitution and child labour. In this view, the local population gains very little from tourism and any extra foreign currency earned is required to meet the increased demand for imported products.[2]

Ecotourism is an idea that has developed out of these concerns. It attempts to take into account the conservation of wildlife and the environment and the needs of the local people. It is an industry that emphasizes the natural environment and the interests of the local community. As with any new endeavor, environmentalists, academics and practitioners have taken a keen interest in debating it, to the extent that there has been a great amount of literature about the term and the different definitions of what constitutes ecotourism.[3]

We contend that ecotourism is more than just a marketing word. It should be understood as a deeper philosophy and practice that considers environmental issues, local community relations, appropriate technology and questions of ecology, economics, and culture. Here, we will discuss ecotourism in a conceptual sense. We define it as follows:

> Ecotourism aims at promoting purposeful travel to natural areas to understand the culture and history of the environment, taking care not to alter the integrity of the ecosystem, while producing economic opportunities that make conservation of natural resources beneficial to local people.[4]

There are three components that we will use to distinguish ecotourism from other forms of tourism. First is the visitation to natural areas, or natural resources, including wildlife. The prefix "eco" is derived from the term "ecology", which is the study of the relationships between the living and non-living organisms in the natural environment.[5] Therefore, one of the distinguishing characteristics of ecotourism is that there should be visitation to the local environment or trips to view wildlife. Second, there is the conservation of the environment. Here, we use the term conservation in a generic sense to include the care and wise (meaning prudent) use of the environment[6] and the management of human activities in it.[7] The third component is the involvement of the local communities to ensure they benefit from the ecotourism industry.

Honey[8] expanded on these components to include respect for cultural values, and visitation to remote areas to enjoy "un-spoilt" environments, while minimizing the impacts on the local environment. To minimize impacts implies the use of renewable and contemporary sources of energy such as solar power, recycling and safe disposal of waste and garbage, and environmentally and culturally sensitive architectural designs. Ecotourism also involves building environmental awareness among the tourists and host communities to promote conservation, to ensure the ecologically sustainable use of invaluable and irreplaceable natural resources.

Ecotourism should also provide direct financial benefits for conservation, and empower local communities to make decisions about if and how natural resources will be used, because un-spoilt areas are becoming increasingly rare and more challenging to maintain. Where communities are partners in ecotourism projects and ecotourism directly benefits communities, there are higher chances that the initiative will be successful. Ecotourism projects frequently lead to better infrastructure and provide benefits to other sectors of society as well. They create new jobs – directly and indirectly, for example, in construction, farming and transport. These benefits help to defuse problems of emigration, since tourist centers usually become established in several regions. Rather than harming cultural values and customs, ecotourism can conversely bring about their revival. Culture may attract just as many tourists as nature. Indigenous and ethnic village communities are encouraged to come to grips with their history and traditions. The visitors often show interest in, and respect for, the residents' knowledge of traditional medicine, for example. This enhances self-confidence and cultural identity, particularly among the youth.[2] Other benefits of ecotourism include the development of markets for local produce and artifacts, the provision of other income-generating activities, and the exposure of local communities to the international community. Some commentators[9] take the ecotourism concept further to link it with support for human rights and good governance practices. The reason for this is partly due to the need to recognize and involve local communities in ecotourism, but also because tourists – especially those from developed countries – tend to avoid tourist locations in developing countries that are characterized by turmoil and bad governance.

Ecotourism, therefore, provides both the means and the opportunities for local people to benefit from the wildlife and natural resources with which they have traditionally lived. But for ecotourism to be successful in the developing world, there must be a change from foreign-owned mass tourism to locally-owned, community-based ecotourism operations.

EVOLUTION OF ECOTOURISM

Conventional mass tourism with its increased commercialization involving photographic wildlife safaris, trophy hunting and coastal holidays is focused on human recreational pursuits without considering the needs for conservation and natural resources management. Groove[10] and Mugisha[11] reviewed the establishment of national parks and other wildlife areas in Africa, and noted that the need for national parks was a result of the concerns about the depletion of wildlife numbers, mostly by European recreational (or sport) hunters, beginning in the late 19th and early 20th centuries. This history helps to explain the inherent conflict between local communities and the establishment of national parks in Africa.[12] With time, however, it has been realized that recreational activities under the banner of tourism represent a great danger and threat to wildlife populations and, consequently, to the very essence of what attracts tourists to a particular destination.

In recent decades, some tourists started longing for "wilderness" in its true form, which prompted the involvement and promotion of conservation ideals within some parts of the tourism industry. Negative impacts of tourism began to be studied and the increasing awareness of them spread among both tourists and conservationists. Mass tourism began to be criticized in favor of nature-based tourism, a term almost synonymous with ecotourism, except for the component of community involvement and benefit that stretches nature-based tourism from the mere application of ecological principles and considerations to a deeper understanding of the distribution of benefits, including provisions for the costs of conservation, and the involvement of the local community.

The issue of the distribution of benefits and costs of conservation and natural resources management has been well studied and documented.[13] It can be argued that the benefit aspect is a key component in natural resources management, and one that originally was ignored at the time of establishing formal conservation areas. Attitudes are changing, however, and it is now asserted that for conservation to succeed tangible benefits must be realized by the local communities that bear the costs of conservation. It is in this context that ecotourism becomes relevant as a tool in wildlife conservation and management.

Whelan[9] examined nature-based tourism in different parts of the world as a management tool for environmental conservation. He cited a number of examples from Kenya as case studies to demonstrate successful nature-based tourism. We argue, however, that nature-based tourism, like mass-tourism, may not necessarily benefit local people and is, therefore, different from ecotourism. Here, we present some case studies from Uganda that, unfortunately, have not yet been well documented, to highlight what we consider to be real ecotourism.

TOURISM IN UGANDA

Uganda lies at the "roof" of the East African plateau, at a high altitude and on the equator. It is the "meeting point" of the larger eco-regions comprised of the tropical rain forests of the Congo Basin and the East African savannah. Its location gives Uganda a unique climate, which is a typical equatorial climate, receiving two rainy seasons and two dry seasons in a year. Temperatures range from a maximum of 30-32 degrees Celsius to a minimum of 16 degrees Celsius. The conditions are quite different from most equatorial countries, which are generally at a lower altitude and are characterized by more humid climatic conditions. Uganda's unique climate has resulted in the country experiencing a special endowment of varied ecological habitats and wildlife species that are unrivalled in the whole of Africa.

Immediately after gaining Independence in 1962, tourism was the second largest foreign exchange earner in the Ugandan economy.[14] At that time, tourism was based largely on the savannah national parks, which offered game viewing as the main product. Unfortunately, civil strife – due to the bad governance that characterized Uganda from the early 1970s to the mid-1980s – led to a dramatic reduction in the size of wildlife populations and seriously affected the tourism products that Uganda could offer. Coupled with the collapse of public services and infrastructure such as roads, tourism in Uganda was more or less non-existent by the 1980s. In 1993, the Ugandan government launched a ten-year integrated tourism master plan to return the industry to its former glory. Persistent banditry activities in the northern parts of Uganda, however, continued to challenge the Ugandan government's ability to insure the necessary security to permit full-blown tourism programs to be developed throughout the entire country.

In 1995, tourism in Uganda underwent a significant shift from the traditional game-viewing products to a primate-based tourism product. This shift was "kick-started" with the introduction of gorilla viewing. Tours were successfully launched in the Bwindi Impenetrable National Park, where habituated mountain gorillas (*Gorilla beringei beringei*) could be visited by tourists on foot. Later, eastern chimpanzees (*Pan tryglodytes schweinfurthii*) were also successfully habituated in Kibale National Park, and tours were conducted on foot to view these primates as well. Following these developments, the traditional savanna parks began to offer guided walks and night-time game drives, under the banner of diversified tourism products, enabling visitors to gain greater understanding about the national parks and their ecological processes.

In 2003, the Ugandan Ministry of Tourism, Trade and Industry launched a new Tourism Policy, which stipulates that ecotourism is the way forward for tourism development in Uganda.[15] This policy recognizes community-based tourism for sustainable development – in this case, development that ensures livelihood security of the local people, without destroying the natural resources base – and poverty alleviation in the rural areas of the country. On the basis of the experience we have had with the ongoing efforts to encourage the private sector to develop environmentally friendly tourism destinations in Uganda, we will now discuss three case studies that may serve as models for ecotourism development in Africa.

The *Nkuringo* Tourism Development Program

The *Nkuringo* tourism development program is an example of how to involve people in tourism development. "*Nkuringo*" is a Rukiga word that originates from the Bakiga pottery artisanship. The word is used to describe a more than 50 meter deep crater valley (which appears as if it were molded by a good potter), found in Nteko parish that borders the Bwindi Impenetrable National Park in southwestern Uganda. The valley was a favorite home range to a group of gorillas that came to be named as *Nkuringo* group, after the name of the valley. This area, however, was owned and used by the local people for cultivation and other land-use practices, thereby creating problems for the Uganda Wildlife Authority and the local people alike. The gorillas not only destroyed crops but also attacked individuals, posing a danger to people's lives. On the other hand, the gorillas were vulnerable to being infected by human diseases, including scabies and internal parasites, such as roundworms. As a consequence, the Uganda Wildlife Authority incurred heavy costs in trying to treat the gorillas for scabies, which was threatening their existence in the region.

In search of a solution to these problems, a dialogue was initiated with the local people. After a long consultative process, involving all the stakeholders concerned, it was agreed that the total of 287 landowners from the local community would be compensated if they left their land in favour of the gorillas. All the landowners were compensated and a total of 233 people who were living on their land moved and found alternative land elsewhere. It was further agreed that part of the land for which compensation was paid would be used for activities – such as tea growing – that were conservation-friendly but did not exacerbate gorilla/human conflicts. In addition, it was agreed to establish a gorilla-based tourism industry in the area, with the community members themselves being the primary beneficiaries. This arrangement resolved all the concerns regarding the management of gorillas and development needs of the people. The concerns about disease transmission were addressed through people relinquishing the use rights of their land through selling, and re-settling away from the land that was being used by the gorillas. Concerns about the loss of crops were also addressed by leaving the land to the gorillas, and practicing agriculture on lands away from the gorilla habitats. To ensure that the gorillas do not extend their home range, buffer crops such as tea are being planned to deter gorillas from further extending their home ranges.

The problems were dealt with, therefore, through the genuine involvement of all stakeholders. The local people were involved right from the beginning of the discussions and they were aware of the benefits that would accrue to them through the ecotourism program that is to be commissioned soon. As a result, the local people were willing to give up the ownership and use rights of their major worldly possession – land. They have also agreed to form an association that will enable them to manage and participate fully in the ecotourism development. With support from the United States Agency for International Development (USAID) and the International Gorilla Conservation Program (IGCP), there are plans to build an eco-lodge to be owned by the community association. The 16-bed eco-lodge will give preference to visitors who have purchased gorilla permits for the *Nkuringo* group. This will ensure that that the lodge owners will be assured of six visitors per night, which is the total number of gorilla permits permissible for the *Nkuringo* group. This is a deliberate effort to ensure that the lodge becomes economically viable and benefits the owners, who are the local people, as well as Uganda Wildlife Authority, the organization responsible for managing wildlife in the country.

The Buhoma community campground

Another case study of an ecotourism project in Uganda is the Buhoma community campground. Like the Nteko example above, Buhoma is one of the parishes neighboring Bwindi Impenetrable National Park, on its eastern side. The campground was started by the Buhoma Women's group in 1994, with the help of U.S. Peace Corps Volunteers, following the commissioning of gorilla-based tourism in Bwindi National Park. The campground was designed to benefit from the gorilla tourism that is based at Buhoma, the Park headquarters for Bwindi National Park. The campground provides simple accommodation and basic meals for tourists who come to track gorillas inside the Park. The proceeds from the camp accrue to the local people who make the decisions about how the funds are to be used. This arrangement has greatly assisted conservation of Bwindi National Park, through creation of local support for the conservation efforts to the National Park. The local people have now developed positive attitudes towards the Park's management, and fully appreciate the conservation of the forest and its natural resources, as they directly benefit from the gorilla tourism. Since the establishment of the campground, tourism revenues accruing to the communities have grown steadily under the management of the community leadership. Also, the ecotourism business undertaken by the community has continued to expand. The latest development is the establishment of a community trail, where members of the community guide tourists through their village, to develop an appreciation of local lifestyles. The tourists are shown how the people live in their village, and how they live in their natural environment, such as how residents make their local beer from plantains.

The Kibale Association for Rural and Economic Development (KAFRED)

The Kibale Association for Rural and Economic Development (KAFRED) is another case study of ecotourism development. This association operates from a small patch of wetland known as the Magombe swamp, which borders Kibale National Park in mid-southwestern Uganda. It is located in Bigodi Parish in the Fort-Portal district. In the local language "Magombe" means "the under-world of death". This, in a way, depicts how the local people have traditionally perceived wetlands – as places of evil or death, or as wastelands.

With the commissioning of chimpanzee-tracking in neighboring Kibale National Park, the local people, again with support from a U.S. Peace Corp Volunteer, took advantage of this development to establish a bird-based ecotourism project in the Magombe swamp.

Kibale National Park is home to numerous chimpanzees. It is also one of the sites of highest species richness and biomass.[16] The Kibale Association, inspired by the high levels of tourist arrivals in the Park, set aside the wetland and their own private land adjacent to it for conservation purposes. Simple tourist infrastructures, such as nature trails and bird-viewing platforms, were established. As well, a number of community members have been trained with support from Uganda Wildlife Authority, non-governmental organizations and other donors, to ensure that professional guiding services are provided.

Tourists who come to visit Kibale National Park spend at least 50% of their time visiting the Magombe wetland. Tourist numbers have risen from a mere 50 tourists per month in 1996, to over 500 visitors per month in 2002.[17] The increase in numbers of tourists should be understood as both a strength and a challenge: a strength in that the revenues will increase for the benefit of the local people, but a challenge in the sense that beyond a certain limit, the principle of ecotourism will be violated with the end result that the environment and the product offered will be spoilt. It is therefore important that continuous monitoring and evaluation are maintained. Since 1994/95, the entire wetland has been managed as a community bird sanctuary, and development programs from the proceeds of this endeavor range from education facilities, such as schools, to health centers in the village of Bigodi. Tourists directly pay the community, which then takes on the responsibility of guiding tourists through their wetland. The community is also responsible for managing the funds that accrue to them from ecotourism. This approach encourages communities to develop positive attitudes towards conservation in general and, in particular, to support the management efforts in Kibale National Park.

There are many other places in Uganda that are modeled on the principles described above. Ultimately, however, it is the issue of ownership and commitment that will determine the success of community-based ecotourism.

SHORTCOMINGS OF ECOTOURISM

Despite the successes and the advantages of ecotourism, there are, nonetheless, limitations and challenges when attempting to use ecotourism as an approach towards successful wildlife conservation and management. Below, we highlight some of the shortcomings, drawing both on our experiences in Uganda and on the experiences of others in different parts of Africa.[9]

The lack of security that characterizes many potential ecotourism destinations in Africa and other developing countries is one of the biggest challenges to the development of successful ecotourism programs. There are incidents of rebellion in different parts of Africa, including Somalia, some West African countries, and in the great lakes regions of Africa. Such uprisings against recognized governments are often planned and executed in wilderness areas, which often happen to be ecotourist destinations.

In Uganda, for example, we witnessed a massacre of tourists in Bwindi National Park in March 1999. There was so much publicity around the world about this unfortunate event that potential visitors cancelled their plans to visit Uganda. This event had a large negative effect on the ecotourism industry, and it took more than a year to restore confidence among tourists from the international community to resume traveling to Uganda.

In addition to the security issues discussed above, health concerns pose another threat as far as tourists are concerned. In rural areas, where poverty conditions dictate people's lifestyles, there is always the potential for disease outbreak that can be a source of risk to tourists. In the Nepal Mountains, and in some pastoral areas in Africa, the use of pit-latrines is culturally shunned, and this creates a problem of water-borne diseases in areas where there are fast flowing rivers. Again in Uganda, we experienced a problem with Ebola viral disease in 2000, and this had a negative effect on tourist arrivals in the country.

Another short-coming of ecotourism is its seasonal nature. There are often high and low seasons, when tourist arrivals go up and down, depending on a number of factors, including the changing of the seasons. During the low season, local communities that become dependent on an ecotourism industry may experience hardships due to a lack of alternative sources of income. Such hardships can lead to the destruction of the very natural resources base that forms the main foundation of the ecotourism industry itself.

The seasonality problem referred to above is exacerbated by a weak domestic market that cannot provide effective demand for ecotourism products during the "off-season" when international tourists are absent. There is little or no interest at either the local or national levels for people to appreciate nature through ecotourism. In the absence of domestic ecotourism, there is almost total dependence on international markets for tourism arrivals.

Lack of local interest also leads to uncertainties in the pricing mechanism. Often, the proprietors of an ecotourism product are concerned about scaring away potential clients by pricing their services too high. As a result they often end up under-pricing their products and losing money. Such under-pricing was detected, for example, in a study of the willingness of international tourists to pay to view chimps at certain locations in Uganda.[18]

Ecotourism destinations are often located in areas remote from city centers. Although this is what makes such a destination more valuable and appealing to ecotourists, it creates a problem of accessibility in terms of reliable transport. Also, many potential ecotourism destinations remain unknown and, hence, are not sufficiently developed. In such circumstances, ecotourism is less applicable as a tool for achieving conservation objectives. Consequently, the local people who would benefit from ecotourism programs still have to live by means of the consumptive use of their natural resources.

As an industry, ecotourism needs investment to ensure standards and competitiveness in the international tourism markets. The nature of the product and services required, coupled with the need to benefit local people, mandates that the local people themselves must invest their own funds in projects. However, there is little investment potential and ability among the local people to do this. Even where such a potential exists, local people are less experienced in, and exposed to, the needs and demands of tourists. In some of the successful cases of ecotourism development, such as in Kenya's Samburu Hills and Tanzania's Koija (near the Serengeti), there are often professional investors who mobilize the communities to pool land resources to develop competitive products. Nevertheless, there are few such investors interested in ecotourism products, considering the level of professionalism required, the often low rate of returns and the patience that is needed to ensure a highly competitive product. There have so far not been any favorable and concerted efforts to create incentives such as soft loans to promote private investment in ecotourism industries.

Ecotourism as a conservation tool has been debated by the elite, professionals and academics, to the extent that the would-be beneficiaries have been left confused about the concept. Coupled with this confusion, the concept has not received adequate publicity and application, especially in remote rural areas, where the concept is supposed to be beneficial and applicable to natural resources management. The concept has remained a subject of debate in the tourism and recreational classrooms, and what has been implemented has been left to the mercy of quick-profit-seeking business communities. There have not been any deliberate efforts to promote, publicize and support the concept by natural resources managers in the relevant rural areas that urgently need to ensure the security of their livelihood, without destroying the natural resources base. Such publicity and support need to target the rural people who would benefit from the industry, as well as the tourists who are interested in, and believe in, ecotourism.

THE WAY FORWARD FOR ECOTOURISM

To ensure that ecotourism is successfully applied in the field of natural resources management, there is a need to take deliberate steps to ensure its applicability. There is also a need to recognize the weaknesses and limitations of ecotourism to ensure its relevance both to conservation and natural resources management, and to local communities. Below we discuss some of the considerations that are needed to make ecotourism a more relevant and effective tool in nature conservation.

First, we should undertake an inventory of all the existing ecotourism destinations all over the world, and prioritize these sites based on the biodiversity that we need to conserve. Second, we need to take stock of the under or not-so-well developed potential ecotourist destinations, on a country-by-country basis. Having determined what there is, there is then a need to sensitize the local communities concerned about the advantages and benefits that could be derived from the identified ecotourism sites in their midst. This sensitization should be accomplished in such a manner that the local people are involved in the planning and development efforts in their communities. The sensitization also should be done concurrently in potential tourist markets, to ensure that the planning and implementation of the project is well coordinated.

Conventional tourism emphasizes the biological and ecological aspects of the environment. These are the values that were guiding principles in establishing national parks and wildlife reserves in Africa and elsewhere. Yet, natural resources have other important values, including cultural and social values, which are left unexploited. Successful ecotourism approaches must of necessity underscore and promote the cultural and social values of an area or region.

To the extent that identified sites are developed, concerned institutions, both governmental and non-governmental, need to provide technical assistance and advice to ecotourism ventures. There is a need to establish an international institutional framework to properly define and support the ecotourism concept and promote it at all levels for the ultimate benefit of the local people and their environment.

Also, there is a need to ensure that ecotourism as a conceptual tool is properly packaged to receive the necessary attention, and to promote private investment in the ecotourism industry. The international donor community should take the initiative to make ecotourism one of the funding components in the efforts to promote the pursuit of ecological sustainability.

NOTES AND SOURCES

[1] World Tourism Organization. Available at http://www.world-tourism.org.

[2] Hausler, N. 2004. Development and Cooperation, 31 (August/September): p. 340.

[3] Roe, D., N. Leader-Williams, and B. Dalal-Clayton. 1997. Take only photographs, leave only footprints. IIED Wildlife

and Development Series No. 10, October 1997, p. 4.

[4] See http://www.devalt.org/da/esb/biodiv/ecotourism.htm.

[5] Odum E.P. 1996. *Ecology and Our Endangered Life-Support System*, 3rd Edition. Sunderland, MA: Sinauer Associates.

[6] See Lavigne, Chapter 1.

[7] See, for example, de la Mare, Chapter 21.

[8] Honey, M. 1999. *Ecotourism and Sustainable Development. Who owns paradise?* Washington DC: Island Press.

[9] Whelan T. (ed.). 1991. *Nature Tourism; Managing for the Environment*. Washington DC, Covelo CA: Island Press.

[10] Grove, R. 1987 Early themes in African Conservation: the cape in the nineteenth century. In D. Anderson and R. Grove (eds.). *People, Policies and Practice*. United Kingdom: Cambridge University Press.

[11] Mugisha A. 2002. Evaluation of community-based conservation approaches: Management of Protected Areas in Uganda. PhD dissertation, University of Florida, Gainsvile FL. USA

[12] See for e.g., Adams, W.M. 2004. *Against Extinction: The Story of Conservation*. London: Earthscan.

[13] e.g. Brandon, K. 1998. Perils to Parks: the social context of threats. pp. 415 - 440. In K. Brandon, K. Redford, and S.E. Sanderson (eds.). *Parks in Peril*. Washington DC: The Nature Conservancy and Island Press; M.W. Murphree. 2000. Community-based Conservation: Old Ways, New Myths and Enduring Challenges. Key Address presented at Conference on African Wildlife Management in the New Millennium. College of African Wildlife Management, Mweka, Tanzania. 13-15, December 2000.

[14] See Key Economic Indicators, published by the Government of Uganda (1995).

[15] Ministry of Tourism, Trade and Industry. 2003. Tourism Policy for Uganda. The Republic of Uganda.

[16] Plumptre, A.J., D. Cox, and S. Mugume. 2003. The status of chimpanzees in Uganda. Albertine Rift Technical Report. Series No. 2. New York: Wildlife Conservation Society.

[17] KAFRED and Kibale National Parks Visitors Statistics Report 2002. Annual Reports of Uganda Wildlife Authority, Kampala, Uganda

[18] Ajarova, L. 2003 Price Estimation and Economic Valuation of Chimpanzee-Based Ecotourism in Uganda. Unpublished MBA dissertation. Eastern and Southern Africa Management Institute (ESAMI), Arusha, Tanzania.

Chapter 11

How Shall We Watch Whales?

Peter J. Corkeron

He maketh a path to shine after him;
one would think the deep to be hoary.
Upon earth there is not his like,
who is made without fear.
Job 41: 32-33[1]

WHALES AS ICONS

Whales – large cetaceans[2] – have held iconic value for some of humanity for centuries. In most parts of the world, maritime technology sufficient to make hunting whales viable did not exist until the early Middle Ages, at the earliest. The incapacity of most early civilizations to manage killing large whales is demonstrated by rhetorical questions in the biblical Book of Job:

Canst thou draw out leviathan with an
hook? or his tongue with a cord
which thou lettest down?

Canst thou put an hook into his nose?
or bore his jaw through with a thorn?
Job 41: 1-2

and

Shall thy companions make a banquet of
him? shall they part him
among the merchants?
Canst thou fill his skin with barbed iron?
or his head with fish spears?

Lay thine hand upon him,
remember the battle, do no more.
Behold, the hope of him is in vain: shall not
one be cast down even at the sight of him?
None is so fierce that dare stir him up: who
then is able to stand before me?
Job 41: 6-10

In this final line, we are brought to the point of this iconic rhetoric: if people could not subdue Leviathan, how could they confront the power of the omnipotent being that created Leviathan? The iconic value of human impotence in the face of whales' power is used to drive home a religious message.

Centuries passed, technology improved and we could indeed "fill his skin with barbed iron", and "part him among the merchants". So we did. At first, this was the work of iron men in wooden ships – whales were lethal quarry pursued beyond the boundaries of the known world. Even so, some whale populations were wiped out,[3] others reduced dramatically in number. Then, through much of the 20th century, industrialized whaling led to rapid near-extinction of most remaining whale populations. [4]

It has been argued that, in general, attitudes to marine mammals through the early days of modern whaling were utilitarian.[5] Although undoubtedly true, there were notable exceptions. In 1922, Sir Sidney Harmer of the British Museum described Norwegian whaling in British subantarctic waters as "insensate slaughter arousing feelings of horror and disgust".[6] D.H. Lawrence's poem "Whales Weep Not", written in 1929, includes allusions to whales' oceanic migrations, to maternal care, equates whales with angels ("great fierce Seraphim"), and depicts male whales protecting females and calves. Not a utilitarian perspective, and prescient (if not always accurate) in its conceptualization of whales

Decades passed, more whales died in the tens of thousands. After visiting a whaling station, Peter Matthiessen observed, "nothing is wasted but the whale itself". This, in his 1971 book, "Blue Meridian", encapsulated the growing view that whales were not just resources to be utilized. Near obliteration of many whale populations provided the initial spur to the "save the whales" movement. But the majestic aspects of whales – their size; the apparent intelligence of some whales; the songs of others – led to rediscovery of the old iconography – whales as magnificent in their own right. The idea grew that viewing such animals as no more than resources cheapened humanity. And from saving whales came the idea of whales as standard bearers of marine environmental issues, so "save the whales" morphed into including "save the whales' environment".

Years passed. The "save the whales" movement had what seemed a great victory. The International Whaling Commission (IWC), the international forum established to manage the whaling industry, adopted a moratorium on commercial whaling. The moratorium ostensibly came into effect by the mid-1980s, but the aim of a complete pause on commercial whaling was thwarted. One tactic used by all three of the major whaling nations – Japan, Norway and Iceland[7] – was to use a loophole in the Convention that established the IWC allowing for whales to be killed for scientific purposes. All three nations have, at some time, continued whaling by using the excuse that research is needed into what whales' "impact" is on other components of marine ecosystems, notably commercially important fish species.

At around the time that the whaling moratorium was being discussed, it became clear that there was good money to be made taking people out on boats to see whales. Whale watching as a commercial use of whales started taking off in the 1960s,[8] but – coincidentally or not – started coming into prominence at around the time of the whaling moratorium.

HOW WE PERCEIVE ANIMALS

Will he make a covenant with thee?
wilt thou take him for a servant for ever?
Job 41:4

Before discussing whale watching, I briefly sketch out some ideas on how the way that we perceive animals affects the way in which we manage our interactions with them.[9] Put simply, animals can be divided into four categories:

- *Nasties* – "we fear or loathe and which we would like to thin out",[10]
- *Lovelies* – "we like, revere or honour and which we wish therefore to conserve";
- *Commodities* – "we use as domesticants or harvest as wild animals"; and
- *Irrelevancies* – "that seldom impinge on us and towards which we direct no strong feelings".

Again, very generally, there are three ways in which we manage our dealings with animal populations: we can control them, use them or cherish them. We tend to control Nasties, cherish Lovelies, and use Commodities.

In many Western countries, whales have gone from being Commodities to Lovelies – the great success of the "save the whales" movement. However, whaling nations now portray whales as competitors for fish,[11] so for many people in these countries, whales have gone from Commodities to Nasties. This is as dramatic a change as that to Lovelies in the non-whaling nations, but does not seem to have been identified as such by the conservation movement. And whale watching uses whales' value as Lovelies to commercialise cherishing.

WHALE WATCHING

Wilt thou play with him as with a bird?
or wilt thou bind him for thy maidens?
Job 41: 5

When talking about "whale watching", I'm referring to commercial boat tours taking people to see free-ranging cetaceans, whether watching from a boat or swimming with animals. I'm focussing on commercial whale watching because that's what conservation NGOs concentrate on when discussing this industry. Personally, I believe that a finer distinction – between viewing large, migratory whales, and inshore dolphins with small home ranges – is needed.[12] This is not yet generally accepted by those engaged in researching or managing nature-based tourism on cetaceans, so I'll conform to generally accepted terminology. I'm not discussing recreational whale watching, when people make their own way to see whales, either by boat, or watching from land. Recreational whale watching raises a set of management issues that can overlap with commercial whale watching, but these are not the focus of this essay.

Whale watching is one of the most rapidly growing forms of nature-based commercial tourism in the world today. Internationally, whale watching has been increasing exponentially over the past two decades, as detailed in a series of reports commissioned by conservation organizations over the past few years.[13] The triumphal tone of these reports seems to suggest that there can be no doubt that this is a "Good Thing". Why is this? There is a lot of good that can be said about whale watching. As a commercial use of whales, it is far less lethal than whaling, although cetaceans have – very occasionally – been killed and injured from being accidentally hit by whale-watching boats. It is not often that conservation groups go out of their way to support the development of an industry, so what are the other arguments used to support whale watching?

There are four main arguments put forward by conservation groups supporting the development of the whale-watching industry.[14] These are: whale-watching boats provide opportunities for research; seeing free-ranging whales is better than seeing captive animals; whale watching is incompatible with whaling, so whale watching provides an economically viable alternative to whaling; and seeing whales while whale watching raises peoples' conservation consciousness. These arguments, or variations of one or more of them, can be seen on the websites of international organizations interested in whale conservation.[15] I concentrate of the latter two of these, as they seem to me to be the most important.

WHALING AND WHALE WATCHING

Whaling nations and whale watching

Is whale watching incompatible with whaling? There are three ways in which incompatibilities could arise: from the behaviour of whales, from the behaviour of whalers, or from the behaviour of tourists. Regarding whale behaviour, there is evidence from the early days of commercial whaling off northern Norway that whales became harder to approach.[16] There are also hints that some Antarctic blue whales are now less likely to avoid ships than in days past.[17] Hunting and watching the same whales in one place – which happened in Iceland in 2003 – seems risky for successful whale watching (not to mention for the whales!).

This overlaps with the issues of the behaviour of whalers. If whalers hunt the inquisitive individuals in a population of whales, animals who can be the mainstay of a whale-watching fleet, the whale-watching experience for tourists could be diminished, as only skittish animals will be left. If whalers deliberately hunt whales in front of whale watchers, the tourism experience might not quite meet tourists' expectations either.[18]

What about the behaviour of tourists? There are whale-watching businesses in Japan, Norway and Iceland, the major whaling nations. In Japan and Norway, the species being watched are different from the species being hunted. At present. In Norway, whale watching is a well-established industry, and tours have been running for nearly two decades. Over this period, commercial whaling re-started in Norway and, in recent years, hunt quotas have increased, with plans to raise quotas substantially in the near future.[19] In Iceland, whaling recently restarted after a hiatus of over a decade. During this decade, whale watching started.

There are no strong data on whether the existence of whaling in some nations puts tourists off going to these places. If it has put some tourists off, more than enough other tourists are ready to take their place to ensure whale watching continues. The risk of lost whale-watch tourism did not stop Icelandic authorities from restarting whaling in 2003. In Japan, "scientific whaling" programmes have expanded, in number of whales killed, and in the number of species killed, over the period since whale watching began.

So whaling not only *occurs* in countries with whale watching, whaling has *increased* in these three countries since whale watching started. Why can whaling and whale watching occur in the same countries? Businesses taking people to watch killer whales or sperm whales in localized areas along the Norwegian coast have no real effect on minke whaling, mostly happening further offshore. These days, whalers and whale watchers rarely even see each other. Most Japanese whaling happens far offshore, well beyond the range of coastal whale watchers, although the recent expansion of Japanese coastal whaling for minke whales may change this. It is only around Iceland where, so far, whaling and whale watching come into close proximity. Will this make minke whales harder for Icelandic whale watchers to approach?

Whaling (in theory, at least) provides food for people to eat – a primary industry. Whale watching provides spectacle, a service (or tertiary) industry. At one level, comparing them is to compare any two industries that use the same natural resource in different ways. But is whaling really a primary industry? Two of the major whaling nations – Japan and Iceland – conduct whaling under the guise of scientific research on marine ecosystems, in theory to provide information useful for managing fisheries. Meat sales are ostensibly just a by-product of the research. Scientific research is not a primary industry, although it could possibly be construed as infrastructure support for primary industry.

More telling, Norway, Japan and Iceland are three of the five Organisation for Economic Co-operation and Development (OECD) countries with the highest subsidies for their agricultural production[20] (Switzerland and Korea are the other two). Norway sets the pace here: about 70% of Norwegian farmers' income is from subsidies, and Norwegian farmers are paid roughly three times the world market prices their produce.[19] Agricultural production in these three nations has less to do with providing food than providing a social (or perhaps political?) service. These days, whale meat is almost entirely for internal consumption in whaling nations. So the internal market for whale meat in whaling nations is distorted, because the food markets in these nations are grossly distorted by their agricultural policies. This being so, the argument that whaling is primarily about providing food seems deeply flawed – even if "scientific whaling" is a ruse to get whale meat to consumers.

Like agricultural production in these countries, whaling is about providing a community service. What is this service? It appears to be pandering to national pride, or at least the national pride of a sector of the community that governments feel obliged to placate. This being so, whale watching can never fulfil the role that whaling does in these countries. Whaling is an expression of national identity that whale watching simply cannot replace.

Other nations and whale watching

Whales are also found in the waters of nations that do not engage in whaling. It may be that in the future, other nations will be "encouraged" to rediscover their whaling heritage, just as nations have been "encouraged" to join the International Whaling Commission.[21] An example might be the Kingdom of Tonga, an island nation in the South Pacific with a per capita gross domestic product (GDP) about 8% of Japan's, and a total GDP of 0.0066% that of Japan's.[22] Tongans engaged in subsistence whaling after learning their whaling skills from open-boat European whalers in the 18th and 19th centuries, and only stopped in 1978. Through the 1990s, there was evidence of foreign nationals encouraging Tongans to recommence whaling.[23]

Tourism is the most important internal source of income in the Tongan economy (remittances from expatriate Tongans are the most important source overall). A study of tourists to Tonga[24] indicated that most of the tourists who visit Tonga now are unlikely to continue to do so if whaling recommences, even if whale meat is used only for local consumption. In this instance, the choice between whaling and whale watching is stark. The whales to be hunted (humpbacks, *Megaptera novaeangliae*) are the same as those watched. Tourists from Western nations – an economic mainstay – are less likely to visit Tonga if whaling recommences. This seems a clear case of both people and whales doing better if whales are seen and not hurt, to paraphrase Fred O'Regan from the International Fund for Animal Welfare (IFAW).[25]

Whaling and whale watching – one size does not fit all

So, there are instances where whaling and whale watching co-exist (Norway, Japan), a case where recommencing whaling may affect the viability of whale watching, although this seems not to matter very much to people of that nation (Iceland) and a case where the resumption of whaling could possibly ruin an existing whale-watching industry (Tonga). What can be drawn from this?

If there is one clear message, it is that the relationship between whaling and whale watching is not simply the case of one replacing the other. This feeds back to whales as Lovelies, Nasties or Commodities. Whaling nations ostensibly view whales as both things to eat (Commodities) and fisheries competitors (Nasties), but more importantly, see the act of killing whales as an expression of national identity. Whale watching thrives by cashing in on cherishing Lovelies. So why should whaling nations care if people view whales along their coastline, as long as it does not stop their own nationals from hunting whales?

What do Japan (the world's third largest economy) and Norway (a non-OPEC oil economy) have to fear if they lose their whale-watching tourism sector? Precious little. For example, the 2004 white paper on Norway's new policy on marine mammals[26] includes discussion of whether an international backlash to the policy will have any negative effect. The paper concludes that it will not.

Iceland, a much smaller economy primarily based on fisheries exports (but with growing importance of nature-based tourism), is far more at risk. However, in this instance, boycotting Icelandic fisheries products *until* whaling stops is surely likely to produce a swifter reaction than pledging to go on holiday in Iceland *if* whaling stops (the current approach adopted by Greenpeace, one NGO with a clear policy relating Icelandic whaling and tourism).

It may be that the risk of losing tourism revenue associated with whale watching will prevent some other nations (e.g. Tonga) from recommencing whaling. How this cost-benefit analysis plays out – and importantly, which individuals will pay the costs and who, individually, will reap the benefits – remains to be seen.

WHALE WATCHING AND ENVIRONMENTAL ENLIGHTENMENT

Does whale watching leave people more enlightened environmentally? And if so, does this make a difference to their environmental impact? Just because people understand that their behaviour affects the environment detrimentally doesn't necessarily mean that they will alter their behaviour.

Both the whale watching and captive display industries claim that they contribute to raising peoples' environmental awareness. There is a case to be made that captive displays helped change peoples' views of cetaceans – particularly killer whales – in years past.[27] However, the same NGOs that dispute the value of captive display in changing peoples' perceptions also firmly assert that whale watching is currently of great benefit, changing peoples' perspectives on whales and hence the marine environment. So how has the marine environment fared in the past two decades,[28] the time when whale watching has been increasing exponentially worldwide?

Marine environmental issues of the late 20th century

Humanity's impacts on marine ecosystems have progressively increased in recent years. Some of this is simply because there are more people on the earth, and they are moving to coastal cities. Some is due to technological developments, particularly in fishing, in military technologies of all sorts and in mass tourism, but also to changes in our technological capacity that have increased the areas from where we can seek, and extract, hydrocarbons. Some impacts (e.g. increases in shipping traffic and hence noise) are exacerbated thanks to increased globalisation of world markets.

Industrial wild-capture fisheries peaked in the 1980s and now appear in decline.[29] Some populations of large predatory fish were dramatically reduced in size in the relatively recent past.[30] As more fisheries are fully exploited or overexploited, industrial fisheries expand their range to new areas and depths. Improved technology of all sorts (vessels, engines, radar, weather forecasting, position fixing, net construction, maritime rescue services) also helps fishermen extract more of the available fish than was possible in the past. Aquaculture businesses, particularly sea cages for fish feedlots and coastal prawn ponds, have boomed to feed markets in the richer countries of the world.

Our carbon dioxide output from using fossil carbon stores remains a problem that refuses to go away. We see increased traffic on our ever-expanding road networks, and greater use of fuel-hungry vehicles over the past two decades. The advent of low-cost airlines, and the growth of peoples' use of air flights for inessential travel (holidays, weekends away) demonstrates that the need for individual self-limitation of our consumption of natural resources remains lost on most people in richer nations. There are indications that global warming caused by excessive consumption of fossil fuels is affecting marine ecosystems in ways that are only now starting to be understood.[31]

Perhaps most importantly, our consumption of resources now outstrips our planet's capacity to regenerate. This change happened some time over the past two decades. And our ecological footprint is not getting any smaller – if anything, the size of our ecological boot just keeps growing.[32] Despite the optimism of the Bjørn Lomborgs of the world,[33] the capacity of our planet's environment to support human life seems in decline. And one nation with around the highest *per capita* consumption of resources (the USA) is also – coincidentally – a place with one of the most valuable whale-watching industries.

But all is not doom and gloom. Some whale populations are increasing in size, as decades of protection from whaling show results.[34] The desire to see whaling – commercial and "scientific" – ended once and for all is a clearly enunciated policy held by the governments of several nations, even if international negotiations to realise this desire seem mired in failure. Enlightened management processes aimed at ensuring the maintenance of populations of marine mammals as viable components of marine ecosystems (such as the application of the Potential Biological Removal [PBR] methods in the USA) are enshrined in law in some countries. Things could always be worse.

What about whale watching?

The rise of whale watching has coincided with a rise in general, and in particular of marine, environmental degradation. There is no simple cause-effect relationship here, but lacking replicate planets, we're not really in a position to assess whether things would be better or worse if we had no whale watching. People in many Western countries may cherish whales, and it may be that whale watching reinforces this emotion. But there seems to be a disconnect between people loving whales and individually making the personal choices, and seeking the societal changes, that are needed to ensure healthy marine environments.

The value of whale watching?

So the important arguments raised by conservation NGOs for supporting whale watching are now looking rather weak. Should NGO support for the whale-watching industry be discarded completely? I suggest not, but support could be refocussed and arguments refined.

WHERE TO WITH WHALE WATCHING?

I will not conceal his parts, nor his power,
nor his comely proportion.
Job 41: 12-13

The whale-watching industry is with us now, and seems unlikely to disappear in the near future. This being so, and given conservation NGOs' continuing support of whale watching, what can be done to minimize the industry's deleterious effects and maximize its positive attributes? Here I sketch out some ideas, by assessing whether blanket support for continued expansion of whale watching is necessarily the best approach for conservation NGOs; and how NGOs can do more to encourage "responsible" whale watching.

The growth of whale watching

Consistently, reports produced by some conservation NGOs glory in the continued exponential increase of the whale-watching industry. But when does enough whale watching become too much? Should every population of whales be watched at every point in their yearly travels? If not, when should they be avoided? Should some populations or individuals be left alone? At present, these questions are speculative as we lack the technical capacity (and the markets) to be with all whales all the time. But these questions can focus attention on the need for appropriate management of whale watching. How do we decide the best way to answer this series of questions?

Managing whale watching needs to account for the detrimental influences that whale watching can have on whales being watched. From the first studies of the effects of whale watching on target animals, research has shown consistently that the behaviour of animals is affected by the presence of whale watch vessels.[35] The question then morphed into demonstrating "long term" behavioural effects and finally "biologically significant" effects.[36] The problem with this is that if the whale-watching industry is increasing exponentially over the period that studies are in progress, by the time an effect can be demonstrated, the industry will be much larger than it was at the start, and so inherently more difficult to control. Research is now beginning to demonstrate effects that are clearly "biologically significant", but this work is based on data collected over many years.[37] In some places now, whale watching can be an important source of human-induced disturbance to whales or dolphins.[38]

Perhaps now is a good time for conservation NGOs to stop actively encouraging further increases in whale watching, except for specific, targeted cases (e.g. the example of Tonga, discussed above). Perhaps seeking a pause in the further growth of whale watching should be considered. An example of such an enlightened approach is that taken by management authorities in Queensland, Australia, where whale watching on humpback whales is allowed only in a few designated whale management areas, all in marine parks. Within the whale management areas, the number of commercial vessels that can operate is limited by permit. In this way, Queensland authorities prevented exponential growth of commercial whale watching.[39]

By ceasing active encouragement of all whale watching, NGOs could acknowledge the intractable relationship between managing to mitigate the effects of a growing whale-watching industry, when we know that effects can take years to demonstrate, and encouraging industry growth world-wide. Supporting the continued exponential increases in whale-watching industries while calling for research demonstrating "biologically significant" effects of whale watching is, in essence, contradictory.

Defining "responsible whale watching"

Conservation NGOs that support whale watching (e.g., IFAW; the Whale and Dolphin Conservation Society [WDCS]; Greenpeace; World Wide Fund for Nature [WWF]) also stress the need for *responsible* whale watching. Adding "responsible" in front of "whale watching" is a start, but without knowing what "responsible" whale watching is, we don't get very far. And having defined *responsible* whale watching, how can it be differentiated from *irresponsible* whale watching? Having established the difference, then, how to support responsible whale watching while discouraging irresponsible whale watching?

It is possible to manoeuvre a boat around cetaceans in a way that does not substantially affect animals' behaviour,[40] a technique used for years by some animal behaviourists studying dolphins.[41] This is obviously the first step for responsible whale watching, but one that is clearly not achieved by most of the commercial whale-watching operations that have been studied. The many guidelines or regulations governing the way people should manoeuvre their vessel in the vicinity of whales are an attempt to help – or coerce – vessel operators to limit their impact on whales.

Commercial whale watching also comes in many guises. An important distinction is between huge industrial enterprises of many vessels working in one area (e.g. whale watching in the waters off the Azores) with small operations where perhaps only one vessel runs whale watches in an area. The extreme is the places where the number of whale-watching vessels approaches the number of individual whales known to exist in the population being watched.[42] Paul Forestell has refined our understanding of this issue by pointing out the important distinction between what he refers to as "subsistence" (a small industry with few boats), "commercial" and "industrial" (huge industry, very many boats) whale watching.[43] "Responsible industrial whale watching" is likely to be an oxymoron.

How much whale watching is too much? Recent research is starting to answer this question quite specifically. David Lusseau[44] has recently shown that if dolphins had commercial tour boats around them for more than 35% of the daylight hours, their overall time spent in rest was reduced. This gives us a way to limit vessel numbers. Until David's study is replicated on populations elsewhere, this proportion of time should be taken as the absolute maximum daily limit that whale-watching vessels can be around cetaceans. Some will argue that as David's study deals with one species at one site with one particular industry, his results cannot be used to infer anything about populations elsewhere. Wearing my geeky scientist hat, I agree that they are absolutely right. However, it is situations like this for which the precautionary principle was devised. Until replicate studies are carried out, David's data are the best available. Perhaps for some species at some sites, a longer period might be possible, but the onus is on those who wish to have vessels around cetaceans for longer to demonstrate how what they are offering can be construed as "responsible".

Supporting responsible whale watching

Some NGOs (WDCS is perhaps the clearest on this) oppose tours that include swimming with cetaceans. A major reason for this distinction is that, in general, swim-with operations are likely to be more intrusive than standard, vessel-based whale watches. This is a very valid point, but what about swim-with operations that are no more intrusive than most vessel-based industries (and may be substantially better than some)? The industry based on swimming with dwarf minke whales within the Great Barrier Reef Marine Park provides an example of what could be considered a "responsible" swim-with operation.[45]

By distinguishing the manner in which they support swim-with and "normal" whale watching, WDCS has started the process of differentiating between "responsible" and "irresponsible" whale watching. Instead of using boat-based = good; swim-with = bad, why not assess what goes into "responsible" whale watching, and see whether operations conform to responsibility?

My list (others may differ) of what is needed to ensure responsible whale watching includes:

- A licensing system for whale-watching tour operations. Without licences, limiting the number of vessels engaged in whale watching in an area is virtually impossible;
- Legally enforceable regulations dealing with the manner in which vessels operate in the vicinity of whales;
- Enforcement of regulations, including a management agency presence on the water. Management staff must have the capacity to fine operators for breaches of regulations and, as a last resort, managers need the capacity to revoke the licence of an operator that consistently flouts regulations;
- For sites where cetaceans have areas where they are most likely to engage in some important behaviour (e.g. resting, feeding), spatially explicit zoning to limit the influence of industry in sensitive areas;[46]
- A scientific monitoring programme to assess the degree of compliance with regulations and the status of the cetaceans being watched. The capacity of the monitoring programme to achieve its stated aims must be demonstrated;
- Formally agreed limits of acceptable change to the aspects of cetaceans' biology that are being monitored. Should these limits be breached, management measures (agreed to in advance) should be implemented. Examples of possible limits include:
  - — An annual, biennial or triennial limit to the number of animals that can be killed or seriously injured by whale-watching vessel strikes,
  - — Limits to the extent to which the population of animals being watched can decline (for whatever reason, not necessarily attributable to whale watching),
  - — Limits to the change in the amount of time that animals spend in particular behavioural states (e.g. if the proportion of time that animals rest is reduced by an agreed amount, management changes are implemented);
- As implementing most of the points above requires spatially explicit regulations, establishing a multiple use marine protected area (MPA), if one does not already exist, will probably be necessary;
- Educating by example with regard to environmental impacts. A couple of possibilities are:
  - — Carbon neutrality for all aspects of the industry (including tourists' air travel), and
  - — On whale watch vessels, serving only seafood caught in a manner that is ecological sustainable;[47]
- An industry-pays policy for the costs of management and monitoring.

Responsible whale-watching enterprises would include all of these points (or whatever list is decided by those willing to take this on). Irresponsible whale watching would include none of them. But definitions are just a start.

Whichever organization (if any) takes on responsibility for developing this idea could consider a star system – five stars for operations that meet the highest standards, no stars for operations that fail to meet any standards. Consumers could then show by their behaviour whether they really care for responsible whale watching or not. Clearly, operations that are directly supported by conservation NGOs (e.g. through WDCS' "out of the blue" holidays) must meet the highest environmental standards.

ICONS REVISITED

whatsoever is under the whole heaven
is mine.
Job 41:11

Once, icons were sacred images. The meaning of icons changed, and came to include people, places or things that stood to represent some set of beliefs or way of life. Now, an icon is an image on our computer screen that we click on to instruct the computer to run some task. Just as the meanings of icons changed, so too has the iconic meaning of whales.

Once, whales were unapproachable, unimaginably powerful giants. They became dangerous animals, sources of wealth to be conquered by brave souls. As technology improved they metamorphosed into big, living tubs of oil. Whales' near disappearance recreated them as symbols of humanity's inappropriate desire for wealth from nature. Now, they are both large nuisances, peceived to be eating too many commercial fish, and big, charismatic animals that are wonderful to see.

The apparent public apathy to current increases in whaling, and to the marine mammal culling currently under way in some countries (e.g. Canada and Norway), suggests that the value of whales (and marine mammals, generally) as environmental icons may be fading. And even within some nations that oppose commercial whaling, there are those who support the idea that whales consume too many fish. The International Coalition of Fisheries Associations (ICFA), whose members include fisheries associations of Australia, New Zealand and the USA, have resolved that whales' consumption of commercial fish is a problem that needs addressing.[48]

The science behind the "whales eat too many fish" idea is poor, but this chapter is not the place for detailed technical analyses of that scientific research. The general question – should whales be able to eat fish (and squid and shellfish) is worth contemplating. One of the several unspoken assumptions regarding whales' "impact" on marine ecosystems is that if the whales weren't there, more fisheries product would be available for people. This further presupposes that we have the technology to wipe out the whales (which we do), the technology to catch the "now unused" fish (which may or may not turn out to be true), that markets exist for these fish (unlikely, as whales eat mostly non-commercial fish[49]), and that our understanding of marine ecosystem processes is so advanced that we know what effects will flow from removing whales. This last presupposition is demonstrably false, as scientists are still wondering over what drives some of the most dramatic and obvious changes we have seen in marine ecosystems in recent years.[50]

A mix of technological development, economics and historical accident has brought us to the point where we humans now exert substantial influence over the biological processes driving marine ecosystems. One of our accidents has been dramatic overfishing of many species, resulting in a general decline in world fisheries production. Scapegoating whales for our mistakes is easy. But scapegoating begs the question of whether whales have any right to exist at all. And if whales are gone, and fisheries still declining, what then? Seals? Seabirds? Sharks? All fish? Shall we leave behind us empty oceans, for our starving grandchildren to squabble over the last jellyfish?

and sorrow is turned into joy before him.
Job 41:22

There is no ocean wilderness any more. Our control over the oceans would seem godlike to our ancestors, but we lack omnipotence. The vessel technologies that allow people to view whales in safety and comfort also bring destruction to their environment. Our desire for new experiences costs fossil fuel, and contributes to the warming planet.

These are the messages that need to come across on whale-watching trips. Conservation groups need to seize the opportunity presented by the current calls to cull marine mammals, and reshape the iconography of whales for this new century.

ACKNOWLEDGEMENTS

My thanks to Dave Lavigne for inviting me to the meeting at Limerick, and for inviting me to write this essay. My previous employer, the Norwegian Institute of Marine Research, funded my trip to Limerick. The manuscript was improved by comments from Sidney Holt, Dave Lavigne, Russell Leaper, Vassili Papastavrou and Sofie Van Parijs. I wrote this while visiting the Bioacoustics Research Program at the Cornell Laboratory of Ornithology, and thank everyone at BRP for their kindness.

NOTES AND SOURCES

[1] Biblical quotations are from the King James Version of the Bible. I'm following St. Thomas Aquinas' view that Leviathan referred to whales. Thoughts for this essay are rooted in Judeo-Christian, Western traditions. A similar review based on Eastern thought would be timely, but beyond me.

[2] Through this essay, when I refer to whales, I'm talking about

cetaceans other than dolphins (Delphinids), porpoises (Phocoenids) and river dolphins.

[3] For example, gray whales in the western North Atlantic, see Mead, J.G. and E.D. Mitchell. 1984. pp. 33-53. In *The Gray whale. Eschrichtius robustus*. Academic Press. New York.

[4] For more on the history of whales and whaling, see Papastavrou and Cooke, Chapter 7.

[5] For example, P. Forestell, 2002. Popular culture and literature. pp. 957-974. In W.F. Perrin, B. Würsig, and H.G.M. Thewissen (eds.). *Encyclopedia of Marine Mammals*. Academic Press, San Diego.

[6] Quoted from page 342 of Tønnessen, J.N. and A.O. Johnsen. 1982. *The History of Modern Whaling*. Translated by R.I. Christophersen. University of California Press, Berkeley, CA.

[7] Nationals of other nations engage in whaling, for example: Canada, Denmark (Greenland, the Faeroe Islands), Indonesia, the Russian Federation, St Vincent and the Grenadines, and the USA. Whaling by these people falls under the "Aboriginal Subsistence" category at the International Whaling Commission – or would, if all were IWC members (Canada and Indonesia are not members). Whale watching occurs in all of these countries, too.

[8] Hoyt, E. 2002. Whale watching. pp. 1305-1310. In W.F. Perrin, B. Würsig, and H.G.M. Thewissen (eds.). *Encyclopedia of Marine Mammals*. Academic Press, San Diego.

[9] Caughley, G. 1985. Problems in wildlife management. pp. 129-135 In H. Messel (ed.). *The Study of Populations*. Pergamon Press, Sydney.

[10] Quotations are from page 129 of Caughley's chapter in *The Study of Populations*.

[11] Reviewed in Kaschner, K. and D. Pauly. 2004. Competition between marine mammals and fisheries: Food for thought. Report from the Fisheries Centre, University of Brtitish Columbia, Canada. Available at http://www.hsus.org/ace/21314.

[12] Corkeron, P.J. 2004. Whalewatching, iconography and marine conservation. Conservation Biology 18: 847-849.

[13] See, for example: Hoyt, E. 2001. Whale watching 2001: worldwide tourism numbers, expenditures, and expanding socioeconomic benefits. International Fund for Animal Welfare, Yarmouth Port, Massachusetts, or Economists@Large & Associates 2004. From whalers to whale watchers. The growth of whale watching tourism in Australia. International Fund for Animal Welfare, Yarmouth Port, Massachusetts.

[14] See also Corkeron, P.J. 2004.

[15] See, for example, http://www.wdcs.org; http://www.ifaw.org: http://www.greenpeace.org: http://www.panda.org.

[16] Tønnessen, J.N. and A.O. Johnsen. 1982.

[17] Corkeron, P.J., P. Ensor, and K. Matsuoka. 1999. Observations of blue whales feeding in Antarctic waters. *Polar Biology* 22:213-215.

[18] Although I've heard an anecdote of tourists disappointed that a "Whale Safari" in Norwegian waters did not include any whale deaths!

[19] Anonymous 2004. *Norsk sjøpattedyrpolitikk*. Stortingsmelding 27 (2003-2004). Available at http://odin.dep.no/filarkiv/207622/STM0304027-TS.pdf. An English a translation of the white paper's summary chapter is available at http://odin.dep.no/filarkiv/202967/marine_mammal_summary_final.pdf.

[20] From OECD 2004. OCED Agricultural Policies 2004. At a glance. Available at: http://www.oecd.org/dataoecd/63/54/32034202.pdf.

[21] Despite dispute regarding allegations that whaling nations have engaged in "vote buying" at the IWC, in 2002 a senior Japanese official admitted that this was the case. See, for example http://www.abc.net.au/pm/s331666.htm.

[22] Country's details are taken from the online version of the CIA World Factbook, http://www.cia.gov/cia/publications/factbook/index.html.

[23] For example, see Orams, M. 2000. Whale watching in Vava'u, Tonga, an important economic resource. Whales Alive 9(2). Available at http://www.csiwhalesalive.org/csi00208.html.

[24] Orams, M.B. 1999. The Economic Benefits of Whale Watching in Vava'u, The Kingdom of Tonga. Centre for Tourism Research, Massey University at Albany, North Shore, New Zealand. 64 pp. + appendices.

[25] O'Regan, F. 2001. Preface. p.1. In E. Hoyt. Whale watching 2001: worldwide tourism numbers, expenditures, and expanding socioeconomic benefits. International Fund for Animal Welfare, Yarmouth Port, Massachusetts.

[26] Anonymous 2004. *Norsk sjøpattedyrpolitikk*. Stortingsmelding 27 (2003-2004). Available at http://odin.dep.no/filarkiv/207622/STM0304027-TS.pdf. An English a translation of the white paper's summary chapter is available at http://odin.dep.no/filarkiv/202967/marine_mammal_summary_final.pdf.

[27] Corkeron P.J. 2002. Captivity. pp. 192-197. In W.F. Perrin, B. Würsig B. and J.G.M. Thewissen (eds.). *The Encyclopedia of Marine Mammals*. Academic Press. San Diego.

[28] The time when whale watching has gone from cottage industry to mass tourism, see Hoyt, E. 2001.

[29] Pauly, D., V. Christensen, S. Guienette, T. J. Pitcher, U. R. Sumaila, C. J. Walters, R. Watson and D. Zeller. 2002. Towards sustainability in world fisheries. Nature 418: 689–695.

[30] Myers, R.A. and B. Worm. 2003. Rapid worldwide depletion of predatory fish communities. Nature. 423: 280-283.

[31] For example: Edwards, M. and A.J. Richardson. 2004. Impact of climate change on marine pelagic phenology and trophic mismatch. Nature 430: 881-884.

[32] For examples, see Wackernagel, M., N.B. Schulz, D. Deumling, A. Callejas Linares, M. Jenkins, V. Kapos, C. Monfreda, J. Loh, N. Myers, R. Norgaard, and J. Randers. 2002. Tracking the ecological overshoot of the human economy. Proceedings of the National Academy of Sciences 99: 9266-9271; Myers, N. and J. Kent. 2003. New consumers: The influence of affluence on the environment. Proceedings of the National Academy of Sciences 100: 4963-4968.

[33] Lomborg, B. 2001. *The Skeptical Environmentalist*. Cambridge University Press. Cambridge, UK.

[34] For an example, see Paterson, R.A., P. Paterson and D. H. Cato. 2001. Status of humpback whales, *Megaptera novaean-*

gliae, in East Australia at the end of the 20th century. Memoirs of the Queensland Museum. 47: 579-586.

[35] See, for example, Baker, C. S. and L. M. Herman. 1989. Behavioral responses of summering humpback whales to vessel traffic: experimental and opportunistic observations. Technical report NPS-NR-TRS-89-01. Final report to the National Park Service, Alaska Regional Office. Anchorage, Alaska.

[36] International Whaling Commission. 2001. Report of the workshop on assessing the long-term effects of whale watching on cetaceans. Annex N. Journal of Cetacean Research and Management 3 (Supplement): 308–315.

[37] For example: Constantine, R., D.H. Brunton, and T. Dennis. 2004. Dolphin-watching tour boats change bottlenose dolphin (*Tursiops truncatus*) behaviour. Biological Conservation 117: 299-307.

[38] Foote, A.D., R.W. Osborne, and A.R. Hoelzel. 2004. Whale call response to masking boat noise. Nature 428: 910.

[39] See Vang, L. 2002. Distribution, abundance and biology of Group V humpback whales, *Megaptera novaeangliae*: A review. The State of Queensland, Environmental Protection Agency. Brisbane.

[40] For example: Corkeron P.J. 1995. Humpback whales (*Megaptera novaeangliae*) in Hervey Bay. Behaviour and interactions with whalewatching vessels. Canadian Journal of Zoology 73: 1290-1299; Lusseau, D. 2003. Male and female bottlenose dolphins (*Tursiops* spp.) have different strategies to avoid interactions with tour boats in Doubtful Sound, New Zealand. Marine Ecology Progress Series 257: 267-274.

[41] Mann, J. 2000. Unraveling the dynamics of social life. Long-term studies and methods. pp. 45-64. In J. Mann, R.C. Connor, P. Tyack, and H. Whitehead (eds.). *Cetacean Societies. Field studies of dolphins and whales*. The University of Chicago Press, Chicago.

[42] See, for example, Figure 2a in Foote, A.D.,R.W. Osborne, and A.R. Hoelzel. 2004. Whale call response to masking boat noise. Nature 428: 910.

[43] P. Forestell, 2002. Popular culture and literature. pp. 957-974. In W.F. Perrin, B. Würsig, and H.G.M. Thewissen (eds.). *Encyclopedia of Marine Mammals*. Academic Press, San Diego.

[44] Lusseau D. 2004. The hidden cost of tourism: Effects of interactions with tour boats on the behavioural budget of two populations of bottlenose dolphins in Fiordland, New Zealand. Ecology and Society 9(1): 2. Available at http://www.ecologyandsociety.org/vol9/iss1/art2.

[45] Birtles, A., P. Arnold, and A. Dunstan. 2002. Commercial swim programs with dwarf minke whales on the northern Great Barrier Reef, Australia: some characteristics of the encounters with management implications. Australian Mammalogy 24: 23-28.

[46] See an example of how this could work in: Lusseau, D. and J.E.S. Higham, 2003. Managing the impacts of dolphin-based tourism through the definition of critical habitats: the case of bottlenose dolphins (*Tursiops* spp.) in Doubtful Sound, New Zealand. Tourism Management 25: 657-667.

[47] Pointers to sustainable seafood can be found through a few organizations now, for example the Marine Stewardship Council, Monterey Bay Aquarium's "Seafood Watch" or the Seafood Choices Alliance.

[48] For example, the ICFA press release of August 14, 2001 "ICFA calls for a scientific approach to whaling", available at the "News" page of the ICFA website: http://www.icfa.net.

[49] Reviewed in Kaschner, K. and D. Pauly 2004.

[50] See, for example: Springer, A.M., J.A. Estes, G.B. van Vliet, T.M. Williams, D.F. Doak, E.M. Danner, K.A. Forney, and B. Pfister. 2003. Sequential megafaunal collapse in the North Pacific Ocean: an ongoing legacy of industrial whaling? Proceedings of the National Academy of Sciences 100: 12223-12228; Yodzis, P. 2001. Must top predators be culled for the sake of fisheries? Trends in Ecology and Evolution 16: 78-83.

Part III:

Factors At Play

CHAPTER 12

ATTITUDES, VALUES AND OBJECTIVES: THE REAL BASIS OF WILDLIFE CONSERVATION

Vivek Menon and David Lavigne

It is widely believed that most conservation philosophies[1] and attitudes toward animals and their natural habitats have utilitarian origins.[2] It is also believed that only in the last two centuries have other views begun to contribute to the conservation debate,[3] e.g. that animals and their habitats have other values beyond the purely utilitarian, that non-human life has intrinsic value, and that our interactions with other species should be guided by animal welfare and other ethical considerations. The latter views are considered by many to be Western mindsets pioneered by European and American naturalists and champions of the wild, including John Muir, Albert Schweitzer, and Aldo Leopold.[4]

There exist, however, in Asian and African cultures, several examples – all of which are over 200 years old, and some dating back more than 2000 years – that show non-utilitarian values being espoused and practiced. In the West, the re-discovery and re-attribution of spiritual, aesthetic, recreational, and sentimental values to wild animals, plants and their habitats was termed "ecological consciousness" by Leopold[5] and "Deep Ecology" by Arne Naess.[6] The former was seen as a transition from scientific ecology, where the relationship of an unbiased observer is separate from the object of the study. In deep ecology, the relationship is one of involved participation that recognizes the role of values and perceptions as they affect our interpretations of the natural world.[7]

The mainstream perception of nature has also changed over time. For centuries, nature and the species comprising it were seen largely as commodities – "natural resources" – of use to humans. This led to a period of over-exploitation, especially in the 19th century.[8] In the early part of the 20th century, there was a renewed celebration of wilderness. In the latter half of the 20th century, we began to view nature from an ecosystem perspective and became concerned about the maintenance of what is now called "biodiversity". The last half of the 20th century also saw calls for the liberation of animals[9] and nature[10] that gave rise to the modern animal rights movement. By the turn of the century, some people began to talk about "culture in nature and nature in culture"[11] and "greater recognition [was] given to the interrelationships between spiritual beliefs, practices of a community and how that community relates to the environment and the world".[12]

Parallel to these developments, there also emerged a "wise use" or "sustainable use" movement, championing the commercial consumptive use and international trade of nature and natural resources, with considerable clout and effect.[13] At the time of writing, the world of nature conservation is split asunder by the "use it or lose it" proponents and the "intrinsic value of nature" proponents, with various intermediate niches occupied by individuals, organizations, pressure lobbies, and governments.

Our changing perception of nature is not merely the result of new facts presented by science (although science has definitely contributed) but reflects the varied and evolving attitudes, values and objectives of individuals, and the societies in which they live.

Wildlife conservation and management is frequently thought of as a scientific undertaking that involves presenting previously unpublished facts in scientific journals (e.g. *Conservation Biology*; *Biological Conservation*) and using such information to inform management decisions. It is sometimes claimed that wildlife management is evolving from an art to a science,[14] or that it is already a science,[15] and that management decisions should be based on the best available science. Topics such as maintenance of biodiversity, protection of endangered species, and the removal of sustainable yields from exploited species, are considered to be in the domain of "science", whereas issues related to animal welfare concerns and animal rights are generally viewed as emotional and non-scientific, and beyond the realm of scientific consideration.

It is our view, however, that virtually all conservation discussions and debates are largely shaped not by the "facts" of science, but rather by the attitudes, values and objectives of individuals, or groups of individuals, in society. The latter include communities, governments and their bureaucracies, non-governmental organizations (NGOs) and transnational corporations, to name a few. In actuality, the decision to remove a sustainable yield from a wildlife population is an ethical choice, as is the choice of a particular sustainable yield and, in theory at least, the choice of the population size in the wild from which to remove such a yield. Similarly, decisions to protect endangered species, to maintain biodiversity, or to implement the precautionary approach, are not dictated by scientific principles, but rather by ethical choices.

Science, of course, has a role to play in describing natural systems and figuring out how they "work," informing debates, providing a range of options for policy makers and managers, and quantifying the risk associated with various options. But, at the end of the day, science cannot make the choice. People do that, and they do it on the basis of their attitudes and values,[16] and on the basis of the objectives that they have for nature and so-called natural resources.

As anyone who works in the field will know, conservation is characterized by controversy and debate. When examined closely, it becomes immediately obvious that the controversies and debates involve disputes over the available "science" (the facts, if you will) and the attitudes, values or objectives of the participants.

Joseph Berry depicts the situation nicely, albeit simplistically, using a conflicts matrix (Figure 12-1).[17] Where facts and values are in agreement, there really is no controversy or debate. In such situations, science can provide a computational "solution". When values are in agreement, but there is some dispute over the "facts", science can test among competing hypotheses, reject those that do not withstand scrutiny and, with luck, move the conflict into the computational box. In many societies, the legal system serves a similar purpose, deciding which "facts" are "true" (or beyond reasonable doubt) and which are not.

In most conservation discussions and debates, however, there is a conflict over values or objectives. In western democracies, disputes over values (or objectives) are the stuff of politics. When the facts are not in dispute but there remains a conflict over values (or objectives), such conflicts are resolved – to the extent that they can be – by the political process.

Finally, there are societal conflicts where both "facts" and values are in dispute. Most conservation debates fit

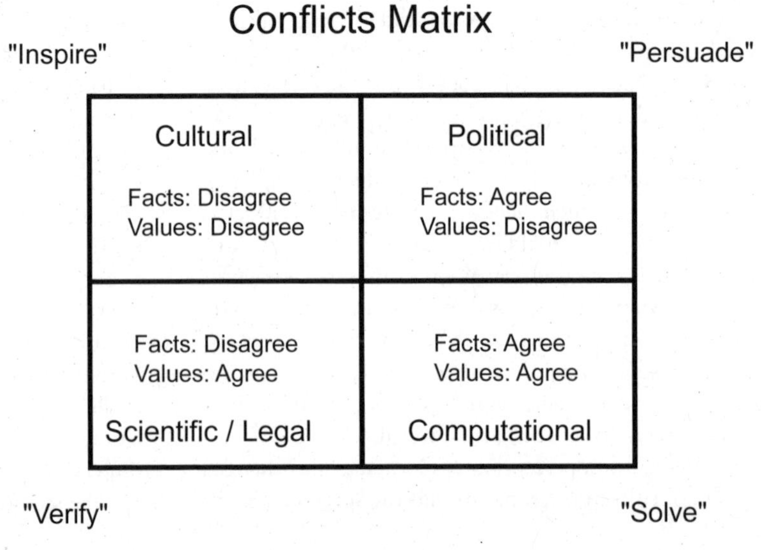

Figure 12–1. Berry's conflicts matrix. Among other things, Berry notes that the contributions of science to conflict resolution vary depending on the nature of the fact/value conflict. Where values are in disagreement, the role of science is limited to persuasion and inspiration, depending on whether or not there is agreement on the facts of the issue.

Conflicts Matrix

"Inspire" "Persuade"

Cultural Facts: Disagree Values: Disagree	Political Facts: Agree Values: Disagree
Facts: Disagree Values: Agree Scientific / Legal	Facts: Agree Values: Agree Computational

"Verify" "Solve"

into this category, which Berry aptly terms a "cultural conflict". According to Berry's scheme, such conflicts remain unresolved unless some agreement can be reached either on the facts, or on the values, thereby moving the dispute into another box in the matrix, where mechanisms do exist to resolve the issue. In the real world, however, management authorities often adopt one set of "facts" as being true and simply reject all others, thereby reducing Berry's cultural conflicts into ones that are dealt with through the political process.

Canada's management of its controversial commercial seal hunt is a classic example. The government simply adopts one set of "facts", neglecting any information that does not suit its purposes, and continues to promote an activity that is not supported by a large proportion of Canadians. The controversy is never resolved, but the hunt continues.[18] Other examples abound, particularly – these days – out of Washington, where the concept of "good science" has come to mean science that supports the policies and objectives of the Bush administration; any science that does not is deemed "bad science".[19]

Regardless, no discussion of the "pursuit of ecological sustainability" – itself a value judgment (or a societal objective) – is complete without some understanding of the ways in which attitudes, values and objectives shape the differing opinions that individuals, organizations, and nations have regarding the interaction between humans, nature and natural resources. Philosophers and thinkers such as Kabilsingh concur, noting that:

> Formal government measures for protection of nature requires acceptance by the people, with the recognition that effective conservation needs to be based on deep value convictions.[20]

Some heads of nations also are veering to this line of thought. As HM King Gyanendra Bir Bikram Shah has recently written,

> Modern conservation is no more about species or ecosystem biology; it is more about the attitudes and behaviour of humans.[21]

In this chapter, we argue that although science provides the facts (as we currently know them) surrounding a particular issue and informs the discussion, it is the attitudes, values and objectives of individual people, societies, and nations and their governing bodies (using facts provided by science to their own ends)[22] that shape conservation policy. It is not "rational" science, therefore, but rather politics, that provides the real basis for conservation in practice.

A FEW WORDS ABOUT ATTITUDES AND VALUES

Attitudes and values are topics discussed by a variety of authors from different fields, including sociology, political science, public opinion polling, and ethics, to mention but four. For that reason, we will begin our discussion by providing definitions that convey the intended meaning in this chapter.

In the present chapter, *attitudes* will be defined simply as, "...learned predispositions to respond in a consistently favourable or unfavourable manner".[23] *Values*, according to Henning "... are individual and collective concepts with emotional, judgmental, and symbolic components that we use to determine what is important, worthwhile, and desirable. Thus, values contain, and at the same time evolve from, judgments and beliefs about what is 'good' or 'bad' and 'right' or 'wrong'...By their very nature values are complex in both interpretation and influence".[24]

That it would be difficult to distinguish between an "attitude" and a "value" using the above definitions is not very surprising. In much of the sociological literature we will cite (particularly that arising from Stephen Kellert's work at Yale), the terms attitudes and values are frequently used as synonyms. Table 12–1 provides Kellert's classification of attitudes (or values) that is widely used in the conservation field, along with descriptions of each term in his typology.[25]

The full spectrum of the attitudes (or values) listed in Kellert's scheme will be observed in society as a whole, but the prevalence of each may change over time.[26] In a similar way, attitudes of individuals may also change throughout their lifetimes. In addition to age, a number of other factors have been shown to affect attitudes, values and objectives. They include: sex, religion, ethnicity, family background, country of residence, location of residence (urban or rural), culture, education, employment, socioeconomic status, political philosophy, and time in history. Some of these factors will be discussed in greater detail later in the chapter.

Another way to frame our discussion is through an examination of the objectives that individuals or societies have for wildlife or wildlife habitats (Table 12–2).[27] It should be pointed out, however, that to hold any objective outlined in each category in Table 12–2 – socioeconomically oriented objectives, ecologically oriented objectives, or ethically oriented objectives – involves value judgments or ethical choices. In other words, the objective of using a wildlife population to provide for scientific uses and increased knowledge is, in and of itself, an ethical choice – a point that is often overlooked or forgotten. It should also be obvious from Table 12–2 that many of the objectives that society might have for wildlife are mutually incompatible and cannot be realized simultane-

Table 12–1. Attitudes toward animals.

Term	Definition
Naturalistic:	Affection for wildlife and the outdoors; satisfaction from direct experience/contact with nature
Humanistic:	Affection for individual animals; emotional attachment, "love" for nature
Aesthetic:	Interest in the physical appeal and beauty of nature
Symbolic:	Interest in the use of nature for metaphorical expression, language, expressive thought
Moralistic:	Concern about the treatment of animals, with strong opposition to exploitation or cruelty toward animals; strong affinity, spiritual reverence, ethical concern for nature
Scientistic:	Interest in the physical attributes and biological functioning of animals
Ecologistic:	Concern for the environment as a system, for interrelationships between wildlife species and natural habitats
Utilitarian:	The practical and material exploitation of natural resources including animals or their habitats
Dominionistic:	The mastery and control of animals and their habitats
Negativistic:	Fear, aversion, and alienation from nature
Neutralistic:	Passive avoidance of animals due to indifference or lack of interest

ously. This, in large measure, explains the controversial nature of the conservation field.

HOW VALUES AND OBJECTIVES ARE THE REAL BASIS OF CONSERVATION

Here, we present five examples that demonstrate how values and objectives affect the conservation of wildlife and their habitats. It is well known that attitudes toward different species or groups of species evoke different value judgments that lead to different conservation strategies and management goals. However, the same species or set of species viewed across 1) geographical regions, and 2) nation states, may show finer variations in attitudes, values and objectives as cultural, economic and political realities weigh in. Furthermore, even within the same region or nation 3) values inculcated by religion and cultural taboos vary as do 4) values dependent on livelihoods and 5) time in history. While, as noted earlier, other factors also affect value systems, these illustrative examples demonstrate the central role that attitudes, values and objectives play in wildlife conservation today.

1. East is East and West is West

In the classic theory of "the clash of civilizations",[28] East and West represent two polarities that seldom meet. Yet, when it comes to attitudes towards Nature, such a view is an oversimplification. Attitudinal surveys indicate that people in the East and West actually exhibit similar attitudes on some issues and dissimilar attitudes on others.[29]

In a survey of attitudes among Japanese (East) and American (West) people to the natural world, Kellert, for example, has found that the "humanistic" attitude[30] toward Nature ranked highest in both regions, exhibiting remarkably similar percentage responses (37 and 38 percent, respectively). Thereafter, however, attitudinal differences began to emerge. While 27.5% of Americans tended to view nature from a "moralistic" perspective, the Japanese response was quite different. Thirty-one percent of Japanese viewed nature in a "negativistic" manner, while a further 28% expressed a "dominionistic" attitude. Not only do Kellert's data reveal marked regional (or cultural) differences in attitudes toward nature, they also demonstrate that the view – held in some quarters – that the East is generally "closer to Nature" than the West is either completely wrong, or a gross oversimplification of a complex topic.

Whether Japan represents a classic Eastern philosophy and/or set of values is, however, a matter of debate, since the concept of interspecies equity first emerged in certain eastern philosophies. Hideo Obara attributes the current disengagement of Japanese culture from nature to the rapid industrialization and militarization of the nation to counter western forces. He says, "the common Japanese view of nature allows the characterization of a mere 'landscape' without wild animals as 'nature'."

This is not, however, a common view throughout the East, and South Asian philosophies and value systems, in particular, are very different from those of the Far East. If South Asian values were analyzed, they might more closely approximate a "moralistic" view.

Table 12–2. Objectives of wildlife management. There are no clear boundaries between the three main heading; various objectives clearly overlap and, in some cases, objectives are either in conflict or mutually exclusive with others under the same or different heading. All objectives must be considered in relation to both long-term sustainable benefits and intermediate or short-term benefits.[27]

Socio-economically Oriented Objectives

1. Provide commodity yields (incl. food, industrial products, luxury items, etc.)
2. Provide for recreation and tourism
   - 2.1 Oriented toward hunting and fishing for sport
   - 2.2 Oriented toward nature observation and tourism
3. Provide employment and cash income
4. Maintain cultural diversity (e.g. survival of traditional and subsistence economies)
5. Provide for distribution of benefits to all levels of society.
   - 5.1 Locally
   - 5.2 Regionally
   - 5.3 Nationally
   - 5.4 Internationally
6. Provide for scientific uses and increased knowledge
7. Provide for education benefits
8. Provide for human health
9. Provide for domestication (e.g. as sources of food and other commodities and as work animals)

Ecologically Oriented Objectives

10. Maintain ecosystem diversity (biodiversity)
11. Maintain ecosystem stability
12. Maintain gene pools (and genetic diversity)
13. Maintain distribution of species and varied environments
14. Maintain the ability of populations to survive fluctuating environmental conditions

Ethically Oriented Objectives

15. Minimize human impacts on populations
16. Avoid inhumane or cruel practices involving wildlife
17. Avoid killing animals at all
18. Enhance survival chances for wildlife populations (especially threatened or endangered species)
19. Maintain options for future generations.

In India, the ethics of nature conservation been characterized as follows:

> Much of India's post-independence wildlife conservation has its roots in a continuous and living tradition that dates back to the Vedic periods... the development of wildlife laws in India from the colonial period to the present day has been inspired by that tradition, and thus India has largely forsaken the route of using wildlife, when it involves taking a life.[31]

Attitudinal differences among geographic regions are also evident between Africa and Asia, using the example of the elephant ivory trade.[32] There can be many different explanations to why eastern Africa, southern Africa, South Asia and eastern Asia have all taken differing positions on trade in ivory and other elephant products. One of the basic differences between African and Asian elephant range states is that the African elephant was valued for its ivory whereas the Asian elephant was valued alive as a "war machine" – a good example where attitudes reflect the different objectives that two regions have for the elephants in their midst. As Sukumar puts it:

> the African elephant was more valuable dead for its ivory, while the Asian had reasons to be left alive as a source of captive stock.[33]

While Africa and Asia both exhibit utilitarian philosophies towards elephants, and both continents have ancient histories of ivory trade, their fundamentally different attitudes toward elephants probably explain why Asian nations put more of their skill and effort into taming elephants and maintaining them in captivity, whereas the equally tamable and tractable African elephant was never made captive for any great length of historical time. Such differences in attitudes may also explain why African range states are today more easily amenable to the idea of the ivory trade than South Asian elephant range states, which have less compunction about catching elephants than killing them.

The Far East of Asia, on the other hand, does not have elephants in the wild and views ivory simply as a commodity that is very high in value. Here,

> What is paramount is the inferred value of the object and the substrate (ivory) ensures that whatever be the object carved out of it, it remains high in value.[34]

Cultural perceptions of ivory in the Far East make it more valuable, therefore, than a live elephant, which is not experienced as part of Japanese Nature.

The difference in the valuation of elephants between eastern and southern Africa is more difficult to fathom culturally. Nonetheless, the governments of Kenya and southern Africa have adopted two quite different utilization models, one involving tourism, and the other, the ivory trade. The success of the tourism model in Kenya may provide one explanation why the country now eschews the commercial consumptive use of its elephant population.

2. National Attitudes

Attitudes toward the conservation of nature may vary dramatically among individual nation states. This is a reflection of the cultural, religious and ethical perspectives of the population, as well as the constitution of the country, its laws, wealth (GDP, natural resources, etc.), not to mention its political system, and current political stability (peace, war, personal security), etc.

India, for example, has a strong cultural and religious basis for the protection of wildlife and the environment. The *Wildlife Protection Act* of 1972 bans the hunting and killing of all wildlife except rats, mice, crows and flying foxes. The list of protected species is not based on any scientific criteria of endangerment but on the principle that all life, with the exception of vermin, should be protected. This strong legislation is supported by several important judgments of the apex and lower courts in India, where spirituality and a national ethic for conservation are prevalent. Dismissing the ivory trade as "pernicious to society" the High Court of Delhi, for example, quoted in equal measure from the *Bhagavad Gita* and Douglas Chadwick's *Fate of the Elephant*,[35] and declared that the ivory trade was a social evil comparable to prostitution and gambling. Supreme Court justice, Arijit Pasayat, expressed an even stronger ethical position, when he declared, "By destroying nature Man is committing matricide, having in a way killed mother Earth".[36] Chief Justice Venkatachaliah was equally clear when he said, "I place Government above big business, individual liberty above government and the environment above all".

While strong ethical realities are the backbone of nature conservation in India, other realities dictate policies as well. This is well reflected in the words of Peritore:

> India's environmental movement has the advantages of Gandhian religion, strong links to native cultural eco-management practices, an excellent intellectual and political infrastructure, and multiple points of access to national and local government. But its sophistication and strength is dissipated by a corrupt and bureaucratically tangled government, by a declining economy, and by an ecological and population crisis that surpasses known techniques of environmental repair and management. The movement, far from being a vanguard, is fighting a rearguard action for cultural and ecological survival.[37]

This mix of culture, ethics, and political and social reality, is what makes up the values of a nation and what sets its societal objectives.

Whales and elephants are charismatic mega-fauna whose conservation has led to polemics and fierce battles on several fronts, all of which involve value judgments and ethical choices. These battles revolve mostly around issues of consumptive and non-consumptive use, the value of "scientific" whaling, and management techniques that involve killing, culling, capture or captivity. The conservation of whales has been among the most contentious and debated of all wildlife conservation issues.[38] Different nations have adopted diametrically opposed views regarding their conservation, and international fora such as the *Convention on International Trade in Endangered Species of Wild Fauna and Flora* (CITES) and the *International Convention for the Regulation of Whaling* have been sharply divided over various moratoria and regulatory systems for commercial whaling and the creation of sanctuaries for the whales. Japan and Norway have been the most vocal proponents of whaling whereas countries like Brazil, India, New Zealand, and the United Kingdom, for example, are now seen as anti-whaling nations.

Such differences among nations are well documented in Freeman and Kellert's comparison of attitudes in six countries: Australia, England, Germany, Japan, Norway,

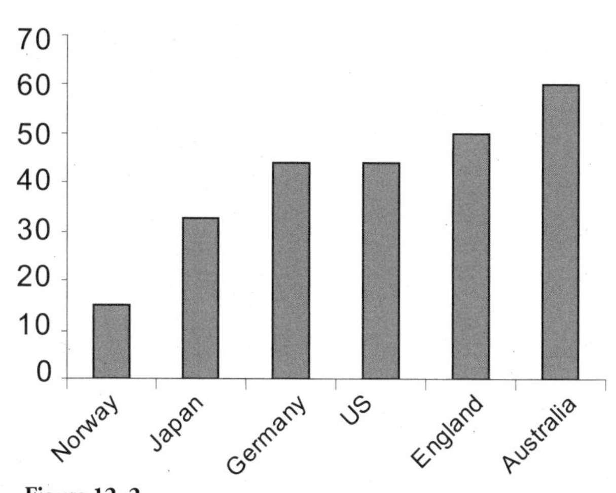

Figure 12–2.
The percentage of people in selected countries opposed to a policy that views whales as a protein food source.

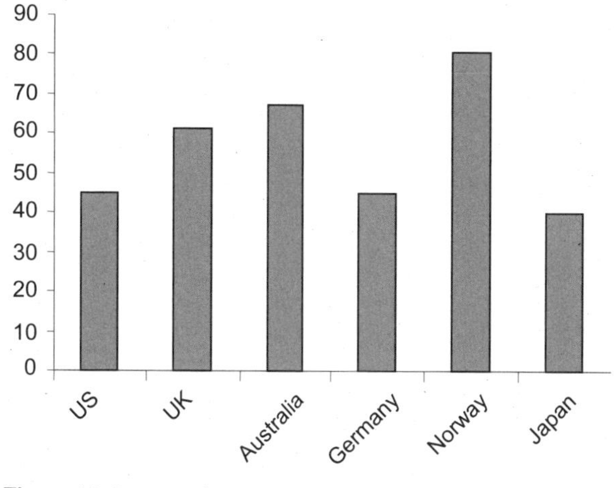

Figure 12–3.
The percentage of people in selected countries which approves of eating lamb.

and the United States. This study dealt specifically with national attitudes toward whaling and the eating of whale meat.[39]

The study found that opposition to viewing whales as a protein source was lowest in the two pro-whaling countries (as expected), and highest in the anti-whaling countries (also expected; see Figure 12-2). What is particularly interesting is that the opposition to whaling in Japan is closer to (although lower than) that of two anti-whaling nations – Germany and the U.S. – than it is to its ideological pro-whaling partner, Norway. This observation may indicate a dichotomy between government policy and public opinion in Japan, or reflect cultural and spiritual differences between Japan and Norway.

However, the moment the debate is taken away from the personal consumption of whale as a protein source to a more "policy-based" question of whether whaling is acceptable under any circumstance, the results followed more expected patterns. Freeman and Kellert write:

> For example, when asked whether they 'opposed the hunting of whales under any circumstances' a majority of respondents in Australia (60%) and Germany (59%) agreed. However, approximately equal-sized majorities in Norway (61%) and Japan (57%) disagreed with the statement that whales should not be hunted under any circumstances. Opposition to whaling under any cirucumstances was more moderated in the U.S. where 48% opposed whaling, and even more evenly divided in England with 43% opposed to whaling 37% not opposed, and a further 19% expressing no strong opinion one way or the other.[40]

Similarly, the different opinions expressed in the six countries regarding the consumption of different kinds of meat – lamb, horse, and kangaroo – is also instructive. While the lowest approval ratings for eating a common domestic animal like lamb was 40% (in Japan), the highest was more than double (81%; in Norway) suggesting some marked cultural differences between the whaling nations (Figure 12–3).

The two countries, however, were more closely allied when the issue was the consumption of kangaroo meat (Figure 12–4) or horseflesh (Figure 12–5), both countries having higher approval ratings than traditionally conservative and protectionist (with respect to whaling) countries, such as the U.S. and the U.K.

Between country differences in attitudes toward the utilization of different species is also telling. Freeman and Kellert report that 79% of German respondents disapproved of whale meat production and sale while 78% disapproved eating wildfowl. Yet the latter scored highly in approval ratings amongst people from Australia, England, Japan, Norway and the U.S.

In short, different countries express different values towards various wildlife species and, often, it is difficult to provide a simple explanation for the differences that are observed.

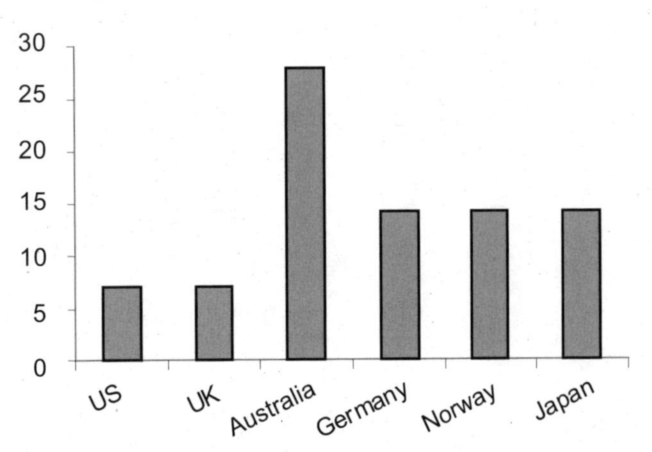

Figure 12–4.
The percentage of people who approve of using kangaroo meat as a source of protein.

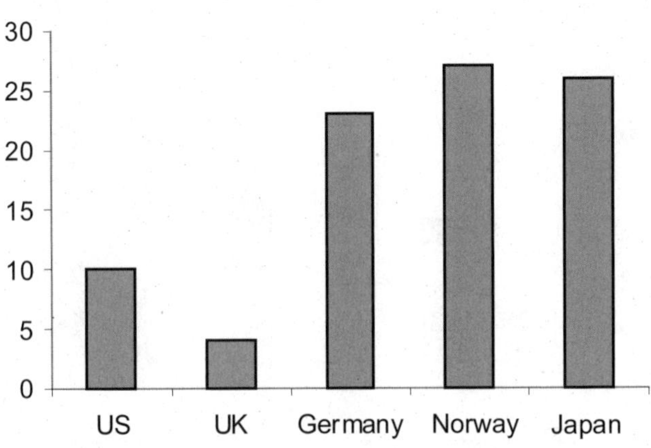

Figure 12–5.
The percentage of people opposed to eating horseflesh.

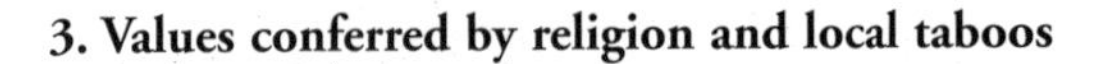

3. Values conferred by religion and local taboos

> *What people do about their ecology depends on what they think about themselves in relation to things around them. Human ecology is deeply conditioned by beliefs about our nature and destiny – that is, by religion.*
>
> Whyte 1967[41]

The values conferred by religion are in many ways the strongest and most immutable of all values. Almost all religions talk of conservation of nature and wildlife. For example:

> That which happens to men also happens to animals
> And one thing happens to them both
> As one dies so dies the other
> For they share the same breath
>
> *Ecclesiastes 3: 19-22*

> There is not an animal on the earth nor a being that flies on its wings, but communities like you.
>
> *Cattle Sura Verse 38, Holy Quran*

> Come back Oh tigers to the woods again.
> For without you the axe will lay it low.
> You, without it, forever homeless go.
>
> *Khuddakapada*

> A right religion is one that teaches respect for the dignity and sanctity of all nature. The wrong religion is one that licenses the indulgence of human greed at the expense of non-human nature.
>
> *Arnold Joseph Toynbee*

For the purposes of our discussion, we will first examine two Asian religions – Buddhism and Jainism – that originated in the same region and are practiced side-by-side in countries such as India. Despite sharing the common principle of *ahimsa*, or non-violence, there are subtle variations and extremes in values and attitudes that are clearly shaped by the particular religion, rather than by region or the state. Then, we will move to another another belief system in a different region of the world and examine the role of cultural taboos in the conservation of an endangered tortoise in Madagascar.

Buddhism and Jainism

The value-oriented approach of certain religions, such as Buddhism with its ecocentric and spiritual approach, bears similarities to Deep Ecology. Henning, for example, says that both Buddhism and Deep Ecology "…use values and perspectives that are based on spiritual and holistic principles for positive change in paradigms, attitudes and practices".[42]

In fact, the very first vow of a Buddhist monk – "I abstain from destroying all life" – represents an extreme expression of a value that has overarching impact on any conservation action contemplated by this particular religious community. The deep reverence for life and the environment is directly linked to the conservation of nature.

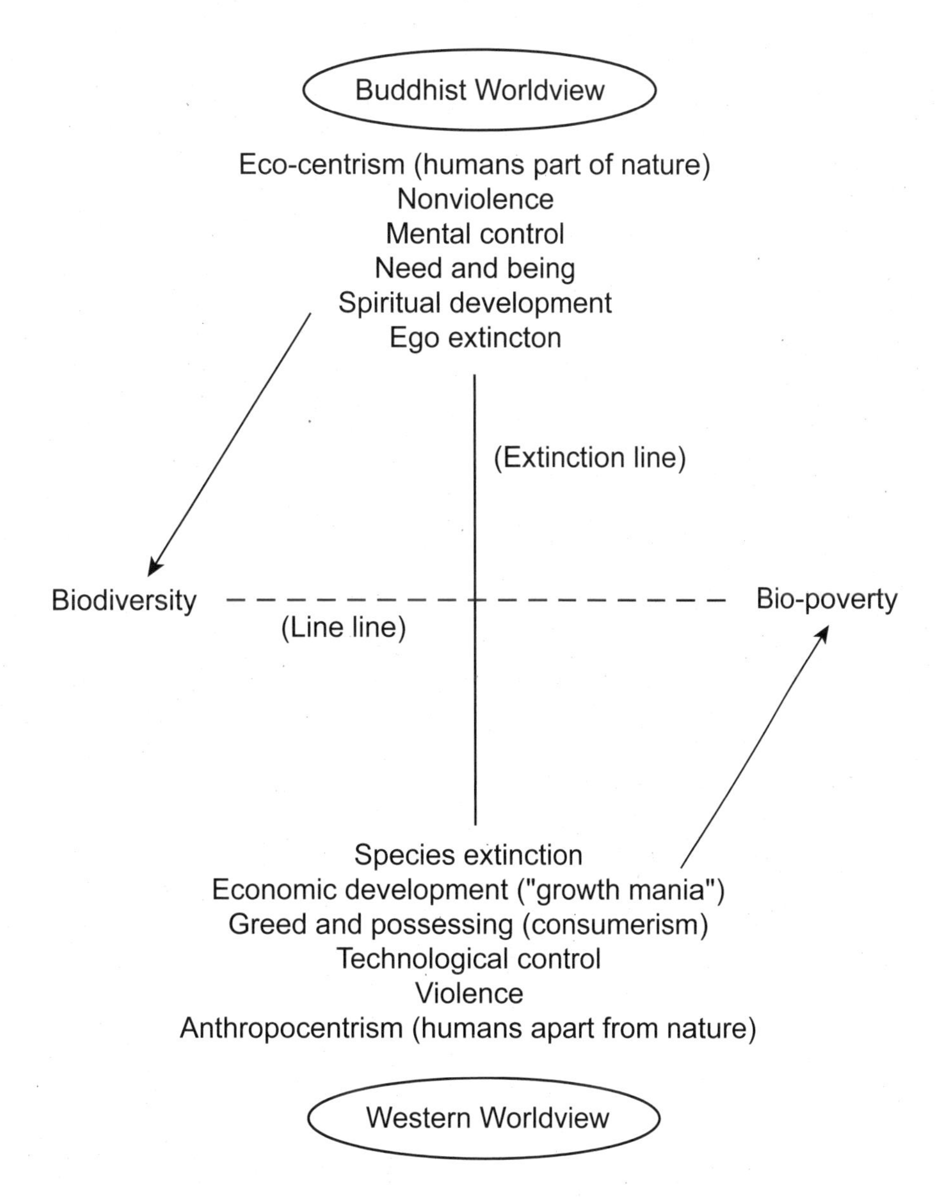

Figure 12–6. A comparison of Western and Buddhist worldviews of the relationship between humans and nature, and the implications for the maintenance of biodiversitiy.

In a landmark conference on culture and environment in Thailand, the assemblage of Thai Buddhists subscribed to a "theory of a moral collapse" as the principle reason for the disequilibrium of ecosystems in Thailand.[43] This view emanates from a Buddhist worldview (as opposed to a Western value system, see Figure 12–6). In the Thai language, the concept of conserving nature means much more than the English words may seem to connote. *Anurak* or 'to conserve' implies a deeper sense of caring than is normally subscribed to the word conservation. Similarly, *thamachat*, or Nature, includes all things in their true and natural state. The two words, put together, translate:

> …as having at the core of one's very being the quality of empathetic caring for all things in the world in their natural conditions; that is to say, to care for them as they really are rather than as I might benefit from them or as I might like them to be.[44]

Buddhist scholars have noted the clear connection between Buddhism and the conservation of nature:

> Wherever Buddhism is influential, studies will usually show some direct benefit for the natural world. In Sri Lanka for example Buddhism has had the largest single

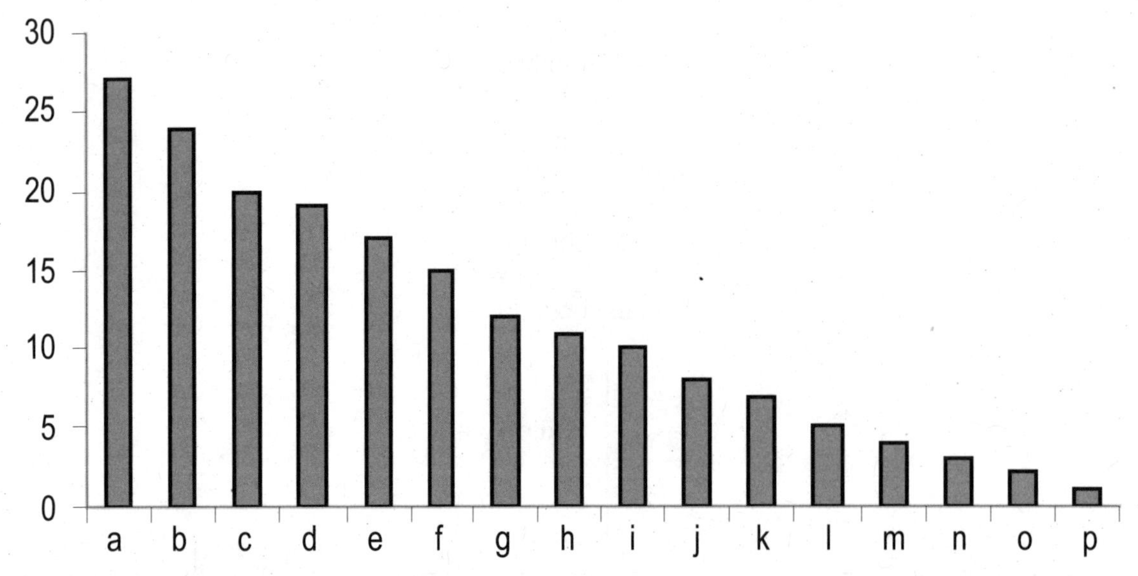

Figure 12–7. Number of times a type of animal appears in the Jataka Tales. Legend: a- Monkey, b- Elephant, c- Jackal, d- Lion, e- Crow, f- Bird, deer, stag, doe, g- Fish, h- Parrot, i- Snake, Tree spirit, j- Horse, k- goose, tiger, tortoise, l- boar, goat quail, m- bull, crocodile, dog, ox, partridge, peacock, rat, vulture, n- cow, crab, crane, cuckoo, lizard, pig, pigeon, serpent, woodpecker, o- antelope, chameleon, chicken, donkey, falcon, osprey, owl, rabbit, rooster, viper, water spirit, p- bear, beetle, buffalo, cat, duck, eagle, elk, fly, fox, frog, grass spirit, hawk, hound, iguana, jay, mongoose, mosquito, mouse, otter, panther, rhinoceros, shrew, wolf.[49]

impact on the conservation of flora and fauna.[45]

One example is the case of tropical forest conservation in Buddhist areas. Tropical forests are arguably the most charismatic of ecosystems: a conglomerate of species, habitats and land-uses. The forests also have within them, the one species that opines and makes value judgments. That species, *Homo sapiens*, attributes both ecocentric and anthropocentric values to the forest.[46] The former include valuing the forest for its biological diversity, including its species and genetic diversity, and its roles in maintaining the web of life, modifying local climate, conserving water and protecting soils. Among the latter, the forest is seen as having agricultural, industrial and medicinal values, e.g. as a source of genetic and other materials. It is valued also as a home to tropical forest people, a source of outdoor recreation, and for its educational and ecotourism benefits. It serves as a source of creative inspiration, and is seen to provide cultural, spiritual and inter-generational values. This diversity of values and the wide cross-section of humanity that "uses" tropical forests undoubtedly have a great impact on how those forests are conserved.

In a predominantly Buddhist state (or if that does not exist in today's world, in forests adjoining a Buddhist-run monastery), the values attributed to the forest may be linked to the fact that *Gautama Buddha* was born in a forest, studied among banyan trees, meditated under Jambo trees, achieved enlightenment under the Bodhi tree, and gained salvation (died) under Sal trees, all symbols of a tropical forest. The fact that Buddhism was born in forests clearly affects the value judgments made by Buddhists about them and their conservation.

An interesting analysis of taxa-specific value systems espoused by a religion is seen in the Buddhist Jataka tales written in Pali.[47] In 50% of the 550 stories (according to the Theravada[48]) of the Jataka, 70 types of animals and 319 animals or groups of animals appear. These animals represent prior life forms of people contemporary to Buddha, and the stories portray animals as auspicious, objectionable, compassionate, cruel, wise, and foolish, among other anthropomorphic manners. Chapple[49] provides an analysis of the number of times that a particular animal appears in the Jataka (Figure 12–7).

In our interpretation of Chapple's work, we note that symbolically powerful and (once) common animals – such as primates, elephants, lions, deer, and crows and other birds – feature most prominently, while smaller animals (e.g. beetles and frogs), and creatures with negativistic attributes (e.g. shrews and wolves), appear much less frequently. While any such analysis is subject to several interpretations, it is sufficient to note here that the animals more commonly mentioned were more suited to convey tales with inherent ecological messages or to illustrate compassion. For example, the lion was viewed as a compassionate yet strong ruler of the forest, the jackal as a scheming realist who lent pragmatism to wisdom and the elephant as both wise and strong. The characters were given strong human traits and the tales were a collection of stories that talked of forest protection, wildlife conservation and nature appreciation along with other messages for daily lives.

Now consider Jainism, which originated more than 2500 years ago and also preaches *ahimsa* as the primary method of attaining eternal liberation from birth or *kevala*.[50] *Ahimsa* leads to several extreme practices related to animal rights and animal welfare that are unrivalled anywhere else in the world. For example, a common expression of *ahimsa* in Jainism is the practice of *pratilekhanao*, or the meticulous inspection of one's attire for any trapped life forms.[51] Jain *munis*, or holy men, wear loose white robes that are shaken clear twice a day to avoid the accidental killing of any life. This custom is taken to a further extreme by the Jain *munis* of the *Digambara* sect who do not wear any clothes at all and cover their mouth and nose with a piece of muslin to prevent insects and other small living beings from being trapped in clothing or inhaled, thus leading to their deaths.

Even more extreme are the alleged practices of the *hastitapasa* or "elephant ascetics"[52] described by the ninth-century commentator Silanka from the 3rd century BC canonical text, *Sutrakrtanga Sutra*. According to Silanka, the elephant ascetics had an unwillingness to overexploit nature and therefore lived off the meat of one large elephant that they killed every year "out of compassion to other life forms".[53] Of course, this tale could be an invention by the author,[54] but it still reflects a religious ethic that guided the practices of a community.

More examples of animal rights, animal welfare, and wildlife preservation can be found in the canons of Jainism. For example, among the 15 occupations or worldly activities that were prohibited for Jains, as early as the 6th century, were the destruction of forests, including the sale of timber; the trade in animal by-products such as ivory, bones, conch shells, pelts and down; the trade in animals; and deriving livelihood from the mutilation of animals.[55] Such prohibitions are part of the *bhogopabhoga-vrata*, the underlying concept of which is the avoidance of injury to animals (including insects) and plant life.[56]

Some of today's most contentious resource management and utilization debates would be rendered differently using the Jain logic. A Jain story illustrates how.

> Once upon a time, six friends went out together. After a while they were hungry and thirsty. They searched for food for some time and finally found a fruit tree. As they ran to the tree, the first man said, "Let's cut the tree down and get the fruit." The second one said, "Don't cut the whole tree down, cut off a whole branch instead." The third friend said, "Why do we need a big branch?" The fourth friend said, "We do not need to cut the branches, let us just climb up and get the bunches of fruit." The fifth friend said, "Why pick that much fruit and waste it? Just pick the fruit that we need to eat." The sixth friend, quietly said,"There is plenty of good fruit on the ground, so let's just eat that first."[57]

Were such communities to whom *ahimsa*, or non-violence, is a basic tenet, to practice conservation or be involved in policy-making, questions such as culling, euthanasia, and other such management practices, would not even arise, because such practices have no place at all in their consideration. Their stance on euthanasia for example has led to several discussions among practitioners of Hinduism and Jainism in India about the right way to dispose of terminally ill and rescued animals. Animal welfare organizations that are funded by Jains – traditionally, a rich community – seldom practice euthanasia or do so covertly, while those funded by Hindus – while holding animal welfare close to their ideals – have fewer ideological problems with the practice.

Taboos

The role of taboos in conservation has been documented, for example, in Madagascar, off Africa's East coast – home to one of the most endangered tortoises in the world, the radiated tortoise, *Geochelone radiata*.[58] Taboos are social prohibitions against things that are holy or unclean, and have a ritualistic connection in most cases. Colding and Falke[59] list a variety of taboos that serve the purpose of nature conservation by regulating human behaviour (Table 12–3).

In Madagascar, approximately half of the radiated tortoise's range is inhabited by the Tandroy people. They regard the species as "dirty" and this perception gives rise to a prohibition on tortoise exploitation, which is expressed as a taboo or *fady*.[60] The potential role of this taboo in the protection and management of the endangered tortoise was studied recently by Lingard *et al.* They found a clear correlation between the following of the taboo in a particular community and the status of the endangered radiated tortoise living there.[61]

At one extreme was the area of Lavanomo, where 93% of people said that they followed the taboo, and up to 99% of the population was estimated to practice it. There, the mean number of *G. radiata* per hectare was found to be 10.8 ± 0.15.

Table 12–3. Resource and habitat taboos and the nature conservation and resource management functions of each category.

Category	Function
Segment taboos	Regulates resource withdrawal
Temporal taboos	Regulates access to resources in time
Method taboos	Regulates methods of resource withdrawal
Life history taboos	Regulates withdrawal of vulnerable life history stages of species
Species-specific taboos	Total protection of species in time and space
Habitat taboos	Restricts access and use of resources in time and space

At the other extreme, was the area of Tsiombe, where only 75% of informants said that they followed the *fady*, and an estimated 80% of them actually did, the mean density of tortoises dropped by an order of magnitude to 0.6 ± 0.04 animals per hectare.

The available evidence indicates that the vastly lower density of tortoises in Tsiombe was a direct effect of illegal killing for consumption and commercial trade, and live capture for the pet trade, resulting from a laxity in the adherence to the taboo. The authors of the study recommended that local taboos attributing value to a particular species in a community be incorporated into official legislation, based on the following rationale (words in brackets are introduced by us):

> environmental regulations that are based on traditional customs and sanctioned by local institutions [read value systems, including taboos] are more likely to be respected than those imposed by external administrative agencies [through legislation, and even if based on scientific fact].[62]

4. Socio-Economics and Values

The inspiration to conserve nature also depends on livelihood issues. People whose livelihoods are directly or indirectly dependent on a particular kind of animal are less likely to have value systems that espouse non-utilization of what they consider to be a "resource". A recent study conducted by TNS Mode for the Wildlife Trust of India (WTI) and the International Fund for Animal Welfare (IFAW) on a conservation issue in India is a case in point. Whale sharks are the largest fish in the world and frequent Indian shores, especially the shores of the maritime state of Gujarat. Gujarat is inhabited by a predominantly vegetarian Hindu community, but there is a minority within the community that includes the Koli fishermen who fish the whale shark as a seasonal activity. In 2002, India protected the whale shark and banned its slaughter. WTI and IFAW planned an awareness campaign in Gujarat to inform the fisher folk of the new ban and the reasons for protecting the fish, and to seek their support for the new conservation measure. At the same time, they wished to celebrate the arrival of the whale shark, a charismatic and gentle giant of the seas, by raising awareness levels among the general populace in the state, especially in the capital city of Ahmedabad. Prior to undertaking this campaign, a survey of attitudes was carried out. Among 425 citizens surveyed throughout the state, endangered animal protection ranked fourth of seven environmental issues, including air, land and water pollution, deforestation, global warming and water table reduction (Table 12-4).

When the data were examined on a regional basis (Table 12-5), it appeared that coastal citizens (Veraval) were more aware (nearly two times more) of the importance of conserving endangered species than those living inland (Ahmedabad). Interestingly, 93% of all citizens of the state said that they would support the cause even if it meant a loss of livelihood for fishermen.

It was also apparent, however, that the importance of endangered species conservation was only an ingrained value, and not necessarily a societal objective, for the fisher folk themselves, when they were confronted with the fact that they had to give up whale-shark fishing. A small qualitative sampling of opinion among the whale-shark fishing community found that many felt that that the whale shark should be killed, not because it is harmful, but because of financial considerations. "It should be killed as it gives livelihood to so many fishermen" was a common refrain.

It is also interesting to note that a common refrain directed by fishermen elsewhere in the world toward cetaceans (whales, dolphins and porpoises) and pinnipeds (fur seals, sea lions, walrus, and true seals) – that they need to be culled to protect fisheries or fish stocks – was also a commonly expressed sentiment directed toward whale sharks. For example, "the whale shark should be killed because they eat the small fish and we don't get any

Table 12–4. Whale shark attitudinal survey (TNS Mode-WTI unpublished). Respondents were asked to identify the three environmental issues (from the list of seven, below) that they considered to be a priority. The numbers in the table represent the percentage of respondents who identified a particular issue.

	All Citizens	Children	Young Adults	Adults
Base sample	425	125	124	176
Air pollution	71	74	69	71
Water pollution	62	70	55	63
Deforestation	58	54	62	58
Specific animal species becoming endangered	**38**	**35**	**40**	**39**
Land pollution	32	34	34	30
Global warming	24	22	29	22
Water table going down	16	13	11	20

Table 12–5. Importance given to each issue in Table 12–5 by the citizens of Ahmedabad (Ahm) and Veraval (Ver).

	All Citizens		Children		Young Adults		Adults	
	Ahm	Ver	Ahm	Ver	Ahm	Ver	Ahm	Ver
Base sample	250	175	75	50	74	50	101	75
Air pollution	71	72	73	76	69	68	70	72
Water pollution	66	57	72	66	65	40	63	61
Deforestation	63	51	63	40	68	54	59	56
Land pollution	33	31	35	32	32	36	32	27
Specific animal species becoming endangered	**28**	**53**	**23**	**54**	**26**	**62**	**34**	**47**
Global warming	26	21	27	14	27	32	26	17
Water table going down	16	15	13	12	12	10	21	20

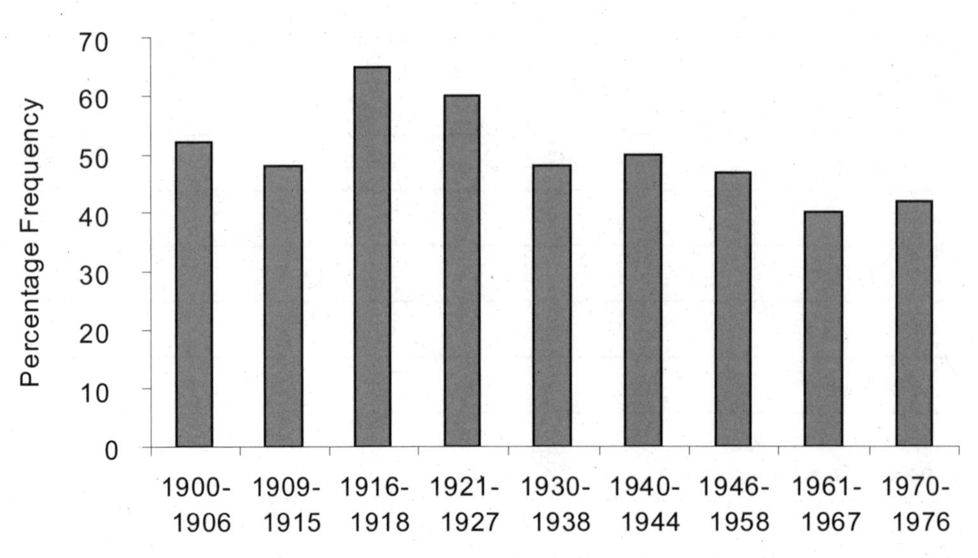

Figure 12–8. Changes in utilitarian attitudes in the United States throughout the 20th century.

small fish," and "small fish are left in the ocean so that fish in general will increase. But the whale shark eats them up, so the fish are decreasing".

Within the same nation, within the single state of Gujarat, and even within a geographically distinct community – the coastal populace – the views of the fishermen who are dependent on the whale shark were dramatically different from the non-dependent citizenry. Indeed, fishermen's views were more similar to those of fishermen from other countries, cultures and religions, who are similarly dependent on fisheries for livelihood.

Subramanian[63] has also demonstrated "the privilege of personal gain over social responsibility and of private wealth over marine wealth" among the artisanal fishermen of Tamil Nadu. He records that:

> I was told several times that as trawler owners grew richer, they contributed less and less to the church fund from which the needs of the village poor were met.

So strong are some of the fishermen's views that even smaller "Kattumaram" or catamaran fisher folk are quoted as saying,

> We didn't need the government or the church to endorse the agreement (of a new fishing rights regime); we had Mary Matha [Mother Mary] as our witness. We know best what is just: where to fish, how to fish and how to protect the sea. [64]

When livelihood issues surface, science plays a tertiary role in conservation. In such circumstances, even religious values and laws based on national value systems are pushed to second place. As Oscar Wilde once observed, "The necessity for a career forces every one to take sides".[65]

5. Time in History

Within the same region, the same nation, and the same community, values and objectives often change over time. In other words, value systems tend to evolve just as species do, acted upon by a multitude of factors.

In the United States, for example there has been a decline over time in public support for commercial whaling. As recently as 1978, 75% of Americans endorsed the hunting of non-endangered whales if it resulted in a useful product.[66] In 1993, only 26% of Americans approved whaling under any circumstance.[67] And, by 1997, only 14% supported (strongly or somewhat) "the killing of whales for their meat, blubber, or other purposes".[68]

The decline in the utilitarian attitude of Americans towards commercial whaling is consistent with the general decline in the utilitarian attitude toward all wildlife species observed in the United States between 1916 and 1976 (Figure 12-8).[69]

Another example of attitudes changing over time can be found in the Japanese Shinto-Buddhist practice of *hojo-e*. *Hojo-e* is a "no-kill ceremony" that involves the symbolic release of living beings.[70] This medieval Buddhist compassion ritual has now evolved into a complex, state-sponsored ritual called *Iwashimizu Hachiman*, allegedly due to quasi-militaristic land grabbing and clan power strategies. This evolution has led to the ironic situation wherein large numbers of fish are now eaten in order to have a ceremony in which fish are released as part of *hojo-e*.

An even more recent change is the paradoxical practice in Japan of memorializing animals that are captured, which includes having elaborate whale banquets to memorialize the whales that are killed. Over a period of time a value based on religion has changed in the same nation state to a value coloured by politics.

DISCUSSION

Much of the technical and academic literature portrays conservation as a scientific undertaking. Claims are frequently made that conservation policies should be scientifically based and that management decisions should be guided by the best available scientific information. In theory, perhaps that is the way things should work but, in practice, that is not the way conservation operates. All conservation issues, from the protection of endangered species to the removal of sustainable yields, to the maintenance of biodiversity, and the adoption of the precautionary approach, all involve value-based choices. Of course, science can inform those choices and, on occasion, the choices that are made may even be consistent with the best scientific advice. But, more often than not, that is not the case.

The driving forces behind any concept of conservation, are the attitudes, values and objectives of individuals, communities, regions or nations. That being the case, it is also self-evident that attitudes, values, and objectives vary among individuals, communities, regions, and nations. There is simply no one view of conservation and no possibility in the foreseeable future, at least, of a single conservation ethic emerging among regions and nations of the world. While a single ethic might be considered ideal, it is precluded by economic realities, and by religious and cultural differences. When controversies arise, as they inevitably will – because of differing visions of what is good or bad, right or wrong – they will continue to be resolved (or at least dealt with) as they usually are: through some form of political process. As Donovan *et al.* wrote years ago,

> ...conflict – the stuff of politics – takes place not only among people and groups and nations, but also among alternative values, or to put it more precisely, competing visions of what is "good".[71]

The situation will be further complicated by the wide disparities among nations in the political systems they employ.

These days we hear much about the promises of globalization for resolving the world's ills. In the case of the conservation field, globalization would require the implementation of a single conservation ethic. But that does not seem realistic or possible. Perhaps a better way to pursue the goal of ecological sustainability, then, is to proceed on a regional or national basis, acknowledging not only religious, cultural, and political differences, but also the current economic disparities and social inequities that exist around the world. It seems obvious that different approaches will be required in different places and at different times. That is the only way we can see to begin to accommodate the different attitudes, values and objectives of individual people and the societies in which they live, even if the over-riding goal remains the pursuit of ecological sustainability on a global basis.

NOTES AND SOURCES

[1] Sukumar, R. 2003. *The Living Elephants: Evolutionary ecology, behaviour and conservation.* Oxford University Press, New York, NY.

[2] Nash, R. F. 1989. *The Rights of Nature: A history of Environmental Ethics.* The University of Wisconsin Press, Madison WI; Worcester, D. 1994. *Nature's Economy: A History of Ecological Ideas*, 2nd Edition, Cambridge University Press, Cambridge, UK.

[3] Lavigne, D.M., V.B. Scheffer, and S.R. Kellert, 1999. The evolution of North American attitudes towards marine mammals. In J.R. Twiss Jr. and R.R. Reeves (Eds.) *Conservation and Management of Marine Mammals.* Washington and London, Smithsonian Institution Press; Nash, R. F. 1989. *The Rights of Nature: A history of Environmental Ethics.* The University of Wisconsin Press, Madison WI; Leopold, A. 1966. *A Sand County Almanac with Essays on Conservation from Round River.* A Sierra Club/Ballantine book, New York, NY.

[4] See Worster 1994; Lavigne *et al.* 1999.

[5] Leopold, A. 1966. *A Sand County Almanac with Essays on Conservation from Round River.* A Sierra Club/Ballantine book, New York, NY.

[6] e.g. Naess, A. 1973. The Shallow and the Deep, Long-Range Ecology Movement: A Summary. Inquiry 16: 95–100.

[7] Henning, D. 2002a. Buddism and Deep Ecology. 1st Books Library, USA; also see Shaw, J.H. 1985. Introduction to Wildlife Management. McGraw-Hill Book Company, New York, NY.

[8] Lavigne, D.M. 2003. Marine Mammals and Fisheries: The Role of Science in the Culling Debate. pp. 31-47. In N. Gales, M. Hindell, and R. Kirkwood (Eds.). *Marine Mammals: Fisheries, Tourism and Management Issues.* CSIRO Publishing, Collingwood, VIC, Australia.

[9] Singer, P. 1975. *Animal Liberation: A New Ethics for our Treatment of Animals.* Avon Books, New York, NY.

[10] Marcuse 1972, cited in Nash, R.F. 1989. *The Rights of Nature: A History of Environmental Ethics.* The University of Washington Press, Madison WI. p. 6.

[11] Jeanrenaud, S. 2002. People-Oriented Approaches to Global Conservation – Is the Leopard Changing its Spots? International Institute of Environment and Development, London, UK.

[12] Henning, D. 2002b. Wildlife, Forests and Spirituality: Deep Ecology in Asia. In V. Menon and M. Sakamoto (Eds.). Heaven and Earth and I: Ethics of Nature Conservation in Asia. Penguin-India.

[13] See Lavigne, Chapter 1.

[14] Giles, R.H. 1978. *Wildlife management.* W. H. Freeman, San Francisco, CA.

[15] Thompson, R. 1992. *The Wildlife Game.* The Nyala Wildlife Publications Trust, Westville, South Africa. Thompson begins Chapter 19, "The principles of wildlife management," with the unequivocal statement "Wildlife management is a *science*...". He goes on to say that the boundaries of wildlife science are "prescribed by legislation and official policy (which reflect a nation's culture)". But the text soon gets very confusing. For example, Thompson also says that "the application of the principles of wildlife management is an art...". Then, he says, "Wildlife management becomes more than just an *art.* It becomes a very specialized *craft*..."). Not all authors are so confused: see Dickinson, N. 1993. *Common Sense Wildlife Management.* Settle Hill Publishing, Altamont, New York. He states (p. 7) that "When practiced in a professional manner, wildlife management is indeed a science...The wildlife manager, as a scientist, must be totally objective and practice the self discipline necessary to make judgements based on facts and truths. There is no room for subjectivity...Dedication to the scientific method is paramount." Hunter, M.L. Jr. (1990. *Wildlife, Forests, and Forestry: Principles of Managing Forests for Biological Diversity.* Prentice-Hall Inc., Englewood Cliffs, New Jersey) also describes "wildlife management as a scientific discipline and profession [that] emerged primarily in the United States through combining the emerging science of ecology with the traditional practices of gameskeepers" (p. 4).

[16] In some fields, the "attitudes" and "values" are treated as synonyms. See Lavigne *et al.* (1999) for a brief discussion. For a different view, see Worcester, Chapter 17.

[17] Berry, J.K. 1993. Distinguishing data from information and understanding GIS World, October. Also see Lavigne, D.M. 2003. Marine mammals and fisheries: The role of science in the culling debate. pp. 31-47. In N. Gales, M. Hindell and R. Kirkwood (Eds.). *Marine Mammals: Fisheries, Tourism and Management Issues.* CSIRO Publishing, Collingwood, Victoria, Australia.

[18] For an example from fisheries, see Hutchings, Chapter 6.

[19] See for example, http://www.ucsusa.org/scientific_integrity/interference/scientific-integrity-in-policy-making-704.html; also see http://www.nrdc.org/bushrecord/science/default.asp; http://www.nrdc.org/bushrecord/science/rfk.asp; http://www.nrdc.org/bushrecord/science/quotations.asp.

[20] Kabilsingh, C. 1987. How Buddhism can help protect Nature, in Nash, N. (Ed.). *Tree of Life: Buddhism and the Protection of Nature.* Buddhist Perception of Nature, Geneva.

[21] G.B.B Shah, His Highness. 2002. Search for a New Value System: Environmental Challenges in Nepal. In V. Menon and M. Sakamoto (Eds.). *Heaven and Earth and I: Ethics of Nature Conservation in Asia*, Penguin-India.

[22] Donovan J.C., R.E. Morgan, and C.P. Potholm. 1981. *People, Power and Politics An introduction to Political Science.* Addison-Wesley Publishing Company, Reading, MA.

[23] Fishbein, M. and I. Ajzen. 1975. *Belief, Attitudes, Intention and Behaviour: An Introduction to Theory and Research.* Addison-Wesley Publishing Company, Reading, MA.

[24] Henning 2002b.

[25] See Lavigne *et al.* 1999. Table 12–1 is adapted from Kellert, S.R. 1980. Activities of the American public relating to animals. Phase II. Results of a U.S. Fish and Wildlife Service funded study of American attitudes, knowledge and behaviors toward wildlife and natural habitats. Superintendent of Documents, U.S. Government Printing Office, Washington, DC; and Kellert, S.R. 1993. The biological basis for human values of nature. pp. 42-69. in S.R. Kellert and E.O. Wilson (eds.). *The Biophilia Hypothesis*, Island Press, Washington, DC.

[26] Kellert, S.R. and M.O. Westervelt. 1982. Historical trends in American animal use and perception. Transactions of the North American Wildlife and Natural Resources Conference 47: 649-664.

[27] Table 12–2 is modified and generalized from Food and Agricultural Organization of the United Nations. 1978. Mammals in the Seas. Vol. 1. Report of the FAO Advisory Committee on Marine Resources Research, Working Party on Marine Mammals. FAO Fisheries Series 5(1) 275 pp. and reprinted from Lavigne, D.M. and V. Geist. 1993. Game ranching: A case study in sustainable utilization. pp. 194-200. In I. Thompson (Ed.). Proceedings of the International Union of Game Biologists XXI Congress. Vol. 2. August 1993. Halifax, Nova Scotia.

[28] Huntington, S. 1996. *The Clash of Civilisations and the remaking of World Order.* Penguin-India.

[29] Kellert, S.R. 1991. Japanese Perceptions of Wildlife, Conservation Biology 5:297-308; Kellert, S.R. 1995. Concepts of Nature East and West. In M.E. Soule and G. Lease (Eds.). *Reinventing Nature? Responses to Postmodern Deconstruction.* Island Press, Washington DC.

[30] Unless otherwise stated, we use Kellert's typology (Table 12–1) throughout to characterize attitudes and values.

[31] Kumar, A. 2002. A History of Compassion: Indian ethics and conservation. In V. Menon and M. Sakamoto (Eds.). *Heaven and Earth and I: Ethics of Nature Conservation in India.* Penguin-India.

[32] For more on the ivory trade, see Kumar and Menon, Chapter 8.

[33] Sukumar, R. 2003. *The Living Elephants: Evolutionary Ecology, Behaviour and Conservation.* Oxford University Press, New York, NY.

[34] Menon, V. 2002 *Tusker: The Story of the Asian Elephant.* Penguin-India.

[35] Chadwick, D.H. 1992. *The Fate of the Elephant.* Viking, Penguin-India Pvt Ltd. New Delhi, India.

[36] Supreme Court, 2002.

[37] Peritore, P.N. 1993. Environmental Attitudes of Indian Elites: Challenging Western Postmodernist Models. Asian Survey 33: 807.

[38] See Papastavrou and Cooke, Chapter 7.

[39] Freeman, M. and S.R. Kellert. 1994. International Attitudes to Whales, whaling and use of whale products: A six-country survey. In M.M.R. Freeman and U. Kreuter (Eds.). *Elephants and Whales: Resources for Whom.* Gordon and Breach Publishers, Basel Switzerland.

[40] Ibid.

[41] White, L.W. 1967. The historical roots of our ecologic crisis. Science 155:1203-1207.

[42] Henning 2002b.

[43] Sponsel, L.E and P.N. Sponsel. 1997. A Theoretical Analysis of the Potential Contribution of the Monastic Community Promoting a Green Society in Thailand. In M.E. Tucker and D.R. Williams (Eds.). *Buddhism and Ecology.* Harvard University Press, Cambridge, MA. Figure 12–6 is redrawn from Tucker and Williams.

[44] Swearer, D.K. 1997. The Hermeneutics of Buddhist Ecology in Contemporary Thailand: Buddhadasa and Dhammapitaka. In M.E. Tucker and D.R. Williams (Eds.). *Buddhism and Ecology.* Harvard University Press, Cambridge, MA.

[45] Kabilsingh 1987.

[46] Henning 2002b.

[47] An ancient language of northern and eastern India that was used in the region where Buddhism originated and took root.

[48] An important Buddhist manuscript.

[49] Chapple, C.K 1997. Animals and Environment in the Buddhist Birth Stories. In M.E. Tucker and D.R. Williams (Eds.). *Buddhism and Ecology.* Harvard University Press, Cambridge, MA.

[50] Jaini, P.S. 1979. The Jaina path of purification. Motilal Banarsidass, Delhi, India.

[51] Vallely, A. 2002. From Liberation to Ecology: Ethical discourses among Orthodox and Diaspora Jains. In C.K. Chapple (Ed.). *Jainism and Ecology.* Harvard University Press, Cambridge, MA.

[52] Generally, an ascetic is one who practices a renunciation of worldly pursuits to achieve spiritual attainment.

[53] Dundas, P. 2002. The limits of a Jain Environmental Ethic. In C.K. Chapple (Ed.). *Jainism and Ecology.* Harvard University Press, Cambridge, MA; Jambuvijaya, M. 1978. *Acarangasutram and Sutrakritangasutra.* Motilal Banarsidass, Delhi, India.

[54] According to Bolle, W.B. 1999. Adda or the Oldest Extant Dispute between Jains and Heretics (Suyagada 2, 6). Journal of Indian Philosophy, 27: 411-437.

[55] Jaini, P.S. 2002. Ecology, Economics and Development in Jainism. In C.K. Chapple (Ed.). *Jainism and Ecology.* Harvard University Press, Cambridge, MA.

[56] Shilapi, S. 2002. The Environmental and Ecological teachings of Tirthankara Mahavira, In C.K. Chapple (Ed.). *Jainism and Ecology.* Harvard University Press, Cambridge, MA.

[57] Ibid.

[58] Lingard, M, N. Raharison, E, Rabakonandrianina, J. Rakotoarisoa, and T. Elmqvist. 2003. The Role of Local Taboos in Conservation and the Management of Species: The Radiated Tortoise in Southern Madagascar. Conservation and Society, Vol. 1, No. 2, New Delhi, India.

[59] Colding, J and C. Folke. 2001. Social Taboos: "Invisible" Systems of Local Resource Management and Biological Conservation, Ecological Applications 11(2): 584-600.

[60] Lingard *et al.* 2003.

[61] Ibid.

[62] Berkes, F and C. Folke. 1998. *Linking Social and Ecological Systems,* Cambridge University Press, Cambridge, UK.

[63] Subramanian, A. 2003. Community, Class and Conservation: Development Politics on the Kanyakumari Coast. Conservation and Society, Vol. 1, No. 2, New Delhi; Vallely, A. 2002.

[64] Subramanian, A. 2003.

[65] Ellmann, R. (Ed.). 1969. *The Artist as Critic.* The University of Chicago Press, Chicago IL. p. 385. The quote is from Wilde's essay, The Critic as Artist, which was first published in 1890. Available at http://www.online-literature.com/wilde/1305/.

[66] Kellert, S.R. 1980. Activities of the American public relating to animals, Phase II, Results of a USFWS funded study of American attitudes, knowledge and behaviors towards wildlife and habitats, Superintendant of Documents, U.S. Government Printing Office, Washington, DC.

[67] Freeman, M. and S.R. Kellert. 1994. International attitudes to whales, whaling and use of whale products: A six-country survey. In M.M.R. Freeman and U. Kreuter (Eds.). *Elephants and Whales: Resources for Whom,* Gordon and Breach Publishers, Basel, Switzerland.

[68] Penn, Schoen and Berland Associates, Inc. 1997. Poll taken for the International Fund for Animal Welfare. Yarmouth Port, MA.

[69] Kellert and Westervelt 1982; Lavigne *et al.* 1999.

[70] Williams, D.R. 1997. Animal Liberation, Death and the State: Rites to Release Animals in Medieval Japan. In M.E. Tucker and D.R. Williams (Eds.). *Buddhism and Ecology,* Harvard University Press, Cambridge, MA.

[71] Donovan *et al.* 1981, p. 4.

A Postmodern Limerick

This limerick stems from the paper that Vivek Menon presented on behalf of himself and David Lavigne. I have cleared it with Vivek.

It needs a little scholarly exegesis. Several of the forum speakers managed to sneak in a little of the vocabulary of the postmodernists – words like deconstruct, discourse, and empowerment, for example. So, I thought I could get away with a postmodern limerick. Specifically, I mean a limerick written with ironic intent – one with two extra lines – and one which ends by deconstructing itself.

An Indian told us that Buddha
Learned to conserve from his Mudha
So whenever it rains
On the North Indian plains
The soils just get gudha and gudha.

Except of course for the vast tonnages that get washed down the Ganges, right through Bangladesh and out into the Bay of Bengal

In amounts that would make Buddha shuddha.

Ward Chesworth 2004

CHAPTER 13

BETWEEN SCIENCE AND ETHICS:
WHAT SCIENCE AND THE SCIENTIFIC METHOD CAN AND CANNOT CONTRIBUTE TO CONSERVATION AND SUSTAINABILITY

William S. Lynn

When David Lavigne first asked that I write this chapter, it was on the heels of a presentation on the ethics of wolf recovery. I have been arguing for years that wolf recovery has little to do with conservation science – the natural and applied sciences contributing to conservation.[1] This form of science and its search for causal knowledge is supposed to be the bedrock of environmental and wildlife policy (e.g. policy principles, laws, regulations, implementation, management and evaluation). I firmly believe that conservation science is necessary for the protection and recovery of wolves in particular, and for the protection of biodiversity more generally. Even so, to expect science to resolve what is at heart a moral conflict is to ask too much of one area of human learning. Conservation science can help inform us about the choices and consequences of our actions. It can even give us insight into what actions produce better or worse outcomes in the world, and these insights may help form better policy. But it cannot resolve for us the deeply rooted ethical conflicts over whether and how we should share the landscape with large carnivores like wolves, with a diverse array of life forms, or with manifestly different ways of life. To answer these questions, we require ethics.[2]

David's charge to me was straightforward. He asked that I answer the following question: "What is it that conservation science and its method can and cannot contribute to wildlife conservation and ecological sustainability? Oh, and please note the place of ethics in this framework". Deceptively straightforward. With this charge in mind, I want to articulate the following argument.

There is no doubt that the world is facing a biodiversity crisis.[3] Responding to this crisis poses a challenge for wildlife conservation and ecological sustainability. The customary response is to emphasize conservation science as the evidenced-based, theory-rich baseline for managing biodiversity, from which research agendas, education and policy follow. Without at all diminishing the importance of conservation science, this response invokes an overly simple understanding of science and the scientific method. Scientific reasoning and evidence does help distinguish better from worse causal explanations. Even so, scientific knowledge is always contingent, and laden with values, worldviews, and vested interests.

Moreover, science alone cannot speak to the origin of the "diversity crisis", a dual crisis impoverishing both humanity and nature. Instead, the origin lies in a deeply rooted cultural conflict over our coexistence with other forms and ways of life. Human land-use and wildlife management has direct consequences for the well-being of all human and non-human life. So too our motivations for learning to live in a more than human world is deeply informed by our moral sensibilities and cultural worldviews. Finally, science and ethics are indispensable to one another.

To answer questions of wildlife conservation and ecological sustainability, we must look between ethics and science. Neither one nor the other alone, nor one above the other, is sufficient to the task, as each has different strengths. Ethics provides insight into the "moral causation" of the diversity crisis, and a vision of a world that respects the diversity of life. Science provides insights into what nature is, how it works, as well as our anthropogenic effects on the geosphere. Like landmarks used during route-finding, both ethics and science help us adjust our compass of environmental and social policy.

I want to make clear at the outset that I am not a conservation scientist, and I will not presume to tell others how to practice natural or applied science *per se*. Doing so violates my deeply held commitment to honouring the insights and skills of colleagues from diverse fields of study. Rather, as someone trained in ethics and the interdisciplinary science of geography, my purpose is to help clarify those elements in the overlapping domains of ethics and science. I also want to make clear that I do not pretend to have the final answer. Science and ethics are contested terrain. My articulations of each will not satisfy every critic.[4] Rather I want to offer a set of comments on how we might think about conservation science and ethics. I mean these comments to be suggestive, not conclusive, and hope they may serve as a point of departure in future discussions.

THE DIVERSITY CRISIS

As Gary Snyder notes, our world is a vast system of flowing energy and cyclic matter, in his poetic vision, a "breathing planet" nested in "sparkling whorls of living light".[5] One of the signature elements of our world is its diversity. Our planet is alive with diverse forms of life and ways of living, human and non-human alike. This diversity is multiform: people,[6] animals, plants, individuals, packs, tribes, populations, societies, ecosystems and cultures (to name but a few). And this diversity is found in a wide variety of places – in the sea, on land, deep underground, and in the air – each interacting with the other at varied scales, from the micro to the macro, from local to regional to global. [7]

Tragically, we also live in a world facing a biodiversity crisis of global proportions. Modern human activity is accelerating the loss of species and ecosystems at a rate and scale unparalleled in natural or human history. With species already disappearing at many times the rate of natural extinction, up to a quarter of the world's land animals and plants may be extinct or endangered by 2050.[8] Conservation scientists have generated a substantial stock of knowledge regarding the factors diminishing biodiversity. The causes of this crisis include habitat degradation, landscape fragmentation, urban sprawl, human population growth, increasing consumption and pollution, and over-exploitation of resources. All of these causes are further complicated by the shifting context of global climate change.[9]

Less understood are the normative[10] values at stake – how humanity should value biodiversity, and how such values should inform our response to the crisis of extinction. Together, these constitute the non-instrumental[11] and normative dimension of biodiversity. Some of these values are about nature itself. One hotly contested issue is whether non-human life is simply a resource for human use, or has a significant value of its own. Another issue is whether our concern for biodiversity should encompass more than wild flora and fauna, and include domestic plants and animals. Other values are about culture and the human interaction with nature. For instance, the cultural diversity of humanity makes cross-cultural norms for ethical decision-making difficult to formulate. Moreover, we have pressing needs to alleviate poverty, advance social justice and defend human rights. These and other particularly human values raise difficult issues about the different range of responsibilities of the world's peoples for protecting biodiversity. Altogether, questions about these broader natural and cultural values go to the heart of "how we ought to live" with non-human life, and how both human communities and the natural world can and ought to flourish together.[12]

What this means is that we are in the midst of a *diversity crisis*. The diversity crisis is really two interrelated crises – a crisis of nature and a crisis of culture. The crisis of nature is driven by humanity's "geographic agency", our power to do good or ill to the living systems of the planet. The effects of this agency are the proximate cause of nature's decline. Using the theories and methods of science, we can hope to measure and model these impacts, gain a measure of prediction and control over them, and thereby alleviate or reverse some of their most deleterious effects. It is for this reason we place a legitimate measure of hope and faith in scientific-technical approaches to wildlife conservation and ecological sustainability.

Unfortunately, our cultural crisis is a bit harder to comprehend. At its heart is a clash of ethics-laden worldviews. These worldviews describe visions of the good life, definitions of moral community, norms of conduct, and attributions of culpability. How we understand and respond to the natural world, not the physical consequences of our actions, is the focus of analysis here.[13] This distinction is crucial. It is our worldviews and ethical sensibilities that not only inform human agency, but characterize the ultimate causes of the diversity crisis. And because the cultural crisis is so morally and socially complex, it spawns ramifications that complicate its resolution. Examples abound. The diversity crisis:

- threatens the cultural survival of the worlds Totemic Peoples whose modes of thought and livelihoods are rooted in the indigenous animals and resources of a region,
- creates a demographic trap whose cycle of spiraling population growth, increasing poverty and degrading habitats besets the so-called developing world, and

- exacerbates the kind of globalization that facilitates irresponsible consumption, the centralization of political-economic power, and the shifting of environmental burdens from areas of wealth to areas of poverty.

Whether humans are a part of the natural world, while at the same time, distinct enough as a species to take moral responsibility for their actions, is the pivotal problem on whose resolution rests any possibility for sustainability. And this has occasioned a global debate over the ecological, social and ethical values that ought to inform our thoughts and practices regarding animal welfare, wildlife conservation and ecological sustainability.[14]

SCIENCE AS THE CUSTOMARY RESPONSE

The customary response to this challenge is to emphasize conservation science, demand science-based environmental policies, and redouble efforts at research and education. This response is conditioned by three presuppositions. First, science provides an objective knowledge of nature, and of our human interactions with nature. Second, this knowledge should be the basis for public policy. Third, education in the methods and facts of science will produce the political and social paradigm shift to motivate and guide a sustainable relationship with the Earth. Thus conservation science becomes the evidenced-based, theory-rich baseline for managing biodiversity in wild and humanized landscapes, from which research agendas, education and policy follow.[15]

Given the interconnected ecological, social and ethical dimensions of the diversity crisis, it may seems incongruous to rely primarily on conservation science as the solution to our problem. Yet the reasoning is simple enough. Ours is a world of facts, and science is best suited to explore, describe and explain these facts. Conservation science therefore provides the data and analyses that inform the purpose and outcome of public policy as it relates to wildlife conservation and ecological sustainability.

In this mindset, the natural sciences tell us the truth about nature, while the social and interdisciplinary sciences model the human interaction with natural systems. If through scientific inquiry we can correctly model the causal interactions within nature, as well as between nature and humanity, then we will be able predict human and natural events, identify and forecast trends, mitigate undesirable outcomes, and manage for certain goals, such as ecological sustainability. This is well expressed in various formulations of the widely adopted "ecosystems management" approach to natural resources. Ecosystems management uses both the natural and social sciences to establish baselines for biodiversity and ecosystems integrity,[16] assess the social and economic needs of a given community or society, and manage the landscape (or seascape) in accordance with desirable social outcomes. Conservation science is a success if it contributes to outcomes that are ecologically sustainable and democratic.[17]

LIMITS OF SCIENCE

Without at all diminishing the importance of conservation science, the customary response invokes an overly simple understanding of science and the scientific method, ignores our historical and philosophical knowledge about science and society, and fails to grasp the tight connection between science and ethics.

Science is not a means for obtaining certain and predictive knowledge about our world, and the hope for a unitary method or conciliant theory is a caricature and pipe-dream respectively. Science is rather a set of theories and methods for seeking causal explanation of natural and/or human phenomena. These theories and methods necessarily differ with respect to the phenomena at hand, and modeling all the sciences on a single model (such as the experimental method) has proved illusory. We can certainly distinguish better from worse causal explanations on the basis of reason and evidence. Even so, scientific knowledge is always contingent, value-laden, informed by larger worldviews, and beholden to systems of power.[18]

Why is this the case?

Modern science was founded on the belief in an orderly universe. Matter and energy conformed to rules that could be measured and modeled. Armed with knowledge of these "laws of nature", scientists could peer into the determinative nature of reality, predict events and control outcomes. Like its ancient and medieval counterparts, science concerned itself with finding causal explanations for the world's phenomena. It differed, however, in its worldview of the universe, and thus the kinds of theories and methods that were needed to produce causal explanations. It no longer focused on an admixture of explanations rooted in theology, human agency, proximate causes and natural processes (the "four causes" of Aristotle), but set its sight squarely on what its precursors labeled "material causation", that is, explanations for natural and social events rooted in the material world. The forces and effects that early modern science both looked for were found in the physical world. Correspondingly, the fields of investigation they praised most were physics, mechanics, optics and chemistry.[19]

At first this worldview was applied to animals and the rest of nature, but not to humankind. Because most scientists of the time believed in a supernatural power (God) that had created an orderly universe (the concept of deism), they also believed that God had invested humankind with a soul. Because they were given souls, humans had "free will" and were exempt from the determinism of natural law. This was especially true of educated white men who most closely resembled the mien of the deity. Very quickly, however, the same concepts of order, determinism, predictivism, measuration, modeling and control were applied to human beings and their societies. Thus the natural and moral philosophy of medieval and early modern Europe was transformed into the natural sciences (e.g. physical, biological), the social sciences (e.g. behavioural, psychological), and the humanities (e.g. history, philosophy) of today. The ultimate goal was the determinative prediction of all natural and human phenomena, a grand theory of nature and society that E. O. Wilson well describes as consilience, or "the unity of knowledge".[20]

Accompanying this faith in consilient theory is a distinctive approach to the method of inquiry. Because it allowed for controlled tests in closed systems that could, with precision, describe chemical and mechanical cause and effect, the "experimental method" became the model for scientific investigation. As our knowledge of the world grows more nuanced, from subatomic particles to complex and chaotic systems – this method is supplemented by others – field trials, statistical techniques, mathematical modeling, etc. Even so, the double blind, controlled experiment remains the touch-stone of scientific methodology to many researchers. It is this basic worldview of science and its method that we teach to schoolchildren, that knocks around inside graduate students heads, and forms the public impression of science. In the philosophy of science we call this the *naturalistic model* of the sciences.[21]

Of course, when we explore the history and philosophy of the natural and human sciences, we quickly learn this worldview is false. It is not false because it is wrong *per se*, but because it conveys only a partial notion of causation. To be fair, the naturalistic model has vastly extended our knowledge of the physical and biological world. It sparked and sustains a technological revolution of unparalleled proportions. It works well at explaining what it was designed to explain – tangible phenomena in relatively closed systems, things we can touch and measure whose causal interactions are bounded and knowable. When scientists say they study facts and produce objective knowledge about the world, they are usually speak from the perspective of the naturalistic model of science.[22]

Yet we live in a world of values as well, which requires an interpretive model of science, one designed to investigate the cultural and social worlds of human and many non-human animals. Values are also facts of life, but they are intangible phenomena that elude capture through quantitative methods. Alongside values, the world of intangible phenomena is large, and includes feelings, thoughts, intentions, reasons and culture (to name a few). These phenomena are just as real, but they cannot be measured and modeled in part or whole. They need to be apprehended if we are to understand their meaning. Questions of meaning, intention and the like are thus qualitatively different phenomena, the sort of facts that the naturalistic sciences were never designed to investigate. What this means in practice is that living beings who are sentient and/or sapient, creatures such as cats and cows and wolves and people, do not completely fit into the theories and methods of the naturalistic model. You cannot describe or explain what people and many animals do, without causal reference to the agency that motivates their behaviour. This is as true with respect to the hunting styles and skills of wolf packs, as it is of the collective decisions of democracies over its economic policy. To study human and animal agency,[23] we require distinctive theories and methods appropriate to the investigation of these phenomena. The development of moral-social theories, cognitive ethology and qualitative methods reflect this distinction between the naturalistic and interpretive models of science.[24]

Most of conservation science is beholden to the naturalistic model, and therein lays it limits. Clear-sighted as it can be when quantifying the functions of ecosystems, its vision is progressively foggy as it passes through animal sociality and on to human agency. Recognizing the limits of conservation science does not imply it is flawed or unimportant. Conservation science is crucial to our efforts at protecting and restoring wildlife, as well as creating greater well-being for people, animals and nature.[25] Even so, the natural sciences cannot on their own circumnavigate the entire landscape of our concerns for diversity.

FACTS AND VALUES

When we discuss questions of facts and values, we quickly run up against the *fact/value dichotomy*. This dichotomy is ingrained in the naturalistic model discussed above, and serves as the justification for making the natural sciences the gold standard of causal explanation. In this dichotomy the natural sciences are associated with facts, reason, empirical truth and objective modes of analysis. The humanities are associated with values, emotions, personal and social preferences, and subjective modes of experience. Science is believed to produce an objective knowledge of the world, while the humanities create subjective states of experience.[26] For some the causal insights of science are so unique, it becomes the only form of true knowledge, a position known as scientism. This scheme marginalizes "non-scientific" modes in knowledge, while valourizing "scientific truth". Sadly, to the degree that we take this dichotomy to heart, we impoverish our understanding of the world in general and our scientific understanding of causation in particular. There are several aspects of this dichotomy we might explore, e.g. its historical development, the logic of the "naturalistic fallacy", and its implications for ontology and epistemology (philosophy of science terms for theories of knowledge and existence, respectively).[27]

Here I want to focus on the most common claim associated with the fact/value dichotomy, the position that science is (or should be) value-free and value-neutral. The reasoning for this runs as follows. Science provides an unbiased outlook on the world, and the scientific method ensures the objectivity of scientific knowledge. Science is therefore free of value claims in itself and neutral with respect to differences over values. In this respect science is like a "tool", neither good nor bad. Its simply provides information. Science may therefore inform the policy process of possible options and likely outcomes, but it cannot choose for us the values we should believe in, much less act upon. In this sense, a science free of values may still be relevant to the public good.[28]

There is something salutary here, especially if we consider objectivity as a commitment to honesty in research. It is even better if we take a broad view of scientific methods, seeing them as systematic practices using reason and evidence in a progressive process of learning. Considered so, objectivity and method can help us expose invidious bias, and avoid errors of fact and interpretation. Based on conversations I have had with scientists, I think this is what most of them really mean when they strive for an "impartial" point of view. And to preserve the appearance and substance of impartiality, many (but not all) avoid taking politically sensitive positions on issues of animal welfare, social justice and environmental protection.

Pushing the notion of objectivity farther, to claim that science is free of values, or that scientists should be neutral with respect to the uses and consequences of science, is to go too far. Science is never value-free and, for better or worse, is laden with value implications. The reason is that a host of moral sensibilities are embedded within the intentions, actions and/or consequences of science.

Scientists are people after all, and their research is unavoidably inflected with values. Many of these values have obvious ethical overtones. We learned to our shame of the scientific abuse of human research subjects through the Nuremberg trials (re: Nazi racism and science) and the Tuskegee experiments (re: American racism and science). We are coming to grips with similar issues involving animal subjects research as well. In addition, science depends on its practitioners telling the truth about their findings, and trusting in the good will of other scientists. Understanding and acting on these values in the form of ethically informed best practices is crucial to maintaining the integrity and credibility of science. Further, science can help us discriminate between better and worse ways of valuing the world. A case in point is the growing recognition that many animals are not automatons, but individuals nested in social groups (e.g. wolves and house cats). This has transformed our beliefs about what responsibilities we have to other animals, whether at home, in the laboratory, in the farm yard, or in the wild.[29]

Science therefore operates in two value-laden *domains of significance*. Both domains are crucial to the integrity and credibility of conservation science. The first is the internal domain, that is, the methods of research and the production of scientific knowledge. We often hear this domain referred to in terms of "professional ethics" or "codes of conduct". Ethics in the internal domain helps ensure the integrity of research. While there are many ways of defining this integrity, it basically serves to uphold two core moral values of science – truth and trust. When speaking of truth, we are referring to such matters as the collection, analysis, interpretation and communication of research. With respect to trust, we are thinking primarily about academic freedom, honesty, transparency, collegiality and conflicts of interest. Along with upholding truth and trust as core values, ethics also helps us define best practices for implementing those values in research. Common examples of best practices include prohibitions against plagiarism, falsification of data, the manipulation of research results, as well as guidelines on avoiding and/or disclosing conflicts of interest, the prior restraint of knowledge, and self-censorship.[30]

The second is the external domain, referring to the uses of scientific knowledge, and the applications of its theories, methods and associated technologies. We often hear this domain referred to in terms of "animal welfare" or "environmental ethics". The reason for this external domain is that science, for better or worse, has direct and indirect impacts on the health and well-being of people, animals and nature. These impacts have consequences at a number of distinct if interconnected scales on individuals, populations, species, and communities, in natural and social systems, and in geographic space and historical time. Ethics helps elucidate the best uses of science by noting how the research practices and knowledge products of science produce more or less well-being in the world.[31]

If we can let go of the fact/value dichotomy and its self-privileging of the naturalistic model in science, then another point swims back into focus: the welfare of animals, the conservation of wildlife, and the integrity of nature are not only, or primarily about, science. The ongoing debate in the United States and elsewhere over the disposition of wildlife in terrestrial and aquatic military zones is a case in point. It raises value-laden questions about our compassion for sentient animals, our commitment to preserving species diversity and the integrity of ecosystems, and our legitimate concerns for national security. The work of scientists will help us determine some of the consequences of military activity on endangered and threatened species, as well as their critical habitats. We could connect this example of value-laden environmental questions into many other areas of wildlife management and landscape protection. Nonetheless, science cannot answer how we ought to balance the well-being of people, animals and nature in a world beset by human violence. Rather it is the ethics-laden discourses of morality, religion, politics, spirituality and the like that generate the moral insights to make discriminating judgments about how we ought to live.[32]

PRACTICAL ETHICS

Up to this point, I have outlined a set of insights and limitation to conservation science. I hope it is now clear why science alone cannot adjudicate (as in justify and choose between) the cultural norms and socio-political structures that generate a constructive or destructive relationship with nature. Having sorted out some of the confusions introduced by the fact/value dichotomy, we are now in a better position to apprehend that ours is a world of both facts and values. To understand and improve our world, the facts of science are never enough. Facts must be complemented by values. Or to put it another way, values are an indispensable kind of intangible fact, much like the tangible facts of the global water and carbon cycles, or trophic energy flows. Both are empirical.

Allow me to caution that adding values to facts does not require changing the way science is done. We gain nothing by side-tracking scientists from what they do well. Even so, conservationists will benefit from a robust ethical analysis that ponders the values that motivate and impact their work. A particularly useful framing of value questions is achieved through ethics. Ethics is not the only way to explore the normative aspects of our relationship to people, animals and nature. Nevertheless, it is an indispensable tool for exploring the meaning and significance of those values that serve the well-being of the human and natural world. Ethics is adept at understanding and adjudicating (i.e. describing, explaining, justifying, choosing) the normative dimensions of human life.

Ethics can be a subject that is difficult to discuss, raising fears that it imposes a rigid or ideological view of the world. There are people who use ethics to shame others, or score debating points. There are also people who justify their dogmatic approach to life with a veneer of ethics. But this is not the main tradition of ethics, or the kind of ethics I am recommending here.

At its best, ethics is an exploration of "how we ought to live".[33] It is a conversation about the moral values that ought to inform our way of life. This search for the moral values we ought to live by involves a twin-fold process of critique and vision. We criticize what detracts from the well-being of ourselves and others, and we envision how we might improve our lives by proactively pressing for positive change. Ethics may be informed and enriched by religion, spirituality, personal experience or social custom, but it is not reducible to these sources. Instead, ethics is a cross-cultural and cross-disciplinary dialogue that uses reason and evidence while discriminating between moral values. Put into practice, ethics helps develop moral principles to guide our thought and action, and improve the well-being of ourselves and others. These others can include different entities (human or non-human) considered at different scales (individuals, social groups, systems). Thus ethics may concern itself with the well-being of individual people and animals, communities, populations and species, as well as cultural, social and ecological systems.[34]

Ethics is also a form of discursive power. Ethics has the power to reveal moral concerns, guide our thought and action in addressing moral problems, and hold people accountable for their (un)ethical actions (irrespective of legalities). Moral critique is the foundation for all movements of social change, whether these are for animal, environmental or social causes. It is for this reason that ethics is indispensable in political life (broadly understood). Moral norms are the foundation of our social customs and laws. This is not to say these norms are always or mostly right. We need only to look at the transformation of norms regarding race, gender and sexual identity for examples of moral progress. Even so, ethics-based arguments motivate the struggle to change and evolve customs and laws. If the process seems a bit unclear, think of it as akin to the development of law. There is much wrangling and many errors, but over time, a trend emerges towards better and deeper understandings. Reason and evidence can do much to contest custom and prejudice. And while there are no external, God-given, *a priori* moral truths to set our sights upon, we can adjust our moral compass to distinguish better from worse ideas and practices based on how they improve or detract from human and non-human well-being.[35]

How then should ethics inform the activities of conservationists? What constitutes a practical ethics for science in the service of wildlife conservation and ecological sustainability? What moral understandings should we consider as we advocate for animal welfare in wild and humanized landscapes? There are several directions we might take to answer this. For the purpose of this chapter, I will focus on the core question of animals and their *moral value*. An analogy may help set the stage for this topic. The moral value of human beings motivates our protection of human subjects in scientific research and civic life. So too, the moral value of animals is motivating our ethical concerns for wildlife conservation and ecological sustainability.

Any practical ethics for conservation must recognize the moral value of animals. While controversial and frequently avoided as a topic of dialogue, one cannot sidestep this issue. Doing so simply undermines the rigour and credibility of one's efforts to inform science with ethics. Discussing the moral value of people, animals and nature can be difficult. It raises a suite of philosophical, religious and social issues that many activists and scientists are uncomfortable discussing. We therefore tend to shy away from the issue completely, or we speak about "social attitudes" with a posture of non-committal objectivity. Being objective as a way to avoid invidious bias in scientific research is a good. Objectivity as an excuse to marginalize value-laden issues in science and society is a mistake. When we misuse objectivity in this way, we simply avoid the root questions that need answering, and this serves conservation poorly.[36]

The justification for protecting human subjects in North America is human rights and civil liberties, both of which are rooted in the dignity and worth of human beings. In ethics-talk, we say that people have *moral value*, that as feeling and thinking creatures we have a responsibility to treat each other with care and respect. It is for moral reasons that we as a society have instituted research rules and human subjects review committees to ensure informed consent, psychological and physical integrity, and justice for vulnerable populations. When it comes to animals we are not so well agreed as a society. The range of species and their differences makes it impossible to simply map human ethics onto animals. Still, virtually all informed ethicists would say that many animals, including most mammals and birds (to name but two evolutionary lineages) are to varying degrees feeling and thinking creatures. Because of their sentience (awareness) and sapience (self-awareness), these creatures have what ethicists call *moral standing:* their moral value demands our attention, consideration and inclusion in a more-than-human moral community. There is broad agreement here based on the facts of biology, ethological studies, and analogies to human welfare. [37]

Please note that neither moral value or moral standing should be equated with "animal rights" *per se.* Animal rights is a broad term, most used to describe an advocacy position, or a philosophical doctrine. Associated with groups like People for the Ethical Treatment of Animals (PETA) and the philosopher Tom Regan, animal rights makes a powerful argument for the recognition of moral rights for sentient animals, rights that are akin to human civil and political rights. Yet to the degree that there is overlap between these arguments, it is more along the lines that there are right and wrong (or better versus worse) ways to treat animals, not an endorsement of particular policies or theories. In addition, also note that the recognition of moral standing does not imply equal treatment between humans and other animals. Moral standing means that being part of a broader moral community, the well-being of animals must be taken into account, not that different individuals or groups need to be treated in exactly the same way.[38]

The ethics of wolf recovery in North America is a good example. For some, wolves are a biological heritage we ought to restore and conserve for our children, citizenry and the world. Future generations will condemn us for failing to take reasonable steps in this regard. Many see in wolves the hand of a Creator for whom the natural world, including wolves, is good. Humans are thereby the stewards of Creation, and wolf recovery is a sacred obligation. Others believe wolves are more than functional units of ecosystems, more than resources for human use. Rather, wolves are self-aware and social beings. This gives wolves, as it does people, a moral standing when it comes to human actions that, for better or worse, have consequences for individuals, packs, populations and species. In this worldview, wolf restoration is an act of restitution for past harms done to creatures with whom we share a common landscape. For still others, wolves are top predators contributing to the health and well-being of the larger community of life. Wolves generate a kind of "natural good" that, while unintentional on their part, is indispensable to ethical adjudications of how we ought to live with the natural world.[39]

In addition, there are a series of value paradigms that stake out distinct positions and implications in this debate – anthropocentrism, biocentrism, ecocentrism and geocentrism. Briefly, *anthropocentrism* gives moral standing to humans alone. *Biocentrism* affords moral standing to living beings, especially the more sentient and sapient. *Ecocentrism* is primarily concerned with the web of life than it is the well-being of individual life-forms. To my way of thinking, articulations of these value paradigms have bogged down in unnecessary and unproductive arguments pitting humans against nature, people against animals, and individuals against systems. For my part, I defend a geocentric approach to moral standing. In *geocentrism*, one values people, animals and the rest of nature, both as individuals and/or collectivities when appropriate. While this paradigm does not eliminate conflict of values, it does recognize the reality of multiple values, as well as our ability to balance those values. For instance, we do this when we balance individual civil liberties against questions of public safety and national defense. Balancing moral-political values is a fact of life in every dimension of human activity, and it should not be seen as an insuperable barrier when it comes to questions of animal-human relationships.[40]

Embracing the moral standing of animals has immediate implications, two of which I will note here. First, it highlights the *moral significance* of conservation. The reason for this is obvious. Our actions have real consequences for the well-being of wildlife and the integrity of their habitats. We can do good or ill to the lives and livelihood of people, animals and nature. Because of this, whether intentional or not, our work is laden with moral implications. In our hearts and minds, the vast majority reading this chapter already will know this. Most of us chose this area of work to make a positive difference in the world. We care for human and non-human individuals, communities and systems. While our reasons and experiences may vary, the caring remains, and it is caring that is perhaps the primary source of our moral sensibilities.[41]

Second, it highlights the *practicality* of ethics in conservation. It reminds us that the stickiest problems in ecosystem and wildlife management are deeply rooted moral conflicts over whether (or how) to coexist with other forms and ways of life. These problems have little to do with a lack of empirical data, quantitative theories or management techniques. Instead, they are a consequence of differing moral visions of how we ought to live with non-human others. To resolve such conflicts, we have to address their moral roots by directly and respectfully engaging the moral differences that divide us. Resolving these differences may be crucial, but doing so is a long process. Fundamentally, it is a matter of instituting an ongoing community dialogue. The dialogue must respect a diversity of opinions, avoid polarization, and focus on best concepts and practices. It is also a process of struggle and the exercise of political power – mobilizing support, negotiating with opponents, and marshalling the agreement of third parties.[42] Overall, it is an effort of cultural transformation that needs to be embraced by advocates and scientists alike. Simply bandaging the wound in our relationship with nature will not heal the dysfunctions. If our efforts at wildlife conservation and ecological sustainability are to succeed, we need to look to the long-term "health" of the geosphere.

Finally, there are different styles of ethics, some of which are more helpful to conservationists than others. One approach to ethics has been to create a body of internally consistent and empirically ungrounded hypotheses (e.g. formal ethics, analytical ethics), and then deductively apply these hypotheses to determine the proper moral outcome in the world (e.g. applied ethics). If this sounds vaguely familiar, it should. It is a model of ethics that mimics the naturalistic model of science.[43] This model of ethics does contribute to the clarification of concepts and systems of thought. I do not mean to undermine its importance. It is, nonetheless, a poor compass in a complex moral landscape. In an article on dueling moralities over feral horses in Australia, I remind readers that:

> Ethics has historically been a form of practical reasoning. Practical reasoning differs markedly from the analytic reasoning that dominates modern moral philosophy (e.g., utilitarianism, deontology, contractarianism). Modern moral philosophers seek a trans-geographical truth, which is to say, deductive axioms of conduct, derived without the benefit of geographic or historical contextualization, and equally applicable to all people, places and circumstances. This not the case, however, for the practical reasoning that is part of alternative traditions of moral philosophy (e.g., casuistry, hermeneutics). Practical reasoning seeks to articulate situationally sensitive principles to guide us in moral and political deliberation. In this view, ethics is not a timeless and placeless code of rules, but the use of moral concepts as rules-of-thumb that help us answer "how we ought to live".[44]

An ethics founded on practical reasoning is what I call *practical ethics*. Practical ethics resonates with Socrates' original question of "how we ought to live". In the contemporary era, this question applies to both humans and non-humans. It focuses on the moral sensibilities that inform (or ought to inform) our individual and collective lives. It seeks to help people make better decisions, incorporating dialogue, democracy and diversity as serious elements in moral deliberation. Taking the failings of analytic and applied ethics to heart, practical ethics looks to diverse moral principles, rooted in the empirical reality of cases, to triangulate on the reasons and resolutions to our moral concerns. [45]

Two insights of practical ethics should be emphasized here. First, ethical concepts cannot be applied by rote, like a grid of latitude and longitude from which we read off the correct "position". Rather, moral understanding is akin to triangulating on the best of several positions, using a plurality of principles (understood as "rules of thumb"). Second, the ethical principles we use should be selected in light of the case at hand. Obviously this relates to the point above, but it is more than a simple statement of conceptual pluralism. It recognizes the interpretive nature of ethics. Our prior understandings condition our current insights. As our presuppositions change, so do our moral insights. Moral principles actively and dynamically reveal the ethical issues at stake, as well as provide guidance on what we ought to do about them. Thus whalers and scientists might argue interminably over the sustainable yield of whale populations.[46] And yet, the insight that whales may have moral value and standing cuts through this debate to reveal fundamental problems in "taking" sapient creatures, especially from their family groups.[47]

The value paradigms mentioned above are also a good case in point. There are times when we may want to take a biocentric point of view, focusing our concern on individual animals and their social groups. This is most useful when the animals in question evince a high degree of sapience. There are other times when we may want to take an ecocentric point of view, shifting the scale of our questions and interpretations to the ecosystems of which individuals are a functional part. Where the creatures in question closely resemble human beings, the arguments for biocentric animal rights carry weight. The great apes fit well in this system of ethics. If the species differs significantly from human beings, the power of biocentric arguments fades. Spiders manifest none of the features that animal rights advocates use to defend primates, and should be thought about in ethically distinct ways. These are not opposing points of view *per se*. They do, however, raise distinct questions and reveal different moral issues.

At the same time, we may wish to take a geocentric point of view when weighing multiple values between different species who share a natural environment. Whales and krill exemplify this point. Using biocentric arguments to defend individual krill against the needs of baleen whales is a concept error in moral understanding. Yet we should still value krill as elements of an ecosystem, an ecocentric insight. Failing to employ biocentric concerns about whales as individuals or social groups is another failure in judgment. Their sapience entitles them to such consideration. Yet we do not want to miss the forest for the trees, valuing whales but not krill. Employing a geocentric paradigm of moral value, we can value both whales and krill, treating them differently in thought and action as is fitting to their distinct natures.

DEEP SUSTAINABILITY

Shortly before the conference that led to this book, *BBC Wildlife* magazine published a cogent article by David Lavigne on wildlife conservation and "wise use".[48] In this article, Lavigne argues that the language and agenda of sustainable development has been hijacked by a "wise use" movement that is hostile to wildlife conservation and ecological integrity. Shielded by a façade of eco-newspeak, this discourse appears to protect animals and their environments while pursuing development, even while their policies and practices are abetting the exploitation of wildlife and wildness. This takes a variety of forms, from creating markets for the natural services of the commons, commodifying wildlife and wild landscapes, dismantling national parks and reserves, and privileging human development over the protection of nature.

Some readers may interpret Lavigne's essay as misanthropic, the perspective of a privileged white Canadian concerned with questions peripheral to the well-being of billions of people from the developing world. For those comfortable with a style of identity politics that accepts or dismisses a perspective based on one's geographic and social location, this will be enough to end the matter. It is not for me.

There are good reasons for worry about the state of nature under the rubric of sustainable development. The steady degradation of ecological integrity through integrated conservation and development strategies is well established.[50] Much of the sustainable development literature is openly anthropocentric, privileging humans above nature, and marginalizing concerns about animal welfare and ecological integrity.[51] Academics provide a mantle of respectability through ill-conceived theories about stages of development, political-ecology, or the social construction of nature.[52] International agencies, transnational corporations and human rights activists further exacerbate this with a myopic focus on the economic, social and technological aspects of human health and welfare.[53]

This is not to deny or diminish the manifest responsibilities we owe human beings who are near or far, or of this and subsequent generations. Basic human rights, the empowerment of women, ridding the world of hunger, malnutrition and disease, ending poverty and illiteracy, creating infrastructures of hope and opportunity for every person and community – these are positive moral goods that we ought to proactively honour, endorse and support.

Yet there is no ethical need or requirement to marginalize the well-being of animals and the rest of nature, while we meet our human responsibilities. And there is a perverse blunder in demanding we sacrifice what little is left of wildlife and wildness, in the vain hope that this will alleviate the burdens of global inequities. Doing so not only dismisses our legitimate moral concerns for the non-human world, it diminishes the sources and resources of cultural diversity and integrity the world over.[54]

Lavigne is not alone in identifying the mendacity of eco-newspeak.[55] Rather, he is cautioning against blind faith in practices that may sound good, but hide a sinister agenda. He is not denying our shared responsibilities to the human world. Nevertheless, what is absent from Lavigne's article, and I mean this in the sense of silence not error, is an articulation of the moral sensibilities that he brings to the table. An ethical framework that will help people understand that his arguments are not misanthropic but geocentric – a positive valuation of people, animals and the rest of nature.

Many people of the world, from all walks and ways of life, have articulated powerful critiques of "sustainable development"[56] and "conservation" as it is envisioned or practiced by the world's elite nations, corporations, social classes or non-governmental organizations. Some of these critiques are motivated by an ethic sensitive to wildlife and wildlands. Those involved in the drafting and dissemination of the Earth Charter are a case in point.

The Charter is "a declaration of fundamental principles for building a just, sustainable and peaceful global society in the 21st Century".[57] Drafted in a 10-year cross-cultural conversation of global reach, the Charter articulates an ethically grounded vision for a sustainable global society that protects and defends its citizens and the earth. The foci of this vision are respect and care for the community of life, ecological integrity, social and economic justice, and democracy, nonviolence and peace. The Charter notes that "environmental protections, human rights, equitable human development, and peace are interdependent and indivisible". The ethical principles of this vision are not meant to be a rule-book or code of conduct (the applied ethics approach), but rules-of-thumb to help civil society, nationalities and international organizations develop best concepts and practices while working towards sustainability (the practical ethics approach).[58]

The genesis of the Charter intersects many other declarations, petitions and social movements committed to environmental protection, social justice and peace. Perhaps the immediate precursor was the report by the World Commission on Environment and Development entitled *Our Common Future*.[59] Shocked by the pace and scale of environmental degradation and inequitable human development, the members of the Commission saw a need for a new moral paradigm centred on the earth and sustainable development – "human survival and well-being could depend on success in elevating sustainable development to a global ethic".[60]

The Charter directly challenges mainstream philosophies of sustainable development, redirecting the concepts and language to the goal of "sustainable communities" and a broader norm of "sustainability". Mainstream discourses of sustainable development emphasized sustaining growth to alleviate poverty. Its commitment to wildlife conservation and ecological integrity could be superficial. The framers of the Charter felt more was needed to properly envision and defend the well-being of humans and nature.[61]

In the Charter, sustainability becomes a moral imperative to remove the cultural, social, economic and political causes of injustice and privation, while simultaneously valuing the creatures and resources on which human flourishing depends. The Charter is thus principally concerned with securing ecological integrity and equitable human community, especially in the face of undemocratic and unegalitarian trends in globalization.[62] As I note elsewhere:

> By broadening the largely economistic and technocratic notion of sustainable development into the moral-political concept of sustainability, the drafters of the Charter were able to emphasize the ethical dimensions that give meaning and direction to human development and environmental protection...[63]

In a similar vein, Denis Goulet notes that:

> The single greatest threat to nature – menacing, irreversible destruction of its regenerative powers – comes from "development". This same development is also the major culprit in perpetuating the underdevelopment of hundreds of millions. The task of eliminating degrading underdevelopment imposes itself with the same urgency, as does the task of safeguarding nature.[64]

I heartily agree, and I refuse to accede to a world made right for people, but at the expense of animals and nature. There is no ethical reason or political-economic necessity that this be so.

It is this geocentric valuing of life — human and non-human, individual and collective — that I believe is at the heart of Lavigne's critique. His anger as a global citizen at eco-newspeak is understandable, especially given its justifications for the swath of destruction our sprawling species cuts through the air, over land and in the seas. More to the point, as a scientist it does not bias his research and writing, it sharpens it. Though implicit, his moral sensitivities intensify his analytic acuity, helping him to espy one cause of the world's biological impoverishment. And is this not one goal of all science, clarity of causal insight?

How then might we express our multiple commitments to people and animals and nature in the context of development? The Charter's norm of sustainability points the way. It simultaneously acknowledges the crying human need that must be met, as well as the responsibility to respect and care for the natural world. Because the Charter is a relatively brief document, it cannot cover each and every contingency in which questions of wildlife conservation, ecological integrity and sustainability intersect. Yet as an open-ended document, it is ripe for further specification. [65]

I suggest we begin to talk and act in terms of *deep sustainability*. By deep sustainability, I mean a way of human life that ethically values, equally respects and proactively cares for the biological and cultural diversity of our world. Obviously, I do not mean sustainable development in the standard anthropocentric and economic mode, as providing a steady supply of material to this and future generations. Nor do I mean ecological sustainability in the sense that Goulet uses it, as preserving the regenerative powers of nature as a whole. Goulet's ecocentric presuppositions are not wrong; preserving the evolutionary-ecological functions of nature is requisite. But a commitment to ecological sustainability does not go far enough in defending the present abundance and distribution of life, or in explicitly recognizing the moral value of non-human forms of life.

Rather, deep sustainability should be an ethics-laden worldview that makes room in our hearts, minds and landscape for the rich diversity of biological and cultural life. It should articulate a true alternative to mainstream as well as "wise-use" notions of sustainable development. It should take a practical approach towards ethics, and emphasize the role of values in defining and creating sustainability. It should be geocentric, and acknowledge the moral value and standing of wild and domestic animals, as well as wild and humanized landscapes. It should embrace both the naturalistic and interpretive sciences as indispensable sources of causal knowledge. In responding to the diversity crisis, it should defend the well-being of people as assiduously as it does wildlife and habitat, but not at the expense of one over the other. It is this kind of sustainability that should envision a world made right for people, animals and the rest of nature.

Aldo Leopold, one icon of North American conservation, says "there are some who can live without wild things and some who cannot".[66] Leopold was someone who could not. He required wildlife and wildness in his life. Please note that he was not stating this solely as a matter of personal preference, but in recognition of his common citizenship and moral responsibilities to what he termed the "land community", a moral community embracing people, animals and places. If we want to live with wildlife and wildness, if we mean to meet our responsibilities for animal welfare and ecological integrity, we must make morally informed choices about "how we ought to live". And if we allow conservation science and ethics to inform one another, then we have not only a better prospect of a broad recovery of wildlife and habitats, but the (re)discovery of a deeper and more sustainable relationship between humanity and the natural world.

ACKNOWLEDGMENTS

I want to thank David Lavigne for inviting me to contribute to this book, as well as its originating conference, "Wildlife Conservation: In Pursuit of Ecological Sustainability", held at the University of Limerick, Ireland from 17-19 June 2004. I also wish to thank his colleagues, Sheryl Fink, Christine Jones and the staff members of IFAW. Their efficient planning, smooth execution and good humour made attending the conference and preparing this chapter a joy.

NOTES AND SOURCES

[1] By conservation science I am referring to the natural and applied sciences in the service of conservation (e.g. ecology, ethology, fisheries and wildlife). As the baseline for developing policies covering wildlife conservation and ecological integrity, conservation science impacts public policy principles, statutory laws, administrative regulations, schemes of implementation, techniques of management, and forms of evaluation. The behavioural and social sciences are represented to a limited degree through the study of "human dimensions" in conservation. Conservation science overlaps with conservation biology and environmental geography, but it emphasizes a larger range of disciplines.

[2] See Lynn, W. S. 2002. Canis Lupus Cosmopolis: Wolves in a Cosmopolitan Worldview, *Worldviews* 6 (3), 300-327.

[3] For more on the biodiversity crisis, see Willison, Chapter 2.

[4] Then again, satisfying every critic is not a concern of mine. I am highly suspicious of assertions of finality in the theory and method of either science or ethics, suspecting that claims to comprehension mask dogmatic assertions. For a recent example of such claims for science, see Wilson, Edward O. 1998. *Consilience: The Unity of Knowledge*, New York: Alfred A. Knopf. A useful critique of Wilson's vision of consilience is articulated by Niles Eldridge and Stephen J. Gould in a dual book review in Civilization magazine. See Eldridge, Niles, and Steven J. Gould (1998) Biology Rules, *Civilization* 5 (Oct/Nov), 86-88.

[5] Snyder, G. 1974. Mother Earth: Her Whales. *Turtle Island*, New York: New Directions, p. 49.

[6] For an explanation of how I conceptualize the relationship between people, animals and nature, please see endnote 25.

[7] Kellert, Stephen R. 1996. *The Value of Life: Biological Diversity and Human Society*, Washington, DC: Island Press; Mayr, Ernst. 1982. *The Growth of Biological Thought: Diversity, Evolution, and Inheritance*, Cambridge: Harvard University Press; Mittermeier, Russell A., and Cristina G. Mittermeier. 1997. *Megadiversity: Earth's Biologically Wealthiest Nations*, Mexico City: Cemex; Wilson, E. O. (ed.). 1988. *Biodiversity*. Washington, DC: National Academy Press; Wilson, Edward O. 1992. *The Diversity of Life*, Cambridge: Harvard University Press.

[8] Thomas, Chris. 2004. Extinction Risk from Climate Change, *Nature* 427, 145-148.

[9] Eldredge, Niles. 1998. *Life in the Balance: Humanity and the Biodiversity Crisis*, Princeton: Princeton University Press;

Meyer, Stephen. 2004. End of the Wild: The Extinction Crisis is Over. We Lost, *Boston Review* (April/May), Available at www.bostonreview.net; Noss, Reed F., and Allen Y. Cooperrider. 1994. *Saving Nature's Legacy: Protecting and Restoring Biodiversity*, Covelo, California: Island Press; Terborgh, John. 1999. *Requiem for Nature*, Washington D.C.: Island Press; Wackernagel, Mathis, and William Rees. 1996. *Our Ecological Footprint: Reducing Human Impact on the Earth*, Gabriola Island, British Columbia: New Society Publishers.

[10] Another way of speaking about ethics is through the concept of "norms", a term that is quite common amongst social scientists. The word norm and its cognates derive from the Latin "*norma*" meaning a carpenter's square, a pattern, or a rule. In modern English, a norm may be many things — a standard or model, the mean value, the average phenomena, or a social custom. In ethics, a norm is a standard for proper conduct. For more on norms in ethical and social theory, see Habermas, Jurgen. 1998. *Between Facts and Norms: Contributions to a Discourse Theory of Law and Democracy*, Cambridge: MIT Press; Selznick, Philip. 1992. *The Moral Commonwealth: Social Theory and the Promise of Community*, Berkeley: University of California Press.

[11] Non-instrumental values are often referred to as "intrinsic values" or "inherent worth", although the concept is more complex than this simple duality reveals. Modern debates over this and related terms stem from Immanuel Kant's distinction between intrinsic and extrinsic values. For Kant, only God and humans have intrinsic value — value in and of themselves. All other things, from rocks to animals to technology, had extrinsic value, that is, value to humans for some instrumental purpose. Thus in Kant's formulation, we refrain from torturing the neighbours dog because we have duties to respect her property, even if we have no duties to the dog himself. For a detailed discussion of intrinsic, instrumental and other values, consult Rolston, Holmes, III. 1994. *Conserving Natural Value*, New York: Columbia University Press.

[12] Engel, J. Ronald, and Joan Gibb Engel (eds.). 1990. *Ethics of Environment and Development: Global Challenge, International Response*. Tucson: University of Arizona Press; Lynn, William S. 2004. Situating the Earth Charter: An Introduction, *Worldviews* 8 (1), 1-15; Rolston, Holmes, III. 1991. Life in Jeopardy on Private Property, in Kohm, Kathryn A. (ed.). *Balancing On the Brink of Extinction: The Endangered Species Act and Lessons for the Future*, Washington, DC: Island Press, 43-61; Rolston, Holmes, III. 1994. *Conserving Natural Value*, New York: Columbia University Press.

[13] Lynn, William S. 2002. Canis Lupus Cosmopolis: Wolves in a Cosmopolitan Worldview, *Worldviews* 6 (3), 300-327; Lynn, William S. 2004. Animals: A More-Than-Human World, in Harrison, Stephan, *et al.*, *Patterned Ground: Entanglements of Nature and Culture*, London: Reaktion Press, 258-260; Sheppard, Eric, and William S. Lynn. 2004. Cities: Imagining Cosmopolis, in Harrison, Stephan, *et al.*, *Patterned Ground: Entanglements of Nature and Culture*, London: Reaktion Press, 52-55.

[14] Other troubles of our world intersect with the diversity crisis. A partial listing should include war, ethnic cleansing, genocide, poverty, malnutrition, hunger, racism, sexism, nationalism, corporatism, the neglect of children, and the abuse of companion, farm and research animals. Again, globalization makes these problems increasingly complex, and terrorism — especially the prospect of bioterrorism — adds yet another illness to burden our social and environmental health. See Dower, Nigel. 1999. *World Ethics: The New Agenda*, Edinburgh: Edinburgh University Press; Harvey, David. 1997. *Justice, Nature and the Geography of Difference*, Cambridge: Blackwell; Harvey, David. 2001. *Spaces of Hope*, Baltimore: Johns Hopkins University Press; Jones, Charles. 1999. *Global Justice: Defending Cosmopolitanism*, New York: Oxford University Press; Lynn, William S. 2003. Act of Ethics: Ethics and Global Activism, *Ethics, Place and Environment* 6 (1), 43-45; Porter, Philip W., and Eric S. Sheppard. 1998. *A World of Difference: Society, Nature, Development*, New York: Guilford Press.

[15] Chicago Wilderness. 2000. *Biodiversity Recovery Plan*, Chicago: Chicago Wilderness. Available at www.chiwild.org/biodiversity.html; Orr, David. 1992. *Ecological Literacy*: Albany: State University of New York Press; Orr, David. 1994. *Earth in Mind: On Education, Environment, and the Human Prospect*, Covelo, California: Island Press; Orr, David. 2002. Four Challenges of Sustainability, *Conservation Biology* 16 (6), 1457-1460; Salafsky, Nick, Richard Margoluis, Kent Redford, and John Robinson. 2002. Improving the Practice of Conservation: A Conceptual Framework and Research Agenda for Conservation Science, *Conservation Biology* 16 (6), 1469-1479; Westley, Frances, and Philip Miller (eds.). 2003. *Experiments in Consilience: Integrating Social and Scientific Responses to Save Endangered Species*. Washington, DC: Island Press.

[16] The definition of ecosystem integrity, like ecosystem health, is slippery. Most often integrity and health are used as a metaphor, juxtaposing the well-being of homeostatic organisms with self-organizing systems, e.g. people and ecosystems, respectively. Health and integrity are also ethics-laden concepts from medicine speaking to the physical flourishing of an organism. In this sense, health is akin to the Greek concept of eudaimonia, e.g. well-being. Bryan Norton attempts to operationally define ecosystem health and integrity by associating health with the autonomous functioning of complex natural systems, and integrity with maintaining the historical diversity of an ecosystem. See Norton, Bryan G. 2003. *Searching for Sustainability: Interdisciplinary Essays in the Philosophy of Conservation Biology*, Cambridge: Cambridge University Press, 176-179. For alternative perspectives on ecological health and integrity, see Aguirre, Alonso, Richard Ostfeld, Gary Tabor, Carol House, and Mary Pearl. 2002. *Conservation Medicine: Ecological Health in Practice*, New York: Oxford University Press; Pimentel, David, Laura Westra, and Reed F. Noss. 2001. *Ecological Integrity: Integrating Environment, Conservation, and Health*, Washington, DC: Island Press; Rockefeller, Steven C., Peter Miller, and Laura Westra ed. 2002. *Just Ecological Integrity: The Ethics of Maintaining Planetary Life*. Lanham: Rowan & Littlefield; Westra, Laura. 1998. *Living in Integrity: A Global Ethic to Restore a Fragmented Earth*, Lanham: Rowan & Littlefield.

[17] See Grumbine, R. Edward. 1994. *Environmental Policy and Biodiversity*, Washington, DC: Island Press; Grumbine, R. Edward. 1996. Reflections on "What is Ecosystem Management", *Conservation Biology* 11 (1), 41-47; Lee, Kai N. 1993. *Compass and Gyroscope: Integrating Science and Politics for the Environment*, Covela: Island Press; Machlis, G. E., J. E. Force, and W. R. Burch. 1997. The Human Ecosystem, Part I: The Human Ecosystem as an Organizing Concept in Ecosystem Management, *Society and Natural Resources* 10, 347-367; Salwasser, Hal. 1994. Ecosystem Management: Can It Sustain Diversity and Productivity?, *Journal of Forestry* August, 6-10.

In public debate, various terms reflect different evaluations of the quality and uses of conservation science. For example, non-governmental organizations frequently describe their advocacy as "science-based". They may even employ staff scientists or consultants to ensure the rigour of their policy recommendations (for an example, see the Annual Reports from Defenders of Wildlife, available at www.defenders.org). Professional societies of scientists, such as the Union of Concerned Scientists (UCS) recommend the use of the "best available science" in the formation of public policy. The UCS also has a "Sound Science Initiative" that tracks the integrity of science used in government reports (see Union of Concerned Scientists. 2004. Scientific Integrity in Policy Making, Cambridge: Union of Concerned Scientists). Yet sound science means different things to different constituencies. It is also a term adopted by political conservative and religious extremists to justify an ideological driven interpretation of scientific inquiry. It is what Paul and Anne Ehrlich term "junk science" in the service of anti-environmental agendas (see Ehrlich, Paul R., and Anne H. Ehrlich. 1996. *Betrayal of Science and Reason: How Anti-Environmental Rhetoric Threatens Our Future*, Covelo, California: Island Press; Lutz, William. 2004. "Sound Science" = Junk Science, *Defenders* (Spring), Available at www.defenders.org.

Interestingly, while members of these three camps would certainly disagree on the character and rigour of science as it relates to environmental questions broadly (e.g. the "reality" of global warming) and wildlife conservation specifically (e.g. the "effectiveness" of the US Endangered Species Act), they all claim science is the basis for their approach to questions of sustainability. Such is the cultural power of science to valourize or marginalize a policy position or a worldview (see Editors. 2004. Cheating Nature: Science and the Bush Administration, *The Economist*, available from www.economist.com; Wakefield, Julie. 2004. Sciences Political Bulldog, *Scientific America*n, available at www.scientificamerican.com.

[18] Lynn, William S. 2004. The Quality of Ethics: Moral Causation in the Interdisciplinary Science of Geography, in Lee, Roger and David M. Smith, *Geographies and Moralities: International Perspectives on Justice, Development and Place*, London: Routledge, 231-244.

[19] Dampier, William Cecil. 1984. *A History of Science, and Its Relations with Philosophy and Religion*, Cambridge: Cambridge University Press; Lindberg, David C. 1982. On the Applicability of Mathematics to Nature: Roger Bacon and His Predecessors, *British Journal for the History of Science* 15, 3-25; Lindberg, David C. 1992. *The Beginnings of Western Science: The European Scientific Tradition in Philosophical, Religious, and Institutional Context, 600 B.C. to A.D. 1450*, Chicago: University of Chicago Press.

[20] Damasio, Antonio, Anne Harrington, Jerome Kagan, Bruce McEwen, Henry Moss, and Rashid Shaikh (eds.). 2001. *Unity of Knowledge: The Convergence of Natural and Human Science*. New York: New York Academy of Sciences; Wilson, Edward O. 1998. *Consilience: The Unity of Knowledge*, New York: Alfred A. Knopf.

[21] Bhaskar, Roy. 1975. *A Realist Theory of Science*, Leeds: Leeds Books; Mill, John Stuart (1987) *The Logic of the Moral Sciences*, La Salle: Open Court. Originally published in 1872; Nagel, Ernest. 1979. *The Structure of Science: Problems In the Logic of Scientific Explanation*, Cambridge: Hackett; Sayer, Andrew. 1984. *Method in Social Science: A Realist Approach*, London, UK: Hutchinson.

[22] Chalmers, Alan. 1978. *What is This Thing Called Science? An Assessment of the Nature and Status of Science and its Methods*, London, UK: Open University Press; Chalmers, Alan. 1990. *Science and Its Fabrication*, Minneapolis: University of Minnesota Press; Wallerstein, Immanuel. 2001. *Unthinking Social Science: The Limits of Nineteenth Century Paradigms*, Second Ed., Philadelphia: Temple University Press.

[23] The definition of agency is usually tied to human beings. An "agent" is the author of his or her own actions, a person who acts out of cultural knowledge and authentic intention. Modern ethology makes it clear that agency is not a feature restricted to people and ethnology, but to many non-human animals. Obviously, human and non-human animals vary greatly in their cognitive capacities. Agency is therefore the ability of a self-conscious being to choose those actions that are within its power to do so. For more on human agency, see Taylor, Charles. 1985. *Human Agency and Language: Philosophical Papers 1*, Cambridge: Cambridge University Press. For animal agency, see Bekoff, Mark, Colin Allen, and Gordon Burghardt ed. 2002. *The Cognitive Animal: Empirical and Theoretical Perspectives on Animal Cognition*. Cambridge: MIT Press.

[24] Bekoff, Mark, Colin Allen, and Gordon Burghardt (eds.). 2002. *The Cognitive Animal: Empirical and Theoretical Perspectives on Animal Cognition*. Cambridge: MIT Press; Bellah, Robert N., Norma Haan, Paul Rabinow, and William Sullivan. 1983. *Social Science as Moral Inquiry*, New York: Columbia University Press; Bernstein, Richard J. 1991. *Beyond Objectivism and Relativism: Science, Hermeneutics and Praxis*, Philadelphia: University of Pennsylvania Press; Crease, Robert (eds.). 1997. *Hermeneutics and the Natural Sciences*. Dordrecht: Kluwer Academic Publishers; Denzin, Norman K., and Yvonna S. Lincoln (eds.). 2000. *Handbook of Qualitative Research*. Second edition, Thousand Oaks, California: Sage.

[25] Throughout this chapter I tend to speak in terms of "people, animals and nature", and it may be helpful to explain why I do so. The usual practice is to think in terms of binary opposites (paired but mutually exclusive categories), with one pole being humans, and the other pole being animals or wildlife or nature or the environment. Thus in various contexts we might hear about humans and nature, or culture and nature, or people and wildlife, or society and natural resources, or

the human-animal bond. What this practice tends to do is centre human beings (and our social institutions) in a moral and conceptual cosmos, orbited by various subjects such as domestic animals, wildlife, or wildlands.

Some believe this makes perfect sense because of humanity's putatively unique status in the universe. Common justifications include our favour in the eyes of a Creator, or our role as the most powerful creatures to evolve on the planet. Others believe these binaries are counterproductive, and would collapse one pole into the other. For example, at one extreme are sociobiologists who say human individuals and society are artifacts of genetic phenomena. At the other are "social construction of nature" theorists who claim nature is the artifact of socioeconomic forces. I cannot explore the details of these perspectives in this endnote or chapter, but one thing is obvious – their self-absorption with humanity.

It is to contest this self-absorption without diminishing the importance of human beings that I speak about people, animals and nature (PAN). PAN is not meant to be a mutually exclusive, empirical categorization of the world. *Homo sapiens* is certainly one kind of animal, with an evolutionary heritage and ecological relationship to a wider natural world. Rather, PAN is my attempt to direct our attention to three equally important spheres of moral concern. In this way, I may share in common with others a concern for the well-being of people, non-human animals, and the rest of the natural world. As importantly, I want to ensure that animals are not collapsed into an all inclusive nature, or that natural systems are not ignored when think about people and other animals. This is too often the case in environmental ethics and animal ethics respectively, where it works mischief in diverting or denying our attention to important moral questions.

[26] The social sciences are situated between the natural sciences and the humanities, and are alternatively praised or damned depending on where a commentator sits in this dichotomy.

[27] Readers interested in these and other fact/value questions may wish to consult Sorell, Tom. 1991. *Scientism: Philosophy and the Infatuation with Science*, London: Routledge.

[28] For a highly influential statement of this position, see Weber, Max. 1946. Science as a Vocation, in Gerth, Hans H. and C. Wright Mills, *From Max Weber: Essays in Sociology*, New York: Oxford University Press, 129-156; Weber, Max. 1978. Value-Judgments in Social Science, in Runciman, W. G., *Weber: Selections in Translation*, Cambridge: Cambridge University Press, 69-98.

[29] Fox, Michael W. 2001. *Bringing Life to Ethics: Global Bioethics for a Humane Society*, Albany: State University of New York Press; Jonsen, Albert R. 1998. *The Birth of Bioethics*, New York: Oxford University Press; Monamy, Vaughan 2000. *Animal Experimentation: A Guide to the Issues*, Cambridge: Cambridge University Press; Rollin, Bernard E. 1999. *An Introduction to Veterinary Medical Ethics: Theories and Cases*, Ames: University of Iowa Press; Rudacille, Deborah 2000. *The Scalpel and the Butterfly: The Conflict Between Animal Research and Animal Protection*, Berkeley: University of California Press.

[30] National Academy of Sciences. 1995. *On Being a Scientist: Responsible Conduct in Research*, Second ed, Washington, DC: National Academy Press. National Academy of Sciences; National Academy of Engineering, and Institute of Medicine. 1992. *Responsible Science: Ensuring the Integrity of the Research Process*, Vol. 1, Washington, DC: National Academy Press; Sigma Xi. 1999. *The Responsible Researcher: Paths and Pitfalls*, Research Triangle Park, NC: Sigma Xi, The Scientific Research Society.

[31] Fox, Michael W. 2001. *Bringing Life to Ethics: Global Bioethics for a Humane Society*, Albany: State University of New York Press; Mighetto, Lisa. 1991. *Wild Animals and American Environmental Ethics*, Tucson: University of Arizona Press; Monamy, V., and M. Gotti. 2001. Practical and Ethical Considerations for Students Conducting Ecological Research Involving Wildlife, *Austral Ecology* 26, 293-300.

[32] Biodiversity Project. 2002. *Ethics for a Small Planet: A Communications Handbook on the Ethical and Theological Reasons for Protecting Biodiversity*, Madison: Biodiversity Project; Orr, David. 2004. *The Last Refuge: Patriotism, Politics and the Environment in an Age of Terror*, Washington DC: Island Press.

[33] Socrates in Plato's Republic, Book 1:352d.

[34] Lynn, William S. 2005. *Practical Ethics: Moral Understanding in a More than Human World*, Book in progress; Weston, Anthony. 1997. *A Practical Companion to Ethics*, New York: Oxford University Press.

[35] Ansbro, John J. (ed.). 2000. Martin Luther King, Jr.: Nonviolent Strategies and Tactics for Social Change. Madison: Madison Books; Gross, Michael L. 1997. Ethics and Activism: The Theory and Practice of Political Morality, New York: Cambridge University Press; Jasper, James. 1997. *The Art of Moral Protest*, Chicago: University of Chicago Press.

[36] A provisional definition of scientific objectivity should centre on fair mindedness, that is, a willingness to change one's mind, an openness to evidence and argument, as well as an absence of prejudice and/or conflicts of interest.

[37] Attfield, Robin. 1999. *The Ethics of the Global Environment*, Purdue: Purdue University Press; Gales, Nick, Andrew Brennan, and Robert Baker. 2003. Ethics and Marine Mammal Research, in Gales, Nick, *et al.* (eds.), *Marine Mammals: Fisheries, Tourism and Management Issues*, Collingwood, Australia: CSIRO; Jamieson, Dale. 2002. *Moralities Progress*, New York: Oxford University Press; Midgley, Mary. 1984. *Animals and Why They Matter*, Athens: University of Georgia Press.

[38] Singer, Peter. 1993. *Practical Ethics*, Second ed., Cambridge: Cambridge University Press.

[39] Lynn, William S. 2002. Canis Lupus Cosmopolis: Wolves in a Cosmopolitan Worldview, *Worldviews* 6 (3), 300-327; Nie, Martin A. 2003. *Beyond Wolves: The Politics of Wolf Recovery and Management*, Mineapolis: University of Minnesota Press; Peterson, Anna L. 2001. *Being Human: Ethics, Environment and Our Place in the World*, Berkeley: University of California Press; Smith, Douglas, Rolf Petersen, and Douglas Houston. 2003. Yellowstone After Wolves, *BioScience* 53 (4), 330-340.

[40] Lynn, William S. 1998. Animals, Ethics and Geography, in Wolch, Jennifer and Jody Emel, *Animal Geographies: Place,*

Politics and Identity in the Nature-Culture Borderlands, London: Verso, 280-298; Lynn, William S. 1998. Contested Moralities: Animals and Moral Value in the Dear/Symanski Debate, *Ethics, Place and Environment* 1, 223-242.

[41] Rabb, George, and Carol Saunders. 2004. The Future of Zoos and Aquariums: Conservation and Caring, *International Zoo Yearbook*, 1-23; Saunders, Carol, and O. E. Myers. 2003. Special Issue: Conservation Psychology, *Human Ecology Review* 10 (2), 87-196.

[42] Sharp, Gene. 1978. *Social Power and Political Freedom*, Boston: Porter Sargent Publishers.

[43] Toulmin, Stephen. 1990. *Cosmopolis: The Hidden Agenda of Modernity*, New York: Free Press.

[44] Lynn, William S. 1998. Contested Moralities: Animals and Moral Value in the Dear/Symanski Debate, *Ethics, Place and Environment* 1, 225.

[45] Examples of practical ethics include Lynn, William S. 2005. *Practical Ethics: Moral Understanding in a More than Human World*, Book in progress; Miller, Richard B. 1996. *Casuistry and Modern Ethics: A Poetics of Practical Reasoning*, Chicago: University of Chicago Press; Toulmin, Stephen, and Albert R. Jonsen. 1988. *The Abuse of Casuistry: A History of Moral Reasoning*, Berkeley: University of California Press.

[46] For more on whales and whaling, and sustainable yields, see Holt, Chapter 4, and Papastavrou and Cooke, Chapter 7.

[47] For more on the history of practical reason (in ancient Greek, *phronesis*), see MacIntyre, Alasdair. 1966. *A Short History of Ethics*, New York: MacMillan; MacIntyre, Alasdair 1984. *After Virtue: A Study in Moral Theory*, Notre Dame: University of Notre Dame Press.

[48] Lavigne, David. 2004. The Return of Big Brother, *BBC Wildlife Magazine* 22 (5), 70-72.

[49] Some of the ideas in that article are reiterated in Lavigne, Chapter 1.

[50] Oates, John. 1999. *Myth and Reality in the Rain Forest: How Conservation Strategies are Failing in West Africa*, Berkeley: University of California Press.

[51] Adams, William M. 1990. *Green Development: Environment and Sustainability in the Third World*, London: Routledge, Chapman and Hall; Nations, James D. 1988. Deep ecology meets the developing world, in Wilson, E. O. (ed.), *Biodiversity*, Washington, DC: National Academy Press, 79-82.

[52] Crist, Eileen. 2004. Against the Social Construction of Nature and Wilderness, *Environmental Ethics* 26, 5-24; Soule, Michael E., and Gary Lease (eds.). 1995. *Reinventing Nature? Responses to Postmodern Deconstruction*. Washington, DC: Island Press.

[53] United Nations. 1992. *Agenda 21: the United Nations Programme of Action From Rio*, New York: United Nations.

[54] Naess, Arne. 1990. Sustainable Development and Deep Ecology, in Engel, J. Ronald and Joan Gibb Engel, *Ethics of Environment and Development: Global Challenge, International Response*, Tucson: University of Arizona Press, 87-96; Rolston, Holmes, III. 1990. Science-Based Versus Traditional Ethics, in Engel, J. Ronald and Joan Gibb Engel, *Ethics of Environment and Development: Global Challenge, International Response*, Tucson: University of Arizona Press, 63-72; Rolston, Holmes, III. 1991. Life in Jeopardy on Private Property, in Kohm, Kathryn A. (ed.), *Balancing On the Brink of Extinction: The Endangered Species Act and Lessons for the Future*, Washington, DC: Island Press, 43-61

[55] Flattau, Edward. 2004. *Peering Through the Bushes*, Philadelphia: Xlibris Publishing; Helvarg, David. 2004. *The War Against the Greens: The 'Wise-Use' Movement, the New Right, and Anti-Environmental Violence*, San Francisco: Sierra Club Books.

[56] See for example, Beder, S. 1996. *The Nature of Sustainable Development.* Second edition. Newham, Australia: Scribe Publications; Lavigne, D.M 2002. Ecological Footprints, Doublespeak, and the Evolution of the Machiavellian Mind, in Chesworth, W., M.R. Moss, and V.G. Thomas (eds.), *Sustainable Development: Mandate or Mantra.* The Kenneth Hammond Lectures on Environment, Energy and Resources 2001 Series. University of Guelph: Faculty of Environmental Sciences. 61-91.

[57] Earth Charter Initiative. 2000. The Earth Charter, San Jose, Costa Rica: Earth Charter Initiative.

[58] Clugston, Richard M. 1997. *The Earth Charter in its Context*, San Jose, Costa Rica: Earth Charter Initiative; Rockefeller, Steven C. 2001. *The Earth Charter: An Ethical Foundation*, San Jose, Costa Rica: Earth Charter Initiative.

[59] World Commission on Environment and Development. 1987. *Our Common Future: A Report by the World Commission on Environment and Development*, Oxford: Oxford University Press.

[60] World Commission on Environment and Development 1987, 308. For critical reviews of sustainable development, see Beder, S. 1996. *The Nature of Sustainable Development.* Second edition. Newham, Australia: Scribe Publications; Lavigne, D.M 2002. Ecological footprints, doublespeak, and the evolution of the Machiavellian mind, in Chesworth, W., M.R. Moss, and V.G. Thomas (eds.), *Sustainable Development: Mandate or Mantra.* The Kenneth Hammond Lectures on Environment, Energy and Resources 2001 Series. Guelph: Faculty of Environmental Sciences. 61-91.

[61] Clugston, Richard M. 2003. The Earth Charter and Good Globalization, Washington, DC: Earth Charter USA Campaign; Rasmussen, Larry. 2001. The Earth Charter, Globalization and Sustainable Community, *The Ecozoic Reader* 2 (1), 37-43.

[62] On the latter, see Tomlinson, John.1999. *Globalization and Culture*, Chicago: University of Chicago Press.

[63] Lynn, William S. 2004. Situating the Earth Charter: An Introduction, *Worldviews* 8 (1), 1-15.

[64] Goulet, Denis. 1990. Development ethics and ecological wisdom, in Engel, J. Ronald and Joan Gibb Engel, *Ethics of environment and development: global challenge, international response*, Tucson, AZ: University of Arizona Press, p. 36. As one of my colleagues points out, this sensibility also applies to the maldevelopment of the so-called developed nations.

[65] For an example of these specifications, see Lynn, William S., and J. Ronald Engel (eds.). 2004. *The Earth Charter and Global Ethics. A Special Edition of Worldviews.* Vol. 8.

[66] Leopold, Aldo (1968) *A Sand County Almanac: And Sketches Here and There*, Oxford: Oxford University Press, vii.

CHAPTER 14

WHY CONVENTIONAL ECONOMIC LOGIC WON'T PROTECT BIODIVERSITY

William E. Rees

FRAMING THE ANALYSIS

Conserving biodiversity is one of the central planks in most policy platforms for achieving ecological sustainability. Remarkably, however, techno-industrial society remains somewhat perplexed about both the relevant core concepts – just what do we mean by "sustainability" and precisely where does so-called "biodiversity" fit in? This paper mostly addresses the second question: just why and how should society act to preserve the diversity of life on Earth? Because economic (monetary) valuation is often held up as essential for conservation, my main focus is on the actual and potential role of economic reasoning in biodiversity conservation. In the course of the analysis, we may also move closer to discovering the meaning of sustainability.

Types of Biodiversity

Biodiversity is a complex and somewhat abstract concept. In general it refers to the range of organic variability found among living organisms in all the habitats and ecosystems on Earth.[1] But to urge the preservation of the great variety of living things is not particularly helpful as a policy direction. For one thing, even assuming biodiversity is central to sustainability, we cannot preserve it all. We therefore need to know what kind of biodiversity, and in what quantity is it needed to "sustain" the global system (including the human enterprise). Most critically, we need to have explicit decision criteria by which to make critical choices among a variety of incompatible values that compete with biodiversity.

I begin, then, by recognizing that "biodiversity" can be subdivided into at least four tightly linked organizational levels:[2]

- *Genetic* diversity refers to within-species variation, to the nuanced information encoded in the DNA and chromosomes – in the genotypes – of individual plants and animals. This genetic information finds expression in the multiple qualities of the individual, in its phenotypic form and function;
- *Species* diversity refers to the total number of species, to the variety of viable life-forms, but even this simple measure presents a problem. While it might be possible to identify all the species making a significant contribution to the structure and function of a particular ecosystem, the aggregate number of species on Earth remains unknown. Fewer than two million have been described scientifically but estimates of the total range up to 30 million.[3] Only a few hundred thousand have been assessed for economic uses.
- *Ecosystems* diversity applies to the structural complexity of biotic communities, to the diversity of relationships and systems-level emergent properties that result from species interacting with each other and with the non-living components of their habitats. It is concerned with the relationship between biotic diversity and community stability and the role of keystone species[4]

whose presence or absence may determine the stability of the whole. This naturally leads to:

- *Functional diversity*, a measure of the capacity of life-support ecosystems to absorb some level of stress, or shock, without flipping from one stability domain to another (particularly one that may not be as amenable to human existence, as has often happened with fisheries collapses[5]). This definition is closely related to Holling's[6] concept of "resilience," the capacity of a system to persist and retain its functional qualities after a disturbance.

While the above classes of biodiversity are the most common foci for debate, Ehrlich and Daily[7] also make a strong case for consideration of "population diversity." Populations are geographic sub-groupings within species that may be defined either ecologically or genetically. An ecological or demographic population is an interbreeding group of conspecific individuals that are sufficiently isolated from other populations that changes in one group does not necessarily induce changes in others. A genetic or Mendelian population is a geographically delimited group that "can evolve independently of other such units; i.e., its evolutionary future is not primarily determined by flows of genetic information from other populations." Ehrlich and Daily[7] observe that "In many parts of the world the extinction of populations, rather than of species, may be the most important facet of the decay of biological diversity. Therefore, consideration only of species extinctions may greatly underestimate the rate of loss of organic diversity as a whole".

Why Care About Biodiversity?

> *Nothing momentous happened to the economy when passenger pigeons went extinct nor will anything momentous happen when/if western Atlantic bluefin tuna are gone. Nothing economically momentous resulted from the extinction of about 20 percent, or 2,000 species, of the world's birds over the past two millennia. As more and more biodiversity is lost, however, the rate of ecological change will accelerate with unpredictable outcome... there is little evidence that these changes will be beneficial to human cultures.*[8]

To many people, especially ecologists, it seems obvious why we must care about biodiversity. Human beings could not exist without biodiversity. We are utterly dependent on other life-forms for food and various less obvious life-support "goods and services." Apart from food, humans derive fibre for clothing, materials for construction and buildings, chemicals for industry, and pharmaceuticals for health, all from other species of plants and animals. Intact forests and other ecosystems moderate local weather patterns (and even the global climate), help mitigate runoff and flooding, purify our water, and provide habitat for fish and wildlife species and therefore the basis for recreational activities and regional economies. Some bio-diverse places may have special heritage or cultural values. And let's not forget that the sheer aesthetic magnificence of many natural ecosystems satisfies deep spiritual needs in many people. All these direct and indirect uses of living nature, each attributable in part to species, ecosystem or functional diversity, fall in the domain of what economists refer to as "use values".[9,10] In addition, some people delight in simply knowing that a particular ecosystem continues to flourish intact somewhere on earth even if they never intend to visit it. Its mere *existence* feeds their souls.

The fact that people need ecosystems services, and use other species to satisfy their own ends, represents "instrumental" and "utilitarian" reasons for biodiversity conservation. From this strictly anthropocentric perspective, other species and ecosystems exist exclusively to satisfy human needs, wants and preferences. We should therefore be concerned about global trends that threaten biodiversity because they may ultimately result in significant economic losses and spiritual bereavement.

Alternative approaches to valuing biodiversity rest on moral and ethical grounds. Other species simply have a right to exist independent of how humans perceive them. They have value unto themselves. Duty-based moral theories therefore argue that humans have moral obligations toward other living things and that this entails morally correct actions.[8] According to Ehrenfeld,[11] "The very existence of diversity is its own warrant for survival". Thus: "If conservation is to succeed the public must come to understand the inherent wrongness of the destruction of biological diversity". In sum, this view holds that humans have a moral obligation not to countenance the wanton destruction of biodiversity.

Further in this vein, some authors have argued that under idealized conditions, certain other species would enjoy something approaching equivalent legal standing to humans and therefore have near equivalent rights.[12] In these circumstances, rights are considered to be enforceable claims that cannot be altered without the consent of affected parties. In the case of biodiversity loss, the affected parties (non-human species) have intrinsic worth, value unto themselves that humans cannot arbitrarily take away. Such a contractarian approach, assuming a kind of "constitutional convention among all kinds of living things," may seem too "far-fetched" for the utilitarian pragmatists in most of us. However, accepting the basic principle would impose an obligation on humans to avoid further biodiversity losses on grounds that it "would violate the rights of the non-human biota under a just constitution".[9] Table 14–1 summarizes many of the ecological functions and human values associated with biodiversity.

There Will Always be Some Loss of Biodiversity: Assessing Value at the Margin

If humans could not exist without biodiversity, does this not mean that the value of biodiversity is effectively infinite? No doubt this is so, but only unambiguously so if we are concerned about the *totality* of biodiversity. Fortunately, that is not what is at issue; we are not facing the simultaneous extinction of all non-human life on Earth.

As noted at the outset, and forgetting the intrinsic values of other species for the moment, the most pressingly pragmatic question for policy in this rapidly developing world is how much biodiversity is enough to ensure the sustainability of the human enterprise? How can we most efficiently and effectively account for the marginal losses – the collapse of a fish stock here, the destruction of an ecosystem there – that seem invariably to accompany human encroachments into nature? To put it in economic terms, we want to know at what point the costs associated with the next increment of development, including the value of biodiversity sacrificed in the process, equal or just exceed any expected benefits of that development. At this stage in the overall development process, the net benefits of economic growth (total benefits minus total costs) will have reached a maximum. This is the point beyond which we should not go. Any further economic growth/development might produce additional dollar income or jobs but these benefits will be more than erased by associated costs, including biodiversity losses. In the best of all possible worlds, rational people would not destroy more value than they would gain in the development process. Instead, having reached the point of optimal economic scale, they would shift their focus from growth and attempt to engineer a steady-state economy.[13]

This sounds simple enough – basic economic reasoning seems to provide both an apparently rational framework (benefit-cost analysis or BCA) and a clear decision criterion by which we can "trade off" biodiversity for other economic benefits in the course of development. All we have to do is add up the total dollar value of the expected losses for each of the value cells in Table 14–1 and include these in our overall decision framework. If the expected costs exceed the expected gains, further development is uneconomic and should not go forward. Problem solved: nature gets her due and development is neatly balanced with biodiversity.

Of course, nothing important to humans is ever quite that simple. As we shall see, BCA is fraught with theoretical and technical problems. Indeed, conflict occurs at the most fundamental level. Many people object to the very notion that biodiversity can or should be "traded off" against other values. As Montgomery and Pollack[14] observe,

> Economists often exasperate environmentalists when they view biodiversity as one desirable goal among many, and insist that it must compete for scarce public and private resources with other goals…

Many environmentalists and biologists feel instinctively with Ehrenfeld[11] that biodiversity ought to trump "other goals" in the development process.

On one central point however, resistance to the economists is futile. The simple fact is that the very existence of humans necessarily compromises the welfare of various other species. Indeed, the history of *Homo sapiens* is one of escalating conflicts with biodiversity.[15] People have always had to make trade-offs between biodiversity and other things they needed or wanted, and mostly they have chosen the "other things".

Table 14–1. The Multiple Functions and Values of Biodiversity

Use Values			Non-Use Values
Direct Values	Indirect Values	Option Value	
Food production (including subsistence) Raw materials Genetic resources Education Recreational enjoyment Aesthetic enjoyment	Flood control Water purification Nutrient recycling Soil formation Wildlife habitat Air pollution mitigation Climate modification	Future use potential	Existence (it's nice just to know it's there) Intrinsic value (species have value unto themselves)

I will detail the economist's approach to making rational choices later. First let's explore the biophysical realities that make deadly conflict between humans and other species inevitable. This is the context that forces us to make economically and ecologically uncomfortable choices.

BIODIVERSITY LOSS: AN INEVITABLE "EMERGENT PROPERTY" OF HUMANKIND-ECOSYSTEM INTERACTION

> For every area of the world that paleontologists have studied and that humans first reached within the last fifty thousand years, human arrival *approximately coincided with massive prehistoric extinctio*ns.[16]

Throughout the history of "modern" humans, whenever people first came to occupy a particular habitat they produced significant effects on the structure and functioning of local ecosystems. I have therefore previously described[17] *H. sapiens* as an archetypal "patch disturbance species". Human patch disturbance is the inevitable consequence of two simple biological realities: first, human beings are large animals with correspondingly large individual energy and material requirements; and second, humans are social beings who universally live in extended groups. The invasion by significant numbers of people of any previously "stable" ecosystem will therefore necessarily produce changes in established energy and material pathways. There will be a reallocation of resources among species in the system to the benefit of some and the detriment of others. To this extent at least, even pre-agricultural hunter-gatherers affected the biodiversity of their world. In North America, South America, and Australia, about 72, 80, and 86 percent respectively of large mammal genera became extinct after human arrival[15]. Pimm *et al.*[18] estimate "that with only Stone Age technology, the Polynesians exterminated >2000 bird species, some ~15% of the world total".

Agriculture, which involves the permanent degradation of entire landscapes increased the impact by orders of magnitude.[19]

To understand better how this historic trend is coming to savage biodiversity in modern industrial times, I draw on recent developments in physics and biology, particularly far-from-equilibrium-thermodynamics as applied to self-organizing holarchic open (SOHO) systems.[20]

SOHO Thermodynamics

The starting point for this interpretation is the second law of thermodynamics. In its simplest form, the second law states that any spontaneous change in an isolated system (one that can exchange neither energy nor matter with its environment) produces an increase in entropy. In simpler terms, this means that when a change occurs in an isolated complex system it becomes less structured, more disordered, and there is less potential for further activity. In short, isolated systems always tend toward a state of maximum entropy, a state in which nothing further can happen.

Now, imagine a homogenized, totally disordered world in which everything is evenly dispersed – there are no distinguishable forms or structures, no gradients of energy or matter. In effect, no finite point in the ecosphere would be distinguishable from any other. For purposes of this discussion I will take this hypothetical randomized distribution of all naturally occurring elements and stable compounds to represent a state of maximum local entropy. This is also, by definition, a state of local thermodynamic equilibrium. This is the state toward which the ecosphere would spontaneously descend over time in the absence of sunlight and life. (The tendency toward increasing entropy can be likened to a relentless form of biophysical gravity.)

Of course, the real world could hardly be different from this randomized primordial soup. The ecosphere is a highly ordered system of mind-boggling complexity, of many-layered structure and steep gradients represented by accumulated energy and differentiated matter. In the course of several billion years, the trend in the ecosphere has been one of increasing order and complexity (even after allowing for occasional catastrophic setbacks). In general, the number of species has climbed from zero to many millions; these highly differentiated life-forms became distributed among hundreds of different ecosystems with multiple trophic levels as organisms adapted to the many physical environments on Earth and co-evolved in response to each other. In short, over geological time, biodiversity at all levels of organization has tended to increase.[21] The ecosphere has clearly been moving ever further from thermodynamic equilibrium. So fundamental is this process that, according to Prigogine[22] "distance from equilibrium becomes an essential parameter in describing nature, much like temperature [is] in [standard] equilibrium thermodynamics".

How is it that the ecosphere can apparently exist and evolve ever greater complexity in the face of the second law? The key is in recognizing that all living systems, from cellular organelles through individual organisms to entire ecosystems are complex, dynamic, *open* systems that can exchange energy and matter with their host "environments." As Erwin Schrödinger[23] observed, organisms are able to maintain themselves and grow "...by eating, drinking, breathing and (in the case of plants) assimilating..." Schrödinger recognized that, like any isolated system, a living organism tends continually to "produce[s] positive entropy – and thus tends to approach the dangerous state of maximum entropy, which is of death. It can only keep aloof from it, i.e. alive, by continually drawing from its environment negative entropy..." ("negative entropy" – also called "negentropy" or "essergy" – is free energy available for work).

Let's now put this in the context of self-organizing holarchic open (SOHO) systems theory. In the past few decades, systems scientists have come to recognize that complex self-producing systems exist in loose nested hierarchies, each component system or "holon" being contained by the next level up and itself comprising a chain of linked sub-systems at lower levels.[24] Each sub-system in the hierarchy maintains itself and grows by "importing" available energy and material (negentropy or essergy) from its host "environment" and processes it internally to generate a more highly organized state. It also exports the resultant degraded energy and material wastes (entropy) back into its host.[25] In short, living systems, as far-from-equilibrium-systems, maintain their local level of organization at the expense of increasing global entropy, particularly the entropy of their immediate host system. Because all such self-organizing systems survive by continuously degrading and dissipating available energy and matter they are called "dissipative structures".[22]

It follows from SOHO systems structure that the integrity and internal diversity of the entire systems complex can be maintained only if the highest level in the hierarchy is resilient and productive enough to support the development and maintenance of all lower level holons and capable of assimilating or dissipating their aggregate entropy production. The highest order dissipative structure on Earth is the ecosphere itself. The ecosphere comprises all the biomes and ecosystems on the planet and maintains itself in a far-from-equilibrium quasi steady-state by assimilating light energy from the sun (the next level up in the thermodynamic systems hierarchy). In effect, using photosynthesis and evapotranspiration, the ecosphere feeds on solar energy to develop and to support all lower order holons – species, populations, individuals, etc. – in the holarchy.

The Ascendancy of the Ecosphere

The fundamental hypothesis of far-from-equilibrium thermodynamics is that life has evolved as a dissipative process in response to the existence of steep gradients of available energy. Certainly, *the existence of essergy gradients is a prerequisite for life.* The theory predicts that whole ecosystems should "organize themselves, in accordance with the second law, to increase the degradation of the [essergy] in incoming energy." A corollary is that "material flow cycles will tend to be closed... to ensure a continued supply of material for the energy-degrading processes".[25]

Over time, therefore, we would expect greater species diversification, more extreme niche specialization, ever more complex structure, and increasingly efficient use of systems resources due to competition and autocatalytic processes, particularly in climatically stable environments. Ulanowicz[26] describes this entire process as the "ascendency" [*sic*] of living systems. His work shows that in the absence of overwhelming external disturbances, the ascendancy – distance from equilibrium – of an ecosystem will exhibit a natural tendency to increase.

Empirical evidence of ecosystem ascendancy is available in the form of infrared scanner surface temperature experiments involving over-flights of terrestrial ecosystems in Oregon. Luvall and Holbo[27] show that surface radiation varies with ecosystem maturity and type. The warmest temperatures were recorded over a rock quarry and a clear-cut; cooler temperatures prevailed above a 25 year-old naturally re-growing forest and a plantation of similar age. The coldest site (26 °K cooler than the clear-cut) was a 400 year-old natural douglas fir old growth forest with a three-tiered canopy. The overall trend reveals that the more developed and diverse the ecosystem, the cooler its surface temperature and the more degraded its re-radiated energy. The clear-cut and quarry were found to have degraded 62%, and the old-growth 90%, of the incoming solar radiation.

Satellite data on outgoing long-wave radiation from the Earth's surface suggest the same phenomenon at a global scale.[24] Deserts, which are biologically impoverished, emit 280 watts/m^2 compared to less than 200 watts/m^2 (29% less) by tropical rain forests. The latter are among the most structurally complex and species-rich ecosystems on Earth. Other biomes fall between these extremes. The low temperatures over the rain-forest are due, in part, to the low temperature of the convective cloud-cover forming over the cool, multi-layered forests below. These low temperatures result from the fact that most of the essergy dissipated by terrestrial ecosystems is degraded into molecular motion, particularly that associated with the evapotranspiration of water.[28] Apart from illustrating the developmental direction of ecosystems, these data suggest that the species, ecosystemic and functional diversity of the ecosphere contribute significantly to the thermodynamics of the global climate.

The Far-from-Equilibrium Thermodynamics of the Human Enterprise

What does all this have to do with the human economy and biodiversity conservation? SOHO theory suggests that today's rapidly accelerating biodiversity losses are a grotesquely amplified version of the patch disturbance activities of early humans. This is because the human enterprise, like the ecosphere, is a self-organizing far-from-equilibrium dissipative structure but with a difference – the human system is also a wholly contained component of the ecosphere. As ecological economist Herman Daly[29] has posited, the human economy is an open, growing dependent sub-system of a materially closed, non-growing finite ecosphere, and this relationship clearly contains the seeds of potential pathology. Just as the ecosphere feeds on the sun, the human enterprise grows and maintains itself by feeding on the ecosphere. The economy therefore has the potential to become dangerously parasitic on the ecosphere.

I have already noted that the ecological integrity and sustainability of the entire SOHO systems complex depends on the capacity of the ecosphere to support the entropic load that lower levels in the holarchy impose upon it. Thus SOHO theory suggests that if the growth-oriented human enterprise comes to demand more useful energy/matter (essergy) than the ecosphere can produce, or discharge more waste (entropy) than the ecosphere can assimilate, then the ascendancy of the human enterprise will necessarily be at the expense of the disordering and potential collapse of the ecosphere (or at least of major host ecosystems).

As noted, the bioproductive capacity of the ecosphere is measured in terms of net primary production (or net photosynthesis) by green plants. It is the accumulation of available energy in plant biomass that provides the low-entropy "fuel" for animal life. This implies that all production by animals is secondary production derived from the consumption and dissipation of the products of primary production by plants. Even most human economic production is secondary production.

Secondary production is fundamentally a *consumptive* process. Vastly more energy and material first produced by nature is dissipated by economic secondary production than is contained in the product. This means that, beyond a certain thermodynamic limit, the accumulation of economic capital – the goal of neoliberal capitalist societies – is *necessarily* at the expense of the "natural capital," including species and ecosystems. Moreover, the law of mass balance ensures that the entire economic throughput of energy and matter – including the portion initially embodied in useful products – is eventually degraded and injected back into the ecosphere as waste. The SOHO model of the economic process thus predicts the pattern of escalating biodiversity losses, resource depletion and pollution that is the stuff of daily headlines today. Remember, in SOHO terms, the expanding human enterprise is structurally positioned to consume and degrade the ecosphere from the inside out.[30]

It is worth noting in passing that most of our economic think-tanks and statistical agencies monitor economic activity using money as the metric. Problematically, many of the material flows to and from nature, as well as the life support services provided by ecosystems remain *invisible* to monetary analyses. In these circumstances, market prices are unreliable indicators of functionally critical forms of ecological scarcity and can have only a limited role in fostering sustainability. I return to this point in a following section.

What Makes Humans Unique?

It might be argued in our defence that the human ecological niche is actually structurally little different from that of any other large consumer organism. After all, many other large social animals from beavers to elephants also qualify as patch disturbers. This reassuring thought is only superficially true. Humans have evolved certain unique qualities that separate us ecologically from all other animals and help to explain the sustainability crisis. Again, we can interpret the result in terms of SOHO theory and far-from-equilibrium thermodynamics.

Ludwig Boltzmann, one of the fathers of thermodynamic theory, was familiar with the concept of Darwinian natural selection. Boltzmann[31] recognized in 1886 that the Darwinian "struggle for existence is a struggle for free energy [essergy] available for work." The reason is simple – the availability of energy makes everything else possible. Drawing on Boltzmann's insight, Lotka[32] later hypothesized that "systems that prevail" (i.e., successful systems) will be "those systems that evolve to maximize their use of the energy [and material] resources available to them." In other words systems (species, ecosystems, etc.) that draw on more resources and use them more efficiently, will eventually competitively displace less effective and efficient systems. This general idea is known today as the "maximum power principle".[33]

Understanding maximum power is central to understanding the contemporary sustainability dilemma. Simply put, *H. sapiens* has evolved unparalleled competitive superiority in appropriating the energy flows and material resources of the ecosphere.[34] Most critical to our success is our capacity for language, particularly written language. This ability has enabled generation after generation of humans to pass on their cumulative knowledge. For thousands of years, the expanding human enterprise has been getting more effective and efficient at exploiting the natural world. Human intelligence and technological prowess have unleashed our species' full expansionary powers.

One result is the remarkable ascendancy of *H. sapiens* relative to all competing species. This general process, which began in earnest with agriculture and the emergence of complex societies eight to ten thousand years ago, has been accelerating rapidly in recent decades.[35] For example, witness the increasing structural complexity and diversification of world's major economies since the beginning of the industrial revolution. Economic ascendancy is exemplified by the sequential layering of economic structure as primary activities (agriculture and resource extraction) gives way to manufacturing, and manufacturing is succeeded by high-end service- and knowledge-based industries. It is also characterized by the emergence of specialty products and niche markets, particularly in larger centres (one doesn't usually find designer boutiques and caviar shops in northern mining towns). Meanwhile, the human population has twice doubled to over six billion people since the beginning of the last century and *per capita* incomes and material standards (material consumption) have been rising exponentially since the beginning of the industrial revolution.

As is the case in ecosystems, the ascendancy of economies is accompanied by a prodigious increase in dissipative capacity. Indeed, the evolution of the economy since the beginning of the industrial revolution has been propelled by the constantly expanding use of "exosomatic" (outside the body) energy supplies, particularly fossil fuels (a cumulative stock of essergy derived from ancient photosynthesis). By the early 1990s, this energy subsidy amounted to 407.5×10^{15} Btu by a population of about 5.5 billion people. "It is as if every [person] in the world had fifty slaves. In a technologial society like the United States, every person has more than 200 such 'ghost slaves'" [at an assumed working level of consumption of 4000 Btu person^{-1} day^{-1}]. So dependent is industrial society on cheap fossil energy that some authors predict the decline and collapse of civilization as supplies run out in coming decades.[36]

Much of this exosomatic energy, along with increasingly sophisticated methods of resource extraction, has been used to exploit the rest of nature, to increase the human "harvest" of everything from fish and logs to ground water and petroleum itself. As a result, humans have become the dominant consumer organism in virtually all the major ecosystems types on earth. In terms of bioenergy and material flows, we are clearly the most significant marine mammal. Fossil energy and modern technology enables the global fishing fleet to appropriate seafood for humans that represents 25-35% of net marine primary productivity from shallow coastal shelves and estuaries, that 10% of the oceans that produces 96% of the catchable fish.[37] Despite diminishing returns to fishing effort, the collapse of several major fisheries, and the unambiguous warnings of fisheries scientists there is no evidence that the pattern of exploitation is changing. Christensen *et al.*[38] and Myers and Worm[39] report that after only fifty years of industrial fishing, the large predatory fish biomass of the world's oceans is only about 10% of pre-industrial levels.

Similarly, humans are the principal consumer in most of the world's significant terrestrial habitats, diverting from grasslands and forests at least 40% of the products of photosynthesis for direct and indirect human use.[40,41] Consequently, by 1988 eleven percent of the 4400 extant mammal species were endangered or critically endangered, and a quarter of all mammal species were on a path of decline which, if not halted, is likely to end in extinction.[42] Meanwhile, increasing human populations and rising consumption levels maintain a subtle but steadily increasing pressure even on long-settled landscapes. McKee *et al.*[43] suggest that human population growth alone will increase the number of threatened species in the average nation 7% by 2020, and 14% by 2050. Their data thus strongly support the view that reducing human population growth is a necessary, if not sufficient, step in the "epic" attempt to conserve biodiversity on the global scale.

The situation in the UK may be typical of densely populated countries. Thomas *et al.*[44] have recently shown that "28% of native plant species have decreased in Britain over the past 40 years, that 54% of native bird species have decreased over 20 years, and that a majority of butterfly species (71% over ~20 years) has declined." Two butterfly species became totally extinct and population extinctions were recorded in all the main ecosystems in the country. The authors suggest that if insects elsewhere are similarly sensitive, then insect extinction rates may well parallel the known extinction rates of vertebrate and plant species "strengthening the hypothesis that the biological world is approaching the sixth major extinction event in its history." This would, however, be the first major extinction caused by the overwhelming domination of a single life-form – humans have become a kind of ecological plague on non-human biodiversity.

Is H. sapiens *Inherently Unsustainable?*

Reasoning that, for sustainability, humans should resemble other "similar" species in key ecological parameters, Fowler and Hobbs[45] tested the hypothesis that *H. sapiens* is "ecologically normal," i.e., that humans fall within the normal range of natural variation observed among such species for a variety of ecologically relevant measures. They found that humans rarely showed normal ecological tendencies. In terms of population size, energy use, carbon dioxide emissions, biomass consumption and geographical range, humans differ from other species by several orders of magnitude. Fisheries collapses and related biodiversity losses are partially explained by the fact that human consumption of biomass was two orders of magnitude greater than the 95% confidence limits for biomass ingestion by 96 other mammal species. In short, Fowler and Hobbs' analysis shows humanity to be an outlier species along many axes in terms of our exploitation of life-support goods and services of nature. They ask whether, in the circumstances, *H. sapiens* is sustainable.

W.M. Hern[46] argues that it presently is not. He likens our species to a kind of planetary disease – the sum of human activities over time "exhibits all four major characteristics of a malignant process: rapid uncontrolled growth; invasion and destruction of adjacent tissues (ecosystems, in the case); metastasis (colonization and urbanization, in this case); and dedifferentiation (loss of distinctiveness in individual components)". Recent reports[47] suggest that should these trends prevail, by 2030, more than half of the world's population will be concentrated mainly in coastal areas and cities, driving resource consumption to even more unsustainable levels. Within a century, the 11 billion people sharing the planet will have encroached on the last vestiges of untouched nature rendering most attempts at habitat rescue futile.[48] Any remaining wildlife preserves will be heavily human-influenced. A distinct species of plant or animal disappears every 20 minutes, and half of all bird and mammal species will be lost within 200-300 years.[49] Hern[46] suggests we do have one theoretical saving grace that might head off a worst-case scenario. We can think and decide not to be a cancer.

All such findings and speculations are actually predictable outcomes of far-from-equilibrium thermodynamics and the fact that the growing human enterprise is a subsystem of the non-growing ecosphere. I have previously argued on these grounds that unsustainability is an inevitable *emergent property* of the systemic interaction of techno-industrial society and the complex systems dynamics of the ecosphere.[34] Like all species, *H. sapiens* has a genetic predisposition to expand to fill all the ecological space and use all the resources available to it. However, unlike other species, humans have been able to mitigate or eliminate many of the systemic negative feedback processes that normally hold populations in check. As a result, our extraordinary evolutionary success in terms of "maximum power," has succeeded not only in competitively displacing less effective and efficient species, but also in depleting many other species and resources upon which we may be dependent for our own long-term survival. The expanding global economy is quite literally dissipating the living ecosphere, putting the future of our own species at risk.[50] Marginal changes such as increased efficiency and improved environmental legislation may buy some time but don't affect the fundamental process.

CAN ECONOMIC LOGIC MITIGATE BIODIVERSITY LOSS AND RESOLVE THE SUSTAINABILITY CRISIS?

Economists argue that one reason for humanity's uninhibited destruction of "the environment" is the very abundance of nature. We naturally tend to undervalue any resource so plentiful that is free for the taking. Economics itself has long treated biophysical goods and life-support services as "free goods" whose contribution to life-quality is not accounted for in market prices. From this perspective, the ecological crisis reflects a fundamental economic axiom: underpricing leads to over-use.

There can be little question that the unpriced contributions of nature to human welfare have historically not been represented in markets and "are [therefore] often given too little weight in policy decisions".[51] Pearce[52] argues that the resultant "asymmetry of valuation" biases the playing field against conservation because, absent prices, the critical role of nature in sustaining the economy is not reflected in either individual or social choices. This example of classic market failure results in destructive policy decisions and inefficient economic performance.

Environmental economists[53] have therefore joined the (un)sustainability debate with a seemingly simple solution. With increasing ecological decay, the time has come to place a dollar value on nature's output – let prices tell the truth. If market prices accurately reflect the true value of the products of nature, then people will adjust their preferences and purchasing patterns accordingly. Just the right amount of current natural income will be consumed and economic efficiency will have been achieved.

The need felt need among economists to price "the environment" implies a sense of impending scarcity, and making sound decisions in a context of scarcity is supposedly what economics is all about. In this context, economists view price as a powerful decision tool – it provides an unambiguous criterion for individuals and communities to choose among mutually exclusive possibilities. We would do well to keep in mind however, that the valuation of nature represents the commodification of global life support.[54] As I have argued elsewhere,[55] "This is worrisomely serious business. For the first time in human history, it seems necessary to some to put a price on the biophysical structures and functions that make higher life possible on Earth".

Most critically, assigning a price to something implies the ability to compress a great deal of information about that thing into a single indicator or metric. In theory, this compacted information should enable us to make "better informed" decisions about the allocation of that thing among competing interests or even determine its fate in the event that it may have to be "traded off" in some economic development decision.[52,56] In this light, and given the risks associated with the destruction of global life support, if we are going to use money value as a decision criterion, the price had better be right!

Putting a Price on Biodiversity

Not long ago I attended an interdisciplinary workshop on strategies for biodiversity conservation. The two economists present couldn't understand what all the fuss was about. For them, biodiversity was simply not an issue. Their paper was founded on two assertions of fact. First, their data from so-called "genetic prospecting" suggested that only about one in 10,000 species of plant/animals show much promise of yielding a pharmaceutically interesting product, i.e., something of economic value. Second, the probability of any species going extinct as a result of any particular development project (e.g., a greenfield industrial park) was virtually nil. The marginal value of biodiversity losses associated with such a development was therefore perishingly small – one ten-thousandth times almost zero is a negligible number. Ergo, biodiversity could be ignored as a factor in most development decisions.

The ecologists' incredulity at these findings was palpable for the rest of the meeting but the form of economists' analysis was paradigmatically true to their discipline. As previously noted, economics is about money value at the margin and about the expression of individual preferences. As Randall[57] observes: "The ethical framework built on this foundation is *utilitarian, anthropocentric* and *instrumentalist* in the way it treats biodiversity [original emphasis]. It is utilitarian in that things count to the extent that people want them; anthropocentric, in that humans are assigning the values; and instrumental, in that biota is regarded as an instrument for human satisfaction".[58]

Environmental economists Pearce and Moran[59] recognize alternative perspectives – that species conservation is a moral issue or that other species may have intrinsic rights – but vigorously defend the utilitarian approach. In their view, debating other perspectives "risks being rather sterile *from the standpoint of getting things done* [original emphasis]...in the real world context of making choices." This argument is based on the view that the moral perspective, or any stand based on the rights of other living organisms (or even future generations of humans), lack criteria for making rational choices among mutually exclusive possibilities, even between different species or biological resources. Is the malarial parasite to be accorded the same inviolable status as the harpy eagle? Are both species ever and always to be granted equivalence in the sustainability debate? And if not, on what basis do we value one over the other?

Pearce and Moran[59] argue that human population growth and economic growth are a fact (and who could dispute this?); that biodiversity losses are inevitable in the competition for limited ecological space (the entire previous section underscored this reality); and that uncomfortable choices must be made. At the very least, then, a ranking criterion is required – if not everything can be saved we need a non-arbitrary way to make policy decisions. Under what circumstances should which species be spared? In a passage likely to chill the blood of any conservationist, they also declare that: "This view is reinforced by the fact that the world is extremely unlikely to devote major resources to biodiversity conservation. We can argue that it should, but we know it will not". (I return to this telling assertion below.)

As already implied, the "ranking criterion" of choice is monetary value. Money is the great leveller, the economists' way of comparing apples to oranges. It assumes that all things are essentially commensurable. From this perspective, *monetary price is a necessary precondition for making economically correct decisions.* The neo-liberal paradigm[60] argues that if we know the money value of some dimension of biodiversity, we have a rational basis for deciding, for example, whether to sacrifice it for the anticipated benefits of some development project. Similarly, if there are several alternative ways of implementing a project, differences among the biodiversity "opportunity costs" associated with the various options can help us to choose among them. In these ways, monetary analysis of biodiversity losses might well contribute to selecting the least social cost option for development.

Contingency Valuation: Who Can Say What Biodiversity is Worth?

Assuming for the moment that we can adequately identify and quantify anticipated biodiversity or losses on their own terms (as might be required for an environmental impact assessment, for example) the problem remains of assigning a monetary value. In rare cases, the market may actually provide a direct indication of perceived value as when drug companies sign contracts with governments for exploration rights and pay royalties, or when we have an accounting of the tourist revenues associated with a particularly attractive natural area.[61] In other cases, it might be possible to assign value based, at least in part, on the market price of near substitutes. Suppose a small number of moose are going to be lost to a micro hydro-electric project. Part of the value of this loss could be based on the price of a similar quantity of dressed beef in local supermarkets. More often than not, however, there is no formal trade in the goods and services derived from biodiversity. In these instances, economists rely on indirect ways of assigning dollar values and determining user preferences such as the travel cost method (e.g., how much do people spend on travel and equipment to go fishing?) or hedonic pricing (e.g., how much do the prices of similar houses differ when one enjoys a spectacular view lot or is close to some other nature-related amenity?). However, by far the most common approach to eliciting consumer values, particularly non-use values is "contingent valuation."

Contingent valuation (CV) usually involves the administration of a questionnaire survey to a statistically valid sample of people. The general objective is to determine how respondents would value a specified hypothetical change in their individual welfare. For example, they might be asked how much they would be willing to pay (WTP) to rescue some specified attribute of the natural environment from the threat of development. Alternatively, they might be asked how much they would be willing to accept (WTA) as compensation for the loss of that attribute. Obviously, the way people respond to such questions will involve elements of their personal perception of use and non-use values of the resource in question. It will also reflect their income or socio-economic status. In any event: "For society, the net value of a proposed change in resource allocation is the interpersonal sum of WTP for those who stand to gain minus the interpersonal sum of WTA for those who stand to lose as a result of the change".[55]

There are, however, many technical and behavioural problems with CV approaches. Fischhoff[62] asks whether there is actually "anything in there". In a particularly detailed critique, Vatn and Bromley[56] identify three major problems associated with valuation, particularly by non-experts:

- The Cognition Problem**:** The valid evaluation of goods and services by individuals assumes that people have perfect knowledge about all the functions of those goods or services. However, in the real world perfect knowledge is unattainable and people are differentially selective in their valuation of different known attributes of a good. Some important qualities may be disregarded out of ignorance.

  Perhaps most significant in the context of valuing biodiversity is the fact that many contributions of species and ecosystems are essentially beneath perception. They are cognitively "invisible." Vatn and Bromley[56] describe such "functionally transparency" to mean that "the precise contribution of a functional element in the ecosystem is not known – indeed is probably unknowable – until it ceases to function". Problem: we cannot value what we cannot know.

  A second cognitive problem is that most people have difficulty in converting ecologically significant attributes into monetary units for comparison with other goods. For this reason, the CV elicitation procedure used "may serve as a means to construct preferences rather than merely uncover them".[56]

- The Incongruity Problem: If the different attributes of an ecologically significant good are "incongruous," or fundamentally at odds with each other in the minds of assessors, then a single measure such as hypothetical price will not reflect all important information. As O'Neill[63] puts it: "Different values are incommensurable; there is no unit through which the different values to which appeal is made ...can be placed upon a common scale." For example, some CV methods may force people to conflate the worth of an ordinary good such as fishing pleasure (use value) with existence value or intrinsic value (the right of salmon to live). Often people are intuitively put off at being asked, in effect, to commodify a moral principle. There are areas of interest where "social norms restrict or reject the commodity fiction".[56]

- The Composition Problem: According to Vatn and Bromley,[56] "the commoditization of environmental goods [as reflected in CV studies] can be looked upon as a product of the felt need to value them. It is not immediately obvious to many – other than economists – why it is necessary to characterize environmental attributes this way". There are several dimensions to this problem. First, in a complex dynamic ecosystem, the whole may actually be dependent on each of its fundamental parts so that the value of any single component cannot be understood independent of the value of the whole. Second, in the same complex systems framework, the value of individual ecosystem components should not be derived from their perceived uniqueness to humans but rather from their functional uniqueness in relation to the integrity of the whole system containing them. Finally, the market or "exchange value" of ordinary commodities depends on the property that their valuable attributes can be appropriated and controlled by the buyer. Biodiversity and related ecological goods and services hardly meet this test. Vatn and Bromley[56] suggest that the economist's disciplinary need to create commodities where they may not be thus encounters the possibility that some ecological goods and services may be technically impossible to price.[64]

Vatn and Bromley[56] summarize their reservations about CV studies in one simple but fundamental point: "Efforts to derive hypothetical values for the complex and interrelated attributes of the environment... result in a non-trivial loss of information". The failure to reflect potentially critical information means that attributes of nature, including biodiversity, will almost invariably be undervalued in any decisions based on the outcome of contingency valuation. Indeed, the inability to assign a valid money value to ecosystems goods and services strips pricing of much of its legitimacy and policy relevance. Contrary to Pearce and Moran,[59] Vatn and Bromley[56] conclude that the economic pricing of nature's goods and services "is neither necessary nor sufficient for *coherent and consistent choices about the environment*" (original emphasis). O'Neill[63] agrees, arguing the very idea money can be used to measure and compare values that are essentially incommensurable is mistaken. "Given the conflicts [among different values] there is no substitute for good practical judgement that is informed by debate amongst practitioners and citizens".

Benefit/Cost Analysis: Light – and Ultimate Darkness

The intractability of the pricing problem as applied to ecosystems values throws a dark shadow over benefit-cost analysis (BCA). And there are other shadows – as Lave and Gruenspecht[65] observe: "The problem with BCA in both theory and practice seems overwhelming yet economists continue to regard BCA as the definitive tool in making public decisions".

As noted earlier, the main strength of the benefit-cost framework is its apparent conceptual simplicity and transparency. However, in recent decades, the theoretical foundations of BCA have been eroded by modern moral philosophers and even some economists reject its economic assumptions. In practice, critical data – such as the monetary value of nature's goods and services – are missing or invalid; the method over-emphasizes economic efficiency while ignoring distributional equity (there may be little overlap between who benefits and who pays), including intergenerational equity (setting appropriate discount rates is problematic); agencies typically provide insufficient time and resources for an exhaustive analysis, and practitioners may be directed to justify a particular outcome or are simply biased by their own beliefs and prejudices. Lave and Gruenspecht[65] conclude that: "The difficulties with missing data, uncertainty, and [too few resources] combine with the theoretical difficulties to make ineffectual any serious claim that an applied study produces an optimal or theoretically justified outcome". In short, BCA is an unreliable tool for determining the true net worth of even isolated projects (and will ever remain a mere idealized construct in contemplating the optimal scale of the economy).

Conceding that the method produces less than optimal results, BCA's defenders argue that society should not make multi-million dollar decisions without having as much information as possible with which to compare policy alternatives. Moreover, the application of BCA may force new questions to the surface, identify significant data gaps that may yield to further research, expose social and institutional flaws in our public decision-making processes, and reveal how people perceive out-of-the-ordinary dimensions of life.

Certainly these incidental benefits of BCA are potentially valuable and can make useful contributions to project decisions. But the real danger lies in getting caught up in the *doing* of the BC analysis, in being blinded to its flaws and ultimately in taking its *results* as accurately reflecting biophysical reality. In these circumstances the analysts would certainly be guilty of what Ehrlich and Daily[7] call "'crackpot rigor' (detailed mathematical analysis of an intractable problem) or 'suboptimization' (doing in the best way possible something that should not be done at all)".[66] For all these reasons, many people will be relieved that at least in the case of publicly-owned natural capital, "the most fundamental environmental choices will continue to be made without prices – and without apologies".[56]

Does It Matter that We Can't Price Nature?

Recall that the purpose of economic valuation is to condense all relevant information about natural assets into a single metric and thus correct the "asymmetry of valuation" that has heretofore biased decision-makers against the environment. Environmentalists hope that better accounting for nature's goods and services will favour conservation. If we know the value of nature's goods and services, we might not destroy them heedlessly. At the least, full cost accounting should favour less ecologically destructive development options. The foregoing discussion shows, however, that despite the best of motives and intentions of economic analysts, it is simply not possible accurately to price biodiversity or many other ecosystems values. But let's stop for a moment: does this really matter over the longer term? Suppose things were different, that we could accurately monetize any attribute of nature. Would it really make much difference to the decision process in a world dedicated to economic growth?

The unavoidable answer in many situations is, "probably not". The purpose of monetization is to provide a common measure of worth that will allow direct comparisons between the value of biodiversity and certain mutually exclusive options. But, as previously noted, humans are necessarily anthropocentric and the world has increasingly adopted a utilitarian and instrumentalist attitude toward nature. In this context, "The sad fact that few conservationists care to face is that many species, perhaps most, do not seem to have any conventional [economic] value at all, even hidden [unpriced] conventional value".[11] This means that for many ordinary development decisions, the present value of anticipated economic gains will exceed the total of all use and intrinsic values of biodiversity sacrificed by the development. Whenever the marginal gain from depleting nature exceeds the value of saving it, people – acting rationally – will choose depletion.[67]

This is particularly true in the case of "open access" (unowned and poorly regulated) resources such as many fish stocks, where the benefits of exploitation accrue to a few individuals but the costs are shared by society at large.[68] In these circumstances, individuals have an incentive to over-exploit natural bio-resources even if the total social costs exceed their private gain. This problem even extends to land reserves ostensibly set aside for conservation purpose. Poaching is a growing problem in game and ecological reserves in Africa and other parts of the world, as sometimes desperate people exploit resident wildlife for their own subsistence or for the bushmeat trade.[69] Even in wealthy Canada, our better known national parks such as Banff and Jasper are under constant pressure from private hotel, ski-hill and related facilities developers wishing to cash in on the revenues generated by the very aesthetic and wildlife attributes that their operations jeopardize.

Some economists have suggested that the solution to the open access problem is the privatization of nature on grounds that the owners of private property have a direct incentive conserve their productive capital. But as Clark[70] famously showed, private ownership of valuable but slowly reproducing species is no guarantee that said "resources" will be carefully husbanded. Consider the choices available to the private owners of a whale stock or perhaps a forested area with a high biodiversity index. Such natural capital assets may have a reproductive / growth rate in the vicinity of two or three percent per year which thus defines the sustainable harvest (natural income) rate. However: "With normal rates of return on investment in the neighbourhood of 10%, the 'optimal' strategy for the whalers [would be] to simply wipe out the whales and invest the proceeds elsewhere".[71]

The problem here is rooted is the discount rate (which measures consumers' "time preference"). The whales or the forest would return only two or three percent of their capital value annually to their owners if harvested sustainably, and much of that return would be absorbed in the cost of the yearly harvest. However, if the owners liquidate the entire stock immediately they could enjoy an annual income of 10% of the capital value (minus a one-time harvest cost) in perpetuity. Clearly any rational self-interested utility maximizer would choose to extinguish his stock of natural capital rather than conserve it intact. So much for market valuation as a shield for biodiversity.

Finally, it might be argued that as "nature" becomes scarcer with human population growth, rising material demand or ecosystem collapse, the existence value and intrinsic worth of remaining biodiversity will increase to its advantage – i.e., the incentive for conservation will increase. To the extent that this slows biodiversity loss it is a point for money pricing. However, at least some of the immediate use or exchange values will also increase with scarcity, perhaps faster than any conservation values.[72] If this happens, the rising scarcity value of ecosystems goods and services is no assurance of protection and may even accelerate resource depletion (see the section on globalization and trade below). Once again, "...extinction may occur no matter what the price response".[8]

To summarize, and to the dismay of all those conservationists who have joined the valuation bandwagon in the hope it would play a preservationist tune, pure economic reasoning generally resonates more with the prevailing symphony of destruction. Recall Pearce and Moran's[59] assertion that "...the world is extremely unlikely to devote major resources to biodiversity conservation". In effect, this is an argument that, absent a crisis, the perceived value of biodiversity is likely always to be less than the measurable value of development. This effectively undermines Pearce and Moran's[59] own argument for pricing biodiversity – the valuation exercise becomes a mere formality that turns against biodiversity by rationalizing its destruction.

From "Safe Minimum Standards" to "Strong Sustainability"

Since econometric approaches do not necessarily conserve nature, some analysts have proposed alternatives compatible with moral or duty-based considerations. For example, Randall[9] argues that if we assume humans should make at least some sacrifice for biodiversity, then we should support a "safe minimum standard" (SMS) for conservation. This would entail preservation of a "sufficient area of habitat... to ensure the survival of each unique species, sub-species or ecosystem, *unless the costs of doing so are intolerably high*" (emphasis added).

The SMS approach clearly shifts the burden of proof "to the case against maintaining the SMS" and accepts the risk "that the costs of preservation may fall disproportionately on present generations and the benefits on future generations".[59] Nevertheless, Randall[9,57] also makes clear that the SMS does not demand unlimited sacrifice by people: human claims trump those of other species if the opportunity costs of conservation are deemed excessive. Once again, in the context of growth-induced scarcity humans will always win at nature's expense.

Many authors, particularly those writing from the perspective of ecological economics, argue that irreducible uncertainty favours a much more cautious approach to the destruction of biodiversity than is provided by the SMS.[73] Ecological economics recognizes the limits to growth[74] and argues the need for humanity to work toward a "steady-state"[75] economy. This implies there is an optimal scale for aggregate human economic activity characterized by a constant safe level of energy and material throughput.

The model argues that before we reach a crisis point of no return, global society should adopt a strong sustainability criterion in which both renewable natural capital and manufactured capital are held intact.[76] Since economic analysis cannot identify the crisis point, we will have to rely on scientific data, debate and seasoned judgment to make the call. The most risk-averse version of the so-called "constant capital stocks" criterion can be stated as follows:

> *Each generation should inherit an adequate per capita physical stock of both manufactured and self-producing natural assets no less than the stock of such assets inherited by the previous generation.*[77]

Note that, far from depleting biodiversity, this criterion suggests that if the human population increases, so too should the natural capital base that produces our natural income of life-support goods and services.

The argument for constant biophysical stocks is conceptually simple: The prevailing system of costs, prices, and market incentives fails absolutely to measure ecological scarcity (particularly functional scarcity) or to determine the appropriate levels of natural capital stocks. Prevailing trends, however, suggest that the economy's "dissipative" activities are already undermining the functional integrity of major ecosystems. Since certain critical natural assets maintain the life-support functions of the ecosphere, the risks associated with their depletion are unacceptable, and there may be no possibility for technological substitution, "conserving [at least] what there is could be a sound risk-averse strategy".[78]

Because its effect would be to preserve remaining biodiversity, the constant natural capital stocks criterion is seemingly more eco-centric than the SMS approach. In fact, however, it is really just a more sophisticated self-serving utilitarian argument, at best a form of ecologically enlightened self-interest. Even so, there is virtually no political or popular support for such a fail-safe approach among contemporary growth-oriented governments or "official" international organizations.

Globalization and the Biodiversity Costs of "Free" Trade

> *During [the 20th] Century, world agriculture has been transformed from a patchwork quilt of nearly independent regions to a global exchange economy. This change in social organization also contributes to the loss of diversity.*[79]

> *Off the remote north-east coast of Borneo, a Malaysian patrol vessel hailed a suspicious looking trawler last week. When marine police boarded, they found a catch of 160 dead giant leatherback turtles, the most endangered of all sea turtles. The poachers, who had poisoned the waters with cyanide, came from China's southern province of Hainan, more than 1,000 miles away.*[80]

> *Myanmar [is] mired in a deforestation crisis: China's appetite for foreign timber has the country's forests disappearing at an alarming rate.*[81]

I have already suggested that economic growth is the proximal cause of much biodiversity loss. Unfortunately, from the perspective of conservation, unimpeded growth is the unchallenged goal of global development theory. Indeed: "In recent years the governing elites of the market democracies have persuaded or cajoled virtually the entire world to adopt a common myth of uncommon power. All major national governments and mainstream international agencies are united in a vision of global development and poverty alleviation centred on unlimited economic expansion fuelled by open markets and more liberalized trade".[34] In this section I advance the hypothesis that globalization and trade may well pose the single greatest threat to biodiversity today.

On an infinite planet untrammelled trade might be an unqualified good. Even on tiny Earth, managed trade in the context of a steady state economy could improve everyone's quality of life without seriously impairing ecological integrity or biodiversity. Unfortunately, the growth ethic and prevailing economic logic give no quarter to the fact that the earth is actually both finite and fragile.

According to conventional trade theory (and common understanding), freer trade should be mutually beneficial to all trading partners.[82] Trade can relieve local shortages – thus seeming to increase local carrying capacity[83] – and catalyze growth at both ends of the trading relationship. Theory suggests that if each country specializes in those few goods or commodities in which it has a comparative advantage and trades for everything else, the world should be able to maximize gross material efficiency. Access to cheap resources and labour, together with economies of scale, results in lower prices. This encourages higher consumption and therefore increases total output. Since trade raises overall production/consumption and also increases the variety of goods and services available to all trading partners, it could potentially raise material standards for everyone. Little wonder that more liberal trade is a mainstay of contemporary globalization strategies.

With all this going for it, what can possibly be wrong with trade? From the perspective of sustainability and bio-conservation, the yellow flags should be obvious.[84] As noted, and consistent with the growth ethic, the objectives of more liberal trade are to relieve resource constraints on local economic expansion and to increase overall economic output. These factors allow population and material growth within all trading regions to be sustained beyond local biophysical limits that would exist in the absence of trade and they stimulate increased global demand for resource commodities of all kinds. To put this in terms of our earlier discussion, unconstrained trade is specifically intended to elevate the human enterprise further from thermodynamic equilibrium. But this both requires even deeper human incursions into remaining patches of natural habitat (displacing non-human species) and permanently increases the overall dissipative pressure on biophysical resources. There is simply no way around this problem – as previously argued, the second law is inviolable and the human enterprise is a dependent wholly contained sub-system of the ecosphere.

The mechanisms by which widening trade circles wreak their impacts on biodiversity can permanently transform both ecological and social systems. Consider the effect of specialization, which, by definition, implies the simplification of regional/national economies. In traditional bio-diverse agro-ecosystems, farmers often intermixed many varieties of the same crop, and different crops, in the same fields, or planted many crops at different times and in different places. This not only provided a natural method of pest control, but helped achieve a dependable food supply – "average production from year to year varied little because of the law of large numbers".[85]

By contrast, the integration of local crop production into producer-competitive global markets generally requires uniform, materially intensive (high-input) production methods. This has many deleterious effects on several levels of biodiversity. The most obvious involves the increasingly universal use of relatively few commercial crop varieties. Traditional varieties – often exquisitely and differentially adapted to local natural conditions – are giving way all over the world to a handful of commercially acceptable cultivars bred or engineered to respond best to increasingly uniform artificial inputs. These chemical inputs then produce their own negative effects. Excess reliance on fertilizers accelerates soil degradation including the destruction of the micro-flora and micro-fauna that comprise the majority of species in most ecosystems; the abuse of chemical biocides reduces the diversity and abundance not only of target pests (usually insects and weeds) but also of non-target organisms in several unrelated taxa, including birds and mammals. "Cropping" natural forests can have similar impacts. For example, clearcut logging in old growth forests may reduce soil biodiversity from thousands to hundreds or tens of species per square metre, with serious consequences for recovery. The resultant loss of vital mycorrhizal fungi is known to limit the success of even low-diversity plantation-oriented reforestation efforts.[86] The bottom line? Genetic, species, and population diversity are all negatively affected by trade-induced ecosystem simplification.

The compulsion to trade can also dramatically alter human behaviour by shifting the perceived value of various goods and services of nature. Economists distinguish between the "use value" and the "exchange value" of any good. Use value is the utility or enjoyment one experiences in using or consuming a good. The exchange value is the money price one could get for that good by selling it in the marketplace. Many local renewable resources such as fish stocks or forests would never be jeopardized if used only to supply the use-related needs of resident populations. However, once people begin to exploit the exchange value of a good in the marketplace, the potential for ecological damage looms large.[87]

The point is that the globalization of trade encourages people everywhere to market "surplus" local resources and use the income to purchase manufactured goods or other things not available (or available only at greater cost) from the local economy. Indeed, international development is frequently based on an "export-led" development model. As described above, countries are encouraged to maximize their incomes by specializing in those crops or products for which they have a natural comparative advantage and use the income to import things they cannot produce at home (Canada produces lumber and wheat; Costa Rica, bananas).

The potential for massive ecological damage – including accelerated biodiversity loss – from trade is truly a modern problem. In pre-industrial times, trade was limited. Most countries and even smaller communities could produce most of their own limited material *needs*. However, the increasing sophistication and diversification of the globalizing economy means that virtually no region or country has the necessary population, human skills or resources to maintain the "good life" on its own. (Consider the explosion of high-tech consumer goods, many unheard of just a few decades ago – autos, computers, cell-phones, home entertainment systems, household appliances, etc. – that are now deemed necessities of modern life.) Most modern countries and regions are therefore essentially forced to trade and, naturally enough, the ecological effects are often greatest on biodiversity-rich developing countries. The latter have to export large quantities of resources to finance their imports of the expensive manufactured goods now required to satisfy their citizens' ever-rising material *wants*.

Both Low Prices *and* High Prices Can Be a Threat to Biodiversity

One result is that trade in agricultural and resource commodities has become a defining characteristic of global techno-industrial society. The problem for many resource-based communities and countries is that in an expanding, increasingly competitive global marketplace, the prices of many resource commodities have been stagnant or falling. (This is partly because there are often many competing suppliers of particular commodities but the few large multi-national trading conglomerates constitute a near-monopoly of buyers.) By contrast, the ever-increasing array of sophisticated manufactured products made from those resources command top dollar in world markets. Of course, for solvency, the value of exports must be at least equal to the value of exports – plus the extra cash needed to pay off the export development loan – but this is made more difficult by deteriorating terms of trade. Faced with heavy debt loads and falling prices, resource exporters must ship an increasing volume of unprocessed resources to finance their loans and pay for imported high-end manufactured goods and services. In many cases, despite rising harvest or depletion rates, economic margins continue to decline and resource exporters have reduced capacity to manage their natural capital stocks sustainably. In the short-term, low commodity prices resulting from unfavourable terms of trade and global competition may delight resource importers and ultimately consumers. However, low prices also encourage wasteful consumption that further accelerates natural capital depletion and biodiversity losses.

Ironically, biodiversity may also be threatened by economic behaviour induced by *rising* prices. The mere possibility of world trade exposes pockets of scarce resources everywhere to the largest possible market and a growing pool of wealthy consumers for whom price is no object. Competition among consumers may bid *up* the market price for highly valued species thus encouraging ever-greater harvest rates as local people respond to rising exchange values. One result is that wildlife poaching for trade is now said to be the world's second most important illegal economic activity after the trade in illicit drugs.

The economic boom in populous China provides several examples of this modern problem. Recent reports reveal that: "Thrilled by the wider choice of food that wealth brings, Chinese people are now consuming the country's beleaguered wildlife at rapid rate… Highly endangered species, such as the Tibetan antelope, called the chiru, have started to appear on Shanghai restaurant menus".[88] Moreover, the opening of trade between China and the countries of Southeast Asia is draining wildlife from the surrounding countries. When trade with China opened up, wildlife "harvesting" boomed in Vietnam. People found they could get high prices for certain animals – as much as $1000 for individuals of one species of turtle, *Cuora trifasciata*, the three-striped box turtle, believed in China to cure cancer. (As many as 12 million turtles may be sold in China every year.) While wildlife still commonly shows up on local restaurant menus in Vietnam, the effects of local commerce in game are nothing like the impact of trade with China. The trade has spread even to remote areas, where local people collect wildlife, then sell it to brokers, who in turn sell the animals to the China trade. This unregulated trade is having a noticeable impact on Vietnam's biodiversity.[89]

Even Canada's wildlife is being impacted by burgeoning global markets for illicit wildlife products. For example, the incidence of poaching on British Columbia's black bear population (along with other bear populations in the Pacific Northwest) has risen sharply in response to Chinese and other Asian markets demand for bear paws and gall bladders for "medicinal" purposes.

The increase in human depredation due to trade opportunities may push some slowly reproducing animals – and plants – closer to population or even species extinction. This is a particular danger if the species in question roams free in the unregulated global commons, accessible to any and all human hunters in pursuit of economic gain. McDaniel and Gowdy[8] describe several cases where prices, even in conventional markets, jeopardize or, at best, provide no protection to "open access" biological resources. In the contemporary case of Atlantic bluefin tuna, demand and market price have increased much faster than the costs of fishing, despite the declining stock. A single fresh bluefin fetches tens of thousands of dollars and the price of tuna just keeps rising. Fishers therefore actually have an escalating economic incentive to continue the hunt, possibly to the last fish.

By contrast, 19th century market prices for North American bison products and for the once super-abundant passenger pigeon rose only insignificantly as their respective populations plunged. Again, markets provided no incentive for conservation even toward the end. The passenger pigeon was extirpated and the few hundred bison that were saved owed their good fortune to the determined efforts of conservationists.

The historic human tendency to ravage the common pool for personal gain led Ophuls and Boyan[90] to observe that we may ultimately be propelled to the brink of ecological chaos "not so much by the evil acts of selfish people as by the everyday acts of ordinary people whose behavior is dominated, usually unconsciously, by the remorseless self-destructive logic of the commons." This reinforces our more general premise that conventional economic reasoning and behaviour reinforce humanity's inherent tendency toward unsustainability. Globalization simply takes humanity's historic conflict with non-human nature to a whole new level of intensity.

EPILOGUE: WHILE WE'RE SPECULATING...

This paper presents a bleak prognosis of prospects for sustainability and biodiversity conservation. The dynamics of the problem are complex but the basic argument is not hard to follow:

- Like all other species, *H. sapiens* has an innate biological imperative to expand into all available habitats. However, our capacity for language and technology make us more successful at this that any other large vertebrate.
- Humans have always had to use other species of animals and plants to survive. Indeed, we could not exist without biodiversity and other forms of "natural capital." Nevertheless:
- Human evolution has endowed us with no inhibitions against extirpating other species or destroying our own habitats. Consistent with this:
- Contemporary economic logic is based on a self-serving utilitarian ethic. All of this is problematic because:
- Material economic growth dissipates energy and material gradients (resources) found in nature. In effect, the economy feeds on the ecosphere.
- The human enterprise is an open growing subsystem of the materially closed non-growing ecosphere.
- The continuous growth of the human enterprise therefore necessarily compromises the integrity of the ecosphere, including biodiversity.
- We are unable to price biodiversity to reflect the full social value of its life-sustaining structure and functions. Markets are therefore incapable of signalling potentially disastrous ecological scarcity.
- In a growing economy, biodiversity is almost invariably traded-off in favour of economic considerations, this despite the fact that:
- Sustainability may depend on maintaining constant stocks of natural capital, including species, populations and ecosystems. Meanwhile:
- Our innate expansionist tendencies are currently being reinforced by the cultural myth of unlimited economic growth currently being abetted by globalization and trade.
- The rate of biodiversity loss is therefore increasing unchecked.

Even this bare-bones summary makes clear that an ever-expanding human economy is fundamentally incompatible with sustainability, particularly the conservation of biodiversity. It is also evident why, in many circumstances, efforts to price biodiversity in the public domain will provide little support for conservation. In any event, price signals in open markets may actually encourage the liquidation of even privately held stocks of "natural capital." In sum, contemporary economic logic and behaviour are simply unreliable allies in the quest for biodiversity conservation and sustainability.

The obvious conclusion is that achieving sustainability will require other than a strictly economic rationale. As McDaniel and Gowdy[8] argue: "Until we move from econocentric resource management to a policy whose [explicit] goal is the preservation of ecosystem integrity, biological resources will continue to be degraded". This argument recognizes that the only way biodiversity can be conserved is if we accord it trump status among moral principles.[91] Biodiversity must stand above all other considerations.

But this presents an immediate problem. As Randall[9] emphasizes: "Surely many would argue that enhancing the life prospects of the worst-off people has moral force at least as powerful as that of protecting biodiversity". In this light, "...pre-eminent value status for biodiversity...is unlikely to survive scrutiny, given the powerful appeal of many other candidates". Are we therefore condemned to a continuous loop, forced back to using faulty evaluation techniques to attempt "least-cost" trade-offs of biodiversity against human interests?

Perhaps for a time we are, but there is a potential way out. The world community would "merely" have to give up on population and economic growth before the pressures on biodiversity are irreversibly fatal. We have seen throughout this discussion that the bane of biodiversity is humanity's innate expansionism as now reflected in neoliberal growth economics. Remove this force and the tension between humans and the rest of nature disappears. Biodiversity would not have to trump other human interests – in the absence of growth pressure the two could coexist (as they still do in some places on the planet).[92] It would be relatively easy to establish a constant natural capital stocks rule. In most of the world, people are already living in remnant ecosystems but there is sufficient biodiversity left on Earth to stabilize the global system and perhaps begin the long-term restoration of biological integrity.

Giving up on growth should not be as difficult as it first appears. In most wealthy countries there is virtually no natural population growth and economic growth is a mere addiction, necessary to neither individual nor population health. Indeed, there is much evidence that human well-being is no longer correlated with income growth in the developed world and in some countries the relationship is negative.[93] Self-interested intelligent people would yield to these data and work to begin the transition to a steady-state economy.[94] Indeed, given present levels of wasteful consumption, steady-state levels of throughput could be substantially lower than current levels with no loss of life-quality. This would free up ecological space to meet the needs of the planet's worst-off and, over-time, reduce the pressure on biodiversity.

While we're contemplating revolutionary changes it wouldn't hurt to wonder whether modern humans can learn to love the non-human world, to come to feel in their bones that the violation of nature is a violation of self. Humans preserve only what they love (and they certainly do not normally subject the things they truly love to market forces). Even an acquired sense of biophilia would greatly reduce the economic pressures on biodiversity particularly if it were part of a generally more life-sustaining and spiritually satisfying life-style than most of us enjoy today.

There seems to be consensus that the human enterprise is currently unsustainable – the debate is about alternative paths to ecological security. Can we get away with fine-tuning the global economy or does the problem demand a more radical "paradigm shift"? David Pearce[10] once argued that the economic valuation approach to biodiversity "does not deny other rationales for [conservation]... Yet it may be unnecessary to resort to such moral arguments. Economic arguments alone could well be sufficient to justify a dramatic reduction in [biodiversity loss]". Well, perhaps in some situations, but not for long in a world committed to continuous material growth.

In these circumstances, we may well be justified in arguing that a shift to steady-state thinking[75,92] and ethics-based approaches are necessary,[95] and possibly even sufficient, for sustainability. This would not deny the economic rationale, yet, with a more balanced system of values, it might not be necessary to resort to crude economic analyses. Ethical and moral arguments alone would be sufficient to halt the arbitrary destruction of biodiversity. (And the question of how to commodify the living world would never come up.)

NOTES AND SOURCES

[1] For more on the topic of biodiversity, see Willison, Chapter 2.

[2] Nunes, PALD. and van den Bergh JCJM. Economic valuation of biodiversity: sense or nonsense? *Ecological Economics* 2001; 39: 203-222.

[3] Wilson, EO. The current state of biological diversity. In Wilson, EO. (ed.). *Biodiversity*. Washington, DC: National Academy Press; 1988.

[4] Keystone species may not be among the dominant or even obvious species in their ecosystems. The determining criterion of a keystone species (such as the sea otter in the inshore-marine kelp forests of western North America) is that their activities determine the structure of the entire community. They can usually be identified only by removal experiments. See Chapter 7, Keystone Species May be Essential to a Community, in Krebs, CJ. *The Message of Ecology*. New York: Harper and Row; 1988 (99-112).

[5] For more on fisheries collapses, see Hutchings, Chapter 6.

[6] Holling, CS. Resilience of Ecosystems: Local Surprise and Global Change. In Clark, W and Munn, T. (eds.), *Sustainable Development of the Biosphere*. Laxenburg, Austria: IIASA and Cambridge: Cambridge University Press; 1985.

[7] Ehrlich, PR and Daily G. Population Extinction and Saving Biodiversity. *Ambio* 1993; XXII (2-3): 64-68.

[8] McDaniel, C and Gowdy JM. Markets and biodiversity loss: some case studies and policy considerations. *International Hournal of Social Economics* 1998; 25 (10): 1454-1465.

[9] Randall, A. The Value of Biodiversity. *Ambio* 1991; 20 (2): 64-68.

[10] Pearce, D. Deforesting the Amazon: Toward an Economic Solution. *EcoDecision* 1991; 1 (1): 40-49.

[11] Eherenfeld, D. Why Put a Value on Biodiversity? Chapter 24 in Wilson, EO. (ed.). *Biodiversity*. Washington, DC: National Academy Press; 1988 (212-217).

[12] Classic early treatments of this subject include Stone, C. Should Trees Have Standing? Toward Legal Rights for Natural Objects. *45 Southern California Law Review 450; 1972* and Tribe, L. Ways Not to Think about Plastic Trees: New Foundations for Environmental Law. *83 Yale Law Journal 1315*; 1974. The ideal conditions for the contractarian approach are set out in Rawls, J. *A Theory of Justice*. Cambridge, MA: Harvard University Press; 1971.

[13] See Czech, Chapter 22.

[14] Montgomery, CA and Pollack, RA. Economics and Biodiversity: Weighing Benefits and Costs of Conservation. *Journal of Forestry* 1996 February; 34-38.

[15] See for example, Ponting, C. *A Green History of the World*. London: Sinclair-Stevenson; 1991.

[16] Diamond, J. *The Third Chimpanzee*. New York: HarperCollins Publishers; 1992. p 355.

[17] Rees, WE. Patch disturbance, Eco-Footprints, and Biological integrity: Revisiting the Limits to Growth. Chapter 8 in Pimentel D, Westra L, and Noss R, (eds.). *Ecological Integrity: Integrating Environment, Conservation and Health*. Washington, DC: Island Press; 2000.

[18] Pimm SL, Russell GJ, Gittleman JL, and Brooks TM. The Future of Biodiversity. *Science* 1995; 296: 347-350.

[19] See Chesworth, Chapter 3.

[20] SOHO theory posits that all living entities exist as self-organizing sub-systems within a loosely over-lapping hierarchy of such systems. The entire structure is termed a holarchy and each recognizable sub-system is a "holon," Holons (e.g., cells, individuals, ecosystems) may be self-producing but they are only quasi-independent. Each holon must draw on higher levels in the holarchy for resources and as sinks for waste. Holons are therefore "open" to both energy and material flows.

[21] See Willison, Chapter 2.

[22] Prigogine, I. *The End of Certainty: Time, Chaos and the New Laws of Nature.* New York: The Free Press; 1997. Chapter 2.

[23] Schrödinger, E. *What is Life: The Physical Aspect of the Living Cell.* Cambridge: Cambridge University Press; 1945.

[24] See, for example, Kay J and Regier H. Uncertainty, complexity, and ecological integrity. In Crabbé P, Holland A, Ryszkowski L, and Westra L, (eds.). *Implementing Ecological Integrity: Restoring Regional and Global Environment and Human Health,* NATO Science Series IV: Earth and Environmental Sciences, Vol 1. Dortrecht: Kluwer Academic Publishers; 2001.

[25] Schneider, ED and Kay, JJ. Complexity and Thermodynamics: Toward a New Ecology. *Futures* 1994; 26: 626-647. p. 364-365. See also: Schneider, ED and Kay, JJ, Order from Disorder: The Thermodynamics of Complexity in Biology. In Murphy, MP and. O'Neill, LAJ (eds.). *What is Life: The Next Fifty Years. Reflections on the Future of Biology.* Cambridge University Press; 1995. Nicholas Georgescu-Roegen famously pioneered the application of these thermodynamic concepts to the modern economy. See, for example, Georgescu-Roegan, N. The Entropy Law and the Economic Problem. *Distinguished Lecture Series no* 1. University of Alabama, Department of Economics. 1971; Georgescu-Roegen, N. *The Entropy Law and the Economic Process.* Cambridge: Harvard University Press; 1991.

[26] Ulanowicz, RE. Ecology, The Ascendent Perspective. New York: Columbia University Press; 1997.

[27] Luvall, JC and Holbo, HR . Measurements of short term thermal responses of coniferous forest canopies using thermal scanner data. *Remote Sens. Environ.* 1989; 27: 1-10.

[28] These coupled cloud-rain forest systems actually have the same long-wave temperature as mid-Canada in February, indicating prodigious essergy dissipating capacity!

[29] Daly, HE. Steady-state economics: concepts, questions, policies. *Gaia* 1992; 6: 333-338.

[30] Rees, WE. Consuming the Earth: The Biophysics of Sustainability. *Ecological Economics* 1999; 29: 23-27.

[31] Boltzmann, L. The second law of thermodynamics (orig. 1886). In *Ludwig Boltzmann, Theoretical Physics and Philosophical Problems,* McGinness B (ed.). New York: D Reidel; 1974. Boltzmann also said that "the struggle for existence of animal beings is... but a struggle for entropy..." which seems contradictory. But he went on: "...the products of the chemical kitchens [of plants] constitute the object of the animal world." These "products" are the result of photosynthetic net entropy production by plants.

[32] Lotka, AJ. Contribution to the Energetics of Evolution. *Proc. Natl. Acad. Sci.* 1922; 8: 147-155.

[33] For a thorough introduction, see: Hall, AS. *Maximum Power: The Ideas and Applications of HT Odum.* Niwot, CO: University Press of Colorado; 1995.

[34] Rees, WE. Globalization and Sustainability: Conflict or Convergence?" *Bulletin of Science, Technology and Society* 2002; 22 (4): 249-268.

[35] See Chesworth, Chapter 3.

[36] Price, D. Energy and Human Evolution. *Population and Environment* 1995; 16: 301-317; Duncan, RC. The Life Expectancy of Industrial Civilization: The Decline to Global Equilibrium. *Population and Environment* 1993; 14: 325-357.

[37] Pauly, D and Christensen V. Primary production required to sustain global fisheries. *Nature* 1995; 374: 255-257. Also see Hutchings, Chapter 6.

[38] Christensen V, Guénette S, Heymans J, Walters C, Watson R, Zeller D, and Pauly D. Hundred-year decline of North Atlantic predatory fishes. *Fish and Fisheries* 2003; 4: 1-24.

[39] Myers RA and Worm B. Rapid worldwide depletion of predatory fish communities. *Nature* 2003; 423: 280–283.

[40] Haberl, H. Human Appropriation of Net Primary Production as An Environmental Indicator: Implications for Sustainable Development. *Ambio* 1997; 26: 143-146.

[41] Vitousek, P, Ehrlich PR, Ehrlich AH, and Matson P. Human appropriation of the products of photosynthesis. *BioScience* 1986; 36: 368-374.

[42] Tuxill, J. *Losing Strands in the web of Life: Vertebrate Declines and the Conservation of Biological Diversity.* Worldwatch Paper 141. Washington, DC: The Worldwatch Institute; 1998.

[43] McKee JK, Sciulli PW, Fooce CD & Waite TA. Forecasting global biodiversity threats associated with human population growth. *Biological Conservation* 2004; 115 (1): 161-164. Also see Willison, Chapter 2.

[44] Thomas, JA (and eight others). Comparative Losses of British butterflies, Birds, and Plants and the Global Extinction Crisis. *Science* 2004; 303: 1879-1881.

[45] Fowler, CW and Hobbs L. Is humanity sustainable? *Proceedings of the Royal Society of London, Series B: Biological Sciences* 2003; 270: 2579-2583.

[46] Hern, WM. Is human culture oncogenic for uncontrolled population growth and ecological destruction? *Human Evolution* 1997; 1-2: 97-105. See also: Hern WM. Why are there so many of us? Description and diagnosis of a planetary ecopathological process. *Population and Environment* 1990; 12: 9-39.

[47] Palmer M. (and twenty others). Ecology for a Crowded Planet. *Science* 2004; 304 (5675): 1251-1252.

[48] There is a good chance, however, that climate change, resource scarcity, systems collapses and geopolitical chaos will intervene well before the human population reaches such heady heights.

[49] University of Texas, Austin Press Release. *Extinction Rate Across the Globe Reaches Historic Proportions.* http://www.sciencedaily.com/releases/2002/01/020109074801.htm. 10 January 2002.

[50] Critics might argue that economic activity is not only dissipative in nature but that it also creates a good deal of order, a kind of order that many prefer to nature. This is true but misses an essential point: the creation of order by the human enterprise requires the transformation of available energy and matter extracted from the ecosphere. Since economic production involves thermodynamic processes that cannot be 100% efficient, the increased order of the human enterprise (the accumulation of manufactured capital) never fully compensates thermodynamically for the disordering of the ecosphere (the depletion of natural capital). In sum, negentropy production in the economy is always less than the negen-

tropy drawn from the ecosphere so the entropy of the total system increases. Continuous economic growth therefore continuously eats away at life-support structures and functions.

[51] Costanza, R. (and twelve others). The value of the world's ecosystem services and natural capital. *Nature* 1997; 387: 252-260.

[52] Pearce, D. *Valuing the Environment: Past Practice, Future Prospect.* CSERGE Working Paper PA 94-02. London: Centre for Social and Economic Research on the Global Environment, University College (London) and University of East Anglia; 1994.

[53] See Czech, Chapter 22.

[54] See Beder, Chapter 5.

[55] Rees, WE. How Should a Parasite Value its Host? *Ecological Economics* 1999; 25: 49-52.

[56] Vatn, A. and Bromley DW. Choices withouth prices without apologies. *Journal of Environmental Economics and Management* 1994; 26: 129-148.

[57] Randall, A. What Mainstream Economists Have to Say about the Value of Biodiversity. Chapter 25 in Wilson EO., (ed.). *Biodiversity.* Washington, DC: National Academy Press; 1988 (217-223).

[58] For more on attitudes and values, and ethics generally, also see Menon & Lavigne, Chapter 12; Lynn, Chapter 13; and Worcester, Chapter 17.

[59] Pearce D and Moran D. *The Economic Value of Biodiversity.* London: Earthscan Publications; 1994. p 30 and 32.

[60] Also see Beder, Chapter 5; especially her note 25 on neoliberalism, economic rationalism, and neoconservatism.

[61] For more on ecotourism, see Mugisha and Ajarova, Chapter 10; and Corkeron, Chapter 11.

[62] Fischhoff, B. Value Elicitation: Is There Anything in There? In Hechter M, Cooper L. Nadel L (eds.). *Values.* Palo Alto, CA: Stanford University Press; 1991.

[63] O'Neill, J. Managing Without Prices: The Monetary Valuation of Biodiversity. *Ambio* 1997; 26 (8): 546-550.

[64] Commoditization is an increasing problem for biodiversity conservation. It reflects the continuous pressure in modern consumer societies "to transform as much of the necessities and pleasures of life as possible into commercial commodities. [The latter] have replaced other forms of need satisfaction in the lives of most people. The more this happens, the more of the earth's resources are turned into commodities, and the more the waste products of consumption are pumped back into the biosphere". (Manno, JP. *Privileged Goods: Commoditization and Its Impact on Environment and Society.* Boca Raton: Lewis Publishers; 2000. p 13.)

[65] Lave, L. and Gruenspecht, H. Increasing the Efficiency and Effectiveness of Environmental Decisions: Benefit-Cost Analysis and Effluent Fees – A Critical Review. *Jour. Air Waste Management Assoc.* 1991; 41 (5): 680-693.

[66] The same criticism applies to various models advanced by economists for the valuation of biodiversity whose seeming analytic rigor and complexity obscures the models' excessive abstraction from real-world ecosystems. (See, for example Weitzman, ML. The Noah's Ark Problem. *Econometrica* 1998; 66 (6): 1279-1298).

[67] To put it another way, if the opportunity costs of conservation (the forgone benefits of development) exceed the perceived value of that which is conserved, development will proceed.

[68] Ecologist Garrett Hardin famously (and erroneously) labelled this "the tragedy of the commons" (Hardin, G. The Tragedy of the Commons. *Science* 1968; 162: 1243-1248.) It is now more accurately known as the "open access problem." (See Ophuls and Boyan, Note 90, for an excellent discussion.)

[69] For more on the bushmeat issue, see Eves, Chapter 9; Oates, Chapter 18; and Milner-Gulland, Chapter 20.

[70] Clark, CW. The Economics of Over-Exploitation. *Science* 1973; 181: 630-634.

[71] Clark, CW. Economic Biases Against Sustainable Development. Chapter 20 in Costanza, R. (ed.). *Ecological Economics: The Science & Management of Sustainability.* New York: Columbia University Press; 1991.

[72] In the extreme, starving farmers will consume their seed grain out of desperation to survive today, even at the expense of a harvest tomorrow. Poor people – and poor countries – have relatively high discount rates.

[73] For the basic argument, see Myers, N. Biodiversity and the Precautionary Principle. *Ambio* 1993; 22 (2-3): 74-79.

[74] For and update on "limits to growth", see Meadows, D., Randers, J. and Meadows, D. *Limits to Growth: The 30-year Update.* White River Junction, VT: Chelsea Green Publishing Company; 2004. 338 pp.

[75] Daly, HE. Steady State Economics (2nd ed.). Washington: Island Press; 1991. Also see Czech, Chapter 22.

[76] Costanza, R and Daly HE. Natural capital and sustainable development. *Conservation Biology* 1992; 1: 37-45. Daly, HE. Sustainable development: from concept and theory towards operational principles. Population and Development Review 1990 (special issue) Reprinted in: Daly, HE. *Steady State Economics* (2nd ed.). Washington: Island Press; 1991; Victor, PA, Hanna E, and Kubursi, A. How strong is weak sustainability? *Economie Appliquée* XLVIII (2): 75-94.

[77] Rees, WE. Achieving Sustainability: Reform or Transformation? *Journal of Planning Literature* 1995; 9 (4) 343-361.

[78] Pearce D., Markandya A and Barbier EB. *Blueprint for a Green Economy.* London: Earthscan Publications; 1989. p. 7.

[79] Norgaard, RB. The Rise of the Global Exchange Economy and the Loss of Biodiversity. In Wilson EO., (ed.). *Biodiversity.* Washington, DC: National Academy Press; 1988. p. 206-207.

[80] Becker, J. Why all the world feels China's growing pains. Independent Digital; 8 May 2004. http://news.independent.co.uk/world/asia/story.jsp?story=519237

[81] York, G. Myanmar mired in a deforestation crisis. Globe and Mail, 13 May 2004; p A14.

[82] Lieth, J. Chapter in T. Schrecker and J. Dalgleish, (eds.). *Growth, Trade And Environmental Values.* Westminster Institute for Ethics and Human Values; London, Ontario: 1994.

[83] That trade acts to increase (local) carrying capacity depends on the fact that each trading region is an open system. However, the world as a whole is effectively closed which

turns things around in ecological terms. Non-renewable resources (e.g., petroleum) imported by and consumed in region "a" are no longer available for future consumption in the exporting region "b." The terms of trade may also lead to the depletion of self-producing resources throughout the trading network (see below). Hence, while exchange can result in a short-term increase in the human population and material standards in each trading region, it also increases global consumption and waste generation. In short, trade tends to accelerate resource dissipation everywhere. It follows from the second law that continuous trade-induced growth of the human enterprise must eventually lead to a *decrease* in global carrying capacity.

[84] I ignore here a significant *economic* problem. The free flow of capital among nations undermines the theory of comparative advantage that confers the presumed universal benefits of trade. Rather than staying home and specializing in those goods for which the domestic economy has a comparative advantage, capital flows out across the world in search of the absolute advantage represented by low-cost labour. China thus becomes the world's manufacturing centre as firms in the developed world "outsource" production at the expense of their domestic employees and economies.

[85] Richards, P. *Indigenous Agricultural Revolution: Ecology and food Production in West Africa.* Boulder, Colo: Westview Press; 1985.

[86] Perry D., Amaranthus M., Borchers J., Borchers S., and Brainerd R. Bootstrapping in ecosystems. *BioScience* 1989; 39 (4): 230-237.

[87] For example, consumption by Canadians alone would have had only a marginal impact on the Northern Cod stocks under their jurisdiction or on the volume of Canada's old-growth forests. Both resources are now greatly trade-depleted – indeed, the industrial cod fishery collapsed in the early 1990s and has remained suspended for over a decade.

[88] Hu Pan. Wealthy Chinese Eating Wildlife into Extinction. *Environment News Service* 2000. 31 January 2000.

[89] Newsletter, Washington: American Museum of Natural History (Center for Biodiversity Conservation). Fall 2000.

[90] Ophuls, W. and Boyan AS. Jr,. *Ecology and the Politics of Scarcity Revisited: The Unraveling of the American Dream.* New York: W.H. Freeman and Company; 1992. p. 199.

[91] For more on the subject of ethics, see Lynn, Chapter 13.

[92] To illustrate, consider contemporary British Columbia, Canada. The province is wealthy, resource rich and enjoys a low average population density compared to most other places on in the developed world. British Columbia boasts one of the world's largest sockeye salmon runs in the Fraser, its greatest river. The Fraser also has enormous potential as a source of hydro-electricity. The potential annual economic value of the energy (particularly for export) is in the hundreds of millions of dollars, vastly more than the economic value of the salmon. Yet any politician who seriously proposed developing that power would face a political storm. The people of BC identify with the salmon; at this stage in the development of the region, they implicitly value the integrity of the salmon run more than the wealth that would be derived from power development. However, excessive population or demand growth, or severe energy shortages in North America, would force a choice between salmon and development of the river and the salmon would almost certainly be sacrificed before the perceived greater need for more energy.

[93] Lane, R. *The Loss of Happiness in Market Democracies.* New Haven: Yale University Press; 2000.

[94] See Czech, Chapter 22.

[95] See Lynn, Chapter 13.

A Triptych of Limerick

On this once verdant planet called Earth
Of big-brained folk there's no dearth
But they talk conservation
Without reservation
While trashing the place of their birth.

The GDPs up in the zillions
While life forms die off by the millions
There's just one conclusion
The money's illusion
(And there's too many people by billions!)

At its meeting in old County Clare
The IFAW did loudly declare
That a world sustainable
Is not attainable
If there's no wildlife there.

William E. Rees, 2004

Chapter 15

Paying for Unsustainable Fisheries: Where the European Union Spends Its Money

Michael Earle

Fish is a vital component of many people's diet, supplying much-needed protein. The average annual consumption of fish in the world is 16.3 kg per person,[1] more than any other source of animal protein.[2] Fish provides more than 2.6 billion people with at least 20% of their animal protein and in some developing countries it is close to half. The fishing and aquaculture industry provides employment for an estimated 38 million people around the world,[3] the vast majority in Asia. In developing countries, most of these jobs are in artisanal fisheries and associated activities. Many developing countries rely upon exports of fish for foreign exchange – more fish is traded in the global marketplace than almost any other food commodity, including coffee and tea, sugar or other types of meat.

Yet there is an anomaly in these figures – trade flows are very one-sided, with the rich nations of the world consuming fish from the waters of the poor ones. Developing countries account for about half of global fish exports,[4] with 82% of the total value of imports going to the developed countries, especially the European Union,[5] Japan and the United States. The Low-Income Food-Deficit Countries[6] provide over 20% of the total value of fish being traded, so that their annual consumption of fish is 14 kg per person, lower than the world average of 16.3 kg; consumption in Africa, at 7.8 kg per person, is the lowest of any region.[7] The hungriest nations are selling their food to the rich nations, an issue that will be discussed later in this chapter.

The status of fish stocks in the world and their ability to support both commercial and artisanal fisheries thus has important consequences not only for the conservation of biodiversity, but also for food security, employment and the balance of payments for many countries.

Yet, according to the most recent analysis by the Food and Agriculture Organization of the United Nations (FAO), there are few grounds for optimism regarding the sustainability of fisheries. Only about one quarter of the world's fish stocks were under-exploited in 2003, that is to say that they might be able to produce more if exploitation were intensified. Over half the stocks were producing catches that are close to their maximum sustainable limits and the final quarter were either over-exploited or depleted.[8] The FAO has observed an increasing trend in the proportion of over-exploited and depleted stocks over the years.[9]

While the total supply of fish from fishing and aquaculture combined has fluctuated between 80 and 90 million metric tonnes since the late 1980s,[10] this level of production has only been maintained by increasing production by aquaculture, to compensate for declines in catches of wild fish. Ironically, the production by aquaculture of certain carnivorous species of fish such as salmon or trout requires large quantities of fish meal and fish oil, exerting further pressure on wild stocks.[11]

There are many proximate reasons for the worrying state of so many fish stocks around the world, but among the most commonly cited are:

- "Too many vessels are chasing too few fish". This simplistic phrase refers to the excess of fishing capacity in the world, in terms of the number of boats as well as their size and technological development compared with the amount of fish that can be caught sustainably.
- Political interference in fisheries management leading directly to over-fishing. One of the most common practices is for politicians to set quotas at levels higher than scientific advice considers to be sustainable.[12]
- Poor control and surveillance programmes, both in waters under the jurisdiction of individual countries but also on the high seas. Without proper enforcement of the rules, over-fishing is encouraged.
- The use of fishing gear that catches too many fish that the fishermen don't want and so throw away. An FAO study in 1994 estimated that 27 million metric tonnes of marine species, fully one quarter of all catches, were discarded, much of it dead or dying. A follow-up study in 2004 concluded that approximately 7 million tonnes were discarded.[13]

While all of these issues are inter-related, the first two act especially strongly and in concert. Ship-owners and fishermen naturally want to catch fish and often resist efforts to restrict their activities through catch limits, seasonal closures, and other measures. The greater the number of fishing vessels relative to the fish that are available to be caught in an ecologically sustainable manner, the more restrictions that are necessary to constrain their activities, and so the greater the resistance from the fishing industry. The obvious targets for their lobbying are the politicians who make the decisions and, frequently, the latter weaken the management measures that are finally imposed.

Another factor that has been proposed as having played a significant role in over-fishing and the decline of fish stocks is the widespread granting by governments of financial aid to the fishing industry – in other words, subsidies. Over the past ten years a discussion has been taking place among governments, non-governmental organizations (NGOs), inter-governmental organizations and representatives of the fishing industry over what impact subsidies have had, if any, in the depletion of so many fish stocks. The World Trade Organization has recently begun negotiations over whether to reduce or eliminate fisheries subsidies. This chapter will review the situation in the European Union (EU) with respect to its programmes of subsidies for the fishing industry, including access agreements negotiated by the EU with other counties, and compare them briefly to levels of subsidies provided elsewhere. Subsidies and their use or misuse, however, are only a small part of the global financial forces that are revolutionizing the way in which fisheries are prosecuted and so a few comments will be included on the trends towards globalization and international trade.

THE EUROPEAN UNION IN THE FISHERIES WORLD

Fisheries are what is known as an "exclusive competence" of the European Union, meaning that all important decisions are made at the level of the EU as a whole – individual governments, such as of Spain, France or Poland, only have limited decision-making ability over the activities of their fleets, their markets or other matters. For instance it is the EU, through its Council of Ministers (the Fisheries Ministers from each of the Member States), that establishes total allowable catches for each fish stock and determines what fishing gear may be used and where, as well as many other fisheries management measures. Member States cannot decide these matters themselves.[14] Similarly, its executive branch, the European Commission, negotiates all fisheries agreements with other countries and represents the EU in all international meetings concerning fisheries. Collectively, this is known as the Common Fisheries Policy, or CFP.

It can be very misleading to examine the activities of any given EU country in isolation. For instance, the statistics on global fish catches published by the FAO suggest that EU countries are of relatively minor importance compared with China, Peru, Japan or the United States – the EU country with the greatest catches is usually Denmark, in 14th place overall for the period 1999-2002. Spain is even lower, at 23rd place in 2002. However, when the combined catches of all 15 Member States[15] are considered, the true importance of the EU becomes evident – it is consistently the third most important fishing power in the world, after China and Peru.

The vast fishing power of the EU fleets, though, is not sufficient to fulfil the demands of the marketplace and there is a large and growing dependence by the EU on fish imports. In 2002, one third of all fish exports went to the EU.

The EU thus has a very large fisheries footprint[16] indeed, in terms of fleet activities, and their impact on fish stocks or markets. The way in which it uses subsidies has a strong influence on the sustainability of this footprint.

THE SCALE OF SUBSIDIES AROUND THE WORLD

In 1993, the FAO published a review paper entitled "Marine Fisheries and Law of the Sea: A Decade of

Change".[17] At that point, the *United Nations Convention on the Law of the Sea* (UNCLoS) had been in force for ten years and the Organization was asking why the state of fish stocks had continued to deteriorate – the extension of national jurisdiction over fisheries from 12 to 200 nautical miles in the late 1970s was supposed to have improved matters but clearly it hadn't. The FAO came to a provocative conclusion – much of the blame lay with the subsidies provided by governments to their fishing industry, which were estimated to be approximately US$54 billion annually.[18] Compared with global revenues from the sale of fish of $70 billion, the figure of $54 billion was massive.

Not surprisingly, a heated debate quickly ensued and a number of studies proposed other estimates.

- A study by the World Bank[19] deduced that subsidies were in the range of US$14 billion to $20.5 billion, about a third of the FAO estimate but still equivalent to a quarter of the sector's total revenues. The author cited a lack of transparency as a major difficulty in the study, suggesting that the figures might be on the low side.
- The Organisation for Economic Co-operation and Development (OECD) came up with another estimate[20] based on the years 1996 and 1997, of between US$6.3 billion and $6.8 billion for its members, which account for about a third of global catches. Since it was not possible for the analysts to obtain reliable numbers on all categories of subsidies, these figures will also be under-estimates.
- Yet another study was done by Asia Pacific Economic Cooperation (APEC)[21] for the Pacific Rim states. The figure for the 21 APEC members was US$12.6 billion; they are responsible for about 70% of global fish production.

The disparities among the above estimates are not simply due to different methodologies and geographical coverage. Any analysis of subsidies is complicated by the fact that governments are remarkably shy about providing information on the type and amount of aid they provide.[22] There are also diverse definitions of what types of financial support should be included in these estimates; so all studies would not include all of the same items.[23] The World Wide Fund for Nature (WWF) attempted to compile the results of the above three studies into a single estimate, bearing in mind their differing characteristics, and concluded that worldwide fisheries subsidies were "at least US$15 billion, if not substantially more".[24]

SUBSIDIES IN THE EUROPEAN UNION

The European Union is correctly perceived to be a generous supporter of its fishing industry, as will be seen from the following discussion. For the sake of simplicity and to avoid theoretical and political discussions over what constitutes a subsidy,[25] this chapter will restrict itself to two specific programmes – direct payments made to the industry and access agreements negotiated with other countries.

The EU's Subsidy Programme

In 1971, the then-six members of the EU[26] decided to establish a common system for the pursuit of fishing activities and the EU has been subsidizing its fishing industry ever since. To begin with, money was only available to encourage the construction and modernization of certain types of fishing vessels, as well as for the processing and marketing of fish. One of the consequences was a rapid expansion in the size of the EU's fleets.

Recognizing the possible negative consequences of an expanding fleet, the Commission took the initiative of introducing the first of a series of programmes to regulate the size of the fishing fleets of the Member States, known as the Multi-Annual Guidance Programmes (MAGPs), which were in effect from 1983-1986. They sought, by means of national objectives for the capacity of each national fleet, to halt, or at least slow down, the surge in increasingly powerful vessels. Although its goals were modest, the first MAGP came to naught: capacity in 1987 was 24% greater than in 1983.[27]

A pattern was established that would last for 20 years, in which there were two programmes acting at cross purposes: on the one hand, regulatory programmes were adopted to limit or, later, to reduce the size of the national fleets, while on the other hand, money was made available for the construction of new vessels. Though the subsidy programmes changed many times over the years, with new measures eligible for support and stricter controls on their implementation, there has until very recently been money to build or enlarge fishing vessels at the same time as the fleets are supposed to be reduced in size and money is made available for scrapping vessels.

To an extent, these programmes can be seen as a poor compromise between the conflicting objectives of the Council of Ministers and the European Commission. The Fisheries Ministers push to have money available to support their domestic fishing industries that want to have bigger boats to be more competitive; this also helps the ship-building industry. The Commission, aware of the deteriorating status of fish stocks, tries to persuade the Member States to reduce the size of their fleets and the intensity of their fishing operations. The result is money

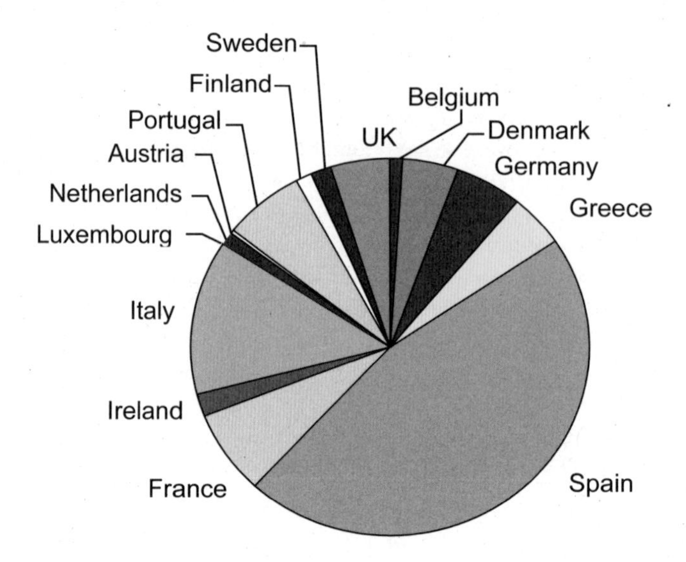

Figure 15–1. FIFG Allocation to Member State (1994-99).

to construct new fishing vessels, (i.e. new capacity) and, simultaneously, money to decrease capacity. In theory, construction of new vessels with public aid was only allowed if the relevant fleet had been reduced to the required level so that there would be no net increase in capacity, but this condition was not always fulfilled. This pattern continued until 2005 (see below).

One cycle of the subsidy programme, known as the Financial Instrument for Fisheries Guidance (FIFG), ran from 1994 through 1999 inclusive.[28] Its objectives were threefold:

- to contribute to achieving a sustainable balance between resources and their exploitation;
- to strengthen the competitiveness of structures and the development of economically viable enterprises in the sector; and
- to improve market supply and the value added to fisheries and aquaculture products.

The Commission has recently published considerable material on the execution of the FIFG[29] that gives insights into the use that Member States made of the money available to them. They were not given total freedom to spend their money as they wished. The FIFG was adopted by the Council of Ministers, so it was an EU-level decision regarding how much money was given to each Member State and what types of activities could be subsidized. Each Member State was then free to spend its money on whichever of these eligible activities it considered to be priorities, so the spending profiles differed considerably.

For the period 1994-1999 the EU contribution to fisheries subsidies via the FIFG was €2.126 billion.[30] As with other EU subsidy programmes, most of the support was reserved for the poorer areas, defined as having a GDP per capita of less than 75% of the EU average. Thus, funding was concentrated in Ireland, northern Scotland, eastern Germany, Spain, Portugal, southern Italy and Greece, while other areas received less support (Figure 15–1). This aid was conditional, though, on contributions from national budgets as well as from private enterprise (the contributions varied according to the activity being subsidized and the region involved). The sum of the EU and national contributions for the period was €3.103 billion. This works out to an average of €517 million per year, or about US$600 million at current exchange rates.

A wide range of programmes was eligible for funding:

- adjustment of fishing effort, including scrapping of vessels, permanent transfer to another country or permanent re-assignment of a vessel to other use;
- re-orientation of fishing vessels by either joint enterprises or temporary joint ventures (associations between EU shipowners and companies in non-EU countries) which were geared to supply the EU market;
- fleet renewal and modernization, including the construction of new vessels provided that they were in conformity with the MAGPs;
- development of aquaculture;
- protection and development of marine resources in coastal areas;
- improvement of port facilities;
- processing and marketing of fishery and aquaculture products;
- promotion of new markets including product labelling, consumer surveys, market studies,

Figure 15–2. FIFG Allocation by Activity (1994-99).

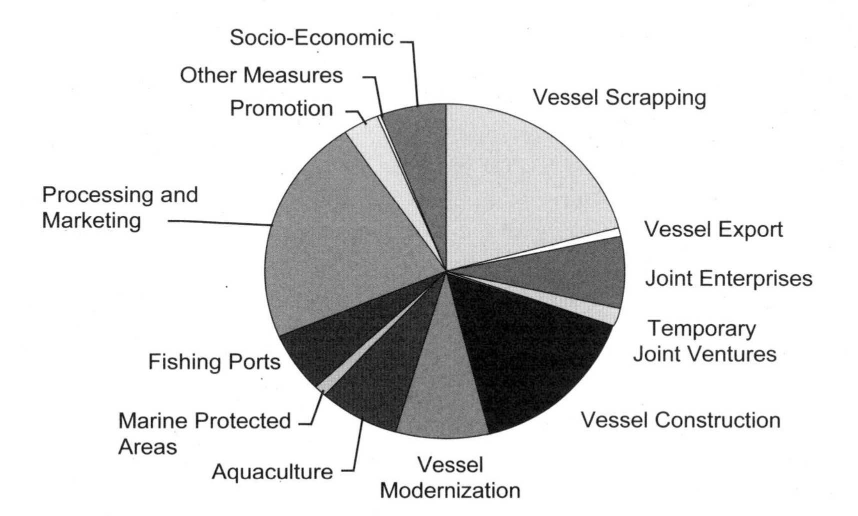

improving marketing conditions, etc.;

- support for temporary cessation of fishing activities under certain conditions;
- measures of a socio-economic nature such as early retirement schemes or compensation for fishermen who lose their jobs following restructuring.

Aid was quite generous in some cases. Construction of new vessels was handsomely supported, with subsidies of several million euros available for the larger vessels. Even a small vessel of 30 Gross Tonnes (GT) was given €300,000 of aid, including the contribution from the national government. Removing vessels from the fleet was encouraged with subsidies of roughly half the amount that was given for construction. A ship-owner could receive €973,000 to scrap a vessel of 350 GT, not a particularly large vessel in today's industrialized climate. He could get the same amount by setting up a joint enterprise to operate in the waters of another country. Even the export of vessels to flags of convenience such as Panama and Belize was subsidized, until that loophole was closed in 2002.

The most heavily funded measures by the EU for 1994-99 were the processing and marketing of fish products, the scrapping of existing fishing vessels, the construction of new vessels, the modernization of existing vessels, and the establishment of joint enterprises in other countries (Figure 15–2). Measures directly linked to the capacity of the fishing fleets received over half of the total amount – €1.776 billion.

The construction of new vessels was supported by €583 million, with a further €275 million for the modernization of other vessels. At the same time, €622 million were used to scrap vessels ten years of age or older. Additional vessels were removed from the fleet registers; many were exported, involving their permanent transfer to the registers of other countries around the world. Others were removed by means of joint enterprises or temporary joint ventures, both of which are associations formed between EU ship-owners and companies in other countries (a total of €296 million was used for these measures).

Such large expenditures on the fishing fleets of the Member States had a considerable impact on their profile. In the late 1990s, the EU fleet comprised approximately 1.8 million GT. The FIFG contributed to the construction of 200,000 GT of new capacity, at the same time as the removal of a further 300,000 GT (demolition, export to other country, joint ventures). In all, more than one quarter of the fleets were either built or removed by these subsidy programmes, not including the projects for the modernization of existing vessels.

These moneys were used to favour increases in the average size of fishing vessel – more money was invested in scrapping smaller vessels of less than 24 metres than in building them, while the converse was true for larger vessels. Consequently, a greater tonnage of smaller boats was removed than was built, with a greater tonnage of larger boats being built than was scrapped. Thus, the net effect of the FIFG as it was used by the Member States was to increase the average size of fishing vessels, though there was considerable diversity among the patterns of the individual Member States.[31]

Despite such large sums being used to build or modernize vessels, the combined fleets of the EU were somewhat smaller in 2000 than they were in 1994. The total tonnage decreased by rather less than 3% while the total engine power decreased by 12%. For a number of reasons, however, the accuracy of these figures is difficult to evaluate.[32]

How much real reduction there was in terms of fishing capacity is more difficult to determine, given that the average size and power of individual vessels increased. Technological advances also resulted in increased fishing efficiency without any change in tonnage. It is thus quite possible that there was no significant reduction in the ability of the EU fleets to catch fish during the period of the first FIFG, from 1994 through 1999.

When the FIFG was renewed for the period 2000 through 2006, it retained more or less the same types of measures concerning the fleet as the previous programme – construction, modernization, decommissioning, export and joint enterprises with other countries (though see below). The only measure that was dropped was the creation of temporary joint ventures with other countries. Since only Spain and Portugal had availed themselves of this possibility during FIFG 1, though, there was little reason to continue. A number of administrative provisions were introduced in order to tighten up the use of the funds to prevent practices deemed undesirable.

The Impact of the Subsidies

Reflecting their divergent views over the need to reduce fleet size in the EU, the Commission and the Council disagree over the contribution that subsidies to the fleet make to over-fishing and the depletion of fish stocks.

There is no doubt that the major stocks fished by the EU fleets have deteriorated over the past 10 to 15 years. Catches of the top ten species taken by the EU fleets in the Northeast Atlantic in 1994, the first year of the FIFG programme, had all declined by 2002, the latest year for which data are available.[33] Some declined precipitously – catches of herring (*Clupea harengus*) dropped by 27%, horse mackerel (*Trachurus trachurus*) by 46%, cod (*Gadus morhua*) by 40%, haddock (*Melanogrammus aeglefinus*) by 33%. It is not that the fleets are catching other species, for nine of the top ten species of 1994 are still in the top ten in 2002 – only blue whiting (*Micromesistius poutassou*) has moved into the top ten, replacing haddock.

These reductions in catches are not simply due to more stringent management measures but reflect serious degradations for these stocks, as well as many other species. Many of the most economically important stocks exploited by the EU have experienced large and long-term declines in abundance.[34]

A detailed analysis of the impact of fleet subsidies on levels of exploitation and stock abundance would require information not only on the size of both fleets and fish stocks, but also on the activity patterns of each vessel and whether or not it benefited from any subsidies. Such data are not available, at least not in the public domain. The widespread decline in the abundance of the most important species fished during and after the infusion of hundreds of millions of euros into the EU fleets, though, suggests that these subsidies did nothing to improve the situation.

That was the view of the European Commission, at least. During the period of debate prior to the reform of the Common Fisheries Policy in 2001-2002, the Commission published a discussion paper in which it analyzed the state of the fish stocks and the fishing industry.[35] It noted the large number of depleted fish stocks and specifically mentioned fleet over-capacity as one of the causes:

> From a biological point of view, the sustainability of a high number of stocks will be threatened if the current levels of exploitation are maintained and, at present, this risk is highest for demersal round fish stocks. Improvement in the state of many stocks is urgent.
>
> Council has fixed some TACs [Total Allowable Catches] systematically at levels higher than those proposed by the Commission on the basis of the indications of scientific advice; over-fishing, discards and fleet over-capacity have also contributed to the current problems.[36]

The paper was very critical of the subsidy policy of the EU:

> Today's subsidies to investment in the fishing industry and certain taxation measures, such as tax-free fuel, do not contribute to this objective [i.e. reduction of fleet capacity]. By artificially reducing the costs and risks of investment in an already over-capitalised industry, they promote over-supply of capital.
>
> The form of aid most favoured by the Community, i.e. aid for capital investment, may have intensified the problem of over-capacity, low profitability and replacement of labour by capital in the catching sector.[37]

In fact, the Commission was declaring its desire to reduce or eliminate certain types of subsidies. When it published its proposals for the CFP reform, one of the changes it sought was to stop, at the end of 2002, the subsidies for vessel construction and modernization, and for the export of vessels and the establishment of joint enterprises in other countries, arguing that:

> Given the urgent need to reduce fishing effort, the use of public aid for new vessels or to make existing vessels more efficient may be counter-productive and can no longer be justified.[38]

Certain Member States, referring to themselves as "Friends of Fishing", resisted such constraints fiercely, but in the end a compromise was reached. Those subsidies that the Commission wanted to stop were indeed terminated, but only at the end of 2004.[39]

The Commission had succeeded in convincing Council that these subsidies had, at the very least, contributed to the over-exploitation of fishery resources in the EU.

FISHERIES AGREEMENTS WITH DEVELOPING COUNTRIES

Europe has had "distant water" fishing fleets for a very long time: vessels from many nations were fishing on the Grand Banks of Newfoundland as early as the 15th century. Most of that activity took place in international waters, where freedom of the high seas was a jealously defended right. In the latter half of the 20th century that changed, as countries began to claim an exclusive economic zone (EEZ) out to a distance of 200 nautical miles from their coastline, well beyond their territorial sea of up to 12 nautical miles. A large number of historical fishing grounds were suddenly closed off for European fleets, as well as other distant-water fishing nations.

The solution was to negotiate bilateral fisheries agreements with individual countries. The legal basis lies in the 1982-UNCLoS:

> **Article 62, 2.** The coastal State shall determine its capacity to harvest the living resources of the exclusive economic zone. Where the coastal State does not have the capacity to harvest the entire allowable catch, it shall, through agreements or other arrangements [....] give other States access to the surplus of the allowable catch [....]

There is thus a clear legal obligation for coastal states to allow other countries access to their waters for fish stocks that they are unable to exploit themselves.

Access rights for European vessels could be obtained in exchange for payments to the government of the other country. The first such agreement that the EU concluded was with Senegal in 1979 – it allowed European trawlers and tuna vessels to fish in Senegalese waters for a period of two years. The number of agreements increased steadily over the next 15 years, especially following the accession of Spain and Portugal in 1985, and by 1995 the EU had concluded a total of 17. Subsequent agreements have extended the fishing tendrils of the EU fleets into the western Pacific, the world's richest tuna grounds. There is also a rather unusual agreement with Greenland. When that island withdrew from the EU in 1985, the EU concluded a fisheries agreement that provided access for EU vessels and also included certain components designed to aid the development of Greenland. The EU also has a series of agreements with countries in the North Atlantic that allow each side to fish in the waters of the other.[40] No money is involved in these agreements and they will not be further considered here.[41]

In 2004, the EU had fisheries agreements with 16 countries in Africa, the Indian Ocean and the western Pacific, and had negotiated a further three to come into force during 2005 (Figure 15–3). Of these, 14 were with Low-Income Food-Deficit countries[42] or Least Developed

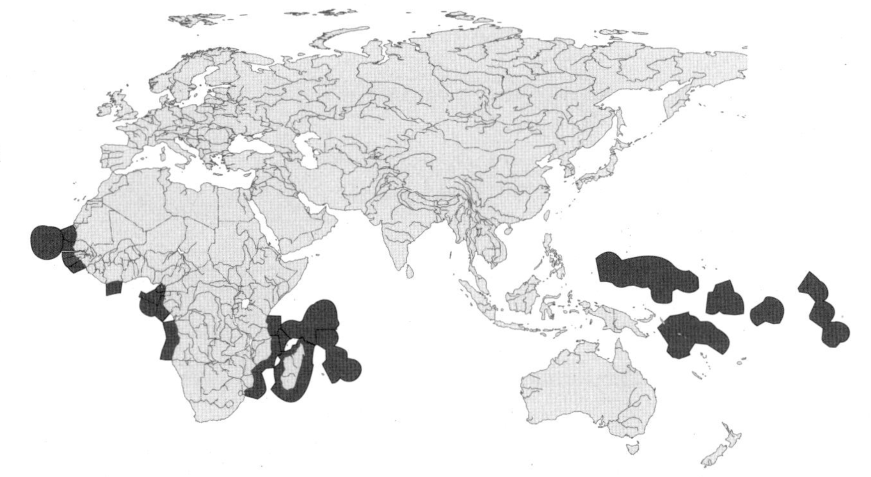

Figure 15–3.

Map showing the EEZs of countries with which the European Union had fisheries access agreements during 2004. EEZ limits are approximate, and based on the map at www.seaaroundus.org. Note that the list varies over time, according to expiration and renewal of the protocols, and an up-to-date list can be found on the webpages of the European Commission (see endnote 41). Some agreements included here were due to enter into force in 2005.

Table 15–1. Table showing status of countries with which the European Union had fisheries access agreements during 2004, as for Figure 15–3. See endnotes 6, 43 and 44 for details of definitions.

	Least Developed	Heavily Indebted	Low-Income Food Deficit
Mauritania	***	***	***
Senegal	***	***	***
Cape Verde	***		***
Guinea-Bissau	***	***	***
Guinea	***	***	***
Ivory Coast		***	***
Sao Tomé and Principe	***	***	***
Gabon			
Equatorial Guinea	***		***
Angola	***		***
Seychelles			
Comoros	***	***	***
Mauritius			
Madagascar	***	***	***
Mozambique	***	***	***
Tanzania	***	***	
Kiribati	***		***
Micronesia			
Solomon Islands	***		***

Countries[43], and ten of them were classified by the World Bank and International Monetary Fund as Heavily Indebted Poor Countries (Table 15–1).[44]

The structure of these agreements is relatively simple. In exchange for a cash payment, the other country grants access to EU fishing vessels to certain fish stocks, usually for a period of two to four years. Payments have been highly variable – the agreement with Morocco from 1995-1999 cost the Commission €125 million per year whereas that with Comoros, a much smaller agreement, cost €350,000 per year (2001-2004).

The agreement with Guinea, first negotiated in 1983, is typical. In exchange for payments totalling €3.4 million per year, the current protocol to the agreement allows EU trawlers and tuna vessels to operate from 1 January 2004 through 31 December 2008, including:

- 34 tuna purse seiners (Spanish and French),
- 14 pole-and-line tuna vessels (Spanish and French),
- 9 surface longliners (Spanish and Portuguese),
- 2,500 Gross Registered Tonnes (GRT) (average per month) of trawlers fishing for finfish and cephalopods[45] (Spanish, Italian and Greek),
- 1,500 GRT (average per month) of trawlers fishing for shrimp (Spanish, Portuguese and Greek).

These are limits on nominal fleet capacity, rather than on the amount of fish that can be caught.[46] Of the total payment, the government of Guinea has committed to spend €1.4 million per year (41% of the total) on "targeted actions": specific activities such as scientific research, control and surveillance, vocational training for young fishermen and others. Though most of the money paid under the agreement comes from Community coffers, additional moneys must be paid to the other-country government by those ship-owners who operate under the protocol, primarily licence fees that vary according to vessel type (e.g. €2,500 per year for a tuna seiner).

The Commission estimates that some 600 EU vessels operate under access agreements, equivalent to 22% of the total Community fleet in terms of tonnage. All Member States of the EU do not participate equally in the agreements, though. Spain is by far the greatest beneficiary, while several other countries fish to lesser extents, including France, Portugal, Italy, Greece, the UK, the Netherlands and Ireland. There has been a trend for more countries to participate in the agreements in recent years, concomitant with the decline in fish stocks in EU waters.

This network of agreements costs the Community approximately €175 million to €200 million per year, depending on the precise protocols in force. This is less than in the late 1990s, when the EU had its agreements with Morocco and Argentina: the annual total then was close to €300 million.

The Community's motivations for paying such sums for the right to fish (in effect, subsidized access for certain of its fishermen) have generally been socio-economic in nature. Among the specific reasons most frequently cited by the Commission are continuing the activity of Community fleets as a source of employment, and securing the supply of fish to the EU market.

As fish stocks in EU waters decrease in abundance and satisfying the increasing demand for fish becomes ever more difficult, the latter reason assumes greater importance. EU vessels operating under other-country agreements catch fish valued at €2 billion, equivalent to 20% of the EU total catch,[47] in exchange for an EU investment of less than €200 million. The increasing dependence of the EU on fish from beyond its own waters, whether obtained through imports or its ever-widening network of fisheries agreements, means that the EU's ecological footprint continues to expand.

Another benefit for the EU is that agreements are a convenient means by which it can export its problems of an excess of fishing capacity. As the Commission stated in 1991:

> ... in the present circumstances where the Community fleet operating in its own waters has surplus capacity, fisheries agreements represent a means in keeping with the Law of the Sea of reducing fishing effort in the Community's home waters.[48]

Impact of the Agreements

Benefits from a European perspective, though, are not always seen as benefits in the country that is on the receiving end of these fleets. During the past 15 years there has been increasing discussion over whether these agreements are in the interests of the other countries. Among the criticisms that have been made by various NGOs, academics or even inter-governmental organizations such as UNEP are:

- They introduce large, technologically advanced fleets into the waters of developing countries, often in competition with local fleets, which can lead to over-exploitation of fish stocks;
- Control and surveillance of fishing activities is frequently poor and occasionally almost non-existent;
- There are limited benefits to the other country in terms of transfer of know-how, technology or employment;
- They make only a limited contribution to the food security of the other country, as much of the catch is landed in the EU;
- Though certain agreements provide large contributions to the national budgets of the government of the other country, the latter can come to rely on them as a source of revenue, thus diminishing their ability to resist the access demands of European negotiators.

An evaluation of the benefits and damages caused by EU-other-country agreements is far beyond the scope of this chapter, however, so only brief descriptions of the situation in West Africa will be provided to illustrate some of the types of problems that can arise.

The fishery resources of the eastern central Atlantic, a region stretching from Morocco to Congo, are heavily over-fished – over 90% of the fish stocks are exploited at or beyond Maximum Sustainable Yield (MSY) levels.[49] Catches in the region have fluctuated between 2.5 and 4 million tonnes annually since the early 1970s, of which about 0.5 million tonnes have been taken by vessels flying European flags, much of it under fisheries agreements with African countries. The EU thus bears at least part of the responsibility for the over-exploited state of the resources.[50] Agreements with Senegal and Guinea are discussed briefly below:

Senegal

The agreement with Senegal dates back to 1979. It has received more scrutiny than other agreements, at least in part because the local fishing industry is so large and dynamic. Two recent reviews of the Senegalese fishing sector included rather negative assessments of the current state of the resources and the sector.

The World Bank is supporting a project to develop sustainable fisheries in Senegal while preserving critical ecosystems. A document describing the project notes:

"All along the West African coast the marine environment is under pressure from excessive fishing, but nowhere is that pressure as severe or are the consequences as far reaching as in Senegal. Excessive fishing has impoverished Senegal's marine environment, which in turn has translated into substantially lower catches per vessel, affecting the volume of Senegal's fish exports (30-40% of Senegal's total exports) and its ability to maintain employment and income for the more than 600,000 Senegalese involved in the fisheries sector".[51]

It attributes this situation to several factors, including "the introduction of a foreign, mainly European fleet, under the umbrella of international fisheries agreements with Senegal". Other causes include a lack of limitation of catches by the Senegalese industrial fleet, expansion of the unregulated artisanal fleet and rising demand in foreign markets for high-value demersal species.

A study by the United Nations Environment Programme (UNEP)[52] provides a detailed analysis of the factors that have led to declining fish stocks in Senegal, resulting in serious problems facing the Senegalese fishing sector. The role played by foreign, especially European, fishing fleets in Senegalese waters is highlighted. Other factors have contributed to the stock depletions, though, including export subsidies for fish awarded by the Senegalese government and a 50% devaluation of the local currency that was recommended by the World Bank, the International Monetary Fund and other donors. These latter two factors encouraged over-exploitation by the Senegalese fleets to produce exports for the European market. The study concludes that there is currently a risk of shortages of fish for local markets.

Guinea

As part of the preparations for negotiating the fisheries protocol for the period 2004-2008, the Commission contracted an evaluation of the impact of the previous protocol (2000-2003). The analysis was quite candid:[53]

- The measures introduced by Guinea for the management of fish stocks have not succeeded in preventing overfishing in the coastal zone;
- The preceding Protocols have not succeeded in improving local market supplies and food security for the population. At the same time, discards are a waste of resources;
- Fisheries protection and control measures in Guinea are inefficient and the previous Protocols have not succeeded in reducing illegal fishing;
- Previous Protocols have also done little to generate local added value;
- Guinean involvement in the fisheries sector has remained largely non-industrial;
- Regional cooperation with other West African countries has been limited;
- The lack of information on fishing in Guinea makes it difficult to carry out an accurate assessment of the activities and results of the fisheries sector.

Considering that the EU has had an agreement with Guinea for 20 years, there had not been much progress at developing sustainable and responsible fishing. The European Parliament was also very critical of the agreement, noting in particular the very poor data available on catches by EU fleets and the over-exploited status of many fish stocks.[54]

The EU considers its fisheries agreements to be commercial in nature, rather than a vehicle to aid the development of the other country. It defends them by insisting that the other country is a sovereign state, free to enter into any agreement it chooses and responsible for the effective management of fisheries in its waters. The degree to which some of the small countries with which the EU has concluded agreements are able to resist the political and economic pressure that the EU is able to exert is a debatable point. Further, if the other country wishes to develop its own capacity to fish, the EU has been known to resist requests to reduce its level of exploitation.

It is perhaps not surprising, given the above, that there has been so much pressure on the EU to change its approach to fisheries agreements.[55] The opportunity came during the reform of the CFP in 2002 and 2003. The Commission admitted that there have been serious weaknesses in the Community's fisheries agreements, such as:

- There is an inability to respond rapidly to emergency circumstances such as decreasing stocks;
- Fishing opportunities are not always based on the real evolution of the resource;
- EU fishing mortality is not always known;
- Often there are insufficient guarantees for the protection of local artisanal fleets.[56]

The solution was to transform what had hitherto been known as "fisheries agreements" into "fisheries partnership agreements". Their primary objectives were to be to:

- maintain the agreements as a means of protecting EU fleet activity and employment, and
- establish sustainable fisheries outside Community waters.[57]

Thus, the agreements were to prevent over-fishing, to improve scientific knowledge of the fisheries, to contribute to the fight against illegal, unregulated and unreported fishing, and to facilitate the integration of the developing coastal state into the global economy.

The terminology has been modernized but it is still too early to tell whether the behaviour of the EU and its fleets will be any more ecologically sustainable. A few agreements have been negotiated since the concept of partnership came into vogue but it will be a few more years before any definitive conclusions can be drawn.

Presumably the primary incentive for the government of another country to sign agreements with the EU is the financial compensation that it receives, which can be substantial in some cases. In exchange, though, some of the fish in their waters disappears into the nets and markets of the European Union. Countries that have only small or undeveloped domestic fishing industries may view this as a convenient way to get money for government coffers with limited regard for the negative consequences for their own people. But even when there is little in the way of domestic fishing industry, if the national government wants to maintain the possibility of developing one later on, it would have to devote at least some of the EU's money to a programme of control and surveillance, to ensure that the EU and other vessels do not irreversibly deplete their fish stocks. For governments with very limited resources, that could be a difficult financial burden to assume with no immediate and obvious gain.

In other cases, there is a very strong and well-developed industry, with the result that there is direct competition between the national fleets and the foreign fleets for fish. Senegal and Mauritania are examples. In the case of Morocco, the government refused to renew its agreement with the EU after it expired in 1999, despite being put under enormous pressure to do so, claiming that it preferred to develop its own industry on its own terms. Morocco has since concluded deals with other countries, such as Japan, which could suggest that the government was given better terms than the Europeans were prepared to offer.[58] This exemplifies the stiff competition that can arise among the various distant-water fishing nations as they seek access to stocks in ever more depleted oceans.

The EU's programmes of direct subsidy to the fishing industry, the FIFG, and the fisheries agreements, do not operate in isolation from one another. Vessels that were built or modernized with public aid could operate under the EU's network of fisheries agreements and, in fact, many were specifically constructed to do so. There are also examples of vessels that received subsidies in order to be re-flagged to countries with which the EU simultaneously had a fisheries agreement, and such re-flagged vessels entered into direct competition with the EU-flagged vessels. Some of these exported vessels may have continued to supply the EU market.

THE ERA OF THE WORLD TRADE ORGANIZATION

The fishing industry of the European Union has spent the past 25 years operating with the support and protection of structural support programmes and fisheries agreements with other countries. All of that is in the throes of changing, as it is being forced to respond to increasing pressure from other members of the World Trade Organization (WTO) to liberalize trade in fish.

In Doha in 2001, the WTO trade ministers agreed to "clarify and improve WTO disciplines" to the use of subsidies in support of the fishing industry.[59] The issue is being debated under the aegis of the Negotiating Group on Rules.[60] The national positions initially expressed differed radically. At one extreme, several countries, often referred to as the Friends of Fish,[61] have argued that most types of support for the industry contribute to over-capacity and so lead to over-exploitation and depletion of resources, including not only money to build and modernize vessels, but also various price-support mechanisms for fish, tax-free fuel, and other measures. They proposed that subsidies promoting over-capacity of the fleets and over-fishing should be banned while other subsidies that reduce costs to producers should be presumed to be harmful and banned unless it can be demonstrated that they have no harmful effects. Other countries, primarily Japan and Korea (later joined by Chinese Taipei), initially maintained that it had not been demonstrated that public aid to the industry leads to over-fishing and so there was no need to treat fisheries subsidies any differently from subsidies to other industrial sectors. Subsequently, though, these countries began to accept the idea of prohibiting certain types of subsidies, such as those related to illegal, unregulated and unreported (IUU) fishing. Developing countries are also questioning the need for them to reduce the relatively modest subsidies they offer, arguing that the developed countries benefited from years of generous subsidies and it would be unfair to deny them the opportunity to develop their own industry so as to enable it to compete in the global fishery.

The EU was initially silent during these discussions, pending the conclusion of the reform of the CFP. Once the outcome of the reform was clear, they began to ally themselves more with the Friends of Fish, supporting the idea of prohibiting subsidies for fleet renewal, and for the transfer of vessels to other countries. Other subsidies should be allowed, according to the EU, including certain types of modernization of vessels that do not increase capacity, payment to fishermen for temporary cessation of fishing activities and for the scrapping of vessels. This position is consistent with the results of the reform of the CFP.

Another point of importance to both the EU and the African, Caribbean, and Pacific (ACP) states is whether or not bilateral agreements for fishing access rights should be considered subsidies and, if so, whether they should be prohibited. These agreements clearly serve to lessen costs for the fishing fleets from the distant water nations, such as the EU, and they can, as outlined above, lead to environmental and socio-economic problems. However, they can also be a significant source of revenue for the other country, including much-needed development assistance in some cases. The simplest way to resolve this incongruity is to de-couple the financial payments for development aid from access to fishing rights – in other words, to provide appropriate assistance to developing countries to fulfil their international responsibilities with respect to fisheries management and to promote the development of their own fisheries without demanding the right to fish in their waters. The EU could also require that those shipowners who benefit from the agreements make a greater contribution to their cost.

Whatever the eventual outcome of these negotiations,[62] it appears that the global fishing industry will not be able to benefit from the very generous support it has had for so many years. The EU admitted as much in 2002 when it agreed to terminate certain types of subsidies it had provided for some 30 years.

The influence of the WTO, however, is not limited to subsidies. Since the 1970s, the EU has had a development cooperation agreement with a group of former colonies among the ACP countries.[63] Article 53 of that agreement provides for the negotiation of fisheries agreements between the EU and ACP countries, as reviewed above. The Agreement also governs access for products from the ACP countries into the EU. Historically, the ACP countries have benefited from duty-free access for a wide range of commodities to the EU market, including most fisheries products, except for certain sensitive items such as sugar, bananas and a few others. Access for these latter products has been subject to a protocol that included both quotas on how much was allowed into the EU as well as a minimum price that the producers would receive.

But such preferential access gives trade advantages to some developing countries that are not extended to others. This has prompted some non-ACP countries to take their complaint before the WTO, most famously in the case of bananas from Latin America. The WTO ruled that the EU could not favour the ACP at the expense of other developing countries and the EU has decided to terminate quota-based access for all products from the ACP.

This will bring to an end the preferential access that the EU has granted the ACP for over 20 years, replacing it with a series of regional free trade associations, known as Economic Partnership Agreements, which will come into effect in 2008. Consistent with the global trend towards trade liberalization, the ACP countries will be obliged to open their markets to EU products, while continuing to receive financial aid. Even the series of commodity protocols, under which the ACP countries benefited from preferential access for fixed quantities of bananas, sugar, beef and veal, are due to be dismantled.[64]

This development could have serious consequences for the economies, and therefore the societies, of many poor countries. If they lose their preferential access to the world's largest market for fish and are forced to compete with fishing industries from other developing countries that may apply less strict environmental and social standards and, at the same time, are also forced to undergo structural economic reforms at home in order to receive debt relief, they could end up in a situation where they see no option but to sell off access to their fishery and other resources to the highest bidder. The case of Senegal described above should serve as a warning – Senegal qualifies under the three categories of Least Developed country, Heavily Indebted country, and Low-Income Food-Deficit country.

CONCLUDING REMARKS

The European Union has been spending large sums of money on its fishing industry for many years and it is difficult to argue that it has contributed to either ecological or economic sustainability: its fleets are too large, its fish stocks are depleted and much of its industry is in decline. Further, not content with depleting its own waters, it is projecting these problems to other regions of the world, often where there is less ability to enforce the rules and so prevent over-exploitation. The EU fisheries footprint is indeed vast, largely unsustainable and, in light of the above, expensive.

What is not easy to prescribe, though, is a solution that would result in ecologically sustainable fishing and social and economic security for fishermen in the EU as well as in other parts of the world.

Over-capacity in the world's fishing fleets has doubtless contributed to the depletion of many fish stocks and, as argued above in the case of the EU, certain types of subsidies have contributed to that over-capacity. A reduction in those subsidies awarded by developed countries to their fleets would thus favour the conservation of fish stocks. The decision by the EU to eliminate certain subsidies for its fleets as part of its reform of the Common Fisheries Policy was important in that regard.[65] However, subsidies are only one factor leading to the depletion of marine resources. Others include poor control and surveillance of fishing activities, the widespread use of flags of convenience, increasing demand for fish especially in the richer countries, poor decision-making on fisheries management measures and the flawed concept of MSY[66] that remains embedded in many fishery agreements and the UNCLoS. Far more comprehensive and radical reforms to the global system of fishery management, including the network of regional fisheries management organizations (RFMOs),[67] are necessary.

Meanwhile, the WTO is the body with a mandate to adopt and enforce rules on subsidies, so the challenge is to use the clout of the WTO to reduce the destructive influence of subsides while not allowing it to exert too much influence over the international system that has been built up for management of fisheries. It should strictly limit its mandate to subsidies and other trade-distorting mechanisms.[68] All other aspects of fisheries are best left to the bodies that are specialized to deal with them, including the FAO and the various RFMOs. Conservation measures that these bodies have so far taken, including catch documentation schemes, trade restrictions and access limits to fish stocks must be preserved, supported and extended, rather than undermined in the name of trade liberalization.

The WTO's "Friends of Fish" should use their enormous political and economic influence to improve the effectiveness of the RFMOs to reduce over-fishing, to provide help to developing countries and to strengthen fisheries management and control and surveillance programmes.

Even if the EU were to transform its domestic fisheries so as to render them sustainable, which would doubtless require subsidies of a nature different to those which have been so popular thus far (though they were available, as described above), it is arguably even more important that the EU act in a responsible manner in its distant water fisheries. Access agreements with developing countries are not necessarily unsustainable: if properly negotiated and implemented, they could be a means for the coastal state to obtain financial and other resources to enable it to improve its own fisheries management regime and contribute to its national development. All too frequently, however, they have tended to lead to more intense exploitation and the decline of resources.

For, after all, the fisheries agreements between the European Union and other countries are but one small, albeit important, part of the complicated relations between the rich industrialized countries and the developing world. They should not be seen in isolation from such issues as third-world debt, poverty and food security.

These issues, obviously, are far beyond the scope of this chapter, or even this book, but at an individual level, consumers can make a contribution by paying more attention to how much fish they eat, how it was caught and by whom. The importance of the role of the market in the pursuit of ecological sustainability cannot be overestimated.

Acknowledgements

My thoughts on many of the matters discussed in this chapter have benefited from intense discussions over many happy years with Hélène Bours, who also read the entire manuscript. Other beneficial comments came from Steve Emmott and Ronald Steenblik. I thank Sheryl Fink for preparing Figure 15–3. Finally, thanks are due to DML for agreeing that an explanation of how governments spend their money shows, more clearly than any policy statement, where their priorities lie.

NOTES AND SOURCES

[1] The Food and Agriculture Organization of the United Nations (FAO) provides statistical analyses of fisheries production, consumption, trade and many other issues. Overviews are published biennially as *The State of World Fisheries and Aquaculture*, the most recent one dating from 2004, referred to here as SOFIA 2004. These are available at http://www.fao.org/sof/sofia/index_en.htm. Detailed statistics can be found in the FAO database FAOSTAT, available at http://faostat.fao.org/faostat/collections. These and other links were verified on 15 August 2005.

[2] Average consumption of other meats for 2002: pork, 15.3 kg per person per year; poultry, 11.7 kg; beef, 9.7 kg and sheep and goat, 1.8 kg. FAOSTAT.

[3] Full and part time. Data for 2002. SOFIA 2004.

[4] In 2002, developing countries provided 55% of exports in terms of quantity and 49% in terms of value. SOFIA 2004.

[5] The EU was the largest importer of fisheries products in 2002, accounting for 35% of all imports (by value). Next came Japan (22%) and the United States (16%). SOFIA 2004.

[6] There were 84 such countries as of May 2005, primarily in Africa (40) and Asia (24) but also in Europe (5), Latin America (4) and the Near East (11). Note that China is classified as a LIFD country. Further details can be found at http://www.fao.org/countryprofiles/lifdc.asp?lang=en.

[7] Other regions – 21.6 kg in North America, 19.8 kg in Europe, 25.6 kg in China or 8.8 kg in South America. SOFIA 2004.

[8] SOFIA 2004. The FAO monitors a large number of stocks for which assessment information is available, comparing each one to its maximum sustainable yield. Since the concept of MSY has been discredited for many years (see Larkin, P. 1977. An epitaph for the concept of maximum sustainable yield. Transactions of the American Fisheries Society: Vol. 106 (1):1–11) the FAO's estimate of fish stocks is probably overly optimistic. See also Holt, Chapter 4.

[9] In the 1993 SOFIA, 69% of stocks were considered to be fully exploited, over–exploited or depleted; in 2003 the figure was 76%.

[10] SOFIA 2004, marine and fresh water combined. This excludes production from China, as doubt has been cast on the reliability of statistics from China. If Chinese data are included, total production has continued to increase almost continually since 1950, reaching 133 million metric tonnes in 2002.

[11] For further discussion of world fisheries, see Holt, Chapter 4, and Hutchings, Chapter 6.

[12] To give but one example, EU Fisheries Ministers set the quota for cod in the Irish Sea an average of 28% higher than the scientific advice throughout the 1990s. See also Hutchings, Chapter 6.

[13] The first study is Dayton Alverson, Mark Freeberg, John Pope, Steven Murawski. *A Global Assessment of Fisheries Bycatch and Discards*. FAO Fisheries Technical Paper 339. Rome. 1994. The second is Kieran Kelleher. *Discards in the World's Marine Fisheries. An Update*. FAO Fisheries Technical Paper 470. Rome. 2005. Kelleher noted that his methodology differed from that of Alverson *et al.*, so the two results are not directly comparable. Nonetheless, he suggested that two trends were evident – reductions in discards due to increased rates of utilization, as well as improvements in gear selectivity. Both reports available at http://www.fao.org/fi/eims_search/publications_form.asp?lang=en.

[14] Individual Member States do have limited power to impose management measures in their EEZ on vessels flying their flag, but, ironically, not on vessels flying other flags.

[15] All figures in this chapter relate to the European Union with 15 Member States, i.e. before the accession of ten new countries in May 2004.

[16] Wackernagel, M. and W. Rees. (1996) *Our Ecological Footprint. Reducing Human Impact on the Earth*. Gabriola Island, BC, Canada: New Society Publishers; Rees, W.E. (2002). Footprint: Our impact on Earth is getting heavier. Nature 420 267-268.

[17] This was published as a Special chapter (revised) of FAO's *The State of Food and Agriculture*, 1992. FAO Fisheries Circular 853.

[18] This figure was derived by comparing estimates of total operating costs for the world's fishing fleet ($124 billion, including vessel repairs, insurance, fuel, gear, capital costs, etc.) and estimates of total revenues ($70 billion). The difference, some $54 billion was thought to be subsidies. (Throughout this chapter, I have used the American convention where 1 billion = 1,000,000,000.)

[19] Mateo Milazzo. *Subsidies in World Fisheries. A Re-Examination*. World Bank Technical Paper No. 406. Washington. 1998. Available at http://www-wds.worldbank.org/.

[20] OECD. *Transition to Responsible Fisheries. Economic and Policy Implications*. Paris. 2000.

[21] APEC. *Study into the Nature and Extent of Subsidies in the Fisheries Sector of APEC Members' Economies*. Document APEC #00-FS-01.1. Singapore. 2000. Available at http://www.apecsec.org.sg/apec/publications/free_downloads/2000.html.

[22] WWF. *Fishing in the Dark. A Symposium on Access to Environmental Information and Government Accountability in Fishing Subsidy Programmes*. Brussels, 2001.

[23] A readable introduction to the literature on fisheries subsidies is William E. Schrank. *Introducing Fisheries Subsidies*. FAO Fisheries Technical Paper 437. Rome. 2003. Available at http://www.fao.org/fi/eims_search/publications_form.asp?lang=en.

[24] WWF. *Hard Facts, Hidden Problems. A Review of Current Data on Fishing Subsidies*. WWF Technical Paper. 2001.

[25] See references in endnotes 19-23, as well as Lena Westlund. *Guide for Identifying, Assessing and Reporting on Subsidies in the Fisheries Sector*. FAO Fisheries Technical Paper 438. Rome. 2004. Available at http://www.fao.org/fi/eims_search/publications_form.asp?lang=en.

[26] Despite the fact that the European Union has previously been known by various names (European Economic Community, European Community), for the sake of simplicity in this chapter it will be referred to as the European Union or EU.

[27] Three other MAGPs followed, with the final one ending on 31 December 2002.

[28] Council Regulation (EC) No. 3699/93. Published in the Official Journal of the European Communities Volume L 346, 31/12/1993 pp. 1-13. Expired.

[29] See the European Commission website at http://europa.eu.int/comm/fisheries/structures/index_en.htm, and http://europa.eu.int/comm/fisheries/news_corner/autres/conf_220504_en.htm.

[30] For the sake of simplicity, all figures are given in euros (€). At the time of these programmes, though, EU budgets were calculated in ECUs, or European Currency Unit, a monetary basket comprising the currencies of all EU Member States. It evolved directly into the euro in 2002. The mean exchange rate in 1993 was US$1.17.

[31] See data published by the European Commission at http://europa.eu.int/comm/fisheries/structures/exec_fin_en.htm.

[32] The units of measurement for tonnage changed during that period as the EU standardized the data in its fleet register. Hitherto, Member States were free to use whatever units they chose, which led to considerable variety. The standardization to Gross Tonnes (GT) was slow and uneven, so that comparisons were very difficult. Two fishing countries, Sweden and Finland, also became members of the EU in 1996. Finally, there were two separate fleet reduction programmes, the MAGPs, one running from 1992 through 1996 and the other from 1997 through 2001. Vessels were classified in different ways in the two programmes.

[33] The species were, in decreasing order of catch volumes, sandeels (*Ammodytes sp.*), Atlantic herring, mackerel (*Scomber scombrus*), sprat (*Sprattus sprattus*), horse mackerel, Atlantic cod, pilchard (*Sardina pilchardus*), Norway pout (*Trisopterus esmarkii*), plaice (*Pleuronectes platessa*), haddock. The FAO publishes annual databases of catches dating back to 1950 and catch data were taken from that database. Available at http://www.fao.org/fi/statist/statist.asp.

[34] European Commission. *Green Paper. The Future of the Common Fisheries Policy*. COM (2001) 135 final. Brussels. 2001. Volume 2 examines in detail the state of fish stocks in the EU. Both volumes available at http://europa.eu.int/comm/fisheries/greenpaper/green1_en.htm#volume1.

[35] *Green Paper*. See endnote 34.

[36] *Green Paper*. See endnote 34, page 6.

[37] *Green Paper*. See endnote 34, pages 14-15.

[38] European Commission. *Communication from the Commission on the Reform of the Common Fisheries Policy - Roadmap*. COM (2002) 181 final. Brussels. 2002. page 11. Available at http://europa.eu.int/comm/fisheries/reform/roadmap_en.htm.

[39] There were some exceptions, for details see Council Regulation (EC) No 2369/2002. Published in the Official Journal of the European Communities Volume L 358, 31/12/2002 pp. 49-56. Available at http://europa.eu.int/comm/fisheries/doc_et_publ/factsheets/legal_texts/regl_en.htm.

[40] Currently there are such agreements with Norway, the Faroes and Iceland. The number of countries has decreased over the years as the EU expands.

[41] Details of current agreements are maintained on the website of the European Commission at http://europa.eu.int/comm/fisheries/doc_et_publ/factsheets/facts/en/pcp4_2.htm. The Commission contracted an exhaustive review of fisheries agreements during the period 1993 through 1997 and a summary of the final report is also available at that site.

[42] See endnote 6.

[43] There are 50 such countries. Further details can be found at http://www.unctad.org/Templates/Page.asp?intItemID=1676&lang=1.

[44] There are 38 such countries. Further details can be found at http://www.worldbank.org/.

[45] Possibly increasing to 3,500 GRT per month if certain conditions are met, including a scientific assessment demonstrating the "sound state of the stocks". The payment would then increase to €4.25 million.

[46] Agreements for tuna fishing do include a reference tonnage that is allocated under the agreement, but if the fleets catch more than the reference tonnage, more money is paid in compensation.

[47] Fish consumed in the EU is increasingly caught outside EU waters. Apart from third-country agreements, the EU also fishes on the high seas and imports large quantities from other countries.

[48] European Commission. La Toja 1991. Document 1 - Fisheries Agreements. Document distributed to a fisheries seminar in the European Parliament 20-21 June 1991.

[49] See endnote 8; also see Holt, Chapter 4.

[50] An international symposium was held in Senegal in 2002 to synthesize and analyze what data were available on the impact of fishing in West Africa during the previous 50 years. The presentations and conclusions are available at http://saup.fisheries.ubc.ca/Dakar/index.htm. The international organization responsible for providing scientific advice for the fisheries in the region is CECAF (Fisheries Committee for the Eastern Central Atlantic). Its documents are available at http://www.fao.org/fi/body/rfb/CECAF/cecaf_home.htm.

[51] World Bank. Senegal. Integrated Marine and Coastal Resource Management. Project Concept Document. Project ID PO58367. 2003.

[52] United Nations Environment Programme. Integrated Assessment of Trade Liberalization and Trade-Related Policies. A Country Study on Fisheries Sector in Senegal. Geneva. 2002. Available at http://www.unep.ch/etu/publications/CSII_Senegal.pdf.

[53] European Commission. Proposal for a Council Regulation on the conclusion of the Protocol defining for the period 1 January 2004 to 31 December 2008 the fishing opportunities and the financial contribution provided for in the Agreement between the European Economic Community and the Government of the Republic of Guinea on fishing off the Guinean coast. COM (2003) 765 final. Available at http://europa.eu.int/eur-lex/en/search/search_lip.htm

[54] Patricia McKenna. Report of the European Parliament on the proposal for a Council regulation on the conclusion of the protocol defining for the period 1 January 2004 to 31 December 2008 the fishing opportunities and the financial contribution provided for in the Agreement between the European Economic Community and the Government of the Republic of Guinea on fishing off the Guinean coast. Document A5-0164/2004. Available at http://www.europarl.eu.int/plenary/default_en.htm.

[55] One clear advantage of the EU's agreements, though, is evident by the very fact that they are so controversial - they are negotiated by a public authority, approved by the European Parliament and published in the EU's Official Journal. Their terms are thus a matter of public knowledge, even if the way in which they are administered is less transparent. Private agreements between EU ship-owners and third countries offer none of these characteristics. They are completely opaque and there is not even a list of them available in the public domain.

[56] *Green Paper*. See endnote 34.

[57] EU Council conclusions, 19 July 2004. Available at http://ue.eu.int/ueDocs/cms_Data/docs/pressData/en/agricult/81505.pdf.

[58] In July 2005, Morocco and the EU agreed to another bilateral fisheries agreement, to last for a period of four years. This one is more modest than the one that existed from 1995-1999. See press release of 28 July 2005 at http://www.europa.eu.int/comm/fisheries/news_corner/press/inf05_38_en.htm.

[59] Articles 28 and 31 of the Doha Ministerial Declaration. Available at http://www.wto.org/english/thewto_e/minist_e/min01_e/mindecl_e.htm.

[60] Information on developments at the WTO on fisheries and other environmental issues may be found in the reports published by the International Centre for Trade and Sustainable Devlopment (ICTSD), including their weekly summary "Bridges" available at http://www.ictsd.org/index.htm. National submissions to the WTO are available at http://docsonline.wto.org/gen_home.asp?language=1&_=1.

[61] The group includes Argentina, Australia, Chile, Ecuador, Iceland, New Zealand, Norway, Peru, the Philipines and the United States.

[62] All negotiations of the Doha Development Round have been extended, given the lack of progress on many fronts; it now looks as if little will be concluded before the end of 2006.

[63] Originally known as the Lomé Convention, signed in 1975, more recently renegotiated as the Cotonou Agreement, valid from 2000 to 2020.

[64] See http://www.acpsec.org/index.htm.

[65] At the time of writing the EU is negotiating its overall budget for the period 2007 through 2013, including its fisheries budget. Certain Member States are urging that fleet subsidies be reinstated, at least to a certain extent. Other Member States, as well as the Commission, are insisting that that cannot be allowed.

[66] For more on this topic, see reference in endnote 8 and Holt, Chapter 4.

[67] Regional Fisheries Management Organizations provide a means for countries fishing in an area of the high seas to discuss and adopt management measures (gear restrictions, closed seasons, capacity limits, etc.) for their vessels. There are many problems with the network of RFMOs, including regions not covered for certain species and the fact that the rules are not applicable to counties that have not joined the RFMO, but this is the means used by the international community. Links to the extant RFMOs can be found at http://www.fao.org/fi/body/rfb/index.htm.

[68] Whether the WTO should even be involved in these matters is open to question by some, who argue that the necessary consideration of environmental criteria to determine whether subsidies should be allowed, such as whether fisheries are over-exploited or not, could lead to the WTO basing its decisions on far wider criteria than its traditional competence of trade. See Roman Grynberg and Natallie Rochester. The emerging architecture of a WTO fisheries subsidies agreement and the interests of Developing Coastal States. Journal of International Trade. *In press.*

CHAPTER 16

THE FREE LUNCH: MYTHS THAT DIRECT CONSERVATION POLICY AND THE NATURAL LAWS THAT CONSTRAIN IT

Ronald J. Brooks

"Thus, from the war of nature, from famine and death, the most exalted object which we are capable of conceiving, namely the production of higher animals, directly follows".[1]

"Everyone has a theory of human nature".[2]

"The human mind is a device for survival and reproduction and reason is just one of its various devices".[3]

"Culture can exert an influence on what kind of survival is targeted – more people or longer life – but there are limits... The major gains of modern medicine have been alleviation of disease in infants, children and young adults... there is nothing to suggest that the function of human minds has changed since our species first appeared – beyond having access to an expanded cultural 'database'".[4]

A PREAMBLE THROUGH THE WASTELAND[5]

In 1859, Darwin[6] unleashed an argument that purported to explain not only the origins of all species but the process by which organisms obtained complex adaptations. Like all great ideas, Darwin's theories of descent with modification and natural (and sexual) selection have aroused a full gamut of intellectual response, from outrage and denial to unfettered admiration and support. Sadly, however, most of the 6.4 billion current human residents of Earth remain largely ignorant of, and indifferent to, Darwinian concepts and their implications for human behaviour.[7] In this chapter, I hope to illustrate that even among those who are alarmed by the potential consequences of the growing human biomass, economy, and consumption on the biotic and abiotic complexity that sustain us, there is widespread failure to account for or acknowledge humanity's origin and creation in Darwinian biology. At every level, humans manufacture myths that negate, deny, and avoid the established role natural selection played in the evolution of our behaviour. We speak fondly and sadly of past golden ages, lost innocence, noble savages, and the interwoven complexity and connectedness between ourselves and the rest of nature. We wax enthusiastically about the inevitability of progress, the lack of limits on humanity's great journey, the infinite capacity of the human mind to solve problems and expand our spirituality, and the necessity and certainty of never-ending economic growth. Such optimistic forecasts and beliefs arise from a deep faith that our pleasures have no real cost, at least in the long view. This certainty is strongly embedded in the myth that there is a free lunch. In this chapter, the central argument will be the claim that social and cultural forces derived from evolutionary history and natural selection have led humans to prefer short-term gains regardless of long-term costs.

My thesis is that the myth of the free lunch underlies all of our approaches to the environment. This myth is natural to all organisms as all are designed by natural selection, which is nothing more than present circumstances. Humans, much more than any other creature, can see into the future but, like weather forecasters, not very well. Nonetheless, we like to believe weather forecasts and, similarly, we express faith in our ability to forecast future environmental scenarios. In truth, we are not good at such prognostication nor at learning from or recalling history. We are the products of a process that records history, does not predict it, and is moulded only by the present.

Central to my argument will be the claim that humans have enormous reluctance to give up their rights and pleasures or reduce their standard of living. Hence, we are highly susceptible to arguments and solutions that let us avoid immediate deprivation regardless of potential long-term consequences. Sustainable developments, clever technologies, "miracle" substances and win-win solutions are eagerly and uncritically embraced, especially when they do not require constraint on our lifestyle. We all adore the free lunch. A key question is whether this adoration has been carved into our psychological framework by our biological history.

In making my arguments, I will try to eschew the sometimes embarrassing penchant of environmentalists to wander piecemeal into grave warnings of planetary collapse and, equally, try to avoid optimistic pronouncements of endless bounty through application of human social skills and ingenuity, or "Popular Science" pronouncements of new technologies that will circumvent the laws of thermodynamics and ecology.

In a recent critique of Rees[8] and the ecological footprint, Fox[9] laments, "I am at a loss to know what Dr. Rees would have us do in a practical way". A good question; but narrow. A better form of the question is what are our "practical" goals in wildlife conservation and ecological sustainability? And we could ask too: Who are "we"? "We" is as important and difficult to define as what we want to do.[10] In this chapter, I ruminate on these questions. My thesis is simple and, in the style of the ancient Chinese mystery, I will lay my essential cards on the table at the outset. "We" are created by the whimsical forces of evolution: natural selection and random effects. These forces have made us to be like all other organisms with the added feature of consciousness, reason, self-awareness, morality, or humanity, whichever of these nouns pleases the reader. "We", in our more thoughtful manifestations have made "humanity" the key feature of our existence and so it may be in selected moments. However, I will vouchsafe that "humanity" is also an outcome of an evolved process and its output is as much dedicated to survival and reproduction as the digestive system is to securing nourishment. Therefore, even when we try to save the environment, we must always do so from our perspective, a perspective honed across 4 billion years of struggle and competition to maximize our genetic survival. As Dawkins[11] has noted, one thing that all our billions of direct ancestors had in common is participation in an unbroken line of successful reproduction. Whether success was relatively wide-ranging as with say, Genghis Khan, or trivial as with an anonymous failure, we will all continue to struggle to a greater or lesser degree to modify the world in our chosen image, an image that maximizes our well-being in the immediate or, at least, short-term sense. There can be no moral position here, only a path chosen by self-absorbed organisms.

In the ensuing argument, I will focus initially on three great self-serving concepts (myths that derive from the free lunch) and try to illustrate their power through their metaphorical and direct connections to our concept of gardening. Gardening is not just an explicit example of our need to manipulate our surroundings, but also is deeply connected with our ideas about conservation and I don't believe we can talk sensibly about conservation goals without examining such connections.[12] Regardless of species, some form of environmental manipulation occurs, in the sense of Dawkins' notion of the extended phenotype.[13] Conservation as the altruistic preservation of other species doesn't exist in any species or human culture that I am aware of.[14] Conservation in an aesthetic sense or a sense of utilitarian self-interest – for example saving rainforests for their potential to produce useful drugs – does exist in many cultures and in some other species; ants and aphids spring to mind. With the huge "success" of agriculture and domestication, the concept of gardening is now expressed in global culture in many forms, but ultimately and transparently as pure self-interest. Whether we grow crops for food, industry, aesthetics or simple pleasure, we recognize that the final and only arbiter is "us". Thus, we must approach conservation similarly to any other survival activity, which is to say as a product of the human social construction of nature.[15] As always, we operate from a position of self-interest and use politics, power, and argument to achieve goals we have rationalized. The details of these goals are culturally derived, but the process of justifying their importance and struggling to achieve them is part of our biological heritage and the drive to extend our phenotype and, more importantly, to raise our genetic fitness.

If the above were the entire argument, then an answer to what we should do could at least be approached. We could, for example, use our vaunted powers of reason to develop criteria for what we want to consume, and then extend this exercise in the context of a definition of ecological sustainability, and then use science to model the Earth as a global garden. In fact, most people approach conservation in precisely this way and almost any discussion of environmental concerns is predicated on optimistic assumptions of our rationality and the prospect of logical solutions that involve merely altering our behaviours and adding a technological miracle or two. But, it seems to me, this approach is too accepting of our ability to use reason and objectivity. As several of the endnotes in this paper indicate, we are far more likely to abandon rational approaches and embrace myths whenever outcomes are unattractive or undesirable, which is virtually always the case.[16] The free lunch is the unshakeable belief of humanity that we can get benefits without cost and that such achievement is part of the natural order of things. As we surge forward with new technology, we get better medical procedures, computers, automobiles, and appliances. We recognize that these have costs but readily accept that these costs can be circumvented or reduced, and we fervently believe that each of these material advances is not only necessary to our well-being, but also must be enshrined in human rights globally. Eventually, we must all be able to own a motor vehicle, have affordable access to organ transplants, palatial housing and unlimited gourmet food that will not lead to obesity. Such dreams typify the mythology of the free lunch.

At the core, although we are Darwinian organisms, we have a penchant for social cooperation. As apes with speech and self-awareness, we have fashioned four great achievements that have elevated our share of the energy pie to great heights compared to expectation. These four accomplishments are control of fire, tools, agriculture, and industrialization. But physically and mentally we are not much different than we were back in late Paleolithic times (circa 50,000 to 15,000 B.P.) and, like all other animals, we still crave the supernormal stimulus.[17] As Dawkins pointed out in his popular explication of the Darwinian paradigm,[18] despite all our cleverness we went through more than 99.99% of our history before someone figured out how we came to be. Before Darwin, the jumbles of myths that accounted for our existence were as numerous as the cultural variations that existed. These myths were usually, I assume, based on common sense, logic, knowledge, culture, reason, hope, and fear of death and the unknown. It's salient that these myths concerning our origins were all also incorrect, usually wildly so. And perhaps none of them proposed a mechanism to explain "us" that is as thoughtless and cruel as natural selection.[19]

Today, we continue to battle in a cultural kaleidoscopic jacuzzi of myths on all subjects, myths upon which the cold blade of science makes only feeble scratches. The first of the aforementioned three great myths that underlie the free lunch and form the interface of our natural and messy entanglement with our surroundings, including all other life forms,[20] is that economic growth is beneficial, progressive, perpetual and necessary.[21] This belief is modern and has spread rapidly in recent decades to infiltrate every aspect of global culture so thoroughly as to be largely accepted as a fundamental truth. This faith in growth is to be expected in a Darwinian organism devoted to maximizing survival, entropy production (energy transformation) and reproductive output, and reinforces the truism that economics is biology.

The second myth is more subtle and less explicitly empirical than the growth myth. This second myth embraces a belief in an ideal past, an age of innocence whose noble savage came to the big city and was corrupted by modern values that reflect the growth ethic, among other evils. Purveyors of this second myth ardently support a no-growth philosophy, in word if not in deed.

These world views clashed until we negotiated a compromise that wonderfully allowed both old myths to be simultaneously denied and maintained – deals made in paradise – and this new (and third) myth was called "sustainable development". Sustainable development allowed humankind not only to continue business as usual, it encouraged us to continue to expand our accomplished expression of Dawkin's "extended phenotype",[22] which describes our manipulation and modification of our surroundings, and other genotypes, to adapt the "environment" to our own needs and desires. And given that the laws of thermodynamics are as obscure to all humanity as the law of natural selection, sustainable development also encourages us to continue to accept the myth that there are things that are highly desirable and have no cost. This notion of the free lunch is more generally referred to as paradise. Where once our high mortality rates and limited technology pushed paradise to the afterlife, our recent scientific and technological success have produced a culture that demands paradise in the here and now. In essence, sustainable development can be translated as a version of our Darwinian belief in the free lunch.[23] And woe betide the Jeremiah who would deny decent people the right to paradise.

The impact (***I***) created by human numbers (***P***) and technology (***T***) varies according to a "real" measure in relation to resources and according to a less concrete measure in relation to "culture", which in the global perspective boils down to wealth or affluence (***A***). Hence the famous equation ***I* = *PAT***.[24] Some cornucopian optimists, like Julian Simon,[25] would like to add reason (***R***). Thus ***I* = *PART***. In the optimistic view, ***R*** and ***T*** will be salvation; in the pessimistic view all four components are our formula for disaster.[26] In a realistic view, the future will not be what we have forecast (to quote baseball legend, Yogi Berra, "The future ain't what it used to be").[27] Regardless, only we humans can judge the outcome as good or bad and accurately predicting the details is largely beyond our capacity. Our biological heritage and the laws of nature will continue to determine the possibilities as they always have. It is indisputable, however, that no outcome will satisfy all of us. The world will bear a major imprint, as it already does, from the vastly extended human organism. Many people despair that it is somehow "unnatural" that humans transform the planet, but this argument does not bear scrutiny. Many times in the past the planet has been transformed by living creatures,[28] most notably by photosythesizing organisms that oxygenated the Earth making large, complex, multicellular life possible. Throughout this chapter, I will be using ideas and pointing out facts that are well established and widely known. The crux of my thesis is that these ideas and facts are ignored when they lead to unpalatable and pessimistic conclusions.

IS THERE A DARWINIAN BASIS FOR CONSERVATION BEHAVIOUR?

As noted at the outset, Charles Darwin outlined two great theories. The first, descent with modification, claimed that all life forms descended from a common ancestor to form the great diversity of organisms occurring today. It is often claimed that more than 99.9% of all species have gone extinct[29] and some would argue that we are entering a new phase of exceptionally high extinction rates, dramatically termed the sixth extinction.[30] In the geological time scale, this potential extinction is interesting but hardly qualifies as a biodiversity disaster, because after other big extinctions, biodiversity bounced back and rose to even greater heights than before.[31] These earlier extinctions were of course the product of natural causes such as comets, volcanic activity, climate change and life forms themselves. According to Turner, "The greatest ecological disaster of all time occurred about two and a half billion years ago, when certain bacteria, blue-green algae [these days referred to as cyanobacteria] specifically, learned photosynthesis" and changed the oxygen content of the atmosphere much to the chagrin of the previously dominant anaerobic microbes which then died off or were relegated to areas where anoxic slime and mud could persist.[32] Calling this event an ecological disaster is not a typical or popular perspective of this particular change, but certainly the parallels to the present are not difficult to see.[33]

Darwin's second theory, or law if you like, was natural selection. He proposed that there was selection of the most adapted forms, by which he meant "nature" favoured some individuals over others on the basis of the possession or lack of particular features. If these features were inherited then those possessing them would be more likely to survive and reproduce, thus passing on these features. Everyone knows this and, as Thomas Huxley very famously pointed out, "How stupid of me not to have thought of it".[34] Regardless, no one apparently thought of it throughout most of human existence, and this absence should convey an important message. There are some important bits to this theory that are germane here. First, Darwin realized that the idea wouldn't work unless there was some means for nature to make choices. Inspired by reading Malthus, he identified overproduction as the ubiquitous explanation for selection and carefully pointed out that not only is there huge mortality in nature, but that those deaths occur mostly in the young. We all know this too, yet like so many unpleasant facts of life, we bury this most fundamental property of ecology in romantic myth using words like integrity, connectedness and interdependence instead of near-genocidal mortality of infants, as key to understanding the secrets of Mother Nature. Thus, most people ignore the point made by Darwin at the start of this essay. Fragile nature is much more palatable than rampant, meaningless death.

Natural selection has properties and effects that should be readily understood by neoconservative thinkers and others who espouse a growth economy.[35] In natural selection, the present is all that really matters. Natural selection clearly does not operate on future or potential long-term gains but only on what is available now.[36] The past experiences of genotypes have created the present forms, so there is an important heritage (adaptation) from the past. Changing environments may render current adaptations useless. To quote the cartoon characters, Calvin and Hobbes, the secret to happiness is short-term, stupid self-interest. From the perspective of natural selection and neoconservative economies, this is an excellent aphorism to describe life.[37]

A second feature of evolution by natural selection is that it purports to explain not just adaptations like the heart, pancreas and giraffes' long necks, but also nervous systems, brains and behaviour. And it is in behaviour that people have a diversity of views that for all but the most hardcore genetic determinists lead to the separation of "man" from beast. Indeed, the objective of most of these views *is* to separate "man" from beast.

In broad terms, our view of human behaviour is often the blank slate, the claim that at birth we have no preformed propensities and that the blanks are filled in by experience. To paraphrase J.B. Watson, the famous behaviourist, we can make any child into anything by appropriate training. The opposite extreme is genetic determinism, the belief that all our behaviour is hardwired via evolution. The blank slate has dominated since about 1920 and only recently have sociobiology[38] and its child, evolutionary psychology,[39] restored to modest credibility the view that some of our behaviour has its roots in evolutionary history. The question I wish to address here is: given that we are organisms and have evolved by natural selection, then from where do we get the "ethic" to be stewards of nature? For example, is conservation behaviour an inherited proclivity or must it be taught; either to a blank slate or to a pre-wired system that has little or no wiring for sacrifice and concern for other species. It seems to me that the answer to this question is central to the development of any strategy that pursues ecological sustainability.

Recently, Paul Ehrlich[40] has argued forcefully that because we have "only 26,000 – 38,000 genes", there is a gene "shortage" in the human genome that he calls the final nail in the coffin of evolutionary psychology. "Details of our ethical evolution cannot be seriously constrained by genetic proclivities".[41] Ehrlich appears to subscribe to some sort of simplistic one-gene one-behaviour model when in fact the relationship of gene combinations and permutations and behaviour output are still very poorly understood but undoubtedly enormously complex. One wonders whether Ehrlich would have a different conclusion if the human genome project had found, say, 130,000 genes or whether he believes there is a threshold value at which genetic constraints force us to rely on good old-fashioned learning and cultural evolution. Regardless, Ehrlich concludes that we can turn anyone into a true and committed steward of nature, as evidenced by the wide diversity of cultures globally and their diverse approaches to nature. For example, at the Limerick forum, the approach to nature by particular religions (Buddhism, Jain for example) was mentioned as a foundation for an environmental ethic.[42] Where Ehrlich and I potentially have agreement is that there is no genetic program for conservation.[43]

A different perspective is given by Dennet[44] who notes that although we may not be genetic robots, we are descended from and built by genetic robots, as mindless as anthrax viruses. Most organisms work on the simple principles that you seek what you need and hide from those who need what you have.[45] Our minds were designed by natural selection to learn, largely through trial and error and then to anticipate when we encounter a familiar pattern. But our mental skills, like those of other organisms, are still limited largely to our experiences, historical and current. We all know we do not learn well from history and that we have poor powers of prognostication and that these powers decline rapidly the further we peer into the future. Prediction of the future fails because of unanticipated and inscrutable complexities of course, but also perhaps more significantly because our needs of the present cloud our judgement.[46] In our evolutionary past, the long-term future was utterly irrelevant.

Let me illustrate our confusion on these points by referring to the Easter Island parable so beloved by those who fear we are heading for ecological disaster.[47] Several authors have pointed out that earlier accounts (e.g. Heyerdahl's)[48] had shown a paternalistic or even racist attitude in explaining the statues on Easter Island as being constructed by a "superior" group (e.g. Incas or even aliens) because they could not accept that the current rabble on the island was capable of either making or moving these huge structures. Modern authors often make it clear that they harbour no such reservations.[49] The statues and the methods for moving them were the product of the Easter Islander.[50] But how was it that, as Shermer[51] has pointed out, these same brilliant builders and seafarers could not see that they were removing all the trees and local food animals when a person could traverse the whole island in a day and see these changes occurring? In other words, why could these people not foresee the catastrophe on the horizon when the evidence was present and plainly visible? Generally, those who use the Easter Island parable are careful to say that the Easter Islanders lacked our science and foresight and thus could not predict what could happen. This is the mandatory, optimistic and hopeful explanation of how we will avoid trashing the planet. Yes, the Islanders were capable of complex construction and cultural sophistication, but they lacked "our" insight into future consequences. I think not. There can be little doubt that the Islanders were quite aware of the repercussions of using all their resources, and went ahead and depleted them anyway.[52] We see similar behaviour in modern societies in regards to disappearing forests or fisheries.[53]

The reality is that if we want a resource and there is significant financial return from it, we will plunder it, at least to commercial extinction. Is this proclivity to plunder to extinction an inherent behaviour pattern? The rationale given to justify plunder is usually complex, but it boils down to self-interest, denial and a naïve belief in the myth of unlimited resources.[54,55] We continue driving our scarce resources to extinction because we still achieve that all-important, short-term gain from such behaviour. We do so because a resource is often worth more as it becomes rarer, because we deny the possibility of using all the resource, because we expect that we will be rescued by the market – Adam Smith's "invisible hand" – or God, because we believe that we are destined to use all the resources or, most incredibly, because resources become commoner the faster we use them.[56] Something will intervene – we will find more or find a substitute.[57] Hence, we find that for declining resources such as timber, oil or fish, the exploiters and cornucopians (fishermen, loggers and corporations) exaggerate supplies[58] or postulate infinite, new, undiscovered sources.[59] Again, the key element in this argument is that future resource limits were irrelevant in the evolution of our ancestors' behaviour.[60]

The basis for the blank slate view that we can make anyone a conservationist seems ironically parallel to the growth economists' denial of our dependence on natural resources.[61] I will give one example of the complexity of directing our behaviour via cultural change. Overpopulation has been a controversial part of the sustainability problem, with many arguing that the population explosion[62] is the bane of ecological sustainability and others arguing that more people are needed to provide markets and solutions to the problems caused by more people (double irony here).[63] Still others argue that trying to reduce population growth or to achieve zero growth is racist and a solution imposed on the developing, coloured races by the developed largely white race.[64] And of course there are the religious arguments against family planning, birth control and abortion. Many laud the so-called demographic transition in which increasing wealth and urbanization of societies is correlated with sharp decline in birth rates.[65] The explanations for this decline are many and varied (empowerment of women, cost-benefit, spread of birth control knowledge and technology, rational response to overpopulation to name a few). However, a recent study suggests that whatever the mechanism, the pattern is universal at least across mammals[66] and is readily predictable from well-known relationships among body size, metabolic rate and reproductive (and other) rates in mammals.[67] These relationships depend on allometric constraints on the structure and function of organisms. Moses and Brown[68] show that the predictive equations for these allometric relationships also predict how fertility rates change as energy consumption changes in human societies. The reason for these parallels can be debated. For example, I don't agree with the explanation offered by Moses and Brown.[69] However, the relevant point here is, "The products of agriculture, industry and technology are commonly thought to have freed modern humans from energetic and biological constraints... but... humans remain organisms constrained by energy. Per capita energy consumption strongly influences the behavioural and economic decisions that ultimately limit the sizes of families and the investment in rearing children".[70] Thus, it appears that we have no more "choice" in this matter than do other mammals, which are likewise constrained in numbers of offspring.[71] The ultimate constraint on all biological and cultural activity is energy. I will return to this subject when I discuss the free lunch.

DO WE HAVE AN INNATE CONSERVATION ETHIC?

Regardless of whether we want to argue that culture can mould us as readily as an advertiser's dream or whether we are hardwired robots, conservation advocates must address the following question: If we evolved as Darwinian organisms and are connected to all other life forms, why do we lack an innate, or at least universal, conservation ethic? The answer in evolution seems obvious. No species expends energy or concern over the conservation or preservation of other species, and indeed, Darwin once threw down the gauntlet to his critics by challenging anyone to find any species showing altruism towards other species.[72] Therefore, we should hardly expect that early humans would display a conservation ethic and a harmonious appreciation of nature with a vision to protect the long-term future. I could go further and say that closeness to nature makes us more Darwinian and less ethical, and even further, and say that any long-term vision, such as those held by most people today, would seek not to curtail our use of resources but to free us from the tyrannies of nature, by enhancing or maintaining our high rate of consumption, our unnaturally high rate of infant survival, our conquest of disease, predators and competitors, and the extension of our abnormally long life expectancy.[73] Whether these goals are ethical or not I am not sure, but it seems crystal clear that the paradox of biodiversity conservation is that it is utterly unnatural. Biodiversity increases and persists despite the frenetic efforts of existing species to exterminate all others by converting them into themselves. Who can pretend this is not what we are doing? In this light, conservation (or the preservation of biodiversity) can be interpreted as a supremely altruistic behaviour[74] but, for most people, it is probably no more ethical than when ants cultivate and defend aphids. We are much more likely to perceive biodiversity and to persuade others to preserve other species in the pragmatic context of self-benefit. Is it ethical to argue that we should reduce contamination in the oceans to maintain the health of those who eat fish. Of course not.

As pointed out in other chapters in this book,[75] conservation/wildlife organizations often start out with idealistic and altruistic objectives to protect wildlife. However, inevitably they gradually transform to institutions that differ little from conventional businesses or corporations. These organizations still have altruistic conservation objectives, but these objectives become infused with concerns for human welfare, confounded by conflicting impacts and increased interest in "practical" goals (e.g. IUCN – The World Conservation Union – had in its 2003 calendar more months with pictures of people than of "wildlife"). And, as Beder[76] discusses, these organizations also develop typical corporate goals and structures and often import corporate, not environmentalist, personnel to leadership positions (e.g. see Royal Botanical Gardens in Ontario, WWF Canada, and Bird Studies Canada). Eventually, the organization, its personnel and profits become the *raisons d'être* while the original objectives and recipients of attention change from being the end to being the means to support the personnel, shareholders and profits. Such scenarios could be predicted from a Darwinian view of conservation.

THE EXTENDED PHENOTYPE

At this point, I want to consider another Darwinian concept that goes a long way to describing not only our interaction with nature, but the character, basis and ineluctability of that interaction. In 1982, Richard Dawkins published *The Extended Phenotype*, a book that, like all his others, effectively popularized Darwinian notions. Dawkins expounded on the reciprocal nature of an organism's interaction with the "environment", a notion widely espoused by environmentalists, although they usually do so from a perspective quite different than classical Darwinism. Dawkins applies his "gene's eye view" (here, I will speak from the organism's view for ease of discussion) and describes how any organism would benefit if it could alter the behaviour of other organisms (e.g. in an intraspecific social context or in interspecific interaction, such as a parasite that manipulates a host to increase the former's chance of transmission) or its physical environment (e.g. an animal such as an earthworm[77] or beaver [78] that alters its physical and biological surroundings to its own advantage). This is a powerful and logical extension of a more passive version of Darwinism in which the organism adapts to the environment it gets and hopes that its adaptations are good enough. In the more dynamic, interactive concept of the extended phenotype, the organism also modifies the environment and, of course, to the modifier's own advantage.

Turner has written a fascinating account of physiological/biochemical examples of the extended organism.[79] Among Turner's examples, he spends an entire chapter on the ways that humble earthworms extend their influence to alter the soil so that the soil becomes a much more earthworm friendly habitat. Turner points out that earthworms are, like other annelids, essentially aquatic but, unlike most other annelids (most leeches and all polychaetes), earthworms live on land, in soil. However, their activities transform the soil in several ways that make the soil more like an aquatic habitat. One major alteration is that earthworms keep soil aggregates large thus slowing soil's inevitable decline toward smaller particle size. When the weathering process occurs without the action of earthworms, the size of the pores between particles declines and the strength with which soil particles hold water increases, such that water becomes less available to worms, plants and other organisms. By increasing aggregate and pore sizes in soils, earthworms can increase the ability of soils to absorb and release water several hundred-fold. Earthworms also speed up the release of nutrients for plants when they consume rotting vegetation. These activities not only "improve" the soil for our crops, they adapt the soil to suit earthworms that, as we recall from undergraduate biology, retain a physiology that requires plenty of easily available water.[81] And most pertinent to our conservation "ethic", earthworms are received favourably because they enhance agricultural output.

In a recent book, Larsen *et al.*[82] present a hypothesis on the origin and evolution of human habitats. They argue that for much of the last million years our ancestors thrived around cliffs and utilized caves for shelter from predators and climate. They present evidence and coherent arguments for their thesis, but I want to dwell on the last part of this story, i.e., the last 20,000 years when humans first began to modify cave dwellings, then expanded out onto the savannah and built structures that resembled caves. According to Larsen *et al.*,[83] "What accounts for our enormous success is that we have managed to propagate our entire habitat, including the species of plants and animals that originally shared the habitat with us... For those wishing to change the trajectory of the conversion of natural habitat into human-modified or human-built (habitats) all over the world, we believe that the task is daunting... If our hypothesis is correct, the past million years of biological evolution has included the evolution of our attitudes toward the building of homes or paradises for ourselves and our families. The form of this construction itself has not changed materially in the 10,000 years since the invention of the brick". This seems to me not only a clear exposition of the extended phenotype but also, yet again, that there is great inertia in our behaviour, cultural variations notwithstanding.

THE GREAT MYTHS OF THE ENVIRONMENT

I wish to consider, briefly, current global views of the environment and how to save and improve it. I will argue that there are two main views[84] – essentially the first two myths described in the "Preamble" – that have been outlined many times before. I will emphasize that whether you look at these views as political, economic or culturally based, they are indefensible from a rational or Darwinian point of view.[85]

The first view I have called the Utilitarian perspective. This is the received or mainstream vision of our global society, generally viewed as optimistic or cornucopian,[86] sometimes more narrowly defined as neoclassical (neoliberal, neoconservative or rational) economics or simply growth economics.[87] Whatever it is called, this approach to our society is based on economics[88] but goes well beyond economics to embrace all society; we could even say society is the free market.[89] Rising from the much revered bible of economics, Adam Smith's *Wealth of Nations*, the modern view of life assumes that humans are like Kant's devils, self-interested, but if properly directed by market forces, these devils can benefit society by wealth creation that will trickle down to society as a whole. Society is essentially a rational market governed by supply and demand, and the success of the creators of wealth is a tide that raises all ships including the poor and ultimately the environment. In this happy world, growth economics has captured the helm.[90] The essence of this received view is that more is better, a growing population is better, resources or their substitutions are as limitless as human creativity, and that all of this expansion leads to progress, whatever that may be.[91]

The received philosophy resides everywhere in modern culture and its information outlets, but especially in advertising where the arguments are made to reinforce the interests of those who control these instruments. But one can look to Lomborg[92] and Simon[93] for the most spirited and data rich popular defences of this philosophy in relation to environmental issues. Apparently, even serious economists take these works as professional.[94]

Growth is necessary and therefore must be perpetual. Even if this is not possible, it has become almost unimaginable for it not to be true. We are so dependent on perpetual growth that we know it must continue. There are two additional reasons for this belief to remain implacably part of our culture. Firstly, it is utterly Darwinian to borrow against the future and so we cleave to this practice with natural ease. Secondly, our adherence to never-ending expansion and increasing consumption and wealth is especially reinforced by the persuasion of advertising and the encouragement of economists who trumpet "In the longer run, the surest way to improve your environment is to become rich".[95] Such endorsements make it difficult to recognize that we are borrowing against the future and closely resemble "buy now, pay later" advertisements that specialize in disguising future costs. But ever-expanding economic prosperity ultimately comes from depleting resources not from adding to them. Yet there can be a clear Darwinian logic to the defence of growth, as expounded in the following quote: "Those who want to stop human (economic) growth will have limited appeal since this solution will banish us all to the tyrannies of nature".[96] This perspective of Darwinian nature raises the frightening spectre of food shortage, lack of shelter and creature comforts, and dangerous insecurity; in other words, the real Darwinian world. Prosperity, however generated, is used to build military, political and economic power to the benefit of those in control who then use this power, albeit not always effectively, to maintain the status quo. What is so hard to understand here?[97]

Contrasting the utilitarian view of the world is the rather less pragmatic perspective I have called Romantic. At its more extreme incantation, the Romantic sees nature as perfection. The legendary Garden of Eden, or paradise occupied by innocent, simple folk and noble savages who represent the Beautiful People Myth (BPM).[98, 99]

Recently, the Romantic view of our interface with the environment has shifted focus from saving wilderness and nature to the alleviation of poverty, misery, gender persecution, and other societal ills as currently perceived.[100] This new emphasis is of course to allow all people to achieve full moral potential (newspeak for the greatest happiness of the greatest number, or "choice" if we use the latest vernacular) and, just as in the Utilitarian's push for full economic potential, the Romantic predicts a tide that will ultimately raise the sinking ship of the beleaguered environment. Suddenly, we see that these two opposing philosophies could seek common ground and, as Beder[101] points out, the corporate mind has spotted the opportunity, the niche so to speak, and moved in. The common ground was staked out in the book *Our Common Future*[102] and, in classic Darwinian style, the proponents of both sides came up with a "win-win" solution. Essentially, they proposed that we could have our cake (development) and not only eat it, but eat it forever (sustainable) and in increasing amounts (still sustainable). There have been many discussions of sustainable development, but I think most would agree that corporate expertise quickly took control[103] and soon it was business as usual, again a Darwinian outcome, as short-term agendas and maintenance of the status quo prevailed.[104] Indeed, the perpetual growth mentality readily attracts advocates from outside mainstream economics (e.g. Bjørn Lomborg, Michael Crichton, and Matt Ridley), who take up the cudgel after seeing the light and, perhaps, the warm acceptance of the rich and powerful.

The Romantic view relies heavily on variations of the BPM.[105] Unfortunately, the available evidence suggests, rather one-sidedly too, that our aboriginal ancestors, whether neolithic, mesolithic, or modern, were not protectors of nature. The data on extinction on islands, of birds, mammals, and reptiles, of the losses of megafauna in North and South America, Australia and Eurasia, and more modern examples such as New Zealand and Easter Island, overwhelmingly show that while our ancestors were in tune with nature, the melody was the usual Darwinian dirge.[106] Our ancestors killed anything they could when they could; their actions were no more conservationist than the proverbial fox in the henhouse.

Indeed, what limited their exploitation was not philosophy, but technology and this statement applies equally to foxes and aboriginals. In the modern context, we see how technology has assisted modern aboriginals (i.e. fishermen, farmers, and other people who live close to nature) in depleting wild populations.[107] I don't think too many conservation biologists today adhere to the aboriginal as conservationist myth, although the BPM still holds great sway in the political landscape and in elementary education in the Romantic tradition. However, many environmentalists adhere to another form of the BPM, that which suggests that certain earlier cultures/civilizations (e.g. Mayan, Sumerian, etc.) or religious doctrines (Tao, Jain, Buddhist, Judaic, Hindu, Shintu, and Christian have all rated mention here)[108] are the keys to a sustainable accommodation with nature. Certainly, some modern religions have beautiful and empathetic views of our relationship to nature, but on inspection I believe it is correct to say that none seem to fare better than any other at saving non-human nature in anything approaching a pristine state.[109] To some extent, this lack of success is a product of population size, but more so it is because of the usual trade-offs that occur so that protection of nature is either overwhelmed by self-interest or reduced to a codified form of the extended phenotype (gardens, small forest patches).

And so we come to the third beautiful myth: sustainable development; the amazing marriage of perpetual growth with stasis, and substantially increased global prosperity. Therefore, we try to fashion modern conservation from the belief that resources are unlimited, either now or in the future, and can best be regulated by the market. This perspective has been welded onto the Romantics' notion that there was a Garden of Eden, occupied by a simple family-oriented culture that lived in harmony in a nature somehow free of predators, disease, food shortages, and even warfare in some cases. How would a Darwinian species resolve these two clearly opposed viewpoints? Well, if as a matter of self-interest neither view can be rejected, then we need growth to sustain our current wealth and to elevate living standards of the billions of humanity living in economic misery and to include the 2 or 3 billion additional persons coming into existence over the next half-century. The only way to do this is to triumphantly, optimistically, and fervently claim that wealth and a perpetual Eden are compatible because it is our right and destiny to be happy, healthy and prosperous (check agendas of all governments, constitutions of nations, most religions and NGOs). A significant basis for acceptance of this view of the future lies in our Darwinian belief in the greatest myth: the free lunch.

THE FREE LUNCH

The term "free lunch" originated some time ago in thermodynamics[110] to describe how one cannot violate the laws of thermodynamics, particularly the first and second laws. However, as many[111] have pointed out, people keep on trying to violate these principles, and keep on believing that it can be/has been done (think of cold fusion, perpetual motion machines, and other gadgets that are claimed to put out more energy than was put in).[112] In a recent book called *The Free Lunch*; science fiction writer, Spider Robinson,[113] created a tale of a place called Dreamworld where there was a free lunch, an antidote to entropy, a return to innocence where "the only sadness is when Dreamworld closes for the night". In other words, this is a novel about paradise, a place where roses have no thorns and don't succumb to black spot. The Dreamworld concept lies at the heart of all optimistic forecasts for our economy, standard of living and environment. After all, as the supreme creation of god/evolution or Intelligent Design, the recent attempt to duplicate sustainable development on the philosophical front, we deserve that future. There is no one who believes there is no free lunch.

The problem faced by sustainability resides in another law; the Law of Supply and Demand that Hardin[114] restated to reflect its Darwinian meaning as: Supply is limited, demand is not. This is a variant of the law that there is no free lunch. The supply/demand law is central to solving the problem of ecological sustainability. It is difficult enough when technological optimists like Bjørn Lomborg or Matt Ridley argue that economic growth is forever, but we also function under political systems that by and large are led by people whose success rests on promises of the free lunch – more money for health care, education, roads and everything else we desire, along with tax cuts/no new taxes and slashing the debt/deficit (see any election platform). The joy of democracy is that it encourages not only our belief in a free lunch, but also our faith that we deserve it as one of our human rights. The sadness of a non-democratic system is that the leaders, once established, don't usually have to persuade the rest of the people that there is a free lunch. Of course, politicians are not the only preachers of something for nothing. CEOs, salespeople of every stripe, and religious preachers, also extol a future free lunch, albeit the latter group tends to emphasize posthumous versions of cold fusion. An outrageous example of the promotion of a free lunch, often promulgated by modern economists, is that a disaster is good for the economy because it creates wealth and jobs. Recent specific examples are the Exxon oil spill in Alaska, the Québec ice storm, the war in Iraq, and the great tsunami of 2004. All of these have been touted as being of benefit to the economy and some "experts" actually argue that such disasters should be encouraged.[115]

An important consequence of the no free lunch law is also a law – the Law of Trophic Decline. This law is well known to ecologists, but not widely known by most other people. It can be stated thus: Every trophic level consumes at a net loss. The result of this loss is the famous ecological pyramid and the essence of why big fierce animals are rare.[116] It is also why these animals are getting rarer. The domestic cat (*Felis cattus*) can show both why lions, tigers, sharks, and other top predators, are vanishing and simultaneously how house cats are affecting global ecology and biodiversity. Consider that, "about 10,000 kg of prey supports about 90 kg of carnivore (mammalian) irrespective of body mass".[117] Thus to support a 5 kg cat requires about 500 kg of prey, all rabbit-sized or smaller. Cats usually get a free lunch, but unfortunately they still continue to hunt. Recent estimates place the number of birds killed annually by domestic and feral cats in the billions in Europe and North America. Not only do cats exert enormous predation pressure, they compete for prey with "natural" predators. And they can win this competition because when prey goes to low levels the cats still get their free lunch. There are still no detailed studies on the ecological impact of cats. This lack is largely because most people like cats and don't think about trophic levels and so songbirds are closely protected in North America, but not from cats. We worry far more about lighted buildings, oil spills, plate glass windows, and giant windmills as killers of birds. And these are serious threats to birds, but cats may exceed all other anthropogenic sources of songbird mortality.

A more direct example of our impact via consideration of trophic levels would be to contemplate having every Chinese person (~1.2 billion) consume an average of one additional chicken per year for lunch. If the chickens were grain-fed and they consumed 10 kg each before slaughter, we would need 10,200,000,000 kg of feed. This is about 40% of the entire Canadian wheat harvest. This lunch would have other expenses as well, but generally we would regard it as free or more likely of significant economic benefit to Canadian farmers and an improvement in the Chinese standard of living. These examples demonstrate our failure to perceive real costs. But, whether either example would have significant effect on wildlife sustainability is unclear.

The concept of the free lunch may be engrained in the behaviour of the other organisms, at least those capable of simple learning. B.F. Skinner found that rare, random reward patterns led to gambling addiction in pigeons, rats and humans at roughly the same rate. In 2002, 32 million Canadians spent $11.3 billion in gambling venues. The percent of Canadian households that gambled that year was weakly correlated with income (<$20,000, 60%, >$80,000, 80%). Rich or poor, we all believe in and seek the free lunch.[118]

How does the free lunch affect ecological sustainability? Well, it seems likely it has a Darwinian basis, perhaps something as simple as: selection doesn't favour the long-term strategy. Given this, then ecological sustainability may be difficult if we humans can actually destroy ecosystems. These statements are pretty straightforward. Where the free lunch gets more subtle and insidious is when environmentalists, being mere Darwinian mammals themselves, try to invoke the concept when advocating strategies for ecological sustainability. Let us consider a recent popular idea to reduce global emissions, and save the environment: biofuels. Neoconservative optimists claim these new fuels are wonderful because they are (or will be) cheap, clean and inexhaustible. We will never run out of oil but, just in case, we should develop a substitute. These fuels are derived from wastes such as cooking oils and crop stubble, or from crops (corn, soybean, sugar cane) grown specifically for fuel. Currently, these fuels are still not as clean or not as cheap as fossil fuel, but with technological improvement soon will be. And, "win-win", we will use up all that "waste" biomass, enhance profits for farmers, expand the economy, and still drive our cars without fear of global warming or running out of fuel. The inherently more pessimistic green view is surprisingly similar but sees biofuel as a necessity because this view has the gloomy prediction that we are running out of fossil fuels and causing global warming in doing so. But "win-win", we can solve these problems using "waste" and following the optimists' argument. And we can still drive our cars, albeit smaller more efficient ones.

It seems to me that these arguments fall perfectly in line with the something-for-nothing machine. There are two sources for biomass. The first is "waste". Waste? What happened to the ecosystem? Where do nutrient and organic matter come from? Indeed, what is oil and gas ultimately? Basically, we are saying the bank reserve is going down, so let's now spend the reserve we used to generate wealth so that we can continue to increase spending. Can one imagine a politician getting elected on such an economic argument? Well yes, if he phrases it differently. In fact, politicians love the biofuel/clean fuel proposals.

The second source of biomass is even more egregious and we call it surplus. We now have initiatives in both developed and developing countries to grow crops not for food, but for fuel, especially transportation fuel. Recently, in Ontario, a farmers' co-op group announced production of biofuel from crops. They gave "win-win" arguments similar to the above and attracted massive financial support.[119] However, life-cycle assessment shows huge environmental costs in erosion[120] and water. But, not only do we add to erosion, at the rate expected from agriculture, we also add further by using the "waste" stubble which would increase the need for fertilizer (which comes from fossil fuel) over that expected in no-till systems. And, we create competition for food prices. Does no one see where this is going? It is hard to envisage a more savagely Darwinian system. The rich maintain their lifestyles at the expense of the poor who will have to pay more for food in an increasingly competitive market as the population grows and food production is curtailed to support fuel production. And what is delicious is that we congratulate ourselves because we are reducing air pollution by using cleaner fuels. Real consumer restraint is not in the Darwinian cards.

THE LAW OF TRANSFORMATION

As an illustration of our adaptations to modify our environment, we can state that no human can resist trying to transform nature to make it "better" and, particularly, to make nature conform to that person's preferences. There are endless examples of this law in action. Suffice it to say that almost nowhere on earth have we not transformed nature. In the southern part of my province of Ontario, no place is more than one kilometre from a road, and it is highly doubtful that a single square meter has not been altered. Nevertheless, this is in Canada, which is often thought to be wilderness or "under populated".

The ubiquity of the law of transformation becomes apparent when we consider environmentalists and so-called restoration ecologists. If anyone should be immune to fiddling with nature it should be outdoor enthusiasts. But instead they are clearly bent on controlling, manipulating and altering nature in myriad ways from restoring it to some earlier state, planting trees, eliminating exotics, controlling fire, culling "unbalanced" predator-prey systems, eliminating disease and so forth. Today, many people argue that our distant ancestors, including Native Americans, altered nature extensively and even created the Amazon as a "garden".[121] Whether this latter claim is true or merely the fantasies of aficionados of the BPM, it seems certain that aboriginals did alter environments with gusto, especially by fire.[122] Even when we go out to experience nature, we cannot resist altering it to make it safer, more convenient or "better" in some way as one can see by the ongoing domestication of vacation and tourist destinations.[123]

A universal example of the law of transformation is gardening. Gardeners often claim to be closest to nature, and like to label themselves true ecologists or even conservationists. Although there is a kernel of truth in such claims, it is also among gardeners that we find behaviour that is antithetical to wildlife conservation. Gardeners attempt to destroy competitors, parasites, herbivores, carnivores, saprophytes and autotrophs from every organic kingdom. As Pollan[124] says, in their fantasies, gardeners use rifles, explosives and the most lethal toxic chemicals to destroy their enemies. This is not conservation, but it is definitely Darwinian. And gardeners go further than mere killing. They also import and foster exotics, transform and import soil, and excavate, unsustainably I might add, ancient peat bogs. They encourage every known form of manipulation of DNA from artificial selection, to hybridization, to cloning, to genetic modification. Monsanto is merely a corporate step in the gardener's pursuit of transformation of nature. And to satisfy our penchant for the supernormal stimulus, the gardeners' mastery of nature is proudly, competitively displayed in grotesque forms of their subject's gigantism or colourful, perfumed and, of course, bloated reproductive organs. As Sidney Holt observed during the Limerick meeting, the future of GM foods and other organisms will either be trivial or will transform our world. Regardless, neither possibility will stop gardeners from exploring all possibilities any more than the horror of nuclear weapons stops their construction and proliferation. To me, the seeming paradox of "love" of nature, which endeavours to alter and destroy natural relationships and forms, is central to the struggle for ecological sustainability.

In extending our phenotype, humans manipulate environments and these environments select humans, and our domestic plants and animals. The natural world evolves with us and changes because of these reciprocal interactions. As David Castle says, "Destruction of the environment is not a paradise lost, but a selective environment changed".[125] I would add that part of this change occurs because of our relentless pursuit of the free lunch. In this quest for a secure, disease-free, fountain of youth, paradisiacal existence, we continue to believe that we can do all these things without associated costs. We even accept implicitly or explicitly that it is our manifest destiny to achieve Eden through god's will or at least through technology. In such a Dreamworld, it becomes hard to define Castle's selective environment. Pollan[126] also develops this argument of exchange by taking the view of non-human organisms, and suggests that one could argue that our domestic plants and animals have manipulated us. For example, he documents changes in marijuana and potatoes as instances of changes not produced by "rational", artificial selection, but by enhancement of the Darwinian success of the plants.[127]

These perspectives raise the question whether our goal in ecological sustainability is subtly becoming nothing more than maximizing biodiversity while optimizing our extended phenotype using the sustainable development "model". The Darwinian choice, the extended phenotype with reduced biodiversity, is by far the more likely scenario of a sustained system because Darwin always trumps the pretenders. When we look at the global picture and imagine all those complex food webs pictured in ecology textbooks, we can now envisage a new simplified global picture with humanity at the top. Already, nearly half of all solar energy captured by life goes to humanity's extended phenotype (direct use and to our domestic species) and half going to the rest of the biosphere. All the stored energy from the past goes to us as well. Given 2-3 billion additional people, expanded technology and expansion of our domestic use of other species for food and medicines, the fraction of the pie for "nature" seems destined to continue shrinking. Given the lack of solid evidence that simpler ecosystems are not as productive nor as stable as complex ones, it seems that expansion of our part of the global energy pie is the choice solution for humankind.[128] Regardless, it is the choice we have made and will continue to make.

CONCLUSION

Energy is everything and all life has evolved in no particular direction other than to maximize the capture and conversion of energy into more life. These activities lead to what we call biodiversity but, from the perspective of the individual organism, the only life that counts are copies of itself, and biodiversity is merely fodder for this self-production. Thus, the recent reverence for the interconnection of all life forms that is revealed by modern evolutionary and genetic theory is based on a mistaken extrapolation of these theories. Connectedness only leads to cooperation with close relatives and even then not all the time. Darwinism is usually red in tooth and claw, and leads to inequalities. Darwin could live with this view of life, but most others ignore it or pretend it isn't so.

Any reasonable consideration of evolution by natural selection encourages the conclusion that conservation, *sensu* altruistic preservation of other life forms, will neither evolve nor be sustained. Therefore, conservation of biodiversity should not be in any sense an innate behaviour, and ultimately it must be defended in pragmatic not altruistic terms. The difficulty with a pragmatic defence is it must argue that long-term benefits are more important than short-term benefits. Such arguments are counterintuitive to organisms evolved from the absolute precedence of immediate payoffs. Hence, we find that economics, which is essentially a branch of biology, is founded on profits and, in particular, immediate profits. Those who make the most profits tend to be those who are in power and who are, of course, loath to relinquish power. Certainly, today we are enmeshed in a political system founded on biological, Darwinian economics, a system that is all gain and no sacrifice. Anyone who suggests real sacrifice, especially for other species, is immediately attacked, or worse, not taken seriously. Conservation that is based on altruism is unnatural and must be viewed as a moral position.

From the above argument and given the unusual wealth of a significant minority, some have tried in the past few decades to engender support for an ethical view of ecological sustainability. But this ethical view is, like any form of altruism, difficult to sell successfully to those who perceive themselves as deprived, a perception that effectively encompasses all members of our species. Hence, the creation of further myths that draw on the evolved responses of humans to non-threatening entities. First, we portray nature as fragile, benign, and nurturing. When that myth is not enough, we play another myth; that nature is necessary to our own self-interest, including that of our descendents. Nevertheless, these myths also largely fail because in the end we need to see the money. Scenery and butterflies look good in calendars, but what is in it for us? Surely a fast car, good wine, and high-class entertainment is better. Once this conflict is evident we graduate to the myth of sustainable development that promises both butterflies and Bentleys, and the free lunch explains how it will work. Paradise is both the here and the hereafter.

Rather than look at myths of perpetual growth, beautiful people, and free lunches, we need to concoct better stories to see where we came from and are likely going. The extended phenotype describes humankind's greatest adaptation. In the past, manipulation of the environment was more constrained by technology. But, with our discovery of the control of fire and later of the control of fossil fuel energy we have become a pervasive influence. The urge to manipulate nature is as ubiquitous as our urge to manipulate one another. Domestication of plants, animals, and ourselves, modification of the landscape, construction of buildings and so on have no limit in human motivation or imagination. Whether we are gardening, farming, or simply occupying an area, we automatically classify and codify nature and always, as best we can judge it, for our own benefit.

There is nothing that I can see to suggest our brains have become more rational or insightful over the past 100,000 years except to have, in the most recent millennia, a greater cultural database. Our power has expanded, but there is no evidence of progress toward purposive ends except wealth, security, and eternal life. All these goals conflict with the pursuit of ecological sustainability.

> *Common sense tells us that life – like the sun – revolves around ourselves.*
>
> *The idea has but one fault, it is wrong.*[129]

NOTES AND SOURCES

[1] Darwin, C.R. 1859. *The Origin of Species*. London: Murray. p. 459.

[2] Pinker, S. 2002. *The Blank Slate: The Modern Denial of Human Nature*. New York: Norton.

[3] Wilson, E.O. 1978. *On Human Nature*. Cambridge MA: Harvard University Press.

[4] Caine, D.B. 1999. *Within Reason: Rationality and Human Behaviour*. New York. Vintage Books: Random House Inc. p. 297.

[5] In *Wasteland*, I am referring to the work of T.S. Eliot, epic yet obscure and complex. The roots of our attitudes and efforts towards conservation are equally convoluted, a witches' brew of myth, biology, social creation, and aesthetics, sometimes

sublime, but more often arcane, maudlin and blatantly narcissistic.

[6] Darwin, C.R. 1859. *The Origin of Species*. London: Murray.

[7] Surveys of attitudes and beliefs concerning evolution have consistently shown that fewer than 10% (in North America) believe that evolution explains human origins. Only a minority of that group have any understanding of evolutionary mechanisms.

[8] Rees, W.E. 2004. Waking the Sleepwalkers – Globalization and Sustainability: Conflict or Convergence. In W. Chesworth, M.R. Moss, and V.G. Thomas (Eds.). *The Human Ecological Footprint.* The Kenneth Hammond Lectures on Environment, Energy and Resources 2002 Series, Faculty of Environmental Sciences, University of Guelph. pp. 1-34.

[9] Fox, G. 2004. Environmentalism, Economics and Social Cooperation: Commentary on the 2002 Hammond Lectures. In W. Chesworth, M.R. Moss, and V.G. Thomas (Eds.). *The Human Ecological Footprint.* The Kenneth Hammond Lectures on Environment, Energy and Resources 2002 Series, Faculty of Environmental Sciences, University of Guelph. pp. 141-167.

[10] Also see Wright, R. 2004. *A Short History of Progress.* Toronto: House of Anansi Press Inc., for a variant of this question.

[11] Dawkins, R. 1995. *River Out of Eden*. London: Weidenfeld and Nicolson.

[12] Pollan, M. 1991. *Second Nature: A Gardener's Education.* New York: Atlantic Monthly Press; Pollan, M. 2001. *The Botany of Desire: A Plant's-Eye View of the World.* New York: Random House Inc.; Freyfogle, E.T. 2004. Conservation and the Lure of the Garden. Conservation Biology 18: 995-1003.

[13] Dawkins, R. 1982. *The Extended Phenotype*. Oxford: W.H. Freeman.

[14] Pinker, S. 2002. *The Blank Slate: The Modern Denial of Human Nature.* New York: Norton; Brown. D.E. 1991. *Human Universals.* New York: McGraw-Hill. Pinker examines our ideas on the origins of our behaviour. He refers to D.E. Brown's list of over 300 behaviours common to all studied human cultures. It is surely noteworthy that conservation and related behaviours are not on these extensive lists.

[15] Evernden, N. 1992. T*he Social Creation of Nature.* Baltimore MD: The Johns Hopkins University Press.

[16] Gardiner, M. 1952. *Fads and Fallacies in the Name of Science.* New York: Dover; Park, R. 2000. *Voodoo Science: The Road from Foolishness to Fraud.* New York: Oxford University Press; Rose, M.R. 1998. *Darwin's Spectre: Evolutionary Biology in the Modern World.* Princeton NJ: Princeton University Press; Sagan, C. 1996. *The Demon Haunted World: Science as a Candle in the Dark.* New York: Random House; Shermer, M. 1997. *Why People Believe Weird Things: Pseudoscience, Superstition, and Other Confusions of our Time.* New York: W.H. Freeman and Co.; Shermer, M. 2002. *Borderlands of Science: Where Sense Meets Nonsense.* New York: Oxford University Press.

[17] Tinbergen, N. 1953. *The Herring Gull's World.* London: Collins.

[18] Dawkins, R. 1976. *The Selfish Gene.* New York: Oxford University Press.

[19] However, our ancestors had ample evidence that the gods played dice with their faiths, as the recent Asian tsunamis have reinforced.

[20] Brooks, R.J. 2001. Earthworms and the Formation of Environmental Ethics and Other Mythologies: A Darwinian Perspective. In W. Chesworth, M.R. Moss, and V.G. Thomas (Eds.). *Malthus and the Third Millennium.* The Kenneth Hammond Lectures on Environment, Energy and Resources 2000 Series, Faculty of Environmental Sciences, University of Guelph. pp. 59-92; Brooks, R.J. 2002. Sustainable Development. In W. Chesworth, M.R. Moss, and V.G. Thomas (Eds.). *Sustainable Development: Mandate or Mantra*? The Kenneth Hammond Lectures on Environment, Energy and Resources 2001 Series, Faculty of Environmental Sciences, University of Guelph. pp. 119-138.

[21] See Czech, Chapter 22.

[22] Dawkins, R. 1982. *The Extended Phenotype.* Oxford: W.H. Freeman; Turner, J.S. 2000. *The Extended Organism: The Physiology of Animal-Built Structures.* Cambridge MA and London UK: Harvard University Press. The concept of the extended phenotype has been elegantly applied by J.S. Turner in his book *The Extended Organism.* I will refer to Turner's ideas several times in this paper as they are central to my perception of human effects and interactions with our biotic and abiotic surroundings, and the constraints of physico-chemical laws on these interactions. I strongly recommend Turner's discussion on earthworms (As the Worm Turns, Chapter 7. But also see: Dawkins, R. 2004. Extended phenotype – but not too extended. A reply to Laland, Turner and Jablonka. Biology and Philosophy 19:377-396. In this paper, Dawkins expresses some reservations about the use of the Extended Phenotype, reservations that may or may not apply to my examples.

[23] Hardin, G. 1993. *Living Within Limits: Ecology, Economics and Population Taboos.* New York: Oxford University Press. "Three great historical paradigms have been identified and labelled. These are the golden age, the endless cycle, and the idea of progress". p. 23.

[24] ibid. I originally encountered this equation in Garrett Hardin's classic book on growth, but it has been around much longer. However, Hardin's discussion of it is incisive and stimulating; also see Willison, Chapter 2.

[25] Simon, J.L. 1996. *The Ultimate Resource.* Princeton NJ: Princeton University Press.

[26] ibid.; Budiansky, S. 2002. How Affluence Could Be Good For the Environment. Nature 416: 581; Ehrlich, P.R., and A. H. Ehrlich. 1990. *The Population Explosion.* New York: Simon and Schuster; Fenech, A., B. Taylor, R. Hansell, and G. Whitelaw. 2001. Major Road Changes In Southern Ontario 1935-1995: Implications for Protected Areas. Report to Environment Canada. Science Horizons Youth Employment Strategy. 14 pp. Available at http://www.utoronto.ca/imap/papers/road_changes.htm; Hardin, G. 1993. *Living Within Limits: Ecology, Economics and Population Taboos.* New York: Oxford University Press; Lomborg, B. 2001. *The Skeptical Environmentalist.* Cambridge UK: Cambridge University Press; Ridley, M. 2001. *Technology and the Environment: The Case for*

Optimism. Prince Philip Lectures. Lecture delivered at the Royal Society of Arts. May 8th, 2001. London UK. In a letter to Nature, Budiansky (2002) argued that environmental impact is not proportional to Population times Affluence as claimed by Hardin (1984) and Ehrlich and Ehrlich (1990), among others. Budiansky pointed out that population has grown as wealth has increased while the area of agricultural land has remained about the same. Therefore, he concluded, perhaps somewhat facetiously, that impact varies with Population divided by Affluence. Like many who have made similar claims that wealth improves the environment (Ridley, 2003; Lomborg, 2001; Simon, 1996), Budiansky ignored the impact of obtaining resources required to increase agricultural productivity. For those who suggest that North American land use is unchanged since 1900, I suggest they look at a few road maps (D.G. Fenech *et al.*, 2000).

[27] Nelson, K. 1982. *Baseball's Greatest Quotes: The Wit, Wisdom, and Wisecracks of America's National Pastime*. New York: Schuster.

[28] Darwin, C.R. 1882. *The Formation of Vegetable Mould Through the Action of Worms: With Observations on their Habits*. London: Murray; Raup, D.M. 1991. *Extinction: Bad Genes or Bad Luck?* New York: W.W. Norton; Turner, J.S. 2000. *The Extended Organism: The Physiology of Animal-Built Structures*. Cambridge MA and London UK: Harvard University Press.

[29] Raup, D.M. 1991. *Extinction: Bad Genes or Bad Luck?* New York: W.W. Norton.

[30] ibid. According to Raup, "A palaeontologist invariably says there have been five mass extinction events known as the 'Big Five'". p. 65. It seems almost a given that extinction rates will gather steam over the 21st century. However, I think it safe to say the term "sixth extinction" is an example of the creation of an environmentalist myth. I am not sure who created the term, but I first encountered it in the title of Leakey and Lewin's book: Leakey, R., and R. Lewin. 1995. *The Sixth Extinction: Patterns of Life and the Future of Humankind*. New York: Doubleday; for more on the sixth extinction, see Willison Chapter 2.

[31] Raup, D.M. 1991. *Extinction: Bad Genes or Bad Luck?* New York: W.W. Norton.

[32] Turner, J.S. 2000. *The Extended Organism: The Physiology of Animal-Built Structures*. Cambridge MA and London UK: Harvard University Press. This is a wonderful statement for two reasons. First, it reminds us that other organisms can profoundly alter the ecosphere and global biodiversity as a "natural" (i.e. Darwinian) process exterminating other species out of pure selfishness but without malice. Second, it places the term "ecological disaster" in proper context as a cultural term.

[33] Ward, P.D. 2004. *Gorgon: Palaeontology, Obsession, and the Greatest Catastrophe in the Earth's History*. New York: Viking Penguin. Peter Ward suggests the anoxic gang got payback during two of the big five extinctions (the end of the Permian and the end of the Triassic). He speculates, on somewhat skimpy evidence, that during these periods, low water levels exposed vast areas of anoxic mud which were then reduced by oxygen which in turn led to a drastic decline in global O_2 levels and asphyxiated most of the world's life forms. My point, however, is that extinction caused by life itself is not unnatural.

[34] Rosenberg, A. 2000. *Darwinism in Philosophy, Social Science and Policy*. Cambridge UK: Cambridge University Press. p. 173.

[35] Dawkins, R. 2003. *A Devil's Chaplain: Reflections on Hope, Lies, Science, and Love*. New York: Houghton Mifflin Co. "As Adam Smith understood long ago, an illusion of harmony and real efficiency will emerge in an economy that is dominated by self-interest at a lower level. A well balanced ecosystem is an economy, not an adaptation" (p.226). It is well known that Darwin read and admired Adam Smith's *Wealth of Nations* (1776; reprinted 1976). Many economists claim that their discipline "discovered" Darwin's theory. Perhaps, the question is: which comes first, neoliberal economic theory or the theory of natural selection?

[36] A second question pursuant to that above might be: Is modern economics any better than natural selection at forecasting the future? Recent studies comparing economists, school children and random guesses in forecasting future stock prices suggest the winner is the random guess. My favourite quote on this topic is: "Economic forecasting is like weather forecasting, but without satellite photographs." (National Post, Dec. 2004).

[37] Vermeij, G.J. 2004. *Nature: An Economic History*. Princeton NJ: Princeton University Press. Vermeij has written perhaps the most interesting book I have read on the similarities between evolutionary theory and economic theory. He notes that to the evolutionist, performance is expressed as fitness and to the economist it is expressed as profit. Both embody the notion of overproduction or surplus. And as Vermeij (p. 2) explains, it is not efficiency that matters most in either case, it is power, the amount of energy produced, consumed or stored per unit time. The reason for this is that what counts for organisms and economists is not what happens two or more generations hence but what is produced now and perhaps into the next generation. In such context, the conservationist and preservationist become losers in the competitive worlds of Darwin and Smith.

[38] Wilson, E.O. 1975. *Sociobiology*. Cambridge MA: Harvard University Press.

[39] Pinker, S. 2002. *The Blank Slate: The Modern Denial of Human Nature*. New York: Norton.

[40] Ehrlich, P.R. 2000. *Human Natures: Genes, Cultures, and the Human Prospect*. Washington, DC: Island Press. Ehrlich, P.R. 2002. *Human Natures, Nature Conservation, and Environmental Ethics*. BioScience 52: 31-43.

[41] Ehrlich, P.R. 2002. *Human Natures, Nature Conservation, and Environmental Ethics*. BioScience 52: 31-43.

[42] See Menon and Lavigne, Chapter 12.

[43] A further point on the "built in" nature of our nature. It is interesting that two of the first countries to have fertility drop below replacement rate were Italy and Ireland, both staunchly Catholic. This apparent contradiction suggests that even strong religious and cultural pressures do not have much effect on our basic responses to economic self-interest.

[44] Dennett, D. 1995. *Darwin's Dangerous Idea: Evolution and the Meaning of Life*. New York: Simon and Schuster.

[45] ibid.

[46] Popper, S.W., R.J. Lempert, and S.C. Bankes. 2005. Shaping the Future. Scientific American 292: 66-71. It is always instructive to discuss these issues with undergraduate students and observe that although they are aware that things used to be different, it is only a vague awareness and it is very difficult for them to see how things could be other that they are at present. They readily believe that our destiny is toward further progress and wealth. For professors, even of economics, the perspective is similar and the time frame is only minutely different.

[47] Diamond, J. 1995. Easter's End. Discover August: 62-69; Ponting, C. 1991. *A Green History of the World.* London: Sinclair-Stevenson; Rees, W.E. 2004. Waking the Sleepwalkers – Globalization and Sustainability: Conflict or Convergence. In W. Chesworth, M.R. Moss, and V.G. Thomas (Eds.). *The Human Ecological Footprint.* The Kenneth Hammond Lectures on Environment, Energy and Resources 2002 Series, Faculty of Environmental Sciences, University of Guelph. pp. 1-34; Wright, R. 2004. *A Short History of Progress.* Toronto: House of Anansi Press Inc.

[48] Heyerdahl, T. 1958. *Aku-Aku: The Secret of Easter Island.* New York: Rand McNally & Company; Heyerdahl, T. 1968. *Sea Routes to Polynesia.* London: Allen and Unwin.

[49] Diamond, J. 1995. Easter's End. Discover August: 62-69.

[50] ibid; Ponting, C. 1991. *A Green History of the World. London: Sinclair-Stevenson.*

[51] Shermer, M. 2002. *Borderlands of Science: Where Sense Meets Nonsense.* New York: Oxford University Press.

[52] ibid.

[53] A similar situation occurred recently in Canada's Queen Charlotte Islands (Haida Gwai) when loggers and environmentalists clashed over whether the trees would soon be depleted. The loggers made it clear that they did not care if the forest ran out in 20 years, what mattered more was to "put food on the table today", etc. In fact the area was pretty much logged out in less than 5 years after those confrontations, surpassing the rapidity of depletion predicted by even the more rabid anti-logging advocates; also see Hutchings, Chapter 6.

[54] Bahn, P., and J. Flenly. 1992. *Easter Island, Earth Island.* London: Thames and Hudson. Bahn and Flenly ask this question too and phrase it in the context of nature and nurture; is the human personality always the same as that of the person who felled the last tree? p. 218.

[55] Keynes, J.M. 1936. *General Theory of Employment, Interest, and Money.* London: MacMillan. "Commerce is rational, but it is driven by irrationality – by risk, gamble, ego, status, ambition and windfall". (p.383).

[56] Frank, T. 2000. *One Market Under God: Extreme Capitalism, Market Populism, and the End of Economic Democracy.* New York: Doubleday; Huber, P., and M.P. Mills. 2005. *The Bottomless Well: The Twilight of Fuel, The Virtue of Waste, and Why We Will Never Run Out of Energy.* New York: Basic Books. In this book, the authors argue that wasting energy is a virtue, not a vice, because the faster we use it, the better we get at finding new supplies, and the less our economy depends on energy: "Energy begets more energy; tomorrow's supply is determined by today's demand. The more energy we seize and use, the more adept we become at finding and seizing more" (p.77). Similar arguments are made by Lomborg, Simon, and other cornucopians. Lomborg, B. 2001. *The Skeptical Environmentalist.* Cambridge UK: Cambridge University Press; Moyers, B. 2004. On Receiving the Harvard Medical School's Global Environment Citizen Award. Available at http://www.commondreams.org/views04/1206-10.html; Simon, J.L. 1996. *The Ultimate Resource.* Princeton NJ: Princeton University Press.

[57] See Czech, B. 2000. *Shoveling Fuel for a Runaway Train: Errant Economists, Shameful Spenders, and Plan to Stop Them All.* Los Angeles: University of California Press, for a detailed exposition and critique of this logic.

[58] Deffeyes, K.S. 2001. *Hubbert's Peak: The Impending World Oil Shortage.* Princeton NJ: Princeton University Press; Unwin, P. 2004. A Brief History of Trees. The Beaver. 84: 21-27. These references make the point that governments and corporations frequently exaggerate reserves of resources to engender various financial advantages such as attracting investors or obtaining lower interest rates. Heinen, J.T., and R.S. Low. 1992. Human Behavioural Ecology and Environmental Conservation. Environmental Conservation 19: 105-116. "We contend that humans, as living organisms, evolved to sequester resources to maximize reproductive success, and that many basic aspects of human behaviour reflect this evolutionary history". p. 113.

[59] Lomborg, B. 2001. *The Skeptical Environmentalist. Cambridge* UK: Cambridge University Press; Simon, J.L. 1996. *The Ultimate Resource.* Princeton NJ: Princeton University Press.

[60] Wright, R. 2004. *A Short History of Progress.* Toronto: House of Anansi Press Inc. Wright (p. 40) states that rule #1 for the prudent parasite is: don't kill off the prey. Many people believe that animals are prudent in this respect, and that is the basis of the beautiful people/noble savage myth. Oddly, Wright spends most of his book trashing this myth about our hunter/gatherer ancestors, yet readily believes that although people aren't prudent, tapeworms are. Of course, this belief in nature's prudence and foresight arises from the widely entrenched myth that organisms evolve adaptations for the "good of the species". I suppose theistic religions also predispose people to accept this view. In fact, Darwinism claims that this never happens and hence my argument that altruistic conservation is unnatural.

[61] I like to think that the Blank Slate is a concept similar to that of infinite substitution in neo-liberal economics. Thus, we can compare J.B. Watson's claims of infinite capacity in child-raising with Robert Solow's (1974, p. 11) claim that through substitution, "the world can, in effect, get along without natural resources." Solow, R. 1974. The Economics of Resources or the Resources of Economics. American Economics Review 64(2): 1-14.

[62] Ehrlich, P.R., and A. H. Ehrlich. 1990. *The Population Explosion.* New York: Simon and Schuster.

[63] Simon, J.L. 1996. *The Ultimate Resource.* Princeton NJ: Princeton University Press.

[64] Abernethy, V.D. 2001. Carrying Capacity: The Tradition and Policy Implications of Limits. Ethics in Science and

Environmental Politics (2001): 9-18; Abernethy, V.D. 2002. Fertility Decline; No Mystery. Ethics in Science and Environmental Politics (2002): 1-11.

[65] ibid.

[66] Moses, M.E., and J.H. Brown. 2003. Allometry of Human Fertility and Energy Use. Ecology Letters 6: 295-300.

[67] For example. Calder III, W. A. 1984. *Size, Function, and Life History*. Cambridge MA: Harvard University Press; Harvey, P.H., and A.F. Read. 1988. How and Why Do Mammalian Life Histories Vary? In Boyce, M.S. (Ed.). *Evolution of Life Histories of Mammals: Theory and Pattern*. New Haven CT: Yale University Press. pp. 213-232; Peters, R.H. 1983. *The Ecological Implications of Body Size*. Cambridge UK: Cambridge University Press.

[68] Moses, M.E., and J.H. Brown. 2003. Allometry of Human Fertility and Energy Use. Ecology Letters 6: 295-300.

[69] ibid.

[70] ibid. p. 299.

[71] Eisenberg, J.F. 1981. *The Mammalian Radiations: An Analysis of Trends in Evolution, Adaptation, and Behaviour*. Chicago: University of Chicago Press; Wright, R. 2004. *A Short History of Progress*. Toronto: House of Anansi Press Inc. Another interesting example of apparent independence of human behaviour from cultural variations is the rise of civilizations. Wright notes several times that civilizations arose independently in different regions at about the same time even though these societies had different cultures, languages, and ecologies. At one point, he expresses a general rule: "Given certain broad conditions, human societies everywhere will move towards greater size, complexity, and environmental demand" (p. 65). Is this genetic, organismal, or cultural determinism?

[72] Darwin, C.R. 1859. *The Origin of Species*. London: Murray.

[73] In the Toronto Star, July 25, 2004, there is a feature article "Evolution's [sic] next stage?" which is devoted to the "Transhuman" (ironically on a page opposite a series on the potentially catastrophic decline in fossil fuel reserves). The article cites "visionary scientists" who seek to improve and perfect humans through DNA manipulation, computer hookups (to improve brain function) and regeneration techniques. These changes will make us "better fit current beauty standards", abolish obesity, make hair thick and glossy, etc. The entire article from its opening sentence on the discovery of tools and fire is a picture of our past and future extended organism. Although the goals seem bizarre, they really are no different in principle than what we and most other species have always done.

[74] Wilson, E.O. 1992. *The Diversity of Life*. Cambridge MA: Harvard University Press.

[75] See Beder, Chapter 5, and Oates, Chapter 18.

[76] See Beder, Chapter 5.

[77] Turner, J.S. 2000. *The Extended Organism: The Physiology of Animal-Built Structures*. Cambridge MA and London UK: Harvard University Press.

[78] Dawkins 1982: but also see Dawkins 2004, and my concluding comment in endnote 22.

[79] Turner, J.S. 2000. *The Extended Organism: The Physiology of Animal-Built Structures*. Cambridge MA and London UK: Harvard University Press. It is interesting that Turner used earthworms because Darwin, still considered one of the world's experts on earthworms, wrote his final book on earthworms ostensibly as part of his long argument (Mayr[80]). Darwin wanted to show how these tiny innocuous friends of agriculture could have huge effects on the soil and on the form of the land itself through prolonged, numerous, and repeated actions, each trivial in itself but having a massive cumulative effect. Given that humans are now moving more soil than all geomorphic forces combined (see Chesworth, Chapter 3), we can be perceived as the modern, and much more powerful, version of earthworms.

[80] Mayr, E. 1991. *One Long Argument: Charles Darwin and the Genesis of Modern Evolutionary Thought*. Cambridge MA: Harvard University Press.

[81] Worms don't come to the surface during rain because they might drown in their burrows; they can live days in an aquarium without any fear of drowning, Rather, they come to the surface for wild, consensual, hermaphroditic sex. Currently, the Canadian Nature Federation is leading a national "Wormwatch" program to involve Canadians in guarding earthworm biodiversity (ironically almost all Canadian earthworms are exotic, introduced alien species, but that is another story). In their literature, the CNF extols the capacity of earthworms to eliminate "waste" ("Imagine all the waste that would accumulate if not for earthworms"). The parallel to humans is almost painful and certainly instructive in showing how even naturalists can be uninformed when thinking about nature. But I am getting ahead of myself. I shall return to waste in a little bit.

[82] Larsen, D.W., U. Matthes, P.E. Kelly, J. Lundholm, and J. Gerrath. 2004. *The Urban Cliff Revolution: New Findings on the Origin and Evolution of Human Habitats*. Markham; Ontario: Fitzhenry & Whiteside.

[83] ibid. p. 178.

[84] Brooks, R.J. 2001. Earthworms and the Formation of Environmental Ethics and Other Mythologies: A Darwinian Perspective. In W. Chesworth, M.R. Moss, and V.G. Thomas (Eds.). *Malthus and the Third Millennium*. The Kenneth Hammond Lectures on Environment, Energy and Resources 2000 Series, Faculty of Environmental Sciences, University of Guelph. pp. 59-92. Robinson, J. 2002. Squaring the Circle?: On the Very Idea of Sustainable Development. In W. Chesworth, M.R. Moss, and V.G. Thomas (Eds.). *Sustainable Development: Mandate or Mantra?* The Kenneth Hammond Lectures on Environment, Energy and Resources 2001 Series, Faculty of Environmental Sciences, University of Guelph. pp. 1-34.

[85] The notion that there are three fundamental "philosophies of the environment" is widespread although they can carry different emphasis (e.g. Wallbank *et al.*). Essentially, they are expansionist (optimistic), static (pessimistic) and ecological (prudent). In my discourse, I am not defending any of these approaches, but merely looking at what people do rather than what they claim to do. Wallbank, T.W., A.M. Taylor, N.M. Bailkey, and G.F. Jewsbury. 1978. *Civilization Past and Present*. Glenview IL: Scott, Foresman and Co.

[86] See Surgeoner for an odd example. Surgeoner, G. 2002. The Challenge of Abundance. In W. Chesworth, M.R. Moss, and

V.G. Thomas (Eds.). *Sustainable Development: Mandate or Mantra?* The Kenneth Hammond Lectures on Environment, Energy and Resources 2001 Series, Faculty of Environmental Sciences, University of Guelph. pp. 51-62.

[87] See Beder, Chapter 5; Rees, Chapter 14; and Czech, Chapter 23. Also see Czech, B. 2000. *Shoveling Fuel for a Runaway Train: Errant Economists, Shameful Spenders, and a Plan to Stop Them All.* Los Angeles: University of California Press. I think the essential element of this viewpoint is that the economy must keep expanding. For example, many others make this claim explicitly noting that society cannot accept a no-growth economy because the perceived costs would be too great. In a recent series in the Toronto Star (e.g. Oct. 2004) on the coming oil "shock", David Olive, after discussing evidence that we are running out of oil and gas, comments that an attack by Al Qaeda on Saudi Arabia's oil facilities could cause a recession, as though this prospect is even more dire than running out of oil. He seems to assume that the depletion of oil reserves will not have this effect, perhaps, although he does not say, because we will have substitutes. In essence, we cannot accept that prosperity will not continue to grow. Recession is apparently even worse than exhaustion of the most important non-renewable resources.

[88] Czech, B. 2000. *Shoveling Fuel for a Runaway Train: Errant Economists, Shameful Spenders, and a Plan to Stop Them All.* Los Angeles: University of California Press.

[89] Frank, T. 2000. *One Market Under God: Extreme Capitalism, Market Populism, and the End of Economic Democracy.* New York: Doubleday; Maynard Smith, J., and E. Szathmary. 1995. *The Major Transitions in Evolution.* New York: W.H. Freeman & Co. Ltd.; Frank (2000) is a great read, as he eviscerates the Neocon trends of the 1990s, when CEOs were "media stars", the stock market was the new form of individual empowerment, and the internet economy, and even the internet itself, were the new democracy. Commerce became a new benevolent force and people predicted the end of government and nations in the new global economy. Thus, Margaret Thatcher pronounced that there is no such thing as society, only economics, and Fukuyama declared the end of history. Unfortunately, commerce, like Darwinism, was not benevolent, and the New World Order quickly proved to be a delusion.

[90] I recommend Czech, B. 2000. *Shoveling Fuel for a Runaway Train: Errant Economists, Shameful Spenders, and a Plan to Stop Them All.* Los Angeles: University of California Press, for a thorough coverage of growth economics and particularly its pervasiveness in our current society in the media, government and every other aspect of our lives. Also see Czech, Chapter 23.

[91] See Wright for some words on progress in this context. Wright, R. 2004. *A Short History of Progress.* Toronto: House of Anansi Press Inc.

[92] Lomborg, B. 2001. *The Skeptical Environmentalist.* Cambridge UK: Cambridge University Press.

[93] Simon, J.L. 1996. *The Ultimate Resource.* Princeton NJ: Princeton University Press.

[94] e.g. Fox, G. 2004. Environmentalism, Economics and Social Cooperation: Commentary on the 2002 Hammond Lectures. In W. Chesworth, M.R. Moss, and V.G. Thomas (Eds.). *The Human Ecological Footprint.* The Kenneth Hammond Lectures on Environment, Energy and Resources 2002 Series, Faculty of Environmental Sciences, University of Guelph. pp. 141-167; Huber, P., and M.P. Mills. 2005. *The Bottomless Well: The Twilight of Fuel, The Virtue of Waste, and Why We Will Never Run Out of Energy.* New York: Basic Books.

[95] Beckermann, W. 1992. Economic Growth and the Environment: How's Growth? How's Environment? World Development 20: 481-496. p. 491.

[96] Saunders, D. Oct. 19, 2002. Globe and Mail.

[97] Czech (2000, p. 44) asks the right question to answer why Neoclassical economists rationalize perpetual growth. "Who benefits?" Czech, however, is too narrow in his answer. He answers, "Any industry and government", because they profit and win elections. I would add that any person benefits or hopes to benefit, and that governments also benefit because growth allows them to perform as Big Men (Harris, 1977), and to embrace unlimited largesse or a free lunch in their campaigns for support and election. Czech, B. 2000. *Shoveling Fuel for a Runaway Train: Errant Economists, Shameful Spenders, and a Plan to Stop Them All.* Los Angeles: University of California Press; Harris, M. 1977. *Cannibals and Kings.* New York: Random House. I can't resist inserting lines here from Leonard Cohen's 1988 song 'Everybody Knows': "Everybody knows that the boat is leaking, everybody knows that the captain lied... Everybody is talking to their pockets, everybody just wants a box of chocolates and a long stem rose. Everybody knows".

[98] Shermer, M. 2002. *Borderlands of Science: Where Sense Meets Nonsense.* New York: Oxford University Press. I refer the reader to Brooks (2001) for a description of the Romantic mindset in full flight. I quoted from a UN document in 1989 called The Vancouver Declaration on Survival in the 21st Century. Suffice to say, the document provides only myths and as an aid to survival one would do as well to chant ancient Celtic poems. Brooks, R.J. 2001. Earthworms and the Formation of Environmental Ethics and Other Mythologies: A Darwinian Perspective. In W. Chesworth, M.R. Moss, and V.G. Thomas (Eds.). *Malthus and the Third Millennium.* The Kenneth Hammond Lectures on Environment, Energy and Resources 2000 Series, Faculty of Environmental Sciences, University of Guelph. pp. 59-92.

[99] A wonderful illustration of this enduring myth is familiar to all Neil Young fans in his song "Cortez the Killer", which lyrically and apocryphally describes the clash of cultures; the corrupt European ("he came dancing across the water") versus the pure indigenous people "and the women all were beautiful and the men stood straight and tall". This confrontation of good and evil, the innocent country bumpkin/noble savage versus the larcenous city slicker/greedy colonialist, reappears in mythology across cultures in literature and in the justification of every tyrant and leader from Stalin to Mao to Osama Bin Laden and countless inspirational tales of achievement and an optimistic future. For example, all political leaders love to tell of their rise from humble origins, Canada's former Prime Minister Jean Chrétien being a good recent example.

[100] Attwell, C.A.M., and F.P.D. Cotterill. 2000. Postmodernism

and African Conservation Science. Biodiversity and Conservation 9: 559-577.

[101] See Beder, Chapter 5

[102] World Commission on Environment and Development. 1990. *Our Common Future*. American edition. New York: Oxford University Press.

[103] See Beder, Chapter 5, and Czech, Chapter 23.

[104] By *status quo*, I mean the hierarchical structure of civilizations (Diamond, 1997, 2005; Fernandez-Armesto; Wright), which always leads to a small number of people with great wealth and a large number with a lot less. Despite the numerous claims that democracy will erase or at least reduce such disparities, it seems to be almost inevitable that eventually the opposite occurs. This trend of increasing wealth disparity is occurring in almost every society on the planet. Czech spends a considerable part of his book lamenting this, but I think he errs in concluding that those at the top (Liquidators) are different than the majority (common folk). In any society of humans or beasts, there are rulers and ruled. These categories depend on genetic and learned differences, but their existence appears pre-ordained. In contrast to Czech's argument that the "common folk" have properly modest needs and wants, the "common folk" usually itch to get their chance at wealth. Regardless, the inertia and maintenance of the *status quo* is primarily the product of those at the top of society, a quintessential expectation of Darwinism. Add to this the evidence that even extremely wealthy people tend to deny their wealth, and if they are unable to accept that reality, it is hard to see how these disparities and high rates of consumption are going to end easily (see also Conniff, 2002). In Conniff's book, he operates on the theory that the rich are like everyone else, just wealthier. But he also argues that the "rich" hate to admit it: "In a recent survey of people with a net worth between 1-4 million, for instance, only 9% would admit to be wealthy" (p. 30). Conniff, R. 2002. *The Natural History of the Rich: A Field Guide*. New York: W.W. Norton & Company; Czech, B. 2000. *Shoveling Fuel for a Runaway Train: Errant Economists, Shameful Spenders, and a Plan to Stop Them All.* Los Angeles: University of California Press; Diamond, J. 1997. *Guns, Germs, and Steel: The Fates of Human Societies*. New York and London: W.W. Norton & Company; Diamond, J. 2005. *Collapse: How Societies Choose to Fail or Succeed*. New York: Viking. The Penguin Group; Fernández-Armesto, F. 2000. *Civilizations*. Toronto; Ontario. Key Porter Books Unlimited; Wright, R. 2004. *A Short History of Progress*. Toronto: House of Anansi Press Inc.

[105] Shermer, M. 2002. *Borderlands of Science: Where Sense Meets Nonsense*. New York: Oxford University Press.

[106] "The Neolithic Revolution, for all its dazzling advances in metallurgy, the arts, writing, politics, and city life, was at its base a matter of the direct control and exploitation of many species for the sake of one: *Homo sapiens*." (Crosby, p. 21); Barnosky, A.D., P.L. Koch, R.S. Feranec, S.L. Wing, and A.B. Shable. 2004. Assessing the Causes of Late Pleistocene Extinctions on the Continents. Science 306: 70-75; Bahn, P., and J. Flenly. 1992. *Easter Island, Earth Island*. London: Thames and Hudson; Crosby, A.W. 1986. *Ecological Imperialism: Biological Expansion of Europe, 900-1900*. Cambridge UK: Cambridge University Press; Flannery, T.F. 1994. *The Future Eaters: An Ecological History of the Australasian Lands and People*. Port Melbourne: Reed Books; Flannery, T.F. 2001. *The Eternal Frontier: An Ecological History of North America and its Peoples*. New York: Atlantic Monthly Press; Jackson, J.B.C. *et al.* 2001. Historical Overfishing and the Recent Collapse of Coastal Ecosystems. Science 293: 629-638; Krech, S. 1999. *The Ecological Indian: Myth and History*. New York: W.W. Norton & Company Inc.; Martin, P.S., and R.G. Klein. (Eds.). 1984. *Quaternary Extinctions: A Prehistoric Revolution*. Tuscon: University of Arizona Press; Wright, R. 2004. *A Short History of Progress*. Toronto: House of Anansi Press Inc.

[107] See Hutchings, Chapter 6; Jackson, J.B.C. *et al.* 2001. Historical Overfishing and the Recent Collapse of Coastal Ecosystems. Science 293: 629-638; Kurlansky, M. 1997. *Cod: A Biography of the Fish that Changed the World*. Toronto: Vintage Canada Edition; Peacock, E., W.R. Haag, and M.L. Warren Jr. 2005. Prehistoric Decline in Freshwater Mussels Coincident with the Advent of Maize Agriculture. Conservation Biology 19: 547-551.

[108] See Menon and Lavigne, Chapter 12.

[109] Hertsgaard, M. 1998. *Earth Odyssey: Around the World in Search of our Environmental Future*. New York: Broadway Books; Kaplan, R.D. 1996. *The Ends of the Earth: From Togo to Turkmenistan, From Iran to Cambodia – A Journey to the Frontiers of Anarchy*. New York: Vintage Departures; Wright, R. 2004. *A Short History of Progress*. Toronto: House of Anansi Press Inc.

[110] Schrödinger, E. 1944. *What is Life?: The Physical Aspect of the Living Cell*. Cambridge UK: Cambridge University Press; also see Rees, Chapter 14.

[111] ibid; Park, R. 2000. *Voodoo Science: The Road from Foolishness to Fraud*. New York: Oxford University Press; Shermer, M. 2002. *Borderlands of Science: Where Sense Meets Nonsense*. New York: Oxford University Press.

[112] Park provides numerous examples of public credulity when presented with apparent examples of unlimited "free" energy (see p. 109 for an example of the hydrogen "economy"). Doty provides a sophisticated critique of the so-called hydrogen economy. Similarly, in recent books (*Eco-Economy*, 2000) and recent talks (University of Guelph 2004), Lester Brown, former head of the World Watch Institute sounds like a Lomborgian cornucopian promising that windmills provide "unlimited, cheap, clean" energy. Oddly, he relies heavily on the "genius" of Kenneth Lay, former CEO of Enron, to bolster his arguments for an economy based on windmills and hydrogen. Brown, L.R. 2001. *Eco-Economy: Building an Economy for the Earth*. New York: W.W. Norton & Co. Inc.; Doty, F.D. 2004. Practical, Clean Energy for Future Transportation. Available at http://www.dotynmr.com /PDF/Doty_H2Price.pdf; Park, R. 2000. *Voodoo Science: The Road from Foolishness to Fraud*. New York: Oxford University Press.

[113] Robinson, S. 2001. *The Free Lunch*. New York: Tom Doherty Associates.

[114] Hardin, G. 1993. *Living Within Limits: Ecology, Economics and Population Taboos*. New York: Oxford University Press.

[115] Czech, B. 2000. *Shoveling Fuel for a Runaway Train: Errant*

Economists, Shameful Spenders, and a Plan to Stop Them All. Los Angeles: University of California Press.

[116] Colinvaux, P. 1978. *Why Big Fierce Animals Are Rare: An Ecologist's Perspective.* Princeton NJ: Princeton University Press.

[117] Carbone, C., and J.L. Gittleman. 2002. A Common Rule for the Scaling of Carnivore Density. Science 295: 2273-2282.

[118] My colleague, Ward Chesworth, points out that this estimate does not include gambling on the stock market, another place where we seek out the free lunch.

[119] Another example of this free lunch concerns so-called "pharma" crops. These crops – mostly corn, the crop that causes the greatest rates of soil erosion and highest use of fertilizer (Jackson 2004) – are grown to produce pharmaceutical and industrial chemicals. A recent report by the Union of Concerned Scientists (Andow *et al.* 2004) notes that although these crops may have substantial commercial and health benefits, they cause risks to the food supply and the environment. For example, as these crops are genetically modified to produce particular compounds, they can injure wildlife, livestock or people that consume them. It is likely they will contaminate these food supplies by accidental mixing of seed or by transfer of pollen. These risks and others associated with high intensity agriculture increase as production increases. Once again, food production will be sacrificed if profits are higher for 'pharma' crops, yet we will perceive such crops as a "win-win" free lunch, because food shortage is not a disease of CEOs. Andow, D., Daniell, H., Gepts, P., Lamkey, K., Nafziger, E. & Strayer, D. 2004. Introduction. In: *A Growing Concern. Protecting the Food Supply in an Era of Pharmaceutical and Industrial Crops.* Ch. 1, 21-32. Union of Concerned Scientists. Available at http://www.ucsusa.org /food_and_environment/biotechnology/page.cfm?pageID=1561; Jackson, W. 2004. Agriculture: The Primary Environmental Challenge of the Century. In W. Chesworth, M.R. Moss, and V.G. Thomas (Eds.). *The Human Ecological Footprint.* The Kenneth Hammond Lectures on Environment, Energy and Resources 2002 Series, Faculty of Environmental Sciences, University of Guelph. pp. 85-100.

[120] ibid.; corn, soybean and sugarcane lead all crops in causing erosion.

[121] Mann, C.C. 2002. 1491. The Atlantic Monthly. March: 41-53. "Before it became the New World, the Western Hemisphere was vastly more populous and sophisticated than has been thought – an altogether more salubrious place to live at the time than, say, Europe. New evidence of both the extent of the population and its agricultural advancement leads to a remarkable conjecture: the Amazon rainforest may be largely a human artefact" (p.41). "Amazonia has become *the* emblem of vanishing wilderness – an admonitory image of untouched Nature. But the rainforest itself may be a cultural artefact – that is, an artificial garden" (p. 50). "If they (modern nations) want to return as much of the American landscape as possible to its 1491 state, they will have to find it within themselves to create the world's largest garden" (p. 53).

[122] Flannery, T.F. 1994. *The Future Eaters: An Ecological History of the Australasian Lands and People.* Port Melbourne: Reed Books; Krech, S. 1999. *The Ecological Indian: Myth and History.* New York: W.W. Norton & Company Inc.

[123] Brooks, R.J. 2004. Loving Nature to Death. Boreal Dip Net: Newsletter of the Canadian Amphibian and Reptile Conservation Network 8: 7-10.

[124] Pollan, M. 2001. *The Botany of Desire: A Plant's-Eye View of the World.* New York: Random House Inc.

[125] Castle, D. 2001. The Ethical Perspective: Environmental Ethics in the Era of Science and Technology. In W. Chesworth, M.R. Moss, and V.G. Thomas (Eds.). *Malthus and the Third Millennium.* The Kenneth Hammond Lectures on Environment, Energy and Resources 2000 Series, Faculty of Environmental Sciences, University of Guelph. pp. 149-161.

[126] Pollan, M. 2001. *The Botany of Desire: A Plant's-Eye View of the World.* New York: Random House Inc.

[127] Ward Chesworth has suggested to me that this argument of Pollan is "ridiculous" because the increase in marijuana and potatoes is an unintended consequence. My response is: of course it is an unintended consequence because natural selection is the "blind watchmaker".

[128] See for example, May, R.M. 1973. *Stability and Complexity in Model Ecosystems.* Princeton, NJ: Princeton University Press.

[129] Jones, S. 2000. *Darwin's Ghost: The Origin of Species Updated.* New York: Random House.

The optimist is always the hero,
While the pessimist scores a big zero.
But when reality strikes,
We all will scream yikes,
As economists emulate Nero.

Ron Brooks 2005

Part IV:

The Way Forward:
Putting Theory into Practice

CHAPTER 17

CHANGING PUBLIC OPINION:
HOW AND WHY SOCIETAL ATTITUDES CHANGE[1]

Robert Worcester

I want to begin with a story: the story of the pig and the chicken.

The chicken went running up to the pig and said,

"*Have you seen the sign? Have you seen the sign"?*

"*Sign? What sign*"? asked the pig.

"*The sign out in front of farmer Brown's*", said the chicken.

"*No, I haven't seen it; what's it got to do with me?*"

"*C'mon, c'mon, I'll show you"! said the chicken.*

She was just dancing around with excitement and went running around to the front of the barnyard. And there was the big sign that said, "Bedrooms to let".

"*What's that got to do with me"?* asked the pig.

"*Don't you see*"? The chicken replied, "*it's a wonderful opportunity for a joint venture*".

The pig, being an Irish pig, pricked up his ears and said, "*Oh? Tell me more*".

And the chicken said, "*Well, people will come and they'll rent the bedrooms and they'll get up in the morning and they'll be hungry*".

And the pig said, "*So*"?

"*So, we'll serve them breakfast*", answered the chicken.

"*Breakfast*"? questioned the pig.

"*Yeah*", she said, "*bacon and eggs*".

"*Bacon and eggs*"?!? exclaimed the pig, as he laid his ears down and started walking away.

"*Where are you going*"? asked the chicken.

"*That's one heck of a joint venture,*" said the pig. "*We'll serve them bacon and eggs. You'll be participating. But I will be involved*"!

You know what caught that pig's attention? He is going to be involved. And that's what I want to talk about: understanding public opinion.

UNDERSTANDING PUBLIC OPINION

Many of you are scientists. Do you know that 50% of the 1500 scientists funded by the Wellcome Trust say they never talk to the public, and that they don't want to talk to the public. Another 40% say they are not equipped to talk to the public, even though they do. And only 10% say they can talk to the public, are trained to do so, and feel that they can do so effectively.

If you work in the field of animal welfare and conservation, you have to engage the public. And to do that, I want you to understand public opinion. A word about public opinion: As a pollster, I don't measure facts; I measure perceptions. As Epictetus said in the 1st century AD, "Perceptions are truth because people believe them".[2]

There are five things I can measure with the tools of my trade. I can measure people's **behaviour**, what they do; I can measure their **knowledge**, what they know, or think they know; and I can measure their **views**. I break down views at three levels. First, there are **opinions**, which I describe – perhaps too poetically for academic adoption – as the ripples on the surface of the public consciousness, easily blown about by the politicians, the media, the pundits, and their talk – not things that people have thought about or care about, not things that affect them and their families, or things that were discussed over breakfast that morning, or things that they are worried about. And they probably didn't even know they had a view until an interviewer said to them, "What do you think about, say, the Metallic Metals Act"? This actually happened. Donald Rugg of Princeton University – in a 1938 PhD thesis – found that 38% of the American people had a view about the Metallic Metals Act, either pro or con. But there wasn't any Metallic Metals Act. He just made it up. Those are what I call opinions.

Then there are **attitudes:** things that people have thought about, have concerned themselves with, things that affect them and their families, things they have discussed with their workmates and schoolmates, things that they care about. And they do have a view, but that view can be changed with one or the other or preferably both of two things: 1) information that is new to that individual, that raises a question about their previous beliefs, and 2) having that information come from a source they respect. If you can get those two things together, you can change people's attitudes.

And then there are **values.** These words – opinions, attitudes and values – are thrown about as if they were synonymous,[3] but in my view they are not. My entire work is based on this model of people's opinions, attitudes, and values.[4] Values are those things that people feel so strongly about, things that they have discussed and debated and come to a conclusion about, perhaps as the Jesuits would say, by the age of five. I would say by the age of 25; very few people change their basic core values after the age of 25. I'm talking about a belief in god, the death penalty, euthanasia, and abortion. And I'm also talking about animal welfare and conservation.

Twenty-five per cent of the British public refuse to countenance animal experimentation that would cause pain in a mouse to help secure a cure for leukemia in children. That's how deeply the animal welfare value runs in the British public. I haven't had an opportunity to do this test in other countries, but I can tell you, that observation explains a lot about IFAW's fundraising, about the protests against veal calves being shipped live over to the continent, and many of the other threads that run through the British public, where I've lived for the past 35 years. So keep those things in mind, because if you take one thing away from my presentation, I want it to be that: to understand this model. I am now going to use this model as a thread, as a division, if you will, to examine some illustrative data.

USING PUBLIC OPINION RESEARCH

I want to talk about using public opinion research. It is really a very simple business. There are three kinds of research involved: desk research, qualitative research, and quantitative research.

Desk research includes things like the IFAW animal welfare data base, where we did as much as we could to pull together all the questions in all the countries of the world where we had access to survey research that was robust – survey research that was from recognized institutes that do good work – not from the fly-by-night companies that come and go – so that we could look at questions on a variety of subjects at any time.

Desk research increasingly involves the use of the World Wide Web. The MORI website,[5] for example, has a mass of data, including most of what I will present here. There you will also find *qualitative research* – focus groups, that sort of thing. Such methods are basically used to test concepts, ideas, and thoughts, and also to get some thoughts back. They are also used to listen to the language that people use when they talk about these concepts. That is the way to use focus groups, rather than the way so many political focus groups are used.

Then there is *quantitative research*, which you are all familiar with, mostly from opinion polls that you read in the newspapers. I'm certainly best known in Great Britain for being Britain's pollster. I'm on radio and television quite a lot, and in other countries as well, Canada and the U.S., particularly, because I come up with these numbers. I know the numbers and follow the numbers. But, this kind of research actually represents less than one percent of our turnover.

The research that we do is in the social field, for governments, charities, and some for corporations on corporate reputations and the like. This research tracks change; you can't do that with focus groups. And it provides the numbers. You have to ask the right sample of people the right questions and add up the figures correctly but, most importantly, and the reason I'm here today, is to use the numbers wisely, and to use them in advocacy.

I'm going to show you a number of examples where we have done surveys for IFAW, Greenpeace, World Wildlife Fund (WWF), The Wildfowl and Wetlands Trust (WWT), and others, where they can use the results in advocating their views, so that people who feel the way the majority of the people feel know they've got the majority of the people with them. That is a very useful use of it.

Survey information is also useful in the management of advocacy campaigns, in the management of what you want to get across. I had a very smart client, nearly 40 years ago now, when I was just starting out in this business in the United States. He called me in and he said we want to find out three things. We want to find out who it is among the American population who already believe what we want them to believe, so we don't spend a penny of our scarce money or a minute of our scarce time to convince them. On the other hand, we want to find out the people who are so opposed to us that we'll never change their minds, so we don't waste a penny of our money or a minute of our time trying to convince them. And where it's really going to pay off, is identifying the people in the middle, the people who can be convinced, will be convinced, or at least will listen to our arguments.

So research can guide semantics, explore concepts, provide insights, obtain information, set the context, and curb claims.

Tony Benn, left wing Labour MP, on the eve of the task force in Great Britain sailing to the Falkland Islands in 1982, held up a sheaf of letters in the House of Commons and said public opinion is swinging massively against the war. The next day in *The Economist* we (MORI) showed that 78% of the British public were in favour of sending the task force to the Falkland Islands. So polling can certainly curb the claims of the demagogues. It can sometimes even clinch arguments.

But polling research doesn't dictate policy or sacrifice your organization's core values. Greenpeace, for example, had me do a 25–country study some years ago. I went down to Tunisia to present the results of that study at eight o'clock on a Sunday morning. I presented the findings, which said that – right across the world – people wanted Greenpeace to accept money from governments and corporations, so long as it didn't carry too high a price tag with it. And Greenpeace said no, our policy is that we will not accept money from governments, and we will not accept money from corporations. And they stuck with it. And I applauded them for doing so, because I don't believe in management by opinion polls. Nonetheless, I do believe that managers make better decisions with the knowledge of what people think, rather than in the ignorance of it.

Polling research should not manage an organization either. It certainly doesn't at MORI. We do staff attitude studies annually, but just because the staff thinks it would be nicer to be down in the south of France than in the south of London, we're not going to move to the south of France, because *we* manage the company.

Research should not add stress or engender conflict. It ought to reduce stress and conflict, by providing the facts to resolve arguments over philosophy or principles.

Research doesn't necessarily have to lead to action either. Some of the best work I have done – shifting to a corporation – was to tell a client not to open a platinum mine – a second platinum mine – because I had done a study in Japan, where half the platinum in the world is consumed, and I was able to prove to them that the market wouldn't sustain the millions and millions of Rand that it was going to take to open that mine.

Finally, research should not confuse the issues; it should clarify them. So let me now give you some examples to illustrate the findings.

EXAMPLES OF PUBLIC OPINION RESEARCH FINDINGS

1. Behaviour

Let's look at the first of the five things that we measure with the tools of our trade: people's behaviour, things like green consumerism. I have developed a typology of a dozen or so things, from writing letters to members of parliament about green issues, to appearing before organized clubs and groups to give a talk on green issues. People who have done four or more (or five or more, I forget which) of the 12 activities are described as "green consumers". Between 1988 and 1990-91, there was a huge increase in green consumerism, post Chernobyl, and as we went up to the global lecture that Margaret Thatcher

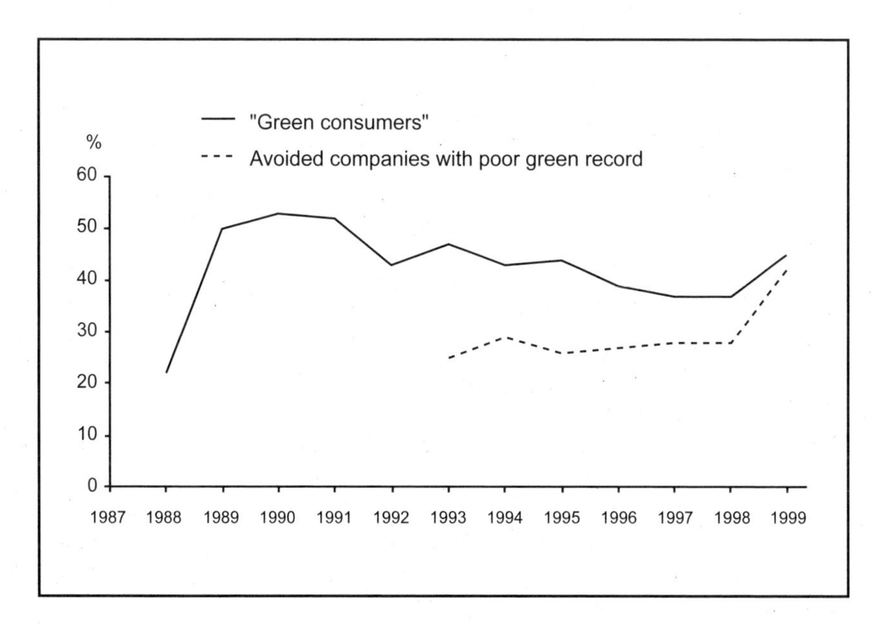

Figure 17–1. Green Consumerism Trends (Britain).

Base: GB Adults. Source: MORI "Business and the Environment" Study.

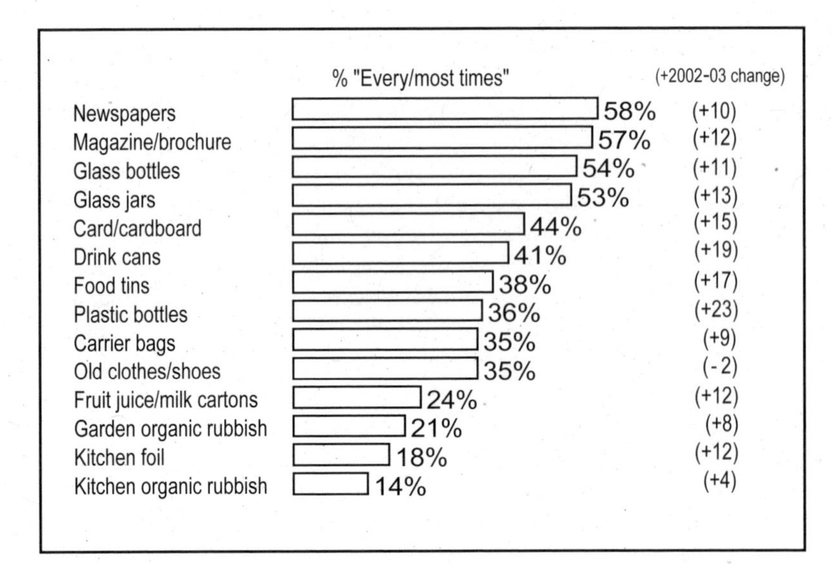

Figure 17–2. Recycling of specific materials.

Response to the question, "How often, if at all, do you recycle?"

Base: 1,314 residents aged 16+ fact-to-face, in-home, London Western Riverside, October-November 2003, Waste Watch / MORI.

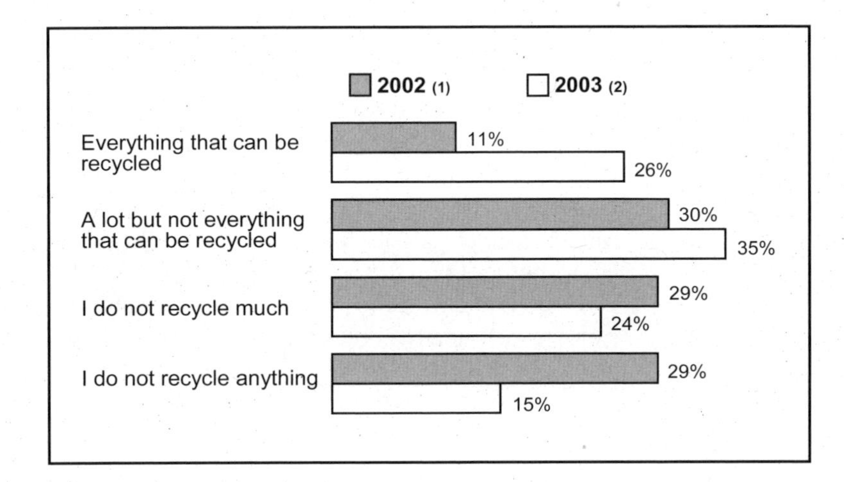

Figure 17–3. Levels of Recycling: Tipping Point.

Response to the question, "Looking at this card, which, if any, of the following statements comes closest to how much you recycle"?

Base: (1) 1,300 residents 16+, face-to-face, in-home, London Western Riverside, Oct-Nov 2002.

(2) Waste Watch / MORI; 1,314 residents aged 16+ face-to-face, in home, London Western Riverside, October-Novemeber 2003, Waste Watch / MORI.

first gave at the Royal Society, and then at the United Nations (Figure 17–1). *Time* magazine's cover – on 1 January 1990, you may remember – was "The Planet." Between 1998 and 1999, there was a sharp rise in people avoiding companies with a poor green record (Figure 17–1).

Figure 17–2 shows another measure of human behavioural change, this time involving the recycling of various materials. Recycling in Great Britain reached a tipping point between 2002 and 2003. By 2003, 58% of people were recycling newspapers and paper products, up 10 points from the year before. Even larger increases were observed in the recycling of magazines and brochures, glass bottles and jars, cardboard, drink cans, and food tins. The largest jump was in plastic bottles (23%). In contrast, when I visit developing countries, plastic bottles are all over the place, littering the countryside.

When people were asked which of the following comes closest to how much they reject and how much they recycle, one in four people claimed in 2003 that they were recycling everything that could be recycled, up from 11% in 2002 (Figure 17–3). And 30-35% recycled a lot, but not everything that could be. Between 2002 and 2003, the number of people who did not recycle anything dropped by half. That's called a scale – it's a Thurston scale technically – and it's a good way to ask questions. So that's behaviour, let's look at knowledge.

2. Knowledge

We conducted a knowledge survey in England and Wales in late 1996 and early 1997. People were asked, "Which of these phrases, if any, had they heard of"? The answers were: climate change, 80%; ozone depletion, 76%, catalytic converter, 75%; and then, sustainable development, 34%; and biodiversity, 22%.

For the World Wildlife Fund, I asked the question, "What do you mean by biodiversity"? I asked that question, I think, in about thirteen countries. Britain was the only country where there was a substantial proportion – I say substantial, but it was less than 5% – who said "Oh, biodiversity, that's homosexuality". So, don't count on people knowing what you know, because a lot of them don't. As recently as May 2004, 45% of the British public said they had never heard about biodiversity and another 39% knew not very much or nothing at all about it. If you don't have awareness, people are not going to listen to your arguments. The minute they hear biodiversity they turn off, and that's nearly half the public in Great Britain.

We also asked people in the United States about the main cause of the greenhouse effect. One person in four didn't know and couldn't provide any answer.

In Britain, in 2004, 51% of people surveyed had never heard of climate change and 50% had never heard of the Kyoto agreement.

So, there is a widespread lack of knowledge on many environmental issues and that is the fight we're fighting.

3. Views: Opinions

As I noted earlier, views and opinions are lightly held and easily blown by the winds. Nonetheless, some of them are very stable.

Most of you are scientists. Not long ago, I had 120 scientists from the British Department for Environment, Food and Rural Affairs (DEFRA) in the audience. I said "hands up those of you who believe that trust in scientists in this country has dropped by 20 points or more in the last five years". Half the hands went up. I said "between ten and twenty percent"; another quarter of the hand went up. "Between zero and ten percent", all the rest of the hands went up, but one. I then said, "stayed the same or increased? One person put his hand up.

And then I showed them Table 17–1: 1999, 63%; 2000, 63%; 2001, 65%; 2002, 64%; 2003, 65%. Dead level; absolutely stable. And yet there wasn't a scientist in that room, but one, who believed it.

But it is not just scientists generically. When you break it down, which of the following scientists do you most trust to give reliable information about the environment: scientists working for universities 50%; for environmental groups, 31%; government, 9%;

Table 17–1. Trust in Types of People. Response to the question, "Now I will read you a list of different types of people. For each would you tell me if you generally trust them to tell the truth, or not?"

Occupations	MORI Veracity Index Trust them to tell the truth (%)									
	'83	'93	'97	'99	'00	'01	'02	'03	'04	'05
Doctors	82	84	86	91	87	89	91	91	92	91
Teachers	79	84	83	89	85	86	85	87	89	88
Professors	n/a	70	70	79	76	78	77	74	80	77
Judges	77	68	72	77	77	78	77	72	75	76
Clergyman / Priests	85	80	71	80	78	78	80	71	75	73
Scientists	n/a	n/a	63	63	63	65	64	65	69	70
Television news readers	63	72	74	74	73	75	71	66	71	63
The Police	61	63	61	61	60	63	59	64	63	58
Ordinary man/woman in street	57	64	56	60	52	52	54	53	55	56
Pollsters	n/a	52	55	49	46	46	47	46	49	50
Civil Servants	25	37	36	47	47	43	45	46	51	44
Trade Union Officials	18	32	27	39	38	38	37	33	39	37
Business Leaders	25	32	29	28	28	27	25	28	30	24
Government Ministers	16	11	12	23	21	20	20	20	23	20
Politicians generally	18	14	15	23	20	17	19	18	22	20
Journalists	19	10	15	15	15	18	13	18	20	16

Base: 2,017 British adults aged 16+, most recent fieldwork 17-21 February 2005.

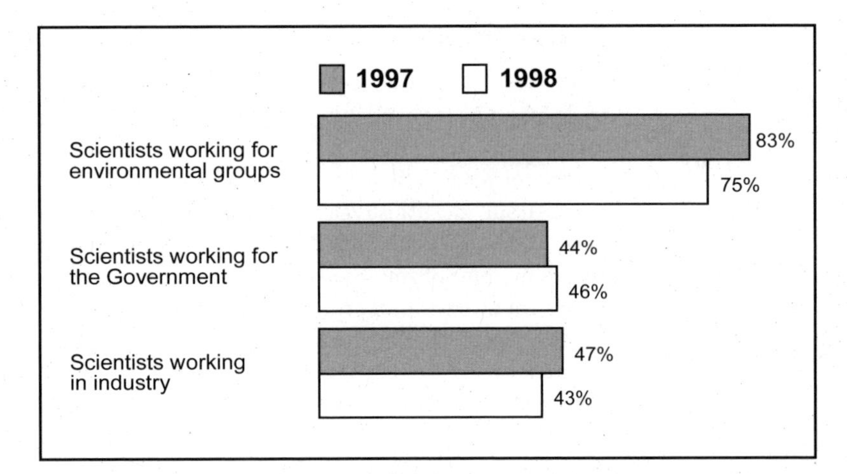

Figure 17–4. Trust In Scientists

Response to the question, "How much confidence would you have in what each of the following have to say about environmental issues?"

Base: c. 2,000 adults 15+ in Great Britain. MORI "Business & The Environment" Study.

industry 6%; and none of these 1%; 3% did not know.

In 1997, I gave the World Environment Day keynote lecture. I noted that only 44% of the people in Britain said that they trust that what scientists working for the government have to say about environmental issues. A senior government scientist came up to me afterwards and said, "But we don't lie to people". I said, "I know you don't lie to people. You don't talk to people." I'm trying to get scientists to talk to people, and you can see where it stands between 1997 and 1998 (Figure 17–4).

We also conducted a seven-country study (France, Germany, Great Britain, Italy, The Netherlands, Poland, and the United States). In those countries, 21, 17, 25, 24, 10, 14, and 29 percent of the people say that the issue of globalization is an extremely important threat to their own countries. Yet, when you ask the same people, with the exception of Poland, twice as many think that global warming is an important threat internationally.

That's important for you to know. And none of you know it.

So, those are opinions.

Views: Attitudes

When people spontaneously talk about the issues that are important in their country, they are speaking of attitudes. Take a look at Figure 17–5. Look closely, the dark line down at the bottom – that's pollution/environment. Respondents were asked about important issues facing Great Britain. It was an open-ended question. They could say anything, and the interviewer coded it. Anything that is solid black – the things that you and I care a lot about – got almost no salience whatsoever. It got quite a lot of attention, however, back in 1989; in fact it was at the top of the poll. Unemployment was on the way down then, but it has fluctuated since then. Trends in unemployment are highly correlated with the pollution/environment figures.

In 1994-1995 we did a European-wide study for Greenpeace International to determine what type of environmental organization people would be willing to support. We found that Europeans wanted an environmental organization that teaches people to live in harmony with nature, one that has no links to political parties and is perceived to be a caring organization. They also wanted their environmental organization to operate on an international level, and to work together with other environmental groups.

I have a son who is a wildlife photographer. When he was at university, he wanted to work for an environmental organization. He was going to go to law school, become an environmental lawyer, and fight the good fight. The summer he graduated, he went to work for Friends of the Earth in Washington. He phoned me in August and said, "I'm not going to work for an environmental organization. I'm not going to become a lawyer. They want to fight each other; I want to help save the planet".

People also want their environmental groups to have realistic aims. Recycling, for example, was considered very worthwhile by 70% of respondents in a 2002 UK poll. I talked earlier about the tipping point. Support for recycling has moved, in my mind, from an opinion to an attitude. Ten years ago, most people didn't have a real attitude about recycling. They'd heard about it, but they didn't care much about it. That's all changed in the last decade.

Now, back to the subject of ecotourism.[6] In 2000, we asked 693 package-holiday takers their top reason for choosing a holiday in relation to age (Figure 17–6). You can see that there is not much of a market for ecotourism compared, for example, with relaxation, enjoying sun, sea, and sand, and of course in Britain, they talk about sex as well!

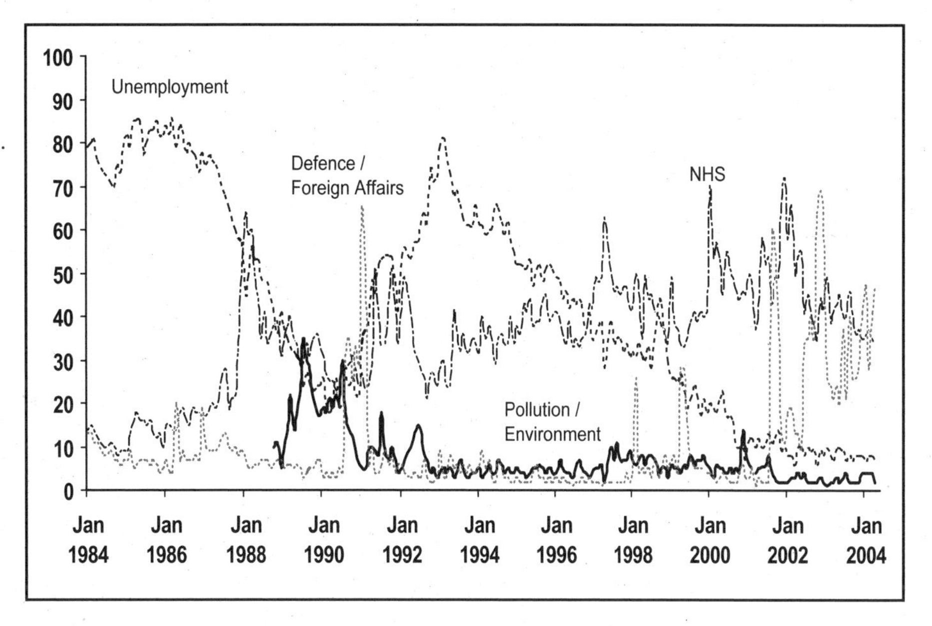

Figure 17–5. Most / Other Important Issues Facing Great Britain.

Base: approx 2,000 British adults.

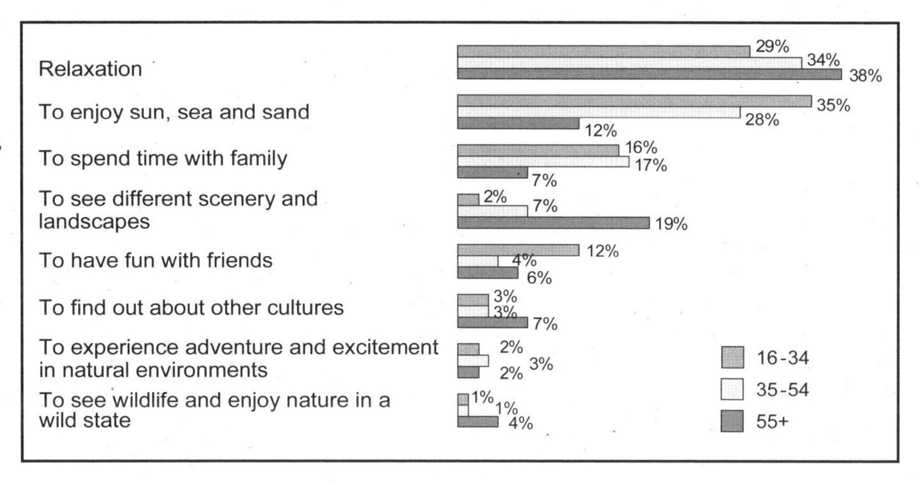

Figure 17–6. Top Reason for Choosing a Holiday – By Age.

Response to the question, "Thinking about your last holiday, what was the main reason for choosing the holiday you went on?"

When you get down to "experience adventure and excitement in natural environments", the figures drop to 2-3%. And, "to see wildlife and enjoy nature in a wild state", the figure is about 1% for all but the 55+ age group, where the number is 4%.

We've just promoted to research director one of my young people – he's been with me I think 15 years now. In his first year we did a study for WWF. We found that – in his words, in the draft report – only 3% of the British public said they would be willing to give a legacy in their will to WWF. I said only? ONLY!? If you could get 3% of the British public to leave a legacy in their will to WWF, they wouldn't have to raise funds in any other way. That would fund them forever. So, while 4% may not sound like much, it sure gives you a good indication of the potential market for ecotourism.

Views: Values

Finally lets get to values, beginning with animal experimentation. The lady I had breakfast this morning with was kind enough to say, spontaneously, not knowing who I was or what I did, that the most interesting MORI study she had seen was the one on animal experimentation. In 2002, we did a study that posed the question, "Provided that all animal welfare regulations were well enforced, please tell us…how acceptable or unacceptable your overall opinions or impressions are of the use of [various animals, including human subjects] in medical research" (Figure 17–7). Eighty-seven percent said, well if humans want to volunteer, it's okay with us. Eleven percent said it was unacceptable to use human subjects in medical research. Twenty-three and 26% said that it was unacceptable to use rats or mice, respectively. Twelve percent

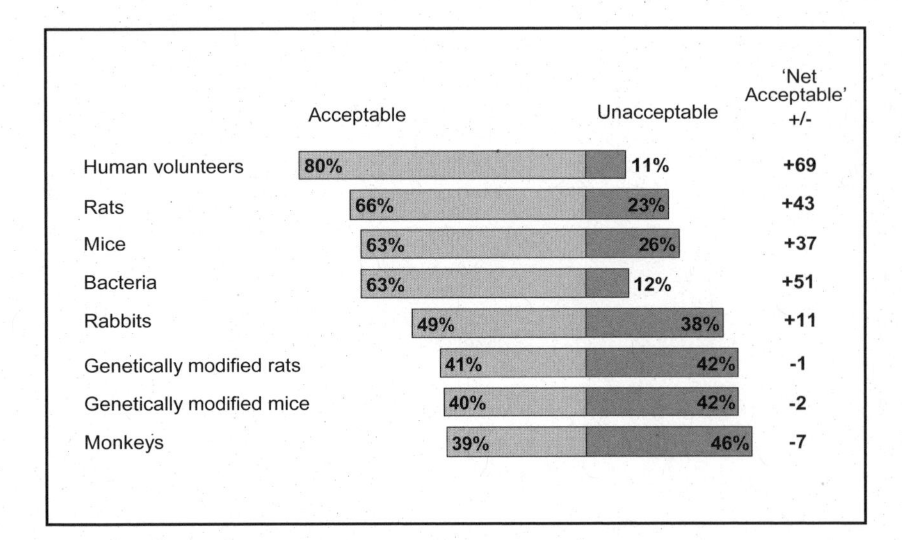

Figure 17–7. Degree of Acceptability.

Response to the question, "Provided that all welfare regulations were well enforced, please tell me, using this card, how acceptable or unacceptable your overall opinions and impressions are of the use of ___ in medical research?"

Base: All GB adults 15+, April/May 2002 (1,125).

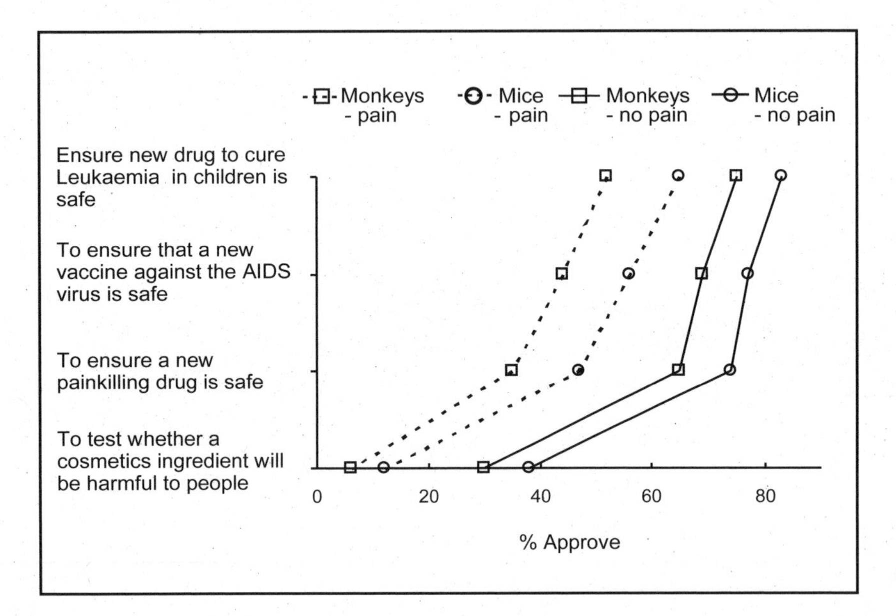

Figure 17 8. Pain vs. Species.

Response to the question, "Please tell me whether you approve or disapprove of the use of live mice/monkeys for each of these goals, if the mice/monkeys are/are not subjected to pain, illness or surgery..."

Base: All respondents (2,009), MORI / New Scientist 1999.

even said it was unacceptable to use bacteria. More people (46%) found experimentation with monkeys unacceptable.

I did a study for *New Scientist* in 1999 that examined whether people approved or disapproved of the use of live monkeys or mice, alternatively, for each of four goals, if the monkeys or mice are, or are not, subjected to pain or surgery (Figure 17–8).[7] It was a very complex research design with over 2,000 people. In that study, almost 40% of people approved of the use of mice as long as no pain was involved. But, at the other extreme, 24% of the British people said that even to find a cure for leukemia in children, they did not find it acceptable to use mice in experiments that cause pain.

In 1996, we asked people in 13 countries (unprompted) "What are the two or three most important problems facing the world as a whole?" On average, 25% of respondents identified "environment, global warming, pollution, and resource depletion", ranking 3rd in a list of 27 issues. Five years later (2001), environment was up six points and ranked 1st among the 27 issues identified.

We have also asked, "What would the effect of the price of fuel doubling have on your use of you car"? Thirty-two percent said that you can double the price of fuel and it wouldn't make any difference in my automobile usage. That is a value. In Germany the figure is 23%; and in Canada, it is 25%. In the United States, it's 18%, but they don't pay very much for gasoline to begin with. In France, the figure is 49%; in Italy, 42%, and to show you what a big job we have in Britain, 43% of respondents say you can double the price but it's not going to make any difference to my driving habits.

Even if we halve the cost of public transport, an average of 60% of respondents still say it would make no difference in their use of the automobile. In Britain 64% say they will not give up their car, 23% in the USA. Nineteen percent of people in Great Britain actually say there is nothing that you can do that would make them give up their car.

Now, the thing that does get people's attention and makes an impact is if you either threaten their health, or promise them better health. Think back to Farmer Brown's pig; you've got involvement there. When you ask people, "How much do you believe environmental problems now affect your health?" 74% of people in China say a great deal, 67% and 50% in India and Nigeria, respectively, whereas the figure drops to 13% in Great Britain, Germany and Japan. When asked "How much if at all do you believe environmental problems will affect the health of our children and grandchildren", 86% in China, 45% in Great Britain, and 50% in the United States, said "a great deal". So there's an emotional tug on which you can pull.

In focus groups, we have gotten statements from people, such as "well really, human activities are in harmony with the environment". Well, we have tested some of these things, and I'll just give you 3 examples.

Concern about the future of the environment. *Which of these opinions comes closest to yours?* The first pie chart (Figure 17–9) is the EU as a total, followed by the Germans, and the British. Only 4% believe that human activity is currently in harmony with the environment, and 44% believe that human activity can lead to irretrievable damage to the environment unless it's changed, 49% in Germany, and 40% in Great Britain.

How about individual action? People often say to me that there's nothing that I as an individual can do; it won't make any difference. Well, we tested the statement, "My actions can make a real difference to the environment" and 43% of individuals polled across Europe, 56% in Germany, but only in 39% in Great Britain said, yes, my actions can make a real difference (Figure 17–10).

The trade off between environment and economic growth. If I could get this across to governments, it would make a difference, but I don't seem to be able to do it. "Please tell me whether you completely agree, mostly agree, mostly disagree, or completely disagree with protecting the environment should be given priority even if it causes slower economic growth and some loss of jobs." Think of all the people in all the governments who are devoted to departments of trade and industry or departments of commerce, to promoting trade, and in employment departments, to promote jobs. Yet there you have 25% who completely agree, and 44% who mostly agree that protecting the environment should be given priority, even if it causes slower economic growth or some loss of jobs. You've even got seven people in ten saying they agree, we want to put the environment first, in the United States! In Great Britain and Germany, the figure is eight people in ten. So help me get that message across, will you?

COMMUNICATION

I will end this chapter on the topic of communication. How do you use this stuff? Consider consumer protests against companies. If you are going to get companies, you've got to scare them. Frankly, I've been in a lot of boardrooms and I do a lot of work for companies, and I try to carry the message. Over the past year, have you considered, or have you actually punished a company you see as not being socially responsible, by refusing to buy their products or speaking critically about them to others. In a global survey conducted by *Environics* and MORI, 23% of respondents to a 1999 survey indicated that they

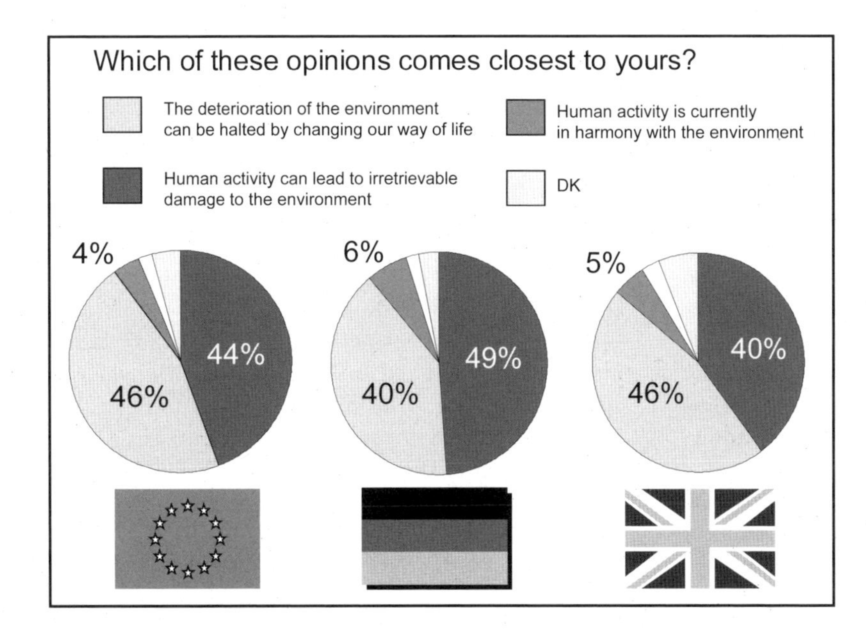

Figure 17–9. Concern About the Future of the Environment.

Base: 16,067 EU, incl. 2,045 German, 1,340 UK adults, Sept-Oct 2002, EORG/Eurobarometer.

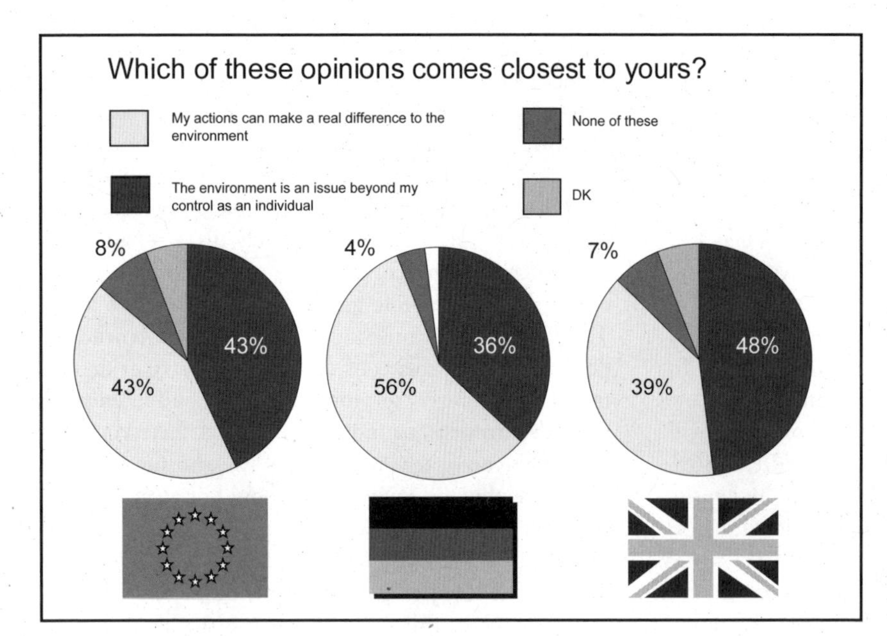

Figure 17–10. Can Individual Actions Make a Difference?

Base: 16,067 EU, incl. 2,045 German, 1,340 UK adults, Sept-Oct 2002, EORG/Eurobarometer.

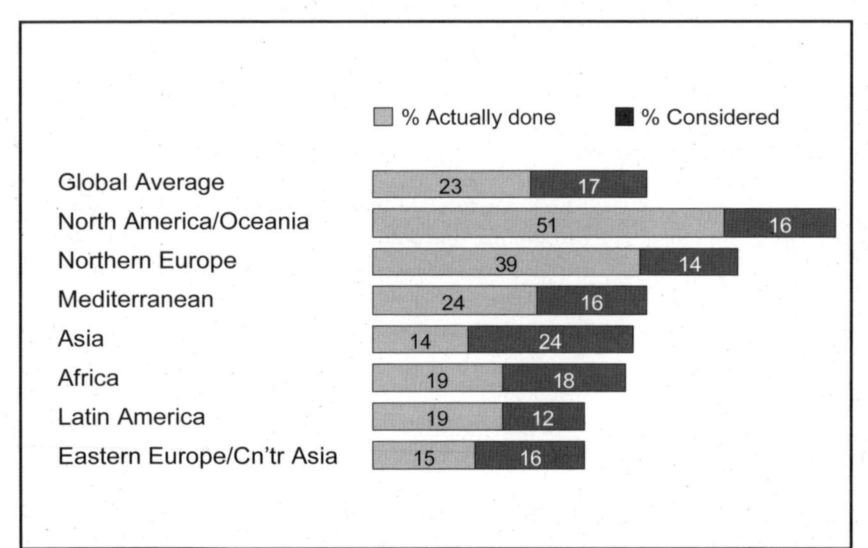

Figure 17–11. Consumer Protests Against Companies.

Response to the question, "Over the past year, have you considered (or have you actually) punished a company you see as not socially responsible – by refusing to buy their products or speaking criticcally about them to others..?"

Base: Adult national populations (1999). Source: Environics.

had changed their minds about buying a product – either quit buying a product or not purchased a product in the first place – because they didn't think the company was socially responsible. In North America/Oceania it was 51%; Northern Europe, 39%; Africa, only 19%; Latin America, only 19%; and Eastern Europe, 15%. But, more have considered it (Figure 17–11).

Another example comes from Dominican support for the South Pacific Sanctuary.[8] Since I saw the whaling presentation earlier,[9] I put in this question: "How strongly do you support or oppose the establishment of a Southern Pacific Sanctuary?" Two-thirds of the people in Dominica say they would like to see their government support a Southern Pacific Sanctuary. Does the government of Dominica know this? I don't know but I certainly hope so. As you can see from Figure 17–12, we broke the results down by gender, by age, by whether respondents were happy or unhappy with their lives, whether they're satisfied or unsatisfied with the performance of their government, and whether the government democracy works well or needs improvement.

One more example: approval for action to stop whaling. In a survey I did in 2003, I asked, "Would you approve or disapprove of the...government taking action to convince Iceland to stop whaling?" Seventy-six, 73, 72, and 58 percent of Dutch, German, British and American respondents said that they would approve of their governments speaking out against Icelandic whaling (Figure 17–13). I hope those governments know that, and apply diplomatic pressure, and apply trade sanctions, as sup-

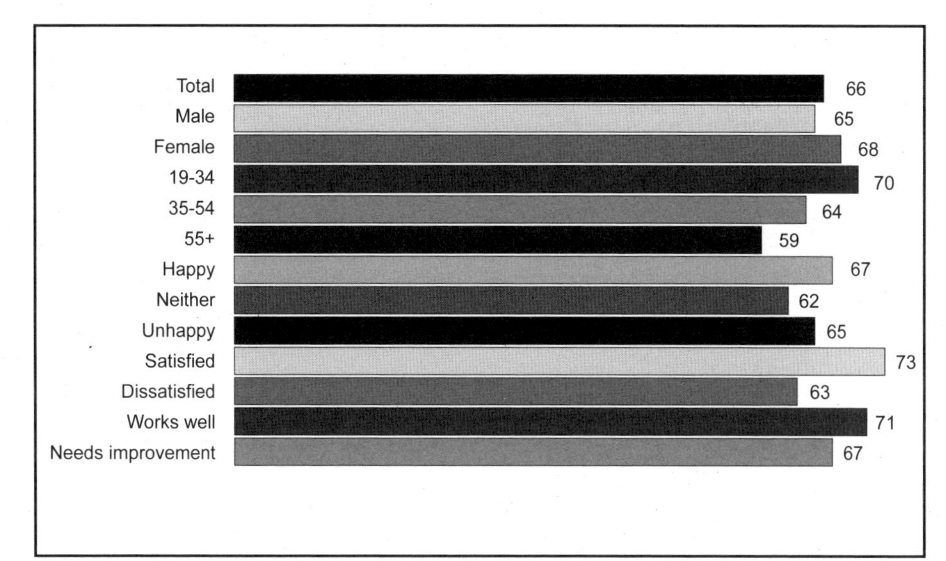

Figure 17–12. Dominican support for a South Pacific Whale Sanctuary.

Response to the question, "How strongly would you want Dominica to support or oppose the establishment of a Southern Pacific Sanctuary?"

Base: 501 adults in Dominica. Source: MORI Caribbean Barometer.

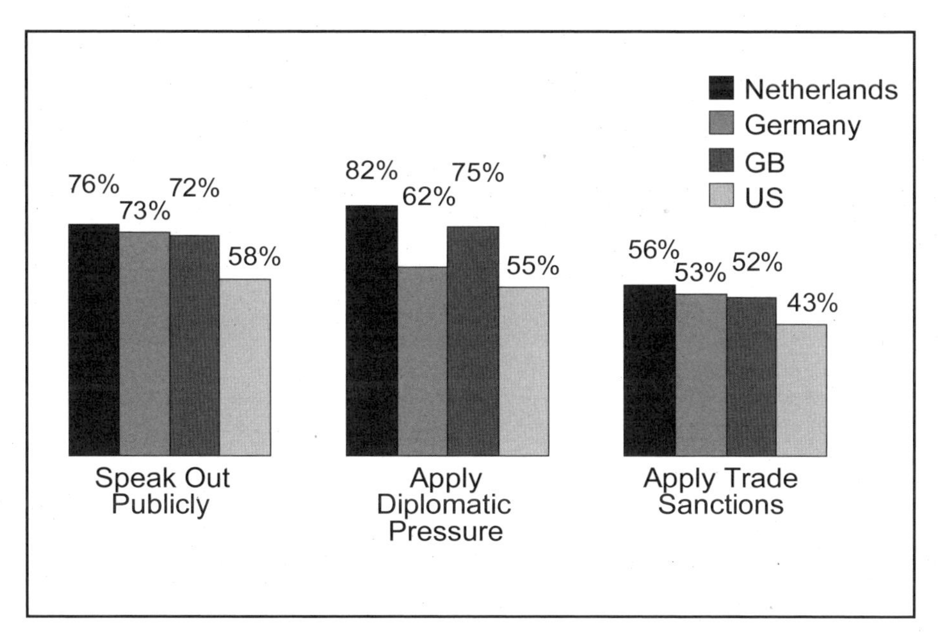

Figure 17–13. Approval for Action to Stop Whaling.

Response to the question, "Would you approve or disapprove of the ___ government taking action to convince Iceland to stop whaling?"

Base: Adults (Netherlands: 502; Germany: 1,000; GB: 1,003; US: 1,012), 2003. Source: IFAW/MORI.

ported by the majority of people in those countries (with the exception that only 43% of Americans supported trade sanctions), if Iceland didn't stop whaling. That would get their attention.

In conclusion, there are four stages of effective communication:

- *Awareness.* Here's who we are. If people don't know that they aren't going to listen to you.
- *Involvement.* Here's what we can do for you. Remember the pig and the chicken.
- *Persuasion.* Here's what we think, because if you've dug the foundation of awareness, and you've built the framework of involvement, then they will pay attention. I can show you advertisements in the newspaper today, ads on the television last night, that are going to fail. They are going to fail because they haven't come from something that people are aware of, something worth listening to, something that's got visibility. Or, they wonder, does it have anything to do with me?
- And finally *action:* here's what we want you to do about it.

NOTES AND SOURCES

[1] This chapter is an edited transcript of Prof. Worcester's presentation at the IFAW Forum.

[2] See Worcester, R. 1999. The British: Reluctant Europeans: British and The Euro – Forecasting the result. Lombard Research, Ltd., London, England. 15 September. Available at http://www.mori.com/pubinfo/rmw/the-british-reluctant-europeans.pdf.

[3] See http://www.mori.com

[4] See Lavigne, D.M., V.B. Scheffer, and S.R. Kellert. 1999. The evolution of North American attitudes toward marine mammals. Pages 10-47 in J.R. Twiss Jr. and R.R. Reeves (eds.). Conservation and Management of Marine Mammals. Smithsonian Institution Press, Washington and London. 471 pp.

[5] For more on attitudes and values, see Menon & Lavigne, Chapter 12.

[6] See Mugisha & Ajarova, Chapter 10, and Corkeron Chapter 11, on ecotourism.

[7] This study is available from the author in the form of a 12,000-word Royal Society discourse.

[8] For more on Dominica, see Martin, Chapter 24.

[9] For a discussion of whales and whaling, see Papastavrou and Cooke, Chapter 7.

Chapter 18

Conservation, Development and Poverty Alleviation: Time for a Change in Attitudes

John F. Oates

In a book published in 1999, I presented my views on the root causes of what I considered to be a crisis for wildlife conservation in the forest zone of West Africa, and argued the need for changes in international conservation policy.[1]

In that book, I argued that the challenges faced by conservation in West Africa (which are not unique to that region) were related in part to large human population increases since World War II, along with economic development and associated road-building, increased forest exploitation, farming, mining and the hunting of wildlife. These forces had led to severe reductions in the extent of forest, and in the numbers of forest animals. Some species that are particularly sensitive to hunting and habitat disturbance (such as red colobus monkeys, *Procolobus badius waldroni*) had been driven to the brink of extinction in many areas.[2]

I suggested that international conservation policies had exacerbated the crisis. Despite evidence that traditional protected areas can conserve tropical nature, new international conservation policies formulated in the 1970s and 1980s de-emphasized the importance of protecting wild nature because of its intrinsic value and, instead, stressed the importance of regarding conservation as a component of economic and social development. In my book, I argued that such policies were based on political considerations and financial expediency, and that they were in part a response to large sums of money being made available by development agencies for projects that included a conservation component. The widely accepted policy of encouraging sustainable development emphasized the management of natural resources to promote human material well-being, rather than the protection of nature.

Projects I witnessed in West Africa that attempted to integrate conservation and development had not evidently improved the status of wild animal populations. Wildlife populations had continued to decline as internationally-sponsored projects, generally managed by highly-paid foreign consultants, emphasized development efforts around protected areas without significantly increasing law-enforcement efforts against poachers. These projects tended to raise the expectations of local people that "development" was being brought to them, only to leave them frustrated when short-term foreign aid grants and contracts ran out. I also saw that conservation-and-development projects located in the surroundings of protected areas could slow emigration, while encouraging immigration by people attracted to the possibilities of short-term material gain.

I noted that in the 1990s, further elaboration of human-centered conservation policy began to stress the idea that integrated conservation-and-development projects were most effectively carried out at the level of loosely-defined human "communities," and community-based natural resource management schemes became popular. Policy-makers argued that local communities of rural people in Africa and elsewhere were inherently "wise users" of nature (in the proper sense of the term) and that, if more power was devolved to them, they would do a better job of conservation than government agencies.

In fact, rural communities in West Africa are rarely egalitarian or especially cooperative, and their members will exploit wildlife populations without regard to long-term sustainability, just as most human populations have done throughout history. Humans are eclectic foragers, and are prepared to exploit any given resource to a point of exhaustion, and then switch to other sources.[3]

Although it was argued in the 1980s and 1990s that traditional conservation, involving the enforcement of hunting laws and the establishment of strictly protected areas, is incompatible with the realities of life for people in poor rural areas in developing countries, I noted that the example of India shows that this is not so. While having one of the densest human populations in the world, and persistent rural poverty, India managed to create and maintain effective nature reserves and to stave off the extinction of endangered species like the tiger.

Based on my personal experiences in West Africa and India, I called for a greater emphasis in international conservation policy on the strict protection, by governments, of nature. This protection, I argued, is of importance not only because of the aesthetic value of nature to people now and in the future, but also for ethical reasons, because of nature's intrinsic value and because of the "right" of other species to exist. And I argued that one mechanism to pay for conservation, in cases where it is not generating revenue from some kind of harvesting, is through the establishment of trust funds, established with money from rich countries.

In the same year that my book appeared, John Terborgh of Duke University published *Requiem for Nature*, in which independently he made many of the same points, and stressed his own concerns.[4] Arguing on the basis of many years of experience in tropical forest ecology and conservation, especially in South America, Terborgh also pointed to a large and rapidly-growing human population as the underlying cause of the rapid disappearance of tropical forests and their fauna, and for general environmental degradation.

Like me, Terborgh argued that nature and biodiversity must be conserved for their own sake, not because of their utilitarian value. No country, he pointed out, has achieved truly sustainable development, and he suggested that, when conservation organizations advocate sustainable use, conservation principles are being abandoned. The best hope, Terborgh argued, for the perpetuation of tropical nature lies in parks and similar protected areas; these should be located in key areas of high species richness and endemism, and they must be large if they are to retain their biodiversity (in part because of the significant role of top predators in ecosystems).

Terborgh agreed that integrated conservation and development projects are flawed by short funding cycles and tend to attract people to the edges of parks, thus increasing destructive pressures. Conservation projects that focus on local people miss the fact, he said, that such people are generally minor players in a larger theater in which powerful figures in central government and business have most influence on environmental policy. Terborgh also warned that the policy of supporting the rights of indigenous people to remain inside forest parks and exploit park resources is likely to lead to the gradual destruction of the parks from within.

Although Terborgh argued that nature conservation works best when it is in the hands of national governments, he also noted that many tropical parks exist only on paper, and that others are generally poorly protected. This is because national institutions are often weak in developing countries, corruption and lawlessness are common, ethnic conflict is widespread, and inequities in power and wealth are the norm.

Based on these observations, Terborgh concluded that, for the foreseeable future, people in rich western countries will probably have to pay for the protection of tropical parks. He suggested that this might best be done through trust fund mechanisms, and perhaps through supporting elite protection forces. He argued that nature protection is of such importance that internationalizing it under the United Nations should be considered, while keeping in mind the key role of national governments.

TRENDS IN INTERNATIONAL CONSERVATION SINCE 1999

Do the arguments presented by John Terborgh and myself in 1999 still hold today? Have there been significant changes in international conservation policy? Have there been obvious gains or losses for conservation, especially in the tropical forest systems on which we focused? It is my perception that there has been some progress, but also many setbacks. I will first summarize what I see as positive trends, then consider what I perceive as less positive developments.

Positive trends

In recent years there has been ever-increasing attention in popular and professional media to "biodiversity" and conservation. For instance, among new professional journals in the field, *Animal Conservation* began publication in 1998, and *Conservation in Practice* (a journal of applied conservation, sister to the more theoretical *Conservation Biology*) appeared in 2000. *Conservation Biology* itself contained 1540 published pages in 1999, and 1884 in 2003. In 2003 The Association for Tropical Biology was renamed The Association for Tropical Biology and Conservation.

More money than ever has flowed to conservation. For instance, the annual operating budgets of three large international conservation NGOs have increased by more than 50 percent from 1999 to 2003, from $332 million to $502 million. Operating expenditures for the World Wide Fund for Nature's (WWF) International Network in its 2003 financial year were CHF 521,631,000 ($386,007,000 at June 30, 2003 exchange rate), compared to CHF 456,010,000 ($291,846,400) in 1999; Conservation International's operating expenditures jumped from $26,576,000 in 1999 to $83,960,000 in 2003; and the international programs division of the Wildlife Conservation Society increased from $13,327,222 in 1999 to $31,843,000 in 2003.[5]

The World Bank reports that in the 12 years prior to 2003, it provided $450 million in biodiversity conservation funding through the Global Environment Facility (GEF).[6] Part of this is presumably accounted for by GEF and World Bank support to the $150 million Critical Ecosystems Partnership Fund (CEPF), launched jointly with Conservation International (CI), the MacArthur Foundation and the Government of Japan in 2001. The CEPF provides funding to "civil society groups" in biodiversity hotspots. Also launched in 2001, with $100 million from the Gordon and Betty Moore Foundation, is the Global Conservation Fund (GCF), which devotes itself to protected area creation and management in hotspots, wilderness areas and important marine regions. Both the CEPF and the GCF are administered through CI.[7]

More meetings and workshops are being held on conservation themes, and these meetings are tending to become larger and more expensive. The World Parks Congress (WPC), convened in Durban, South Africa over 10 days in September 2003, attracted over 2,500 participants, and was so large that it took participants hours just to register.[8]

More attention is being given to the importance of protected areas in conservation. At the WPC, it was announced that 11.5% of the planet's land area is now within protected areas. In 2002, the Wildlife Conservation Society and other organizations helped Gabon to establish a network of 13 national parks, where there had been none before; these parks cover 11% of Gabon's land area. Conservation International has sponsored a study that found that in a sample of 93 parks and other protected areas in 21 countries, the protected areas showed less land clearance, higher wildlife populations, less loss of commercial trees, less burning and less grazing than surrounding areas.[9]

"Conservation concessions" have also been introduced as a new protection tool. Under such schemes, a conservation organization arranges to pay government or other landholders for the right to use an area of land for conservation rather than for exploitative use, such as logging. This approach has been developed by CI and others, and has already been put into effect in a 340,000 acre (137,600 ha) area of the Peruvian Amazon.[10]

In 2001 the World Bank introduced a new environment strategy, which includes a revised safeguard policy designed to minimize the adverse environmental impacts of projects supported by the bank.[11] The bank has also conducted a third comprehensive review of its environmental impact policies.[12] This review emphasizes the need to make safeguard policies uniform, transparent and effective, and to incorporate them at an early stage of project planning.

Less Positive Trends

Despite the large number of words and relatively very large sums of money now being devoted to conservation, the status of wild places and of wild plants and animals continues to deteriorate, at least over much of the tropics. Although the world appears to be paying more attention than ever to conservation, and although a good deal of this attention pays at least lip service to the importance of protecting nature, there still exists a large constituency for the ideas that (a) conservation should be seen as intimately linked to development, and (b) that people's perceived needs should be given high priority in the planning of conservation areas and projects. Some examples of these less positive trends follow.

Continuing losses of wildlife and habitat

Examples of continued loss of tropical habitat and wildlife come from the three countries with the largest areas of tropical forest: Brazil, Democratic Republic of Congo (DRC), and Indonesia.

In April 2004, the government of Brazil announced that the rate of Amazonian deforestation rose 2 percent in 2003, to produce a loss of 9,169 square miles (23,750 km^2) of forest, the second highest annual loss recorded since 1988.[13]

Conflict and anarchy in the eastern Democratic Republic of Congo (especially pronounced from 1997-2003) led to increased bushmeat hunting, and some villages had to abandon agriculture and depend almost solely on bushmeat; in the remote Okapi Faunal Reserve, pygmies are reported to have been forced to hunt at greater and greater distances from the road to assure a catch.[14] Assuming a more stable future, over the next decade the government of DRC plans to increase timber production from $44{,}000m^3/yr$ to around $10{,}000{,}000m^3/yr$.[15] In the forests of central Africa as a whole, the scale of the bushmeat trade has continued to increase, as forests in previously inaccessible areas are opened up by logging operations.[16]

A study of deforestation across Indonesia has predicted that, if current rates of loss continue, all the lowland forest on Sumatra will be gone by 2005, and on Kalimantan by 2010.[17] A separate analysis of remote sensing data from Kalimantan alone has shown that from 1985 through 2001, forest in West Kalimantan's lowland protected areas was reduced by 63 percent, while in other parts of Kalimantan 48 percent of forest cover was lost in protected areas, and much of the remaining forest was reduced to small fragments.[18]

Continuing emphasis on integrating conservation with development

Although more money is being spent by international conservation organizations and donor agencies, this has not necessarily translated to better nature protection on the ground.

Conservation organizations have indeed become richer, but a significant part of the increase in their expenditures is accounted for by larger administrative offices (often in neighborhoods with high rents) and larger numbers of personnel in their headquarters. Meanwhile, there is a tendency for field programs to be run with short-term grants from foundations and aid agencies, such as the U.S. Agency for International Development (USAID), the European Union Development Fund, the Canadian International Development Agency (CIDA), Germany's Gesellschaft für Technische Zusammenarbeit (GTZ) and the U.K. Department for International Development (DFID, formerly Overseas Development Administration, ODA). To receive funds from such donors, projects must often satisfy criteria that can include paying attention to such humanitarian issues as community development, poverty alleviation, gender equity and the curbing of HIV/AIDS.

Aid-agency funded projects, which typically have short-term budgets of one million dollars or more, not only promote the fallacy of development as a means to better conservation, they also directly encourage behaviors that are the antithesis of nature conservation. The potential availability of large sums of money over a short period of time incites a clamor from consultants and non-governmental organizations (NGOs) eager not to miss their share. Such funding also tends to encourage corruption. The problem of corruption in externally-funded conservation-and-development projects has previously been raised by John MacKinnon, who has suggested that donor agencies have been naïve about this problem.[19]

In the area of West Africa where my own research and conservation efforts are concentrated, and where wild nature is under very great threat, much of the money and attention in conservation is still being devoted to projects that stress development, poverty alleviation and community-based management, rather than the basic protection of threatened wildlife.

For example, USAID is currently funding a consultancy group to consider agricultural intensification and community-based natural resources management in a zone around Cross River National Park in Nigeria. Local conservation NGOs are participating in this project, as they are with a program funded by CIDA. The CIDA-funded program has as its stated aim the strengthening of NGO capacity to support communities around the Cross River National Park and the Obudu Plateau, and includes projects looking at issues such as community resource mapping, gender equity, and HIV/AIDS awareness. The Nigerian Conservation Foundation, one of the partner NGOs in this program, and usually considered to be Nigeria's foremost non-governmental wildlife conservation organization, now states that it "focuses attention on building a harmonious relationship between the natural resources it seeks to conserve and the people who depend directly on these resources for their food, shelter and clothing. NCF's field projects try to integrate conservation with rural development in order to achieve long-term sustainable development of our natural resources for the benefit of all".[20]

In Cameroon, GTZ has sponsored studies in the Takamanda Forest Reserve, looking at ways to give villagers living inside the forest greater rights to exploit forest resources. There are recent signs, however, that the government of Cameroon may resist this proposal, and support more stringent protection of the area, one of the last homes of the critically endangered Cross River gorilla (*Gorilla gorilla diehli*).

At Cameroon's Korup National Park, WWF, ODA, GTZ and other agencies have been involved for many years with the Ministry of the Environment and Forestry (MINEF) in a conservation-and-development project.[21] Major funding for this Korup Project has come from the European Union and the German government, but much of this funding has gone not to nature protection but to a socioeconomic development program in a zone around the park.[22] The U.K.'s ODA withdrew from Korup in 1998, and GTZ's component of the project ended in late 2003, to be replaced by an organization called Conservation and Development Service (CODEV), which has a much reduced level of funding and technical assistance compared to the earlier Korup Project. The national park, covering an area of 1,259 km^2, has only 26 guards, and only two of these have long-term employment contracts with MINEF. The remaining guards are employed by WWF with short-term funding, and their motivation is low. Expectations of development assistance to the villages in and around the park have been only partly met, creating resentment and making the villagers uncooperative with park management. Not surprisingly, hunting is reported still to be rife over most of the park away from park headquarters, from which small-scale ecotourism and research efforts operate.

Increasing emphasis on relationships between poverty alleviation and conservation

The United Nations Millennium Declaration in 2000 stressed poverty eradication as a major concern in development policy (U. N. General Assembly Resolution 55/2, 2000), and poverty eradication or alleviation is now incorporated into most development-agency policies. These policies have a strong influence on conservation strategies because of the linkages that have arisen between conservation organizations and development agencies.

The United Nations Development Programme, in particular, has argued that poverty reduction can play a pivotal role in environmental protection; the United Nations Development Programme's (UNDP) Human Development Report 2003 states:

> ...many environmental problems stem from poverty – often contributing to a downward spiral in which poverty exacerbates environmental degradation and environmental degradation exacerbates poverty. In poor rural areas, for example, there are close links among high infant mortality, high fertility, high population growth and extensive deforestation, as peasants fell tropical forests for firewood and new farmland... And when poor people degrade their environment, it is often because they have been denied their rights to natural resources by wealthy elites.[23]

Meanwhile, on its web page describing its policy on biodiversity, the World Bank states that all the activities it supports (including establishing and strengthening protected areas) have important links to poverty alleviation initiatives.[24]

In the same way that United Nations policy formulations played a major role in driving the 1980s combination of conservation with sustainable development, the new stress on poverty, and the funds made available for its alleviation, have led to the design of increasing numbers of projects claiming that they will solve conservation problems by reducing rural poverty. Indeed, in a recent commentary, Sanderson & Redford suggest that poverty alleviation is supplanting biodiversity conservation on the international aid agenda. [25]

Sanderson & Redford express their disagreement with the argument that poverty alleviation on its own will achieve conservation, even when no specific biodiversity protection measures are attempted. I share their skepticism. Poverty eradication is, of course, an important humanitarian goal; and some of the relationships between poverty and environmental degradation laid out by the UNDP may exist in some form. On the other hand, many years of efforts to promote sustainable development in the tropics, and to link development to conservation, have met with little success, or downright failure. I am not aware of good evidence that the linking of poverty-alleviation with conservation will be more successful, either for reducing poverty or for protecting nature.

In the field of environmental economics, it is still a matter of debate whether alleviating poverty (and therefore increasing material wealth) automatically translates into gains for the environment and nature conservation. For instance, the Environmental Kuznets Curve suggests that as per capita income rises, environmental degradation initially increases, only to decrease when people become really wealthy.[26]

Support for the idea that increasing wealth can actually lead to increasing pressure on the environment comes from an article by Naidoo & Adamowicz in *Conservation Biology*.[27] They tested relationships between wealth and wildlife conservation by comparing data from over 100 countries on land use and gross national product, with the numbers of threatened plant, invertebrate, fish, amphibian, reptile, bird and mammal species. With the exception of birds and, in some cases, mammals, all taxonomic groups showed increasing numbers of threatened species as the log of per-capita GNP increased. Only bird data conform to the Kuznets curve, with numbers of threatened species first rising with rising GNP, then declining, perhaps related to a public liking for birds in some rich countries.

Such analyses conform to commonplace observations. In general, as people get wealthier they consume more and more products from their environment and, with the use of more technology, they increase their extractive and destructive powers exponentially. Only at levels of high wealth do societies show increasing concern for environmental quality, at which point they may then act selectively to protect preferred species and sites. To quote from John MacKinnon:

> Every step of the development ladder is accompanied by an overall increase in resource use levels. People never have 'enough.' When a man graduates from a bicycle to a motor bike he suddenly needs money to pay for the fuel. This need forces him to cut even more forest or sell more wildlife than when he was a subsistence farmer.[28]

Rural poverty in tropical countries is surely much more susceptible to alleviation by large-scale changes in governance and culture than by small-scale NGO conservation projects run on short-term grants. And as with development projects generally, poverty-alleviation efforts conducted close to conservation target areas are likely, if successful, to increase rather than decrease pressures on nature.

Calls for greater rights for indigenous people

Along with poverty alleviation, another issue that has gained salience in conservation policy formulations in recent years, particularly for tropical regions, is the rights of indigenous people in relation to protected areas. This was one of the core issues addressed at the World Parks Congress in Durban in September 2003, and it is reported that the congress gave increased recognition to indigenous peoples' rights.[29] In a declaration to the congress, a coalition of indigenous peoples noted that they should be regarded as "rights-holders, not merely stakeholders," and called for restitution for lands expropriated as protected areas, for compensation to be provided, and for indigenous peoples to be able to participate in all aspects of protected area administration and management.[30]

This complex and vexing issue cannot be addressed adequately in a short paper of general scope. A particularly useful discussion on the issue can be found in the pages of the journal *Oryx* in 1998 and 1999. In this exchange, John Burton pointed out the difficulty in defining which people are indigenous in any given area.[31]

In searching for a definition of indigenous people, I found that the United Nations refers to Article 1 of the International Labour Organization's Convention on Indigenous and Tribal Peoples, which states that people may be regarded as indigenous:

> ...on account of their descent from the populations which inhabited the country, or a geographical region to which the country belongs, at the time of conquest or colonisation or the establishment of present state boundaries and who, irrespective of their legal status, retain some or all of their own social, economic, cultural and political institutions.[32]

This is obviously a very broad definition, that can be interpreted to give very many people indigenous status, especially since Article 2 of the convention states that "Self-identification as indigenous or tribal shall be regarded as a fundamental criterion for determining the groups to which the provisions of this Convention apply." With regard to the declaration made to the World Parks Congress, this interpretation of "indigenous" could allow almost any group of people to claim rights to occupy and/or manage a protected area. As John Henshaw pointed out in the discussion published in *Oryx*, this highlights the inadequacy of approaching wildlife and habitat conservation, and its necessary biological underpinnings, from the standpoint of political doctrines.[33]

Social scientists writing about conservation often make the argument that the creation of protected areas involves removing indigenous or local people from their land.[34] This argument ignores the fact that many tropical parks have been established in precisely those areas that have contained few or no resident humans in recent times, and that it is the low densities of people in these areas that has led to the survival of the rich biodiversity which the parks have been created to protect. Almost all regional and national land-use planning that is designed to produce long-term benefits for a majority of people has some adverse consequences for a small number of people. Plans to create new roads, railways, airports and reservoirs will generally displace some local residents, and it is well-established policy that such residents receive appropriate compensation when they relocate. I see no obvious reason why national protected-area planning should proceed according to a different philosophy.

More conservation workshops – but do they help?

Projects that emphasize linkages between conservation and development, and between conservation and poverty alleviation, often have a series of workshops as a major component. In places like West Africa, frequent workshops of this kind over the last decade have been a powerful mechanism for inculcating in local conservationists the notion that wildlife conservation is part and parcel of sustainable development and poverty alleviation. To me, there are parallels with missionary activity. Frequent preaching of the same message often does lead to its widespread adoption, especially when rewards are attached. Workshop attendees expect to be paid relatively handsome per diems to attend workshops and to be put up in good hotels, and these are strong incentives in countries where the average daily income is $1-$2/day. Well-funded, comfortable meetings also reinforce the idea that large funds are potentially available for projects that combine conservation with development. Workshops also seem to be favored by their sponsors, because they are a discrete activity, easier to administer and report on than are the difficult long-term realities of trying to make a tropical protected area work better. Most of those involved in a workshop probably feel that something worthwhile has occurred, when nothing has changed actually on the ground, and may not change.

THE NEED TO CHANGE ATTITUDES

Nature conservation continues to face huge challenges, especially in the tropics. Since 1999, tropical conservation has received plenty of attention (at least on paper and in digital form) and there has been increased acknowledgement of the importance of protected areas, but the balance of evidence suggests that actual wildlife and habitat conservation in the tropics continues to lose ground. Not only in general public discourse, but also within conservation circles, the rights and needs of one species, *Homo sapiens*, continue to be stressed relative to those of the millions of other species with which we share the planet. There is widespread acceptance of the propositions that conservation is part of something called sustainable development, that community-based conservation works (despite much evidence to the contrary), and that conservation can be achieved through projects to alleviate poverty. This dogma has been adopted by conservation practitioners in tropical countries, to the general detriment of effective nature conservation.

In my view, wildlife in West Africa and other similar parts of the tropics will continue to lose ground unless we can do a better job of changing attitudes as to what conservation is about. If we continue to regard conservation as just another aspect of efforts to increase the material well-being of humans, we will not retain many other species or the complex ecosystems and processes which sustain them. Rather than continuing to follow this inappropriate course, we need to go back to basics and examine why it is that people interested in wildlife and nature have the concern they do for conservation. Frequently, the root cause of this concern is the aesthetic value of nature; finding animals, plants and wild places beautiful and inspiring. This valuation is often combined with a belief that nature has an intrinsic value, and that other species have some right to exist, to find some corner of the planet where they can live freely.

We know that aesthetic and ethical values towards nature can be nurtured in young people. As I said in my book, such distinguished individuals as Richard Leakey, George Schaller and Edward Wilson, who have all thought deeply about conservation, have independently made the point that one can find an inherent appreciation of nature in people everywhere.[35] This is widely seen in the fascination of young people everywhere with animals.

The need to reduce poverty and to increase human well-being, especially in tropical countries, should not be minimized. But this issue has become too closely entangled with nature conservation. Of course there are some linkages between development and conservation, but conservation also has to be seen as very important in its own right. Many people today value nature for aesthetic and ethical, rather than economic, reasons, and this is likely to be the case in the future. Quite apart from the right of other species to exist, preserving nature now can therefore be seen as a benefit for future generations of humans.

For these reasons, I think that conservationists should not be afraid of trying to persuade others of the importance of their core values. Changing the attitudes of those who shape policies will not be an easy task, however, especially because these policy-makers reflect the increasingly materialistic attitudes of the constituents they represent.

It is my perception that in the last 30-40 years, people in many parts of the world have become more rather than less selfish and materialistic, and selfish materialism is not a sturdy foundation on which to build real conservation. The emphasis of a link between conservation and development, and the widespread use of professional consultancy companies to implement conservation-and-development projects, have tended to promote a materialistic approach to conservation at the expense of an aesthetic and ethical one.

These realities, the fact that it can take a long time to change attitudes, and the fact that nature faces a crisis in many parts of the tropics, suggest to me that, in the short term, we are probably going to have to rely on parks as a last refuge for much of tropical nature, and employ a rather hard-nosed monetary approach to park protection. As John Terborgh and others have suggested, those people in rich western countries who are concerned about tropical nature will probably have to pay for its protection, at least for the time being.[36] Money is needed both to compensate people whose livelihoods may be disrupted by the existence of a park (fair compensation is more compatible with conservation than user rights), and to cover the direct costs of protection (including staff salaries, equipment and infrastructure). Payments by people and organizations in western countries for the protection of tropical parks are probably best achieved by using long-term trust-fund mechanisms rather than by short-term grants.[37] Such payments, and any necessary outside technical assistance, require the full cooperation of national governments, who can be resentful of too direct a western role in the management of their natural resources.

This approach is essentially short-term crisis management. In the long-run, conservation seems likely to fail unless there is a strong constituency of people urging that nature be protected because of its aesthetic and intrinsic values.

NOTES AND SOURCES

[1] Oates, J.F. 1999. *Myth and Reality in the Rain Forest: How Conservation Strategies are Failing in West Africa*. Berkeley: University of California Press.

[2] Oates, J.F., Abedi-Lartey, M., McGraw, W.S., Struhsaker, T.T. & Whitesides, G.H. 2000. Extinction of a West African red colobus monkey. *Conservation Biology* 14: 1526-1532.

[3] See Brooks, Chapter 16, for a Darwinian explanation of such behaviour.

[4] Terborgh, J. 1999. *Requiem for Nature*. Washington, D.C.:

Island Press.

[5] According to financial statements on the organizations' web sites and personal communications from staff of CI, WCS and WWF. The money spent by international conservation organizations is still modest compared with the revenues and expenditures of major international business corporations.

[6] Available at: http://lnweb18.worldbank.org/ESSD/envext.nsf /48ByDocName/Biodiversity.

[7] From the website of Conservation International, http://www.conservation.org.

[8] Salafsky, N. 2003. The ghost of SCB future. *Society for Conservation Biology Newsletter* 10 (4): 1,12.

[9] Bruner, A.G., Gullison, R.E., Rice, R.E. & da Fonseca, G.A.B. 2001. Effectiveness of parks in protecting tropical biodiversity. *Science* 291: 125-128.

[10] Rice, R. 2002. *Conservation Concessions – Concept Description.* Center for Applied Biodiversity Science, Conservation International, Washington, D.C. 5 pp.

[11] World Bank. 2001. *Making Sustainable Commitments: An Environment Strategy for the World Bank.* Washington, D.C.: The World Bank.

[12] World Bank. 2002. *Third Environmental Assessment Review (FY96-00).* Washington, D.C.: The World Bank.

[13] Downie, A. 2004. Amazon destruction rising fast. *Christian Science Monitor*, 22 April 2004.

[14] Wildlife Conservation Society. 2003. *Democratic Republic of Congo Environmental Analysis. Final Report.* Report to USAID, Washington, D.C.; for more on the bushmeat issue, see Eves, Chapter 9, and Milner-Gulland, Chapter 20.

[15] Wildlife Conservation Society. 2003. *Democratic Republic of Congo Environmental Analysis. Final Report.* Report to USAID, Washington, D.C.

[16] Fa, J.E., Currie, D. & Meeuwig, J. 2003. Bushmeat and food security in the Congo Basin: linkages between wildlife and people's future. *Environmental Conservation* 30: 71-78.

[17] Jepson, P., Jarvie, J.K., MacKinnon, K. & Monk, K.A. 2001. The end for Indonesia's lowland forests? *Science* 292: 859-861.

[18] Curran, L.M., Trigg, S.N., McDonald, A.K., Astiani, D., Hardiono, Y.M., Siregar, P., Caniago, I. & Kasischke, E. 2004. Lowland forest loss in protected areas of Indonesian Borneo. *Science* 303: 1000-1003.

[19] MacKinnon, J. 2002. "Avenues of futility in conservation." Unpublished presentation to annual meeting of Society for Conservation Biology, Canterbury, England.

[20] From http://www.onesky.ca/Nigeria/partners.html.

[21] In December 2004, the Cameroon Ministry of Environment and Forestry (MINEF) was split into two new ministries, with the Ministry of Forests and Wildlife (MINFOF)now responsible for Korup National Park.

[22] MINEF, 2003. *A Management Plan for Korup National Park and its Peripheral Zone.* Ministry of the Environment and Forestry, Yaounde, Cameroon; Oates, J.F. 1999. *Myth and Reality in the Rain Forest: How Conservation Strategies are Failing in West Africa.* Berkeley: University of California Press.

[23] UNDP. 2003. *Human Development Report 2003.* New York: Oxford University Press.

[24] Available at: http://lnweb18.worldbank.org/ESSD/ envext.nsf/48BYDocName/Biodiversity.

[25] Sanderson, S.E. & Redford, K.H. 2003. Contested relationships between biodiversity conservation and poverty alleviation. *Oryx* 37: 389-390.

[26] Barbier, E.B. 1997. Introduction to Environmental Kuznets Curve special issue. *Environment and Development Economics* 2: 369-382.

[27] Naidoo, R. & Adamowicz, W.L. 2001. Effects of economic prosperity on numbers of threatened species. *Conservation Biology* 15: 1021-1029.

[28] MacKinnon, J. 2002. "Avenues of futility in conservation." Unpublished presentation to annual meeting of Society for Conservation Biology, Canterbury, England.

[29] DeRose, A.M. 2003. *Fifth IUCN World Parks Congress. Special Bulletin on Global Process, no. 5.* Washington, D.C.: World Resources Institute. Available at: http://governance.wri.org/project_description2.cfm?ProjectID=148.

[30] "The Indigenous Peoples' Declaration to the World Parks Congress." September 2003. Available at http://www.treatycouncil.org/section_211812142.htm.

[31] Burton, J.A. 1999. Traditional rights – what do they mean? *Oryx* 33: 2-3.

[32] International Labour Organization. 1989. C169 Indigenous and Tribal Peoples Convention. Geneva, Switzerland: International Labour Organization. Available at http://www.ilo.org/ilolex/english/convdisp1.htm.

[33] Henshaw, J. 1999. Indigenous people and conservation. *Oryx* 33: 4-5.

[34] Brockington, D. & Schmidt-Soltau, K. 2004. The social and environmental impacts of wilderness and development. *Oryx* 38: 140-142.

[35] Oates, J.F. 1999. *Myth and Reality in the Rain Forest: How Conservation Strategies are Failing in West Africa.* Berkeley: University of California Press. p. 247.

[36] Terborgh, J. 1999. *Requiem for Nature.* Washington, D.C.: Island Press.

[37] Global Environment Facility. 1999. *Experience with Conservation Trust Funds.* GEF Evaluation Report no. 1-99. Washington D.C.: UNDP, UNEP, World Bank, 80 pp.

Chapter 19

The North American Model of Wildlife Conservation: A Means of Creating Wealth and Protecting Public Health While Generating Biodiversity[1]

Valerius Geist

The North American Wildlife Conservation Model has evolved over nearly a century in response to the near elimination of wildlife from most of the continent by the end of the 19th century. Garrett Hardin's "Tragedy of the Commons"[2] had run its course to the bitter end, followed by the extermination of "vermin", including grizzly bear, wolf, and even cougar, over wide areas of their range where they interfered with cattle and sheep production. Several once spectacularly abundant species went extinct, foremost the passenger pigeon and the Eskimo curlew. Waterfowl, shore birds, and even songbirds were then severely depleted by market hunting and uncontrolled pot-hunting, while the habitat of wildlife was being converted to plowed fields for corn, wheat or cotton, livestock pastures and urban sprawl.

Yet in these dark hours for wildlife there arose a unique system of wildlife conservation and management that restored wildlife to the North American continent and made it a source of wealth and employment. This restoration of wildlife and biodiversity to North America is probably the greatest environmental achievement of the 20th century, and the North American Model of Wildlife Conservation may be one of the great achievements of North American culture. It is most significant that it turned Garrett Hardin's "Tragedy of the Commons" into a "Triumph of the Commons"[3] and, contrary to advocates for private wildlife, it showed that private ownership of wildlife is, in the long term, incompatible with conservation.

The North American approach to wildlife conservation has been examined by a number of symposia,[4] as well as in discussions in the popular press,[5] and on the Internet. It encompasses both the United States and Canada, and was formed in close cooperation between leading individuals from both nations. Canada, a loyal colony of Great Britain, opted not for the manner of wildlife conservation of its European mother country, but chose instead to unite under new common policies with the United States. It is a model based on raw grassroots democracy, and is thus the product of innumerable political discussions and decades of hands-on experience. Consequently, it is not the product of a single mind, but expresses the collective wisdom of nearly a century of continent-wide debate and hard bargaining. It has retained what has worked. It has thus a deep wisdom that could not have been invented by any single mind.

We have before us, therefore, a successful conservation model, and one worthy of scrutiny regardless of one's political philosophy. And yet, ironically, this model of wildlife conservation has only recently been recognized as such.[6] It is poorly known or understood even in North America; it is politically incorrect for much of the urban electorate; and it is opposed by various special interests, including some agricultural and environmental organizations. You will not hear about it on radio or TV, and even among wildlife managers, there are a good many that must plead ignorance when asked about the North American Wildlife Conservation Model.

Nonetheless, a close examination of that model is most illuminating as it is pregnant with tested ideas about how to manage a renewable resource in a sustainable manner. However, it requires certain pre-conditions to flourish, such as a tradition of grassroots democracy, the acceptance of wildlife as food, ready access by all citizens in good standing to wildlife harvest, and the availability of the requisite tools, including weapons, which raises questions about its universality and transferability.

THE SUCCESSES OF THE NORTH AMERICAN MODEL OF WILDLIFE CONSERVATION

In reviewing the major achievements of the North American Wildlife Conservation model I am following primarily two publications.[7] Briefly, the achievements are as follows:

- **The recovery of wildlife and biodiversity continent-wide**. This includes the recovery of species that were at the brink of extinction a century ago, which means most species of wildlife. Some conservation efforts went so well that, in the case of the buffalo, the society dedicated to saving it –The American Bison Society – voted itself out of existence, considering its mandate fulfilled. Between 1974 and 1999, wild sheep in North America increased in number by almost 50 percent.[8] There are once again millions of white-tailed deer in North America as well as other big game, but the recovery also included waterfowl, shorebirds and songbirds. Where recovery remains wanting, concentrated efforts are at work to restore species, including the much-publicized efforts to restore grey wolves and whooping cranes. The plight of a few forms, however, has not been addressed by wildlife conservation groups, most notably the woodland caribou.[9]
- **It generated a novel economic use of wildlife** so that great wealth and employment are created. Wildlife is not merely sustained; it continues to grow and prosper! In 1996 some 77 million U.S. citizens spent in excess of 100 billion dollars on wildlife-related activities.[10] They created about 50,000 jobs per billion dollars (U.S.) in throughput. There are similar trends for Canada.[11] (The following may help visualizing the sheer size of the U.S. wildlife economy: if one divides the total first-time expenditure of 101 billion dollars into the area of the United States then one obtains an annual expenditure of about $27,500 per square mile.) Here we can also study the distinction between markets that destroy wildlife, such as markets in *dead wildlife*, and markets that increase wildlife abundance, such as markets based on encountering *living wildlife*. An example of the worth of wildlife is documented by the annual auctions for special big game hunting permits, such as the "governor's or premier's permits" for mountain sheep, but also for elk, moose and deer. These auctions, open to all, are limited to one permit for one trophy animal per year. For the less affluent, raffles have been established for a similar permit. In 1998, a record $405,000 was bid to hunt one bighorn ram in Alberta, Canada.[12] Hunting also creates public benefits such as the "freedom of the woods" that results from keeping large and potentially dangerous carnivores timid and afraid of humans, as without this we could not use our woods and campgrounds safely. In addition, once wildlife populations expand, hunting keeps in check such populations which otherwise could expand to cause damage to agriculture, forestry or the environment at large.
- **It led to a new uniquely North American profession: the university-trained wildlife biologist or manager.** The first notable practitioner among these was Aldo Leopold.[13] He rose to be an idol of not only wildlife biologists, but of the environmental movement at large with his inspirational writing.[14] It insured that North America's wildlife received well-qualified, professional attention and care in its conservation and management.
- **Public participation.** One of the greatest achievements of North American wildlife conservation is public involvement with wildlife. This includes the whole-hearted participation of the blue-collar segment of society in contrast to the primary involvement of the elite in European societies. This makes for a large volunteer force willing to act on behalf of wildlife. Outwardly, public involvement takes the form of a large number of conservation organizations, formed at the federal, provincial or state, and local levels. Notable among these are sportsmen's organizations supporting single species or related groups of wildlife, such as the Rocky Mountain Elk Foundation, Mule Deer Foundation, Ducks Unlimited, Foundation for North American Wild Sheep, and Wild Turkey Foundation. There are also effective conservation societies such as the venerable Boone & Crockett Club, the Campfire Club and the Audubon Society. The volunteers have great

achievements to their credit. The Rocky Mountain Elk foundation has conserved over 3.8 million acres of elk habitat since its inception. A volunteer force of less than 6,000 Americans and Canadians, uniting biologists, managers, hunters, guides, outfitters, and interested parties in a common cause under the Foundation for North American Wild Sheep, increased the mountain sheep population by almost 50% in the last 25 years. Yet this is a small foundation![8] These are examples – and there are many others – of what volunteers, irrespective of nationality, in free association, without call for legislation or government funding, can achieve under existing legislation. The genius of North America's system of wildlife conservation is that it captured the enthusiasm and support of all strata of society.

- **Taxing for wildlife.** North Americans generated a secure funding base for wildlife conservation when the user-pay principle was adopted as policy by the American Game Conference in 1930. Ever since, North Americans have taxed themselves on behalf of wildlife (Migratory Bird Stamp Act 1934; Pitman-Robertson, Dingell-Johnson and Fish & Wildlife Conservation Act, Alberta's Buck for Wildlife Fund, etc.).[15]
- **Habitat conservation.** North Americans created an extensive public system of protected areas for wildlife, including great national parks and monuments, wildlife refuges, provincial parks and ecological reserves. Habitat conservation on agricultural land results from initiatives such as the U.S. Conservation Reserve Program. In addition, there are significant ongoing private efforts to acquire habitat such as those by the Nature Conservancy, or the many foundations dedicated to wildlife. They act continentally, continually acquiring habitat by purchase or gift, or by securing habitat protection through liens on the land. In addition, military reserves, by long tradition, respect wildlife's presence and contain some of the finest wildlife habitats and populations.
- **International treaties.** North Americans recognised early on the need to protect and manage wildlife that crossed national borders in its migrations. They negotiated the first and effective international wildlife treaties, such as the 1911 Fur Seal Treaty, but above all the famous 1916 Convention for the Protection of Migratory Birds.
- **Conservation of large predators.** Despite early and continuing sentiments against large predators, they were nevertheless retained or reintroduced as a functioning entity of ecosystems. They are controlled, or protected or reintroduced, depending on circumstances. Also, predators are better off under hunting regulations, because the kill is very closely controlled, is under constant public scrutiny and persons are held accountable for each kill. Not so in Canada's national parks, where bears have a very high chance of being killed due to concerns for public safety.[16]
- **Preservation of non-game species.** From the very outset, the out-of-doors was considered an integrated whole. That is to say, very early on, under the so-called Roosevelt Doctrine,[17] conservation was considered broadly. Consequently, the history of bringing non-game species under the same umbrella as game species has a very long history. However, not all conservation was altruistic; rather, it was usually motivated by utility. This included songbirds that were considered early in this century effective allies against various crop insect pests.[18] Moreover, the focus on particularly desirable game species casts a broad halo effect from which non-game species benefit. Although specific legislation to save endangered species has been in effect across the continent, such legislation could not succeed in the absence of a hunting culture, which had practiced broadly based habitat conservation that simultaneously conserved biodiversity.
- **Law enforcement.** Enforcing conservation law in North America is normally a remarkably civil affair, although it can be as dangerous as in Europe[19] when commercial poaching is involved.[20] Because wildlife conservation is broad-based and an exercise in participatory democracy, there is much self-policing involved. The North American approach differs from European models in which wildlife is private property and its protection is pursued accordingly.

FOUNDATION POLICIES

The seven primary or root policies, the foundation values on which the North American Wildlife Conservation Model is built, were best summarized in a collaborative paper that included the insights of Shane Mahoney, then chief of Research of the Newfoundland and Labrador Wildlife Division, and John F. Organ, Wildlife Program

Chief of the U.S. Fish and Wildlife Service. This paper is of primary importance.[21]

1. Wildlife as public trust resources

Wildlife in North America is public property, not merely *de jure*, but also *de facto*. Wildlife may be held privately, but only as a trust for the public and at the discretion of the sovereign. The Public Trust doctrine has a long history in the U.S.

Why is public ownership of wildlife so important for wildlife conservation?

- Public ownership prevents the inevitable consequence of private ownership, such as the domestication of wildlife as well as its genetic alteration to fit market whims. Domestication systematically diminishes the anti-predator adaptations of a species by making it more tractable and easier to control under conditions of captivity. Domestication also has led to severely reduced brain-size.[22] Domestication is done so as to serve specific markets and therefore leads to genetic alteration of a species to produce desirable products. Gigantic antlers in deer or horns in buffalo are examples, as well as the restructuring of bison to assume the carcass conformation of cattle. The latter is done to increase the carcass value, as the carcass of domestic cattle has a higher proportion of high-priced cuts compared to those of wild bison. Selecting for antler size in deer selects for social incompetence. Domestication is thus the systematic genetic alteration of innate adaptations. Such altered stock can escape into the public domain and pollute public wildlife irreversibly.
- Public ownership of wildlife largely prevents the mixing in captivity of many species and thereby prevents what parasitologists have labelled "transporting the zoo" (of pathogens and parasites). Each species carries its contingent of pathogens and parasites which, transferred to another species, may mutate into strains dangerous to public health. Transferring wildlife into domestication increases the risk of pathogens escaping into human populations. Private ownership of wildlife generates a disease bridge across which may pass diseases affecting livestock and public health. Retaining wildlife in strict public trust prevents wildlife farming and the building of a disease bridge between wildlife, livestock and people. It is good public health policy. The recent SARS epidemic originated in farmed wildlife, namely in farmed palm civet cats in China.[23] In any confrontation of private agricultural and public wildlife interests, wildlife is inevitably the loser.[24]
- Wildlife in public ownership insures the ecological basis for native cultures to continue. One way to diminish native cultures is to make wildlife and their habitat private property.[25]
- Because wildlife is in the public domain, it is possible to consider national systems of wildlife sanctuaries and wildlife treaties.[26]
- Because the state is ultimately responsible for wildlife, it is possible to hire professionals to practice conservation and management on behalf of the public. Here lies the origin of the North American profession of wildlife biologists.
- Wildlife in the public domain is subject to public scrutiny and concern. The public has a say in how wildlife is to be treated. When grizzly bears become private property, *de jure* – or *de facto* by virtue of being turned over to owners of private or leased land, their fate is no longer the public's business.
- Once wildlife is made private it pits private wildlife against public wildlife, a battle in which public wildlife is the inevitable loser.[27]

2. Elimination of markets for wildlife

The elimination of trafficking in dead game animals, or their parts and products derived from them, is one of the most effective and important policies of wildlife conservation. Its introduction was revolutionary, as North Americans were avid consumers and traders of wildlife at the turn of the 20th century.

Why is the elimination of markets in wildlife and its parts and products so important to conservation?

- The elimination of markets in dead wildlife eliminates a financial incentive for the illegal taking and selling of public wildlife. Where such incentive exists it promotes illegal markets and encourages the criminal element to enter and ruthlessly exploit wildlife. Law enforcement under such circumstances is hazardous in the extreme and of questionable efficiency.[28]
- Eliminating monetary value from wildlife encourages the public to enjoy wildlife for its own sake. A grizzly bear is no longer a walking bank account.
- The acquisition of wildlife outside the market place is bound to significant private effort. The

resulting individual efforts and exertions, the "sweat equity", as well as the significant monetary expenses incurred, act as deterrents to killing wildlife. So does the inability to sell legally killed wildlife.

3. Allocation of wildlife by law

Allocation of surplus wildlife for consumption by law, and not by the market place, insured an equal allocation of wildlife to citizens irrespective of wealth, social standing or land ownership. Every citizen in good standing is able to participate in the annual harvest of wildlife within the laws set by legislatures. In this instance, aboriginal people are an exception because treaty rights also govern their wildlife harvest.

Why is allocation by law so important to wildlife conservation?

- This policy generates a sense of propriety and ownership by those participating in the wildlife harvest and is fundamental to the public participation in wildlife conservation, be it directly as volunteers or indirectly via the legislatures.
- This policy, by encouraging citizens to regard wildlife as their own, generates large national and continental organizations of citizens who join together into societies on behalf of wildlife. Large foundations dedicated to single species or species cluster are a North American phenomenon. These non-government organizations channel funds and the efforts of volunteers towards the maintenance and spread of such wildlife as well as the acquisition of their habitat.
- Because all citizens in good standing have access to wildlife as prescribed by law, it removes wildlife from any image of elitism, or as the plaything of the filthy rich, a symbol of privilege. Wildlife controlled privately by an elite can become a symbol of the hated elite and suffer the consequences. This can be particularly tragic when public sentiment against the elite and their symbols are unleashed in revolutions.[29]
- Egalitarian allocation provides the basis for an equitable cost of conserving wildlife through a user-pays principle. Because enough of the public avail themselves of the opportunity to obtain wildlife for private consumption, there is enough funding for conservation. User-pay means that hunters are footing most of the bill for wildlife conservation and, in so doing, provide a benefit to society at large – the maintenance of wildlife and the continent's biodiversity.
- An egalitarian distribution of opportunities to acquire wildlife also generates indirect public benefits. One of these is the "freedom of the woods". In this case, the harassment of bears – through inefficient hunting – conditions them to avoid humans, allowing for safe camping and hiking. Clearly, this depends on reasonably large numbers of hunters going into bear habitat.

4. Wildlife can only be killed for a legitimate purpose

Wildlife can be killed only for cause. That is, it can be killed for food, for fur, in self-defence or in the protection of property. Wanton waste of hunted wildlife may be considered a felony in some jurisdictions. This policy obliges all hunters to properly make use of animals killed.

Why is killing wildlife only for cause a desirable conservation policy?

- This policy outlaws wanton slaughter that, in the days of market hunting, was commonly practised, and viewed as a mark of prowess among so-called hunters. Killing wildlife only for cause reduces mortality and questions all killing.
- Allocation plus regulation of the taking of wildlife by law is *enforced inefficiency*. This is a very important point, as it is the enforced inefficiency of harvest that generates wealth and employment. Efficient harvest, by contrast, eliminates wildlife without generating public wealth. Since an animal taken in hunting must not be wasted, it insures that the hunter spends a fair sum of money in transporting, processing, storing and consuming the animal. This generates a demand for services.
- Enforced inefficiency also triggers the invention of gadgetry, a consequence of ingenuity rewarded by the marketplace. Ironically, North America's wildlife economy is thus comparable to the economy inherent in the automobile industry, where the unending multiplications of a product that generates some convenience at best, or, at worst merely enhances the owner's status, produce huge wealth. Such gadgetry in no way enhances transportation efficiency.

5. Wildlife is considered an international resource

Wildlife is considered an international resource to be managed co-operatively by sovereign states. This policy is basic to international wildlife treaties as well as the broad based, continental co-operation between professionals and conservation organisations.

Why is wildlife considered formally as an international resource conducive to conservation?

- This policy brings wildlife to the highest political level as a public good. It insures federal involvement in all nations affected.
- This forces – by law – all federal, provincial, state and municipal jurisdictions affected into active cooperation.
- This generates a lasting federal attention to wildlife crossing the borders.
- Treaty law is considered strong law that supersedes that of lower national jurisdictions. Thus treaties are effective conservation and management tools.

6. Science is the proper tool for the discharge of wildlife policy

Science is considered to be the proper tool for discharging management responsibilities. This is the Roosevelt Doctrine.[30] This is another basic policy that gave rise to science-based wildlife professionals hired by the state to perform wildlife conservation.

Why is science important?

- Science is by and large our best tool to formulate appropriate management and policy options, because it is based on a disinterested pursuit of understanding. It stands apart from political considerations and favours a hands-off policy by elected representatives.[31]
- This policy assures that public wildlife is in the hands of exceedingly well-educated individuals and that it is scrutinized continuously.

7. The democracy of hunting

To paraphrase from Geist *et al.*[3] the concept of "sport hunting" has origins in Europe.[32] The term "sport" as applied to hunting referred originally to a code of honor rather than to a frivolous recreational pursuit. It was subsequently adopted to distinguish hunting under codes of fair chase from market hunting, and is not an appropriate descriptor of either the modern European or North American hunting.[33] The European archetype was dramatically different than what emerged as "sport hunting" during the 20th century in North America. The European model allocated wildlife by land ownership, privilege or income, whereas in North America, all citizens in good standing can participate. The European model, a manifestation of class conflict between aristocracy and commoners, led to wildlife poaching as a means for inflicting revenge on the ruling class.[34] Indeed, in Africa today efforts to combat poaching have led to development of programs designed to direct economic returns on hunting fees to the rural indigenous peoples who otherwise would have no reason to stop poachers.[35] In North America, where all citizens have the opportunity to participate, everyone is a stakeholder, not just the privileged. This has been termed by Leopold the "democracy of sport".[36] The foremost spokesman for egalitarian allocation, and participation of the common man in hunting was Theodore Roosevelt. He wrote eloquently of the societal gain to be made by keeping land available for hunting by the common people.[37] Hunting as a deep-rooted passion is thus fundamental to wildlife conservation,[38] but only within a framework of honorable, ethical conduct.[39] By adopting a code of "fair chase", North Americans explicitly opposed the excesses of wildlife slaughter, particularly in enclosures, as practiced in Europe at the turn of the 20th century, as well as historically.[40]

WHAT CAN WE LEARN FROM THE NORTH AMERICAN WILDLIFE CONSERVATION MODEL?

- Hunters support wildlife conservation because there is something in it for them: a payoff in their annual allocation of wildlife. The motive is selfish, not idealistic. As a profit motive drives a capitalistic economy, so a profit motive drives the North American system of wildlife conservation: the hope for a richer harvest and a richer experience in hunting. Consequently, with self-interests in wildlife, hunters become concerned, active spokespersons for, and supporters of, wildlife, and experience shows that wildlife will then flourish. A ruling elite that elevates wildlife against the self-interests of the common man causes wildlife to suffer and be destroyed by the common man, if and when the opportunity to grasp power arises in revolutions. This lesson goes back to medieval forest laws (which were in essence animal rights legislation), and is valid for today's top-down animal rights legislation. Our only hope to retain a thriving biodiversity is to embrace a human-centred view for the use of the biosphere, in which wildlife provides for human needs and aspirations and is therefore valued by a broad segment of society. A romantic, purely eco-centric view, that is, an impersonal and unselfish view of biosphere management that excludes broadly held aspirations to use resources by common people cannot but fail. How much wildlife can mean emotionally is illustrated by the novelist William Faulkner's response to being informed that he had won the

1949 Nobel Prize for Literature and would have to go to Sweden to receive it. Faulkner said "I can't get away. I'm going deer hunting!" And he so informed the Nobel prize officials by mail.[41]

- Wildlife must remain a harvestable resource, supplying in the first instance food for our tables. It is an alternative to agriculture generating utility from the land. It must not be viewed as a purely recreational resource, as a source of "sport" or entertainment. Its first order of utility is the provision of a harvest of unusual food of exceptionally high nutritional value.[42] Wildlife thrives with attention and dies from neglect. Utility fosters attention.
- We must, therefore, retain the utility of wildlife. For instance, songbirds were historically protected not for moral or ethical reasons and not because they are cute and entertaining, but because they were valued as destroyers of insect pests in fields, forests and gardens.[43] Today songbirds have no utility in North America, and enjoy little organised public support such as is enjoyed by native game birds, including turkey, ruffed grouse and waterfowl. Songbirds may have the protection of the law, but little in the form of tangible popular support – even from bird watchers.
- We must examine for retention the seven basic conservation policies that have served us so well in bringing back wildlife and retaining continental biodiversity in North America. These contain many counterintuitive lessons about how to maintain and foster a public resource. Would we but dare manage forests the way we – cheerfully – manage wildlife. Would we but manage marine fisheries the way we manage wildlife in North America – with an open, transparent and accountable system.
- One must point to the awesome power of the democratic process, in which we set aside willingly our differences and unite in a public cause – fostering the welfare of wildlife and, through it, the biosphere as well. One should recognize the power of volunteers as social equalizers, as reciprocal carriers of information and power. In this way, one retains accountability and openness that has characterised to date the relationship between wildlife managers and the public in North America. Establishing a partnership between managers and the public, and unlocking the spirit to act in the public good, is an essential component to achieve wildlife conservation.
- Today wildlife conservation in North America is beginning to suffer from an ignorance of the past, be it an uninformed judiciary or through uninformed managers of wildlife unable to defend the system. As Immanuel Kant once quipped: *We learn from history that we do not learn from history.* We must buck that trend!

The universality of the North American Wildlife Conservation Model is in doubt, however, because it is built on some fundamental assumptions, the primary one being that all citizens may participate in both the harvest of wildlife and its management. And those assumptions entail the availability of firearms to all citizens and not merely the country's elite. An armed citizenry, one practiced in the art of grassroots democracy and accepting of decisions reached by public debate and compromise, is fundamental. Therefore, there has to be an acceptance of responsibility for a public resource, despite embracing a capitalistic economy and values. Citizens must see wildlife as a common good and must accept sharing on trust. Even the country's elite must participate in the processes of wildlife conservation and must not be exempt from such. There must be willingness by the public to privately support wildlife, accepting public efforts at conservation as minimal at best.

Some agricultural interests would like to tie wildlife ownership to land ownership and make wildlife a private resource to be managed according to market demands and sold to the highest bidder. Such interests openly oppose the North American Wildlife Conservation model. The same goes for corporations who, for whatever reasons, control large land areas and are interested in generating revenue by leasing out hunting rights to the highest bidder. Support for these efforts comes from a significant sector of urban-based affluent hunters who chaff at bag limits, short seasons and crowded hunting grounds. Such individuals are effectively supported by gun control advocates who lobby for a disarmed public. In practice, gun control means disarming the blue-collar segment of society leaving the elite well armed. Without effective, egalitarian public hunting there will be little opposition to the privatization of wildlife, making it a plaything of the elite as it has been so often in the past. Canada's most unfortunate gun control legislation is well on the way to doing just that and it is thus in opposition to the North American Wildlife Conservation Model. It is self-evident, however, that in dictatorships this model is unlikely to be accepted, based as it is on armed civilians who practice effective grassroots democracy.

NOTES AND SOURCES

[1] An earlier version of this chapter may be found at: http://www.albertawilderness.ca/Issues/WL/Archive/AR0411 WL.pdf. An excerpted version later appeared in Wild Lands Advocate 12(6), December 2004, published by Alberta Wilderness Association, Calgary, Alberta. Available at: http://www.albertawilderness.ca/Events/Lectures/AL2004AR.pdf.

[2] Hardin, G. 1968. The Tragedy of the Commons. *Science* 162:1243-1248.

[3] See the misuse of the concept of the *Tragedy of the Commons* in relation to North American wildlife in *The Economist,* 22 October 1988.

[4] Geist, V. 1994. Wildlife conservation as wealth. *Nature* (London) 368:491-492; Geist, V. 1995. North American policies of wildlife conservation. pp. 77-129 in V. Geist and I. McTaggart Cowan (eds.) *Wildlife Conservation Policy.* Detselig, Calgary. Geist, V. 2000. A century of wildlife conservation successes and how to repeat it. pp. 17-22 in W. D. Mansell (ed.) *Proceedings of the 2000 Premier's Symposium on North America's Hunting Heritage.* Wildlife Forever, Eden Prairie, Minnesota. Geist, V., S.P. Mahoney, and J. F. Organ. 2001. Why hunting has defined the North American model of wildlife conservation. *Transactions of the North American Wildlife and Natural Resources Conference.* 66:175-183.

[5] Posewitz, J. 2004. *Rifle in Hand.* Riverbend Publishing, Helena Montana.

[6] Geist, V. 1988. How markets in wildlife meat and parts, and the sale of hunting privileges, jeopardizes conservation. *Conservation Biology* 2(1): 1-12.

[7] Geist, V. 1995 ibid.; Geist, V., Mahoney S. and J. F. Organ 2001. ibid.

[8] You can read all about it in D. Toweill and V. Geist. 1999. *Return of Royalty.* Foundation for North American Wild Sheep and Boone & Crockett Club, Missoula, Montana.

[9] Geist, V. 2003. Of history and man, modern and Neanderthal: A creation story founded on a historic perspective on how to conserve wildlife, woodland caribou, in particular. *Rangifer,* Special issue No. 14, pp. 57-63.

[10] U.S. Department of the Interior, Fish and Wildlife Service and U.S. Department of Commerce, Bureau of the Census. *1996 National Survey of Fishing, Hunting and Wildlife – Associated Recreation.* Available at http://www.census.gov/prod/3/97pubs/fhw96nat.pdf.

[11] Wilkerson, O. 2002. Then and Now: Human-Wildlife Interactions in Simon Fraser's New Caledonia. *Western Geography* 12:319-368. Available at: http://office.geog.uvic.ca/dept/wcag/wilkerson1.pdf.

[12] Trophy hunting has developed historically a number of times to excess in the Occident, in particular in late medieval times in Europe, but also during the 19th and 20th centuries. These auctions and raffles are an outgrowth of this. However, in North America trophy hunting was advanced by the Boone and Crockett Club originally as a means of reducing wildlife slaughter, shifting the emphasis from numbers of big game shot to quality of animals taken. Sportsmen gained standing not from the number of animals shot, but by the quality of trophies taken. In a similar vein the club advances "fair chase" in order to oppose imitation in North America of the European elite's killing of big game in enclosures, or "canned shoots" as popularly described. In Europe the quest for superior trophies led to detailed investigations into how to grow massive antlers in deer, leading to the finding that, following luxurious feeding, it was essential to discourage males from rutting. Males that abstained from rutting saved body resources. These were then available for inflated body and antler growth. The exceptionally rare "trophy" male in nature was, consequently, a male that did not participate in breeding and was thus of no consequences genetically. That is, trophy males are likely to be of low fitness (see Geist V. 2000. Under what system of wildlife management are ungulates least domesticated? pp. 310-319 in Elisabeth S. Vrba and George B. Schaller (eds.) *Antelopes, Deer and their Relatives.* Yale University Press, New Haven, CT).

[13] Leopold, A. 1933. *Game management.* C. Scribner's Sons, New York, NY. 481 pp.

[14] Leopold, A. 1949. *A Sand County Almanac: and Sketches Here and There.* Oxford University Press, New York.

[15] For an excellent review of legislative achievements see the *Foreword* written by Laurence R. Jahn to Aldo Leopold's 1986 edition of his 1933 *Game Management,* The University of Wisconsin Press, Madison, WI.

[16] McLellan, B.N., F.W. Hovey, R.D. Mace, J.G. Woods, D.W. Carney, M.L. Gibeau, W.L. Wakkinen, and W.F. Kasworm, 1999. Rates and Causes of Grizzly Bear Mortality in the Interior Mountains of British Columbia, Alberta, Montana, Washington and Idaho. *Journal of Wildlife Management* 83(3):911-920.

[17] Leopold, A. 1933. ibid. pp. 17-18.

[18] Hewitt, C. G. 1921. *The Conservation of Wildlife in Canada.* C. Scribner's Sons, New York.

[19] Busdorf, O. 1954. *Wilddieberei und Förstermorde.* Gersbach und Sohn Verlag, Braunschweig. 311 pp.

[20] Grosz, T. 1999. *Wildlife Wars.* Johnson Books, Denver; 2000. *For the Love of Wildness.* Johnson Books, Denver; 2001. *A Sword for Mother Nature.* Johnson Books, Denver; 2002. *Defending our Wildlife Heritage.* Johnson Books, Denver.

[21] Geist, V., S.P. Mahoney, and J. F. Organ. 2001. Why hunting has defined the North American model of wildlife conservation. *Transactions of the North American Wildlife and Natural Resources Conference.* 66:175-183.

[22] This is old information, see p. 368 in Geist, V. 1978. *Life Strategies, Human Evolution, Environmental Design.* Springer Verlag, New York.

[23] See World Chelonian Trust, http://www.chelonia.org/Articles/Palm_Civits_SARS.htm; National Geographic News, http://news.nationalgeographic.com/news/2004/01/0109_04 0109_SARS.html.

[24] See discussion in Geist 1995.

[25] See discussion in Geist 199.5.

[26] See Hewitt 1921.

[27] See discussion in Geist 1995.

[28] See the most illuminating book series on wildlife law enforcement by Terry Grosz (endnote 20). Available at http://www.johnsonbooks.com/catalog/author.php?mode=book&id=12.

[29] See discussion in Geist 1995.

[30] Leopold, A. 1933; ibid. pp. 17-18.

[31] But also see Hutchings, Chapter 6.

[32] Herbert, H.W. 1849. *Frank Forester's field sports of the United States and British provinces of North America.* Stringer and Townsend, New York, NY.

[33] Organ, J.F., R.M. Muth, J.E. Dizard, S.J. Williamson, and T.A. Decker. 1998. Fair chase and humane treatment: Balancing the ethics of hunting and trapping. *Trans. No. Am. Wildl. and Natur. Resour. Conf.* 63:528-543.

[34] Manning, R.B. 1993. *Hunters and poachers: A social and cultural history of unlawful hunting in England, 1485-1630.* Clarendon Press, Oxford, U.K. 255 pp.; Girtler, R. 2003. *Wilderer. Rebellen in den Bergen.* Bohlau, Vienna. Threlfall, W. 1995 Conservation and wildlife management in Britain. pp. 27-76 in V. Geist and I. McTaggart Cowan. 1995. ibid.

[35] Kock, M.D. 1996. Zimbabwe: a model for the sustainable use of wildlife and the development of innovative wildlife management practices. pp. 229-249 in V.J. Taylor and N.Dunstone, eds., *The exploitation of mammal populations.* Chapman and Hall, London, U.K. 415 pp.

[36] Meine, C. 1988. Aldo Leopold: His Life and Work. Univ. Wisconsin Press, Madison. 638 pp. See p. 169.

[37] Roosevelt, T. Van Dyke T. S. Eliot, D. G. and A. J. Stone 1902. *The Deer Family.* MacMillan Company, New York. See pp. 18-20.

[38] Mahoney, S. 2000. The enduring relevance of man's first try at life: Hunting viewed from the escarpment of history. pp. 13-17 in W. D. Mansell (ed.) *Proceedings of the 2000 Premier's Symposium on North America's Hunting Heritage.* Wildlife Forever, Eden Prairie, Minnesota; Shepard, Florence R. 2001. Bringing the sacred game home. pp. 22-28. in W. D. Mansell (ed.) *Proceedings of the 2000 Premier's Symposium on North America's Hunting Heritage.* Wildlife Forever, Eden Prairie, Minnesota.

[39] Posewitz, J. 1994. *Beyond Fair Chase.* Falcon Press, Helena and Billings, Montana.

[40] See Geist 1995, as well as Threlfall, W. 1995. Conservation and wildlife management in Britain. pp. 27-76 in V. Geist and I. McTaggart Cowan 1995 ibid.; Stahl, D. 1979. *Wild, Lebendige Umwelt.* K. Alber, Munich/Freiburg, 349 pp.

[41] Wegner, R. 2001. *Legendary Deer Camps.* Krause Publications, Iola, Wisconsin. p. 158.

[42] Medeiros, L.C., Busboon, J. R., Field, R. A., Williams, Janet C., Miller, G. L. and Betty Holmes. 2002. *Nutritional content of game meat.* College of Agriculture, University of Wyoming. Laramie, WY; Cordain, L., Watkins, B., Kehler, M. and L. Rogers. 2002. Fatty acid analysis of wild ruminant tissues: evolutionary implications for reducing diet-related chronic disease. *European Journal of Clinical Nutrition* 56(3):181-91.

[43] Hewitt, C. G. 1921; ibid.

CHAPTER 20

DEVELOPING A FRAMEWORK FOR ASSESSING THE SUSTAINABILITY OF BUSHMEAT HUNTING

E.J. Milner-Gulland

The hunting of animals in tropical forests for food ("bushmeat hunting") has emerged as a major conservation issue over the last few years, with concerns about the "bushmeat crisis" focusing on the forests of West/Central Africa.[1] Eating wildlife is not new, and reasons for our current preoccupation with this issue are complex.[2] However it is undeniable that bushmeat species are being severely overexploited in many areas, and that vulnerable species are threatened with extinction.[3,4] In other places, however, the situation seems relatively stable, with vulnerable species already extirpated, and the remaining resilient species apparently supplying markets on a relatively sustainable basis.[5] In order to prioritise conservation action effectively, we need to assess the relative sustainability of bushmeat hunting in different areas. We can also use sustainability assessments as the foundation for examining options for intervention to conserve species endangered by over-exploitation.

Bushmeat is a high-profile example of the issues facing conservation more widely. It involves balancing the imperatives of improving the livelihoods of desperately poor people and conservation of endangered species.[6] It also highlights the issues of controlling profitable commercial trade, and of battling against weak or non-existent governance.[7] Because bushmeat hunting is relatively non-selective, vulnerable species are killed along with resilient ones, and species-specific conservation actions are less straightforward to implement. Finally, the widespread and diffuse nature of the problem makes monitoring and intervention difficult to target. The closest analogue to bushmeat is artisanal reef-based fishing, but the separation of the resource from human habitation, the visibility of users and the limitation in landing sites makes fishing easier to monitor and control than bushmeat hunting. Hence bushmeat is a useful, if extreme, model for examining approaches to assessing sustainability, and more importantly, implementing effective and scientifically sound management. If conservation action for bushmeat species is successful, a significant step forward will have been taken for the conservation of natural resources more generally.

In this chapter, I will discuss the kinds of data that are required for a full assessment of the sustainability of bushmeat hunting, and compare these requirements to the data that are usually available. I will then address some of the issues that arise out of the weaknesses in the available data, and particularly the difficulty of inferring lack of sustainability from market data alone. There are already some methods in use for assessing the sustainability of bushmeat hunting. Simple indices of the ratio of extraction to production are the most widespread of these. I will examine the potential of these indices to provide a robust estimate of sustainability under realistic field conditions, and identify the key components that are needed in a valid sustainability index. New frameworks for assessing sustainability include Bayesian networks and simulation modelling. Finally I consider the prospects for conserving bushmeat species and identify priority situations for conservation intervention.

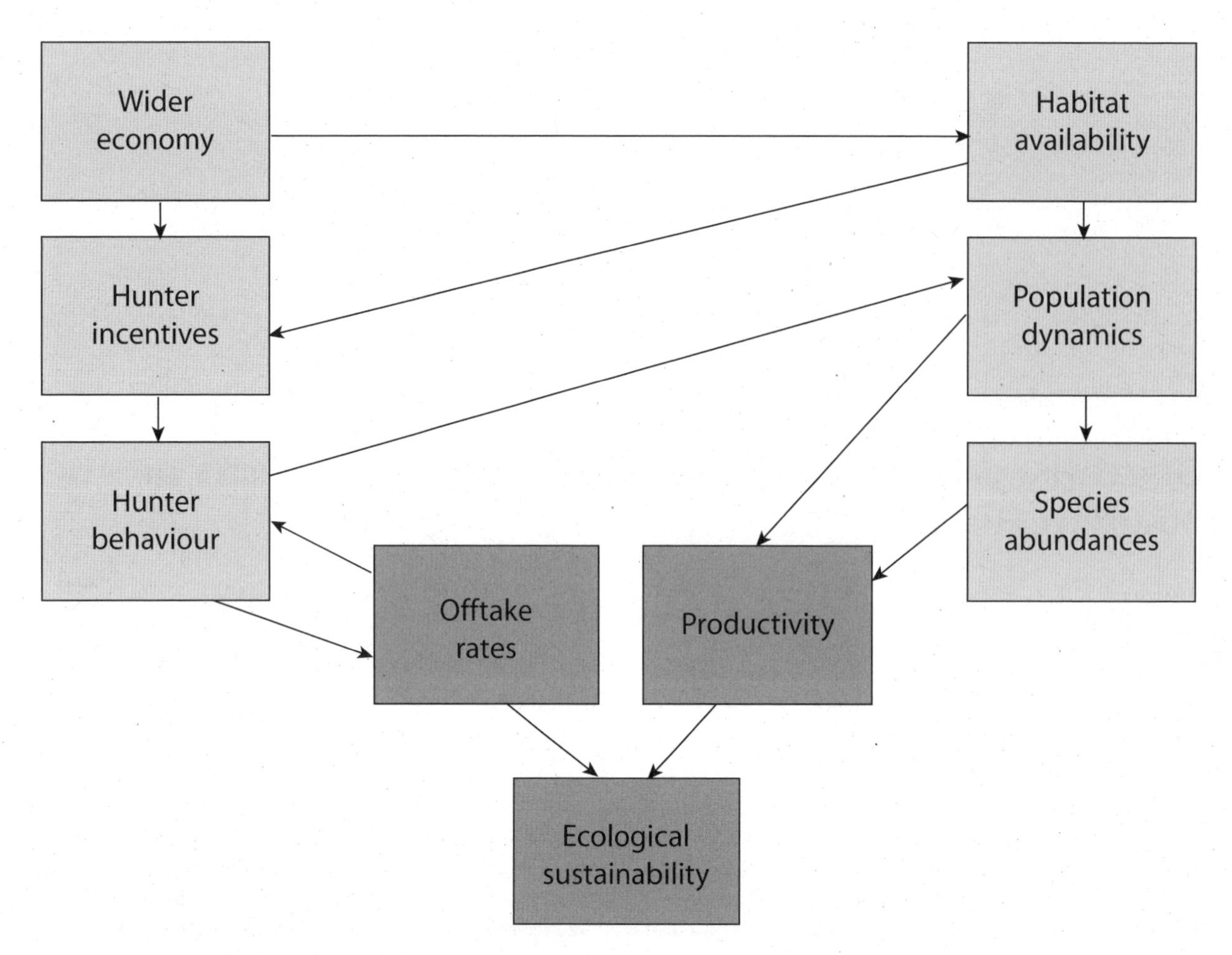

Figure 20–1. A schematic representation of bushmeat as a bioeconomic system. Ecological sustainability depends on the balance between offtake and productivity. The increasingly large-scale economic and biological influences are shown, together with arrows representing possible interactions between the biological and economic components. These interactions are fundamental to bioeconomic systems, and include the effects of macroeconomic policy on habitat availability for hunted species.

DATA REQUIREMENTS

Like most bio-economic systems, bushmeat hunting is complex, multi-scale, dynamic and heterogeneous (Figure 20–1). At the smallest scale, sustainability ultimately depends on the behaviour of individual hunters and their prey. This in turn is affected by the hunter's household economy, determining his investment in different activites. Whether someone chooses to hunt or farm depends on the relative profitability of each activity, which is influenced by consumer demand for the goods produced and costs of inputs. A consumer's choice between foodstuffs depends on the interaction between preferences, affordability and availability. The incentives felt by individual consumers and producers are products of the wider economy within which they are embedded.

On the biological side, prey availability is determined by the distribution and abundance of individuals (which may be influenced by avoidance of disturbance caused by hunting). A key determinant of prey distribution is habitat distribution, which is determined both locally by the hunting/farming tradeoff (more farming often means more habitat conversion) and regionally by government decisions on land clearance and logging rights. Finally, the behaviour of individuals and institutions is influenced by the regulatory regime and the effectiveness with which it is implemented and enforced on the local and regional scales.

The Individual Hunter

The idea of hunters as optimal foragers, maximising some fitness proxy such as profit per hunt, was explored by Alvard.[8] Hunters in his study villages in Amazonia did not act as "ecologically noble savages", conserving resources, but as rational utility-maximisers. This chimes with the predictions of both behavioural ecology and economics.[9] The interactions between individual hunters and their prey can be complex, and vary with the gear type used. Rowcliffe *et al.*[10] extended standard foraging theory to consider the difference between snare hunting, when the gear is assumed static and the animals mobile, and gun hunting, when the gear is assumed mobile and the prey static. The encounter rate of prey and the approach distance needed to trigger the gear vary substantially between these two extremes.

A number of empirical studies have followed hunters and obtained useful data on their behaviour and movement patterns.[11] However there has been virtually no work done on the rules of thumb which hunters use to determine when to change their hunting behaviour. For example, how low must catches be, and for how long, before they decide to act, and what steps do they take to increase catches: add more snares, move snares locally, or leave the area completely? These decision rules are critical to prediction of the level of depletion reached in hunted areas, but are not well understood.

The individual hunter is embedded in a household with various productive activities. Barrett & Arcese[12] used an economic model to suggest that agricultural productivity is a critical determinant of poaching rates in the Serengeti, while Damania *et al.*[13] looked at the decision to invest in bushmeat hunting. Damania *et al.*'s model showed that improved agricultural productivity does not necessarily reduce hunting rates, because the consequent improvement in income leads to greater demand for meat by the household and also gives hunters the ability to replace less efficient, cheaper technology, such as snares, with guns. Given that guns target vulnerable species such as primates as well as the more resilient species, improvements in agricultural incomes can exacerbate conservation problems.

In order to model the offtake of meat by individual hunters, a range of data on the livelihoods available to actual and potential hunting households is required, as well as specific information about behaviour while out hunting. Needless to say, very little of this kind of information is collected, and virtually never as part of an integrated project that also collects information on prey densities and consumer behaviour.

The Market

Once an animal is caught by the hunter, there is a long way to go before it reaches the market, where most data are collected. Along the way, individual animals drop out of the commodity chain (Figure 20–2). The animals reaching the end-market are a highly biassed selection of those killed, chiefly because low value individuals killed by non-professional hunters are generally eaten or bartered in the local villages, and only higher-valued animals are worth transporting to market. If law enforcement is effective, protected species are also likely not to appear in the end market.

The main determinant of whether it is worth selling bushmeat in the market is the price it fetches compared to the costs of sale. Transport costs are often a large proportion of total hunter costs,[14] and these depend on factors in the wider economy such as the price of petrol. Bushmeat prices are determined by consumer demand. Often bushmeat is sold in open markets, where it is competing with many other foodstuffs; hence the price and availability of other meats is an important determinant of bushmeat

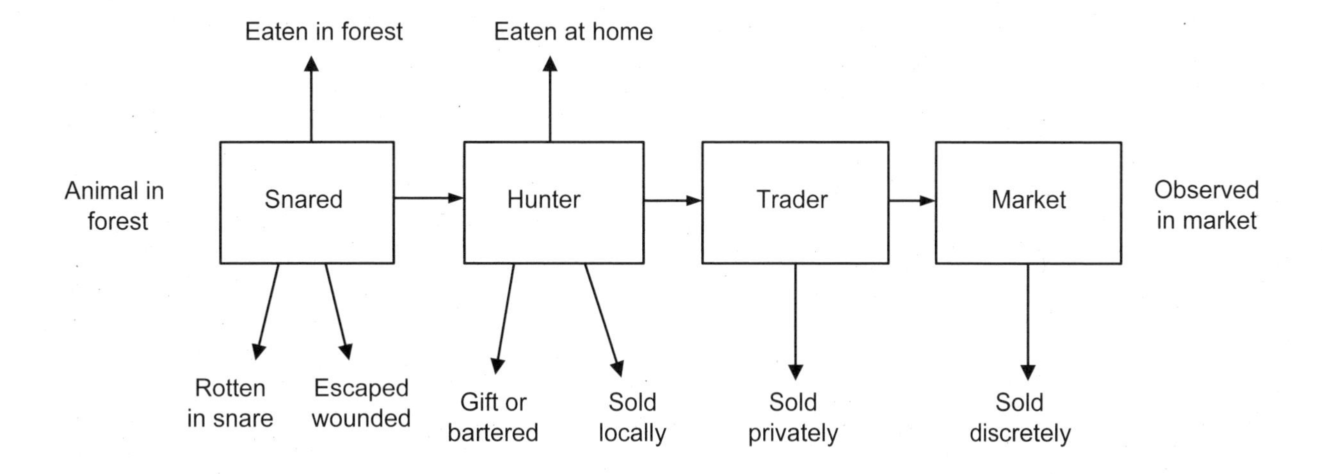

Figure 20–2. A stylised commodity chain, showing the journey of a carcass from being trapped in the forest to being observed in the end market. The actual structure of the commodity chain will vary from place to place, but this gives an idea of some of the main ways in which meat is disposed of. Different species will be more or less likely to reach the end of the chain, depending on their value and protected status.

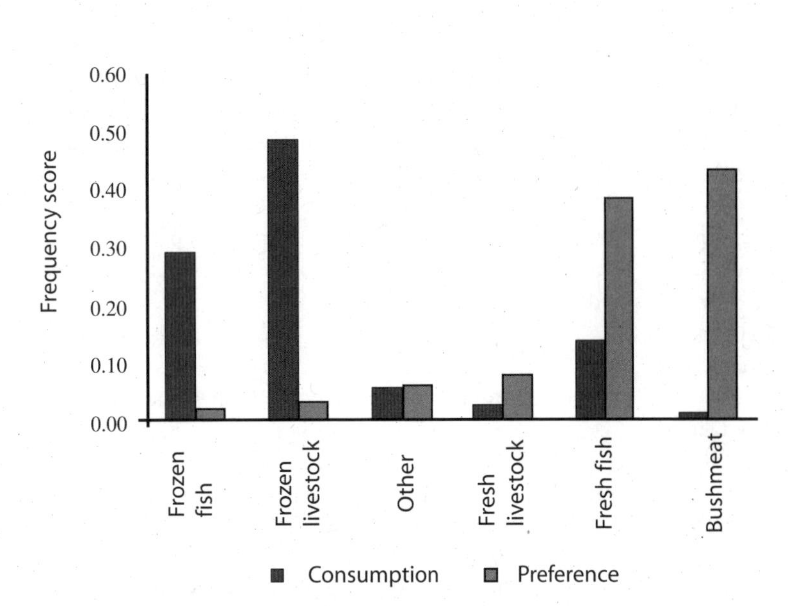

Figure 20–3. Consumption and preference scores for urban consumers in Bata, Equatorial Guinea.[17] Data are taken from a survey of 100 randomly selected households, in which individual meats were ranked 1-3 in order of frequency of consumption and preference. Scores are the normalised sums, for all meats in a category, of the number of people ranking an item 1st-3rd, multiplied by 3 for 1st place, 2 for 2nd place or 1 for 3rd place.

demand. Brashares *et al.* [15] have shown that poaching rates in Ghanaian National Parks increase in years with poor fishing yields, and that the effect is stronger the nearer the Park is to the coast. Consumer choice is a complex process, and the few published studies of bushmeat consumption suggest that consumer preferences, consumption and availability are all strongly inter-related. Often the person's nationality or tribal origin and wealth class also play a role.[16] East *et al.*[17] show that urban consumers in Equatorial Guinea have a strong preference for fresh fish and bushmeat, but that they mostly consume much less preferred frozen fish and livestock (Figure 20–3). This is because frozen produce is much cheaper; preferences are followed more closely by wealthier consumers. This is particularly significant because Equatorial Guinea is currently experiencing an oil boom that is increasing both the urban population and the average income of cities such as Bata; this is likely to enable people to satisfy their preferences more easily. Hence it seems likely that demand for bushmeat in Bata will rise in the near future, increasing pressure on wildlife populations.

The effects of income, substitute goods and price on demand are captured in elasticities. Elasticities measure the percentage by which the quantity of a good demanded changes with a 1% change in the explantory variable. As yet there is only one published study of elasticities of demand for bushmeat, by Wilkie & Godoy,[18] and this is for Amerindian households in South America rather than for an African site. The lack of understanding of the effects of income and prices on consumer demand is a major impediment to policy-making, because the commercial bushmeat trade is founded on consumer demand. Many policies proposed for controlling the commercial bushmeat trade rely on altering prices or incomes, and the effects of these changes on demand and offtake rates are not necessarily obvious.[13]

Biological Data

Good data on the abundance of forest mammals are very difficult to obtain, often relying on indirect methods such as dung counts, which are notoriously unreliable.[19] However, much effort has been put into obtaining estimates of population densities of hunted species, often comparing protected and exploited areas.[20] There is usually signficant variation between sites, partly explained by differences in factors such as vegetation and soil type, but also linked to hunting. As populations get lower and hence less observable due to hunting, either more effort must be put into surveys or the variation around estimates of population size will be higher. For example, Peres[21] estimated the biomass of large-bodied species at 18 forest sites in Amazonia as 663±50 kg/km^2 in unhunted areas, and 232±85 kg/km^2 in heavily hunted sites.

The biology of many hunted forest species is very poorly known; without long-term studies, data on reproductive rates and mortality schedules can be difficult to obtain. Hence the assumptions used in sustainability analyses for many commonly traded species are based on allometric relationships between measurable variables such as mean body mass and the variables of interest, such as intrinsic rate of increase, group size or carrying capacity.[22] These relationships are adequate for crude estimation in the absence of other data, but often have substantial variation around them. This also contributes to the uncertainty surrounding any estimates of the productivity of exploited populations.

Bio-economic Interactions

The interactions between biological and economic processes occur at all scales. At the small scale, changes in habitat type from forest to farm-bush as villagers expand agriculture (and back as agriculture declines) lead to changes in local species composition. At the larger scale, bushmeat hunting and logging are often linked and devastating in their synergistic effects on wildlife and habitats.[23] One key side-effect of logging is improvements in roads and infrastructure, which can promote overexploitation by opening up access to markets,[24] although there are also examples where access to markets reduces hunting by allowing people to purchase domestic meat and encouraging them to concentrate on farming.[25]

In Sulawesi, road improvements in the early 1990s led to a dramatic increase in wild pig hunting as traders could drive further into the forest and still return to market with saleable produce.[26] This led to increased travel times for a case study dealer as nearby forests were depleted of pigs (Figure 20–4). However, habitat conversion was also occurring along the road, reducing the amount of primary forest available to the endemic babirusa wild pig (*Babyrousa babyrussa*, Figure 20–5). The babirusa is one of two species of endemic wild pigs traded together in North Sulawesi; the other, the Sulawesi wild pig (*Sus celebensis*) is able to persist in secondary forest, and hence is probably less affected by habitat loss. The Sulawesi wild pig is also substantially more resilient to hunting than the babirusa, such that the combination of hunting and habitat loss impacts far more severely on one species than the other. This example illustrates the complexity of bioeconomic interactions affecting hunting sustainability.

Although there is a general realisation that there are wider influences on hunting sustainability and that they

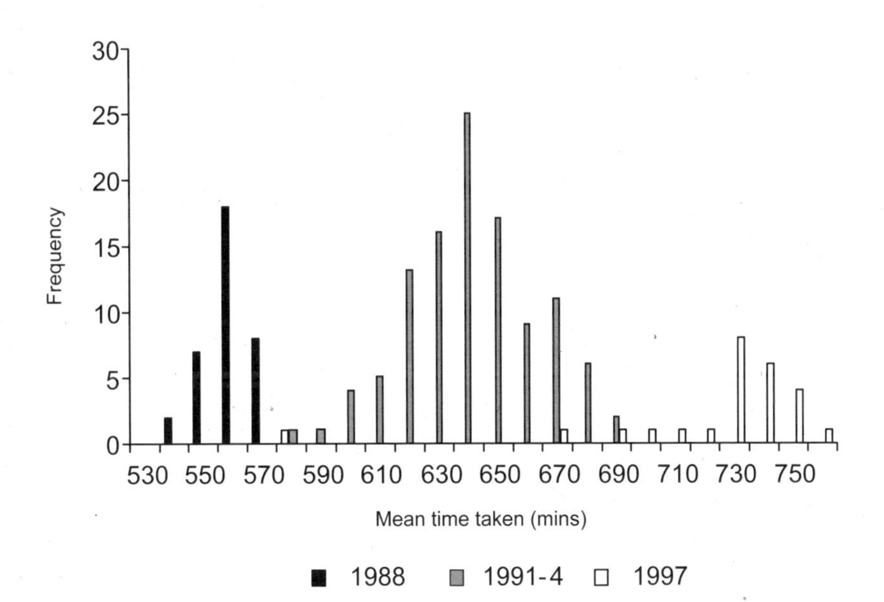

Figure 20–4. Travel times for a case study wild pig dealer in North Sulawesi, 1988-1997. The travel times are for single journeys, weighted by the number of pigs purchased in each location along the way.[28]

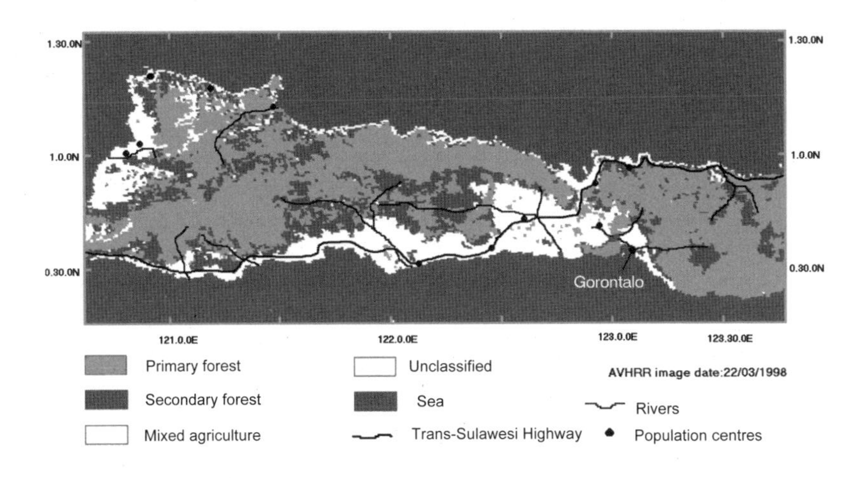

Figure 20–5. Vegetation of Gorontalo region, North Sulawesi, in 1998. Classifications are taken from a ground-truthed Advanced Very High Resolution Radiometer (AVHRR) image, and divided into primary and secondary forest or mixed agriculture. The major road along which pig dealers travel is shown. This is the region from which the majority of the wild pigs for sale were being hunted at the time. The image shows substantial forest conversion emanating from roads and rivers.[51]

have synergistic effects, these dynamic interactions are difficult to take into account in sustainability assessments, and so are often ignored.

THE EFFECT OF DATA DEFICIENCIES

Because population estimates are so uncertain, and ridden with unquantifiable biases, it is extremely difficult to estimate the sustainable level of production from a given area. The detailed surveys that would allow us to quantify sustainable offtakes to an adequate degree of accuracy are few and far between in tropical forests worldwide. Examples such as Peres,[21] in which multiple sites are examined, such that the effects of hunting can be separated from other factors, are very rare. Abundance data alone cannot be used for sustainability assessments because abundance is only one component of productivity; data on species life histories, population growth rates and catchabilities are also needed. This is not to say that a crude estimate of productivity cannot be obtained based on abundance, allometric relationships and inferences from offtake data, but this cannot form a robust basis for managed hunting.

Market data are easier to obtain than biological data, and in theory could provide signals of lack of sustainability. For example, a decline in the number of animals on sale or an increase in their price might indicate depletion of the wildlife stock. In a multi-species system, it would be expected that larger, slower-growing species would disappear from the market first, a phenomenon observed in fisheries.[27] Finally, as observed in North Sulawesi, an increase in the distance animals travel to the market might be a strong signal of depletion.[28] However, a more thorough examination of market data shows how misleading such trends can be.

An Example of Problems with Market Data

To illustrate these points, Crookes *et al.*[29] examined trends in a dataset collected by the Ghanaian Wildlife Department in Atwemonom market, Kumasi, Ghana. The dataset comprised 36,099 animals, observed on 2,446 market days from 1987 to 2002. This is a mature market, in which 7 relatively common and resilient species make up 95% of the open season trade; vulnerable species are hunted, and do appear occasionally in the market,[30] but are not in this dataset at levels that permit statistical examination of trends.

The data show no evidence for a decline in the number or biomass of traded species, nor in the proportion of slower-growing species. However, there was a sharp increase in price from 1999 to 2002 (Figure 20–6a), together with a significant increase in the proportion of the trade that came from more distant sources. This increase was almost entirely fuelled by an increase in the proportion of grasscutters (*Thryonomys swinderianus*) on sale (Figure 20–6b). Most market datasets do not include data on the location where animals were killed or the method of killing. However, this information is extremely useful, and is included in the Atwemonom dataset. The proportion of individuals killed by cutlass rather than gun increased for all species at the time of the price increase (Figure 20–7). Cutlass-killed animals are often trapped first, and this gear combination represents cheap technology that might be used by new entrants and non-professional hunters, compared to the professionals who tend to use expensive but more efficient gear such as guns.

The trends in this dataset can be interpreted in very different ways. They are compatible with an explanation based on unsustainable use, or with hunting becoming more profitable due to extraneous circumstances:

Explanation 1. Unsustainable use. Consumer demand for bushmeat started to grow rapidly in the late 1990s, due to some extrinsic factor such as urban population increase, taste changes or scarcity of alternative foodstuffs. Bushmeat hunters are struggling to keep up with demand, hence prices are rising but quantities supplied are relatively constant. In order to maintain supply, traders are having to recruit hunters from further and further afield. Wildlife depletion in the forests makes gun hunting less viable economically, so hunters are using cheaper technology, trapping near their fields. Grasscutters are the most valuable animal that is routinely trapped, and so are now sold more than other species.

Explanation 2. A profitable profession. Increasing consumer demand is leading to rising prices. The higher prices are tempting people to sell in the market, including hunters from distant locations who did not previously sell to Kumasi due to prohibitive transport costs. New entrants to hunting start cheaply with trapping, expecting to invest in a gun if their trade continues to be profitable. Similarly, non-professional hunters who previously only trapped around their fields for subsistence are tempted now to sell their produce (particularly the more valuable grasscutters).

Under explanation 1, the expectation is that prices will continue to rise unless extrinsic factors change, and that local offtakes will decline as produce is transported increasingly farther to the market. Under explanation 2, the impact on sustainability depends on whether effort is increasing due to new entrants, or whether people are simply choosing to sell more of their catch. However with time, hunting is likely to become less sustainable as profits encourage people to invest in more efficient technologies.

This example illustrates how difficult it is to make inferences about trends in market data without support-

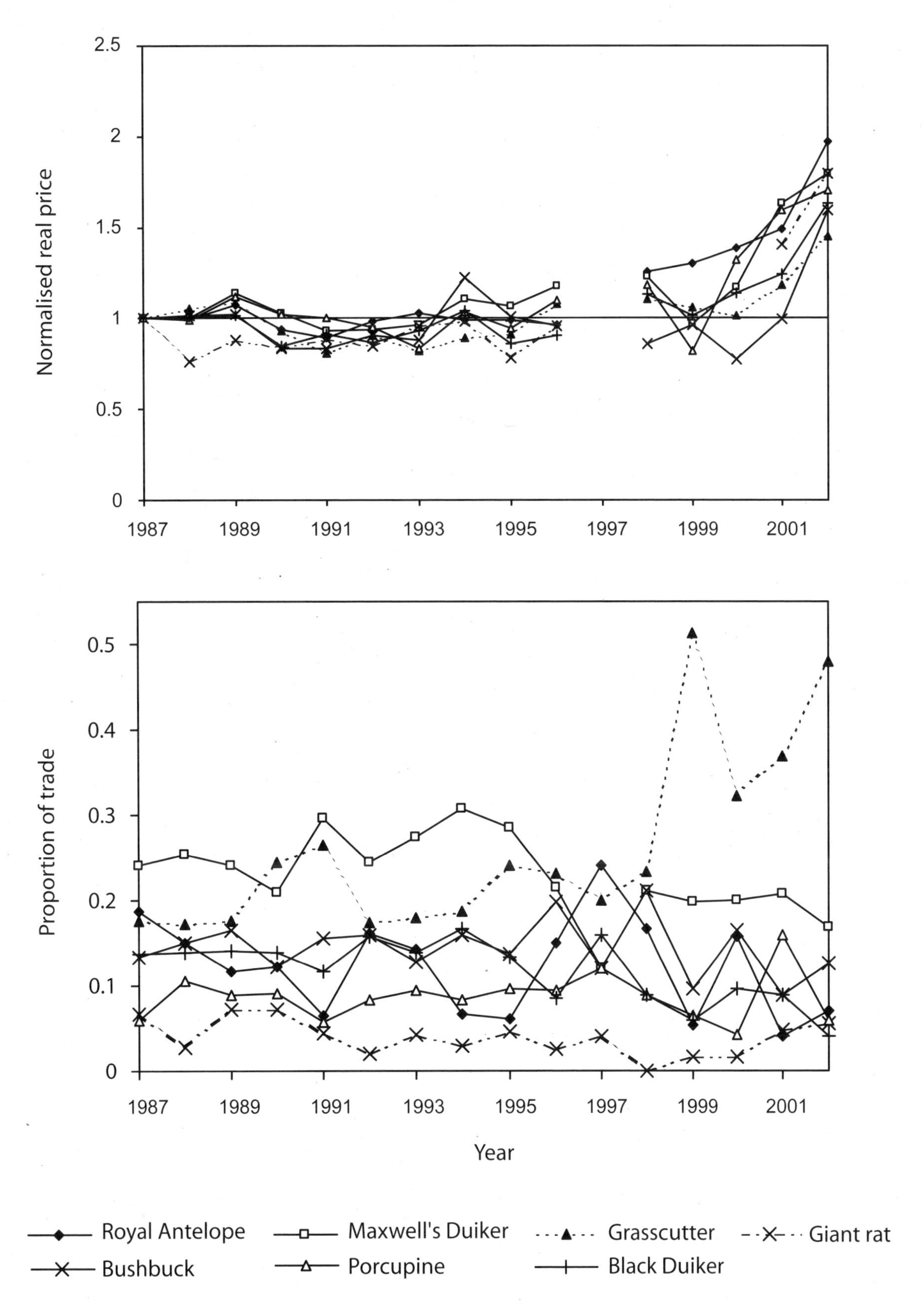

Figure 20–6. a) Normalised real price of the seven most commonly traded bushmeat species at Atwemonom market, Kumasi, Ghana, based on a dataset collected by the Ghanaian Wildlife Department. **b)** Changes in the proportion of the Wildlife Department dataset represented by the 7 species over the same time period, using data from the open season when all these species were legally traded.[29]

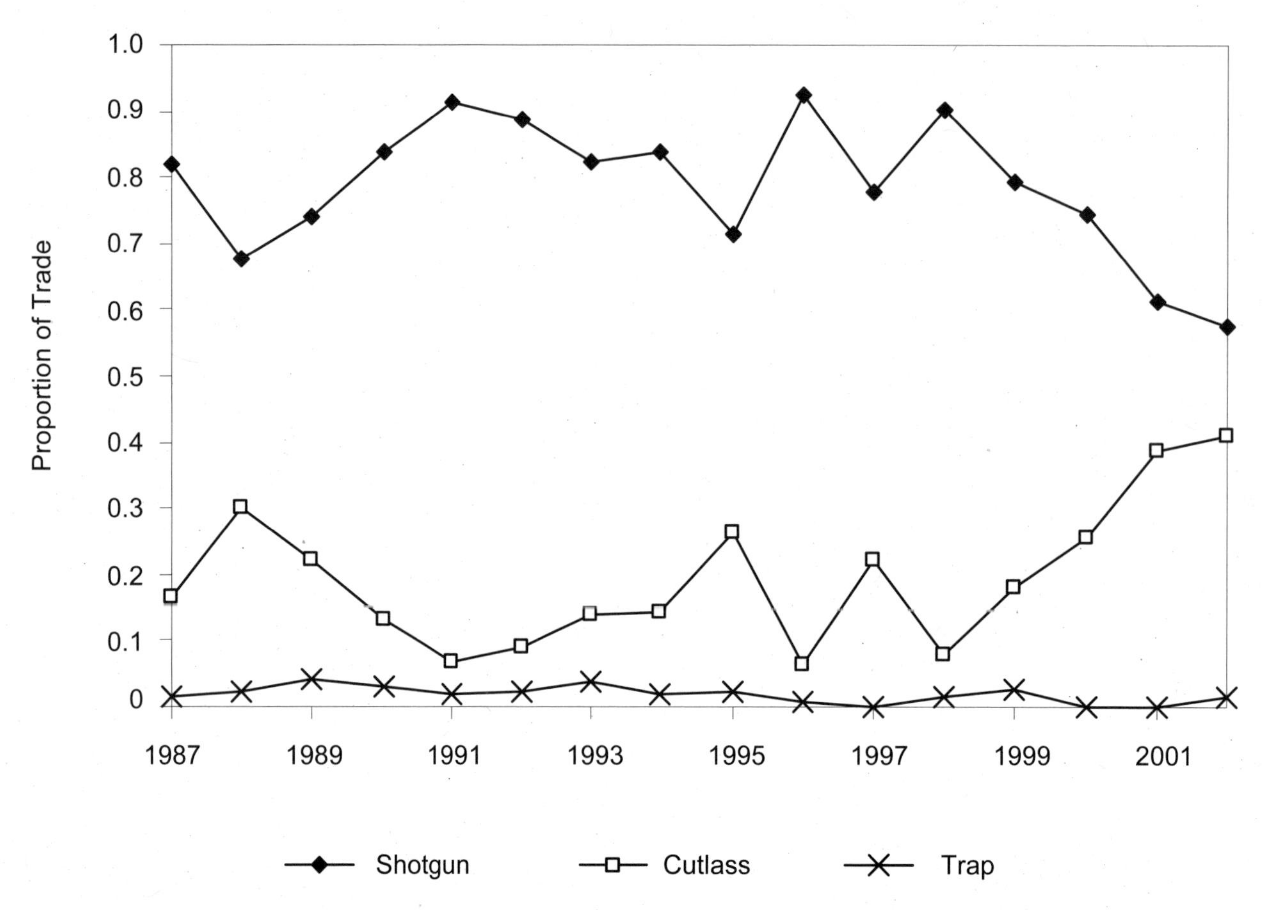

Figure 20–7. The proportion of animals in the Atwemonom dataset that were killed by guns, cutlasses or directly in traps over the period 1987-2002.[29]

ing information on changes in species' abundance or on hunter behaviour. In particular, the dynamics of hunter effort, including movement of individuals into and out of the profession, are important but rarely considered in the literature. Market data represent an amalgamation of influences, including wildlife stock depletion and external factors, and hence should be treated with caution unless accompanied by supporting data.

APPROACHES TO THE PROBLEM

Given that high-quality data for sustainability assessments of bushmeat hunting will never be available from more than a handful of sites, how best can we approach the problem, using data that are realistically obtainable? One response to the poor coverage of bushmeat data is the Bushmeat Crisis Task Force's Information Management and Analysis Project,[31] which aims to collect spatially referenced information on the intensity and location of bushmeat hunting and markets. If widely used, this could go a long way towards building a more integrated view of the dynamics of the bushmeat trade. On the more local level, superficially the most attractive way forward is to try to find a robust index of sustainability, based on the limited data that can feasibly be obtained.

Sustainability Indices

Detailed analyses of the sustainability of bushmeat hunting are virtually non-existent. Generally, sustainability assessments for local areas involve obtaining estimates of animal abundance and comparing them to offtake data from hunters.[32] There is a lot of effort and scientific rigour involved in getting these data. By contrast, and despite being the point of the exercise, the sustainability analysis is usually extremely crude. The almost universal method for calculating sustainability in bushmeat studies is Robinson & Redford's[33] index:

$$\boldsymbol{P} = 0.6\,\boldsymbol{K}\,(\boldsymbol{R_{max}} - 1)\,\boldsymbol{F}$$

where $\boldsymbol{P}$ = production (animals/km^2/year), $\boldsymbol{K}$ = carrying capacity (animals/km^2), $\boldsymbol{R_{max}}$ = arithmetic intrinsic rate of population increase (animals/year), $\boldsymbol{F}$ = natural mortality correction factor.

Milner-Gulland & Akcakaya[34] showed that if this index were used as a basis for management, it would rapidly send the population to extinction. This is because it does not take uncertainty and bias or current abundance into account. However it has been widely applied in bushmeat research because it is simple and does not require large quantities of data. One similarly simple sustainability index that does perform well is the National Marine Fisheries Service's index of the sustainability of cetacean bycatch:[35]

$$P = 0.5\,N\,(Rmax - 1)\,F$$

where the notation is as above with the exception of: F = correction factor for bias and uncertainty between 0 and 1, and N = a minimum estimate of abundance (animals/km^2).

This takes uncertainty and bias into account partly through F and partly because N is a minimum estimate, while the use of N in place of K ensures current abundance is included in the analysis. Given the robustness of this index, and the fact that in general, simple harvesting strategies are more robust than complex ones,[36] this suggests that the use of sustainability indices may be worth exploring further. However it is also clear that simple indices based on flawed models can be extremely misleading.

Another type of index that has been proposed for bushmeat is prey profiles.[10] This builds on the idea that in a multi-species system, the prey composition should change as species are depleted. Reductions in the proportions of vulnerable species in the offtake could act as a warning signal of depletion. Prey profiles are as yet untested in the field, but could potentially be useful if applied at the village scale to situations in which effort and technology remained constant over time and market dynamics were not relevant. This is likely to be a relatively limited subset of bushmeat hunting situations.

Bayesian Network Approaches

One way of including a realistic representation of uncertainty into a simple model is use Bayesian Belief Networks (BBNs). BBNs provide an intuitive analytical framework for combining various types of information of differing degrees of certainty, so that all the available information can be used. The system is represented in an "influence diagram", which is a network of influences between variables (such as the population of a bushmeat species, the number taken by hunters, the number sold on in a market, and so on), each of which has a probability distribution attached to it. This probability distribution can be derived from actual survey data, expert opinion or from the literature. Several user-friendly packages exist allowing a model to be constructed rapidly and straightforwardly.[37]

Figure 20–8 is a simple example of an influence diagram for assessing the sustainability of bushmeat hunting. It shows that indicators of the sustainability of use can be obtained, but more importantly that they are accompanied by a realistic visualisation of their uncertainty. As new information is obtained for a particular variable, its probability distribution can be updated. This updating affects all the other variables in the network, both in the direction of causality and against it. For example, if several individuals of a particular species appear in a market, this observation updates probability distributions throughout the network, ultimately affecting the probability distribution for the population size of the species prior to the offtake. Policy interventions can easily be modelled within this framework. For example, one could predict the effects of a ban on market sales of a particularly vulnerable species on all other species in the market, because changes in the marketability of the banned species would propagate through the network to affect the dynamics of all other species.

One powerful use of BBNs is in identifying data needs. By examining the sensitivity of results to improvements in knowledge about each variable, the model shows where reducing uncertainty is most critical. For example, the model may show that the proportion of offtake which goes to commercial markets as compared to subsistence use and wastage is a variable with a high degree of uncertainty which is also critical to the sustainability of offtake. This facility allows researchers and managers to target resources most effectively.

There is a growing interest in the potential of BBNs for conservation. Recently, a BBN model was developed for assessing the impacts of land use changes on bull trout populations in the USA.[38] Another recent application of BBNs is in modelling uncertainties in fish stock assessment and the impact of seal culling on fish stocks.[39] Marcot *et al.*[40] have used BBNs for evaluating population viability under different land management alternatives, while Wisdom *et al.*[41] used BBNs in conservation planning for the greater sage-grouse (*Centrocercus urophasianus*). However, these are isolated cases, and BBNs are still not widely used in conservation planning. They could make an important impact, if only as a way of improving the transparency of conservationists' assumptions about uncertainty and causal linkages between variables in the system, and improving the effectiveness with which data of various types and qualities are used. Although complex to develop, the final interface could be straightforward enough for non-modellers to use, which is the main attraction of sustainability indices such as Robinson & Redford's.

BBNs do have drawbacks, and are not the only modelling tool that could be used to represent uncertainty. The attractive graphical representation soon becomes unwieldy under more realistic conditions, particularly when a dynamic network is required, as it would be for bushmeat. It is also important to realise that the BBN's representation of uncertainty is built on point estimates,

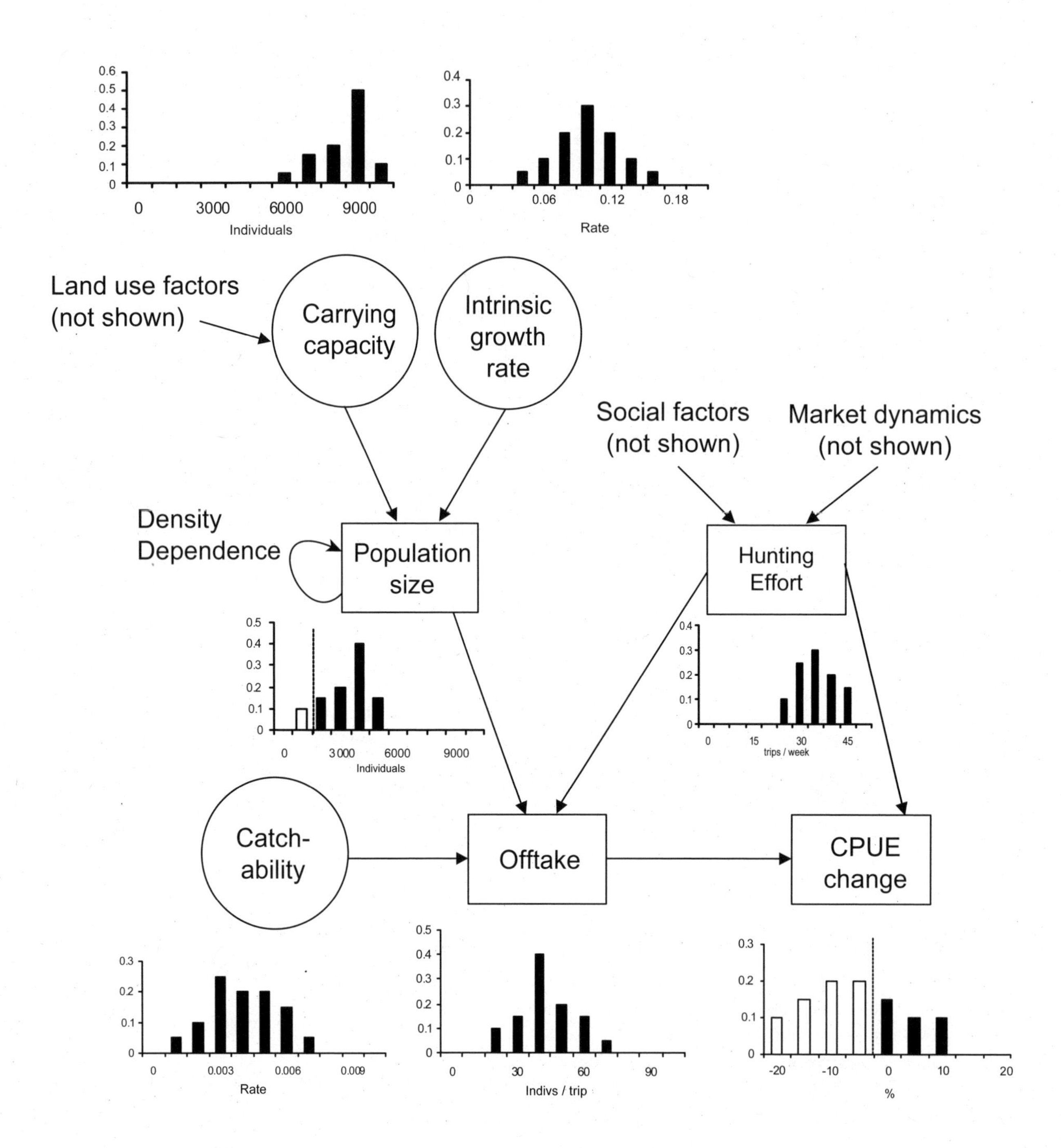

Figure 20–8. An example of a simple Bayesian Belief Network. The influence diagram shows how variables relate to each other, and for each variable a probability distribution of values is given. Note that this example is simplified for demonstration purposes, showing one location, one point in time, and one species. Two measures of sustainability are included in this example: i) Is the population size below a threshold level of 10% of carrying capacity? The probability of being below the threshold (dashed line) is shown by the white column. ii) The change in catch per unit effort (from previous time periods). A declining Catch Per Unit Effort (CPUE) is an unreliable indicator of unsustainable use, which would not be used in practice, but is shown here for demonstration purposes. The probability of a declining CPUE is the sum of the white columns in the probability distribution.

rather than distributions. A Bayesian model, in the normal sense of the word, provides a better representation of uncertainty based on probability distributions, with the loss of some of the strengths of BBNs, such as backwards updating. Bayesian models are already widely used in fisheries management,[42] but the relative lack of quantitative analysis in bushmeat research and management, and the lack of capacity for policy implementation mean that they are not likely to be used for bushmeat in the foreseeable future.

Using Virtual Worlds

As computing power becomes less of a limitation, simulation modelling has become a much more powerful tool in understanding the dynamics of complex systems. The advantage of this is that policy interventions can be tested by simulation rather than in the real world. As the real world is effectively one realisation of a stochastic simulation, predictive power should be improved by running simulation models repeatedly. A particularly powerful feature of the simulation approach is the ability to model the observation process itself. Thus the effect of observation uncertainty can be included as well as the process uncertainty inherent in the system[43] (e.g. climate-driven population fluctuations). Model uncertainty can be addressed by running the simulation under a range of possible assumptions about system structure, and using a model selection procedure such as Akaike's Information Criterion.[44]

There is an increasing ecological literature using simulation modelling to address observation uncertainty.[45] One of the early applications to natural resource management was the Revised Management Procedure developed by the scientific committee of the International Whaling Commission[46] (Figure 20–9). This is an example of the broader "operating model" approach to fisheries management[47], in which management plans are developed on the basis of flawed observational data, but the performance of the system can be monitored using variables based on actual values of the system such as true population size. Fisheries models of this type tend not to include individual human behaviour, but instead concentrate on the options for a single management authority; this is not realistic for most bushmeat hunting, in which the management authority has little control over individual behaviour.

Within bushmeat research, the simulation modelling approach has yet to catch on. Bousquet *et al.*[48] use a individual-based model of individual hunters and duikers (*Cephalophus monticola*) to show how individual agents

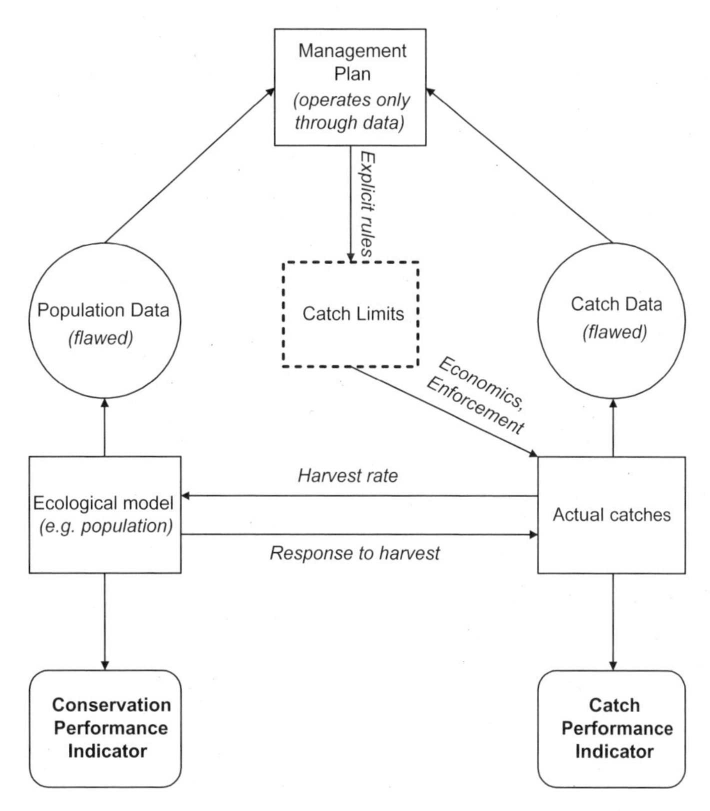

Figure 20–9. The conceptual basis of the International Whaling Commission's revised management procedure, showing the relationship between the true system dynamics and the management plan operating through data. The performance of management rules can be assessed in the simulation using the performance indicators, which represent the true status of stocks and catches.[46,52]

act when hunting a common property resource, but do not address observational uncertainty. Some of the drawbacks of the simulation approach include the need for computer programming expertise, and the problem that complex models can lose heuristic power. Simple models are far more enlightening than complex ones so long as the simplifying assumptions are valid. Although limited in some ways, simulation modelling is potentially a very useful tool in bushmeat research, particularly in testing frameworks for managing uncertainty.

PROSPECTS FOR BUSHMEAT

Bushmeat hunting is complex ecologically, socially and economically. Conservationists will not be able to obtain data for full sustainability assessments in any but a handful of locations. Under these circumstances, action still has to be taken to safeguard vulnerable species and to ensure that wildlife is not extirpated over wide swathes of tropical forest. Fa *et al.*[3] estimate that bushmeat hunting is currently about 2.8 times the sustainable level in the Congo Basin as a whole; despite the broadbrush nature of this estimate, it is clear that the wildlife resource in this area is not able to sustain current offtake levels. However, there is heterogeneity in offtake rates, and some way to capture this is required, in order to prioritise interventions.

The key to moving forward in this situation is to acknowledge that sustainability assessments for bushmeat are plagued with uncertainty, and ensure that this uncertainty is realistically represented in assessments. We need to strengthen our understanding of the socio-economic context of bushmeat hunting, so that we can be sure that our interventions are having the desired effect on people's decision-making.[49] And crucially, effective protected areas are required in order to ensure the persistence of unhunted ecological communities.

Bushmeat hunting in Africa can be crudely divided into three scenarios: mature markets, frontier bonanzas and declining source-sinks. In some areas it seems that hunting has continued for decades to supply mature bushmeat markets. The animals are coming from the farm-bush matrix rather than from forested areas, and are all resilient, fast-growing species. The vulnerable species are already extirpated. Cowlishaw *et al.*[5] describe such a scenario for Takoradi market in Ghana. These areas are likely to be of low priority for conservation intervention, but research is still required on the potential of these areas to produce a sustainable bushmeat supply, and on the livelihood importance of bushmeat to people living there.[50] There is also no clear understanding of the extent or importance of these areas as a component of overall bushmeat supply.

The frontier bonanza scenario is the one of most conservation concern. Here, primary forest is being opened up for logging or other commercial or development activities. Wildlife communities are rapidly extirpated as immigrants move into the area, and intervention is required to safeguard both hunted animals and often also the livelihoods of vulnerable people already living in the area.[23,24] This is where pressure on governments or multinationals and urgent action to protect habitat are crucial.

Perhaps the most widespread and difficult form of bushmeat hunting is the declining source-sink system. In this scenario a build-up of hunting effort is reducing wildlife populations over a wide area, but with spatial variation caused by differences in habitat, travel costs or other factors. The causes for this build-up are complex and interacting, as discussed above. Here the problem is at its most intractable, and imaginative interventions such as no-take zones, community-based management and alternative livelihoods are most often tried. It is to this scenario that the suggestions for future management approaches made in this chapter are most likely to be relevant. Sustainable bushmeat hunting in these areas is only going to be achieved, if at all, when it is addressed in the context of the wider economy, and with a full understanding of the inherent uncertainties.

ACKNOWLEDGEMENTS

I am very grateful to Conservation International for funding some of the work that is presented in this chapter. I thank Bob Burn, Jim Cannon, Guy Cowlishaw, Doug Crookes, Noelle Kümpel, Stephen Ling, Marcus Rowcliffe & David Wilkie for useful discussions that helped form the ideas presented here, and Tom Evans for permission to use his data in Figure 20–5.

NOTES AND SOURCES

[1] See Eves, Chapter 9, for more information on the bushmeat crisis.

[2] Milner-Gulland, E.J., Bennett, E.L. and the SCB 2002 Annual Conference Wild Meat Group (2003) Wild meat - the bigger picture. Trends in Ecology and Evolution 18, 351-357.

[3] Fa, J.E., Currie, D., Meeuwig, J. (2003) Bushmeat and food security in the Congo basin: linkages between wildlife and people's future. Environmental Conservation 30, 71-78.

[4] Bowen-Jones, E. (1998) A review of the commercial bushmeat trade with emphasis on Central/West Africa and the Great Apes. Report to the Ape Alliance, obtainable from: http://www.4apes.com/bushmeat/report/bushmeat.pdf.

[5] Cowlishaw, G., Mendelson, S., Rowcliffe, J.M. (2005) Evidence for post-depletion sustainability in a mature bushmeat market. Journal of Applied Ecology 42, 460-468.

[6] Bennett, E.L. (2002) Is there a link between bushmeat and

food security? Conservation Biology 16, 590-592.

[7] Smith, R.J., Muir, R.D.J., Walpole, M.J., Balmford, A., Leader-Williams, N. (2003) Governance and the loss of biodiversity. Nature 426, 67-70.

[8] Alvard, M.S. (1993) Testing the "ecologically noble savage" hypothesis: interspecific prey choice by Piro hunters of Amazonian Peru. Human Ecology 21, 355-387.

[9] But see the work of E. Ostrom and others on how communities can govern natural resource use for community rather than solely individual benefit, e.g. Dietz, T., Ostrom, E., Stern, P.C. (2003) The struggle to govern the commons. Science 302, 1907-1912; Pretty, J. (2003) Social capital and the collective management of resources. Science 302, 1912-1915; for more on human behaviour, see Brooks, Chapter 16.

[10] Rowcliffe, J.M., Cowlishaw, G.C. Long, J. (2003) A Model of Human Hunting Impacts in Multi-Prey Communities. Journal of Applied Ecology 40, 872-89.

[11] For example, see Muchaal, P. K., Ngandjui, G. (1999) Impact of village hunting on wildlife populations in the western Dia Reserve, Cameroon. Conservation Biology 13, 385-396.

[12] Barrett, C. B., Arcese, P. (1998) Wildlife Harvesting in Integrated Conservation and Development Projects: Linking Harvest to Household Demand, Agricultural Production and Environmental Shocks in the Serengeti. Land Economics 74, 449-65.

[13] Damania, R., Milner-Gulland, E.J., Crookes, D.J. (2005) A bioeconomic model of bushmeat hunting. Proceedings of the Royal Society of London B 272, 259-266.

[14] Mendelson S, Cowlishaw G, Rowcliffe J.M. (2003) Anatomy of a bushmeat commodity chain in Takoradi, Ghana. The Journal of Peasant Studies 31, 73-100.

[15] Brashares, J.S., Arcese, P., Sam, M.K., Coppolillo, P.B., Sinclair, A.R.E., Balmford, A. (2002) Bushmeat hunting, wildlife declines, and fish supply in West Africa. Science 306, 1180-1183.

[16] Fa, J.E., Juste, J., Burn, R.W., Broad, G. (2002) Bushmeat consumption and preferences of two ethnic groups in Bioko Island, West Africa. Human Ecology 30, 397-416.

[17] East, T., Kumpel, N., Milner-Gulland, E.J., Rowcliffe, J.M. (2005) Determinants of urban bushmeat consumption in Rio Muni, Equatorial Guinea. Biological Conservation 126, 206-215.

[18] Wilkie, D. Godoy, R.A. (2001) Income and price elasticities of bushmeat demand in lowland Amerindian societies. Conservation Biology 15, 761-769.

[19] Smart, J.R.C., Ward, A.I., White, P.C.L. (2004) Monitoring woodland deer populations in the UK: an imprecise science. Mammal Review 34, 99-114.

[20] For example, studies such as Peres, Hart and Fitzgibbon *et al.*, all in: Robinson, J.G. & Bennett, E.L. (2000) *Hunting for sustainability in tropical forests.* New York: Columbia University Press.

[21] Peres, C.A. (2000) Evaluating the impact and sustainability of subsistence hunting at multiple Amazonian forest sites. Pages 31-56 of: Robinson, J.G. & Bennett, E.L. (2000) *Hunting for sustainability in tropical forests.* New York: Columbia University Press.

[22] See endnote 10 for an example.

[23] Robinson, J.G., Redford, K.H., Bennett, E.L. (1999) Wildlife harvest in logged tropical forests. Science 284, 595-596.

[24] For example Wilkie, D.S., Sidle, J.G., Boundzanga, G.C. (1992) Mechanised logging, market hunting and a bank loan in Congo. Conservation Biology 6, 570-580.

[25] Ayres, J.M., de Magalhaes Lima, D., de Souza Martins, E., Barreiros. J.L.K. (1991) On the track of the road: changes in subsistence hunting in a Brazilian Amazonian village. Pages 82-92 in Robinson, J.G. & Redford, K.H. *Neotropical Wildlife Use and Conservation.* Chicago: University of Chicago Press.

[26] Clayton, L., Keeling, M., Milner-Gulland, E.J. (1997) Bringing home the bacon: A spatial model of wild pig harvesting in Sulawesi, Indonesia. Ecological Applications 7, 642-652.

[27] Roberts, C.M. (1997) Ecological advice for the global fisheries crisis, Trends in Ecology & Evolution 12, 35-38.

[28] Milner-Gulland, E.J., Clayton, L.M. (2002) The trade in wild pigs in North Sulawesi, Indonesia. Ecological Economics 42, 165-183.

[29] Crookes, D.J., Milner-Gulland, E.J., Ankudey, N. (submitted) What can long-term market data tell us? A case study of Atwemonom market, Ghana.

[30] Ntiamoa-Baidu, Y. 1998 *Wildlife development plan. Volume 6-Sustainable harvesting, production and use of bushmeat.* Accra, Ghana: Wildlife Department.

[31] See http://www.bushmeat.org/imap/index.html.

[32] For example endnote 11 and Fitzgibbon, C.D., Mogaka, H., Fanshawe, J.H. (2000) Threatened mammals, subsistence harvesting and high human population densities: a recipe for disaster? Pages 154-167 in Robinson, J.G. & Bennett, E.L. (2000) *Hunting for sustainability in tropical forests.* New York: Columbia University Press.

[33] Robinson, J.G., Redford, K.H. (1991) Sustainable harvest of neo-tropical mammals. Pages 415-429 in Robinson, J.G. and Redford, K.H. (1991) *Neo-tropical Wildlife Use and Conservation.* Chicago: University of Chicago Press.

[34] Milner-Gulland, E.J., Akcakaya, H.R. (2001) Sustainability indices for exploited populations. Trends in Ecology and Evolution 16, 686-692.

[35] Wade, P.R. (1998) Calculating limits to the allowable human-caused mortality of cetaceans and pinnipeds. Marine Mammal Science 14, 1-37.

[36] Ludwig, D., Walters, C.J. (1985) Are age-structured models appropriate for catch-effort data? Canadian journal of fisheries & aquatic science 42, 1066-1072.

[37] For example Netica, www.norsys.com, or Hugin, www.hugin.com.

[38] Lee D.C. (2000) Assessing land-use impacts on bull trout using Bayesian belief networks, in Ferson, F., Burgman M. *Quantitative Methods in Conservation Biology.* New York: Springer.

[39] Hammond T.R., O'Brien, C.M. (2001) An application of the Bayesian approach to stock assessment model uncertainty. ICES Journal of Marine Science 58, 648-656.

[40] Marcot, B. G., Holthausen, R.S., Raphael, M.G., Rowland, M. and Wisdom, M. (2001) Using Bayesian belief networks to evaluate fish and wildlife population viability under land management alternatives from an environmental impact statement. Forest Ecology and Management 153, 29-42.

[41] Wisdom, M.J., Wales, B.C., Rowland, M.M., Raphael, M.G., Holthausen, R.S., Rich, T.D., Saab, V.A. (2002) Performance of Greater Sage-Grouse models for conservation assessment in the Interior Columbia Basin, USA. Conservation Biology 16, 1232-1242.

[42] For example McAllister, M.K., Starr, P.J., Restrepo, V.R., Kirkwood, G.P. (1999) Formulating quantitative methods to evaluate fishery management systems: what fishery processes should we model and what trade-offs do we make. ICES Journal of Marine Science, 56, 900-916.

[43] Shea, K. & the NCEAS Working Group (1998) Management of populations in conservation, harvesting & control. Trends in Ecology and Evolution 13, 371-374.

[44] Hilborn, R. and Mangel, M. (1997) *The Ecological Detective: Confronting Models with Data.* Princeton: Princeton University Press.

[45] For example Berger, U., Wagner, G., Wolff, W.F. (1999) Virtual biologists observe virtual grasshoppers: an assessment of different mobility parameters for the analysis of movement patterns. Ecological Modelling 115, 119-127.

[46] Cooke, J.G. (1995) The International Whaling Commission's Revised Management Procedure as an example of a new approach to fishery management. Pages 647-657 In Blix A.S. *et al. Whales, seals, fish and man.* Amsterdam: Elsevier Science.

[47] Hilborn, R. and Walters, C. (1992) *Quantitative Fisheries Stock Assessment: Choice, Dynamics and Uncertainty.* London: Chapman & Hall.

[48] Bousquet, F., LePage, Ch., Bakam, I., Takforyan, A. 2001. A spatially-explicit individual-based model of blue duikers population dynamics: multi-agent simulations of bushmeat hunting in an eastern Cameroonian village. Ecological Modelling 138, 31-346.

[49] c.f. endnote 12.

[50] For an excellent example of the kind of research that is required, see de Merode, E., Homewood, K., Cowlishaw, G. (2004) The value of bushmeat and other wild foods to rural households living in extreme poverty in Democratic Republic of Congo. Biological Conservation 118, 573-581.

[51] Evans, T.E. (1998) Forest mapping in North Sulawesi. MSc thesis, University of Warwick, UK.

[52] Milner-Gulland, E.J. & Mace, R. (1998) *Conservation of Biological Resources.* Oxford: Blackwell Science.

Chapter 21

What is Wrong with our Approaches to Fisheries and Wildlife Management? – An Engineering Perspective.

William K. de la Mare

When we think of the subject *Wildlife Conservation – in pursuit of ecological sustainability* our thoughts usually tend towards public awareness campaigns backed by scientific studies of the state of the world and how to make these generate the political case for conservation. In my essay, I am going to take a different perspective because many jurisdictions have already adopted laws and policies that give great weight to conservation, and yet many of the benefits we expected to follow from these changes have fallen short of our hopes. Poor conservation outcomes continue despite public commitments to "sustainable development" and "ecosystem-based management". Of course, I do not mean that the need for public awareness, backed by good science and directed towards political action has diminished, but we need not only calls for action but concrete ideas as to what those actions should be. The perception is that even where we have good laws and policies they are failing to deliver. This shows that we need to identify the sources of the failures and hence promote corrective actions.

There are three broad reasons why we, as societies, continue to achieve poor conservation outcomes:

1. We are not serious about ecological sustainability.
2. We do not know how to achieve it.
3. Our institutions are not up to the task.

There is little doubt that (1) plays a major part in the failures of fisheries and wildlife management (FWM). I doubt that (2) is a major contributor; we have developed a range of approaches to FWM that have the potential to be effective. However, one problem is that these approaches are not applied as frequently or as thoroughly as required. This means that (3) must be a substantial part of the problem. If FWM practitioners are unaware of, or unable to implement, sound approaches to FWM, then our institutions are at fault. I argue that even if we were to solve problem (1), we would still fail to achieve ecological sustainability unless we also reform the institutional framework of FWM.

In this essay I will tend to concentrate on fisheries because that is where I have the greatest experience. However, I have had some experience with other wildlife management issues, from which I can suppose that the development of a coherent management paradigm does not appear to be significantly closer than that developed so far for fisheries. I have also had another life away from the science of fisheries management where I worked in electronic, mechanical, chemical and control systems engineering. This has often had me pondering why engineering is so much more successful in achieving its objectives than we seem to be in the field conventionally known as fisheries and wildlife management.

An important qualification is that none of our human institutions are perfect or even perfectible. However, that should not stop us from enquiring into what seems to work well and what does not. In contrasting engineering with FWM, I am not supposing that the practice of engineering is perfect, or that many of the principles used in engineering are not evolving in FWM. My premise is that instead of rediscovering the lessons of engineering slowly and painfully by trial and error, we should compare directly engineering and FWM to see where the experience of the engineering profession can guide us more quickly to our goal of ecological sustainability.

There are many famous engineering disasters; no doubt there are more to come. However, engineering is a profession in which disasters are a rarity; thousands of mundane engineering projects are completed everyday, millions of technological products are in use continuously; catastrophic failure is rare. In contrast, fisheries mismanagement is common, for example, 58% of some 232 well studied fish populations had declined by 80% or more.[1] In many cases, it is unclear when, or even if, the species or the ecosystems that supported them will ever return to the conditions that prevailed before their collapse. The human hardship and economic losses from our mismanagement of the exploitation of fisheries and wildlife resources are also catastrophes. Many commentators now refer to the crisis in world fisheries. Wildlife and remnant natural ecosystems are under pressure globally from the inexorable greed of the developed world and from the economically marginalised who are displaced onto shrinking areas of previously little-used and marginal habitat.

What are the differences that allow us to be confident that engineering enterprises have a very high probability of success, while our fisheries and wildlife management systems are now widely recognised as being highly failure prone? That we can be enormously successful in the field of engineering suggests that we should examine the processes that are responsible for that success. Then we should examine whether those same processes are at work in FWM and, if they are, why then do they fail? If some of the processes are absent, we should determine whether they could be applied or adapted to FWM. Of course, many of our failures in FWM arise from institutional or political causes, but engineering also operates within the same institutional and political framework and still attains a high level of success. So could there be something about the practice of engineering and its interactions with our institutional and political systems that accounts for the difference?

We could suppose that FWM is intrinsically much more difficult than engineering, for example, that engineering has to deal with less complex systems than the natural world. This may be true to some extent, but engineers deal successfully with very complex systems. A commercial airliner consists of millions of parts, and no one engineer understands how they all work. Such a piece of technology is exceedingly complex. Nor can it be that aircraft engineers have complete knowledge; the flow of air over an aircraft wing can be chaotic and unpredictable. Engineering successfully deals with other bewilderingly complex systems that are expected to remain serviceable in the face of unexpected events. The recent (2003) blackout in the Northeast of the USA and Eastern Canada shows that catastrophic failures can occur but, equally importantly, that such failures are surprisingly rare – failures of such magnitude occur less than once a decade even though hundreds of equipment failures occur every day. Moreover, engineers do have to deal with the vagaries of the natural world. For example, extreme weather and unknown properties of rocks and soils that support structures are routinely allowed for in construction. Power grids are hit by the unpredictable effects of solar flares and lightning. In many cases, failures occur not because of faulty engineering but because of failures in our other institutions: the systems are not used as designed or we fail to enforce adequate maintenance and training standards (two factors that contributed to the mega-blackout).[2]

From a historical perspective, engineering and FWM have both long been key human activities. We probably tend to think of technology as a recent activity, but tool making and construction date back to the hunter-gatherer period of human history. Our most successful "fisheries and wildlife management" activities came to be known as agriculture – one of our first and most transformative and engineering approaches to the environment. That which was left outside of agriculture is what we now consider to be FWM. The key characteristic that leads to certain species and ecosystems remaining outside agriculture is low controllability, in the sense of both controlling the animals and excluding other humans from their use. Interestingly, the relatively recent trend of establishing property rights over fish by means of individual transferable quotas (ITQ) is a step towards establishing a form of ranching in the marine environment.

On the face of it, we might suppose that our collective human experience in FWM is, on evolutionary time scales, much longer than our experience in engineering, so that it should be engineers who are studying the long and successful history of FWM in order to improve their discipline. However, despite its apparently shorter history, engineering is more highly evolved than FWM. Part of this is due to historical attitudes. There was no perceived need for FWM for much of human history because there was always the frontier, with new stocks of wildlife to be had just over the next hill. We usually found substitutes for products from wildlife, and we did not value wildlife because it was one of nature's "free" services. When we lose a natural service it is usually gradually, and often imperceptibly. By contrast, when your bridge falls, the loss is immediate and obvious. If you need the bridge you have to replace it, and you do not want the new one to fall. We value the bridge because we expended effort and treasure to build it. So engineering is more mature because historically we have demanded more of it and more ruefully lamented its failures.

That engineering is a more successful and trusted professional discipline than FWM is starkly indicated by:

Engineers

- Trusted to deal within areas of expertise
- High levels of technical success
- Users trust the solutions offered
- Politicians defer to technical expertise
- Profession is self regulating
- Able to designate who is an engineer

Fisheries and Wildlife Managers

- Areas of expertise not well defined
- Low levels of technical success
- Users dispute the solutions offered
- Politicians interfere
- Not recognised as a profession
- Does not have recognised membership

A key point about the way ahead in FWM is that, although there are many things yet to learn, we already have appropriate paradigms[3,4] with which to pursue ecological sustainability. The main problem is that the paradigms are not widely understood, accepted or applied. The paradigms are closely related to an engineering approach to problem solving. The fact that good ideas for improving FWM are around but resisted, ignored or overlooked suggests that current FWM institutions are part of the problem, and that institutional reform must be part of the solution. Therefore, a major task is to incorporate the paradigms into social, political and educational processes, a process that is much more complete with engineering paradigms and the engineering profession. Improvements in FWM will depend on the transformation of Fisheries and Wildlife Management into a professional discipline with similar attributes to that of Engineering.

A COMPARATIVE ANALYSIS OF ENGINEERING AND FISHERIES AND WILDLIFE MANAGEMENT

The broad objectives of engineering and FWM

Engineering and FWM both contribute to our aspirations, and that entails the full complexity of our mix of humanist and ecocentric philosophies. In contemporary societies both engineering and FWM are expected to increase human well-being in the terms encapsulated in "sustainable development". However, in this essay I am going to consider our objectives at a more operational level and, in particular, what the two disciplines imply in terms of professional attributes and objectives. In both activities the objectives are to provide for meeting society's needs, including protecting the environment. Both activities have ethical practice as objectives. If we accept that both the objectives and the constraints imposed by society are broadly similar for each, then differences in broad objectives cannot explain why engineering is more successful than FWM. The key differences must therefore be primarily institutional.

Both FWM and engineering are scientifically based

Engineering has a long history of empirical knowledge that has been incorporated into its body of knowledge. Engineering is supported by engineering science, but engineering science is not the practice of engineering. The other contributor to the scientific base of engineering is research and development. We can think of the "pure" end of engineering science as including fields such as physics, chemistry and materials science – developing basic theories about the nature of energy and matter and how we might be able to create new or improve existing materials. Microbiology and genetics are now looking much more like engineering sciences than biological sciences. Research and development usually relates to how to turn engineering science into better engineering outcomes, how to improve processes or turn discoveries into practical technologies.

Engineering is outcome oriented. Engineering depends on applying what we know – not on a wish list of what we would like to know. Engineering theories face tough tests; they are tried out in practice and modified or discarded. Engineering failures are the subject of intense scrutiny. A major engineering failure will lead to judicial review or Commissions of Inquiry. Considerable scientific and investigative effort is expended to determine why an engineering product failed, particularly if there is injury and loss of life or significant property or economic loss. Aeroplane crash or bridge collapse investigations are well known examples. Much of the investigation is scientifically based and one of the principle objectives is to add to the body of engineering knowledge to ensure that faulty design is avoided in future, and that other instances of the defective design are corrected. The science is combined with empirical experience to build a body of knowledge that becomes codified in the form of engineering standards. Engineering is committed to learning from its mistakes. This is because we hold engineers liable for their mistakes, and they usually have little choice when it comes to acknowledging their errors because the consequences of engineering blunders are usually obvious. It is also very hard to shift the blame for engineering errors onto others or the vagaries of nature. Even though engineers err, we trust in the engineering profession that the errors will be confronted and corrected.

In contrast to engineering, FWM tends to be process oriented. We spend more time on the data and methods we are going to use to assess the state of exploited fish and wildlife than we do on how the results can be translated into required outcomes of ecological sustainability. Similarly we put more effort into setting regulations than we do in measuring whether they have been effective.

In the early days, fisheries and wildlife science drew on the academic interests of "gentleman naturalists" who tended to ignore the empirical experience of "working class" harvesters, and so failed to absorb, or at least consciously evaluate, empirical knowledge. Historically, a great error was made when the need for FWM began to become apparent in the 19th and 20th centuries, because this was also the age of unbridled optimism about our capacity to find scientific solutions for every problem. The supposition that if we studied the animals we would be able to "manage" them is an enduring fallacy. This supposition created a vast enterprise of studying the detailed life history of commercially important species and, although many important things were learned, it is also now clear that this cannot be enough to predict the effects of exploitation on a community of species. In a nutshell, biology is necessary, but not sufficient for FWM, in much the same way that physics is necessary, but not sufficient, for engineering.

Engineers do not study molecules when designing structures; it is sufficient to study the bulk properties of the materials by direct experiment. Engineering has a much greater experience of focusing on the relevant. If something breaks, engineers study why it broke; they rarely study why it did not break (unless it was supposed to). Once the properties of a material are understood, engineers do not have to repeat the studies unless a new material or application is to be developed. Information on the properties of various materials is available to the practicing engineer in the form of specifications expressed in a standard way.

When it comes to empirical experience, most engineers listen when an artisan says that there is an easier way to fabricate something or that he or she is concerned that something is not right. This is for two reasons; the artisan could well be right or, if there are good reasons for sticking to the plan, it is important that those implementing it understand those reasons if the project is to succeed. The practice of FWM still has a paternalistic and arrogant undercurrent, where any form of knowledge other than science (in fact, scientific orthodoxy) is discounted. FWM is long on science and pure research and short on empiricism and known good practice. Research and development in the sense analogous to engineering is underrepresented in FWM.

Pure scientists address questions such as, how does this system work? They aspire to work on open-ended questions where theories are incomplete. So practitioners in FWM with training in pure science tend towards reductionism; if we understand all the pieces we will be able to predict how species will respond to exploitation. Reductionism usually leads to a large list of things we wish we knew. Unlike the application of engineering theories, many of these predictive FWM theories have been tried, have failed and, despite their failures, they have then been retained. This may be in part due to the predominance of scientists with pure science training in FWM. Pure sciences have a relatively poor record in discarding flawed theories; we cling to our favourite theories long after they are broken. Ecology is particularly prone to this pathology.[5] In engineering, a broken theory may mean a failed project with serious losses; in FWM it may mean broken populations, human, non-human animal or both; in pure science it is an inconvenience.

A good example of a failed FWM theory is the use of catch per unit effort data in stock assessment. It has been known for at least 30 years that such data are unreliable indicators of the status of exploited populations[6] and yet these data are still widely used in fisheries management. One may suppose that this reflects a dearth of alternatives, but I do not believe this to be the case. We should be consciously compiling those methods that seem to work and those experiences that seem not to work so as to undertake the research and development phase of our nascent discipline.

What engineers and FW managers do is different when the best available science is poor. Engineers first try to design around the uncertainty, and if this is not cost efficient they undertake further, highly targeted research. Traditionally FWM has ignored the uncertainty. Now FW managers are supposed to act cautiously in the face of uncertainty in accordance with the precautionary approach.[7,8] However, this rarely takes the form of a design either to acquire the required knowledge or to evaluate whether the extant management strategy is sufficiently robust to work adequately despite the uncertainty. When we do undertake further research, it is usually without evaluating the management system to find out which uncertainties are the most important. We tend to rely on intuition as to the best piece of research to carry out, and without evaluating the effect of reducing a few of probably many uncertainties on the overall management result.

An interesting example is the relatively recent enthusiasm for "ecosystem management". Many people have come to the conclusion that single species and sectoral management has failed and so the obvious way ahead is to capitalise on our failures to solve the difficult problems of single-species management by making management even more difficult. Many scientists are enthusiastic about this idea because they can use it to justify even more and broader scientific research. Environmental NGOs like it because it saves having to worry about the humdrum of small decisions about whether a particular form of human activity is sustainable or not, because they can always say that we are not looking at the big picture. Cynics like it because they can use it to justify continuing with business as usual while everyone tries to figure out just what ecosystem-based management means.

This is management based on what we wish we knew rather than what we know. I am not for a moment advocating that we abandon ideas that we need more holistic approaches to the management of human activities.[9] My concern is that we are always eager to hitch ourselves to the next bandwagon, when actually making progress involves attention to the humdrum. The problem with ecosystem management is that it easily leads us into the trap of saying we don't know enough, the best available science is not sufficient, we need more research. This is not to say that we don't need more research – we clearly do – but we must not fall into the trap that we need more research **before** we take any action, nor should we rush off and do just any research in the **hope** that it will solve some poorly specified problem. Scientists are forever justifying their predilection for studying various interesting scientific problems because they claim it will be of use to conservation and management, claims that they are virtually never called on to demonstrate. Just as classically we perpetrated the fallacy that to manage fish and wildlife it was sufficient to study animals, we are now in danger of creating a new fallacy that to manage ecosystems we have to understand them. First, ecosystems only exist in our heads[6,10] and, because they are rather subjective, we will spend many fruitless hours attempting to decide whether this patch of turf is an ecosystem or not. Second we do not manage ecosystems (they manage us); our goal is to manage human activities so that our use of ecosystem services is sustainable.[9]

In contrast, engineering conquers complexity by appropriate division of a project into sub-projects. It does this without falling into the trap of reductionism, because it employs a top-down approach, but with iteration. An engineering project has to have an overall design, but as a designer one usually finds that some aspect of the grand design cannot be finalised until it is clear what the properties are of one or more of its components. Once clarified, the grand design can be revised to the next iteration. Engineers cannot afford to lose sight of the requirement that all of the sub-projects have to work together, and each has to be completed. The worst engineering foul-ups in terms of delays and cost overruns occur when this principle is overlooked, but sooner or later, engineers are reminded when they forget to deal with reality.

There is a structural problem, a partial vacuum in the centre of FWM. In engineering there is a clear gradation from pure science to engineering science to engineering practice. The greater numbers of engineers are involved in practice, but engineering scientists and practicing engineers see themselves as participants in the same profession. In contrast, FWM science is dominated by biologists, whose traditional training in pure science has not suited them to the tasks they face. They see themselves as distinct from FW managers, who have heterogeneous professional training which in most cases will not have suited them to the tasks they face either. Communication between these groups is often poor.[11] Each tends to blame the other for the failures of their enterprise, which they do not see as being shared. They lack a common basis of technical understanding because they see themselves as separate professions, and they receive little, if any, training that bridges the gap. Hence, we need to fill the gap between the pure science training of biologists and the social and political process by developing a new profession of FWM that develops the applied science of FWM, including a research and development function and the accumulation of empirical experience.

Part of our problem is that we suppose our solutions to FWM problems have to be "discovered by the application of science"; actually our solutions have to be tailored to our existing scientific knowledge, and that brings me to the next important set of differences between engineering and FWM.

Role of design

Engineering projects are designed and evaluated before being put into practice. In most countries this is a legal requirement. The essence of the engineering professions is to design things. In contrast, FWM is dominated by scientists who were trained to see their role as finding out how nature works. They have no historical tradition in designing systems, they see the world as an extant system:"we cannot design nature, our job is to discover and understand nature's design".

Some twenty years ago I began to demonstrate the idea that the management system for fishery or wildlife exploitation was a problem in control theory and that such systems were in fact "designs" and that they could be evaluated.[12] Reactions to these ideas included scepticism and indifference through to outright hostility. Fortunately a small number of people had been developing similar thoughts[4] and a few more embraced these ideas. At the recent (2004) World Fisheries Congress in Vancouver, these ideas now seem much closer to becoming generally accepted at least by some FW scientists. That being said, the number of cases where FWM systems have been designed or evaluated to see if they can work even in principle is still pitifully small. Even so there remains considerable resistance among FW managers, who always seem to be saying that they need "flexibility", which usually means having no specific objectives, or re-inventing them whenever tough decisions loom. At least several major fisheries collapses have been the result of promising to be good later – but we need to give the industry time to adjust. The natural systems that we exploit are indifferent to pleas for flexibility; they have their own rules, and we have to develop our rules accordingly.

If you set out to design something you need a clear idea of what it is you are trying to achieve. Engineering relies on (and has the advantage of) achieving practical outcomes. No engineering project starts from the premise that we need a bridge or an airport but we will decide which a bit later down the track. Engineering projects are expected to work as intended. Building a bridge that gets you 90% of the way across a river is no bridge at all. And it will be too small to use as an airport.

FWM should achieve practical outcomes; where we fail is that we are often too coy to say what they are, or too craven go beyond the vague generalities that allow everyone to interpret the objectives in the ways that suit them. While this may be convenient politics, it is lousy management. Engineers are not permitted to say they are going to build something and have half the population expecting a bridge and the other half an airport.

We ruthlessly require that engineering designs be efficient. We do not usually require bridges to be made of brick or masonry to keep lots of people in work – we require that the most cost-effective solution be found. This means that we are used to evaluating alternative designs and costing them. After a while we don't need to do this every time because engineers develop formulae that give sufficiently accurate forecasts of the costs of implementing a project in different ways and so narrow the field of possible candidates. This becomes part of the practical empirical knowledge drawn on by professional engineers.

However, there is one depressing similarity between engineering and FWM, and that is in the transfer of solutions designed for use in highly developed countries to less developed countries. There is little point in foisting inappropriate designs (either for engineering or FWM) on societies that lack the resources to support them. The ideas of appropriate technology are just as important for FWM as they are for any other form of infrastructure development.

Decisions, decisions, decisions…

When you design something like a structure or a chemical process you have to make lots of decisions. Engineers are used to making decisions. If you are to complete a project you do not have the option of deferring decisions indefinitely. Many engineering decisions are a consequence of having a clear idea of the objectives of the project. You cannot design a thing unless you know what it is to achieve. I wrote earlier that we already have a suitable paradigm for the development of FWM. It is simply the standard management paradigm for any project; define objectives and evaluate the means for achieving them. Some years ago I suggested that we could think of this as a MOP (Management Oriented Paradigm) for tidying fisheries management.[3] A MOP consists of the following elements:

1. management objectives which are measurable;
2. a management procedure based on decision rules;
3. assessments based on specific data and methods;
4. a prospective evaluation of the management procedure using performance measures.

The first three elements need to be set up in such a way that the properties of the resultant system can be evaluated. The evaluation is primarily carried out using simulation methods, in which the physical world is replaced by hypothetical simulated worlds. Different simulated worlds are used to characterise our uncertainties about the real world. Models are also used that reflect how fisheries and fisheries investment respond to the physical and regulatory environments, because our management system is not about managing fish and wildlife populations, it is about managing us.

Like any other system or structure, FW management systems have to be designed. There are many control measures, assessment methods and tools available. The choices about which to include are taken by evaluating their ability to meet the objectives at a cost that is commensurate with the benefits. The development process is iterative, with various approaches to management being simulated, modified and re-evaluated. After a recent lecture in which I discussed the MOP, I was asked by a student, what if you cannot afford to evaluate management policies? My answer was, if a society cannot afford to design its management policies in terms of deciding what it is that it is trying to achieve, determining what means are at its disposal and then evaluating whether the policy can be made to work, then it is not serious about management. That is the nub of many of our problems, when you look at how we tackle FWM issues, it is starkly clear that in very many cases we do not take them seriously. Deciding not to decide on some issue is a decision to continue the *status quo*. If that involves the continuation of the exploitation of a fish or wildlife population or the degradation of their habitat then we are deciding that we are indifferent to the outcome; in essence we are deciding not to manage our activities. Delay in taking decisions has been a common factor in the collapse of many fisheries.

Sometimes we hear the view that fisheries and wildlife systems may be so unpredictable as to be "unmanageable". This view arises from the faulty premise that we manage fish or wildlife. A MOP is not broken by this observation because in saying that something is unmanageable we are really saying that we have not organised ourselves to take into account that we are unable to predict or rely upon a particular system conforming to our wishes of a steady state.[13] We face the choice of transforming the world into a more predictable form, as we have with irrigation and agriculture, or we should reform our activities so that we know what to do when the unpredictable downturn begins. We should not continue to flog a declining resource because we failed to plan for any alternatives. This is one of our commonest failings, recognising in good time that we need to reduce exploitation because the abundance of an exploited population has taken a downturn. This failure means that we make a bad state of affairs worse, and that consequently we will have exposed the population to even greater risks of extirpation, or at least substantially delayed its recovery to productive levels,[1] while reducing the circumstances of the people dependent on it. This means that we should have some form of insurance, in the form of alternative employment of people and capital when the inevitable downturn occurs in some unpredictable species. Prudent levels of insurance are something we can evaluate using a MOP.

Engineers tackle "predictable failure" by redundancy and monitoring. By predictable failure, I mean that a failure is inevitable, but when it will occur cannot be accurately predicted. A key engineering strategy is redundancy; in a typical industrial process any piece of equipment prone to predictable failure will have a backup. Critical pieces of equipment might have two backups. Engineers monitor the performance of equipment to identify incipient failure so that the backup can be brought into service without any serious disruption.

In FWM we do not plan for failure; we hardly ever consciously set aside some region (e.g., a protected area) or species so that there is a backup when things turn sour. We usually allow exploitation on everything simultaneously. Even when alternatives emerge serendipitously we may fail to organise ourselves so that those people who were worst affected are the first in line when an alternative emerges. A recent example is the collapse of the northern cod fishery in Canada[11] that left more than 10,000 small-scale inshore fishermen without substantial employment. The paradox is that the fisheries in the region are now worth more than before the cod collapse because they are now dominated by a high-value shrimp fishery.[14] However, most of the 10,000 unemployed cod fishermen are still there, but now we have a relatively small number of shrimp millionaires as well.[14]

Trust me, I'm a professional

Engineering is a well established example of a profession since it has all of the attributes set out in Table 21–1. We can also use Table 21–1 to determine if FWM has the characteristics of a profession. Clearly FWM in pursuit of ecological sustainability is an essential service (1) and is concerned with an identified area of need or function (2). It also satisfies (4) – involvement in decision-making in the service of clients – and usually (10) strong motivation and lifetime commitment. These are sufficient conditions for asserting that FWM should be a profession.

How does FWM stack up against the other requirements and can this help explain its poor standing as a profession? Although (3) is true, the body of knowledge and repertoire of skills can be identified, they are not sufficiently known by many of the practitioners currently employed in FWM, nor are our successes and failures being sufficiently and systematically accumulated to add to that body. Another problem for FWM is (5); although we can identify the under-girding disciplines, concentration is still too narrowly focused on biological sciences, with insufficient attention being paid to mathematics and statistics, control and management theory, sociology and economics. In terms of (6), FWM has relatively little in the way of professional associations, certainly none in the same class as engineering professional associations, and this is probably a key factor in the lack of recognition of FWM as a profession. Both (7), standards, and (8), training, are great weaknesses in the development of FWM as a profession. A high level of public trust (9) is one of our most serious problems. The keys to progress as a profession are to establish a track record of success and to continue the trend of working with public and industry groups to increase public understanding of the nature of FWM. Both of these will require that we address the other weaknesses in relation to particularly (3), (5), (7) and (8). Then we should be able to achieve (11) and (12), which will then further help with reducing the effects of the political weaknesses of FWM.

"Politics is the art of looking for trouble, finding it everywhere, diagnosing it incorrectly and applying the wrong remedies."[16]

A society deciding what it wants to achieve is politics. However, societies do not have unbounded choices but usually consider a series of options, and these inevitably involve tradeoffs. The role of professions is of course not to make those decisions on behalf of society, but that does not mean that professions do not express views and advocate various solutions (however, both engineering and

Table 21–1. Defining characteristics of professions[15]

1. Professions are occupationally related social institutions established and maintained as a means of providing essential services to the individual and the society.
2. Each profession is concerned with an identified area of need or function (for example, maintenance of physical and emotional health, preservation of rights and freedom, enhancing the opportunity to learn).
3. The profession collectively, and the professional individually, possesses a body of knowledge and a repertoire of behaviors and skills (professional culture) needed in the practice of the profession; such knowledge, behavior, and skills normally are not possessed by the nonprofessional.
4. Members of the profession are involved in decision-making in the service of the client. These decisions are made in accordance with the most valid knowledge available, against a background of principles and theories, and within the context of possible impact on other related conditions or decisions.
5. The profession is based on one or more undergirding disciplines from which it builds its own applied knowledge and skills.
6. The profession is organized into one or more professional associations, which, within broad limits of social accountability, are granted autonomy in control of the actual work of the profession and the conditions that surround it (admissions, educational standards, examination and licensing, career line, ethical and performance standards, professional discipline).
7. The profession has agreed-upon performance standards for admission to the profession and for continuance within it.
8. Preparation for and induction into the profession is provided through a protracted preparation program, usually in a professional school on a college or university campus.
9. There is a high level of public trust and confidence in the profession and in individual practitioners, based upon the profession's demonstrated capacity to provide service markedly beyond that which would otherwise be available.
10. Individual practitioners are characterized by a strong service motivation and lifetime commitment to competence.
11. Authority to practice in any individual case derives from the client or the employing organization; accountability for the competence of professional practice within the particular case is to the profession itself.
12. There is relative freedom from direct on-the-job supervision and from direct public evaluation of the individual practitioner. The professional accepts responsibility in the name of his or her profession and is accountable through his or her profession to the society.

FWM are usually at their worst when engineers and scientists make political decisions in policy vacuums, e.g. flood control by the US Army Corps of Engineers for much of the 20th century). Moreover, we rely on professions to delineate what is possible given the current state of resources, what is technically feasible, what the various options are expected to cost and what sort of benefits will result, including what are the trade-offs between conflicting objectives.

We are used to deciding on practical tradeoffs in engineering projects. Once we have decided to implement an engineering project we trust engineers to design it and supervise its implementation. The decision, say, to build a dam is a political one; what sort of dam to build is subject to the advice of engineers. It would be a very brave politician indeed who would override the advice of the engineering profession. Although mega-projects are not immune to political interference, engineers are in any case well able to defend their expert turf against political interference. Even when engineers work for governments, Professional Engineering Associations provide protection in the form of agreed best practice and ethical standards that enable its practitioners to resist political pressures at the technical level.

Engineering fares well in the political process because it has earned society's trust through its:

- Track record of success
- Commitment to learning
- Commitment to ethical practice[17]
- Standards and regulations
- Accountability
- Resistance to political expedience
- Distinction between technical and political processes

FWM does not fare so well in the political process because of its:

- Track record of frequent failure
- Reluctance to learn from mistakes
- Lack of guidance on ethical practice
- Few standards and enforceable practices
- Low levels of accountability
- Susceptibility to political interference
- Confusion between technical and political processes

For politicians, credit for the sustainable use of natural systems is much more nebulous than it is for an engineering mega-project. One hardly ever receives credit for things that do not go wrong – whereas fixing things is one method for achieving fame and adulation. This is one reason, coupled with the history of regarding natural services as free, that conservation has a low rating in terms of political relevance. However, politicians usually share the blame when things go awry, so they have some incentive for reforming the practice of FWM. Opportunities for reform usually only arise when things are seen to be broken. However, politicians are not reluctant to override the day-to-day advice of FW managers.[11] When doing so they enjoy the benefits of popularity with whatever constituency lobbied them, and they do not incur much in the way of opprobrium from FW managers. In most countries, FW managers are civil servants who can be instructed or pressured by politicians to adjust their advice. Even without instructions, most civil servants know "what will go down well with the Minister". The lack of a professional culture with accepted standards increases the vulnerability of FW managers to such pressures; they lack the protective mantle of an organised professional association. The absence of an independent professional body also decreases the likelihood that politicians will be held to account. For a time, environmental NGOs were successful in this role, but this success is declining because NGOs are now less likely to be regarded as impartial,[18] and indeed the rewards for a politician may well arise from showing environmental do-gooders just who is in charge.

In many jurisdictions, a politician is, in legal fact, the Fisheries and Wildlife Manager, and there is a confusion of delegation and technical responsibilities between the politician and the bureaucracy. A better system would look more like engineering, where management objectives are political decisions. However, once the political decisions are reached, we should be able to rely on professional FW managers to implement the decisions using methods in which there is widely shared trust. Such a system reduces opportunities for direct political intervention at the behest of vested interests.

Even when engineers work in government bureaucracies they remain accountable to their profession. When an engineer designs something, he or she has to sign their name to it, as does his or her supervisor. Signing your name to something has a salutary effect on your sense of responsibility for what you do. Although currently many bureaucratic systems expose FW managers to political pressure, the system also shields managers from responsibility by its ability to spread the blame around. So, while professional reform will involve the reduction of bureaucratic anonymity, increasing accountability has to be tempered by increased protection for individual managers, so long as they have discharged their responsibilities according to defined professional and ethical standards.

The role of standards

An Engineering Standard is a book of rules compiled by a Standards Committee. The rules define acceptable engineering practice and engineers apply those standards in the routine exercise of their profession. The rules usually specify matters such as how a structure is to be designed, allowable stresses in various materials, how metals are to be welded, what sort of documentation is to be produced and so on. There are standards for almost every aspect of practical engineering. They codify much of the theoretical and empirical knowledge used by engineers and they are continually revised as engineering knowledge and practice evolve. Standards usually have backing in law and failure to apply the Standards is professional misconduct for which an engineer is legally and financially liable. However, Standards also limit the liability of engineers in the event of an engineering failure so long as they have been adhered to. Reading and applying Standards is fundamental in the training that engineers receive.

It is not that standards are completely absent in FWM. In fisheries we have a number of "standards" set by international bodies, particularly the Food and Agricultural Organisation of the United Nations (FAO). These include the application of precautionary management[7] and the Code of Conduct for Responsible Fisheries.[19] Other standards are emerging through the Marine Stewardship Council.[20] But the compilation of standards is not a core activity of FW managers in the way it is for engineers. The lack of accepted standards makes it much more difficult for FW managers to defend the decisions they make against outside pressures. We should be actively compiling our theory and our practical experience of what works and what does not in the form of standards so that professional FW managers can find and apply sound practice.

Role of professional associations

The peer group for engineers is other engineers. Engineers can point to a set of accepted professional standards that guide their conduct. This enables them to resist cooption, and they have a professional association that can support them in the event of coercion. Ethical standards and codes of conduct are an essential component of the engineering profession. Engineers delineate the professional duties they owe to their employers, those they owe to their profession and those they owe to society. Legislation on engineering regulations is based on advice from Engineering professional associations and usually results in the force of law backing up the rights and responsibilities of the profession. The law and professional standards reinforce each other. If an engineer follows the rules and Engineering Standards but nonetheless something goes wrong, the professional association can provide technical and, often, legal support. This assists the engineering professions to confront mistakes in the practice of engineering and to learn from them.

In the absence of a professional association, the peer group for FW managers is more likely to be the bureaucracy, its political supervisors and those that it regulates. This increases the proneness of FW managers to capture by those that they regulate, both by direct cooption from daily contact and indirectly through political influence. The absence of profession-specific ethical standards and codes of conduct make it difficult for FW managers to delineate the bounds of practice within which they can comply with political demands. The normal perception of the duty of a FW manager is to their employer. Their duties to society and to the species and habitats on which they manage human impacts are either absent or vaguely specified in legislation. Legislation is often vague because of the lack of FWM professional standards and codes of practice that carry their imprint into the law. If a FW manager makes an error, he or she lacks an institutional professional support system, and either the bureaucracy finds him or her a convenient repository for all blame, or has a system where blame is so diffused that no one has to accept any. In either case, mistakes are not really confronted and the opportunity for institutional learning is stunted.

In many jurisdictions, a key role of an Engineering Professional Association is that they define who is legally able to practice as an engineer. They designate the professional skills that an engineer requires, and membership of the association is a sufficient qualification to practice. The principal method that engineers qualify for membership is by completing a degree in engineering, accredited by the Engineering Professional Association, at a University or College.

In some places FWM is moving towards establishing professional associations, for example, Registered Professional Biologists in Canada, and the certification schemes in the USA initiated by the Wildlife Society and by the American Fisheries Society. These are certainly steps in the right direction. However, the focus of these schemes is predominantly on fisheries and wildlife science, although the American Fisheries Society does recognize non-scientific training and experience in fisheries as grounds for certification. These steps need to be emulated more widely, but also expanded to recognise that FWM is a profession that involves a range of technical skills that link policy and science to ecological sustainability as well as social and economic outcomes. Thus, these schemes also require further development to reflect better the breadth of education required for certification and then in accrediting University and College Degree courses that provide the required training. Further, these emerging schemes will need to be strengthened through legal recognition that confer rights and responsibilities similar to those accorded to professional associations such as in Engineering and Medicine.

Role of education

An interesting insight into our professional development as FW managers occurred at the recent (2004) World Fisheries Congress in Vancouver. During the panel discussion in the closing session a panelist exhorted the attendees to produce statements about how fisheries management can be improved, directly implying that such material is lacking. In a response to this exhortation a participant asked the audience for a show of hands as to how many had not read the (1995) FAO document on the "Precautionary Approach to the Management of Capture Fisheries".[7] More than half the audience admitted to having not read this fundamental piece of literature. This document and similar documents has now been around for sufficient time for at least FW managers that have graduated in the last ten years to have seen them. That they have not shows that they are not receiving the breadth of training they need for the tasks they face in FWM.

Engineers require high-level skills in quantitative methods and a sound knowledge of mathematics. A typical engineering course has high-level mathematics subjects in every year, and consequently engineering graduates are virtually applied mathematicians. Much of engineering training is directed towards avoiding the mistakes of the past. However, this is not achieved by dwelling on the study of mistakes but by studying the known good solutions that were devised to avoid their repetition. However, we should always study some mistakes because, besides combating hubris, it helps us to recognise patterns of practice where mistakes become more likely.

FWM requires high-level skills in quantitative methods and a sound knowledge of mathematics. Most graduates who end up employed in fisheries science and management have backgrounds in biology or even non-scientific training. Many biology programs are still producing graduates with limited mathematical and statistical training. Moreover, typical statistics courses concentrate on hypothesis testing – a pure science preoccupation. The important statistical subject in FWM is estimation. The key analyses are not about rejecting hypotheses that observed samples are similar between two groups of animals, but of the type; how many animals are there in a given area? The Masters students who enter the fisheries program at my university usually have Bachelors Degrees in Biology. Although we require our students to have received good grades in calculus and statistics, many of them took these subjects early in their degrees. In their higher level undergraduate studies they have had little or no further exposure to mathematics, to the extent that one of my teaching tasks is to refresh students' knowledge of basic mathematics. Very few students have any undergraduate experience in computer programming or mathematical modeling, skills of enormous utility in FWM.

If we are to fill the gap between pure science and FW management as it is practiced today we need new educational courses designed to equip practitioners with the mix of skills they require. Since FWM is largely about managing human activities, FW managers need education in this area and in economics. Not only do we often do a poor job at a technical level in designing our management systems, we have too frequently failed to consult on what is socially and operationally feasible, and how to ensure that the regulatory systems we set up will be observed. The body of knowledge relevant to our nascent discipline is a unique combination of the applications of:

- Biology and ecology
- Economics
- Mathematics, statistics and estimation
- Control systems theory
- Sociology
- Management theory
- Politics and policy formation

CONCLUDING REMARK

I often get the impression that the lack of trust accorded to FW managers from both industry and environmental NGOs is now bordering on contempt. Failure to arrest this trend will be a serious error if it means that we turn good people away from careers that will contribute to a sustainable future. We do not turn our backs on engineering even when we have calamities; we should not turn away from the development of a scientifically based profession of FWM because of its failures either. Thus, the foregoing critique of FWM as a profession is not intended to focus criticism on the people that currently practice it. The focus of my criticism is on our societal failure to recognise FWM as a profession that requires a unique body of knowledge and to ensure that the people taking up careers in FWM have the skills, training and institutional support they need to be effective. If we succeed in that, then we should also begin to achieve more surely our objectives of *Wildlife Conservation – in pursuit of ecological sustainability.*

NOTES AND SOURCES

[1] J. A. Hutchings and J. D. Reynolds, "Marine Fish Population Collapses: Consequences for Recovery and Extinction Risk". *Bioscience*, Vol. 54 No. 4, 2004. For more on fisheries, see Holt, Chapter 4 and Hutchings, Chapter 6.

[2] C. Biever, "'Preventable' failures caused US power blackout". *New Scientist*, London, 2003.

[3] W. K. de la Mare, "Tidier fisheries management requires a new MOP (management oriented paradigm)". *Reviews in Fish Biology and Fisheries*, Vol. 8, 1998, pp 349-356.

[4] C. J. Walters, *Adaptive management of renewable resources.* McGraw Hill, New York, 1986.

[5] R. H. Peters, *A critique for ecology*. Cambridge University Press, Cambridge, 1991.

[6] R. Hilborn and C. J. Walters, *Quantitative fisheries stock assessment: choice, dynamics, & Uncertainty*. Chapman & Hall, 1992.

[7] FAO, *Precautionary Approach to Fisheries, Part 1: Guidelines on the precautionary approach to capture fisheries and species introductions.* Food and Agriculture Organization (FAO) of the United Nations, Fisheries Technical Paper No. 350/1, FAO, Rome, 1995.

[8] S. R. Dovers and J. W. Handmer, "Ignorance, the Precautionary Principle, and sustainability". *Ambio*, Vol 24, No. 2, 1995, pp 92-97.

[9] See W. K. de la Mare, "Marine ecosystem based management as a hierarchical control system". *Marine Policy* Vol 29, No. 1, 2005, pp 57-68.

[10] F. B. Golley, *A history of the ecosystem concept in ecology: more than the sum of the parts.* Yale University Press, New Haven. 1993.

[11] See Hutchings, Chapter 6.

[12] W. K. de la Mare, "Simulation studies on management procedures". *Report of the International Whaling Commission*, Vol. 36, 1986, pp 429-450.

[13] See Czech, Chapter 22.

[14] R. Hilborn, "Examples of implementation of management procedures and how they help reconcile maintaining fisheries with conservation of resources". Oral presentation to the World Fisheries Congress, Vancouver, 2004.

[15] Internet document attributed to the University of Virginia http://www.adprima.com/profession.htm

[16] A quotation from Marx, Groucho not Karl!

[17] See Lynn, Chapter 13

[18] S. Beder, *Global spin: the corporate assault on environmentalism* Totnes, UK : Green Books, 2002. Rev. ed.

[19] FAO, *Code of Conduct for Responsible Fisheries*, Food and Agricultural Organisation of the United Nations, Rome, 1995.

[20] MSC. "Certification Guidelines", Marine Stewardship Council, London, 1997.

ADDENDUM

During the discussions at the Forum some participants expressed the view that natural systems are more complex than engineering systems and this explains why engineering is relatively more successful than FWM. As I assert in my paper, I do not believe this to be a substantial cause of the differences in the success rates of engineering and FWM. Consider a petro-chemical plant; this is an extremely complex system. If viewed at a molecular level as a physical and chemical system, it is chaotic and unpredictable. Fluid flows are turbulent; the chemical compositions of the feedstocks and their physical and chemical properties are variable and incompletely known. The plant is operated by erratic and unreliable humans and hundreds of unobserved variables can affect the outcome. The design of the system would consider literally thousands of variables. No one could predict where to set the hundreds of control valves, pump speeds, furnace thermostats, cooling water flows and so on to produce the required outputs. The plant changes over time; catalysts degrade, pipes erode, heat exchangers become fouled, and equipment wears or fails. Yet engineering is able to deal with this complexity to a high degree of reliability. By a blend of empiricism and theory, engineers have developed experience on which classes of complexity can be ignored. They design the plant to deal with foreseeable uncertainties so that its operation can be adjusted using negative feedback. Engineers approach complexity hierarchically by dividing complex systems into assemblies of less complex subsystems. Many engineering components are designed using simplified and empirical models. Even chaotic systems such as turbulent fluid flow have predictable emergent properties. It is these approaches that engineers exploit to deal with complex systems. We should apply similar ideas in FWM.

Gro promoted sustainable development
Used by corporations for global envelopment
It cut up the pie
So the rich can go high
And the poor get enviro-dishevelment.

William de la Mare 2004

CHAPTER 22

IMPLEMENTING THE PRECAUTIONARY APPROACH: TOWARDS ENABLING LEGISLATION FOR MARINE MAMMAL CONSERVATION IN CANADA

Michelle Campbell & Vernon G. Thomas

There are at least 33 cetacean and 11 pinniped species that inhabit the waters of Canada's three oceans, including a number of resident and migratory populations.[1] Twenty-four populations of these marine mammals are considered endangered, threatened or at risk by the Committee on the Status of Endangered Wildlife in Canada (COSEWIC).[2] Threats affecting marine mammals include habitat degradation, noise and chemical pollution, accidental strikes by ships, tourism activities, and incidental by-catch by the commercial fishing industry,[3] hunting and culling. Despite the number of populations endangered or threatened, and the numerous threats to their survival, there are no laws in Canada for the deliberate protection and conservation of marine mammals.[4] Their use is regulated currently under the federal *Fisheries Act*.[5] Marine mammals are managed by Fisheries and Oceans Canada – the same government department responsible for managing the fishing industry – and under the Act marine mammals are defined as fish. This biologically invalid definition has been a barrier to the development of comprehensive research and conservation programs for marine mammals in Canada.[6]

In 1996, the federal government enacted the *Oceans Act*, the purpose of which was, among others, to delineate Canada's exclusive economic zone, promote the understanding of oceans ecosystems, provide for the conservation of marine resources and the maintenance of biodiversity, promote the wide application of precaution in oceans management decision-making, and promote integrated management of ocean resources.[7] As a requirement under the *Oceans Act*, in 2002 Fisheries and Oceans Canada released the *Oceans Strategy*, which is the government's plan for the integrated management of Canada's oceans.[8] The Strategy lists the international principle of precaution (variously known as the precautionary approach or precautionary principle[9]) as one of the principles to guide all ocean management decision-making. Canada has stated that by implementing the Strategy, it intends to become a world leader in ocean management.

The precautionary principle or approach means that action to protect the environment may be required even in situations where there is scientific uncertainty as to cause and effect relationships. The concept evolved from a shift in thinking about environmental management and regulation brought about by the inadequacies of environmental policies under the traditional permissive approach. The permissive approach is based on the rationale that science can accurately predict threats and provide solutions to mitigate harms after they have been identified.[10] Precaution is characterized by uncertainty and, as such, it differs from preventive measures, which are undertaken when the risks are known.[11] The precautionary principle first appeared explicitly in international law in 1987 when it was included in the Second North Sea Ministerial Conference in relation to marine pollution,[12] and has appeared in virtually all international environmental treaties and conventions since the early 1990s. Precaution was initially applied in the context of pollution, especially marine pollution, but since has also been included in a variety of treaties and instruments relating to the ozone layer, biodiversity, and fisheries conservation.[13] The precautionary principle finds its most complete application in international fisheries regulation.[14] Nevertheless, the use of the precautionary approach to address biodiversity conservation and ecosystem-based management more generally, is weak. For example, the precautionary principle appears only in the preamble to the *Convention on Biological Diversity* (CBD).[15] Thus, it may be useful as a guiding principle, but it does not create legally binding obligations on States under international law. This means that the principle may carry little weight when States are designing national strategies to implement the CBD.

The precautionary approach could be applied to decisions relating to the conservation of marine mammals, especially because of the uncertain risks of numerous ocean activities, such as shipping, oil and gas exploration, fishing, and whale watching.[16] The consequences of these activities include population reduction from incidental bycatch in the fishing industry, accidental strikes by ships, pollution, and destruction of habitat. Additionally, sounds in the ocean (e.g. from seismic exploration and other sources, such as low and medium frequency SONAR),[17] whale watching,[18] shipping, and fishing activities[19] have the potential to alter behaviour and have an impact on normal breeding and foraging activities of individual animals.[20]

The purpose of this chapter is to outline the status of the precautionary principle in international biodiversity and marine conservation law, and to discuss the application of precaution in Canadian federal biodiversity and marine conservation law and policy, including how precaution has been dealt with in Canada's new *Oceans Strategy*. After outlining the commitments to precaution made in the Strategy, recommendations on how to implement these commitments to enhance marine mammal conservation will be presented.

PRECAUTION IN INTERNATIONAL BIODIVERSITY AND MARINE CONSERVATION LAW

A major question regarding the precautionary principle is whether it has crystallized into a norm of customary international law.[21] If it has not yet attained the status of customary law, then it is only binding on States according to the ratification of international conventions, or to the extent it has been included in national law. Some commentators consider that it is "emerging as a principle of customary international law",[22] while others state more definitively that its presence in recent international environmental instruments suggests that it has already attained that status.[23] Others, however, are more cautious about declaring that precaution has become customary international law, stating that it is not specific enough to be translated into a binding norm.[24] The latter argument is that since there is no consensus in international law as to how precaution should be defined, then there is no way of enforcing its implementation as a binding rule of law. Nevertheless, nations that have ratified international agreements that require the application of precaution would have certain good faith obligations to follow it in national decision-making.

The precautionary principle has been included in numerous treaties and agreements relating to biodiversity and marine conservation. For example, it is part of the Bergen Ministerial Declaration on Sustainable Development, the Rio Declaration on Environment and Development, the *Convention on Biological Diversity* (CBD), the United Nations Straddling Fish Stocks Agreement, and the Fort Lauderdale resolution of the *Convention on the International Trade of Endangered Species* (CITES). Additionally, certain international moratoriums and management approaches have been described as precautionary actions. The most notable are the International Whaling Commission's (IWC) 1986 moratorium on commercial whaling,[25] the 1994 IWC Revised Management Procedure for determining quota sizes,[26] and the 1991 ban on the use of large-scale drift nets in high seas fisheries.[27]

Despite its appearance in these international instruments relating to biodiversity and marine conservation, the precautionary principle has not been defined consistently. For example, in the 1990 Bergen Ministerial Declaration on Sustainable Development in the European Commission for Europe (ECE), it was stated that:

> In order to achieve sustainable development, policies must be based on the precautionary principle. Environmental measures must anticipate, prevent and attack the causes of environmental degradation. Where there are threats of serious or irreversible damage, lack of full scientific certainty should not be used as a reason for postponing measures to prevent environmental degradation.[28]

Just two years later, the precautionary principle appeared in a similar format in Principle 15 of the 1992 Rio Declaration, where it was stated as follows:

> In order to protect the environment, the precautionary approach shall be widely applied by States according to their capabilities. Where there are threats of serious or irreversible damage, lack of full scientific certainty shall not be used as a reason for

> postponing *cost-effective* measures to prevent environmental degradation [emphasis added].[29]

In the Preamble of the 1992 CBD, the precautionary principle was expressed as: "where there is a threat of significant reduction or loss of biological diversity, lack of full scientific certainty should not be used as a reason for postponing measures to avoid or minimize such a threat".

In 1994, the Fort Lauderdale Resolution was agreed upon by parties to CITES.[30] It requires Parties to apply precaution and act in the best interest of the conservation of the species when considering proposals for the amendment of the Appendices to the Convention. The States resolved to "apply the precautionary principle so that scientific uncertainty should not be used as a reason for failing to act in the best interest of the conservation of the species".

The 1995 statement on the precautionary principle in the United Nations Straddling Fish Stocks Agreement is the most comprehensive statement on the precautionary principle.[31] This is because Article 6 and Annex II provide guidelines for the application of precaution, in addition to a definition. For example, under Article 6.1 of the Agreement, Parties are required to apply a precautionary approach to conservation, management, and exploitation of straddling and highly migratory fish stocks on the high seas. Specifically, Article 6.2 provides that, "States shall be more cautious when information is uncertain, unreliable, or inadequate. The absence of adequate scientific information shall not be used as a reason for postponing or failing to take conservation and management measures". Article 6 of the Agreement also requires States to collect information and develop research programmes to assess the impact of fishing on non-target species, undertake monitoring, and adopt cautious conservation measures for new or exploratory fisheries.

Despite the definitional differences among international instruments relating to biodiversity and marine conservation, a number of key elements have emerged. First, there must be scientific uncertainty before the principle will be invoked. Second, there is normally some level of evidentiary threshold that triggers application. This is stated as "serious or irreversible damage" in its strictest form in the Rio and Bergen Declarations on sustainable development, whereas a "threat of significant reduction" is enough to warrant action in the context of biodiversity conservation. Third, most definitions provide that where there is uncertainty and some sort of risk of harm, there is a duty to act to reduce or minimize that harm. What remains uncertain is the precise threshold that must be met and the scope of the duty to act. It has been suggested that the application of precaution will vary among and within the various management regimes, and as such should necessarily be applied differently in the context of pollution management, fisheries management, or biodiversity conservation.[32]

Additionally, some definitions, such as in the Rio Declaration, include a provision that actions be cost-effective, indicating that application of the precautionary principle should be linked to economic concerns, and be subject to a cost-benefit analysis.[33] This is a problem because economic concerns could be used as an excuse to avoid or stop regulatory action. Also, a costs provision could be manipulated because the definitions do not state whose costs and whose benefits should be considered.[34] The CBD includes no requirement that measures be cost-effective. This suggests that economic interests – for example, an opportunistic expansion of Canada's commercial seal hunt to accommodate a growing market demand – should not predominate when considering precautionary measures to protect biodiversity.

Some commentators have stated that the correct interpretation of the precautionary principle is that the normal burden of proof is shifted, such that it is the responsibility of the proponent of an activity to show that an activity will not harm the environment.[35] However, the precise extent of such a reversal of the normal burden of proof is the subject of controversy. Another issue that has not been resolved is whether the precautionary principle creates a positive legal duty to anticipate problems (as suggested in the Bergen Declaration definition), thus placing an obligation on States to apply it not only to specific science-based decisions involving risk, but also to legislative frameworks and institutional structures if these might lead to harmful decisions or omissions. Thus, it is possible that regulatory *inaction* could invoke liability under the precautionary principle.

Despite the variety of definitions, and the unresolved issues regarding its application, it is clear that the precautionary principle involves value-laden choices as to what level the appropriate threshold should be set, and what level of risk society is willing to accept.[36]

PRECAUTION IN CANADIAN NATIONAL BIODIVERSITY AND MARINE CONSERVATION LAW AND POLICY

Canada has ratified the 1992 CBD and the 1995 UN Fish Stocks Agreement.[37] However, unlike some nations, before the provisions of ratified treaties will be considered to have legal effect in Canada they must be transformed into domestic law, i.e. internal law can only be altered by the enactment of domestic statutes.[38] Thus, while the ratification of a treaty advocating the use of the precautionary principle may bring with it a good faith obligation to comply,[39] no legal liability will be attached unless it is incorporated into federal (or provincial) law.

VanderZwaag has described the embrace of the precautionary approach in Canadian law as hesitant, as it has only been expressly included in a limited number of federal statutes.[40] For example, it has been included in the *Canadian Environmental Protection Act, 1999* (CEPA), which regulates the release of toxic substances.[41] The language of precaution mirrors that found in the Rio Declaration and, as such, contains the high threshold of "serious or irreversible damage" as well as the cost-effective provision.[42]

The concept of precaution has been included in two statutes relating to biodiversity conservation. However, its inclusion has generally been weak. For example, the recent *Species at Risk Act* (SARA) includes the language of precaution although does not refer explicitly to the precautionary principle.[43] The Preamble states that,

> the Government of Canada is committed to conserving biological diversity and to the principle that, if there are threats of serious or irreversible damage to a wildlife species, cost-effective measures to prevent the reduction or loss of the species should not be postponed for a lack of full scientific certainty.

Section 38 of SARA also requires that the "principle", as stated in the Preamble, must be "considered" by the competent minister in the preparation of a species recovery strategy, action plan, or management plan.[44] The precautionary language in SARA has legal effect because it is included in a substantive provision with respect to particular actions required under the Act. However, the provision is relatively weak because the inclusion of the term "considered", without qualification, creates broad ministerial discretion. The commitment is also potentially crippled because of the inclusion of the cost-effective provision, which could be used as a justification for stalling precautionary action.

The new *Canada National Marine Conservation Areas Act*, which enables the designation of marine protected areas under the management of Parks Canada, also recognizes the precautionary principle both in the preamble and in a substantive provision.[45] The preambular statement creates a lower threshold than SARA, only requiring a "threat" of damage to trigger the application of precaution. It states:

> Whereas the Government of Canada is committed to adopting the precautionary principle in the conservation and management of the marine environment so that, where there are threats of environmental damage, lack of scientific certainty is not used as a reason for postponing preventive measures.

In the context of broader ecosystem conservation, precaution has been included in the *Oceans Act*, an Act that provides for the management of the ocean and the setting up of marine protected areas under the management of Fisheries and Oceans Canada.[46] The Preamble states that, "Canada promotes the wide application of the precautionary approach to the conservation, management and exploitation of marine resources in order to protect these resources and preserve the marine environment". The precautionary approach is not defined in the Preamble, but in a substantive section requiring the development of a national oceans strategy, the Act provides that such a strategy is to be guided by "the precautionary approach, that is erring on the side of caution".[47] Despite its inclusion in the Act, no meaningful legally binding obligations or duties have been created. This is because the preamble is meant merely to guide the decisions made under the Act, and as such can only be used by courts as an interpretive aid when construing the substantive clauses. The substantive clause is weak because even though it requires government to be guided by the precautionary approach in the development of its national ocean strategy, determining the scope of that guidance is left completely to the discretion of the Minister of Fisheries and Oceans.

Precaution has also been embraced in certain federal fisheries policy.[48] For example, the *Policy for Selective Fishing in Canada's Pacific Fisheries* cites conservation as its primary objective, and commits to not allowing a resource to be compromised by short-term considerations.[49] Additionally, the *New Fisheries Policy* recognizes precaution in that before a new fishery will be permitted, the burden is on the proponent to carry out stock assessments and collect scientific information to provide a reasonable scientific basis for the management of the new fishery.[50] This policy also states long-term conservation as a guiding principle, and explicitly provides that a precautionary approach to new fisheries will be applied.

In July 2003, the Government of Canada released *A Framework for the Application of Precaution in Science-based Decision Making about Risk*.[51] This top-level national policy document was developed over a two-year period through a multi-departmental process and with stakeholder participation.[52] The Framework is meant to provide the "guiding principles for the application of precaution to science-based decision making in areas of federal regulatory activity for the protection of health and safety and the environment and the conservation of natural resources".[53] It outlines five general principles of application, which describe distinguishing features of precautionary decision making, and five principles for precautionary measures, which describe characteristics that apply once a decision that measures are warranted has been made. Since the Framework is meant to provide guiding principles regarding the application of precaution to all relevant government departments, it is worthwhile discussing some of its limitations. Three main limitations have been highlighted by VanderZwaag.[54] First, the third principle of application requires that "sound scientific information and its evaluation must be the basis for applying precaution...". This is problematic because sound scientific information is not necessary to justify applying precaution, and its inclusion attempts to limit the principle by hinging it on such criteria.[55] Second, the Framework does not explicitly provide for a general reversal of the burden of proof. Instead, it states that the responsibility for providing the sound scientific basis should be on the proponent of a project that may carry a risk of serious harm, but that this may vary depending on the particular situation.[56] This is also problematic because it allows too much discretion. A third limitation is that the Framework does not establish a principle that decision-makers should assess alternatives to identify options that are least environmentally harmful.[57] Coupled with the recommendation in the Framework that measures be cost-effective, this is a problem because it provides an easy way for regulators to justify not applying precaution.

In addition to the limitations described by VanderZwaag, we have identified two additional limitations. First, the Framework defines the precautionary approach as one that recognizes that "the absence of full scientific certainty shall not be used as a reason for postponing decisions where there is a risk of serious or irreversible harm".[58] Although subtle, this formulation of precaution is even less protective than the Rio definition, cited above. Note that in Canada's formulation it is "decisions" that are not to be postponed, instead of "measures to prevent environmental degradation" that are not to be postponed. In practice, this difference could amount to semantics, or it could amount to a different philosophical expression of the precautionary approach – one that is more concerned with decision-making, rather than protecting the environment.

Second, it is stated in the Framework that Canada's application of precaution is "flexible and responsive to particular circumstances". While this may be the case with respect to aspects of the principles of precaution, it does not seem to be the case with respect to the threshold for its very application. The threshold that will trigger the application of precaution is set rigidly at "serious and irreversible harm". In fact, this threshold is described as a tenet of precaution.[59] Setting such a high threshold could pose a problem in cases where a lower trigger threshold is more appropriate. For example, recall that the trigger threshold for biodiversity conservation at the international level is a mere "threat" of harm.[60] Despite this, when Canadian legislators and managers are developing measures for precautionary action for the conservation of biodiversity nationally, they may feel obligated to use the stricter threshold of "serious and irreversible" harm because this is stated in the Framework and there is no language indicating that any other trigger might be acceptable. As noted above, commentators have recognized that in different contexts, precaution will need to be defined and applied in different ways. Additionally, the fourth principle for precautionary measures states that such measures should be cost-effective.[61] This requirement may not be flexible enough, however, to accommodate all the various situations where precaution may be necessary. For example, recall that the CBD does not mention cost-effectiveness when describing precaution, which may indicate that in the case of biodiversity conservation this factor should not be considered particularly relevant.

OCEANS STRATEGY AND MARINE MAMMALS

The *Oceans Act* requires Fisheries and Oceans Canada to develop a national strategy for the management of Canada's oceans.[62] *Canada's Oceans Strategy* is the government's policy response and is meant to set the overall strategic direction for the management of coastal and marine ecosystems.[63] The stated overarching goal of the Strategy is to "ensure healthy, safe, and prosperous oceans for the benefit of current and future generations of Canadians".[64] The Strategy is based on the three principles set out in the *Oceans Act*, that is, sustainable development, integrated management and the precautionary approach. All three of these principles are supposed to guide all ocean management decision-making. Sustainable development is at the core of the Strategy, and is described as recognizing the need for an approach that integrates social, economic and environmental aspects of decision-making,[65] such that ocean resource development is carefully considered so as to not compromise the ability of future generations to meet their needs. Integrated management is central to the Strategy and is described as "a commitment to planning and managing human activities in a comprehensive manner while considering all factors necessary for the conservation and sustainable use[66] of marine resources and the shared use of ocean spaces".[67] The development of long-term and large-scale integrated management plans for all of Canada's oceans are envisioned in the Strategy. The precautionary approach is described in the Strategy as a key principle to be applied to ocean management. To that end, the Strategy commits to protecting marine resources and preserving the marine environment by promoting the wide application of precaution to the conservation, management and exploitation of marine resources.[68] The Strategy adopts the *Oceans Act* definition of precaution – that is, "erring on the side of caution" – but at the same time states that the Government of Canada Framework will guide the application of precaution. Despite this potential conflict, it is also expected that further commitments in the Strategy, such as promoting an ecosystem-based approach to management, applying conservation measures to maintain biodiversity and productivity (including establishment of marine protected areas), promoting a greater understanding of the marine environment, and maintaining the community structure of ecosystems, will clarify and guide the application of precaution from an oceans perspective.[69]

Three main policy objectives to advance oceans management activities have been identified in the Strategy, for which the three principles outlined should act as a guide. These are: understanding and protecting the marine environment;[70] supporting sustainable economic opportunities;[71] and international leadership.[72] The objective of understanding and protecting the marine environment is the most relevant to the protection and conservation of marine mammals. A number of activities to help implement this aspect of the Strategy over a four-year period are outlined. It is recognized that some activities will require establishing committees or conducting research; some will require policy approval and new financial resources; and others may require legislative or regulatory change.[73] The three categories of activities under this objective are: improving the scientific knowledge base for estuarine, coastal, and marine ecosystems; developing policies and programs to prevent marine pollution; and undertaking activities relating to the conservation and protection of the marine environment.[74] Overall, the specific activities listed in these three areas do not address biodiversity and ecosystem concerns adequately, nor do they seem to provide the necessary direction to make decisions with precaution. There is a very general statement that scientific knowledge will be improved by collecting, monitoring and disseminating "information",[75] but no indication as to what information will be collected. Thus, it is uncertain whether the collection of species and ecosystem biodiversity information will be a priority at all. Although the development of a strategy for a national network of marine protected areas is listed as an activity, there is nothing addressing the effects of industry and fisheries on biodiversity and habitat, and no general direction on species conservation. However, the Strategy does provide that government will support new legislation, regulations and policies to protect marine species at risk.

Another main section of the Strategy sets out commitments to improve ocean governance. First, it is stated that the government will develop and support activities to establish institutional governance mechanisms to enhance decision-making that is coordinated and collaborative. Commitments in this area include, among others, strengthening national and regional institutional arrangements, supporting an Advisory Council on Oceans, and supporting an Oceans Task Group of fisheries ministers. Second, stewardship and public awareness will be promoted to help Canadians engage in ocean management activities. Third, the Strategy provides for the implementation of an integrated management planning program, which is touted as the cornerstone of the oceans governance approach. The program will establish advisory bodies for the conservation and protection of ecosystems, as well as the promotion of economic opportunities.[76] One major activity in this area is to support the implementation of the *Policy and Operational Framework for Integrated Management of Estuarine, Coastal and Marine Environments in Canada.*[77]

The Operational Framework discusses how Canada will achieve its legislative and policy commitment to integrated oceans management. A number of principles to guide integrated management are listed in the Operational Framework. The main ones are sustainable development, ecosystem-based management, the precautionary approach, and conservation. Ecosystem-based management is not defined in the Operational Framework. Briefly, it can be thought of as integrating scientific knowledge of ecological relationships within a complex socio-political and values framework toward the general goal of protecting native ecosystem structure and function over the long-term.[78] In this policy document, the *Oceans Act* definition of precaution is re-iterated (that is, erring on the side of caution). However, the Government of Canada Framework language is also adopted, such that it appears that the threshold trigger of "serious or irreversible harm" has been accepted.[79] The Operational Framework's "principle of conservation" recognizes that "the protection, maintenance, and rehabilitation of living marine resources, their habitats and supporting ecosystems are important".

The government intends to establish a system of Large Ocean Management Areas (LOMA) and smaller Coastal Management Areas (CMA) over the long term through a system of integrated management planning. In light of the "principle of conservation", a truly integrated management planning process would necessarily dictate that marine mammals must be considered explicitly in the conservation of biodiversity. Indeed, regarding LOMAs, the Operational Framework provides that consideration must include the conservation and protection of commercial and non-commercial fishery resources, including marine mammals, endangered or threatened marine species and areas of high biodiversity or biological productivity.[80] Other aspects to consider include an assessment of ecosystem characteristics, and the identification of ecologically sensitive habitat, marine species and features in need of special protection. This would include an identification of marine protected areas.

Given the specific commitments outlined above, it can be said that with respect to species and ecosystem biodiversity conservation, the Operational Framework is precautionary in and of itself, and the potential exists for government to approach biodiversity conservation in general, and marine mammal conservation specifically, in a new, more precautionary, manner. What remains to be seen is whether precautionary measures will actually be undertaken given that the strict threshold of "serious or irreversible harm" has been adopted.

PROGRESS AND LIMITATIONS IN CANADA'S CONSERVATION OF MARINE MAMMALS

In 2003, Canada finally ratified the *United Nations Convention on the Law of the Sea* (UNCLOS), and with that assumed a greater responsibility for managing all of its resources within the Exclusive Economic Zone (EEZ), especially the exploitation of its marine biota. The passage of the *Canada National Marine Conservation Areas Act* (NMCAA) in 2002 also attests to a federal commitment to protect marine areas of natural and national significance.[81] The NMCAA can be seen to support many of the same objectives as the *Oceans Act*, despite the fact that the two pieces of legislation emanate from two very different government departments (Heritage Canada and Fisheries and Oceans, respectively). Such redundancy is unusual in Canadian federal law,[82] but reflects a growing inter-agency interest in marine conservation.

The first Marine Protected Area in Canada (the Gully) was created in 2004 under the *Oceans Act*, and 11 other areas in Canada are being considered for similar status. The Gully, off the Nova Scotia coast was selected, in part, because of its numerous marine mammal species.[83] The Department of Fisheries and Oceans has also implemented the Eastern Scotian Shelf Integrated Management Initiative under the *Oceans Act* to attempt to apply an ecosystem-based approach to this region of high productivity and biodiversity.[84] Thus, insofar as the *Oceans Act* is starting to be used for the better conservation and management of marine mammals' habitats, its use also could be extended to ensure better conservation of the marine mammal species.

A dichotomy of federal interests exists between marine mammal species of commercial value and those of little commercial value. Canada has ceased being a nation of non-native whalers, but continues to be the largest commercial sealing nation in the world. Canada has not been a party to the International Whaling Commission (IWC) since 1982, despite its diversity of marine mammals, and so is not at the international table when decisions are being made about their protection and exploitation. Canada left the IWC declaring that it no longer had a direct interest in whaling. Presumably, Canada's continued failure to rejoin the IWC relates to the aboriginal co-management agreements Canada has entered into with its aboriginal people.[85]

Similarly, a dichotomy exists between the single-species style of management that Fisheries and Oceans uses to determine quotas for seals and whales and the ecosystem-based approach to management that is espoused in Canada's *Ocean Strategy*. Single-species management and ecosystem-based management can be regarded as occupying the extreme positions along a management continuum.[86] Ecosystem-based management integrates large spatial scales, larger time frames, more parameters, uncertainty, and more species interactions than simple single-species approaches. It realizes the direct and indirect economic services provided by the system, and endeavors to ensure that the diversity, productivity and persistence of the ecosystem are maintained. This dichotomy in management is consistent with neither the precautionary approach nor the integrated management approach advocated in the *Oceans Act*.

The single-species approach to seals (especially harp and hooded seals, *Pagophilus groenlandicus*, and *Cystophora cristata*, respectively), is based on a too simplistic notion of marine food web structure and predator-prey interactions,[87] as when the killing of seals is used to justify the recovery of Atlantic cod (*Gadus morhua*) stocks. However, to those charged with the task of protecting commercial fish stocks, the culling of seals is viewed as a justifiable and precautionary approach to fisheries management. Recent quotas for harp seals issued by Fisheries and Oceans have been criticized by ecologists as being far too high, and based upon uncertain population size and age structure estimates.[88] These authors regarded the setting of those quotas as being inconsistent with the precautionary approach and insensitive to risk-averse considerations because the quotas do not account for uncertainty. Moreover, the Canadian quotas were set without regard to a more recent and precautionary approach to assessing biological removal levels based upon precaution and the inclusion of uncertainty in assessment models, developed in the U.S.[89] These concerns must now be even greater following the higher harp seal quotas announced for the 2003-2005 seasons.[90] Fisheries and Oceans justifies the higher quota on a larger population size of the species. However, critics of the quota attribute the high quota to enhanced economic opportunities for seal products in Asia and some European countries, coupled with the need to provide more economic opportunities to financially-challenged parts of Atlantic Canada.[91] Should the critics of the Fisheries and Oceans seal management process be correct, then this aspect of marine mammal management is markedly disconnected from the vision of ecosystem-based management presented in Canada's *Ocean Strategy*, with its deference to the precautionary approach and integrated management. It is also possible that the current management of seals contravenes Canada's obligations under the United Nations *Convention on the Law of the Sea* (i.e. that harvesting should not over-exploit stocks). Johnston *et al.* also remarked that although the harp seal is a migratory species, and subject to harvesting by Greenlanders, Canada has not developed an international, bilateral, protocol that recognizes the exploitation that this population receives at different stages of its annual cycle.[92] This is the antithesis of precaution.

The Inuit of Canada have Constitutional rights to kill marine mammals for purposes of subsistence and ceremony, and beluga whales (*Delphinapterus leucas*) and narwhals (*Monodon monoceros*) are routinely taken on a community-quota basis assigned by Fisheries and Oceans. A recent accession to this species list is the bowhead whale (*Balaena mysticetus*), a severely endangered, and otherwise protected species in Canada.[93] The exploitation of this species by the Inuit runs counter to the provisions of SARA, and it is not endorsed by the IWC. From a biological perspective, it is not advisable to harvest from such a small ***K***-selected population, and so the issued quota may not reflect precaution and risk aversion, and be incompatible with the aim of ecosystem-based management. However, in this management scenario, it is difficult to reconcile the biological need to persist of bowhead whales with the human cultural desire for bowhead whale hunting among the Inuit, because proponents for both sides could invoke the precautionary approach in their arguments.

THE WAY FORWARD

Precaution is characterized by uncertainty, not ignorance, and cannot be applied in the absence of knowledge.[94] It has been recognized that conservation programs (i.e. all programs designed to maintain stocks of species at levels consistent with their being used by humans in consumptive and non-consumptive ways), inventories, monitoring, and protected areas are all precautionary measures.[95] Thus, with respect to marine mammal biodiversity, it will be essential to collect information and undertake monitoring regarding the various threats to populations so that the evidentiary burden can be addressed appropriately before risk-based decisions are made. Given the current lack of research and conservation programs in Canada for marine mammals, if the commitments in the *Oceans Strategy* are to be realized, a functional and appropriate governance structure, i.e. one that is not dominated by commercial fishing interests, will be necessary. It has been noted by several commentators that an appropriate legislative framework is necessary to give effect to the precautionary principle[96] because it is people and their activities that are being managed, not marine mammals.[97]

To enable the collection of information and research to make the commitment to apply precaution meaningfully, three main reforms are recommended. Firstly, responsibility for the management and conservation of marine mammals should be removed from the *Fisheries Act* and moved to the *Oceans Act*.[98] This is necessary because a progressive approach to marine mammal conservation requires appropriate enabling legislation. The *Fisheries Act* is inappropriate because it is primarily concerned with commercial interests. Enhanced marine mammal conservation has been realized in Australia, New Zealand, and the U.S. where separate marine mammal legislation has been enacted.[99] The Australian example is illustrative. In 1999, its government enacted the *Environment Protection and Biodiversity Conservation Act*, which provides for specific conservation measures relating to marine mammals. Also, such a change would better complement other Canadian conservation legislation. This is important in the case of the federal species at risk legislation (SARA), which is inadequate on its own to deal with marine mammals, because the ultimate listing of species as endangered, threatened or of special concern has been left to the discretion of politicians, rather than assigned to the relevant scientific experts.[100] As a result, of the 24 marine mammal populations listed as endangered, threatened or of special concern by the scientific Committee on the Status of Endangered Wildlife in Canada (COSEWIC),[101] to date only four populations have been added to the species list under SARA.[102]

The *Oceans Strategy* and the accompanying Operational Framework state that institutional and legislative reform will be necessary to achieve integrated management and goals set under the new ocean governance regime.[103] Thus, we argue secondly that to implement an enhanced legislative regime as recommended above, it would be necessary to create new institutional structures to deal with marine mammals generally, and at-risk marine mammals specifically.[104] Such governance structures have been successfully implemented in the U.S., New Zealand, and Australia.[105] For example, in the U.S., marine mammal conservation is the responsibility of the Marine Mammal Conservation Division, which is part of the Office of Protected Resources, a separate agency within the National Marine Fisheries Service. In Australia and New Zealand, authority over marine mammal conservation has been transferred to environment or conservation departments that do not deal with fisheries concerns. New Zealand provides an instructive example

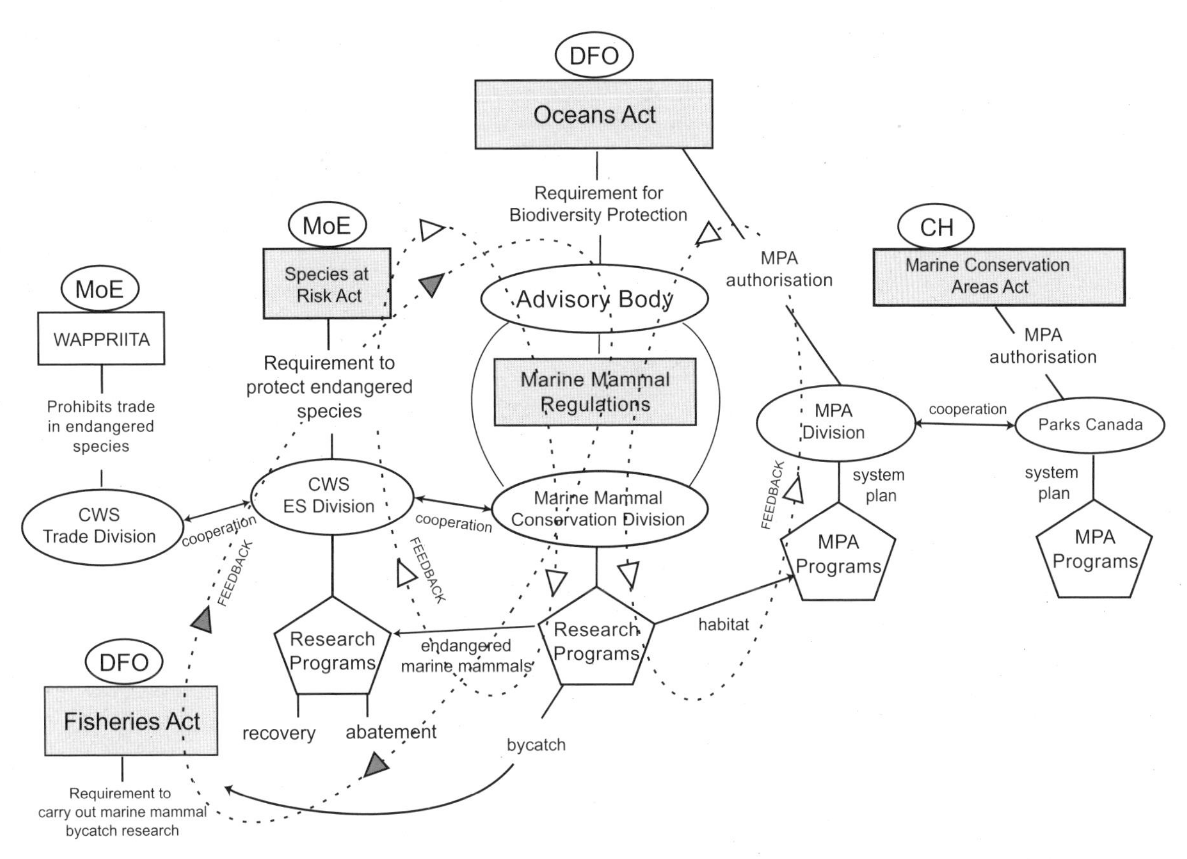

Figure 22–1. An improved Canadian legal and institutional framework for the protection and conservation of marine mammals. Note the degree of feedback built into this framework, both within and between the various Ministries and Divisions.

Legend: CH – Canadian Heritage, CWS - Canadian Wildlife Service; DFO – Department of Fisheries and Oceans (Fisheries and Oceans Canada), ES - Endangered Species, MoE – Ministry of the Environment, MPA –Marine Protected Areas, WAPPRIITA – Wild Animal and Plant Protection and Regulation of International and Interprovincial Trade Act.

of how marine mammal conservation can improve after such a shift in management authority. In that country, marine mammal conservation was transferred from the Ministry of Agriculture and Fisheries to the Department of Conservation in 1987. Since that time, marine mammal conservation initiatives have become more prominent, funding has increased, and previous conflicts of interest that arose when the fisheries department managed marine mammal conservation have disappeared.[106]

In Canada, a separate agency within Fisheries and Oceans Canada could be set up to deal with marine biodiversity conservation (see Figure 22–1 for an example of a revised institutional framework).[107] Such a change is needed to complement the goals identified in the policies and ensure these goals can be delivered. A change in the institutional structure would also help to ensure that economic and commercial interests do not overshadow non-consumptive conservation interests, and ensure that particular cultural interests do not override ecological concerns.[108] For example, adherence to a precautionary approach would require that seal hunt quotas for harp and hooded seals be reconsidered and set on a basis other than replacement yield (the current measure). This is because setting quotas based on the concept of replacement yield is not considered precautionary. Further, in the "market-driven" hunts of recent years, quotas have been set above replacement yield and annual catches have often exceeded the quotas.[109] Under a precautionary approach, increasing yields to accommodate higher market demands would not be justified.

Thirdly, substantive sections should be added to the *Oceans Act* requiring specific application of the precautionary approach. It is important to create legally binding and enforceable obligations on government to apply precaution to override the discretionary aspects often found in Canadian law. It would also be helpful for government to develop new standards or thresholds, either through policy or law, for the application of precaution in relation to biodiversity concerns that are separate from the standards that are applied in relation to pollution concerns.[110]

CONCLUSION

If Canada is to realize its lofty objective of becoming a world leader in ocean management through the implementation of integrated management, and if it is to give effect to its stated guiding principles of conservation and precaution in ocean management, then it will need to make the governance changes such as those suggested above. As long as marine mammals are defined and managed as fish, their concerns will be overshadowed by the economic interests of the fishing industry, and their future will remain uncertain.

NOTES AND SOURCES

[1] "Cetaceans" include the large whales, dolphins and porpoises; "pinnipeds" include fur seals, sea lions, the walrus and true seals.

[2] Committee on the Status of Endangered Wildlife in Canada (COSEWIC), *Canadian Species at Risk* (Ottawa: Environment Canada, November 2003).

[3] See generally Randall R. Reeves, Stephen Leatherwood & IUCN/SSC Cetacean Specialist Group, *Dolphins, Porpoises and Whales: 1994-1998 Action Plan for the Conservation of Cetaceans* (Gland, Switzerland: IUCN, 1994); Gary K. Meffe, William F. Perrin, & Paul K. Dayton, "Marine Mammal Conservation: Guiding Principles and their Implementation" in John R. Twiss Jr. & Randall R. Reeves eds., *Conservation and Management of Marine Mammals* (Washington: Smithsonian Institution Press, 1999) 437.

[4] The term "conservation" is used in this paper, unless otherwise stated, where use by humans is implied, whether that use is consumptive or non-consumptive. See Lavigne Chapter 1 for further discussion.

[5] R.S.C. 1985, c. F-14.

[6] M.L. Campbell & V.G. Thomas, "Protection and Conservation of Marine Mammals in Canada: A Case for Legislative Reform" (2002) 7(2) Ocean & Coastal L.J. 221.

[7] S.C. 1996, c. 31.

[8] Government of Canada, *Canada's Oceans Strategy: Our Oceans, Our Future* (Ottawa: Fisheries and Oceans Canada, 2002) [*Strategy*].

[9] The terms "precautionary approach" and "precautionary principle" are often used interchangeably. In some cases, the precautionary principle has been applied so as to prohibit completely a certain type of industry or activity (e.g. the UN high seas driftnet fishing ban); thus some nations prefer the use of the term "approach", which is thought to be more flexible in that it incorporates socio-economic concerns as well as environmental sustainability concerns (see Pamela M. Mace & Wendy L. Gabriel, "Evolution, Scope, and Current Applications of the Precautionary Approach in Fisheries" Proceedings, 5th NMFS NSAW, 1999 NOAA Tech. Memo NMFS-F/SPO 40, 65-73). It is also thought that the term "approach" may suggest that it is less than a rule and more a way of considering or handling something (see Ellen Hey, "The Precautionary Concept in Environmental Policy and Law: Institutionalizing Caution" (1992) 4 Geo. Int'l. Envtl. L. Rev. 303); also see Mace & Gabriel, *ibid.*, at 65 and John M. MacDonald, "Appreciating the Precautionary Principle as an Ethical Evolution in Ocean Management" (1994) 26 Ocean Devel. & Int'l L. 256. Further, it is thought that the term "approach" may carry with it less legal obligations. It is interesting to note that the European Union has decided to consider precaution in a wide range of decisions, and has specifically decided to use the term "principle" to describe the application of precaution. At the same time, Canada has insisted on using the term "precautionary approach" in the majority of its documents and policies.

[10] Owen McIntyre & Thomas Mosedale, "The Precautionary Principle as a Norm of Customary International Law" (1997) 9(2) J. Envtl. L. 221, at 222; Charmain Barton, "The Status of the Precautionary Principle in Australia: Its Emergence in Legislation and as a Common Law Doctrine" (1998) 22 Harv. Envtl. L. Rev. 509, at 512. The concept originated in the 1970s in West Germany in relation to German water protection laws, and was known as "Vorsorgeprinzip", or the "foresight" principle. The idea was that environmental damage should be avoided by implementing a forward-looking planning process to avoid harmful activities; see Carolyn Raffensperger and Joel Tickner, eds., *Protecting Public Health & the Environment: Implementing the Precautionary Principle* (Washington, D.C.: Island Press, 1999).

[11] Barton, *ibid.*, at 513; Alan Stewart, "Scientific Uncertainty, Ecologically Sustainable Development and the Precautionary Principle" (1999) 8 Griffith L.R. 350, at 368; Poul Harremoës, *The Precautionary Principle in the 20th Century: Late Lessons From Early Warnings* (London: Earthscan Publications, 2001), at 189.

[12] Barton, *supra* note 10, at 515.

[13] For an overview, see McIntyre & Mosedale, *supra* note 10. Examples include the 1987 London Ministerial Declaration of the Second International Conference on the Protection of the North Sea, the 1987 Montreal Protocol for the Protection of the Ozone Layer, the 1991 *Convention on the Ban of Import into Africa and the Control of Transboundary Movement and Management of Hazardous Wastes within Africa* (the Bamako Convention), the 1992 *Paris Convention for the Protection of the Marine Environment of the North-East Atlantic* (OSPAR), the 1992 *Baltic Sea Convention*, and the 1992 *Convention on Biological Diversity*.

[14] Sonia Boutillon, "The Precautionary Principle: Development of an International Standard" (2002) 23 Mich. J. Int'l. L. 429, at 436.

[15] 31 ILM (1992) 822.

[16] For a discussion on the use of the precautionary principle in the regulation of whale watching, see Alexandre Gillespie, "Whale-Watching and the Precautionary Principle: The Difficulties of the New Zealand Domestic Response in the Whaling Debate" (1997) 17 N.Z.U.L. Rev. 254; and, Jon Lien, "The Conservation Basis for the Regulation of Whale Watching in Canada by the Department of Fisheries and Oceans: A Precautionary Approach" (2001) Canadian Technical Report of Fisheries and Aquatic Sciences 2363.

[17] M Jasny, *Sounding the Depths: Supertankers, Sonar, and the Rise of Undersea Noise* (New York: National Defences

Research Council, 1999); Committee on Potential Impacts of Ambient Noise in the Ocean on Marine Mammals. Ocean Noise and Marine Mammals. National Research Council of the National Academies, The National Academies Press, Washington, D.C. (2003) see http://www.nap.edu.

[18] See Corkeron, Chapter 11.

[19] For more on fishing, see Holt, Chapter 4, and Hutchings, Chapter 6.

[20] See, for example, B. Würsig, and W.J. Richardson, "Noise, Effects of" in W.F. Perrin, B. Würsig, and J.G.M. Thewissen eds., *Encyclopedia of Marine Mammals* (San Diego: Academic Press, 2002) at 794-802.

[21] Before something can be considered a binding rule of customary law, two elements must be proved: 1) evidence of consistent state practice; 2) *opinio juris*, that is, that a state has acted because it believes the law requires it to do so (see Phillip Sands QC, *Principles of International Environmental Law*, 2nd Ed. (Cambridge: Cambridge University Press, 2003).

[22] James Cameron & Juli Abouchar, "The Precautionary Principle: A Fundamental Principle of Law and Policy for the Protection of the Global Environment" (1991) 14(1) B.C. Int'l. & Comp. L. Rev. 1, at 21.

[23] McIntyre & Mosedale, supra note 10, at 235.

[24] Boutillon, *supra* note 14, at 442.

[25] The IWC is the management body set up under the 1946 *International Convention for the Regulation of Whaling* (Dec. 2, 1946, 161 U.N.T.S. 72).

[26] IWC [1994] "Annex H: The Revised Management Procedure [RMP] for Baleen Whales" *Rep Int Whal Commn* 44, at 145-52; see also Mace & Gabriel, *supra* note 9, at 70.

[27] UNGA Res 46/215 on Large-Scale Pelagic Drift-Net Fishing and its Impact on the Living Marine Resources of the World's Oceans and Seas, 20 December 1991, (UN Doc A/46/645/Add 6); see also Grant J. Hewison, "The Precautionary Approach to Fisheries Management: An Environmental Perspective" (1996) 11(3) Int'l J. Marine & Coastal L. 301, at 305-8.

[28] Adopted May 16, 1990, at Art. 7. Conference on "Action for a Common Future", Bergen, Norway, May 8-16, 1990.

[29] 13 June 1992, A/CONF.151/5/REV.1.

[30] Resolution of the Conference of the Parties: Criteria for Amendment of Appendices I and II, agreed at the Ninth Meeting of the Conference of the Parties, Fort Lauderdale (US), 7-18 November 1994; 1973 *Convention on International Trade in Endangered Species of Wild Fauna and Flora*, 993 UNTS 243.

[31] Agreement for the Implementation of the Provisions of the UN Conference of the Law of the Sea of 10 December 1982 relating to the Conservation and Management of Straddling Fish Stocks and Highly Migratory Fish Stocks, UN Conference on Straddling Stocks and Highly Migratory Fish Stocks, Sixth Session, New York, 24 July – 4 August 1995, A/CONF. 164/37 [the "Agreement"].

[32] MacDonald, *supra* note 9, at 277.

[33] For further reference see: K. Kuntz-Duriseti, "Evaluating the economic value of the precautionary principle: using cost benefit analysis to place a value on precaution" (2004) 7 Env'tl Science & Pol. 291-301. Also see Rees, Chapter 14.

[34] Nancy Myers, "Debating the Precautionary Principle" in Allan Greenbaum, Alex Wellington, & Ron Pushchak eds., *Environmental Law in Social Context: A Canadian Perspective* (Concord: Captus Press, 2002) 389.

[35] Sands, *supra* note 21, at 273.

[36] Stephen R. Dovers & John W. Handmer, "Ignorance, Sustainability, and the Precautionary Principle: Towards an Analytical Framework" in Ronnie Harding & Elizabeth Fisher eds., *Perspectives on the Precautionary Principle* (Annandale, N.S.W.: Federation Press, 1999) at 173; Stewart, *supra* note 11, at 369; MacDonald, *supra* note 9 at 277.

[37] Canada became a party to the CBD in 1992 and the Fish Stocks Agreement in 1999.

[38] Peter W. Hogg *Constitutional Law of Canada*, looseleaf (Scarborough, ON: Carswell, 1997) vol. 1 at para. 11.4(a). This position has its roots in the 1932 *Labour Conventions* case, which said that while the power to ratify an agreement rests with the Executive branch of Parliament, implementation requires legislation to bring its obligations into law (*A.-G. Can. v. A.-G. Ont. (Labour Conventions)*, [1937] A.C. 326, 1 D.L.R. 673 (P.C.)). Further, implementation must follow the division of powers between the federal and provincial governments as set out in the *Constitution Act, 1867*.

[39] Simon Lyster, *International Wildlife Law: An analysis of international treaties concerned with the conservation of wildlife.* (Cambridge UK: Grotius Publications Limited, 1985).

[40] David VanderZwaag, "The Precautionary Principle in Environmental Law and Policy: Elusive Rhetoric and First Embraces" (1998) 8 J. Env. L. & Prac. 355, at 369.

[41] S.C. 1999, c. 33.

[42] In addition to appearing in the Preamble, an administrative duty is imposed by section 2, which states that, "in the administration of the Act, the Government of Canada shall... exercise its powers in a manner that protects the environment and human health, [and] applies the precautionary principle". Despite requiring that the precautionary principle must be followed, CEPA does not include provisions on how it is to be implemented (David VanderZwaag, "Canada and the Precautionary Principle/Approach in Ocean and Coastal Management: Wading and Wandering in Tricky Currents" (2003) 34(1) Ottawa Law Review 117, at 136) [VanderZwaag (2003)].

[43] S.C. 2002, c. 29 [SARA].

[44] *Ibid.*, at s. 38.

[45] S.C. 2002, c. 18 [NMCAA].

[46] *Supra* note 7.

[47] *Supra* note 7, at s.30. Recall that the other two principles are sustainable development and integrated management.

[48] See VanderZwaag (2003), *supra* note 42.

[49] Fisheries and Oceans Canada, *A Policy for Selective Fishing in Canada's Pacific Fisheries* (Ottawa: Fisheries and Oceans Canada, 2001), online: Fisheries and Oceans Canada, http://www-comm.pac.dfo-mpo.gc.ca/publications/selectivep_e.pdf, at 8.

[50] Fisheries and Oceans Canada, *New Emerging Fisheries Policy* (Ottawa: Fisheries and Oceans Canada, 2001), online: Fisheries and Oceans Canada http://www.dfo-

mpo.gc.ca/communic/fish_man/nefp_e.htm.

[51] Government of Canada, *A Framework for the Application of Precaution in Science-based Decision Making about Risk* (Ottawa: Privy Council Office, 2003) [Framework].

[52] The departments involved were: Agriculture and Agri-Food Canada, Canadian Environmental Assessment Agency, Canadian Food Inspection Agency, Fisheries and Oceans Canada, Department of Foreign Affairs and International Trade, Environment Canada, Finance Canada, Health Canada, Industry Canada, Justice Canada, Natural Resources Canada, Privy Council Office, Transport Canada and the Treasury Board Secretariat.

[53] Framework, *supra* note 51, at s. 1.0.

[54] VanderZwaag (2003), *supra* note 42.

[55] *Ibid.*, at 126.

[56] Principle 4.3; see *ibid.*

[57] *Ibid.*

[58] Framework, *supra* note 51.

[59] Framework, *supra* note 51, at section 1.0.

[60] See CBD, *supra* note 15 and accompanying text.

[61] Framework, *supra* note 51, at s. 4.9.

[62] *Supra* note 8, at s.29.

[63] *Strategy*, *supra* note 8.

[64] *Ibid.*, at 10.

[65] For critical commentaries on the subject of sustainable development, see for e.g. S. Beder, *Global Spin: The Corporate Assault on Environmentalism* (Melbourne: Scribe Publications, 1997). Also see Rees, Chapter 14, and Brooks, Chapter 16.

[66] The term "sustainable use" is not accompanied by a list of procedures that would ensure that the use of marine resources can be maintained at stated levels over time. Sustainable use is, however, a highly controversial concept. See for e.g. D. Lavigne, Ecological Footprints, Doublespeak, and the Evolution of the Machiavellian Mind, at 63, and R.J., Brooks, "Sustainable Development", in W. Chesworth, M. Moss, and V.G. Thomas eds., *Sustainable Development: Mandate or Mantra?* (Guelph: Faculty of Environmental Sciences, University of Guelph, 2002) at 119. Also see Lavigne, Chapter 1, and Holt, Chapter 4.

[67] *Strategy*, *supra* note 8., at 11.

[68] *Ibid.*

[69] *Ibid.,* at 12. Note that an "ecosystem-based approach to management" is not defined specifically in the Strategy. The use of the terms "health and integrity" together in reference to the maintenance of ecosystems is problematic. Each term carries different philosophic connotations regarding ecosystems, and cannot be equated. We prefer the term "ecosystem integrity". The ecosystem health metaphor is poor because it does not accurately reflect ecology or health science – this is because health is not a property of ecosystems (for a full critique of the ecosystem health metaphor, see Glenn W. Suter II, "A Critique of Ecosystem Health Concepts and Indexes" (1993) 12 Envt'l Toxicology & Chem 1533).

[70] The *Strategy* provides that understanding the marine environment depends on, *inter alia,* undertaking rigorous scientific research, identifying key ecosystem functions, and developing indicators. At the same time, protecting the marine environment can only occur from a greater understanding of the ocean ecosystem and the effects of pollutants and unique, sensitive and ecologically significant marine areas and species.

[71] The Strategy provides that the conservation and sustainable use of fisheries is a key goal for this objective. Also important are offshore energy and mineral resource development, shipping, shipbuilding and industrial marine industry, seabed mapping, marine communications and data management, eco-tourism operations, and cruise-ships and waterfront development operations.

[72] The Strategy provides that Canada intends to be an international leader in ocean management, in part by adhering to and encouraging compliance with the *United Nations Convention on the Law of the Sea* (UNCLOS), the Straddling Stocks Agreement, Agenda 21 to the 1992 United Nations Conference on Environment and Development.

[73] Strategy, *supra* note 8, at 22.

[74] *Ibid.*, at 23.

[75] *Ibid.*, at 22.

[76] *Ibid.*, at 19.

[77] Government of Canada, *Policy and Operational Framework for Integrated Management of Estuarine, Coastal and Marine Environments in Canada* (Ottawa: Fisheries and Oceans Canada, 2002) [Operational Framework]. Note that the development and implementation of integrated management plans is also required by the *Oceans Act* (ss. 31, 32).

[78] Edward R. Grumbine, "What is Ecosystem Management?" (1994) 8(1) *Conservation Biology* 27, at 30.

[79] Operational Framework, *supra* note 77, at 9.

[80] *Ibid.*, at 17.

[81] *Supra* note 45.

[82] M. Campbell & V.G. Thomas, "Constitutional Impacts on Conservation – Effects of Federalism on Biodiversity Protection" (2002) 32(5) Env'tl Pol. & L. 223.

[83] Glen Harrison and Derek G. Fenton eds., *The Gully: A Scientific Review of its Environment and Ecosystem* (Ottawa: Fisheries and Oceans Canada, 1998).

[84] R.J. Rutherford, G.J. Herbert & S.S. Coffen-Smout, "Integrated ocean management and the collaborative planning process: the Eastern Scotian Shelf Integrated Management (ESSIM) Initiative." (2005) 29 Marine Policy 75.

[85] See John R. Twiss Jr. & Randall R. Reeves eds., *Conservation and Management of Marine Mammals* (Washington D.C.: Smithsonian Institution Press, 1999) at 14 & 31.

[86] W.K. de la Mare, "Marine ecosystem-based management as a hierarchical control system" (2005) 29 Marine Policy 57.

[87] Peter Yodzis, "Must top predators be culled for the sake of fisheries?" (2001) 16 Trends in Ecology and Evolution 78.

[88] D.W. Johnston, P. Meisenheimer, & D.M. Lavigne, "An evaluation of management objectives for Canada's commercial harp seal hunt, 1996-1998" (2000) 14 Conservation Biology 729.

[89] P.R. Wade, "Calculating limits to the allowable human-caused mortality of cetaceans and pinnipeds" (1998) 14 Marine Mammal Science 1.

[90] The quota was 975,000 harp seals over three years, resulting in the largest annual quotas since the introduction of quota management in 1971.

[91] Also see Hutchings, Chapter 6, for the notion that marine management decisions, particularly fishery management decisions, have long been linked to politics.

[92] *Supra* note 88.

[93] R.W. Moshenko, S.E. Cosens & T.A. Thomas, *Conservation strategy for bowhead whales (*Balaena mysticetus*) in the Eastern Canadian Arctic*. Recovery Plan No. 24. (Ottawa: Recovery of Nationally Endangered Wildlife (RENEW), 2003).

[94] J. Harwood & K . Stokes, "Coping with uncertainty in ecological advice: lessons from fisheries." (2003) 18 Trends in Ecology and Evolution 617.

[95] See Malcolm MacGarvin, "The Precautionary Principle, Science and Policy" in Ronnie Harding and Elizabeth Fisher, eds., *Perspectives on the Precautionary Principle* (Annandale, N.S.W.: Federation Press, 1999) 225, at 226.

[96] See Cameron & Abouchar, *supra* note 22, at 23; McIntyre & Mosedale, *supra* note 10, at 236; Felicity Nagorcka, "Saying what you mean and meaning what you say: precaution, science and the importance of language" (2003) 20 E.P.L.J. 211, at 219; Karla Sperling, "If Caution Really Mattered" (1999) 16(5) E.P.L.J. 425, at 434 & 437; Elizabeth C. Fisher, "The precautionary principle as a legal standard for public decision-making: The role of judicial and merits review in ensuring reasoned deliberation" in Ronnie Harding and Elizabeth Fisher, eds., *Perspectives on the Precautionary Principle* (Annandale, N.S.W.: The Federation Press, 1999) 83, at 84 & 96.

[97] See, for e.g. Holt, Chapter 4, and de la Mare, Chapter 21.

[98] For a full discussion of this point see Campbell & Thomas, *supra* note 6.

[99] *Ibid.*, at 226-34.

[100] SARA, *supra* note 43, at s.27. For a further discussion of the problems with the SARA listing process, refer to Hutchings, Chapter 6.

[101] COSEWIC, *supra* note 2.

[102] SARA, *supra* note 43, at Schedule 1. These are: Killer Whale (*Orcinus orca*) Northeast Pacific southern resident population (endangered); Killer Whale (*Orcinus orca*) Northeast Pacific northern resident population (threatened); Killer Whale (*Orcinus orca*) Northeast Pacific transient population (threatened); and, Killer Whale (*Orcinus orca*) Northeast Pacific offshore population (special concern).

[103] See, respectively, *supra* note 8, at 22 and *supra* note 77, at 11 & 18.

[104] See also, Hutchings, *supra*, where he states that policy can be strengthened by the work of an independent, expert science advisory body.

[105] Campbell & Thomas, *supra* note 6, at 227-8.

[106] *Ibid.*, at 228.

[107] *Ibid.*, at 257.

[108] A change in institutional structure would require moving an amended form of the *Marine Mammal Regulations*, SOR/93-56, to the *Oceans Act*. However, this would not mean that quotas for aboriginal hunting could not be set. Hunting quotas would still be set; however, the quotas would be required to adhere to the precautionary approach as well as the criteria for ecosystem management.

[109] See Colin W. Clark, "Marine Reserves and the Precautionary Management of Fisheries" (1996) 6(2) Ecological Applications 369, at 369; and MacDonald, *supra* note 9 at 271. Also see Holt, Chapter 4.

[110] Nagorka, *supra* note 96, at 221.

The Canadian Department of Fishes
Throws caution to the wind as it wishes
That the demise of cod
Was all a mirage
Its policies are as clear as my Guinness.

Jeff Hutchings 2004

CHAPTER 23

THE STEADY STATE REVOLUTION AS A PREREQUISITE FOR WILDLIFE CONSERVATION AND ECOLOGICAL SUSTAINABILITY[1]

Brian Czech

The biggest threat to the welfare of wild animals is habitat destruction. Habitat destruction, meanwhile, is a function of human economic activity. Therefore, economic growth is another way of describing the biggest threat to the welfare of wild animals, and with much more relevant policy implications.

Economic growth is an increase in the production and consumption of goods and services. It entails increasing human populations, *per capita* consumption, or both. The size of the economy is generally indicated by gross domestic product (GDP).

The relationship between economic growth and habitat destruction is readily ascertained by looking at the list of causes of species endangerment. For example, in the United States these causes include agriculture, domestic livestock production, mining, logging, and other extractive sectors of the economy.[2]

Another major cause of species endangerment is urbanization, which represents the growth of much of the labor force and the consumer population as well as a variety of industrial and services sectors.[3] Closely related economic infrastructure (such as roads, reservoirs, pipelines, power lines, telecommunications facilities, and wind farms) constitutes another major cause of species endangerment.[4]

Outdoor recreation is another major cause, and is also cited as a prominent economic sector.[5] Pollution, a byproduct of economic production, is an insidious, widespread, and pervasive threat to animal welfare and biodiversity conservation. Non-native invasive species, which disperse as a function of international trade, constitute one of the biggest and most rapidly growing threats to ecological integrity.

Global warming is becoming recognized as another threat to species,[6] although the mechanisms are less direct. Temperature is a key variable in ecological function and species composition. To put it in somewhat simplified terms, global warming is likely to "push" polar species (such as polar bears) off the ends of the earth and create unprecedented niches near the equator that will only be filled through the slow process of evolution. It has also been implicated in increased incidences of human and wildlife diseases.[7] Global warming is largely a function of greenhouse gas emissions from the burning of fossil fuels.[8] The large, industrialized economies are mostly fossil-fueled, so global warming is also a function of economic growth.

In other words, the threats to biodiversity conservation are a Who's Who of the human economy. This is readily explained with basic principles of ecology. One such principle is "competitive exclusion", whereby no species proliferates except at the expense of other species with overlapping niches.[9] Due to the tremendous breadth of the human niche, which expands as a function of technological progress, the human economy grows at the competitive exclusion of wildlife in the aggregate.

Another relevant aspect of ecology is trophic theory.[10] The entire "economy of nature" (that is, the production and consumption activities of nonhuman species) is founded upon the producers, or plants, which produce their own food via photosynthesis. Primary consumers, or animals that eat plants, constitute the next trophic level. Secondary consumers prey upon primary consumers, and so forth. In some ecosystems there may be six or seven trophic levels and, in all ecosystems, the top trophic level is called the "super-carnivores". Mixed throughout this trophic system are "service providers" which are not readily categorized in trophic levels. These include decomposers, scavengers, and symbiotic parasites. In addition, many species that fit neatly into a particular trophic level also provide incidental services, such as pollination, soil aeration, and nutrient cycling.

Perhaps the most important thing gleaned from trophic theory is that the size of the entire enterprise depends upon the size of the producer trophic level. Growth in the economy of nature requires growth of the producer trophic level or, in other words, on an increase in primary production, i.e., photosynthesis. There is a limit to the size of the economy of nature imposed by primary production, which in turn is limited by solar energy and the availability of physical resources such as nutrients, minerals, and water.

Philosophically, some prefer to classify humans as part of the economy of nature, in which case humans are clearly super-carnivores. They can eat virtually anything edible to them and are rarely threatened themselves by predators, especially in developed nations. As the trophic level comprised by humans expands in biomass, it exerts "trophic compression" upon the rest of the economy of nature. This is another way of illustrating the principle of competitive exclusion that makes it even clearer that there is a limit to human economic growth imposed by the other trophic levels and ultimately by primary production.

ECONOMIC GROWTH AS A NATIONAL GOAL

In most industrialized nations, economic growth is the highest priority in the domestic policy arena.[11] In the United States, for example, economic growth has been a primary, perennial, and bipartisan goal of the public and polity since the Great Depression. During much of that time, a major stimulus was the Cold War. Defeating Soviet communism without precipitating nuclear war entailed a strategy of economic growth for the sake of showing the world that a capitalist system was more conducive to economic efficiency and an enjoyable lifestyle. Just as important was the need to stay ahead of the Soviets militarily, which required economic growth to finance the war chest. The Cold War was basically a GDP race.

With the collapse of the Soviet Union in 1988, greedier goals have continued to drive economic growth in the United States. There is still a significant populace in the United States living in poverty but, instead of instituting progressive reforms for redistributing wealth, the American government has adopted supply-side economics[12] and the logic that "a rising tide lifts all boats". Little thought is given to the supply of "water" or the number of boats that can be accommodated in the "ocean".

American economic philosophy and policy is extremely important to international economics for several reasons. First, the American government and society remain the standard of capitalist democracy in many parts of the world, although its image has been tarnished in recent years. Perhaps more importantly from the standpoint of animal welfare and conservation, the United States is by far the largest consumer in the world. In terms of GDP, the United States comprises approximately one fourth of the world's production and consumption, with an annual GDP currently over $10 trillion.[13] Finally, whether or not the American model of political economy is emulated by other nations, the economic might of the United States equates to tremendous political power and influence over international economic affairs and agreements. For example, the United States controls the biggest levers in the World Bank, International Monetary Fund, and World Trade Organization.[14]

There are many critics of economic growth as a national goal in the United States.[15] However, their arguments are suppressed and get very little media attention. Therefore, the American public seldom hears about the threats posed by economic growth. Polls indicate that 63% of Americans believe there is no limit to economic growth; a natural corollary is that there is no conflict between economic growth and long-term economic or, indeed, ecological sustainability.

NEOCLASSICAL ECONOMICS AND ECONOMIC GROWTH THEORY

Economics, both theoretical and applied, has a long history of being swayed, manipulated, and corrupted by vested interests.[16] An episode of corruption particularly relevant to economic growth and conservation occurred in response to the powerful, populist economics of Henry George (1839-1897). Mason Gaffney has described this episode in depth in *The Corruption of Economics*.[17]

In *Progress and Poverty* (1879), George argued the greatest source of economic inefficiency and inequity was the tendency of "rent" (i.e., all income associated with location and exploitation of natural resources) to accrue to landowners who did little to earn it.[19] This phenomenon led to speculation in land markets, causing booms and busts, especially during the settling of the American West. Great tracts of land were commandeered by timber, railroad, cattle, and other land barons. As farmers, laborers, and capitalists worked to produce goods and services, the value of nearby lands naturally increased, whether or not the land baron did any work. The rents they received, therefore, were unfairly disproportionate to the wages of labor and the profits of capital.

George struck a nerve in the American body politic and engendered many supporters in the United States and abroad. While Marxists were promoting revolution and the "expropriation" of the capitalists in Europe,[19] George and the populists argued for a tax on land in the United States, and this at the time when the American tax code was being developed. Frightened, the land barons fought George on every front they could.[20]

This period in history corresponded with the development of economics as a distinct discipline, i.e., what came to be called "neoclassical" economics. The classical economics of the 19th century was really a combination of what we classify today as political science and economics. Indeed, its practitioners referred to their subject matter as "political economy".[21] In contrast, neoclassical economics, especially in Europe, took on a sheen of independence from political affairs. Supply and demand, prices, and principles of marginality became more prominent than the older debates about class structure and models of political economy, such as capitalist democracy versus socialism or communism

In the United States, economics departments in academia were in their formative stages. Gaffney documented how land barons often established or patronized these departments and hired economists to refute George. Examples included Columbia University, the University of Chicago, and Johns Hopkins University; all bellwether institutions of higher learning in economics. Led by John Bates Clark at Columbia and Frank Knight at the University of Chicago, economists denied the importance of land (including the natural resources comprising it) as a distinct factor of production, thereby hiding it from prominence as the tax code was being developed. The old "land, labor, and capital" of the classical economists rapidly became "labor and capital", where land was either ignored or considered a form of capital.

This corruption of economics had tremendous implications for taxation and the distribution of wealth. However, for our purposes, it is more relevant to note the effect on the economic production function. Today, when we open a typical macroeconomics textbook, we find that "$\boldsymbol{Y} = \boldsymbol{f}(\boldsymbol{K},\boldsymbol{L})$", or "production (***Y***) is a function of capital (***K***) and labor" (***L***). If land is mentioned, it is considered substitutable by manufactured capital. The corrupted production function is amenable to a theory of economic growth that fails to recognize any limits to economic growth.[22]

Economic growth theory went through several major stages, with roots in the work of John Maynard Keynes[23] and Sir Roy Harrod.[24] The major stages are associated with the early work of Robert Solow (1950s),[25] Robert Lucas (1980s),[26] and David Romer (1990s).[27] The Romer model serves as the foundation of current economic growth theory.

The most important aspect of the latter model, for our purposes, is Romer's slant on technological progress, which in economic terms refers to increasing output per unit input. Romer rightly pointed out that labor, the "***L***" in the production function, includes a subset engaged in research and development, or "R&D", which gives rise to technological progress. Research and development, and the resulting technological progress, is required for increasing *per capita* GDP growth and, therefore (as economists generally assume), increasing human welfare.

It doesn't take long to ascertain one of the astounding implications of the Romer model, when we consider that only a certain fraction of the labor force may be engaged in R&D. The only sure way to get more R&D is to have more people conducting it. In other words, a common interpretation of the Romer model is that population growth is required for *per capita* GDP growth![28]

This hypothesis is essentially the same argument made by the late Julian Simon for a decade preceding Romer's work.[29] Simon simply said there was no limit to population growth because, as population growth caused environmental problems, more human brains were available to solve those problems. The Romer model is embellished by higher mathematics and a great deal more sophistication, especially pertaining to the role of patent law (which is beyond the scope of the present essay). However, it is just as ecologically unsound as Simon's "pop economics". At its core is the landless production function and an assumption of unlimited economic growth.

To say there is no limit to economic growth on a finite land mass is mathematically equivalent to saying we can have a stable, steady state economy on a perpetually diminishing land mass. For example, with technological progress, we could have the $40 trillion global economy contained first on a continent, then in a nation, and ultimately in a nanotube. This is as ludicrous as saying there is no limit to economic growth on Earth. Yet, we continually hear, "There is no conflict between economic growth and environmental protection." Why?

THE IRON TRIANGLE

Political scientists have a concept called the "iron triangle".[30] An iron triangle consists of a special interest group, a supportive political faction, and a professional society (usually manifest in a government agency), which dominates a policy arena and fends off all comers. Iron triangles are not necessarily conspiratorial. They can simply materialize when interest groups, politicians, and professionals have similar backgrounds, perspectives, and mutual economic and political interests.[31]

Let us consider the iron triangle most relevant to the conflict between economic growth and conservation in the United States (Figure 23–1). This iron triangle is a virtual juggernaut in the policy arena, because the "special interest" is the corporate community at large, and the political "faction" is the political community at large! Corporations are concerned primarily with profits and, therefore, are served by a national policy of economic growth. Meanwhile, the American campaign financing system ensures political fealty to the corporate community.[32] Most Americans have some sense of this impure aspect of American politics; witness the sweeping support for campaign finance reform. Similar iron triangles exist in other capitalist governments as well.

While many people are aware of the tight relationships between corporations and politicians, most are oblivious to the third side of the iron triangle surrounding the economic growth policy table. The third side is comprised of neoclassical economics. Neoclassical economics has become such an abstract, mathematical discipline that many scholars claim it has lost much of its relevance to ecological, social, and political issues.[33] Nonetheless, neoclassical economics is the mainstream school of economics throughout the modern world.[34] It feeds the politicians the politically expedient theory of unlimited economic growth and the corollary that there is no conflict between economic growth and environmental protection.[35] The neoclassical theory of unlimited growth also helps to maintain "consumer confidence", which is conducive to corporate profit. The influence of neoclassical economic growth theory has dire implications for wildlife conservation and animal welfare.[36]

In response to growing discontent with neoclassical economics, various academic reform movements, societies, and schools of thought have arisen.[37] Examples include the International Society for Ecological Economics, the South African New Economics Network (SANE), the Post-Autistic Economics Movement (formed by French university students), various new schools of Marxism, and the gradual resurrection of political economy in the academic literature. Mason Gaffney and the Georgists were mentioned earlier.

Those concerned with wildlife conservation, however, should use discretion in their critiques of neoclassical economics. Neoclassical economics has given us much, especially in the realm of microeconomics. For example, cost-benefit analysis coupled with contingent valuation of wildlife have helped wildlife managers make better decisions and illustrate the value of wild animals to American society.[38] Our critique should be targeted primarily toward conventional macroeconomics, especially its theory of unlimited economic growth. The pursuit of ecological sustainability requires us to weigh in at the economic policy table, but the iron triangle blocks our path.

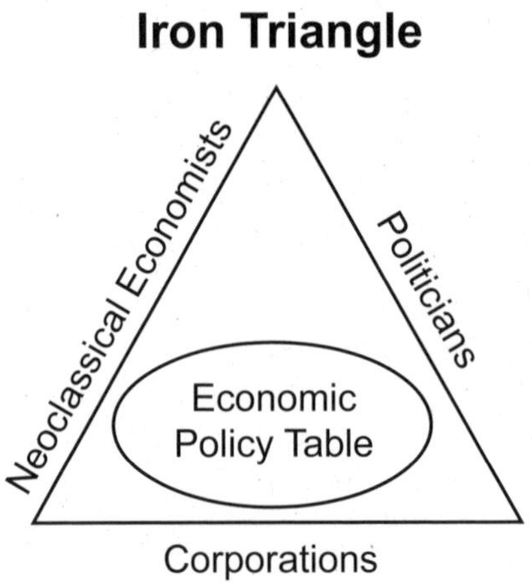

Figure 23–1. The Iron Triangle of Economic Policy.

If there is a weakness in the iron triangle, it is clearly not the corporate community with its vast resources.[39] Nor is it the political community, connected to corporate resources as it is.[40] The iron triangle's weakness is neoclassical economics, partly because of its somewhat weaker attachment to the corporate community and partly because of the duress it is currently under from so many different directions.

For those concerned about the future of wildlife conservation, a major ally is the ecological economics movement (represented by the International Society for Ecological Economics and various national chapters). Other allies include the natural resources professions, which are beginning to scrutinize neoclassical economics and the implications of economic growth for conservation.[41] A good example is The Wildlife Society, which published a technical review on economic growth in 2003 that described a "fundamental conflict between economic growth and wildlife conservation" and adopted a position on economic growth in 2004. Other professional societies that have taken positions describing the conflict between economic growth and ecological integrity include the United States Society for Ecological Economics and the North America Section of the Society for Conservation Biology. At the time of writing, similar efforts are underway in the American Fisheries Society, Ecological Society of America, and American Society of Mammalogists. The Center for the Advancement of the Steady State Economy (CASSE), a non-profit organization based in Arlington, Virginia, has been instrumental in these efforts, and it's own position on economic growth is often used as a template from which economic growth positions are developed. The CASSE position on economic growth has also been endorsed by several environmental organizations.

THE STEADY STATE REVOLUTION

"Revolution" is perhaps an overused and cheapened word, here applied to a new kind of toothbrush, there to a shifted preference in entertainment. It is commonly used in the rhetoric of marketing, politics, and propaganda due to its powerful connotations of history-changing events. The American Revolution, the French Revolution, and the communist revolutions of the 20th century, in particular, have created these connotations.

Yet there is nothing magical about most revolutions. A revolution is simply an episode of change distinguished by its magnitude and pace. A revolution in political economy typically also includes an element of revolt. It happens when evolutionary change is deemed unacceptably slow by one or more classes in the public or polity.

The Steady State Revolution I sketched out in *Shoveling Fuel for a Runaway Train* would replace the national goal of economic growth with the goal of a steady-state economy in the United States. It has an academic and a social phase, both of which are required for the establishment of a steady-state economy.

The academic phase of the Steady State Revolution entails the replacement or usurpation of neoclassical economics by ecological economics.[42] While much of the neoclassical work on the allocation of resources among the factors of production would be retained, macroeconomics would develop a focus on the issue of scale: that is, the size of the economy relative to the ecosystem that supports it. The distribution of wealth, so important to ecological sustainability and, therefore, to animal welfare and conservation, would receive much more attention than it does in neoclassical economics. The study of the allocation of resources would also be marked by a return to its classical roots, where land was emphasized as a distinct factor of production, and the production function would be reconstructed accordingly.

The academic phase of the Steady State Revolution entails much more, however, than just the reform of economics courses and rewriting of economics textbooks. It also entails curriculum reform in such areas as history, business, and ecology. For example, history students will learn about the corruption of economics by the iron triangle; business students will be taught the principles of sustainability; and ecology students will learn how to apply ecological principles in the economic policy arena. The academic phase also entails "extension" work by faculty members to get the principles of ecological economics into the public domain.

Much of the groundwork has already been done to facilitate the academic phase of the Steady State Revolution. For example, we now have an International Society for Ecological Economics and various national chapters, the journal *Ecological Economics*, and a bona fide textbook on ecological economics.[43] Frankly, albeit simplistically, the rest of the academic phase will depend upon the insights, energy, and commitment of faculty and students. In a very real sense, it will be a test of the human will, or at least the subset in academia, to achieve ecological and economic sustainability and equity.

The social phase of the Steady State Revolution will be dependent to a large degree upon the success of the academic phase. However, it may be possible for the social phase to transpire in the absence of an overwhelming academic phase. Talented policy entrepreneurs and energetic political activists might be able to lead the social phase with only the basics of ecological economics under their belts. The prospects for that depend as much on political context, major events pertaining to sustainability (for example, grid blackouts, oil spills, and water wars), and media coverage, as on the revolutionaries themselves.

While the academic phase of the Steady State Revolution has to some extent already begun, it will never constitute a sufficient condition for the establishment of a steady-state economy. The social phase and its political results are clearly prerequisites. The rest of this chapter will be focused on the social phase.

THE SOCIAL PHASE OF THE STEADY STATE REVOLUTION

As with most revolutions involving wholesale transformations in political economy, the Steady State Revolution has a class structure. However, unlike capitalist and communist revolutions, in which classes are identified based upon ownership of the means of production, classes in the Steady State Revolution are identified based upon their members' propensities to consume. The simplest and, in many ways, most useful metric for gauging consumption is called "personal consumption expenditure", or the amount of money spent by an individual annually. In the United States, average personal consumption expenditures (broken down by quintiles) are tracked by the Bureau of Labor Statistics.[44]

In the Steady State Revolution, the uppermost percentile in personal consumption expenditures is called the "liquidating class" or "liquidators"; the lowest 80th percentile is called the "steady state class" or "steady staters"; and the intermediate 19th percentile is called the "amorphic class" or "amorphs". The quintessential feature of the Steady State Revolution is the social interaction between the steady state class and the liquidating class.

The steady state class, having been sufficiently informed via academia and policy entrepreneurs about the perils of economic growth (reconstructed in the Steady State Revolution as "economic bloating") develops an attitude[45] toward the liquidating class conducive to changing the latter's propensity to consume. This attitude and the actions it engenders among steady staters are concisely captured in the word "castigation".

In other words, the liquidating class is castigated by the steady state class. Castigation is not a violent process, but rather takes the form of social ridicule, censure, and shunning. As we will see, this is a potent force for engendering social change. The amorphic class is relatively uninvolved in this process, but observes it unfolding and, in turn, seeks to avoid identification with the liquidators. The result of this process, to put a twist on supply-side theory, is "trickle-down consumption" leading to the cessation of economic bloating, at least in the context of a stabilized population. Trickle-down consumption and related economic, political, and psychological features of the Steady State Revolution will be treated in more detail in the following sections. First, however, a few words about population are in order.

The establishment of the steady state economy clearly requires a stabilized population as well as stabilized *per capita* consumption. The focus of the Steady State Revolution is on *per capita* consumption for several reasons. First, much has been said and is known about the problems posed by population growth.[46] Despite Romer's model of economic growth and the abuse of it by the iron triangle for the sake of assuaging public concerns about population growth, it is reasonable to expect that most citizens have enough common sense to recognize the unsustainability of perpetual population growth. *Per capita* consumption, on the other hand, is a subtler phenomenon requiring more than common sense for ascertaining its implications for sustainability.

Second, there is no population policy arena in the United States (the most important nation for the fruition of the Steady State Revolution). There is, however, a huge policy arena devoted to economic growth.[46] Once economic growth is recognized as economic bloating and a major threat to the nation, policy analysts and policy makers will be forced to scrutinize the components of economic growth. Population and *per capita* consumption will readily come to light. Of the two, however, *per capita* consumption will be more ripe for addressing with an assortment of policies, if only because population control remains a relatively taboo topic for religious and ideological reasons. That will have to change, of course, but in the interim *per capita* consumption may be addressed; there is nothing sacred about conspicuous consumption.

Yet another reason for focusing on *per capita* consumption, especially in the United States, is because the impact associated with the consumption of an American citizen is much greater than the impact associated with the consumption of the average world citizen.[47] In other words, Americans need consumption reform even more than population reform.

Nothing in this chapter should be construed to suggest that stabilizing population is not absolutely essential to establishing a sustainable, steady state economy. Rather, the *focus* of this chapter is simply on *per capita* consumption for the reasons given above.

ECONOMIC RATIONALE FOR THE STEADY STATE REVOLUTION

Let us assume – until the next section, at least – that the Steady State Revolution is politically viable. In other words, the steady state class develops opprobrium toward conspicuous consumption and castigates the liquidating class. Let us also assume – until a later section – that this castigation of the liquidating class has enough of a psychological effect to alter the liquidators' propensity to consume. How could such a minor development in consumer affairs, which seemingly affects only the behaviour of one percent of consumers, lead to the establishment of a steady-state economy?

First, a reduction in the consumption by the liquidating class may not be such a minor effect after all. Liquidators spend lavishly, and their consumption behaviour results in the liquidation of a tremendous amount of natural capital.

However, there is a much more compelling reason for the significance of the castigation of the liquidating class. Castigated by 80% of citizens, the behaviour of the liquidators will change. Most notably, the conspicuousness of their consumption will decline. (*Ceteris paribus*, their actual consumption will decline.) For those at the margin constituting the boundary between the amorphic class and the liquidating class, even a small amount of behavioural change will result in their vacating the liquidating class and joining the amorphic class. The most liquidating of the liquidators would have to change their consumption behaviour dramatically to vacate the liquidating class and avoid the ire of the steady state class.

In any event, erstwhile liquidators will become amorphs, but since the liquidating class by definition always consists of the upper one percentile in personal consumption expenditures, there is always a new liquidator to take the place of the old. This is what is meant by the phrase "trickle-down consumption" (Figure 23–2). Liquidators trickle down from the most conspicuous levels of consumption, but as they do, the liquidating class is reconstituted by ex-amorphs. In other words, the propensity to consume is ratcheted downward, with the greatest

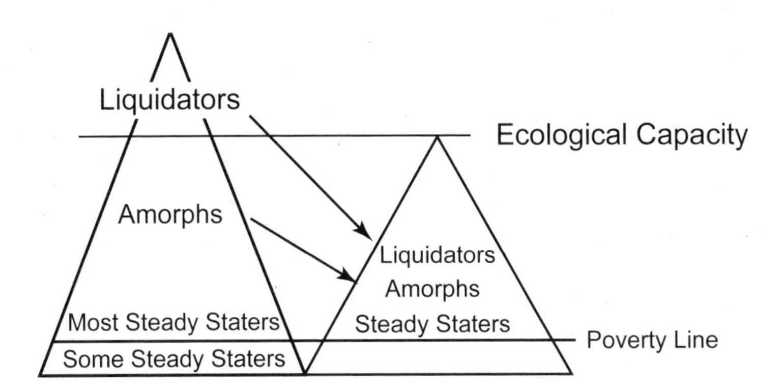

Figure 23–2. Trickle-down consumption.

reductions occurring in the highest levels of personal consumption expenditures.

When does the ratcheting stop? This is the least of our worries in a full-world, bloated economy, but few people will willingly reduce their consumption to a level of abject poverty. In other words, the ratcheting will proceed until a much more appropriate level of consumption is achieved across the board, whereupon the castigation will cease. The effect will be similar to recent reductions in smoking in public places. Smoking has been dramatically reduced largely due to the stigma attached to it in a society where non-smokers have finally put their feet down and said, in effect, "That's enough of that." Now that much less smoking is occurring, there is much less behaviour left to stigmatize.

There is another major economic benefit to trickle-down consumption. While the conspicuous consumption of the liquidating class is diminished, incomes of the erstwhile liquidators do not just collapse. Liquidators who consume less have more surplus income, and the income must go somewhere. Liquidators will seek to find other ways and means to make their money "work" for them in terms of social standing. By the nature of the Steady State Revolution, one of the most likely alternatives to conspicuous consumption is philanthropy, including charity for the poor. The beneficiaries include those who may have been living below the poverty line. In other words, poverty, another major economic problem of the United States (and many other nations) is alleviated, and without the inefficiencies and political difficulties of taxation designed to redistribute wealth.

POLITICAL RATIONALE FOR THE STEADY STATE REVOLUTION

The seeds of revolution may be sown by a small minority, but the viability of any revolution in political economy in a generally democratic system depends upon the solidarity of a majority class that demands reform.[48] In the Steady State Revolution, solidarity among the steady staters results from the sowing of ecological economics in academia, which then bears fruit in society with the help of policy entrepreneurs and activists. The propensity of citizens that leads to solidarity among steady staters is the concern for posterity. When 80% of citizens find that another 1% is having a devastating impact on their children's and grandchildren's long-term ecological and economic prospects, they will be moved to action. In this sense, the Steady State Revolution is similar to a Marxist revolution in which the proletariat develops a unified front to modify the behaviour of the capitalists.[49] A major difference is that the Steady State Revolution is peaceable due to the psychological power of social castigation.

The amorphs in the Steady State Revolution are roughly analogous to the bourgeoisie in a Marxist revolution. Some of the bourgeoisie, prior to a Marxist revolution, are simply budding capitalists, and many are aspiring capitalists. Similarly, prior to the Steady State Revolution, some amorphs are almost in the liquidating class, and many aspire to be. The aspirations of the amorphs change dramatically, however, when the steady state class castigates the liquidating class. Watching this unfold, amorphs will temper their consumption to avoid being deemed liquidators.

Another important political prerequisite to a revolution in political economy is that the classes are readily identifiable. Such is the case in the Steady State Revolution due to the very conspicuousness of the liquidators' consumption. In early 21st century American society, Hummers, mansions, fur coats and related trappings are the "uniforms" of the liquidating class. Conversely, hybrid cars, small houses, and modest organic cotton clothing readily identify the steady staters.

Eventually, of course, consumption behaviour will need to change across a much wider swathe of society than the upper 1% in personal consumption expenditures. Again, this broader change will be effected through the process of trickle-down consumption. It is important to begin the revolution with a large, relatively responsible majority identifying a small, reckless minority for the sake of developing solidarity and to add to the psychological viability of the revolution.

PSYCHOLOGICAL RATIONALE FOR THE STEADY STATE REVOLUTION

What makes us think an attitude developed by a "lower" class will have an effect on the behaviour of the wealthiest? This is readily answered using the theory of motivation developed by Abraham Maslow (1908-1970). Maslow developed a "hierarchy of needs" that explains much of human behaviour.[50] He identified five basic levels of needs, increasing in complexity, that tend to dominate human activities in succession. In the first level, people are motivated by the need for nutrition. When people are short on food and water, they lose interest in other things and behave in whatever way is required to obtain nutrition. In the second level, the need for safety motivates behaviour. Using a simple but illustrative metaphor of an animal at a watering hole, the first concern is drinking, but as thirst is slaked the animal becomes more wary of predators and this wariness affects the animal's behaviour accordingly. In the third level, behaviour is motivated by the need for love, affection, and sexual reproduction. At this level things become more complicated. People are "hard-wired" genetically to mate and have a host of affectionate emotions that may also have genetic origins. In any event, the behaviour of the well-fed, safe and secure person will tend to be oriented toward mating and affection.

At first glance, this aspect of Maslow's hierarchy may seem at odds with the notion of a "demographic transition" in which wealthier societies tend to have lower birthrates. However, demographic transition theorists do not simplistically argue, "the poorer, the more kids", all across the spectrum of wealth. That would be tantamount to arguing that literally starving societies have the highest birthrates. Rather, the demographic transition is focused on the intermediate interface between poverty and moderate wealth.

Nor did Maslow simplistically deny that people need love and affection, and do reproduce, even when struggling to put food on the table and a roof overhead. Maslow's point was that people need a certain amount of nutrition and security before love, affection, and reproductive needs become the dominant motive. That amount of nutrition and security may be quite minimal, appearing as poverty by most modern standards, yet if it is not attained, the love and affection needs will not be the driving force in the society.

Demographic transition theory posits that, as wealth increases from a relatively poor, typically agrarian level, - things other than reproduction become more important in driving behavior, and birth rates decline.[51] This is quite similar to what Maslow posited by identifying his fourth and fifth levels in the hierarchy of needs.

The fourth level is where things become even more complex. Behaviour is motivated in the fourth level by self-esteem, which may be garnered in a wide variety of ways. However, it has long been known that conspicuous consumption is often conducted for the sake of displaying one's prowess in worldly affairs. As such, it clearly is a behaviour subject to the motives of self-esteem. This is the primary level at which the Steady State Revolution will be engaged.

When liquidators are castigated for conspicuous consumption rather than emulated (as is largely the case in current affairs), self-esteem will tend to be lost rather than gained. Liquidators will therefore be motivated to consume less and regain their self-esteem. This may not work smoothly across the board because some liquidators may be sociopathic or may not interact enough with mainstream society to experience the castigation. In general, however, the Steady State Revolution will turn the tables of self-esteem based on consumption 180 degrees. Steady-statish consumption will be upheld as responsible, unselfish, exemplary behaviour conducive to the welfare of the poor and especially of posterity. Steady staters will be "selected for" in the evolution of a steady state economy.

The castigation of the liquidating class will also operate more powerfully at the third level in Maslow's hierarchy of needs, especially if young steady staters develop a preference for mates who are not liquidators. In fact, the third and fourth levels overlap to the extent that many people's self-esteem is based partly on their mates.

The fifth level in Maslow's hierarchy is called "self-actualization" and is the most complex concept in Maslow's theory. Self-actualization is beyond the scope of this chapter, but it should be noted that Maslow was adamant about materialistic behaviour (such as conspicuous consumption) falling short of self-actualization. In the Steady State Revolution, then, the liquidator is perceived not only as problematic for posterity but as falling short of the higher standards of human experience. Those driven to conspicuously consume for the sake of self-esteem may be perceived as suffering from "liquidator syndrome", a sort of psychological retardation that denies them the finer experiences in life and lowers their attractiveness to potential mates, friends, and fellow citizens.

INTERNATIONALIZING THE STEADY STATE REVOLUTION

As originally set forth in *Shoveling Fuel for a Runaway Train*,[52] the Steady State Revolution was intended as an American movement. The transition to a steady state economy in the United States is perhaps the most important development for the sake of a sustainable world economy and international stability. However, the Steady

State Revolution will eventually be needed in all nations with the goal of economic growth. Some of these nations, however, deserve and need a certain amount of economic growth, especially in *per capita* terms, before the establishment of a steady state economy becomes politically feasible or ethically appropriate.

Due to the importance of an American Steady State Revolution, we can also consider another means to that end, i.e. international diplomacy. In an international Steady State Revolution of diplomacy, the principles of ecological economics and limits to economic growth are recognized by a "steady state class of nations" who then find especially problematic the "liquidating class among nations" including the United States, Japan, some of the western European nations, and perhaps the richest of the Middle East OPEC[53] nations. Certainly the United States is the primary subject of this international Steady State Revolution, however.

The analogy to social castigation on the international stage is diplomatic sanction, and the United Nations is probably the most important venue in which this may transpire. The steady state nations develop solidarity based upon ethical and responsible consumption patterns. They express their opprobrium toward the consumption patterns of the United States in the form of economic sanctions, especially sanctions affecting international trade. The effect on the United States polity would be similar to the effect that castigation by American steady staters will have on American liquidators; i.e., its consumption patterns will become much more conservation oriented, leading toward the establishment of an American steady state economy. The American government, upon establishing a steady state economy, may then turn some of its considerable resources toward assisting other nations in the establishment of their own steady state economies rather than putting its greatest efforts into economic growth, as is currently and unsustainably the case.

NOTES AND SOURCES

[1] Much of the source material for this essay may be found in: Czech, B. 2000. *Shoveling Fuel for a Runaway Train.* University of California Press, Berkeley and Los Angeles. 210 pp. Also see http://www.steadystate.org.

[2] Czech, B., P. R. Krausman, and P. K. Devers. 2000. Economic associations among causes of species endangerment in the United States. Bioscience 50(7):593-601.

[3] Hanink, D. M. 1997. *Principles and applications of economic geography: economy, policy, environment.* John Wiley and Sons, New York. 512 pp.

[4] See endnote 2 above.

[5] Czech, B., and P. R. Krausman. 2001. *The Endangered Species Act: History, Conservation Biology, and Public Policy.* Johns Hopkins University Press, Baltimore, MD.

[6] Brown, P. 1998. *Climate, Biodiversity, and Forests: Issues and Opportunities Emerging from the Kyoto Protocol.* World Resources Institute, Washington, DC. 36pp. Available at: http://www.wri.org/ffi/climate/index.html.

[7] Harvell, C.D., Mitchell, C.E., Ward, J.R., Altizer, S., Dobson, A., Ostfeld, R.S., and Samuel, M.D. 2002. Climate warming and disease risks for terrestrial and marine biota. Science 296:2158-2162.

[8] Intergovernmental Panel on Climate Change. 1996. *Climate Change 1995: The Science of Climate Change.* Cambridge University Press, Cambridge, UK. 572 pp.

[9] Pianka, E. R. 1974. *Evolutionary ecology.* Harper and Row, New York. 356pp.

[10] Begon, M., J. L. Harper, and C. R. Townsend. 1996. *Ecology: individuals, populations and communities.* Third edition. Blackwell Science, Oxford, UK. 1068 pp. For more on trophic interactions and implications, see Brooks, Chapter 16.

[11] Ayres, R. U. 1999. *Turning point: an end to the growth paradigm.* Earthscan, London, UK. 258pp.

[12] Collins, R. M. 2000. *More: the politics of economic growth in postwar America.* Oxford University Press, Oxford, U.K. 299 pp.

[13] The World Factbook, U.S. Central Intelligence Agency. Available at: http://www.cia.gov/cia/publications/factbook/geos/us.html#Econ.

[14] Sardar, Z., and M. W. Davies. 2003. *Why do people hate America?* Disinformation Company, New York. 240 pp.

[15] Daly, H. E., and J. Farley. 2003. *Ecological economics: principles and applications.* Island Press, Washington, DC. 450 pp.

[16] See endnote 14. Also see Beder, S. 2002. Global spin: the corporate assault on environmentalism (revised edition). Chelsea Green, White River Junction, VT. 336 pp.

[17] Gaffney, M., and F. Harrison. 1994. *The corruption of economics.* Shepheard-Walwyn, London, UK. 271 pp.

[18] George, H. 1929. *Progress and poverty.* Vanguard Press, New York. 214 pp.

[19] Heilbroner, R. L. 1992. *The worldly philosophers: the lives, times, and ideas of the great economic thinkers.* Sixth edition. Simon and Schuster, New York. 365 pp.

[20] See endnote 17.

[21] See endnote 19.

[22] See endnote 17.

[23] Meadows, D., J. Randers, and D. Meadows. 2004. *Limits to Growth: The 30-Year Update.* Chelsea Green Publishing Company, White River Junction, Vermont. 338 pp. For more on limits to growth, see Rees, Chapter 14.

[24] Besomi, D. 2001. Harrod's dynamics and the theory of growth: the story of a mistaken attribution. Cambridge Journal of Economics 25:79-96.

[25] See especially: Solow, R. M. A contribution to the theory of economic growth. Quarterly Journal of Economics 70 (February):65-94.

[26] See especially: Lucas, R. E. Jr. 1988. On the mechanics of economic development. Journal of Monetary Economics 22 (July):3-42.

[27] See especially: Romer, P. M. 1990. Endogenous technological change. Journal of Political Economy 98(October):S71-S102.

[28] See, for example: Jones, C. I. 1998. *Introduction to economic growth.* W. W. Norton, New York. 200pp.

[29] Simon, J. L. 1981. *The ultimate resource.* Princeton University, Princeton, New Jersey. 415 pp.

[30] Adams, G. 1981. *The politics of defense contracting : the Iron Triangle.* Transaction Books, New Brunswick, NJ. 465pp.

[31] See, for example: Browne, W. P. 1992. *Sacred cows and hot potatoes: agrarian myths and agricultural policy.* Westview Press, Boulder, Colorado. 151 pp.; Miroff, B., R. Seidelman, and T. Swanstrom. 2002. *The democratic debate: an introduction to American politics.* Third edition. Houghton Mifflin, Boston, Massachusetts. 548 pp.

[32] Korten, D. 2001. *When corporations rule the world.* Second edition. Kumarian Press, Bloomfield, Connecticut. 384 pp.; Browne, W. P. 1992. Sacred cows and hot potatoes: agrarian myths and agricultural policy. Westview Press, Boulder, Colorado. 151 pp.

[33] Ormerod, P. 1997. *The death of economics.* John Wiley and Sons, New York. 230 pp.

[34] Ibid.

[35] Czech, B. , E. Allen, D. Batker, P. Beier, H. Daly, J. Erickson, P. Garrettson, V. Geist, J. Gowdy, L. Greenwalt, H. Hands, P. Krausman, P. Magee, C. Miller, K. Novak, G. Pullis, C. Robinson, J. Santa-Barbara, J. Teer, D. Trauger, and C. Willer. 2003. The iron triangle: why The Wildlife Society needs to take a position on economic growth. Wildlife Society Bulletin 31(2):574-577.

[36] Hall, C. A. S., P. W. Jones, T. M. Donovan, and J. P. Gibbs. 2000. The implications of mainstream economics for wildlife conservation. Wildlife Society Bulletin 28:16-25.

[37] See endnote 33. Also see Beder, Chapter 5; Rees, Chapter 14; and Brooks, Chapter 16.

[38] Loomis, J. B. 2000. Can environmental economic valuations techniques aid ecological economics and wildlife conservation? Wildlife Society Bulletin 28:52-60.

[39] Czech, B., P. Angermeier, H. Daly, P. Pister, and R. Hughes. 2004. Fish conservation , sustainable fisheries, and economic growth: no more fish stories. Fisheries 29(8):36-37. For other thoughts on cost-benefit analyses, see Rees, Chapter 14.

[40] See endnote 32 (Korten 2001).

[41] Greider, W. 1992. Who will tell the people? Simon and Schuster, New York. 464 pp.

[42] Trauger, D. L., B. Czech, J. D. Erickson, P. R. Garrettson, B. J. Kernohan, C. A. Miller. 2003. The relationship of economic growth to wildlife conservation. Wildlife Society Technical Review 03-1. The Wildlife Society, Bethesda, Maryland. 22 pp.

[43] Ecological economics differs from conventional or "neoclassical" economics by incorporating the natural sciences as the conceptual foundation for economic theory, especially economic growth theory. Unlike neoclassical economic growth theory, ecological economics posits limits to economic growth based upon laws of thermodynamics and ecological principles. Also see endnotes 1 and 17.

[44] ftp://ftp.bls.gov/pub/special.requests/ce/share/2002/quintile.txt

[45] See Menon and Lavigne, Chapter 12, and Worcester, Chapter 17, on the subject of attitudes and values.

[46] Brown, L. R., G. Gardner, and B. Halweil. 1999. *Beyond Malthus: nineteen dimensions of the population challenge.* W. W. Norton and Company, New York. 167pp.

[47] Hall, C. A. S., R. G. Pontius, J. Y. Ko, and L. Coleman. 1994. The environmental impact of having a baby in the United States. Population and Environment 15:505-524.

[48] Brown, M. B. 1995. *Models in political economy: a guide to the arguments.* Second edition. Penguin Books, New York. 418 pp. Also see Best, Chapter 25.

[49] Marx, K., and F. Engels. 1998. *The Communist manifesto.* Monthly Review Press, New York. 112 pp.

[50] Maslow, A.H. 1943. A theory of human motivation. Psychological Review 50:370-396.

[51] For more on "demographic transition, see http://www.uwmc.uwc.edu/geography/Demotrans/demtran.htm. Also see Lockwood, M. Development Policy and the African Demographic Transition: Issues and Questions. Journal of International Development 7, no. 1 (1995): 1-23.

[52] See endnote 1.

[53] The Organization of the Petroleum Exporting Countries. See http://www.opec.org.

CHAPTER 24

TOWARDS A NEW ARCHITECTURE OF WILDLIFE CONSERVATION IN THE DEVELOPING WORLD:
AN INTEGRATED DEVELOPMENT PLANNING APPROACH

Atherton Martin

I have been asked to focus on mechanisms for wildlife conservation, the "how to" component of our discussions. In so doing, I will draw on my experience as an environmental activist, minister of government, private development consultant, and small eco-hotelier.

A major question before us is whether we need a new approach to wildlife conservation. Let me begin by signalling several of the challenges that have to be confronted and for which practical and effective mechanisms need to be found. First among these is to be aware that the conservation, protection and management of human activities in the biosphere that is so critical to plant and animal species is also critical to the conservation, protection and management of our own species. Strategies of conservation that fail to make that link are destined to fail. If nothing else, in an increasingly threatened global biosphere, the powerful drive for survival that is in all life forms should fuel the need for coordinated and integrated approaches to conservation.

Dominica's natural history clearly reveals those links. The present population is indeed fortunate to have been preceded by several generations of the indigenous people of the Caribbean, the Carib or Karifuna People, who for centuries tended the flora and fauna of Waitukubuli (the Carib name for Dominica). When Christopher Columbus lost his way to India and ended up in the Caribbean, it was the Carib people who welcomed him, until he violated their hospitality by attempting to claim the island for his European monarch.

For the next three hundred years (1492 to 1800) the Carib people defended and protected the entire island ecosystem against European settlement. This allowed the world community to inherit one of the most intact and healthy island ecologies where, in 1998, the United Nations Educational, Scientific and Cultural Organization (UNESCO) declared a 17,000 acre National Park the island's first World Heritage Site.

There is a lesson here for the development of mechanisms for conservation that simultaneously protect ecologies in the name of "human development". This lesson is particularly urgent today, given the deteriorating human condition in so many developing countries, and given the advance in technologies for natural resource extraction that have been refined and honed to the point where:

- Thousands of acres of natural, mostly tropical, forests are destroyed every hour;
- Species of plants and animals are brought closer to extinction with each passing day; and
- Millions of human beings do not have sufficient food, water or shelter, education or health care that would allow them to live beyond their thirtieth birthday.

Sensible strategies for conservation have to be built, therefore, around connecting environment and development issues from the design to the implementation stage.

MAKING THE LINKS: FOOD PRODUCTION AND CONSERVATION

As someone trained in the science of soil management, I discovered very early in my professional career as an agronomist that breaking the link between environment and development can be disastrous both for people and the biosphere. My experience in Haiti in 1989 made this painfully clear.

I had been recruited by an international relief agency to train farmers in a remote area of the country in the production of vegetables. A diagnostic study by the agency had determined that the poor supply of vegetables in the area was the result of limited technological capacity and the unavailability of fertilizer and seed. After two weeks in the village, I had visited small farms that could only be described as models of mixed farming. The local knowledge about what to plant, when and how to plant, how to cultivate, store and process the farm products was well established and was being passed on effectively to the youth of the village. Seed was not only available but was being produced by some farmers for sale to their colleagues. Composting was well developed and the mixture of crops and livestock ensured a steady supply of that input.

I telephoned the national office of the agency in the capital, Port-au-Prince (eight hours by lorry from the village) to report on the situation and to inquire about the diagnostic study that led to my being recruited. I received a fax copy of the report and discovered that the conclusion on low production from the area was based on observations that farmers from the village did not deliver much produce to sell at the marketplace in the capital.

When I presented this finding to the farmers, they all laughed and made jokes about those "parachute people" who come to Haiti and never take the time to talk with them, but always seem to have answers to their problems. To make a long story short, I was invited to go with them on the Friday trip to the capital so that I could see the problem for myself. In a nutshell, there were twelve security checkpoints between the village and the capital. At every checkpoint, the soldiers required the 10-ton lorry to be off-loaded. Each soldier took a share of the best produce and the farmers loaded what was left back on to the lorry as they continued the trip to market. This happened twelve times! By the time the lorry got to the capital, there was very little produce left for sale.

The moral of the story: whether it is wildlife conservation or vegetable production, before designing a response to the problem, be sure of what the problem is. Talk with the people; they probably know what is wrong and what needs to be done to fix it.

The problem in this case was that people like me were trained at university to promote the notion of technical answers to the "problem" of developing-country agriculture. I later discovered that the agribusiness corporations, which financed much of the research at major universities in the West, were promoting a worldview of hunger caused by deficient technology as part of a strategy to expand the markets for their own technologies.

Today, the food industry, the aid industry, the trade talk industry (the series of conferences and seminars that talks about trade as if people were not involved or impacted), and most of the media industry, all have vested interests in perpetuating this myth. The Food First Institute for Food and Development Policy and others have made this point for decades. The ever-expanding impact of the rings of concentric circles of technological responses to food and agriculture are increasingly based on the culturally imperialistic notion that people in developing countries are lazy, dumb, and incapable of managing 'complex' systems of anything.

One of the lessons to be drawn here for wildlife conservation is that the barriers to conservation are often cultural and social, so that appropriate mechanisms need to be culturally sensitive and to take account of such issues as communication, family and community traditions, community governance and administrative structures, and local knowledge.

Another challenge that I raise is that there is often a wide gap between the reality of human survival, production and conservation at the village level and the language of international environmental agreements. The mechanisms that we seek and select must help bridge that gap; they must help ensure that the ground level reality is accurately reflected in the national, regional and international agreements and, in fact, drives the global agenda.

WILDLIFE CONSERVATION TODAY

Earlier chapters have examined many of the issues confronting the modern conservation movement. I simply wish to extract those that, for me, are critical to informing the rationale for a new architecture of wildlife conservation.

Poverty

We are challenged to redefine our strategies by the burgeoning numbers of human beings living in damaged and destroyed environments and for whom poverty has become "a way of life". There is the need to connect approaches to the protection of natural environments, and the plant and animal species that inhabit them, with the needs of the people who share the same spaces.

There is also the need to find ways to demonstrate, practically, the links between investing in the protection of natural systems of streams, reefs, forests and savannahs, and investing in helping people put food on their tables, roofs over their heads, and jobs in their hands.

War and Social Strife

Given the prevalence of domestic conflict played out in the theatre of the developing world for many decades since the establishment of the United Nations, and given the clear links between international political and corporate decisions relating to the extraction and use of natural resources, how do strategies for wildlife conservation address the fact that:

- in countries containing the most important natural resources needed for sustaining life on Earth (drinking water and forests as carbon sinks), millions of people exist under the most miserable conditions;
- the countries with the largest ecological footprints (consuming most of the earth's resources) are the richest, and those with the smallest footprints are the poorest and most at risk;
- the real weapons of mass destruction (of the biosphere) are economic, political and military mechanisms deployed not according to the precautionary principle, or considerations of carrying capacity, but according to one universal edict – MIGHT IS RIGHT – or, as we used to put it in the 60s, "money talks and everything else walks/bows/scrapes/does what they are told".

What is responsible for this grossly ecologically and people unfriendly situation? Do our strategies for wildlife conservation address the root causes of this imbalance?

What is the mechanism for dealing with local-level conservation and protection of habitat that is capable of stopping, say, a World Bank-financed dam or mine or timber project, projects that use your tax dollars and the profits from the stock market to finance these "weapons of mass destruction"?

Was Mr. Ghandi right? Is peace the only weapon that can defeat poverty? What is the link between peace and conservation?

Local Knowledge

After the land, the first resource of a people that is tackled by an invader is the mind. Fortunately, the mind is often the last thing of a people to be defeated.

Local knowledge is born out of centuries of working with and understanding natural systems, plants and animals, and developing the capacity to live within the limits of biosphere. That is the main legacy of systems of traditional local knowledge. Have we developed systematic mechanisms of wildlife conservation that allow us to understand how these communities have protected and passed on that knowledge through generations?

Have we found ways to inform researched science with practical science? Where are those mechanisms, and how do we protect and pass these on? Remember the story of the Haitian farmers.

I recall visiting the National Seed Bank of Ethiopia in Addis Ababa and walking through centuries of genetic information and food security. I visited farms around Addis that were dedicated to the annual reproduction of seed that was initially over five thousand years old and re-supplying the bank. Then it was described to me how scientists from major agricultural institutes were given seed material from the bank in Addis as part of the process of creating the new hybrid varieties of grains that triggered the "green revolution". That well-financed innovation in agriculture employed mechanisms of modern agribusiness that led to the exponential expansion in the use of agro-chemicals, mono-crop agriculture in the developing world, export agriculture at the expense of local food production, hunger in the face of food surpluses, endemic poverty where there had been none for centuries, and habitat destruction after only decades of "modernization", in contrast to centuries of protection through local knowledge.

Taken all together, there is a message here for those engaged in the search for mechanisms of wildlife conservation in today's world. The message is about the need for those mechanisms to attend to such issues as valuing and listening to local knowledge, and protecting and passing on local knowledge to the next generation. Implementing this approach would signal a shift from the World Trade Organization (WTO) practice of treating intellectual property rights as a commercial item to be traded. It would require international financial institutions to recognize the value of local knowledge, and it would buttress the growing realization on the part of international non-governmental organizations (NGOs), that best practices are not always those documented in studies and annual reports.

One important issue that we must confront in our search for improved approaches to wildlife conservation is the following: In the face of powerful global forces engaged with these issues, how do local communities, regions and nations take action in their own interest to protect their natural systems and wildlife?

The Power To Make Change

In the 1970s, one of the Caribbean's leading economists, Dr. George Beckford of Jamaica, wrote a book entitled *The Poor and the Powerless*, in which he coined the phrase "Plantation Economy" to describe the fact that one hundred and fifty years after emancipation, the people of the region were still enslaved. He submitted that the processes and institutions of decision-making had remained the same after colonial rule was ended so that the minds, behaviour, awareness, control and capacity to influence the direction of development were still firmly in the hands of institutions and persons sitting in the metropolitan capitals.

Dr. Beckford submitted that the exercise of power to decide to protect, expand, trade, invest, or extract, needed to shift to the local, national, and regional arenas in order to establish balance, order and direction that reflected the wishes of the local population.

Later that same decade, another Jamaican, Robert Nestor Marley, wrote and sang a song that included the following lines:

> Emancipate yourself from mental slavery,
> none but our selves can free our minds...

Life itself is this miraculous circular journey that we share with everyone else on the planet. We may not be able to measure it but this is undoubtedly how life is, circular.

Nature teaches us nothing, if not about the notion of cycles, interdependence and the power of self-sustainability, and the power of local knowledge as the platform from which global knowledge is perceived and filtered. Some years ago, a few of us explored the possibilities of illustrating the flow of knowledge as the fuel of the development process. We evolved something rather ominously called the "***CIA***" theory of the power circle, where the cliché, "knowledge is power", is applied to the journey that we all make from first becoming aware (***A***), to beginning to influence (***I***), then to become a part of the control (***C***) mechanisms in our society. The Power Circle (Figure 24–1) is a simple graphic illustration of that "***CIA***" theory of change.

Where we are in that circle as individuals, communities, organizations, nations or species, helps shape our response to most issues including the respect for, use, and conservation of natural resources and systems for living. Are we in control? Who is? Are we able to influence others? Who is influencing us? Are we aware of what is going on locally, nationally, and globally? Are others aware of our situation?

The question is, where are you in the Power Circle? Where are you in relation to other individuals, organizations, communities, governments, global corporations and institutions like the International Monetary Fund (IMF), the World Bank, the International Fund for Animal Welfare (IFAW), and Caribbean Conservation Association (CCA)?

OUR RESPONSE

These challenges notwithstanding, and before we conclude that we need a new approach to wildlife conservation complete with new mechanisms, we need to assess how we have responded thus far. Have decisions been made and strategic corporate plans developed based on science, conscience, funding, politics, partnership, respect, rights, responsibilities, ethics or knowledge, on other considerations such as the Power Circle analysis, or on all of the above?

In many cases, the response has been to attempt to graft sustainable approaches and mechanisms on to the existing systems of economic management and governance. Special Offices have been created. Ministries of Environment have been established, but the Planning and Finance Ministries still retain control over the formulation of policy and the allocation of financial resources.

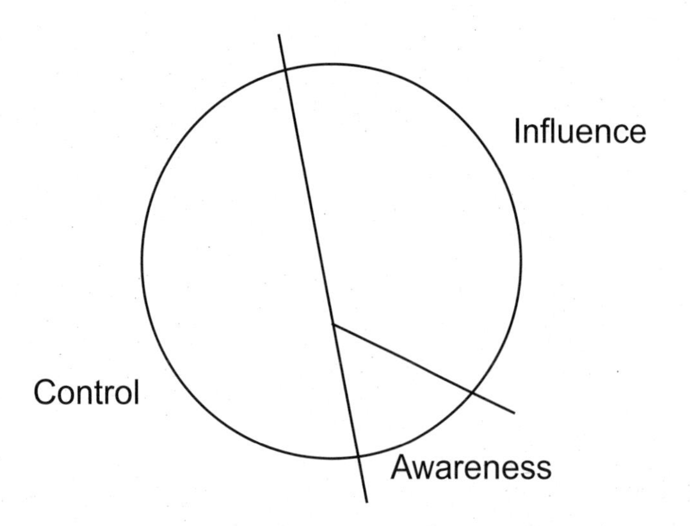

Figure 24–1.

The Power Circle

The standard indicators of "growth" (GDP, fiscal deficits, debt ratios, etc.) are still used to evaluate economic performance to the exclusion of indicators such as the ecological footprint, wildlife populations, biodiversity, the recharge rate of watersheds, stream flow, soil and water quality, etc.

Issues of participation, transparency, accountability and governance remain "soft" issues that always lose out to the "hard" numbers of the economic statistics. After decades of conferences, summits and agreements, ecological sustainability has definitely not become mainstream. The concept of holistic and integrated planning remains threatening to the traditional planners and policy makers, and so it is "business as usual".

At the International Level

The last few decades have seen the proliferation of multilateral organizations, agreements, conventions and summits focused on measures for arresting the rapid deterioration of global ecosystems. These include: the *Convention on Biological Diversity* (CBD), the *Framework Convention on Climate Change*, the International Water Agreement, Agenda 21, the *United Nations Convention on the Law of the Sea* (UNCLoS), the *Convention on International Trade in Endangered Species* (CITES), the Specially Protected Areas and Wildlife (SPAW) Protocol, the *International Convention for the Regulation of Whaling*, and many others.

One aim of these global initiatives has been to draw attention to the fact that the uncontrolled extraction of resources from natural systems (in the name of economic growth) has in many instances led to the irreversible destruction of habitats that are vital to the continued existence of both humans and non-human animals alike.

The call has been for the international community to assemble around strategies and work plans that would continue to build widespread awareness of the crisis, and assemble institutions and resources that could be put to work together to stop and reverse this mad rush to global ecological disaster.

In some circles it was realized that the global response needed to be more than scientific and environmental, and linked to the shaping of economic and social policies that had often instigated the human assault on wildlife and their habitats in the name of "feeding" the world's people and, more recently, in the name of "globalization". In response, the idea of sustainable development emerged at the 1992 Earth Summit in Rio, as being central to the strategy for connecting the care and management of the environment to the care and management of human material needs.

While the international community has largely endorsed the concept of sustainable development there has been precious little movement from the core of the major world economic institutions that would suggest an institutionalization and internationalization of the principles of sustainable development as the guide for environmental and economic planning.[1]

The evolution of the *General Agreement on Tariffs and Trade* (GATT) into the WTO gave new focus to the linkages between trade issues and environmental stewardship and sustainable development. But the environment continues to be traded off in favour of unfettered economic growth.

The two most influential international financial institutions, the IMF and the World Bank, in spite of fascinating internal literature to the contrary, have continued to insist on application of the tried and failed mechanisms of economic management. The deterioration of human and ecological conditions is living testimony to this officially sanctioned tragedy.

When added to the new threats of pandemics, intense climatic events, heightened personal security and conflict, with the resulting migrations of millions of people and animals across regions, it is nothing short of criminal for us to accept "business as usual" as the way forward.

At the Regional and National Levels: Lessons from the Caribbean

With the advent of UNCLoS and the definition of an Exclusive Economic Zone (EEZ), small islands have experienced the sudden expansion of their domestic domain. With the inclusion of significant areas of marine territory in the resources to be stewarded, these islands have come face-to-face with the challenges of wildlife management and conservation on a scale well beyond their capacities. This has made these island states even more vulnerable to the influence and control of larger more powerful states with vested commercial interests in the animal, mineral and energy resources that are now part of the EEZs of these islands.

Several small islands states of the Caribbean have in recent years found themselves at the forefront of the attempts by other nations, international agencies and NGOs, to influence the approach to and mechanisms for marine wildlife conservation. The case of Dominica is instructive and pertinent to our search for new mechanisms for wildlife conservation.

Whales and Wildlife Conservation in Dominica

For the past 12 years, Dominica has taken positions at the International Whaling Commission (IWC) that have been most closely associated with the positions held by countries interested in the recommencement of commercial whaling, stopping the establishment of whale sanctuaries, and downplaying the importance of whale-watching as the most ecologically sustainable use of whales.[2] Two of the countries that have taken the lead on these positions at the IWC are Japan and Norway.

Since 1992, three different Dominican governments have taken the same position on whaling. Each administration has claimed that the country's votes at the IWC were in the best interest of Dominica and were based on the scientific data and not on the promise of aid from Japan. The construction of a US$18.6 million Fisheries Complex with Japanese Government aid in the capital city, Roseau, and the construction of a second such complex in the coastal village of Marigot, remain the most tangible evidence of possible benefits related to Dominica's voting pattern at the IWC.

In the context of this discussion regarding the implications of Dominica's involvement with the IWC for the country's strategy on conservation, we recall that the IWC is an organization that was set up by whaling countries to regulate that industry in the face of depleted whale stocks. In light of this, an obvious question that arises is: If Dominica has never been engaged in commercial whaling, why do we belong to, and participate in, that organization?

Issues of Control, Influence and Awareness

The first time around, we were coaxed into joining the IWC by international NGOs intent on getting the IWC to support the protection of the world's whale populations through a ban on commercial whaling. The second time around, Dominica was coaxed into rejoining the IWC by pro-whaling nations intent on gathering votes that would block the establishment of whale sanctuaries.

The first time around, Dominica's vote seemed consistent with what by then was the country's unique tourism image as a "nature island" arising from a commitment to the protection and conservation of habitats and wildlife, and its ability to offer a different experience to the visitor. The second time around, the vote seemed to depart from that "nature island" image and legacy.

All independent nations enjoy the privilege and responsibility of participating in international fora of their choosing. There are dozens of international organizations that a country can choose to join. For small, developing countries, one of the most vital considerations in making that decision, is the extent to which the country's international association will advance its domestic agenda in such areas as trade, aid, investment, and training. In that regard, since Dominica has never been a whaling country, a legitimate question can be: What is the domestic agenda driving our participation in a forum dealing with issues of the regulation of commercial whaling?

Local Knowledge

The answer to that question has its beginning in 1989, thanks to the pioneering work of local hotelier and dive shop operator, Fitzroy Armour, who started recording whale sightings in Dominica's waters noting the frequency, location and species. Soon, scuba divers in Dominica were being exposed to whales as an added attraction to their planned diving trip. In 1993, with the assistance of IFAW, special techniques for whale watching in Dominica were developed and scuba diving personnel were trained to maximize boat-based sightings of whales in ways that were sensitive to the behaviour of these mammals.[3]

As a direct result of this initiative by IFAW and the Dominica Conservation Association (DCA), several local dive shops started marketing the activity of whale watching, which then became a new and unique element of the Dominica tourism experience. In fact, in just seven years (1992 to 1999), this activity moved from earning EC (Eastern Caribbean) $40,000 per year to EC$1.2 million per year.

It seemed clear that if participation in the IWC had any sustained material relevance for Dominica, it would have to be in promoting the development of whale watching, a fast-growing segment of the tourism sector that could create more jobs, earn foreign exchange, expand local investment in the economy and diversify the economic options for a small island state.

The Path to Sustainability

On the question of sanctuaries, Dominica has always been and still is, the logical proponent of this measure for the protection of endangered species and their habitats. The reason is very simple. Dominica is the one country in the Caribbean where the establishment of sanctuaries for water, trees and wildlife has been standard practice for over 50 years. Our extensive system of forest reserves and national parks covers close to 30% of the island's total landmass, one of the highest in the region and, certainly, one of the highest in the world. The success of this strategy of establishing sanctuaries on land has recently been extended to the establishment of the country's first marine reserve in the south of the island.

Dominica is no stranger, therefore, to the benefits of sanctuaries as part of the strategy for the conservation of vital natural resources such as soil, trees, water and the animal species associated with these habitats. Through the development of nature tourism, and the emerging prospects for organic agriculture, Dominica is also aware of crucial links between conservation and prospects for agriculture, health, housing, tourism and recreation.

Substantial work has been done to establish the direct links between the protection and conservation of natural resources and the prospects for economic development. Notable among these initiatives are: the 1967 Shankland and Cox Report; the 1992 United Nations Conference on Environment and Development (UNCED) Program of Action, Agenda 21; the 1994 National Environmental

Action Plan; the 1994 Program of Action of the UN Conference on the Sustainable Development of Small Island Developing States; the National Biodiversity Strategy now being prepared by the Environmental Coordinating Unit of the Ministry of Agriculture; the Organization of Eastern Caribbean States (OECS) Environmental Charter; and others.

The successful promotion of Dominica for over 30 years, as "The Nature Island of the Caribbean" is due directly to the vision and persistence of the foresters, scientists, conservationists and others who have pioneered this strategy of sustainable development (meaning "development that is ecologically sustainable") through the conservation and management of natural resources locally and globally.

In fact, Dominica has strongly signalled its commitment to environmental stewardship locally and globally, by signing on to important international agreements such as the *Convention on Biological Diversity* and the *Framework Convention on Climate Change*. The clear message that we have sent to the world is that while we will continue to care for and manage our national resources, we are also committed to playing our part internationally for the proper conservation and management of the Earth's resources for the benefit of all of humanity now and in the future. There is, therefore, every reason for Dominica to be one of the leading proponents at the IWC for the extension of this successful strategy to the South Atlantic, Southern Pacific and to other parts of the Earth's oceans and seas.

SCIENCE, SANCTUARIES AND SOLIDARITY

There are some who tell us that the scientific data supports the killing of whales while not supporting the establishment of sanctuaries. We should also be reminded that the majority of countries participating in the IWC have supported the establishment of sanctuaries. In fact at the IWC meeting in 2000, 18 countries supported the establishment of the South Pacific Sanctuary. At that same meeting 11 countries (including 6 OECS countries), voted against the establishment of the sanctuary. Why did the majority vote not prevail? Simply because, under the rules of the IWC, there must be a 75% majority for the establishment of a sanctuary. In spite of the majority in support of the Sanctuary, these 6 OECS countries together were able to block its adoption.

Who are these countries that support and oppose the sanctuary? Well, included among the countries supporting the establishment of the Pacific Sanctuary are: the United Kingdom, France, the United States of America, New Zealand and Australia, to name a few. Included in the list of countries opposing the sanctuary are Japan and Norway. All these countries have substantial national scientific capability in the relevant areas of competence for the IWC deliberations.

It can be submitted that the scientific capability of each of these countries (all former whaling countries) is at par with the others. It must mean therefore, that Dominica and the other OECS States participating in the IWC, have taken our cue from an assessment of scientific data by the significant minority at the IWC, i.e. the five nations besides the six of us who voted against the establishment of a whale sanctuary in the South Pacific.

In this regard, as Dominica establishes its independent position at the IWC as in other fora, we remind ourselves and others that it was outstanding science and outstanding local scientists who were instrumental in establishing the strategy of terrestrial sanctuaries as an effective means of maintaining a healthy natural resource base for agriculture, tourism and the general development and health of the people of this island. That admirable track record makes a resounding call for an approach to the management of natural systems that is based on conservation, protection and proper management of fragile resources by the people, for the people.

In light of this, at the IWC, Dominica should be a logical supporter of the establishment of protected areas in keeping with the island's enviable track record as a place where nature is conserved as the basis for economic and social development.

It is worth noting that one of the cardinal principles of conservation is the precautionary principle,[4] which states that in the face of a lack of scientific data decisions should favour caution, discretion and non-extraction.

GLOBALIZATION AND THE NEW SOVEREIGNTY

The sovereignty of our nation is under attack, not by NGOs, but by the WTO, the Free Trade Area of the Americas (FTAA), the Financial Action Task Force (FATF), and the Organization for Economic Cooperation and Development (OECD). Governments that give in to these institutions give up their sovereignty. Governments that vote one way in order to get the promise of aid, violate their sovereignty. In contrast, Governments that have the vision and courage to prepare and persist with proven development strategies that work for their people are protecting and consolidating their dignity and self-determination.

In this regard, what about the views and development aspirations of the people of the South Pacific? Dominica would have taken serious objection to any effort at preventing us from establishing the Morne Trois Piton National Park, the Morne Diabolitin National Park, the Cabrits National Park or the Soufriere, Scottshead, Marine

Reserve. We would have considered that an affront to our national pride and right to self-determination.

The Dominica Labour Party (currently forming the government) has called for the establishment of the Caribbean as a zone of environmental protection. The Governments of Caribbean Community (CARICOM) have for years, called for the Caribbean to be a nuclear-free zone. The people of the South Pacific ask for no less than respect for their right to determine how to protect global resources that are directly vital to their economic well being. Many of these states have declared their EEZs as sanctuaries but because whales are highly migratory they needed the IWC to go the next stage and adopt a large area of the South Pacific in order for these measures to be truly effective.

It is worth noting that the Caribbean has, for many years now, condemned the passage through the Caribbean Sea of ships carrying spent and re-processed nuclear materials (plutonium) moving between Japan and Europe. Our right to protect our seas and people from major nuclear accidents has been championed by all the Governments of the Region. We as a people are violated every time those ships transit the region, in open defiance of our Governments' position and our peoples' safety.

DEVELOPMENT AND CONSERVATION: THE MEANING OF SUSTAINABILITY

The issues raised by the IWC are primarily related to how we as a people and a nation view ourselves; how we view the resources of the Earth (*Apres Bondie c'est la Ter*, "After God, the Earth"; from the Dominica Coat of Arms); how we respect and build upon our history and tradition of care and management of these resources; how we strengthen our ability to develop patiently economic opportunities from the resources that we have learned to conserve; how we deepen and extend the successful promotion of our island as the quintessential nature destination of the Caribbean; how we convert that unique image into a major economic breakthrough in tourism, health, nutrition, organic agriculture, renewable energy, and a genuine culture of democracy, that guarantees both ecological and economic sustainability of, for, and by the people.

The real meaning of these issues is not that they define how we feel about whales, or Japan, or Greenpeace, or IFAW, or anybody else. The real meaning of these issues is that we seize the opportunity to convert the notion of "nature island" into a comprehensive development strategy for Dominica that gives us the competitive and comparative edge that secures our future for generations to come.

Dominica is one of the region's leaders in conservation and management of specially protected areas. In more recent times, Dominicans have established some of the best-managed and regulated terrestrial parks and marine scuba diving businesses anywhere in the world. Dominica has recently had one of its National Parks declared a World Heritage Site by UNESCO and the island is consistently listed as one of the top five dive destinations in the world. This is no accident. Instead, it is the direct result and continuation of a long and proud tradition of sensible and sensitive management of our natural resources by the residents of the island. It is not an accident that over the last few years when the actions of the United States Government and the WTO have traumatized our banana industry and, more recently, our fledgling off-shore sector, it is nature-based tourism that has earned over EC$100 million (US$37 million) per year, employing over 2,000 persons full-time and contributing EC$35 million (US$13 million) to direct public revenues in 1999.

THE MESSAGE

Conservation and the effective management of natural systems do result in direct economic benefit to people. Dominica's nature-based tourism demonstrates that environmentally friendly development is possible and that the same approach can be extended to agriculture, health and other sectors.

In the context of the IWC, this tradition of environmentally friendly economic sustainability can continue to find expression if we remain consistent and support the creation of those specially protected areas called sanctuaries and stand strongly for the protection of whales and other endangered species that inhabit those areas in Dominica as in other parts of the world.

This approach to economic development is not short term. It demonstrates vision, requires planning, demands leadership, and it works. Dominica's tourism is living proof of that. The challenge and the task are to apply these same principles of vision, leadership and planning to the other sectors of the economy.

MANAGING THE ECONOMIC TRADE-OFFS: HOW WE CARE FOR OUR FUTURE BY MANAGING OUR PRESENT

Sometimes, the economic argument for conservation is clouded by the more spectacular action associated with the extraction of resources as in mining and logging, or the construction of large buildings as opposed to the more careful management of parks and other whole systems for the generation of sustainable income and livelihoods for present and future generations.

Year	Japan's trade balance with Dominica	Dominican exports to United Kingdom
1995	EC$14,509,991	EC$46,533,187
1996	EC$19,386,724	EC$48,211,518
1997	EC$15,845,291	EC$45,488,873
1998	EC$16,889,318	EC$41,365,645
1999	EC$10,314,371	EC$40,770,465
2000	EC$ 5,315,435	EC$35,862,752
Total	EC$82,261,130	EC$259,232,440

Table 24–1. Japan's trade balance with Dominica compared with Dominica's exports to the United Kingdom, 1995-2000, inclusive.

Over the years, Dominica has quietly demonstrated that the sensible management of natural systems has allowed us to generate significant economic benefit. In 1999, EC$131 million was earned from nature tourism as people came to Dominica from many other countries primarily to hike, dive, snorkel and just relax in a setting of exquisite, natural peace and quiet.

Incidentally, many of these visitors came from the same countries that are supporting the establishment of the South Pacific Sanctuary for whales (Britain, France, Germany, and the United States). It is also true that the people who buy and consume our main export, bananas, come from Britain, a country that believes that there are strong arguments in favour of the establishment of a whale sanctuary in the South Pacific.

On the other hand, Japan and Norway purchase precious little from Dominica (in the year 2000, EC$15,356 for Japan and EC$0 for Norway as compared, for example, with EC$36,063,973 for UK). In that same year, we transferred EC$24,987,126 to Japan to purchase cars and other depreciating assets. If we allowed ourselves for a moment, to determine our position on this and other international issues, based on raw economic benefit, the trade data suggest that our interests lie in cooperation and collaboration with countries like the UK with whom we have favourable trade balances and not with the manufacturers and exporters of Japan.

Data from the Central Statistical Office of the Government of Dominica indicate that Japan has enjoyed a favourable trade balance with Dominica over the period 1995-2000 (Table 24–1). Dominican exports to the United Kingdom, obtained from the same source, also prove an indication as to where Dominica obtains the money to purchase Japanese products (Table 24–1). What then is the explanation for our position on whale sanctuaries and marine conservation at the IWC?

THE POWER CIRCLE REVISITED

I return to an earlier point. Some argue that Dominica has voted the way we have because it has been in the country's economic interest. The trade data with Japan does not support this claim. The logic behind this claim seems to be that in the new globalized world economy, small, poor countries must be prepared to "take what we can get even if it means trading off the one thing which gives us equal rank with other countries of the world, our vote at international fora".

Not only does this make a mockery of the "sovereignty" argument, but, we can ask, "What have we got for 12 years of voting at the IWC?" How has it benefited Dominica to take positions at the IWC that are directly counter to the island's trading interests and our tradition and track record of the development and management of protected areas for people, plants and animals?

Let us examine the facts. In the mid-nineties, Japanese aid allowed the construction of a fisheries complex in the constituency held by the then sitting Prime Minister, Hon. Mary Eugenia Charles. Japanese aid for the construction of a second fisheries complex was promised to Prime Minister Edison James in the late nineties. That facility was to be built in Mr. James' constituency. That facility was never built. Japanese aid was then promised to the late Prime Minister Roosevelt Douglas. The facility was to be built in Mr. Douglas' constituency.

There is a pattern here of aid and the promise of aid, for projects that move around depending on the location of the Prime Minister's constituency and not according to any reasoned plan for the development of the fisheries sector. I recall that in 2000 when the Japanese Ambassador to Dominica came to see me as then Minister of Agriculture, I shared with him the details of the plans for development of Dominica's fisheries sector. That plan had been the subject of years of preparation by communities, NGOs, technicians and planners at the national and OECS regional level and was a full section of the proposed OECS Development Strategy. During the time that I was sharing this information, the Ambassador was looking out the window of my office. When I was through, he looked at me and said, "Fisheries Complex!" He had come to me having already decided what it was that Dominica needed. That was the last I saw of the Ambassador from Japan. Quite apart from the obvious political implications of a foreign Government wanting to decide for another Government what were its development needs, let us examine this pattern of aid giving from the point of view of net economic impact on the recipient country.

The one fisheries complex that has been built in Roseau cost US$18.6 million (approximately EC$45 million. Half of the building is devoted to office and meeting space for the 20 members of staff of the Fisheries Division. The other half is devoted to the provision of cold storage space that has been rented out to private importers of frozen products and, retail space for the sale of fish. The value of that complex is approximately equivalent to half of Dominica's trade deficit with Japan over the period 1995 to 2000. Put another way, that is the cost of 1800 vehicles at an average price of EC$25,000. Where does the money come from to purchase these goods from Japan? We have already established that it comes from trade with Britain, a country that was convinced that the scientific data available at the IWC supported the establishment of the whale sanctuary in the South Pacific.

The question is: how has Dominica benefited from persisting with a voting pattern at the IWC that is in direct contradiction to our clear trading and economic interests and in contradiction to our tradition and strategy of protection, conservation and management of natural areas on land and in the sea?

One can surmise that prior to the emergence of the WTO, special agreements between Britain and her former colonies facilitated special protected trade relations. Since the WTO, these protected arrangements are no longer sacred. The fate of bananas and the challenges to the offshore financial sector reveal the impact of the new world economic order on small economies like Dominica. It is not reasonable to expect that Britain or any other traditional trading partner will risk further WTO-driven trade retaliation in order to protect us.

An equally important question is: how has Dominica earned the foreign exchange that has been used to purchase exports from Japan and other countries? The answer is by using the island's natural resources to produce agricultural goods and a unique tourism experience.

CONSERVATION, AGRICULTURE AND TOURISM

In recent years, the largest share of Dominica's foreign exchange has been earned from agriculture and tourism, the latter most directly the result of the practice of creating and managing parks, reserves and protected areas. To put these economic issues in an even more current perspective for Dominica, in 1999, the country earned EC$45,533,187 from export of bananas and other commodities and, EC$131 million from tourism. That is, in one year, banana earnings equalled the value of the Roseau fisheries complex and tourism earnings were three times the value of the fisheries complex.

Eighty-five percent of the money earned by tourism was directly attributable to stay-over tourism, or the visitors who come to Dominica precisely to enjoy the natural environment of the island that is directly linked to more than 50 years of conservation and protection policies and programs. Some 2000 Dominicans were directly employed by the tourism sector while close to 2000 farmers were on record as selling bananas to the Dominica Banana Marketing Corporation (DBMC) in that year.

Where do the economic interests of the people of Dominica and the Dominican economy lie? Is it in more fisheries complexes or in the diversifying and strengthening of the agriculture and tourism sectors, both of which will generate demand for the goods and services of all other sectors?

Who benefits from the nature tourism product? The tour companies, taxi operators, dive and whale watch operators, tour guides, hotel and restaurant owners and workers, musicians, vendors, club and bar owners as well as the Government Treasury are all beneficiaries of this nature tourism. If we can resolve the difficulties facing the banana sector and the manufacturing sector, while we strengthen and extend the tourism sector, Dominica would be well on its way to sustainable prosperity.

SILENT TRUTH

There is a silent truth that needs to be shared. There are many thousands of Dominicans who, without being aware of it, benefit today from the practice of protection and conservation started over 50 years ago. Witness the fact that as many other Caribbean countries have suffered from the severe effects of drought this year, Dominica though affected, has not experienced the acute crisis that has plagued some of our neighbours.

This legacy of being concerned about our environment and protecting our plants and wildlife habitats has served Dominica well. As Dominicans, we have to think long and hard about how much aid it would take to make us turn away from that path of sustainability. The legacy of the Nature Island is not for sale.

There is another silent truth that we must face up to. The European Union (EU), not Japan, is the largest external aid donor to Dominica, the African, Caribbean and Pacific States. Between 1991 and 2001 Dominica received EC$134,686,700 from the EU, compared with EC$40,842,600 from Japan.

EU assistance, like all aid, is also tied to conditions. These conditions, however, do not require us to turn away from policies and practices that are tried and proven. The aid is not given on condition that we support the policies of the countries of the European Union. The aid is available under terms and conditions that Dominica was a part of the negotiating. Although there is growing evidence that this situation with the EU is changing, this a very different arrangement compared to what we are required to do at the IWC in order to get Japanese aid.

HARSH REALITIES

On the 27th July 2001, the closing day of the IWC meeting for that year, Dominica participated in a major three-day Caribbean trade, investment and tourism promotion activity – CARIBBEAN EXPO 2001 – organized by CARICOM and the Commonwealth Secretariat. The theme selected for the Dominica booth at that event was "Dominica, the Nature Island".

The same travelers, tour operators, international agencies and media that had just witnessed our vote at the IWC against wildlife conservation, were now encouraged to continue to trade with us, to invest in our economy, and to book their next holiday in Dominica, the Nature Island.

In business, one of the cardinal rules is that the customer is always right. Are we ready for customers who are convinced that the right thing to do is to support the establishment of sanctuaries for whales and other species?

TOUGH CHOICES

Ultimately, the choice is ours. We can abandon our proud legacy of nature protection and conservation that has left us with a real chance for a sustainable future, or we can find partners who identify with and respect our own traditions of natural resource conservation through protection, to strengthen our nature-based tourism that has strong links with a new, clean agriculture, health care, education, research, and recreation, for Dominicans and visitors alike.

As we do all this, we can really test the extent of external support for Dominica and determine once and for all if the promise of aid is linked to our vote or if support for our development is based on mutual respect and a genuine desire to help Dominicans develop our island as we determine.

Remember, even as we have voted against Britain, France and others at the IWC, they have not stopped aid to Dominica. Remember too that the changed world economic order suggests that this arrangement is not likely to continue. By the same token, can we expect Japanese aid to Dominica to arrive if Dominica votes with the majority at the IWC in support of the South Pacific Whale Sanctuary? This is a test of the basis for the relations between our two nations.

VISION AND LEGACY

These are hard times but they cannot be hard enough to reduce the legacy of this great little country to a vote at the IWC in return for two more fisheries buildings. Dominica is bigger than that. Dominica is better than that. Dominica deserves more than that. Dominica is not for sale, not now, not ever!

LESSONS LEARNED

Notwithstanding our accession to the St. Georges Declaration[5] or the OECS Development Charter, or the many other Regional and International accords to which we are signatory, the definition and implementation of an effective strategy for wildlife conservation depends ultimately on the national resolve at the household, community, enterprise and government levels.

The clearest means of expression of that resolve is within the framework of an Integrated Development Plan (IDP) that has been formulated with the full engagement of all segments of the local population, reflects the aspirations and commitment of the people, and is theirs. An Integrated Development Plan is defined in Chapter 8 of Agenda 21 as: "a policy/action framework that reflects a long term perspective and cross-sectoral approach as the basis for decisions, taking account of the linkages between and within the various political, economic, social and environmental issues involved in the development process" (Section 8.4(b)).

The IDP is also defined as "providing strategic guidance to lower order plans, such as economic and physical plans as well as sector/facet plans... thus ensuring that the broad goals and objectives of each plan, at whatever level in the tiered system, will be in conformity with the guiding principles enunciated in the Integrated Development Plan". This means that economic plans no longer give direction to the management of natural systems, for example, but respond to the holistic analysis of the value of natural systems to the environmental, social, and economic well-being of the people of the nation.

In January 2003 Dominica completed the first Integrated Development Plan prepared for an entire country. Along with the recently completed National Biodiversity Strategy, the IDP places Dominica in the forefront of countries with an overarching framework for broad-based conservation and development. The Principles applied in the preparation of the IDP were:

- **Ownership**: Developing commitment to the policy/programme/project based on opportunities for active involvement in the process of shaping the vision, setting the goals, designing the mechanisms, timing and sequencing the actions for implementation of decisions.
- **Decentralization**: Opening up opportunities and providing the capacity for local and national organizations to share the responsibility for management of social and economic programmes as well as geographic spaces.
- **Governance**: Initiating processes and procedures of public administration that are responsive to citizens and residents.
- **Transparency**: Ensuring openness in public and private administration so that all information that is vital to decision-making is reliable and available in a timely manner and in a form that is useable by all stakeholders.
- **Participation**: Providing an active role within the planning and decision-making process for all stakeholders including the opportunities for women, youth, indigenous people, and other marginalized groups.
- **Partnership**: Engaging informal and formal mechanisms of cooperation, coordination and collaboration that focus on agreed goals in ways that reduce suspicion and distrust between stakeholders and enhance progress towards sustainable development.
- **Accountability**: Ensuring that persons in positions of trust and responsibility are required to account for the decisions they make and the resources under their care.
- **Cross-sectoral**: Taking account of impact of policies, plans and actions in one sector on other sectors.
- **Cross-cutting**: Addressing stakeholders' interests which cut across the conventional definitions of economic and social sectors and which require an holistic approach.

These principles are not new and like most things of real value, they have been around for such a long time that most people have forgotten how to use them. These principles are, nonetheless, the core of the package of mechanisms that are needed to make a difference in our approach to wildlife conservation and to development in general. Put another way, the changes that are needed are not in the science or the technology but in the attitude, and approach to the design and management of our programs.

A recent study on integrated coastal zone management in the Caribbean[6] drew the following lessons for the design and implementation of mechanisms of integrated planning to wildlife conservation:

- When planning is not participatory or has been separated from management and vision, strong partnerships among co-management stakeholders is less likely.
- An IDP approach builds trust, capacity, respect and legitimacy of the process and content of the plan.
- Once engaged, the IDP must genuinely consider and use the input of stakeholders in order to be credible.
- The plan should be endorsed politically and legally.
- Prior to implementation, the IDP approach should be widely publicized and disseminated for it to be actively adopted by all concerned.
- The IDP process must become standard operating procedure if it is to be institutionalized.
- Building collaboration for co-management takes time, is complex, is possible and is essential for sustainability and success.

These lessons are echoed in many of the experiences of integrated and participatory approaches to development actions taken by communities in the Caribbean and other parts of the developing world.

In preparing the Integrated Development Plan for Dominica, the TDI Team found it helpful to consolidate the IDP approach into five cross-cutting clusters for action:

1. *Link social and economic development with the environment*
   - Production and consumption activities must allow for maintenance of natural life support systems.
   - Utilizing information based on a Natural Resource Inventory.
   - Utilizing more solar, wind, water and geothermal energy sources.
   - Popularizing new skills in preventive health care.

- Opening up centers of "Sustainable Living" for residents and visitors.

2. *Address inter-generational poverty*
   - Promote job creation in agriculture, health, tourism and other sectors.
   - Sustain domestic demand by targeting poverty in rural and urban areas.
   - Build coalitions to develop neighbourhood and local area concept plans and provide a basis for local enterprises.
   - Promote social investment funding.

3. *Focus on human development*
   - Human development cuts across the concepts of wellness, skills, capabilities, employment and incomes.
   - Utilize participatory and consultative approaches to focus health services on geographic equity.
   - Use improvements in technology to support student learning.
   - Allow local communities to adjust local school curriculum.
   - Establish a "Rural Industry Support Fund" to support physical asset development in communities.

4. *Decentralization*
   - IDP defines decentralization as the transfer of responsibility to democratically independent local levels of governance giving people more managerial discretion.
   - Strengthen partnerships to share responsibility for education curriculum and health facilities.
   - Achieve accountability through inter-agency and department cooperation in Ministries.
   - Create multi-sectoral and multi-enterprise zones for agricultural expansion.
   - Provide information and incentives for community-based tourism.

5. *Manage information for local empowerment*

The IDP identifies 5 important information systems that require immediate attention.

- Information organized into a "Sustainability Frame of Reference".
- Information organized into a "Structural and Functional Frames of Reference".
- Information System to support Human Development Assessment.
- Natural Resource Inventory System.
- Disaster Management Information System.

CONCLUSION

In the opening section of their remarkable book, *Sharing Nature's Interest: Ecological Footprints as an Indicator of Sustainability*, Chambers, Simmons and Wackernagle[7] further distil the advice on guidelines and mechanisms for tackling barriers to sustainable living as follows:

1. Make the links between human quality of life and a healthy environment.
2. Ensure that the needs of all people are met.
3. Do not cheat on your children.

I end this chapter with the following advice from C.S. Lewis on how we can continue this journey for a life of equity and quality for all God's creatures:

> Progress means getting nearer to the place you want to be. And if you take a wrong turning, then to go forward does not get you any nearer. If you are on the wrong road, progress means doing an about face and walking back to the right road, and in that case the man who turns back the soonest is the most progressive man.

ACKNOWLEDGEMENTS

On behalf of the members and the Board of the Caribbean Conservation Association (CCA), I wish to thank IFAW for conceiving of this Forum. I know IFAW well enough to appreciate that it took long months of preparation to get us all to Limerick in the hope that one of us would say something that was sensible. At least sensible enough to give IFAW and others a sharpened focus on the increasingly challenging issue of conserving plant and animal species at a time when the lives of millions of the human species are endangered. It is a tribute to the animals and plants that the human species is not as endangered by them as they are by us, *Homo sapiens*, a real misnomer in the circumstances.

In addition to the personal feeling of partnership that I feel with IFAW's CEO, Fred O'Regan, and with Kelvin Alie and the many others at IFAW, as President of the CCA, the largest environmental organization in the Caribbean, I convey to our partner, IFAW, warm greetings from the Board and Members of the CCA. The CCA is a unique organization in which Governments, NGOs, and individuals sit together as members, on Committees, on the Board and in the Plenary at the Annual General

Meeting determining ways to conserve, protect and manage the fragile natural resources of the Caribbean.

IFAW and CCA have, over the last few years, nurtured a functional partnership into a formal relationship that I believe is a living example of a holistic, collaborative approach to conservation. This is one of the elements of the new architecture for wildlife conservation that I believe deserves to be explored, expanded and institutionalized as part of the toolkit for wildlife conservation in this century.

NOTES AND SOURCES

[1] For further discussions and critiques of the concept of sustainable development, see Beder, S. 1996. *The Nature of Sustainable Development.* Second Edition. Newham, Australia: Scribe Publications; Chesworth, W., M.R. Moss, and V.G. Thomas. (eds.). 2002. *Sustainable Development: Mandate or Mantra.* The Kenneth Hammond Lectures on Environment, Energy and Resources 2001 Series.Guelph: Faculty of Environmental Sciences, University of Guelph. Also see Brooks, Chapter 16.

[2] For more on whales, whaling, and the International Whaling Commission, see Papastavrou and Cooke, Chapter 7.

[3] For more on whale watching, see Corkeron, Chapter 11.

[4] Agenda 21, Principle 15.

[5] OECS. 2001. The St. George's Declaration of Principles for Environmental Sustainability in the OECS. OECS Secretariat. Castries, St. Lucia.

[6] McConney, P., Pomeroy, R., Mahon, R. 2003. Guidelines for coastal resource co-management in the Caribbean: communicating the concepts and conditions that favour success. CCA, CERMES, MRAG, NRSP.

[7] Chambers, N., C. Simmons, and M. Wackernagel. 2000. Sharing Nature's Interests: ecological footprints as an indicator of sustainability. London, UK: Earthscan, UK.

CHAPTER 25

SAVING THE PLANET TO DEATH: THE NEED TO REFORM THE MOST IMPORTANT BARRIER TO ECOLOGICAL SUSTAINABLITY

Stephen Best

CONCLUSION

> *To will implies delay, therefore now do.*
>
> John Donne

The deadliest issue facing humankind today is our failure to achieve an ecologically sustainable global economy. That failure is killing more of us now than any other threat, and is wrecking any hopes or dreams we might wish for our children, our grandchildren, and the generations of children to come. It is burying our future.

The first and most significant barrier to our achieving ecological sustainability is not the "usual suspects" often identified by the environmental and animal protection movement: industry, government, globalization, nationalism, capitalism, consumerism, nor any of the other "isms". The barrier now is those who would save our planet: the major environmental and animal protection non-governmental organizations (ENGOs) – or, "the Greens" – and, in particular, the caring, dedicated, and knowledgeable men and women who manage and govern them.

The revolutionary political, economic, and social changes necessary to achieve goals on the order demanded by ecological sustainability happen only when sufficient coercive power is applied to those who benefit from and defend the *status quo*. The necessary power exists in the environmental and animal protection movement. However, of the hundreds of thousands of organizations struggling to protect the environment and animals, fewer than a thousand control the efforts of almost three-quarters of the hundreds of millions of people who actively support environmental and animal protection causes. These major ENGOs are in turn managed and governed by just a few thousand men and women. Thus, a few thousand people control the political effectiveness of hundreds of millions.

The millions of environmental and animal protection supporters controlled by the major ENGOs are the key to humankind achieving ecological sustainability. These supporters are the only people on the planet who might be made to appreciate and respond effectively to the dire urgency of the current situation *and* who have the necessary resources – the necessary power – to compel corporations and governments to make the crucial reforms needed for us to achieve ecological sustainability. However, because of the major ENGO's self-interest, corporate conservativeness, and comfortable complacency, this power has never been used to its fullest. Even when the major ENGOs have gingerly and gently applied the power of these many millions of people, it has had no adequate effect on achieving ecological sustainability.

Consequently, if the people who govern and manage the Greens cannot act with the urgency that the current crisis demands, if they cannot reform themselves, and if they cannot apply all the power they hold hostage within their hundreds of millions of supporters, then ecological sustainability cannot be achieved; human suffering and dying will escalate. If the men and women who manage and govern the major ENGOs will not themselves make the revolutionary changes within their own organizations necessary to achieve ecological sustainability, then there is little reason to expect that anyone else will.

To conclude, the few thousand men and women who manage and govern the major Green organizations are doing little more today than saving the planet to death. They are now the first and most important barrier to achieving ecological sustainability. It is they – not the public, corporations or governments – who must be petitioned, pleaded with, and convinced to overcome themselves and change. It is they who must be convinced to ask themselves, "If not us, who? If not now, when?" It is they who must be convinced to act. They are the first barrier and our last hope for achieving ecological sustainability.

INTRODUCTION

> *Apparent agreement masks the fight over what exactly "sustainable development" should mean – a fight in which the stakes are very high.*
>
> Herman E. Daly

Today, the overarching threat to our lives and civilization is our persistent failure to establish an ecologically sustainable global economy and culture. Ecological sustainability is – and always has been – the foundation issue for human communities and civilizations. Regardless of one's hopes, without the necessary resources and security to realize them, they are moot. Everything begins with the environment. Exceed the carrying capacity of the environment and, ulitmately, civilization collapses. Yet because of wonderful technologies and fossil fuel-based energy, the *Zeitgeist* is that our civilization – unlike that of the Mayans and others[1] – is immune from collapse. Nothing could be further from the truth.

More people suffer and die from the direct and indirect causes of our continuing failure to establish an ecologically sustainable global human community than from any other cause.[2] This threat, and its dire consequences, diminish every person's life. It is likely that you, the reader, are either personally suffering or have a friend or family member who is suffering or who has died because we – as individuals, as a species, and as a civilization – refuse or are unable to do what is necessary to achieve ecological sustainability. It is also likely that your personal response to this assault on you has not been sufficient to decrease the threat. The reason is our evolutionary shortcoming as a species inhabiting a modern, fuel-based, and technological environment of our own creation.[3] If you or your family is killed slowly enough, you are incapable of a commensurate emotional response. Killing you slowly – over years rather than a day – does not invoke in you the fight or flight response you need to survive. Perversely, not only will you suffer or die, but also it is likely you will aid and abet your pain or premature demise in order to achieve your short-term desires. If we want to survive with some semblance of civilization, we will have to rely on our intellect, not our emotions. That will be very difficult. We are not – as we like to delude ourselves – a thinking animal that feels, we are a feeling animal that thinks. As one political operative observed, "Elections are won and lost on emotion, not logic".[4]

Here is the reality check on our attempts to achieve ecological sustainability. Every year, governments around the world spend billions of dollars implementing policies to protect animals and the environment. Thousands of responsible corporations invest heavily to reform their environmentally harmful practices. Thousands of scien tists work to find new answers to ecological questions. Hundreds of thousands of ENGOs manage untold numbers of projects to save the environment. Hundreds of millions of people contribute tens of billions of dollars to the ENGOs, and sign thousands of petitions and mail millions of protest letters.

The result? For the last forty years, this annual activity has failed to reverse the decline in environmental quality that continues to harm and kill more and more men, women, and children in every country every year. Despite all the evidence, despite the suffering, and despite the deaths, the dire consequences of our failure to achieve ecological sustainability continue to increase. As the United Nations World Summit on Sustainable Development concluded in 2002,

> The global environment continues to suffer. Loss of biodiversity continues, fish stocks continue to be depleted, desertification claims more and more fertile land, the adverse effects of climate change are already evident, natural disasters are more frequent and more devastating, and developing countries more vulnerable, and air, water and marine pollution continue to rob millions of a decent life.[5]

The suffering and deaths of humans and non-humans alike increase despite the fact that we are aware of what we must do to save our lives and the ecological foundation that supports us. Not only do we know what must be done, we also have the necessary scientific and technical skills, and economic and natural resources to do it. Some of us – no more than a few thousand – even have the political and economic power to make it happen.

Given our widely acknowledged and well-documented failure to achieve – or even begin to approach – ecological sustainability, the only rational conclusion is that what we have been doing to protect animals and the environment these many decades simply does not work. If it did work, environmental quality would be improving. The only rational course is to abandon what we have spent four decades proving fails, and often exacerbates the situation, and use strategies that history has repeatedly shown will actually make a difference.

In *The Challenge of Sustainability*, Mohamed T. El-Ashry, the Chief Executive Officer and Chairman of the Global Environment Facility (GEF) asks, "So, what will it take to protect our biological heritage, avoid the devastation that climate change could bring, sustain the soil and water that give us life, protect human health, and reduce the scourge of poverty and hunger?" Answering his own question El-Ashry writes, "It will take leaders from all walks of life who are willing to think and act differently and lead the way".[6] El-Ashry's answer is the stuff of social revolution, which is exactly what is required to achieve ecological sustainability. Nothing less will do.

Think differently. Act differently. Lead the way. The most important is "Lead the way". Leading the way does not mean courting public opinion; educating the public, the media, or government officials; presenting scientific findings; leveling criticism; suggesting what must be done and when; peacefully demonstrating; forming partnerships with stakeholders; rescuing a few wild animals; restoring a "unique" ecosystem of special interest; signing petitions; or mailing protest letters. We know that none of this is enough. It never will be enough to cause the political, social, and economic change on the scale necessary for humankind to achieve ecological sustainability. Leading the way means making revolutionary changes happen by acquiring and applying power. As Bertrand Russell noted, "Those whose love of power is not strong are unlikely to have much influence on the course of events. The men who cause social changes are, as a rule, men who strongly desire to do so[7]...The ultimate aim of those who have power (and we all have some) should be to promote social cooperation, not in one group against another, but in the whole human race".[8]

To achieve an ecologically sustainable global community a qualitatively different type – a revolutionary type – of leadership is required. We need leaders who understand the challenges of ecological sustainability, who are willing and have the competency to acquire or to access the necessary power, and who have the skill to apply power effectively in order to compel social and economic change of a revolutionary order. It is unlikely they will come "from all walks of life". Where these leaders might be found is among those who populate the boardrooms and the executive suites of the major environmental and animal protection ENGOs. There are only a few of them, numbering in the hundreds or low thousands. Their names and mailing addresses are a matter of public record. These people already have the power necessary to influence the course of events. Among them are some of the most knowledgeable about the need to achieve ecological sustainability. However, achieving ecological sustainability – without incurring vast human suffering and massive environmental degradation – will not be possible unless among those who now govern and manage the major ENGOs, leaders are found who "are willing to think and act differently and lead the way," something the major ENGOs have failed to do, with few sporadic exceptions, for decades.[9]

This chapter began with its conclusion so that it will remain firmly top of mind as the premises and arguments that support it are presented. The reason is that the conclusion may seem simplistic and absurd, even immoral, and perhaps, for some of those implicated, disrespectful and offensive. The conclusion implies that most of the people now governing, managing and employed by the ENGOs must make changes in their own thinking and actions of a magnitude no less than that which they have been scolding the rest of the world to make.

The conclusion – and what it entails for the major ENGOs – also serves as a predictor of our odds of achieving ecological sustainability. If the people who govern and manage the Greens are themselves unable or unwilling to make the necessary changes to achieve ecological sustainability, then it is highly unlikely that anyone else will, or even can. Therefore, how the Greens respond to what is actually required of them will be a strong indicator of whether or not ecological sustainability can be achieved at all.

The conclusion is predicated on three fundamental premises. All are contentious, as are most aspects of the pursuit of ecological sustainability. The first is the urgent necessity for the global human community to achieve an ecologically sustainable economy[10] to avoid a future plagued by suffering, despair, and death. Responsible independent authorities and agencies, and most ENGOs generally accept this premise. However, others are campaigning vehemently and, unfortunately, successfully against this view.[11] The others are winning.

The second premise is the optimistic assumption that it is possible for the global human community to make the necessary political, social, and economic changes necessary to achieve ecological sustainability. Economist Herman Daly suggests that the "technical and economic problems involved in achieving sustainability are not that difficult".[12] Nonetheless, the necessary changes are of such a revolutionary nature[13] that many informed and thoughtful people doubt that we, as a species, are ethically and intellectually capable of making the changes.[14] Many also doubt there is sufficient time – even if the changes are made – to reverse the "sixth great extinction".[15]

The third premise is that the kind of revolutionary social, economic and political changes necessary to achieve ecological sustainability can only come about by the deliberate political actions of a few individuals, not by some hoped-for spontaneous elevation of global environmental consciousness and responsibility. Five centuries ago Niccolò Machiavelli observed that "One should take it as a general rule that rarely, if ever, does it happen that a state... is either well-ordered at the outset or radically transformed vis-à-vis its old institutions unless this is done by one person".[16] Machiavelli's observation was based on the history of the Roman Empire, a millennium before him. We have seen this general rule borne out time and time again since. The human catalyst for revolution is, if not "one," always just a few.

Saving the Planet to Death concludes that the revolutionary changes needed to achieve ecological sustainability will come about only if out of the ranks of those people who manage and govern the major ENGOs there emerge leaders who will be the catalyst for what must be done. However, in order for effective leaders to emerge and force the necessary changes, the management and governance practices, policies, and programs of the major environmental and animal protection ENGOs must change first. Only then can the organizations do what they have failed to do for these many decades: effectively exert to the maximum effect possible all of the political and economic power latent in their supporter bases. Indeed, so egregious and culpable are the past and continuing failures of the governors and managers of the major ENGOs to exert the full force of their latent political and economic power that they, as a group, now stand as the leading impediment to achieving ecological sustainability. They are more to blame for our failure to move towards ecological sustainability than are industry and government.

If the powerful, highly informed people who control the major ENGOs do not believe that the danger of failing to achieve ecological sustainability is serious enough to warrant themselves changing – and exerting all of the power in their memberships to save and rehabilitate the ecological systems that support humankind – then one might forgive anyone else who mimics their timidity and complacency and refuses to radically change his or her ways.

THINKING DIFFERENTLY

After all else has failed, men turn to reason.

Abba Eban, 1967

Think differently. Act differently. Lead the way. Acting differently and leading the way begins with thinking differently. Thinking differently does not require novel ideas. Human history is long, broad, and deep with experience. Humanity's collected wisdom fills libraries where one can find practical answers – political, economic, ethical, scientific, and technical – to the dilemmas we face, including the problem of achieving ecological sustainability. New thinking is not required. What is required, first, is that we reject current thinking that, while self-serving, comfortable, and entrenched, demonstrably is not producing the necessary solutions. Second, our thinking about ecological sustainability must be commensurate with the scale of the issue. And third, with the magnitude of the issue in mind, we must embrace ideas and concepts that have proven their efficacy on the scale required to develop and implement solutions. As to the question, "Who must think differently?" the answer is not the "public" or the royal "we" or even the ubiquitous "they," but rather those who have the power to compel the necessary changes, can choose to act differently, and could lead the way: the "Who" is the men and women who govern and manage the major ENGOs.

To think differently about achieving ecological sustainability, it is first necessary to admit that what is being done now by government, industry, non-governmental organizations, and individuals – while some of the efforts are undoubtedly necessary – is insufficient to achieve ecological sustainability. Just a few examples of the current thinking that has proven inadequate to answer the challenge of achieving ecological sustainability will illustrate this fact.

But first, a caveat. The environmental movement, including the major ENGOs, deserves much credit for some notable successes in a few countries: for example, the Montreal Protocol on acid rain, DDT bans, emission controls on motor vehicles, the elimination of lead in gasoline, and the protection of marine mammals. As laudable and necessary as these successes were, tragically, they and all the other environmental victories since have not offset the general decline in global environmental quality.

The 2003 World Wildlife Fund-Canada (WWF-Canada) annual report, *Our first 35 years were a dress rehearsal*, is the first example of old and deficient thinking about the ecological threats facing us. WWF-Canada is one of the major and most prominent conservation groups in Canada. They enjoy close, financially beneficial relationships with industry and Canada's various governments. Their 2003 annual report is, in fact, a fund raising vehicle, so it may seem unfair to present it as evidence of WWF-Canada's faulty thinking. However, a comparison between WWF-Canada's history and the projects it funds, and what is argued in *Our first 35 years were a dress rehearsal,* suggests that the publication is an accurate statement of how the executive and board of WWF-Canada think about solving the issue of ecological sustainability. Consequently, *Our first 35 years...* must be what the WWF-Canada's executive and board want their contributors to believe about saving the planet.

Page two and three of the report is a four-colour double spread. The background is a picture of a caribou migration. Overlaid on the picture is the bold headline "Now It's Showtime". A text line across the bottom of the page reads,

> Saving nature one hectare, one river, one inlet at a time isn't fast enough. So WWF is scaling up. We're taking the skills we've honed over the last four decades and we're going after the big stuff. Conservation can't wait. Neither can we.

The annual report then goes on to describe new campaigns that are not qualitatively different from what the WWF has been doing over the course of its history in Canada. Under the headline on page four, "Center Stage," three senior corporate officers write,

> Our sights are set on something breathtaking. It's conservation on a grand scale. It's conservation that will make the world sit up and take notice. It's what is truly needed.
>
> What makes us think we can do it? WWF knows how to collaborate. We know how to work with Aboriginal people, business, government, and other conservation groups to save nature. We've proven it time and time again.

The facts contradict WWF-Canada's claim, notwithstanding that they tell their supporters that they have proven "time and time again" the efficacy of their approach to saving nature. Canada's environmental quality is getting worse, not better. It has been getting worse since at least 1970. In 1995, the Washington-based National Center for Economic Alternatives issued a report that documented a 38.1% decline in Canada's environmental quality between 1970 and 1990[17] – two decades during WWF-Canada's self-proclaimed "dress rehearsal".

More recently, in 2004, Canada's Commissioner of the Environment and Sustainable Development, Johanne Gélinas, reported,

> ...I am concerned at signs that Canada's environmental status and reputation may be slipping. For example, the Conference Board of Canada rated the relative performance of 23 member countries of the Organisation for Economic Cooperation and Development (OECD) on a range of environmental issues, using OECD data. On that basis, Canada's overall environmental performance was downgraded from an already disappointing twelfth-place ranking in 2002 to sixteenth in 2003. Pollsters have also noted a decline in Canadians' confidence that their country is showing strong leadership on world environmental issues.[18]

From 1970 to 2004, the empirical evidence is that whatever the WWF-Canada and all of the other Canadian conservation, environmental protection, and animal groups had been doing for 35 years to save nature was not sufficient to halt, let alone reverse, the degradation of Canada's environment. Yet WWF-Canada is proposing doing nothing qualitatively different for the next thirty-five years, just more of the same. Thirty-five years of environmental degradation has not changed how WWF-Canada thinks about achieving ecological sustainability. The best that it can suggest is a quantitative change in its activities. Since 1970 the environmental movement has grown by 5000%.[19] Yet, despite this quantitative growth in members and revenue, WWF-Canada's solution to Canada's degrading environment is more of its own growth. There is no hint in *Our first 35 years were a dress rehearsal* of what is truly required: a revolutionary change in thinking.

"WWF-Think" is not unique. Most of the world's major ENGOs have not progressed beyond the 1980s in how they think about solving environmental and animal protection and ecological sustainability issues.

The Environmental Defence Fund (EDF), a Washington-based NGO, provides a second example of poor thinking. The EDF published a calendar in 1996 that offered a 10-point program to "Save the Earth" (EDF's phrase):

1. Visit and help support our national parks.
2. Recycle newspapers, glass, plastic and aluminum.
3. Conserve energy and use energy efficient lighting.
4. Keep tires properly inflated to improve gas

mileage and extend tire life.

5. Plant trees.
6. Organize a Christmas tree recycling program in your community.
7. Find an alternative to chemical pesticides for your lawn.
8. Purchase only those brands of tuna marked 'dolphin-safe'.
9. Organize a community group to clean up a local stream, highway, park, or beach.
10. Become a member of EDF.

As Peter Montague, the editor of *Rachel's Environment & Health News*, commented:

> What I notice here is the complete absence of any ideas commensurate with the size and nature of the problems faced by the world's environment. I'm not against recycling Christmas trees – if you MUST have one – but who can believe that recycling Christmas trees – or supporting EDF as it works overtime to amend and re-amend the Clean Air Act – is part of any serious effort to 'save the Earth?' I am forced to conclude once again that the mainstream environmental movement in the U.S. has run out of ideas and has no worthy vision.[20]

The EDF is not the only group to issue a ten-point program to save the Earth. The David Suzuki Foundation of Canada, which gives us a third example of inadequate strategic thinking, promotes its ten point "Nature Challenge: ten simple things you can do to protect the environment". Even if every Canadian took the Nature Challenge, it would be woefully insufficient to "protect the environment". Nevertheless, the David Suzuki Foundation says that its ten-point list comprises the "10 most effective ways we can help conserve nature and improve our quality of life". [21]

Lastly, and once again, the World Wildlife Fund – this time its international body based in Gland, Switzerland, now named the World Wide Fund for Nature (WWF) – provides a final example of thinking that cannot lead to solutions. In time for Christmas 2004, WWF-International earned the ridicule of Colin Isaacs, the editor of *The Gallon Environment Letter*, for a list it published called "Ten things not to buy for Christmas".[22] The planet-saving list included such items as Beluga caviar in the number one spot and a Shahtoosh at number five. A Shahtoosh is a high fashion scarf woven from the hair of the Tibetan antelope. As Isaacs laments, "Christmas in Panda Land... Maybe WWF should get its head out of Harrods and start looking at some of the environmentally damaging things that ordinary people buy! Shahtoosh, indeed!"[23]

More troubling is the quality of information that the EDF, the David Suzuki Foundation, the WWF and others are giving their supporters: information that implicitly trivializes the issues and understates by orders of magnitude what is, in fact, required to "save the Earth". What people believe – which forms the basis for their actions – is wholly decided by the information they receive.[24] When environmental and animal protection groups give their members misleading information – no matter how conducive to marketing their issue-based wares, raising funds, or sheltering their members from the full enormity of an unpalatable truth – they reduce their supporters' political effectiveness.

The consequence of this poor thinking is that today in North America the major environmental and animal protection groups, despite their size and the support they enjoy, are incapable of securing the passage of any major environmental legislation, such as that which was passed in the 1960s and 1970s. The movement is stronger in terms of supporters and donations, but weaker in effectiveness. This picture, with only a few notable exceptions such as, possibly, the German Greens, is much the same in the rest of world.

If the obvious can be admitted – which is that the way most of the major ENGOs have been thinking about achieving ecological sustainability for the last twenty-five years has not moved human civilization closer to the goal, although it may, at best, have retarded the rate of degradation – we can begin to look at the scale of the problem which, in turn, entails the scale of the necessary solutions. Only after that, can a discussion about acting differently actually begin.

It will not be necessary in this chapter, given its intended audience, to devote many words to the urgency of achieving ecological sustainability. However, it is worth reiterating the dire scale, the seeming intractableness, and the insidiousness of the problem to reemphasize that the current way of thinking about it has not been – and never can be – sufficient to implement the known solutions.

Consider, first, how long we have been aware of the need to achieve ecological sustainability by means other than suffering through uncontrolled catastrophic events. The horrific life and death consequences of human civilization not achieving ecological sustainability are well understood. In 1948, William Vogt warned in *Road to Survival* that:

> ...unless... man adjusts his way of living, in its fullest sense, to the imperatives imposed by the *limited* resources of his environment

> – we may as well give up all hope of continuing civilized life. Like Gadarene swine, we shall rush down a war-torn slope of barbarian existence in the blackened rubble.[25]

A year later, Aldo Leopold raised the alarm in *A Sand County Almanac*.[26] In 1962, Rachel Carson echoed Vogt, Leopold and others in *Silent Spring*.[27] In 1992, the Union of Concerned Scientists issued a *World's Scientists' Warning to Humanity*. Signed by "some 1,700 of the world's leading scientists, including the majority of Nobel laureates in the sciences," the warning said, in part:

> ... a great change in our stewardship of the Earth and the life on it is required, if vast human misery is to be avoided and our global home on this planet is not to be irretrievably mutilated.[28]

In 1997, the scientists reiterated their warning, and noted that since 1992 "progress has been woefully inadequate. Some of the most serious problems have worsened. Invaluable time has been squandered because so few *leaders* [my italics] have risen to the challenge".[29] In 2002, the Global Environment Facility (GEF) reported in *Challenge to Sustainability* that:

> We have, in the last decade, seen environmental problems mount – from extreme weather patterns and melting glaciers that point to a changing climate, to air and water pollution that threatens human health, to deforestation and land degradation that are undermining the Earth's capacity to sustain humanity.[30]

And in 2004, *The Living Planet Index 2004*,[31] an annual publication of the WWF documented the same bad news. In the foreword to *The Living Planet Index 2004*, the Director General of the WWF, Dr. Claude Martin, reported:

> Unfortunately, the news is not good. The [Living Planet Index] declined by about 40 per cent from 1970 to 2000, which represents a critical blow to the vitality and resilience of the world's natural systems. During the same period, humanity's Ecological Footprint grew to exceed the Earth's biological carrying capacity by 20 per cent... When we compare the current Ecological Footprint with the capacity of the Earth's life-supporting ecosystems, we must conclude that we no longer live within the sustainable limits of the planet.
>
> Ecosystems are suffering, the global climate is changing, and the further we continue down this path of unsustainable consumption and exploitation, the more difficult it will become to protect and restore the biodiversity that remains.

What is evident from this half-century long litany of warnings, failure, and continued environmental degradation is that whatever the various governmental and non-governmental environmental and animal protection organizations and industry were doing for the last half century to protect the environment, while possibly necessary, was not sufficient to move the global human community towards ecological sustainability. At best, as E. O. Wilson has observed,

> You could say that the rate at which [the environment is] being degraded has maybe been slowed a little bit as a result of the environmental awareness that we have. But it's continuing downhill.[32]

Why is that we can research and intellectualize the problem of ecological sustainability – even endure the catastrophic consequences – and not respond appropriately? For the human species it is a matter of time. Human beings can have their health destroyed and even be grotesquely killed with little complaint as long as it is done slowly. Kill them quickly – such as happened with the September 11, 2001 attack on the World Trade Center – and the response is emotionally charged, disproportionate, and worthy of overwhelming public support. Not to diminish the tragedy of September 11th, but the number of deaths was insignificant compared to the premature and preventable deaths in New York City caused every year by environmental problems.

We know that the deadly consequences of, and the solution to, achieving ecological sustainability occurs over a time span that tends not to excite an emotional response in most individuals and, therefore, not in government and industry. In April 2000, a representative from Environment Voters made an oral presentation to the Canadian government's Standing Senate Committee on Legal and Constitutional Affairs. The witness's testimony included:

> Imagine, if you will, that tomorrow a local chemical producing company accidentally releases a toxic plume into the air that settles on Ottawa and kills – over a period of a few days – 2,000 people! This would be a Bhopal scale event. The sheer horror of the disaster would trigger a massive and immediate response from emergency, medical,

law enforcement, and news services, and in the aftermath from political and legal agencies.

The deaths would be devastating for the families involved, and severe health consequences would be suffered by tens of thousands for years to come. The economic effects would be in the hundreds of millions, perhaps billions of dollars. Governments at all levels would take measures to insure that such a tragedy would never happen again. The offending company would likely never resume operations.

Contrast that with this.

In Ontario this year, almost 2,000 people *will die prematurely* because of poor air quality. The only differences from the fictional scenario described above will be that these people won't die over one weekend, they won't die in one place, and the blame won't fall on just one company. But these 2,000 real people will die just as painfully and their families will suffer just as much. Another difference between the reality and the fiction is that as of yet no government is [prepared] to take measures to insure that the real [deaths are not repeated year after year.]

The crime is that these 2,000 people are going to die needlessly. It's too late for them now, their fate has already been sealed, despite the fact that Canadian governments – particularly the federal government – have always had the legislative powers they needed to prevent their deaths. The reality is these people are going to die because our elected politicians and political parties have been compelled for perfectly [understandable] political reasons not to exercise their powers in a way that would have saved these peoples' lives.[33]

As it turned out the witness was dead wrong when he told the Senators about the need to kill 2,000 people to "trigger a massive and immediate response". Just one month later, the tainted-water tragedy in Walkerton, Ontario, showed that killing seven people quickly would be sufficient.[34]

While killing people quickly is a crime, killing them slowly is tolerated as a cost of doing business. Moreover, as long as people – including environmentalists – are killed slowly enough, they tend not to respond as if their lives are in danger, as if they were going to die tomorrow. Thinking differently – despite the fact that being killed slowly tends not to invoke in us an intense emotional response – requires that we force ourselves to respond appropriately, commensurate with the suffering and deaths that failure to achieve ecological sustainability is causing. Part of thinking differently, and therefore acting differently, is to consider solutions for achieving ecological sustainability in the context of a time frame that is relevant to the issue and not to the genetically hard-wired, short-term emotions of the human animal.

If we can accept that the way we have been thinking about the challenge and the enormity of achieving ecological sustainability has not been adequate to implement solutions, and if we can acknowledge and incorporate into our thinking the true scale of the problem, we can begin to understand the scale of the solutions and what must be done to realize them.

In considering the scale of the solutions, it is worth reiterating that novel approaches are not required. Just as some have been thinking clearly about the size of the problem, others – for half a century – have been suggesting strategies at the corresponding scale. In 1948, William Vogt wrote:

> Conservation is not going to save the world. Nor is control of populations. Economic, political, educational, and other measures are indispensable; but unless population control and conservation are included, other means are certain to fail.[35]
>
> Drastic measures are inescapable. Above everything else, we must reorganize our thinking. If we are to escape the crash we must abandon all thought of living unto ourselves. We form an Earth-company, and the lot of the Indiana farmer can no longer be isolated from that of the Bantu... An eroding hillside in Mexico or Yugoslavia affects the living standard and probability of survival of the American people.[36]

In 1980, André Gorz was writing:

> As long as we remain within the framework of a civilization based on inequality, growth will necessarily appear to the mass of people as the promise – albeit entirely illusory – that they will one day cease being

> 'under-privileged,' and the limitation of growth as the threat of permanent mediocrity. It is not so much growth that must be attacked as the illusions which it sustains, the dynamic ever-growing and ever-frustrated needs on which it is based, and the competition which it institutionalizes by inciting each individual to seek to rise 'above' all others. The motto of our society could be: *That which is good for everyone is without value; to be respectable you must have something 'better' than the next person.*
>
> Now it is the very opposite which must be affirmed in order to break with the ideology of growth: *The only things worthy of each are those which are good for all; the only things worthy of being produced are those which neither privilege nor diminish anyone; it is possible to be happier with less affluence, for in a society without privilege no one will be poor.*[37]

And, half a century after Vogt, in 2002, the authors of the *Report of the World Summit on Sustainable Development* wrote,

> ...we assume a collective responsibility to advance and strengthen the interdependent and mutually reinforcing pillars of sustainable development – economic development, social development and environmental protection – at the local, national, regional and global levels.
>
> We recognize that poverty eradication, changing consumption and production patterns and protecting and managing the natural resource base for economic and social development are overarching objectives of and essential requirements for sustainable development.
>
> The deep fault line that divides human society between the rich and the poor and the ever-increasing gap between the developed and developing worlds pose a major threat to global prosperity, security and stability.
>
> We welcome the focus of the Johannesburg Summit on the indivisibility of human dignity and are resolved, through decisions on targets, timetables and partnerships, to speedily increase access to such basic requirements as clean water, sanitation, adequate shelter, energy, health care, food security and the protection of biodiversity. At the same time, we will work together to help one another gain access to financial resources, benefit from the opening of markets, ensure capacity-building, use modern technology to bring about development and make sure that there is technology transfer, human resource development, education and training to banish underdevelopment forever.
>
> We reaffirm our pledge to place particular focus on, and give priority attention to, the fight against the worldwide conditions that pose severe threats to the sustainable development of our people, which include: chronic hunger; malnutrition; foreign occupation; armed conflict; illicit drug problems; organized crime; corruption; natural disasters; illicit arms trafficking; trafficking in persons; terrorism; intolerance and incitement to racial, ethnic, religious and other hatreds; xenophobia; and endemic, communicable and chronic diseases, in particular HIV/AIDS, malaria and tuberculosis.[38]

The only adequate characterization of what the "representatives of the peoples of the world, assembled at the World Summit on Sustainable Development in Johannesburg, South Africa, from 2 to 4 September 2002"[39] are calling for is "revolution" – a political revolution, a social revolution, and an economic revolution. Unfortunately, the assembled representatives also extolled pure nonsense when they suggested that the means to implement the necessary revolution is "sustained economic growth and sustainable development".[40] "Sustained economic growth" *and* "sustainable development" are as feasible on a finite planet with finite resources and a finite sink for pollution as a perpetual motion machine.[41] Daly argues, for example, that "Sustainable development... necessarily means a radical shift from a growth economy and all it entails to a steady-state economy,[42] certainly in the North, and eventually in the South as well"[43] – something the representatives of the people were not ready to suggest in South Africa.

Nonetheless, despite the fact that parties to the World Summit on Sustainable Development had their heads in an economic "cloud cuckooland,"[44] they can see no other solution to achieving ecological sustainability except a revolutionary change in the global political and economic systems under which the richer nations prosper, and to which most less developed nations aspire. Revolutionary change is the necessary solution to the problem of achieving ecological sustainability. If revolutionary change is implemented, it will dramatically affect the lives and the livelihoods of every person on the planet. We know from experience that it is a revolution that those who benefit most from the current system will aggressively resist.[45] Ironically, it will also be resisted by those who would benefit most from such a revolution for the simple reason that human beings – the conservative, social animals that we are – tend to resist change. This necessary revolution will not occur – if it is to occur at all – because it is what we *ought* to do, nor will it occur because it is what must be done to achieve ecological sustainability and avoid environmental catastrophe. It will only occur if it is made to happen by revolutionaries.

Unfortunately, as the WWF-Canada annual report, the 1996 EDF calendar, the David Suzuki Nature Challenge, the WWF-International Christmas shopping list demonstrate, few organizations and individuals in a position to be agents, catalysts, and instigators for the revolutionary changes needed are thinking in this way. As David Pimentel noted:

> Historically, decisions to protect the environment have been based on isolated crises and are usually made only when catastrophes strike. Instead of examining the problem in a holistic manner, such *ad hoc* decisions have been designed to protect and/or promote a particular resource or aspect of human well-being in the short-term. Our concern, based on past experience, is that these urgent issues concerning human carrying capacity of the world may not be addressed until the situation becomes intolerable or, possibly, irreversible.[46]

Revolutions are, by definition, upheavals and, therefore, inherently difficult. For that reason, most people argue and hope that one is not necessary to achieve ecological sustainability. They have faith that we can just slightly modify the present geo-political, economic system or, even better, make it give us more but with less pollution and more equity. This is in fact the United Nation's view of sustainable development. The hope is that any plan to achieve ecological sustainability will not unduly disturb those who profit most from present practices, and who often contribute large sums to environmental causes. Those who doubt the need for a revolution should ask themselves how many more years of failure, how much more environmental degradation, how many more corpses will it take before they consider something other than the continued comfortable and often self-serving, appeasement of the most powerful beneficiaries of the current system?

Before leaving this discussion of thinking differently, one final matter should be raised. It also introduces the next section of this chapter: *Acting Differently*. It is an omission in the *Report of the World Summit on Sustainable Development*: an omission that is endemic to these kinds of reports whether produced by international bodies, national governments and agencies, or ENGOs. The omission is practical and proven strategies and tactics or even insights about how individuals or ENGOs might go about actually making governments-of-the-day change public policy, or about how they might make other political and economic entities acquiesce to the revolutionary solutions required. Like so many reports before it, when it comes to the problem of actually making its suggested solutions happen, the *Report of the World Summit on Sustainable Development* is silent, and offers instead generic platitudes and the always futile appeals to cooperation and good will.

ACTING DIFFERENTLY

> *Let us in the name of radical pragmatism not forget that in our system with all its repressions we can still speak out and denounce the administration, attack its policies, work to build an opposition political base.*
>
> Saul Alinsky, 1971

Earlier in this chapter, a quote from William Vogt included a reference to Gadarene Swine. The Gadarene Swine Fallacy[47] is the erroneous assumption that a majority group, moving herd-like in one direction, is going the right way. And, conversely, that an individual who is not in step with the herd is going the wrong way. To the majority group, the odd sheep may appear to be traveling in the wrong direction or hold the wrong beliefs, but not to an ideal observer. To an ideal observer – notwithstanding the arguments of the minority "Wise Use" movement and their sympathizers[48] – the characterization earlier of the problems, consequences, and solutions associated with attaining ecological sustainability fairly states what confronts us. Yet, like Gadarene swine, responsible governments, industries, and the major Greens continue down a path of principles, practices, policies, projects, and campaigns – including the shibboleth of "sustainable development" – that has proven it neither leads to ecological sustainability nor to the social equity and justice necessary to achieve that goal.

The major ENGOs are good at unearthing problems. They are very good at suggesting solutions. They are excellent at telling and chastising everyone, but themselves, what should be done to correct the problems and implement the solutions. But, they are hopeless at compelling governments, industries, and institutions to do what must be done to solve the problems. Advocates can often decide what is necessary to achieve a goal, but it is the opponent or the goal that determines what is sufficient. To achieve ecological sustainability, we must act differently. We must do what is both necessary *and* sufficient.

What the last twenty-five years of environmental degradation should have taught the major ENGOs is that, as big as they are, they never had, do not have now, and never will have the human and financial resources needed to clean up – let alone repair or reverse – the on-going damage caused to the environment by the global economy and culture. Nevertheless, they are working tirelessly to clean up ecosystems and rehabilitate endangered species. The Greens never had, do not have now, and never will have the research resources needed to uncover the full extent of environmental degradation and its effects on people and ecosystems, nor do all the studies necessary to understand how best to repair and enhance the environment. Yet many groups spend a large portion of their budgets on scientific research. The major groups never had, do not have now, and never will have the educational resources needed to compete with the multi-billion dollar, world advertising and marketing industry working to convince us that reckless, unsustainable over-consumption is the route to happiness. Yet groups continue to spend donor's funds on public awareness and educational campaigns. The major groups never had, do not have now, and never will have the legal resources necessary to prosecute enough offenders of environmental law – or sue enough governments that fail to enforce the law – to make any long-term difference. When environmental groups lose in court, precedents are often set which increase environmental degradation. When they win, governments often change the law to permit the environmental damage to continue. Yet environmental groups still rely on the courts to address environmental problems. Add fund raising and overhead to the list of activities above, and you have the budget items of most major ENGOs. These activities – along with demonstrations and media stunts – have been the strategic and tactical bulwark of the environmental and animal protection movement for the last three decades. A comparison of the environmental movement in the early 1980s, when it began to lose its effectiveness,[49] with the environmental movement of the early 21st century, shows almost no change in strategy or tactics. The same cannot be said of those who oppose the environmental and animal protection movements.[50] If Benjamin Franklin's definition of insanity, "doing the same thing over and over and expecting different results," is correct, then the environmental and animal protection movements are surely insane.

The problem – which has been obvious, at least since the 1992 Earth Summit in Rio de Janeiro – is that the Green movement never had, does not have now, and never will have the necessary resources for its traditional strategies and tactics to protect the environment, let alone address the necessary social justice and development issues. What is apparent is that the movement, in particular the major ENGOs, does have the financial resources and the public support to be a decisive political and economic force, and that it has sufficient power to make governments – that do have the necessary resources and the responsibility – promulgate, fund, and enforce responsible environmental and social justice policies that will, over time, not only save and protect our environment, but also facilitate its repair and enhancement.

Given that all else has failed and holds no promise for the future, acting differently means employing concepts, strategies, and tactics that have proven their efficacy throughout human history at changing the course of human affairs: specifically, the acquisition and the judicious application of sufficient negotiable power. Acting differently implies acting effectively and making progress. To do that, proponents of ecological sustainability must understand and accept, no matter how personally distasteful or immoral it may seem, that the "fundamental concept in social science is Power, in the same sense in which Energy is the functional concept in physics".[51] The exercise of power – in another word "politics" – determines everything that is decidable by people. Like termites and chimpanzees, we are a social animal. Without the support of a group, our chances of staying alive approach nil; our chances of building a civilization *are* nil. Politics is how we control and direct ourselves in groups – no matter how small or primitive, no matter how large or civilized. "What drives everything in our society is not facts, but politics".[52]

Power is the capacity and willingness of an individual, organization, institution, or state either to provide a benefit or to exact a cost. Depending on whom the wielder of power is, the benefits and costs can range from the extremes of a kiss, happiness, economic windfall, political success, and prestige to murder, bankruptcy, incarceration, torture, and death. Power is absolutely necessary, and nothing else will do, when one group wants to succeed at imposing its values on another, which is exactly what those who are advocating the need for ecological sustainability are trying to do.[53] Alexander Hamilton wrote in the Federalist Papers, "What is power, but the ability or faculty of doing a thing? What is the ability to do a thing, but the power of employing the means necessary to its execution?"[54] Unfortunately – and this is why progress is not being made – the majority of those involved in promoting ecological sustainability are either frightened, reluctant, willfully ignorant of the necessity, or do not understand how to use power to achieve their ends. Many morally abhor the exercise of power, blaming it for our troubles. Some erroneously believe that public awareness campaigns, educating school children, the law, or scientific research will eventually produce on their own enough spontaneous global behavioural changes in enough individuals who influence institutions, industries, and governments that ecological sustainability will result.

From the perspective of a political activist, the most significant forms of power that the major ENGOs enjoy are political and economic. Political power is the capacity to influence who wins and who loses an election. Economic power is the capacity to influence what people buy or do not buy – in other words, influence profit and loss. Both these forms of power flow from the fact that the major ENGOs enjoy the trust and support of millions of people whose contributions and, to a lesser extent, their votes and buying habits can be used to provide a benefit for or exact a cost from politicians and businesses.

The importance of the major ENGOs using their power to the maximum extent possible is beyond exaggeration. Given that achieving ecological sustainability is about life or death, to do anything less is culpable. All public policy and all business decisions are negotiated by relevant actors from positions of power. Those with the most power (i.e. those with the capacity and willingness to provide benefits or exact costs) decide what course of action will be followed. The more power they use and the more routinely they use it, the more relevant they become, and the more progress they make. And, power begets power. The more it is used the stronger it becomes – no different than a muscle or a mind. Those without power – which includes those who choose not to use their power, like most of the major ENGOs today – are irrelevant, no matter how compelling or just their cause or how much public opinion they have on their side.

Self-interest is usually a major consideration when politicians and business people make important public policy and economic decisions. Therefore to achieve ecological sustainability, it will be necessary for the major ENGOs to use their political and economic power to make it in the short-term political best interests of politicians and economic best interests of business to make the necessary and sufficient changes required.

"Politics," Bismarck said, "is the art of the possible". Politics is, in fact – as Václav Havel observed – "the art of the impossible".[55] Within the lifetime of many of the readers of this chapter, politics – the exercise of power – has accomplished much good that was once thought impossible: the unification of Europe, female political and economic parity in many of the Western Democracies, the end of the Cold War and the liberation and democratization of many of the former countries of the Soviet Union, the end of apartheid in South Africa, the reduction of commercial whaling, labor and industrial safety improvements, the end of state-sanctioned segregation in the United States, universal health care in most of the Western Democracies, and so on. On the environmental front, where successes have been achieved it was usually when power – political and/or economic – was applied effectively. For example, in the 1980s, the International Fund for Animal Welfare (IFAW) was able to reduce the Canadian seal hunt by 90% as a consequence of applying political power in the European Community, and economic power in the form of a Canadian fish products boycott in the United Kingdom and the United States.[56] In the 1990s, Greenpeace, using its considerable power to influence consumers, was able to compel the British Columbia forest industry to improve its forestry practices, something no amount of scientific study, public protest, or appeals to the British Columbia government had been able to accomplish.[57] People for the Ethical Treatment of Animals (PETA) has been able to marshal consumers to force much of the American fast food industry to buy animal products only from suppliers who meet the highest standards of animal care.

In the United States, much of the current environmental legislation came from the politically charged radical movements of the 1960s and the 1970s that "represented a vast bipartisan voting constituency". Candidates discovered that their commitment to the environment had political relevance – which is power – and could influence whether they were elected or not.[58] By 1995, however, the major U.S. environmental groups had abandoned the effective use of their political and economic power and devolved into "paper tigers". As a result, with no reason to be concerned about the major environmental groups, the U.S. Senate suspended funding for new prospective listings under the Endangered Species Act, and President Clinton and Vice President Gore, realizing the "vaunted organizing and lobbying power of the mainstream environmental movement had turned out to be a sham" began to abandon their environmental campaign promises.[59]

For the most part, the major Green groups have abandoned the use of power – economic and political – in favour of non-confrontational or uncontroversial clean-ups and rehabilitation projects, scientific research, public awareness and children's education campaigns, and law suits defending legislation that is generally inadequate to protect the environment or animals. In the political realm, where public policy is decided, they tend to confuse achievement with access and appearing reasonable to politicians.[60]

The ways and means to clean up our environment, protect it from further degradation, and enhance it are well known. Almost every environmental issue or social justice issue facing the world today has an off-the-shelf solution. All environmental and social justice issues are being effectively dealt with somewhere, but not everywhere. True, in the area of the environment, some of the damage already done cannot be repaired. Many coral reefs, for example, are lost. We may never again see vast schools of cod off the east coast of Canada again. Nonetheless, where there were once coral reefs or old growth forests or cod,[61] new, but different and robust, ecosystems can evolve. An environment can be saved. All is not lost.

Given the daunting, revolutionary scope of the requirements necessary to achieve ecological sustainability, we should ask the question, "Is it realistically possible for us to do what will be required to achieve ecological sustainability?" The answer is a qualified "yes," because we have never been in a better position – in theory, at least – to compel and implement the necessary revolutionary changes. "We" does not mean humanity in general. Most people are not engaged in the struggle to achieve ecological sustainability; they simply suffer for our failures. "We" means those who have the power – whether they choose to use it or not – to contribute to the solution.

Globalization has exacerbated environmental degradation. However, some of the very factors that make globalization possible make implementing an ecological sustainability revolution more possible now than at any time since Vogt wrote *Road to Survival.* These factors include more open and accountable governments in the western democracies, supra-national institutions like the World Trade Organization and the various trade conventions, increased wealth in the western democracies, and excellent, low cost global communications. All of these factors can be successfully exploited by the economically thriving major ENGOs based in the western democracies. We also know – and have qualified and experienced practitioners of – the proven strategies and tactics that can be used to wield political and economic power effectively. All of these factors offer the major ENGOs the opportunity not just to influence but also to control much public policy. One recent example of social justice groups using modern organization and communication tools was the successful campaign to defeat the Multilateral Agreement on Investment.[62] And, in the 2004 U.S. elections, the Internet as an organizing tool – albeit not necessarily a decisive one – was again proven effective by www.MoveOn.org and www.DemocracyForAmerica.com, the political progeny of the campaign to elect Howard Dean for President. Since the 2004 U.S. elections, both of these virtual organizations have continued to grow, and are steadily increasing their capacity to influence public policy.

Earlier it was said that people – including environmentalists – will tolerate being killed, and even make common cause with their killers, if it is done slowly. Happily, the reverse is also true: people will tolerate being saved slowly. When we look back on 1948, it becomes apparent that had our fathers and mothers and their elected officials been compelled – through the judicious and sufficient use of power – to heed people like Aldo Leopold and William Vogt and been persuaded or compelled by law to make consistent, modest improvements and concessions to ecological sustainability on the order of a one or two percent improvement every year (the rate of decline since 1970 reported in *Living Planet Index 2004*) we would be much better off today. Indeed, it is likely that many of our current ecological crises would have been averted or at least ameliorated, and along with them, many national and global political and economic tensions and social injustices.

Fully implementing solutions to achieve ecological sustainability in a short time frame – a decade or less, or even one generation – is impossible. Human societies, unless devastated by tragedy, simply do not and will not respond that quickly. What can be accomplished, safely and securely without any undue hardship caused by the transition from one set of global, economic standards to another – from an economy based on growth to a steady-state one,[63] for example – is an annual net improvement of one or two percent. In many countries, with some indicators (air and water quality, and waste management), this rate of improvement has been exceeded.[64] If this rate of improvement is achieved overall, we can expect that the global environment will recover. It will be different from what it is now and from some imagined Eden-like past, but it will likely not only be able to sustain itself and us, but also thrive. Life, as we know, has amazing recuperative powers.

A one or two percent positive rate of change is possible, and has been achieved in the past for peaceful undertakings of the magnitude being contemplated here. Roosevelt managed it when he implemented the New Deal and lifted the people of the United States out of the Depression.[65] Winston Churchill, George Marshall, René Courtin and others managed it when they laid the foundation, built upon by many since, for the European Union, arguably – given the historic rivalries of the nations involved – a more difficult task than achieving ecological sustainability. Achieving ecological sustainability is on the same scale as these undertakings. It should be cautioned that the noun phrase "one or two percent positive rate of change" unintentionally trivializes the effort required. This will be an enormous and challenging undertaking.

Most executives and governors of the major ENGOs will be immediately aware that if they chose to exercise power as described above they would have to restructure and reorient their organizations. Building the capacity for political action, particularly electoral politics, and for economic action, in the form of consumer boycotts and endorsements, would become their fundamental organizing strategic principle, because this is from where their negotiable power would flow. Charitable status – which carries with its many benefits debilitating restrictions that render ENGOs politically impotent – would have to be abandoned. For the same reason, so would reliance on funds from many foundations and large donors. Projects that ought to be done by government because they are in the broad public interest and are now done by ENGOs – for example, land acquisition, wildlife rescue programs, and companion animal shelters – would have to be triaged, and the liberated funds used to pressure governments (who can be held to a higher level of accountability and transparency than non-profit organizations) to take on those responsibilities. Changing the law and making new laws through politics – pressure and electoral – would replace challenging laws in court. The governors of the major ENGOs would have to require their executives and staffs to design and implement campaigns that would produce measurable results that would demonstrably move us toward the goal of ecological sustainability.

Too many NGOs confuse activity with achievement. Activity is less demanding and less measurable than achievement so it becomes the default. Activity is a meeting with a legislative aide, an audience with a Senator, testimony before a Parliamentary committee, ordering a public opinion poll, running a full page ad, and protests. Achievement is a law passed that increases funding for public transit, a national building code amendment that improves energy efficiency, thirty percent of a national coastline being declared a non-consumptive marine sanctuary, and an end to Canada's commercial seal hunt. Activity takes work. Achievement takes more work *and* the application of power. Activity without achievement weakens an ENGO. Activity and achievement strengthens it. Activity requires scientists, lawyers, educators, and middle managers. Achievement demands political operatives, social activists, and community organizers. To insure the power in their supporter bases is used effectively, the governors of the major ENGOs would have to reward and promote those with a record of achievement, and treat less handsomely those whose record is merely activity and the avoidance of corporate problems.

It is worth cautioning that just as societies and nations cannot be reformed too quickly, neither can large ENGOs, no matter how urgent the need. The total reformation of a multi-million dollar ENGO with hundreds of thousands of supporters and hundreds of employees could take a decade. In some cases, because of the internal culture of an ENGO, total reformation might never be achieved. The major ENGOs are like societies in that if progress is to be made they must be reformed cautiously, incrementally, and fairly – with due regard for everyone's self-interest. Anything less, as with nations, will result in failure.

The managers and governors of the major ENGOs can take comfort from the fact that to reform themselves (like achieving ecological sustainability) novel ideas and practices are not necessary. There are many examples for the major ENGOs to follow. Many smaller, local ENGOs are very efficient at applying power effectively, and are having the kinds of results that are necessary if ecological sustainability is to be achieved. These ENGOs also tend to be very well-managed. Because of their limited supporter bases and reliance on volunteer staffs to be successful, the smaller ENGOs are required to use optimum methods to achieve their ends. Indeed, so successful are these smaller groups that it is common for major ENGOs to hitch a low-cost ride from them for fund raising purposes. These smaller ENGOs have much to teach the majors. The most important is that what the best of the smaller ENGOs are doing – particularly in the areas of governance, staff and volunteer relations, fund raising and donor relations, coalition building, government relations, and the application of power – can be scaled up for use by the major ENGOs. Rather than retain very expensive consultants (whose prime objective is usually maximizing billable hours and not the welfare of their clients) to help them deal with management and campaign problems, the governors and managers of the major ENGOs would learn much by spending some volunteer time with these smaller groups. At the very least, the governors of the major ENGOs should recruit most of their executives from the ranks of the smaller, successful ENGOs and not from the scientific community, the corporate world, fund raising and non-profit management schools, law schools, or government agencies.

Yes, it might be possible to implement the social revolution necessary to achieve ecological sustainability. We know what the objectives are and the necessary reforms to achieve them. There is good evidence that the changes needed can be realized through small, incremental, economically and socially tolerable steps. There are excellent, low-cost global communication systems to facilitate political organizing and information dissemination. There are the national and international political, economic, and legal institutions that do respond, if sometimes reluctantly, to pressure, and that are competent to implement and enforce the changes. The western democracies have the necessary economic wherewithal to finance the changes. Lastly, there is a large, well-informed, affluent constituency that has already demonstrated its commitment and willingness to support initiatives to compel reluctant governments and recalcitrant corporate entities to implement the necessary public policy and economic changes that would lead to ecological sustainability.

It is important to realize, however, that actually achieving it will be extremely difficult. Most of those who now profit from or who are comfortable in the present system will aggressively resist every change – no matter how trivial or incremental – that would adversely affect them however slightly. To achieve ecological sustainability, we must expect and plan for every advance to be a bitter and hard fought battle. Winning those fights will require the use of strategies and tactics more aggressive than those used by the advocates for those interests and individuals who benefit from economic inequality, lack of opportunity, social injustice, and a degraded environment.

To take advantage of our knowledge of the problems and solutions associated with achieving ecological sustainability, the enhanced global organizing environment, and the untapped economic and political power in the major ENGOs, leaders are required – leaders, as El-Ashry pointed out, who are willing to think and act differently and lead the way.

LEADING THE WAY

> *Those who believe in social change must fully accept their own leadership role in the process and recognize that neither politicians nor political parties are the prime movers of progressive change.*
>
> Randy Shaw, 1996.

A discussion about leaders involves at least three elements: the need for leaders, where they might come from, and whether it is likely they will emerge. The first two elements are relatively easy to dispose of. The third is so problematic that it may doom any hope of humankind achieving ecological sustainability without suffering Vogt's "rush down a war-torn slope of barbarian existence in the blackened rubble".

Achieving ecological sustainability demands a political movement based on the judicious and sufficient application of coercive power, rather than protest and persuasion. Strong and judicious leaders are a fundamental prerequisite to such movements. It is rare for organizations that place excessive importance on decentralization, managerial consensus, and internal harmony to develop strategies and tactics that produce the most effective application of power. In organizations, democratic or collegial consensus decision-making tends to produce not the best decision, but usually the least objectionable and most self-serving one. As Machiavelli observed, "though the many are incompetent to draw up the constitution since diversity of opinion will prevent them from discovering how best to do it, yet when they realize it has been done, they will not agree to abandon it".[66] Leadership is a "vital ingredient in the effectiveness of organizations".[67] Saul Alinsky has argued that the "building of many mass power organizations to merge into a national popular power force cannot come without many organizers".[68] Leaders are necessary, and little can be accomplished without them: "the greater the need for economic and social change, the greater the need for leadership to guide the process".[69]

El-Ashry suggests that "leaders from all walks of life" are needed. Perhaps, but in the beginning it is likely that the potential leaders of a power-based, ecological sustainability movement are currently active in the ENGOs. It is within the ENGOs that people with the necessary leadership, political, and management skills, and who have the greatest understanding and appreciation of the urgency to achieve ecological sustainability,[70] might be found. However, a closer look at the Green movement reveals that the necessary, sufficiently funded and supported leaders may remain still-born, unless the men and women who manage and govern the major ENGOs can think and act differently, as El-Ashry asks, and reform their own organizations first.

The necessity to reform the major ENGOs stems from the fact that they control most of the financial resources and access to the supporters of the Green movement. There are over 33,000 environmental and animal protection charities in the United States.[71] That number does not include tens of thousands of non-profit organizations that do not qualify for charitable "501(c)3" status. A small sampling of United Kingdom statistics[72] produces similar results. However, in the United States, the "twenty-four organizations that comprise the mainstream sector of the [environmental] movement" receive 70% of the donations.[73] The fact is that, worldwide, just a few major ENGOs essentially control the millions of supporters, and therefore the negotiable power of the environmental and animal protection movement, and by implication whether or not we have any hope of achieving an ecologically sustainable global society.

It is important to appreciate that this concentration of resources and supporters in a small number of ENGOs is more insidious and debilitating than may appear at first glance, and yet, ironically offers more reason for hope. Most of the people who actively support environmental and animal protection contribute to and are loyal to the major environmental and animal protection groups. This can best be understood in terms of brand loyalty, by the enormous marketing power of the major groups, and by the fact that most, if not all, of the information that supporters receive about their organizations and the issues is controlled by the organizations themselves. Organizations do not criticize themselves. As James Q. Wilson suggested "the behavior of persons who lead or speak for an organization can best be understood in terms of their efforts to maintain and enhance the organization and their position in it".[74] Consequently, the enormous latent political and economic power that the environmental and animal protection movement enjoys is, for all practical purposes, inaccessible without the permission of the men and women who manage and govern the major ENGOs. They are the gatekeepers to the political and economic power needed to achieve ecological sustainability. The probability of leaders – no matter how competent – emerging independent of the major ENGOs, and then somehow acquiring the necessary resources to influence public and economic policy, is very low to impossible.

Nonetheless, there is hope. The major ENGOs are all sympathetic to the need to achieve ecological sustainability, and their numbers are very small, indeed. In the United States, for example, there are 107 environmental and animal protection charities that raised more the $20 million, and 306 which raised between $5 and $20 million.[75] What these statistics demonstrate is that the actual number of men and women – the executives and governors – who control the environmental and animal protection movement and can decide, for all practical purposes, whether or not it will ever unleash all its latent coercive power and achieve ecological sustainability is quite small, a few thousand at most. Moreover, among these few thousand are many men and women who fully understand exactly what is at stake in the global pursuit of ecological sustainability.

Unfortunately, no matter how determined a trustee or executive of a major group might be, reform of any large organization is difficult, and often proves impossible.

> Broaching questions about the future with almost any mainstream [environmental organization] leader will draw one into a discussion of federal rulemaking, organizational development, or fund raising strategies – likely items on the agenda of the next board meeting. When leaders do take time to reconsider their mission and explore a vision of the future... they find themselves restrained by the imperatives of their benefactors and the sheer size of their organizations. They tend to continue on the track that has kept them alive thus far, whether it protects the environment or not.[76]

Non-profit organizations – particularly when they grow and commoditize their issues – suffer from a mutually-reinforcing range of internal and external impediments that prevent them from being as effective as their budgets and memberships would allow. These impediments result in an *Alice in Wonderland* situation: non-profits – unlike business, labor, or political organizations – do not need to actually achieve their objectives in order to thrive. The positive feedback mechanisms that relate to achieving their stated goals – i.e. saving harp seals, panda bears, or the Earth – do not function to keep non-profit and charitable organizations on track. As a result, for non-profits – particularly environmental non-profits – losing is an option, and activity tends to become a substitute for achievement. In contrast, if businesses fail to earn a profit, they go bankrupt; if labor unions fail to improve wages and working conditions, their leaders are voted out of office; and if politicians fail to benefit their donors and relevant voting constituencies, they lose elections. The managers and governors of ENGOs are not held accountable with comparable feedback mechanisms when they fail to "save the Earth". All of the important economic and career feedback mechanisms that influence the staffs and managers of ENGOs relate to internal politics, fund raising, and supporter acquisition, not to achieving the purposes of the organization. Not infrequently, those who are best at achieving an ENGO's stated goals are first to be shown the door. Further exacerbating this bizarre environment is the fact that the managers of the ENGOs control most of the information that their "customers" receive about their performance and successes. All of the third party watchdog organizations that monitor ENGOs evaluate them on their management structures and accounting and audit practices not on their achievements. For the donor, this is as absurd as deciding to buy a car based on which accounting firm an automobile manufacturer retains rather than the quality of the product. The situation is so egregious that it is often better financially for a non-profit to lose than to win. Charitable status, which is cherished by many groups in order to attract tax deductible donations, compounds the problem by restricting ENGOs to exercising only an insignificant fraction of the power in their supporter bases – a fact fully understood by politicians.[77]

Indeed, the *Through the Looking Glass* regulatory framework and topsy-turvy financial environment in which non-profits operate is so antithetical to achieving their public interest objectives that the larger – and potentially more influential – a non-profit becomes the less effective and aggressive it tends to be. A truly depressing comparison of trends underlines this situation. The first trend is that since 1970, the environmental movement has increased in funds and membership by about 5,000%.[78] Today, the movement is a multi-billion dollar industry – worth $6.95 billion[79] in the United States in 2003 alone, where "chief executives at nine of the nation's ten largest environmental groups earned $200,000 and up, and one topped $300,000".[80] In 2002, one group, The Nature Conservancy, Inc., raised almost a billion dollars.[81] Tragically, the other trend is that while the movement was enjoying 5,000% growth, the Earth it was supposed to be saving suffered a 40% decline in environmental quality and in its capacity to sustain life.[82]

Within the Green movement are the leaders and resources necessary to create the kind of global power-based movement needed to achieve ecological sustainability. But those leaders will never emerge, and those resources and powers will never be unleashed, unless the gatekeepers – the senior executives and governors of the major ENGOs – liberate them. These gatekeepers have the power to prevent the awful escalation of the human tragedies so many are needlessly suffering now, and that so many authorities have been warning us about for so many years. It may turn out that it is beyond the will and abilities of enough of the few thousand men and women who control the major ENGOs to reform their organizations, to begin to use their power as effectively as their predecessors used theirs in the 1960s and 1970s, and to lead the necessary wave of power-based environmental activism that might result in the social changes needed for us to achieve ecological sustainability. As Lester Brown observed,

> Achieving a sustainable society will not be possible without a massive reordering of priorities. This is in turn dependent on political action by individuals and by public interest groups... If we fail, it will not be because we did not know what needed to be done. Unlike the Mayans, we know what must be done. What we will soon discover is whether we have the vision and the will to do it.[83]

About the environmental movement, sociologist Robert Nisbet wrote, "When the history of the twentieth century is written, the single most important social movement will be judged to be environmentalism".[84] His prediction could not have been more wrong. Because of the impediments discussed above, the environmental movement – a globe-girdling, multi-billion dollar movement, financially supported and trusted by hundreds of millions of people – has been a spectacular failure, despite the best efforts of some of the most dedicated, self-sacrificing people one might ever hope to meet. Perhaps it is true that things would be worse without the environmental and animal protection movements, but the promise – the implied contract with supporters – made to this day, and the reason people still donate and protest, was not to slow environmental degradation, but to end it and reverse it.

Analyzing why environmentalism failed to become the "the single most important social movement" of the twentieth century has become a political science and publishing cottage industry, as has contriving fixes for the environmental movement. Whether it is Mark Dowie's *Losing Ground*, Pulitzer Prize winner Tom Knudson's series of articles for the *Sacramento Bee*, Peter Montague's excellent Environmental Research Foundation, or Michael Shellenberger and Ted Nordhaus's *The Death of Environmentalism*, they all miss the point. The environmental movement cannot save the environment. All it can do – which is what is necessary and sufficient – is use its awesome political and economic power to leverage and force governments and industry to do it. This is what it has failed to do since the mid-1980s. This is why the environmental movement has failed – and continues to fail – to save the environment.

Whether or not we "Save the Earth" depends on those few men and women who control the major ENGOs. Many of them are sensitive to the urgency of achieving ecological sustainability. They have the power necessary to achieve it. But, if they do not have the vision or will to reform their own organizations (a less challenging task than what they are asking the rest of us to do) in order to save even their own children and grandchildren, there is little hope that anyone else, or any organization, will emerge that will be able to slowly compel the necessary global economic and political reforms.

Since Biblical times, two questions have nagged those men and women whom fate has burdened with the opportunity and means to do great good and prevent great harm. The questions that the men and women who are executives and trustees or directors of the major ENGOs need to answer are: "If not us, who? If not now, when?"[85] They should address their answers to themselves, to their children and grandchildren, and, if they worship one, to their deity.

NOTES AND SOURCES

[1] Brown, Lester R. *Building a Sustainable Society*. New York: Norton, 1981. p. 3.

[2] Pimentel, David, *et al.* "Ecology of Increasing Disease: Population growth and environmental degradation." *Bioscience* Vol. 48. No. 10 October, 1998.

[3] For more on this topic, see Brooks, Chapter 16.

[4] Napolitan, Joseph. *The Election Game and How to Win It.* New York: Doubleday, 1972.

[5] *Report of the World Summit on Sustainable Development.* Johannesburg, South Africa, 26 August - 4 September 2002. New York: United Nations, 2002. p. 3. Also see Willison, Chapter 2, on the loss of biodiversity; Holt, Chapter 4 and Hutchings, Chapter 6, on the state of world fisheries; and Chesworth, Chapter 3, on the impacts of agriculture on the planet.

[6] Livernash, Robert. *The Challenge of Sustainability.* Washington: Global Environment Facility, 2002. p. 93.

[7] Russell, Bertrand. *Power: The role of man's will to power in the world's economic and political affairs.* 1938. New York: Norton, 1969. pp. 14-15.

[8] Russell, Bertrand. Power: *The role of man's will to power in the world's economic and political affairs.* 1938. New York: Norton, 1969. p. 271.

[9] Dowie, Mark. *Losing Ground: American Environmentalism at the Close of the Twentieth Century.* Cambridge: MIT Press, 1996. Knudson, Tom. "Environment Inc." *Sacramento Bee* 22-26 April 2001. 25 October 2004. Available at: http://www.sacbee.com/static/archive/news/projects/environment/index02.html; Montague, Peter. "#744-Part 4: Rebuilding the Movement to Win." *Rachel's Environment & Health News.* February 13, 2002; Shaw, Randy. *The Activists Handbook.* Berkeley: Univ. of California Press, 1996. Wilson, Jeremy. "Green Lobbies: Pressure Groups and Environmental Policy." *Canadian Environmental Policy: Ecosystems, Politics, and Process.* Robert Boardman, ed. Toronto: Oxford UP, 1992.

[10] See Czech, Chapter 23, for more on a steady state economy.

[11] Helvarg, David. *The War Against the Greens: The "Wise Use" Movement, the New Right, and Anti-Environmental Violence.* San Francisco: Sierra Club, 1994; Dowie, Mark. *Losing Ground: American Environmentalism at the Close of the Twentieth Century.* Cambridge: MIT Press, 1996. pp. 83-104.

[12] Daley, Herman E. Beyond Growth: The Economics of Sustainable Development. Boston: Beacon Press, 1996. p. 224.

[13] *Report of the World Summit on Sustainable Development.* Johannesburg, South Africa, 26 August - 4 September 2002. New York: United Nations, 2002. p. 1-5.

[14] Kemp, David D. *Global Environmental Issues: A Climatological Approach.* London: Routledge, 1990. pp. 193-196.

[15] Wilson, Edward O. *The Diversity of Life.* Cambridge: Harvard UP, 1992. Also see Willison, Chapter 3.

[16] Machiavelli, Niccolò. *The Discourses.* 1531. New York: Penguin, 1970. p. 132.

[17] Alperovitz, Gar, *et al. Index of Environmental Trends: An Assessment of Twenty-One Key Environmental Indicators in Nine Industrialized Countries over the Past Two Decades.* Washington: National Center for Economic Alternatives, 1995. pp. 20-21.

[18] Gélinas, Johanne. *Report of the Commissioner of the Environment and Sustainable Development to the House of Commons.* Ottawa: Office of the Auditor General of Canada, 2004. http://www.oag-bvg.gc.ca. p. 2.

[19] Dowie, Mark. *Losing Ground: American Environmentalism at the Close of the Twentieth Century.* Cambridge: MIT Press, 1996. pp. 40-41.

[20] Montague, Peter. "1996 in Review: More Straight Talk." *Rachel's Environment & Health News #525.* December 19, 1996. Available at http://www.rachel.org/home_eng.htm.

[21] David Suzuki Foundation, 2005. Available at http://www.davidsuzuki.org/WOL/Challenge.

[22] "Ten things not to buy for Christmas." World Wildlife Fund. Gland: 21 December 2004. Available at: http://www.panda.org/news_facts/newsroom/other_news/news.cfm?uNewsID=17351&uLangID=1.

[23] Issacs, Colin, ed. "Christmas in Panda Land." *The Gallon Environment Letter.* Fisherville, Ontario: Vol. 9, No. 24, December 22, 2004.

[24] Ellul, Jacques. *Propaganda: The Formation of Men's Attitudes.* New York: Vintage, 1965. Also see Worcester, Chapter 17. Lippman, Walter. *Public Opinion.* New York: Free Press (Macmillan), 1965.

[25] Vogt, William. *Road to Survival.* New York: William Sloane, 1948.

[26] Leopold, Aldo. *A Sand County Almanac.* 1949. New York: Oxford UP, 1966.

[27] Carson, Rachel. *Silent Spring.* 1962. New York: Fawcett, 1970.

[28] "World Scientists' Warning to Humanity." *Union of Concerned Scientists.* 1992. Available at http://www.ucsusa.org/ucs/about/page.cfm?pageID=1009.

[29] "World Scientists' Call for Action (1997)." *Union of Concerned Scientists.* 1997. Available at: http://www.ucsusa.org/ucs/about/page.cfm?pageID=1007.

[30] Livernash, Robert. *The Challenge of Sustainability.* Washington: Global Environment Facility, 2002.

[31] Loh, Jonathan and Wackernagel, Mathis, eds. *Living Planet Index 2004.* Gland: WWF, 2004.

[32] Branfman, Fred. "Living in shimmering disequilibrium." Salon.com (22 April 2000). 25 October 2004. Available at http://www.salon.com/people/feature/2000/04/22/eowilson.

[33] Best, Stephen. Oral Presentation to the Standing Senate Committee on Legal and Constitutional Affairs. Toronto: Environment Voters, 2000. Available at http://www.environmentvoters.org/C2ReporttoStandingSenateCommittee.html.

[34] "Ontario's rural heartland in shock." CBC News, May 2000, updated Oct. 18, 2004. Available at: http://www.cbc.ca/news/background/walkerton/index.html.

[35] Vogt, William. *Road to Survival.* New York: William Sloane, 1948. p. 264.

[36] Vogt, William. *Road to Survival.* New York: William Sloane, 1948. p. 285.

[37] Gorz, André. *Ecology as Politics.* New York: Black Rose, 1980. p. 8. Also see Czech, Chapter 23.

[38] *Report of the World Summit on Sustainable Development.* Johannesburg, South Africa, 26 August - 4 September 2002. New York: United Nations, 2002. pp. 1-3.

[39] *Report of the World Summit on Sustainable Development.* Johannesburg, South Africa, 26 August - 4 September 2002. New York: United Nations, 2002. p. 1.

[40] *Report of the World Summit on Sustainable Development.* Johannesburg, South Africa, 26 August - 4 September 2002. New York: United Nations, 2002. p. 52.

[41] Also see Brooks, Chapter 16.

[42] See Czech, Chapter 23.

[43] Daley, Herman E. *Beyond Growth: The Economics of Sustainable Development.* Boston: Beacon Press, 1996. p. 31.

[44] Lavigne, David. "In My View." *BBC Wildlife.* September 2002. p. 65.

[45] Helvarg, David. *The War Against the Greens: The "Wise Use" Movement, the New Right, and Anti-Environmental Violence.*

San Francisco: Sierra Club, 1994.; Daley, Herman E. Beyond Growth: The Economics of Sustainable Development. Boston: Beacon Press, 1996.; Milloy, Steven J. JunkScience.com. November 2004. Available at http://www.junkscience.com.

[46] Pimentel, David, *et al. Will Limits of the Earth's Resources Control Human Numbers*? Ithaca: Cornell University, 1999.

[47] Gadara was the ancient city of Palestine southeast of the Sea of Galilee and subsequently destroyed. The name was later adopted by a district east of Jordan and called Gadarenes, or Gergesenes. It was the site of the famous miracle of the swine, in which Jesus conjured demonic spirits into the body of swine and let them perish in the sea. The story is recounted in the Synoptic Gospels, Luke 8:26-33. Available at http://www.philosophicalsociety.com.

[48] Helvarg, David. *The War Against the Greens: The "Wise Use" Movement, the New Right, and Anti-Environmental Violence.* San Francisco: Sierra Club, 1994.

[49] Dowie, Mark. *Losing Ground: American Environmentalism at the Close of the Twentieth Century.* Cambridge: MIT Press, 1996. p. ix.

[50] Grefe, Edward A. *Fighting to Win: Business Political Power.* New York: Harcourt Brace Jovanovich, 1981.

[51] Russell, Bertrand. *Power: The role of man's will to power in the world's economic and political affairs.* 1938. New York: Norton, 1969.

[52] Lavigne, David. Television interview for Environment Voters documentary. Taped March 2000 in Guelph, Ontario.

[53] Ellis, Richard J. and Thompson, Fred. *The Culture Wars by Other Means: Environmental Attitudes and Cultural Biases in the Pacific Northwest.* Vancouver: SFU-UBC Centre for the Study of Government and Business, 2000.

[54] Alinsky, Saul D. *Rules for Radicals.* 1971. New York: Vintage, 1989. p. 52.

[55] Havel, Václav. *The Art of the Impossible.* New York: Knopf., 1997.

[56] Mowat, Farley. *Sea of Slaughter.* Toronto: McClelland and Stewart, 1984. pp. 388-401.

[57] Stanbury, W. T. *Environmental Groups and the International Conflict Over the Forests of British Columbia, 1990 to 2000.* Vancouver: SFU-UBC Centre for the Study of Government and Business, 2000.

[58] Dowie, Mark. *Losing Ground: American Environmentalism at the Close of the Twentieth Century.* Cambridge: MIT Press, 1996. pp. 33-34.

[59] Cockburn, Andrew. "Big Green Faces Time of Reckoning." *Los Angles Times* 23 March 1995.

[60] Shaw, Randy. *The Activists Handbook.* Berkeley: Univ. of California Press, 1996. pp. 60-61.

[61] See Hutchings, Chapter 6.

[62] Shah, Anup. "Multilateral Agreement on Investment." *Global Issues.* December 20, 2000. Available at http://www.globalissues.org/TradeRelated/MAI.asp.

[63] Daley, Herman E. *Beyond Growth: The Economics of Sustainable Development*. Boston: Beacon Press, 1996. Also see Czech, Chapter 23.

[64] Alperovitz, Gar, *et al. Index of Environmental Trends: An Assessment of Twenty-One Key Environmental Indicators in Nine Industrialized Countries over the Past Two Decades.* Washington: National Center for Economic Alternatives, 1995.

[65] Black, Conrad. *Franklin Delano Roosevelt: Champion of Freedom.* New York: Public Affairs, 2003.

[66] Machiavelli, Niccolò. *The Discourses.* 1531. New York: Penguin, 1970. p. 132.

[67] Handy, Charles B. *Understanding Organizations.* New York: Penguin, 1976. p. 107.

[68] Alinsky, Saul D. *Rules for Radicals.* 1971. New York: Vintage, 1989. p. 63.

[69] Brown, Lester R. *Building a Sustainable Society.* New York: Norton, 1981. p. 311.

[70] Ellis, Richard J. and Thompson, Fred. *The Culture Wars by Other Means: Environmental Attitudes and Cultural Biases in the Pacific Northwest.* Vancouver: SFU-UBC Centre for the Study of Government and Business, 2000. pp. 33-41.

[71] GuideStar: The National Database of Nonprofit Organizations. Available at http://www.guidestar.org/index.jsp.

[72] Charity Commission for England and Wales. Available at http://www.charity-commission.gov.uk.

[73] Dowie, Mark. *Losing Ground: American Environmentalism at the Close of the Twentieth Century.* Cambridge: MIT Press, 1996. p. 41.

[74] Wilson, James Q. *Political Organizations.* Princeton: Princeton UP, 1995. p. 9.

[75] GuideStar: The National Database of Nonprofit Organizations. Available at http://www.guidestar.org/index.jsp.

[76] Dowie, Mark. *Losing Ground: American Environmentalism at the Close of the Twentieth Century.* Cambridge: MIT Press, 1996. p. 256.

[77] Webb, Kernaghan. *Cinderella's Slippers? The Role of Charitable Tax Status in Financing Canadian Interest Groups.* Vancouver: SFU-UBC Centre for the Study of Government and Business, 2000.

[78] Dowie, Mark. *Losing Ground: American Environmentalism at the Close of the Twentieth Century.* Cambridge: MIT Press, 1996. pp. 40-41.

[79] *Giving USA 2004.* Giving USA Foundation-AAFRC Trust for Philanthropy.

[80] Knudson, Tom. "Environment Inc." *Sacramento Bee* 22-26 April 2001. 25 October 2004. Available at: http://www.sacbee.com/static/archive/news/projects/environment/index02.html.

[81] GuideStar: The National Database of Nonprofit Organizations. Available at: http://www.guidestar.org/index.jsp.

[82] Loh, Jonathan and Wackernagel, Mathis, eds. *Living Planet Index 2004.* Gland: WWF, 2004. p. 2. For further discussion of the Nature Conservancy, see Beder, Chapter 5.

[83] Brown, Lester R. *Building a Sustainable Society.* New York: Norton, 1981. p. 372.

[84] Caldwell, Lynton. Quoted in "Globalizing Environmentalism: Thresholds of a New Phase in International Relations," *Society and Natural Resources*, 4, p.259. Available at: http://www.antiwar.com/stromberg/s090500.html.

[85] Rabbi Hillel the elder, ca 70BC - 10 AD.

CHAPTER 26

REINVENTING WILDLIFE CONSERVATION FOR THE 21ST CENTURY

David Lavigne, Rosamund Kidman Cox, Vivek Menon, and Michael Wamithi

"Never has it been so urgent to ensure the ethical and ecological integrity of this world for generations to come. It is a moral obligation".

M. Jean 2005[1]

"Perhaps the 21st century will be the age of miracles. If it is not, the human race as we know it is unlikely to see the 22nd".

J.R. Whitehead 1995[2]

"Do you believe in miracles?"

A. Michaels 1980[3]

Planet Earth is in the midst of a biodiversity crisis.[4] More species are being lost more rapidly than ever before. Many view it as the sixth mass extinction, but this one is quite different from the earlier ones. Unlike previous events, extinctions today are being caused not by some unavoidable catastrophe, but rather by the activities and behaviour of one superabundant, virtually omnipresent and dominant species – *Homo sapiens*. If maintenance of biodiversity is a primary goal of the conservation movement, then the movement is failing.

In principle, the ongoing loss of species can still be greatly reduced or curtailed. But in order for that to happen, we will need a new conservation paradigm. That paradigm must acknowledge and understand the lessons of history, the realities of the present, and what can be anticipated with reasonable certainty in the coming decades. It also must deal with the inevitable and inescapable uncertainties in a prudent and precautionary manner. That, in a nutshell, was the conclusion arising out of the IFAW Forum in Limerick, and it is a conclusion implicit and, in some places, explicit, throughout this book.

Every reader will have his or her own interpretations of the messages contained in this volume. Those interpretations will depend on individual circumstances, life experiences, education and training, and numerous other factors. They will depend, ultimately, on how the information presented is filtered through the readers "genes, values and belief systems".[5] Which is another way of saying, upfront, that not all readers – and likely not all the authors – will agree with our interpretations or recommendations, filtered as they are through our own "genes, values and belief systems". With that in mind, we will attempt to place our discussions into the context of history[6] and, most importantly, attempt to plot a way forward, if there is to be any hope of gaining ground in the pursuit of ecological sustainability, as the 21st century unfolds.

SPEAKING OF HISTORY

There is a tendency these days to think of conservation as largely a 20th century phenomenon. Nonetheless, the idea of conservation has been around for much longer than that. Mahesh Rangarajan, in a recent book about the history of India's wildlife, for example, writes:

> The protection of elephants became serious business by the time of the Mauryan rulers such as Ashoka (4th and 3rd Centuries BC). The *Arthashastra*, containing maxims of ancient statecraft, lays down the duties of the Protector of the Elephant Forests with no room for doubt: On the border of the forest, he should establish a forest for elephants guarded by foresters. The Superintendent should with the help of the guards... protect the elephants whether on the mountain, along a river, along lakes or in marshy tracts... they should kill anyone slaying an elephant.[7]

The comment sounds remarkably modern, describing laws to protect elephants, together with a means of enforcement, and penalties for those who were found to be breaking the law.

In the West, the philosophical roots of modern conservation can be traced back at least to the 16th century (Figure 1–1). The distinct eras of conservation identified in North America, dating back to 1600, generally describe the trends in much of the world, at least since about 1850.[8]

For the first 60 years of 20th century conservation, attention still remained focussed almost exclusively on "resources" of utility to humans. Beginning in the early 1960s, however, both progressive and protectionist schools of conservation[9] were subsumed into a larger, more encompassing environmental movement that began to realize that humans were very much part of the equation (Figure 1–1). People started to talk not just about the environment, but also about the human environment, e.g. the 1972 United Nations Conference on the Human Environment. By 1980, we entered a new phase in which humans (and human activities) became the central focus. Now it wasn't just about sustaining humans, other species, and their environments, but about sustaining economic growth and human development, under the mantra of something called "sustainable development",[10] e.g. the 1992 United Nations Conference on Environment and Development. Economics and development moved to centre stage, and steps were taken to roll back the environmental gains of the 20th century in order that economic growth and development could continue unabated. The pendulum has now swung so far that a Canadian environment minister recently felt comfortable suggesting that his Department of the Environment should actually be renamed the Department of Sustainable Economy.[11]

SOME TAKE-HOME MESSAGES FROM LIMERICK

The current state of the planet was briefly reviewed in Chapter 1, and the subsequent chapters provide additional information and detail, and so there is no need to repeat that material here. The message is, quite simply, that the environment continues to lose ground,[12] many species with which we share the planet continue to lose ground, and the human species continues to lose ground, on many different fronts. While those are obviously important take-home messages, they are not for us the most interesting ones. The reason for this is simple. Our failure to achieve ecological sustainability and the resulting problems facing the planet (or, more precisely, the human species) have been documented and discussed *ad nauseam* in the conservation and wider literature for decades.[13] Consider the following excerpt:

> Current discussion of the environmental crisis often centers on the pollution problems of the industrial world... Both food production and economic development prospects in Africa, Asia, and Latin America are now dimmed by accelerating destruction of the land's productivity... Rampant deforestation of mountain slopes results in the erosion of precious topsoil and a disastrous increase in flooding on the plains below. In drier zones, burgeoning populations create new deserts. While oil prices dominate headlines, the fuel crisis of the third of mankind dependent on increasingly scarce firewood for cooking intensifies.

The words sound familiar and ring true, and they could have been written in late summer and early fall of 2005. The fact is, however, they appear on the back cover of Eric Eckholm's 1976 book, *Losing Ground*, not to be confused with Mark Dowie's book of the same title, referred to in Chapter 1.[14] Not only do we keep repeating the same dismal facts – updated, of course – decade after decade, we even recycle book titles to characterize the enduring trends.

Another example:

> ...the time for decision is the present. With the consumption of each additional barrel of oil and ton of coal, with the addition of each new mouth to be fed, with the loss of each additional inch of topsoil, the situation becomes more inflexible and difficult to resolve. Man is rapidly creating a

> situation from which he will have increasing difficulty extracting himself.

Those words, very similar to words spoken at the IFAW Forum, were written by Harrison Brown, author of a book with the (almost) modern sounding title *The Challenge of Man's Future*. The year was 1954.[15]

One can go back even further. Aldo Leopold's *Sand County Almanac*, published posthumously in 1949, was all about the failure to achieve ecological sustainability, particularly in the US. "Conservation is getting nowhere," he wrote, "because it is incompatible with our Abrahamic concept of the land. We abuse land," he continued, "because we regard it as a commodity belonging to us".[16]

For those who, in 2006, promote the commercial consumptive use of nature and natural resources (the "wise users" referred to in Chapter 1 and discussed further by Sharon Beder in Chapter 5), Leopold had this to say,

> A system of conservation based only on economic self-interest is hopelessly lopsided. It tends to ignore, and thus eventually to eliminate, many elements in the land community that lack commercial value, but that are (as far as we know) essential to its healthy functioning. It assumes, falsely, I think, that the economic parts of the biotic clock will function without the uneconomic parts.[17]

We are reminded of another of Leopold's observations:

> Despite nearly a century of propaganda, conservation still proceeds at a snail's pace; progress still consists largely of letterhead pieties and convention oratory. On the back forty[18] we still slip two steps backward for every forward stride.

Those are sobering words for participants in yet another convention on the topic of ecological sustainability. But they also underscore another message that we took home from Limerick. That message is that although we have known about the problems for decades, we have had very little success in doing anything much to resolve them. Things continue to go from bad to worse, although even that reality tends to get lost through the problem of "shifting baselines".[19] Our individual and collective memories tend to be very short and each successive generation interprets present circumstances in relation to its memory of the past, rather than in relation to some earlier baseline, such as the memories of past generations.

That particular message raises a couple of questions. Knowing about the problems confronting the human condition, why have we continually failed to mitigate or solve them? And, how do we begin to chart a way forward that just might allow us finally to gain ground in the pursuit of ecological sustainability? That prescription – how to move forward – is the ultimate take-home message from the IFAW Forum and from this particular book. But before we attempt to pull together that prescription, let us first make some observations about what has transpired in the world since we departed Limerick in June 2004.

POST-FORUM HAPPENINGS

Two years have passed since the contributing authors to this volume gathered in Limerick to discuss conservation, and the pursuit of ecological sustainability. During that time, numerous articles, books and reports have appeared that are relevant to the discussion. A few of these are highlighted below.

The State of the Planet

More reports have appeared on the state of the environment, most notably, the *Millennium Ecosystem Assessment Reports*.[20] The findings will be of no surprise to the readers of this book. They include:

- Some 60 per cent of the planet's ecosystem services are currently being degraded by human activities.
- These activities include polluting the atmosphere with excess greenhouse gases, draining freshwater aquifers, overharvesting our forests and fisheries, polluting our oceans and introducing alien species to new regions. As a result, 20 per cent of the world's coral reefs have been lost, 40 per cent of the planet's rivers have been fragmented, and our climate has been disrupted...
- Conservation is essential to maintain and enhance humanity's quality of life...
- Simply put, we must find sustainability.

Among the numerous new books on the state of the planet was Ronald Wright's *A Short History of Progress*, published in late 2004, in which he reinforced the view that "our modern predicament is as old as civilization, a 10,000 year experiment we have unleashed but seldom controlled".[21] In early 2005, Jared Diamond's *Collapse: How Societies Choose to Fail or Succeed* covered some of the same ground, and more.[22] Both authors – like the participants in the IFAW Forum – called for reform but, tellingly, Diamond's modest suggestions for dealing with the human predicament are buried in the "further readings" section at the end of his massive tome, rather than being featured prominently in the actual text.

In September 2005, *Scientific American* published a special issue entitled "Crossroads for Planet Earth".[23] The cover carried the optimistic and promising banner, "A plan for a bright future beyond 2050". But when readers finally reach the article entitled "How should we set priorities?" the answer seems to be that mythological "well functioning market" discussed earlier in this book[24] – further confirmation, perhaps, that the "wise users" really are the dominant player in 21st century conservation.

Meanwhile, reports about the disappearance of the polar ice caps and the dire consequences for polar bears and other northern inhabitants, including humans, became increasingly common during 2005.[25] Elsewhere, the residents of the remote settlement of Tegua, on the South Pacific Island of Vanuatu, became possibly the first climate change refugees when they were relocated to higher ground because of coastal flooding problems.[26] The US space agency NASA now reports that 2005 was the warmest year on record.[27]

The Human Population "Crisis"

Since the IFAW Forum, the human population crisis has been in the news on several occasions. For decades it has been recognized that failures to achieve ecological sustainability on a global scale are ultimately related to too many people consuming too many resources too quickly. "We are eating away at our natural capital rather than living off the interest" was the way David Suzuki put it in his overview of the *Millennium Ecosystem Assessment.* One obvious thing we could do to address this particular problem would be to lower the human birth rate.

But wait. Overpopulation was *not* the "crisis" that was being discussed in the news over the past few months. The "crisis" being proclaimed was the *declining* birth rate in various parts of the world. Just before the IFAW Forum, the Australian government announced subsidies to encourage Australians to produce more children. Since 1 July 2004, it has paid new mothers A$3000 for each baby born. A year later, a triumphant news story heralded a new baby boom in Australia.[28]

Since Limerick, the push for increased human population growth has become a recurrent theme, particularly in developed countries. Just after his inauguration in the summer of 2005, Pope Benedict XVI lamented the declining birth rates across western Europe and urged the faithful to have more children.[29]

More recently, a front-page story in one of Canada's national newspapers carried the headline "The growing problem of shrinking population".[30] "There is a real crisis looming" was the header as the story continued on the inside pages. The "crisis" is that without immigration Canada's population could start falling immediately, that China's population is expected to start falling by 2025, and that the US, India, and parts of Africa will, according to the United Nations, see their populations start to drop after 2050. The Prime Minister of France was quoted as saying "We must do more to allow French families to have as many children as they want" and, predictably, the article reported that the French government was paying generous subsidies of Can$1400 (approximately €1000) per month to parents on the birth of their third child, with additional incentives to reproduce even more children. "Baby deficit" was the title on the world map included with the article.

Taming the Oceans and "Re-wilding" North America

In the summer of 2005, *Nature* published two telling commentaries offering putative solutions to specific failures to achieve ecological sustainability. The first, written by John Mara, addressed our failure "to manage the ocean's fisheries".[31] Neglecting a message repeated throughout this book that humans *can't* manage wild ecosystems,[32] and the historical evidence that agriculture has been an unsustainable activity throughout its 10,000-year history,[33] Mara unabashedly recommended that "Following the cultivation of the land for food, society must take the next step: domestication of the oceans".

A month later, an even more startling article appeared. Entitled "Re-wilding North America", it was co-authored by 13 individuals including, remarkably, Michael Soulé, sometimes called the "father of conservation biology". The paper suggested transplanting endangered Eurasian and African wildlife – including, among other species, cheetahs, elephants and lions – to the North American plains, not only to prevent their extinction but also to "reinvigorate wild places".[34] This "alternative conservation strategy for the twenty-first century" entirely neglects the observation (repeated in the *Millennium Ecosystem Assessment Reports* cited above) that the introduction of alien species is a major factor in the loss of biodiversity. It neglects long-held concerns about disease transmission from exotic wildlife to endemic and domestic animals, and the current worry about pandemics arising from novel diseases jumping from wildlife or domestic animals to people. It also neglects the fact that some North Americans, including many farmers and ranchers, are already troubled by the modest success of wolf reintroduction programs.[35] If a few indigenous predators can cause so much concern, one can only imagine the wildlife management problems that might arise if exotic and hungry predators – such as lions and cheetahs – began to appear on the North American prairies.

Political Interference in Science

Conservation science continued to take a back seat to politics over the past 18 months. In Canada, the government rejected the advice of its Committee on the Status of Endangered Wildlife in Canada (COSEWIC) to list a number of marine species as Endangered. Among the species the government chose not to list was Atlantic cod, *Gadus morhua*, despite the fact that it has been reduced to 1 per cent or less of its former abundance.[36] This turn of events was not unexpected. At the 1996–IUCN meeting in Montreal, the Canadian government, led by its Department of Fisheries and Oceans, vigorously disputed the addition of cod and other commercial fish species to the Red List of Threatened and Endangered Species. Background information provided by Jeff Hutchings – a COSEWIC member – in Chapter 6, gives reason to believe that Canada is unlikely to list cod as Endangered under its new *Species at Risk Act*. In Canada, adding species to the Endangered list is now the prerogative of politicians, regardless of the recommendations of the expert committee established to conduct the status assessments.

Another example comes from the US, where the Bush administration's continued disdain for science moved the Union of Concerned Scientists to issue a report in July 2004 entitled "Scientific Integrity in Policy Making: Further Investigation of the Bush Administration's Misuse of Science", an update of an earlier report that appeared in February 2004.[37] In 2005, Chris Mooney's book, *The Republican War on Science*, added another chapter to the wider story of *Science Under Siege: The Politicians' War on Nature and Truth*.[38]

For the purposes of our discussion, it is sufficient to note that in any debate over the pursuit of ecological sustainability or related topics – most notably global warming – science is increasingly a pawn in the political process. While this is not a new phenomenon, it seems to be an increasingly common problem. These days, when governments distinguish between "good" science and "bad" science, the former usually refers to science that agrees with or supports their policies whereas the latter is science that does not. We will return to the urgent need to separate science from the political process later in the chapter.

Intelligent Design

The science of evolutionary biology, which has so much to contribute to our understanding of the problems associated with gaining ground in the pursuit of ecological sustainability – from the characterization of biodiversity to its influences on the philosophical underpinnings of conservation, not to mention its contributions to understanding human behaviour[39] – was also singled out for renewed attack over the past two years. A largely American phenomenon, the most recent pseudoscientific attack by creationists on Darwin's theory of evolution brought back memories of the Scopes Monkey Trial of the 1920s[40] and the creation science debate of the 1980s.[41] For the 21st century, the creationists' argument has been reframed as "intelligent design".

The site of the most recent spectacle was the small town of Dover, Pennsylvania, where in September 2004 the local school board instructed biology teachers to teach "intelligent design"[42] as one alternative to Darwin's theory of natural selection. A number of parents objected and sued the board, claiming that intelligent design amounts to a religious belief that has no place in a biology course, because it breaches the separation of church and state mandated by the US Constitution.[43] In a remarkably insightful decision, Judge John E. Jones III (incidentally, a Republican appointee of President George Bush) ruled that intelligent design is "nothing less than the progeny of creationism" and should not be taught in public schools.[44] In early 2006, the Dover School Board rescinded its policy of presenting intelligent design as an alternative to evolution in high school biology classes.

History tells us, however, that we have not heard the end of this particular challenge to evolutionary biology. Indeed, even before the end of 2005, a Colorado senator announced that he was considering introducing legislation that could put intelligent design into science classrooms in that state.[45] And Kentucky governor Ernie Fletcher announced his support for adding intelligent design to the public school science curriculum,[46] a position he reiterated in his State of the Commonwealth address in early 2006.[47] The creationists' worldview seems likely to remain both a distraction and an impediment to the pursuit of ecological sustainability for a while yet.[48]

Eroding the Conservation Gains of the Past

One stated goal of the "wise-use" initiative is to roll back the conservation and environmental gains of the past forty years, in particular, environmental legislation. Less regulation, not more, is a key plank in the anti-environmentalists' platform. The most recent evidence that they are continuing to gain ground in their relentless push occurred in the fall of 2005, when US Congressman Richard Pombo, Chairman of the House Resources Committee, introduced the *Threatened and Endangered Species Recovery Act of 2005*. The bill, passed by the House in September, would compensate landowners if species-protection requirements hindered plans for development. It would also put political appointees in charge of making some scientific determinations and block the designation of "critical habitat" in some areas. While conservatives argued that the bill did not go far enough to protect private property rights,[49] environmentalists claimed that it gutted the *Endangered Species Act*.[50] A Senate version of the bill, introduced in December 2005, contained further provisions designed to please landowners. Secretary of the Interior Gale Norton is backing the efforts of Congress to rewrite the *Endangered Species Act* to give more power to political appointees and to provide landowners with additional tax breaks for "helping" endangered species.[51]

Renewed attempts by the Republican administration to open up Alaska's Arctic National Wildlife Refuge (ANWR) to oil drilling also continued in 2005. Senate Democrats, along with a few Republicans, managed to block the move, orchestrated by Alaskan Republican Senator Ted Stevens.[52] The battle, which pits environmentalists and native people in the US and Canada against proponents of development and the oil industry, will undoubtedly continue.[53]

In Africa, meanwhile, Kenya is in the process of redrafting its wildlife policy and amending its wildlife legislation. The government is under considerable pressure from lobby groups to follow many other African countries and promote the commercial consumptive use of wildlife. An early indication that the Kenyan government might be moving in a new direction occurred in early 2005 when plans were announced to export live animals to a zoo in Thailand. The planned shipment has, for now, been stalled by a court decision. Nonetheless, the debate continues, as does the pressure on the government to promote the commercial consumptive use of its wildlife under the mantra of "sustainable utilization". Until now, Kenya has distinguished itself among African countries by supporting non-consumptive uses of its diverse wildlife fauna.[54]

Some Sobering Reminders

Late 2004 and 2005 will be remembered for a number of environmental catastrophes. The tsunami in Asia, hurricane Katrina in the southern US,[55] a massive earthquake in Pakistan, and the spread of avian influenza reminded us that we do not and can never manage or control nature and "natural" phenomena.

The Current State of Conservation and Environmentalism

Finally, suggestions that the conservation movement (or environmentalism, more generally) needs to be revised and reinvigorated or be abandoned and replaced by a new paradigm[56] have become increasingly common since the IFAW Forum. Bill Vitek, writing in *Conservation Biology* in December 2004, argued that we need to "abandon environmentalism for the sake of the revolution".[57] His suggested approach not only throws "politics and spirituality into the scientific mix", but also "evolutionary biology, the second law of thermodynamics, and ecosystem dynamics". The "death of environmentalism" was also the subject of an article that stimulated wide debate in late 2004 and 2005.[58] Although the motivation behind the various calls for the abandonment or death of either conservation or environmentalism was apparently different in each case, it is notable that the idea was, and remains, "in the air". The time has come to move forward.

THE CHOICES TODAY

The choices available to us today are the same choices we have had for decades. We can remain in denial and argue that there is no problem, following in the tradition, for example, of Julian Simon and, more recently, Bjorn Lomborg.[59] Or, we can finally accept that there are problems, and look for solutions, as Harrison Brown recommended half a century ago.[60]

The contributing authors to this book have chosen the latter path. They have demonstrated that we humans do indeed have serious problems; they also demonstrated that we have a sufficient understanding of the factors involved to make rational decisions to solve many of the identified problems. And, most importantly, through their insights into the problems and possible solutions, they provide some hope for a better and more secure future, a hope that provides the essential motivation to work towards finding the solutions we seek.

Nonetheless, it also seems obvious that it really will take a revolution[61] – an economic, cultural and political revolution – to gain ground in the pursuit of ecological sustainability, let alone to achieve it and, even then, it will take decades to get there. So, what is needed now is a new conservation paradigm (or a New Conservation Movement) to head us in the right direction. The good news is we know how to do it. The lingering question, of course, is "Will we"?

REINVENTING WILDLIFE CONSERVATION IN THE 21ST CENTURY

The suggestion that it is time for the conservation movement to reinvent itself is neither radical nor new. More than half a century ago, Gifford Pinchot insisted that,

> conservation must be reinvigorated, revived, remanned, revitalized by each successive generation, its implications, its urgencies, its logistics translated in terms of the present of each of them.[62]

Nor is the need for occasional renewal unique to the conservation and environmental movements. As Valerie Curtis recently noted, even public-health campaigns have to reinvent themselves to grab people's attention.[63] She also offered some words of wisdom that apply to any movement contemplating renewal. As anyone who has campaigned for social change will know, getting people to do something, even something that is good for them, is never easy. Attempting to change human behaviour means understanding people's deepest desires, and campaigners need to target their audience's emotions, because emotions are the ultimate decision-makers. An essential step in that process, as Robert Worcester emphasized in

Chapter 17, is to get people "involved".

We must be frank, however. There are no magic bullets for facilitating social change – in this case, for implementing a new conservation paradigm and gaining ground in the pursuit of ecological sustainability. But this much is also obvious: change will not happen on its own, or overnight. It is a long-term proposition, one that if started today, might take 40 to 50 years to implement.[64] Nonetheless, much has been learned over the past 100 years to provide guidance if global society really wishes to pursue such a goal.

If individuals in society decide to make the ethical choice to pursue ecological sustainability, and to maintain biodiversity, and that in itself is a big IF, it is worth reiterating here that the pursuit of ecological sustainability, including the maintenance of biodiversity, is not about whether any exploitation or use is consumptive or non-consumptive,[65] but rather about whether a particular use is, indeed, biologically sustainable. In other words, ecological sustainability is not something that happens automatically as a result of commercial consumptive use and free trade in wildlife. In fact, history indicates, more often than not, that commercial use and trade are detrimental to wildlife populations and their habitats and efforts to make them "sustainable" are all too easily overwhelmed by market forces and exacerbated by lack of political will.

Scepticism aside, however, there are reasons for hope. Another world really is possible. It will just take a major social, economic, and political revolution to create it. One step in that direction involves the creation of a new and revitalized conservation movement, designed specifically to tackle the problems that confront the human condition today and into the foreseeable future. The question is, how do we proceed from here? We begin with some suggestions.

1. Establish clear goals

For present purposes, we will assume that the goal of the New Conservation Movement is to halt and reverse current population declines, and the loss of species and habitats, in pursuit of the ultimate goal: the achievement of biological and ecological sustainability.

2. Recognize that "conservation" is value-based

For the past 75 years or more, conservation has more often than not been portrayed as a scientific undertaking. Such a portrayal, however, is misleading for the simple reason that the decision "to conserve" depends not on science nor on scientific evidence, but rather on societal attitudes, values, and objectives. Decisions to remove a particular sustainable yield from a wild population, to declare a species endangered, to exterminate a "pest", or to maintain biodiversity – like concerns for the welfare of individual animals – are dictated by ethical choices and not by science.[66] Further, societal attitudes, values and objectives – not to mention needs – vary from place to place and from time to time, depending on a plethora of factors mentioned earlier in this book.[67]

How any individual, organization, or sovereign state, defines conservation will depend entirely on the values of the individual, organization or state. Each of the different schools of conservation identified in this book – preservationist conservation, progressive conservation, and "wise use" – have, for example, defined conservation in different ways, reflecting their different values or their different views of what is right or wrong, good or bad.

If different schools of conservation represent different (and, thus, competing) sets of values, and politics involves making choices between competing values[68], then it follows that conservation is ultimately about politics and political choices. If "all politics is local" – a common refrain amongst those who play in the political game[69] – so too must conservation ultimately be viewed as a local issue. Which begs for a definition of what is appropriately meant by the term local – a topic we will return to under the title of "Scale Issues" below.

Within the value-laden field of conservation, science has a vitally important role to play, but that role is very specific and somewhat "peripheral"[70] to mainstream conservation discussions, debates, and decisions. As David Orr recently put it,

> ...science on its own can give no reason for sustaining humankind [or anything else]. It can...[however] ...create the knowledge that will cause our demise or that will allow us to live at peace with one another and nature.[71]

We will discuss the role of science in conservation shortly.

3. Adopt a "geocentric" conservation ethic

Local, regional, and national differences in values notwithstanding, there is arguably still a need for a widely adopted conservation ethic characterizing the relationship between all humans and nature. This too is an old idea, but the urgency is greater now than ever before. It was central to Aldo Leopold's *Land Ethic,* in which he argued that we must adopt a more ecological and ecocentric[57] approach to our dealings with the rest of nature. What he seems to have meant is that we must abandon our anthropocentric worldview, where we are the centre of the universe and nature exists, and is used, solely for our benefit;[58] and we must recognize and accept that we – both as individuals and as a species – really are an integral part of the biosphere.

In some fields, however, the term *ecocentrism* has more precise connotations. Among some ethicists, for example, *ecocentrism* emphasizes species and ecosystems but, unlike *biocentrism*, does not explicitly include individual animals as a locus of moral concern. But, as several authors have already noted, and we will return to this idea shortly, there are good reasons – and numerous precedents – to recognize that individual animals (e.g. elephants, as Kumar and Menon argue in Chapter 8) have intrinsic value and, therefore, deserve moral consideration as well. The addition of individual animals to the mix suggests that the sort of conservation ethic we are searching for would best be described as geocentric.[74] *Geocentrism* (Earth-centred) assigns moral value to both the parts and the wholes of the Earth. In other words, individual animals, species, and ecosystems all have concurrent moral value – i.e. they are intrinsic ends in themselves, as well as being instrumental means to other ends.[75]

Traditionally, progressive conservation has been concerned primarily with the welfare of populations and species, leaving concern for individual animals to humane societies and animal welfare organizations such as IFAW. The adoption of a geocentric conservation ethic removes the artificial separation of individual animals and populations (which, of course, are simply collections of individuals belonging to the same species) and puts animal welfare where it naturally belongs – on the modern conservation agenda.

4. Recognize the central role of values in the formation of public policy

Promoting and implementing a new conservation paradigm is essentially an advocacy activity designed to influence public policy and legislation and, thereby, to facilitate social change. The recognition and acceptance that conservation is all about values is fundamental to understanding how best to proceed. Those who advocate the pursuit of ecological sustainability are, of course, promoting a particular set of values. Public opinion, one target of the advocacy, is also based on values.[76] When values are in dispute, as they frequently are in the field of conservation, society chooses among competing values (or visions of what is right or wrong, good or bad) – at least in the democratic world – through a process called politics. Politicians, who are sometimes referred to as "policy-makers" or "law-makers", are reactive creatures and they usually implement public policy[77] that reflects the values of those who "win" the political argument. Legislation has long been recognized as "the record... or the moral sense of the community".[78] More accurately, legislation reflects the moral sense of the "winners" of the political debate, and it is important to recognize here that the winners often represent a minority or a plurality of the population and, very often, not the majority.

It is from such an understanding of how "the system" works that the "radical right" has become, arguably, the dominant force in US politics today. Systematically, over some 40 years or more, conservatives set out their core values, set up organizations to promote those values, supported political candidates who stood for those values, and enacted legislation that reflected those values.[79] If progressive conservationists want to re-invent themselves and reverse the recent successes of the "wise-use" initiative, they have much to learn from the strategies and tactics employed by US conservatives, including the "wise-users", over the past few decades. We provide the outline of some of the things we have gleaned from that approach later in the chapter.

5. Clarify the role of science in public-policy formation

Some readers may have wondered about the lack of any mention of science in the previous section. After all, we frequently hear the view that natural resources policy should be based on the best available science. Academics, for example, often argue that policy should always be based on sound, relevant science. It wasn't so many years ago that three wildlife biologists published a paper in the mainstream wildlife literature entitled, "Adaptive Resource Management: Policy as Hypothesis, Management by Experiment".[80] The naïve implication was that natural resources management is actually a science experiment, with policy representing some sort of testable hypothesis. Governments frequently play a similarly misleading game when they justify particular policy decisions by arguing that they are "based on the best available science". Gro Harlem Brundtland – the former Prime Minister of Norway, former advocate of "scientific" whaling, and former chair of the World Commission on Environment and Development, mentioned elsewhere in this book[81] – reinforced such a view when she wrote in the pages of *Science* that,

> ... there is no other basis for sound political decisions than the best available scientific evidence ... science must underpin our policies.[82]

While an argument can be made that, in an ideal world, science should be the basis of many public-policy decisions, particularly in a field such as conservation, Brundtland and the others cited in the preceding paragraph must surely know that, in practice, it does not. It is values – the values of the politically relevant – and not science, that "are at the center of the natural resource policy process, and thus problem solvers must consider people's perspectives [rather than just science] in order to find solutions".[83] As Christopher Bernabo has pointed out, "Science is mute on the values that underlie the decisions societies make".[84] And it does not "provide the answers to the policy makers' ultimate questions". In the political world, subjective value judgments and political expediency determine which policy is "right".

Tim Clark put it this way in his policy textbook,

> Science alone offers no methods for integrating the broad range of relevant data ... required for intelligent management and policy decisions.[85]

In short, policy is not a "hypothesis", and management is not an experiment. Policy decisions are value-based and, ultimately, policies are determined by weighing the political costs and benefits of alternative policy options.

Nonetheless, science and scientists have an important role to play in the policy-formation process.[86] Scientists can provide the facts as they are understood, and they can reject certain claims as being false. In short, they can generally inform and inspire the debate and sometimes persuade policy-makers that one alternative is more desirable than another. Politicians will of course eagerly use the science when it happens to coincide with their own policy objectives.[87] When it doesn't, however, the scientific advice will be usually ignored or rejected. Regardless, the scientist's main role should be one of providing a range of options for policy-makers to consider, outlining the potential risks and benefits associated with each option. Obviously, informed policy decisions are better than uninformed decisions.

In order for science to maintain its role in the policy-formation process, however, there must be some changes made in how science is conducted in the conservation field. Scientific research, by its very nature, depends on the free and open exchange of ideas and information and the freedom to reject hypotheses in the face of new evidence.[88] Yet government scientists, in particular, increasingly face situations where bureaucrats or politicians attempt to influence their work and manipulate their findings – the sort of misuse and abuse of science that is on-going in the US (as noted above) and elsewhere today. When this happens, government scientists often find themselves unable to comment on management decisions or policies, and certainly they are well advised not to criticize their employer, even when decisions run contrary to their scientific advice.[89]

Similar pressures exist for scientists in academia and in the private sector who depend on governments, foundations, non-governmental organizations, or corporations for research grants and contract funding.[90] When pressed, many scientists defend their silence by arguing that science and, therefore, scientists, are unbiased and objective, and that commenting on policy issues, or becoming involved in public controversies, is somehow inappropriate. Such a defense neglects the obvious, that scientists are human and, like everyone else, have their own values and built-in biases. Even their selection of research topics involves value judgments, as does the determination of which research proposals get funded, and which do not.

For these and other reasons, conservationists must consider incorporating ethics increasingly into science and technology, as Bill Lynn recommended in Chapter 13. As he has written elsewhere,

> ... ethics is indispensable in any science or practice involving human beings. We cannot explain what we do, much less justify our actions, without some reference to the ethics-based norms that inform our individual and collective activity. For all these reasons, we cannot pretend that science and ethics are separate fields of endeavor. Rather a sound science... must be complemented by an equally sound ethic...[91]

There are other ways to reduce the bureaucratic and political interference that has become so rampant in the scientific world in recent years. One suggestion, made by Jeff Hutchings and colleagues a few years ago, is that government science should not be funded by governments, but rather by some politically independent institution,[92] operating at "arms length" from the political process. In other words, to function effectively, scientists and science must be given the same freedoms enjoyed by the judiciary in democratic societies.[93] In this way, science and the work of scientists would no longer be put in compromising positions when politicians make decisions that are contrary to their scientific advice.[94]

Conservation in the 21st century also requires "better models, metaphors and measures to describe the human enterprise relative to the biosphere".[95] Considerable progress has already been made on the development of a new generation of management models that facilitate decision-making, despite the ever present and inevitable gaps in knowledge.[96] E.J. Milner-Gulland's work on assessing the sustainability of the bushmeat trade in Chapter 20 is a recent example.

Finally, conservation biologists and other scientists who work in this field must become more professional and more accountable for the advice they provide to policy makers and management authorities. As Bill de la Mare notes in Chapter 21, when a bridge collapses, the engineers are held responsible; when a fish stock collapses, we still search for scapegoats, like seals or whales.[97]

6. Establish fundamental principles for 21st century conservation

Once it is acknowledged that conservation is shaped by attitudes, values and objectives, it is possible to draft a set of principles to guide the development and implementation of a new conservation paradigm. One way to design such a paradigm was outlined by Fred O'Regan in the Foreword of this book. He argued that we need to "deconstruct" the existing conservation movement and rebuild a new paradigm, saving the elements of the old approaches that seemed to work while abandoning those ideas and approaches that did not – a good example of learning from the lessons of history. To start this sort of process, we have begun to compile a list of some fundamental principles for 21st century conservation (Table 26-1). Most of these principles are drawn from the literature of progressive conservation. Some focus on individual animals, as noted above, others on populations, species, and ecosystems, providing further justification for adopting the term "geocentric" to describe conservation in the 21st century.

7. Identify real conservation problems and work towards finding real solutions

One of the questions raised earlier in this chapter was why, when we know about the problems confronting the human condition, have we so often failed to mitigate or solve them. One answer is that conservationists and environmentalists often become sidetracked and embark on actions that, in reality, do not actually address the problems at hand. Both government agencies and non-governmental organizations, for example, frequently confuse process with outcome.[116] Enormous amounts of time, energy, and resources are invested in holding meetings, developing management plans or strategic plans, doing research and monitoring, writing reports, and conducting the extensive bureaucratic business associated with conservation. Such activities – however necessary – are processes that do not directly promote conservation goals. They only do that if they are linked to the implementation of some direct action and often that simply never happens. Success can only be measured by progress toward the ultimate goal, such as the recovery of depleted and endangered populations.

Innumerable other examples exist where humans simply have failed to tackle real problems with initiatives that actually address the identified problems. For example, where there are problems with over-exploited wildlife populations leading to an increasing number of endangered species, the solution involves providing increased protection with a view to halting their declines and promoting their recovery, not rolling back protective legislation to facilitate the commercial consumptive use and free trade in wildlife, endangered species included.

In the case of over-fishing, it is well understood that the major issue is that too many boats and too many fishermen are chasing too few fish. The solution is to reduce the amount of fishing, and develop and implement ecologically sustainable fishing practices. The problem will not be resolved by subsidizing the fishing industry further to deplete the oceans[117] or by culling marine predators – particularly seals and whales – on the fallacious grounds that they are draining the oceans of fish.[118]

Similarly conservationists, environmentalists, and others, have known about the human over-population crisis for decades (some might say centuries). The logical way of dealing with it would be to lower the population growth rate, and to applaud those nations where that is now happening. The solution is not for governments (or others) to pretend that the real problem is a declining birth rate and to subsidize the production of more people, as is currently happening, in parts of the developed world.

Another good example relates to the current human food shortage in many parts of the developing world. One putative solution would be to solve the distribution problems that prevent so many people from receiving the food that is already available. That approach, which would deal directly with the problem, seems more likely to provide relief than recommending the development of genetically modified foods (a popular alternative suggestion) because – among other things –the same distribution and affordability issues that plague the delivery of existing food products also plague the delivery of GM technology.

A related example involves the problems of economic disparity and social inequity mentioned elsewhere in this book. Concerned countries could consider implementing real debt-reduction schemes and fostering good governance to ameliorate the situation, rather than promoting something called "sustainable development" to maintain the *status quo*.[119]

In conclusion, if we are to gain ground in the pursuit of ecological sustainability, and solve other problems confronting the human condition, conservation in the 21st century will have to do a much better job of finding solutions that are likely to produce measurable results.

8. Put old myths to rest

Discussions of what conservationists must do in the future can get somewhat esoteric, and concerned people sometimes ask, "But what can we do, beginning tomorrow morning?" One place to start would be for the progressive conservation movement – as part of its continuing evolution – to abandon forthwith a number of widely held myths that simply get in the way of moving forward. Myths such as the "balance of nature", which renowned British ecologist Charles Elton tried to put to rest more than 70 years ago; the myth of the free lunch, which Ron Brooks discusses in Chapter 16;[120] and the myth that creationism and its offspring, intelligent design, are legitimate scientific alternatives to evolution by means of natural selection. We must also abandon the

Table 26–1. Principles of Wildlife Conservation for the 21st Century.

- Wildlife[98] has aesthetic, cultural, ecological, economic, intrinsic,[99] recreational, scientific, social, and spiritual values that need to be acknowledged, preserved and passed on to future generations.
- Recognizing that wildlife has intrinsic value broadens conservation to include both individual animals and the populations they comprise. In other words, 21st century "geocentric" conservation must be concerned about the welfare of individual animals (traditionally termed animal welfare, or animal protection) and the welfare of wild populations (the traditional focus of progressive conservation).[100]
- As a consequence, "people should treat all creatures decently, and protect them from cruelty, avoidable suffering, and unnecessary killing".[101] This principle reaffirms that animal welfare is an integral part of modern conservation.
- Conservation measures that compromise the welfare of individual animals to achieve goals at the level of the population should not be the preferred means of addressing wildlife conservation issues.
- Capturing wild animals for live trade and captivity should not be permitted. Bringing individual wild animals into captivity for short periods may be permitted, however, to deal with animal welfare issues such as disease, injury or estrangement.
- All attempts to domesticate wild animals should be discouraged.
- Efforts to protect and conserve wildlife and wildlife habitat should begin long before species become rare and more costly to protect.[102]
- The maintenance of viable wildlife populations and functioning ecosystems should take precedence over their use by people.[103]
- Recreational and other uses of wilderness must not compromise the very essence of "wilderness" as untrammelled wild lands, and any such uses should be compatible with this basic principle.
- Wildlife belongs to everyone and no one. It is protected and held in trust for society by governments or appropriate intergovernmental conventions (i.e. central management authorities at an appropriate scale).[104]
- "Highly migratory species" belong to all nations and not just those that wish to exploit them.[105]
- Effective conservation of wildlife relies upon a well-informed and involved public.[106]
- Any material benefits derived from wildlife must be allocated by law – following consultation with the public, who collectively "own" the resource – and not by the marketplace, birthright, land ownership, or social position.[107]
- All individuals share the costs of conserving wildlife. Those whose actions result in additional costs should bear them.[108]
- The onus must be on those who wish to use nature and natural resources, e.g. exploiters and developers, to demonstrate that their actions will not be detrimental to the goal of achieving of biological and ecological sustainability.
- The exploiter/developer pays. Those exploiting wildlife should bear the full costs of ensuring that any exploitation is ecologically sustainable, including the cost of enforcing any catch limits, and any scientific research required to determine those catch limits.[109] Exploitation (and depletion) of wild living resources should no longer be subsidized by governments.[110]
- The use of wildlife for subsistence[111] purposes by human populations should not be equated with their commercial consumptive use.
- Use of wildlife and ecosystems should be frugal (parsimonious) and efficient (not wasteful), ensuring that any use is biologically and ecologically sustainable.[112] (Parsimonious use means taking as little as you need, rather than as much as deemed possible as implied, for example, by the idea of maximum sustainable yield.)
- Human development should not threaten the integrity of nature or the survival of other species.[113]
- Development of one society or generation should not limit the opportunities of other societies or generations.
- Each generation should leave to the future a world that is at least as biologically and ecologically diverse and productive as the one it inherited, and – given the current state of the planet – a world in which the physical environment (including the land, water, and atmosphere) is less polluted than the one it inherited.
- The establishment of protected areas – where human impacts, including exploitation and development, on wildlife and their habitats are reduced to an absolute minimum – is an essential component of any plan to achieve biological and ecological sustainability.[114]
- Ultimately, human use of nature must be guided by humility, prudence, and precaution.[115]

dangerously deceptive myth that we know how to *manage* wildlife, habitats or ecosystems. In short, we don't. As Sidney Holt noted more than two decades ago and reiterated in Chapter 3, all we can really do is attempt to manage human activities, to our own ultimate benefit or harm. Instead of pursuing the maximum sustainable yield – Peter Larkin wrote that obituary in 1973 – as the objective of fisheries management, we must look toward more "parsimonious use" (Table 26–1), with a view to reducing human impacts and avoiding irreversible ecological damage.

Then there are the various myths arising from neo-classical economics discussed by Bill Rees in Chapter 14. These include the view that we can have continually increasing growth in a finite world and the myth that the conservation of wildlife is best achieved through the marketplace. The conservation community must reject these myths – confident in the knowledge that money is not the common currency of ecosystems[121] – and join Brian Czech (Chapter 23) in promoting the transition to a steady state economy.[122]

The New Conservation Movement must also reject the oxymoronic idea of sustainable development, especially the three-legged stool model that depicts sustainability as being perched on legs of environment, economy and social equity.[123] In such a model, economic considerations always take priority over environmental and social concerns. It makes for a very unsteady stool. As John Oates argued in Chapter 18, we need another model, one in which society and economics (including development) are embedded within the environment. The reality is that we cannot have a healthy economy or ever hope to enhance social equity unless we have, first and foremost, a sustainable environment.[124] Using such a model, it will become axiomatic that when we use nature – as we inevitably will – we must strive to reduce the risks of causing irreversible damage to the biosphere and its component parts through a more vigorous application of the precautionary approach.

9. Issues of scale

A major challenge for a revitalized conservation movement will be to deal with problems of scale. Simply arguing that all conservation should be "local" – a natural conclusion arising earlier in this chapter from the recognition that conservation, ultimately, is a political activity – or that all conservation should be "community-based", or "decentralized", is insufficient without operational definitions of what these terms mean. In the future, conservation actions will have to be designed for the appropriate ecological scale – whether local, regional, national or international – depending on the nature of the problem being addressed. Providing definitions – or new words – to describe the appropriate scale for conservation decisions is complicated, for biological, ecological, social, and political reasons.

On the biological and ecological front, animals do not recognize political boundaries. The situation appears relatively simple when a population or species is confined to a single jurisdiction. In many countries, the appropriate scale for conservation action is enshrined in law. In Canada, for example, marine mammals are a federal responsibility, so the "community" of concern is the entire country and the management authority is the federal government. White-tailed deer, on the other hand, are a provincial responsibility, thereby defining a different "local" community and a different management authority.

Dealing with widely dispersed species that live in more than one jurisdiction, and highly migratory species that cross international boundaries, is a more difficult proposition. Indeed, it has been observed that the habit of crossing international borders can increase the probability that a species will become endangered.[125] Such examples suggest, on biological grounds alone, that the appropriate scale for conservation must be expanded to include those countries where the species in question lives. Now the relevant "community" is defined by "range states". There are already many long-established examples of bilateral and multilateral agreements that acknowledge this reality, most notably the *Convention on Migratory Species* and its daughter agreements. We need more of them, however, and we need them to become more effective in implementing modern conservation principles (Table 26–1). This is especially true at a time when we can predict that current species' ranges and ecosystem boundaries will continue to be modified as plants and animals respond to changing global temperatures in the coming decades.

For species living on the high seas – such as the great whales discussed in Chapter 7 – that are owned by everyone and no one, international agreements, such as the *International Convention for the Regulation of Whaling* and the *United Nations Convention on the Law of the Sea*, usually define the relevant "community" of concern and specify the responsible management authority.

On the social and political front, scale issues are important because people living in different regions and countries of the world, with their different cultures, religions, and political systems, have different attitudes, values, objectives, and needs.[126] On these grounds alone, 21st century conservation must – in the words of Char Miller – become "a more elastic concept, stretching to meet the distinct social contexts, cultural matrices, and political environments in which it must function…".[127] This, of course, is the antithesis of globalization – the road down which we currently gallop. As Gifford Pinchot – one of the fathers of modern conservation – noted more than 50 years ago, we must

> ...see to it that the rights of people to govern themselves shall not be controlled by great monopolies [or trans-national corporations, oligopolies, etc.] through their power over natural resources. [128]

Atherton Martin emphasized this point from the perspective of small island nations in Chapter 24.

The appropriate scale for conservation in the 21st century must be determined, therefore, on a case-by-case basis, taking into consideration the ecological, social, cultural, and political realities. On pragmatic grounds, we anticipate that there will be relatively few situations where the globalization of conservation philosophies and initiatives will be deemed the appropriate scale for pursuing the goal of ecological sustainability. And yet, we are reminded of the first law of ecology: everything is connected to everything else (Figure 1-2). All of "the distinct social contexts, cultural matrices, and political environments" mentioned previously are open subsystems within the global system: the biosphere.[129] Conservation decisions that happen in one subsystem usually have implications elsewhere. So, to be successful, a New Conservation Movement must also retain a global dimension that brings all the various subsystems together.

British MP Michael Meacher[130] – a former Secretary of State for Environment – suggests that this might be accomplished through a strengthened United Nations Environment Programme (UNEP), structured to function as a global environmental agency, backed by a World Environmental Court. The Court would function as "the supreme legal authority for settling issues regarding harm to the environment, in whatever form, by land or water, or in the air or upper atmosphere". "The basis of its legal operations," Meacher proposes, "would be a global environmental charter specifying the ecological conditions which must be met if the biosphere is to be allowed to operate within tolerable elasticities from its natural functioning". If Meacher's ideas seem fanciful, it is only because they underscore the magnitude of the social, economic and political revolution that is necessary if we are ever to achieve the goal of ecological sustainability.

PROMOTING AND IMPLEMENTING A NEW CONSERVATION PARADIGM

A major reason why the traditional progressive conservation movement has been losing ground for more than two decades is that it has been caught up in a "radical right-wing transformation" that has been developing, especially in the US, but elsewhere as well, over the past 40-50 years.[131] The anti-environmentalist "wise-use" initiative, which has essentially supplanted traditional progressive conservation over the past two decades, is itself a child of the conservative right.[132] It has essentially commandeered the language of conservation,[133] and as Don Hazen – among others – has noted, "When you control the language, you control the message, and the corporate media does the rest".[134] In response, parts of the old 20th century conservation movement have continued with their old methods and approaches, despite the fact that they no longer seem to produce the desired results, whereas others – including even some of the old mainstream conservation organizations – seem largely to have abandoned ship, allowing themselves to be co-opted either by the political establishment[135] or by the "wise-use" initiative itself.[136] Given these realities, what do progressive conservationists need to do if they really want to reinvent themselves and gain ground in the pursuit of ecological sustainability?

The paradigm shift required in order to stem the loss of species in the on-going extinction crisis will not happen on its own or overnight. It will require a more realistic appraisal of our current circumstances, and the public and political will to deal with the obvious problems that confront us. We can find little guidance in the traditional conservation literature on how specifically to resurrect a movement and implement a new paradigm. Some suggestions emerged, however, during the IFAW Forum, and there is a growing body of literature that provides additional ideas and guidance. The latter comes largely from two other movements that are, coincidentally and simultaneously, trying to re-invent themselves: the progressive or liberal movement in the US[137] and, somewhat ironically, the conservative movement in Canada.[138]

Like progressive conservationists, these two "movements" have had successes in the past but, in recent decades, have fallen on tough times. The two political movements are now studying the lessons of history and drawing heavily on the "textbook" written by the conservative right in the US as it rose to its present position of political power over the past 40 to 50 years. That "textbook" provides a useful starting point also for progressive conservationists who wish to promote a new conservation paradigm for the 21st century. The message is that if liberals (or Democrats) in the US, conservatives in Canada, or progressive conservationists worldwide are ever to resurrect themselves they must re-learn what it takes to gain the political power required to implement their policies and to achieve their goals.[139]

It would take an entire book to outline in detail what progressive conservationists must do to revitalize their movement over the next few decades. Here, we can only begin to provide an outline of some of the things that need to be considered and possibly implemented if we are to move forward in the years to come.

To begin

From the outset, conservationists and environmentalists must learn from the lessons of history, keeping in mind the old adage that those who do not learn from history are doomed to repeat it.[140] For wildlife conservationists, at least two particular histories are important: the history of the conservation movement itself, including an understanding of what has worked in the past and what has not; and the history of social and political change, which provides a "cookbook" on how to gain ground in the pursuit of ecological sustainability.

Today, conservationists and environmentalists are often viewed as purveyors of doom and gloom. But if people are convinced that they are doomed, it will be impossible to rally their support for positive change. It is essential, therefore, that individuals are provided with reasons for hope.[141] One way to do this is to celebrate past victories. They demonstrate that success has happened in the past and provide hope that it can happen again in the future. Without such hope, there really is no future for the movement. Conservationists must also be vigilant and ready to protect past gains, especially today when so many of them – as noted earlier – are under attack.

Another message is that conservationists must also begin to think bigger and over the longer term. We must abandon our emphasis on projects and single-issue campaigns and learn to think strategically about large moral goals.[142] While it remains essential to know where the other side is coming from, it is also important for the conservation movement to forsake its negative and reactive tendencies from the past, and to become a more positive and proactive promoter of a new vision. It must abandon its past defensiveness and be constantly on the offensive, "creating waves" [143] and raising its profile, in both the public and policy arenas.[144] In so doing, we have to keep in mind that human decision-making is driven not by rationality,[145] but rather by emotions.[146] Voters in democratic countries, for example, tend to cast their ballots according to their identity, i.e. in keeping with their values, and not necessarily even in keeping with their own self-interest.[147] As a consequence, we need to remember that facts and rational arguments (or even being "right") are rarely sufficient to win in the political arena. And let us remember that conservation decisions, ultimately, are political decisions.

A few words about long-term planning

Long-term planning is easier said than done. It is extremely difficult to set priorities, or to develop strategies and tactics, far in advance, simply because we don't know precisely how the world will change along the way. This is particularly true when it comes to conservation and the problems facing the human condition over the next 50 to 100 years.

Future uncertainty notwithstanding, there are some changes which can be anticipated with varying degrees of certainty that will have an impact on any attempt to implement a new conservation paradigm. Human population size, for example, is projected to increase from about 6.5 billion in 2005 to 6.8, 7.2, and 9.2 billion in 2010, 2015 and 2050 respectively.[148] Human population growth will be accompanied by increased demand for natural resources, including oil and natural gas, water, land (including wildlife habitat, for housing, agriculture, and other developments), forests, fisheries and wildlife, and wildlife products. As a result of this increasing demand, we can anticipate a further depletion of natural resources, including biodiversity, a further loss of productive soils, more interference with evolutionary processes (through selective exploitation, biotechnology and, likely, ill-conceived "restoration" projects[149]), and increased pollution (including increased greenhouse gas emissions), exacerbating the problem of climate change (likely, global warming). The average temperature of the Earth's surface has risen by 0.6 °C since the late 1800s; it is currently predicted to increase by another 1.4 – 5.8 °C by 2100.[150]

Increasing global temperatures have serious implications for the distribution and abundance of both plants and animals, which tend to be highly correlated not only with each other, but also with environmental factors such as ambient temperature.[151] We are currently anticipating the appearance and spread of new and emerging infectious diseases, e.g. an influenza pandemic is expected within the next two decades, and we are already dealing with the threat that avian influenza will be transmitted to humans on a global scale in the near future.

As far as the political environment is concerned, there are few signs that the conservatives, including the "wise-use" initiative, will back off anytime soon. There are some indications, however, that progressives are beginning to think about how to regain lost ground,[152] and hopefully, a revitalized conservation/environmental movement will be part of that initiative.

In the longer term, numerous prognosticators have suggested that we are moving toward a new hierarchy of global economic and military power, with the emergence over the next 50 years of China and India[153] as global powers, the rise of countries like Brazil and Indonesia,[154] the re-emergence of Russia,[155] the rise of Islam, and a concomitant decline in European and American influence. Given the central role of values in conservation, it goes without saying that any changes in the world power structure will be accompanied by shifts in prevailing values, in both space and time, that will have profound implications for conservation later in the 21st century.

A New Conservation Movement will also have to deal with the issue of shifting baselines referred to earlier in this chapter,[156] specifically, the incremental lowering of standards with respect to nature, where each successive generation lacks knowledge of how things used to be, redefines what is "natural" according to its personal experience, and sets the stage for the next generation's shifting baseline. In other words, a New Conservation Movement will have to overcome the progressive lowering of expectations that occurs naturally with each successive generation. More books and other educational initiatives on the history of the biosphere and the history of conservation might be one place to start.

Develop a coherent ideology[157]

A New Conservation Movement, like any social or political movement, must develop a coherent ideology. In George Lakoff's terms, this translates into knowing one's values. Since conservationists exist as part of a larger progressive movement, we might start by adopting established progressive values of particular relevance to a New Conservation Movement. Such values include empathy (or compassion), responsibility, protection (or security), fairness, fulfillment, community, cooperation, honesty, and open communication (transparency). We might add to this list respect for the law, to distinguish a New Conservation Movement from those who resort to unlawful activities to promote their causes.

While progressive conservationists, much like the "wise-use" initiative, have always focussed on the utilitarian values of nature, a New Conservation Movement must emphasize to a greater extent its humanistic, aesthetic, symbolic (Table 12–1), and intrinsic values (Table 26–1) as well. Most human societies, for example, value great works of art, not for their utilitarian value but because "they can be uplifting, make connections, engender feeling states that benefit our psyche or just make us wonder and engage in life".[158] So why don't we include similar arguments to safeguard biodiversity and the integrity of ecosystems?

Nigel Collar, a Senior Research Associate in the Conservation Science Group at the University of Cambridge, has taken this idea perhaps one step further, arguing that,

> A[n]... honest admission that the natural world is an inalienable component of the human capacity to experience freedom ... would transform the way we treat the natural environment and hence the prospects for long-term survival of biological diversity...[159]

That is certainly food for thought as we develop a set of values for a New Conservation Movement. So too is Bill de la Mare's more pragmatic and selfish utilitarian argument, in Chapter 21, that humans must come to value ecological sustainability as much as they value safe bridges and safe buildings. And at the risk of raising the odd eyebrow among some of our colleagues, we might also suggest adding love and respect to our list of putative values for 21st century conservation.

Numerous authors from a variety of fields have argued the need to foster a love and respect for nature, reasoning that we will only protect that which we love. Bill Rees raised the topic in Chapter 14, as he has done on previous occasions[160] to mixed reviews, even from some of the contributors to this book.[161] The idea, of course, is not new. Aldo Leopold himself made a similar suggestion in his *Land Ethic*, more than 50 years ago. There, he wrote,

> We abuse land because we regard it as a commodity belonging to us. When we see land as a community to which we belong, we may begin to use it with love and respect. [162]

The American nature writer, Joseph Wood Krutch extended Leopold's message, particularly in an essay entitled "Conservation is Not Enough".[163] Quoting Henry David Thoreau, he wrote,

> ...without some realization that "this curious world" is at least beautiful as well as useful, "conservation" is doomed. We must live for something besides making a living.

A bit later, he concluded,

> Without some "love of nature" for itself there is no possibility of solving "the problem of conservation"

Similar sentiments have been attributed to Baba Dioum, a Senegalese conservationist, in a 1968 speech to the general assembly of the International Union for the Conservation of Nature and Natural Resources, New Delhi, India. In that speech, Dioum argued,

> In the end we will conserve only what we love.[164]

More recently, David Orr, a regular contributor to *Conservation Biology*, has written:

> A growing number of scientists now believe, with Stephen J. Gould, that "we cannot win this battle to save [objectively measurable] species and environments without forging an [entirely subjective] emotional bond between ourselves and nature as well – for we will not fight to save what we do not love".[165]

That "love" is increasingly "in the air" was also in evidence at the 2003 World Wolf Congress, in Banff, Alberta, Canada. One prominent member of the scientific community even used the word during his plenary lecture. In our experience, such a thing would not have been observed at "proper" scientific meetings only a few years ago, and the fact that it was mentioned in Banff, with little obvious reaction from the audience, must be viewed as evidence that perhaps attitudes are indeed changing, even within parts of the scientific community. One begins to wonder if a New Conservation Movement could turn planet Earth into one of Kevin Roberts' "lovemarks". Roberts defines lovemarks as,

> ...a new way of thinking about the things we love...they are about love and respect: they speak to us as thinking and feeling human beings. Lovemarks embody mystery, sensuality and intimacy.[166]

Such a change in how humans view the planet would require, among other things, the kind of spiritual renewal that David Orr has characterized as one of the four challenges of sustainability.[167]

Frame the debate

Lakoff[168] devotes an entire chapter to the topic of framing, which we won't attempt to repeat here. Suffice it to say that frames are mental structures that shape the way people see the world".[169] A New Conservation Movement must frame the issues to reflect its values, vision, and mission in its own language. It must avoid attacking the values of the opposition because all that does is keep their views at the forefront of the debate.[170] It must re-gain control of the language of conservation, remembering that whoever controls the language controls the debate.[171] It must no longer allow others to co-opt the language of conservation and give new meanings to old words, as they have been doing for the past two decades.[172]

It is also imperative that conservationists frame their issues in ways that resonate with target audiences. Facts presented in ways that do not fit an audience's frame will have little or no impact. The same facts, when framed to appeal to human emotions, can have a very different outcome. A recent – if somewhat accidental – example relates to the campaign against the international trade in live animals.

For decades people and organizations concerned with animal welfare and conservation have campaigned against the international trade in live wildlife, including both captive and wild birds. Such campaigns have largely been based on animal welfare concerns (facts) about individual animals in trade, or on the very real conservation concern that such trade will threaten wild populations (another fact). They have failed, however, to stop international trade in captive and wild animals, whether they be tropical fish, amphibians, reptiles, birds, or mammals.

In the latter half of 2005, there was considerable concern about the spread of bird flu (avian influenza) through Asia to Europe, and much speculation about the possibility that the virus might "jump" to humans and mutate, causing a pandemic that one "expert" suggested might kill from 5 million to 150 million people. The issue became reframed like this: the international trade in wild birds is a threat to human health. Within weeks the European Union imposed a ban – albeit a temporary one – on the international trade in live birds.

What happened? The threat of a pandemic that could kill millions suddenly got people emotionally "involved" in the issue, much like Farmer Brown's pig was "involved" in Worcester's parable in Chapter 17. Once the link with humans was made and journalists and politicians became involved, governments soon got the message and responded. The response was exactly what animal welfarists and conservationists had been working towards for decades: a cessation, however temporary, in the live-bird trade.

If the idea that the trade in wild birds (and, indeed, the trade in wild animals) is undesirable becomes reinforced frequently enough, we can possibly envision the day when all such trade might be banned completely. The immediate take-home lesson, however, is this. Once the international trade in live birds was framed in terms of human health, and humans became emotionally engaged and involved, governments suddenly responded in the desired way. This example underscores the messages of Worcester and Lakoff, mentioned earlier. And, it makes Kevin Roberts' point that, "Our reason must have emotion to engage it".[173]

The message for geocentric conservationists in the 21st century is that they need to frame their debates in ways that appeal to people's emotions,[174] and in ways that involve them directly, if they have any hope in making progress in the pursuit of ecological sustainability.

Build infrastructure

Historically, the conservation movement has been comprised largely of a plethora of non-governmental organizations functioning on local, regional, national and international scales. While such organizations frequently get together and form coalitions on specific issues, they remain, nonetheless, independent entities – essentially competitors in the non-governmental organization "industry" – that more often than not end up vying with each other for public funding, media attention, and the kudos when victories are achieved.

To build a New Conservation Movement[175] will require a different approach. Essentially, it will demand a new infrastructure to establish a critical mass of progressive conservationists dedicated to implementing the movement's principles (e.g. Table 26–1). We discuss some aspects of such an infrastructure below.[176]

Set up think tanks

Following the example of successful political movements, a New Conservation Movement might begin by creating a network of think tanks, strategically located in both the developed and the developing world.[177] Such an initiative would require a major investment of capital that could only come, initially at least, from the major conservation and environmental organizations referred to by Steve Best in Chapter 25.

These think tanks would function as "idea factories" for a New Conservation Movement. They would study public-policy issues in conservation, think strategically about how most effectively to influence the media, the public, and the policy-makers, and figure out how to market their ideas to various target audiences, and get people who share their values into public office. In short, such think tanks would undertake the strategic thinking required to facilitate the kind of societal revolution that would be needed to achieve, decades down the road, something approaching ecological sustainability.

Such an undertaking would require a dazzling range of expertise. That expertise would include not only the usual advocates (campaigners, lawyers), scientists (conservation biologists, ecologists, evolutionary biologists, modellers, statisticians, taxonomists, etc.), educators, environmental economists, policy wonks, and pollsters, but also wildlife veterinarians and other animal-welfare specialists, cognitive psychologists (who study how people think and make decisions), earth scientists, engineers, historians (mindful of the lessons of history), media and communications experts, linguists (who understand the art of language), philosophers (including, especially, ethicists), political scientists, political strategists, social scientists (studying various human societies, past and present, and the inter-relationships between individuals and society), theologians, and, some would argue, convincingly, poets.[178]

In addition to hiring the best and brightest thinkers who share the values of the New Conservation Movement to staff the think tanks, long-term planning requires the education and training of the next generation of geocentric conservationists. There is, therefore, a need to identify proactively promising young conservationists and environmentalists, and provide them with scholarships, fellowships, internships, professorships and research grants, not only to get them into the movement but also to keep them there for the long haul. It is also vital to place the brightest young progressive conservationists in various professions, including law, the media, the civil service, politics, and the various branches of academia mentioned above, much as the radical right (and the "wise-use" initiative) has already managed to do, throughout much of the modern world.

Develop grassroots initiatives

Part of any successful political campaign involves grassroots initiatives. It wasn't so many years ago that business consultant Edward Grefe borrowed approaches from a variety of single-issue groups – including the environmental movement – in order to advise and teach corporations "how to become involved in the political process".[179] Part of that process involved "grassroots" participation. More than a decade ago, Mark Dowie argued that grassroots activism – local people acting locally to protect local places – represents the new civil and moral authority, the "fourth wave" of the environmental movement.[179] Bearing in mind our earlier comments about "scale", a New Conservation Movement needs to develop support for its values and initiatives on scales appropriate to the issues at hand, and such initiatives will often benefit greatly from grassroots' support.

Form alliances and build coalitions

Existing conservation and environmental groups not only need to form alliances with each other and build coalitions to promote common values, they also need to form alliances and build coalitions with other progressive organizations who share their core values. One only has to look at the "informal amalgamation of individuals and groups" involved in the "wise-use" initiative to appreciate the importance of this recommendation. In short, "United we stand, divided we fall".

Market the Ideology

A New Conservation Movement can have the best ideas in the world, but they will have no impact at all unless the message gets out to the media and the public – one of the things that conservative think tanks have accomplished. These think tanks provide a constant stream of information and deliver their specific messages through the proliferation of books, magazines (both in hard copy and electronic formats), opinion pieces, press releases, and the like. A New Conservation Movement needs to redouble its existing efforts, using the traditional methods as well as placing increasing emphasis on modern technologies, including the Internet (everything from informative web sites, to chat forums and blogs) and podcasting, to get its ideas and principles into the public discourse. It needs not only to promote the ideals of geocentric conservation, but also to repeat them over and over again until they become an accepted part of the public discourse. And, of course, it will have to fend off any attempts by the opposition to undermine those efforts.

In an interview during the Limerick Forum, pollster Bob Worcester noted that in Britain, only about one person in four was even aware of the problems of wildlife conservation, and that in other places in the world the number decreases to one in ten, fifteen, or twenty. Only half of all Britons have even heard of the *Kyoto Protocol*, despite all the coverage and debate it has generated in recent years. And, as he noted in Chapter 17, 45 per cent of the British public in 2004 had never heard of "biodiversity"; some even confused it with homosexuality. Even more sobering, perhaps, is the recent news that Britons say they do know about intelligent design, and that 41 per cent of respondents in a recent Ipsos MORI poll would like to see it taught in science lessons in British schools.[181]

Obviously, there is an urgent need for the public to become better informed on conservation and environmental issues. A New Conservation Movement must, therefore, revise its communications (or public relations) strategies to include mechanisms for better informing the public, e.g. improved education programs, and better communication with target audiences.[182] It must not only generate awareness, but also persuade people to take on its values, to get involved and, ultimately, to be moved into action.[183]

The communications strategy must include mechanisms for getting the message out through the media, and so a media strategy will have to be developed as well.[184] If we follow the example of the conservatives in the US, a New Conservation Movement might want to think about establishing media outlets in radio and television and on the Internet, and to organize speaking engagements in schools and on college campuses around the world.

At every stage, political strategists, using polling data and focus groups, will be required to make the messages most appealing to the target audiences. That process will involve a continual monitoring and assessment of past actions, and societal trends, and will require constant reframing of the issues to keep the public and the politicians engaged as various baselines shift with the passage of time.

Efforts must be concentrated on the uninformed and undecided. There is, as Worcester reminded us, little point in preaching to the converted or attempting to get the other side to change its core values. Even among those "objective and unbiased scientists" we mentioned earlier, it is well known that paradigm shifts are never fully completed until the old guard dies out. In short, strongly held opinions and values are virtually impossible to change in the face of new evidence, even amongst those trained to reject hypotheses in the course of their career work.

Funding

All of the above initiatives will require significant inputs of funding. The most likely source of such funding, initially at least, will be the major Green organizations discussed in Chapter 25. Major foundations that share the values of the New Conservation Movement are another obvious source of support. It is sobering to remember, however, that the anti-environmentalists (the "wise users") have a 25-year head-start on any New Conservation Movement. They already have their infrastructure in place and their pockets are lined with contributions from big business, corporations and, indeed, governments. While any New Conservation Movement will initially have to depend on private funding sources, it should gain increased support from individuals and, if it begins to gain ground on political fronts, from government sources as well, much as the liberal political movement has done in Canada in recent years.[185]

Provide inspiring leadership

Like all social movements, a New Conservation Movement will require dynamic and charismatic leadership. Leaders must have vision and idealism, and they must inspire other leaders and not just followers. Without inspiring leadership, one can virtually guarantee that the planet will continue to lose biodiversity at increasing rates throughout the remainder of the 21st century.

The question is, where will that leadership come from? In theory, it could come from within governments, international institutions, non-governmental organizations, or even trans-national corporations. Best argues convincingly in Chapter 25 that initially, at least, such leadership is most likely to come from within the existing large and powerful conservation and environmental organizations that, collectively, he calls "the Greens." But, he argues, to assume such a leadership role, the Greens will have to reinvent themselves, before they attempt to reinvent the entire movement. Provocative, yes. A huge challenge, yes. But also a tremendous opportunity, should the Greens wish to rise to the challenge and take it on.

One can envision, perhaps somewhere down the road, that leaders may also emerge from within green corporations and, if the movement is successful, from green political parties jumping reactively on the bandwagon, as is their usual custom.

Parting thought

Promoting and implementing a new conservation paradigm (or movement) in pursuit of ecological sustainability requires the same skills and approaches as any other political or value-based movement. But in order for any political movement to gain ground, it is necessary to gain political power and to exercise that power to the maximum. It has been done before, and it can be done again. And that alone, perhaps, is another reason for hope.

ONE MORE MOUNTAIN TO CLIMB

While the way to build a successful social and political movement is reasonably clear, we would be remiss if we did not acknowledge that there is one huge mountain to climb in order to re-invent a conservation movement that just might be successful in gaining ground in the pursuit of ecological sustainability. And that mountain is us – *Homo sapiens*.

Humans are good Darwinian animals. We are concerned primarily with selfish self-interest, and we are addicted to consumerism.[186] The altruistic behaviour required to solve so many of our global problems does not come easily to Darwinian animals. And, as Ron Brooks noted in Chapter 16, there is nothing in our evolutionary legacy that predisposes us (or any other animal) to undertake long-term planning, let alone the sorts of long-term conservation planning discussed earlier in this chapter.

In addition, while all life forms seem to practice deception in one form or another, humans appear to have elevated this art to include self-deception. Indeed, we seem to have evolved what some people call "Machiavellian intelligence".[187] One of the unfortunate spin-offs, especially in the present context, is that we have "considerable capacity for self-delusion when the truth is unpalatable".[188]

Let us give one example of self-delusion playing a role in the current situation. As a species, we have difficulty coming to grips with our individual mortality. Rather than confront our limited life spans, most human societies have developed myths to get around the issue. These myths take a variety of forms but almost invariably involve a "life-after-death."[189] How can a species in which individuals deny their own mortality even begin to contemplate the death (i.e. extinction) of our entire species?[190] We doubt that we can. But even if we could, it is unlikely that we will. The possibility of our extinction – well, actually, its inevitability – is simply too far down the road, i.e., beyond our own lifetimes and those of our children and grandchildren, to disturb us very deeply or keep our attention for very long.

The nature of individual human beings notwithstanding, there is an added complication: the behaviour of humans in groups. It has been argued that "the evolution of [human] intellect [including Machiavellian intelligence] was primarily driven by selection for manipulative, social expertise within groups, where the most challenging problem faced by individuals was dealing with their companions."[191] It is not surprising, then, that further evidence of deception and self-deception becomes apparent when one examines even superficially the behaviour of humans in groups. We discuss two examples below: the behaviour of governments and the behaviour of corporations, the two most powerful institutions in the modern world.

Let us begin with governments and examine the practice of politics. Politics is "bloodless conflict among individuals, groups, and nations… among alternative values, or … competing visions of what is 'good'".[192] Politics is also "the father of lies. In political arenas … the participants will distort the advantages of their positions and the disadvantages of their opponents." Fair enough, but "they will [also] shade the truth – first for their audiences; then in many cases, for themselves".[193] Shading the truth for their audiences is deception;[194] some would call it "tactical deception".[195] Shading the truth "for themselves" requires self-deception.

Of course, this sort of behaviour is to be expected. Machiavelli (1469-1527) long ago described the need for such deceptive behaviour among political leaders in his classic work *The Prince*.[196] But what perhaps is less well understood, are the consequences that often emerge from such group behaviour.

The late historian, Barbara Tuchman, for example, described a "phenomenon … noticeable throughout history: the pursuit by governments of policies contrary to … the self-interest of the constituency or state involved".[197] She termed this phenomenon "wooden-headedness." "Wooden-headedness," she wrote, "plays a large role in government … acting according to wish while not allowing oneself to be deflected by facts".[198] The mindless pursuit of continued economic growth, sustainable development, and an unsustainable agriculture in a finite world, may one day be recognized as examples of 21st century wooden-headedness.

Like governments, corporations are made up of human beings and so provide another opportunity to examine human group behaviour.[199] Corporations, like individuals, are characterized primarily by selfish self-interest. They are concerned, first and foremost with their shareholders and with profit-maximization. The bottom line is more important than the public interest. Generally, corporations exhibit no moral conscience, as a number of high-profile recent events attest. Indeed, if corporations were people, their behaviour would be seen to exhibit all the traits of a prototypical "psychopath".[200] And, if that were not bad enough, *The Economist* informs us that the "infinitely more powerful … modern state has the capacity to behave … as a more dangerous psychopath than any corporation can ever hope to become".[201]

Either way, when you put a number of selfish, self-interested individual Darwinian human beings into a group situation (e.g. have them form a government, or work together in a bureaucracy or a corporation) what typically emerges is something that appears quite un-Darwinian: decisions that ultimately act against (rather than promote) the collective self-interest of the group.

It is considerations such as these that lead Ron Brooks, the quintessential cynic, to remark in Chapter 16, that "any reasonable consideration of evolution by natural selection encourages the conclusion that conservation, *sensu* altruistic preservation of other life forms, will neither evolve nor be sustained." But then, even he leaves the door slightly ajar. "Conservation", he says, "... must be viewed as a moral position" and, ultimately, "be defended in pragmatic... terms". Which is essentially what we've been arguing throughout.

Discussions such as this usually end on an optimistic note. As Donald Johansen and James Shreeve pointed out in their book, *Lucy's Child: The Discovery of a Human Ancestor*, even if our large brains and "superior" intelligence evolved as a result of the advantages gained by deception and self-deception, there were spin-offs.[202] Our intelligence – they argued – permits us to ponder the consequences of our actions and to take appropriate steps to ameliorate the situation. It gives us the "flexibility" to shake off the shackles of self-deception. According to this scenario, we could all wake up one day, smell the coffee (or the flowers) and, drawing on all of our ingenuity, find real solutions to the human population problem, the depletion of natural resources, the loss of biodiversity, environmental pollution, and global warming, not to mention the disparities among rich and poor. And, perhaps that will happen. Our concern is, however, that such optimism may simply be symptomatic of more self-deception.

WHEN WILL THINGS CHANGE?[103]

At first glance, the prospect that humans anywhere will overcome their self-delusional tendencies and take real steps to gain ground in the pursuit of ecological sustainability seems unlikely at this juncture in history.

The world's dominant institutions – governments and multi- and trans-national corporations – continue their blind pursuit of increasing economic growth and increased profits. Today, it is difficult to imagine how individuals and nongovernmental organizations that recognize the folly in such policies can really do anything to change things in the future. But, as several authors have noted recently, the fact is that they probably can, if only they have the will to do so. And, while governments and corporations may represent the two most powerful institutions in the world today, there is a third potential power broker to consider, and that power broker is "people".

According to Nicanor Perlas, modern society can be viewed as having three realms: the economic, the political and the cultural.[204] On the world stage, the economic realm is the purview of international corporations and three major international organizations concerned with development: the International Monetary Fund, the World Bank – both established by the West following World War II – and the World Trade Organization – which emerged out of the General Agreement on Tariffs and Trade (GATT) in the mid-1990s.[205]

Governments, of course, dominate in the political realm. These days, corruption within government is one of the very serious threats to biodiversity conservation. Yet, as David Orr has pointed out, only governments moved by an ethically robust and organized citizenry (see below) can act to ensure a fair distribution of wealth within and between generations.[206] Good governance is also required to produce the necessary conservation legislation, and to ensure compliance through enforcement, locally, regionally, nationally, and internationally. As William Laurance recently noted,

> Determined efforts will be needed by all nations to fight corruption and slow rampant overexploitation of natural resources. For their part, wealthy nations must strive to become part of the solution, rather than part of the problem.[207]

And, as Nigel Collar has concluded,

> ... it is only when the political institutions on this planet legally enshrine a permanent, quasi-absolute value for nature – a non-economic, non-negotiable value based on our shared sense of the importance of nature to the human mind – that we will have the framework and basis for a new era of ethically driven global governance.[208]

That leaves the cultural realm, and it is occupied in Perlas's scheme by civil society, which is comprised of individual humans. While it is obvious that both corporations and governments currently have the power and are in control of the world situation, individual humans also have power, should they choose – or, in some parts of the world, be allowed – to exercise it. Noam Chomsky recently went as far as to suggest that public opinion today is the world's "second superpower".[209]

In democratic societies, at least, people have power in the political realm because they cast the votes that elect the politicians. Governments (not to mention political parties and individual politicians) really have only one over-riding goal and that is to get elected or re-elected. Consequently, they are, as we have noted elsewhere, reactive – as opposed to being proactive – which explains why they spend so much of the people's money monitoring public opinion. And therein lies the answer to the question that opened this section. Al Gore – before he became the US Vice President – put it as succinctly as anyone:

> When enough people insist upon change to embolden the politicians to break away from the short-term perspective ... the political system will fall over itself to respond to this just demand that we save the environment for future generations.[210]

Corporations are just as vulnerable as governments to public pressure but, in the economic realm, it is not votes that count but rather consumer behaviour in the marketplace. If no one buys their products, they lose their market share, their profits drop and their shareholders get anxious. Eventually, they respond in predictable and understandable ways and bow to public pressure.

There are an increasing number of examples where the power of civil society (or public opinion) has shaped events on local,[211] regional and global scales. Examples include the civil rights and women's movements of the 1960s[212] and the environmental movement during its heyday of the 1960s to the early 1980s.[213] A more recent example was the derailment of the *Multilateral Agreement on Investment* (MAI), perhaps one of the first examples of the power of the people being mounted using the Internet.[214]

Chomsky also recently wrote,

> One can discern two trajectories in current history: one aiming toward hegemony [i.e., power], acting rationally within a lunatic doctrinal framework as it threatens survival; the other dedicated to the belief that 'another world is possible'...[215]

Our parting question is whether civil society will remain uninvolved, complacent, and silent (remember Richard Nixon's "silent majority"[216]) and accept the "lunatic doctrinal framework" that currently threatens human survival. Or will it say, enough is enough, exercise its vast political power, and demand change, in the belief that "another world" really is still "possible"?

Either way, it will provide a test of James Suroweicki's hypothesis about the "Wisdom of Crowds".[217] He argues that "Large groups of people [and here, we're thinking of Perlas's civil society] are smarter than an elite few [governments and corporations], no matter how brilliant – better at solving problems, fostering innovation, coming to wise decisions, even predicting the future". Our earlier observations about human nature and the emergent behaviour of humans in groups notwithstanding, we can only hope he is right.

FINAL THOUGHTS

The fathers of the conservation movement – including Gifford Pinchot and Aldo Leopold to mention but two – have been dead now for more than 50 years. The conservation and environmental movements to which they contributed so much have been losing ground for the past 30. We are now in the midst of a life-and-death crisis – an extinction crisis, an environmental crisis, and a human crisis. People who care about biodiversity, the environment, and other people can no longer afford to "sleepwalk" into the future.[218] To effect change 50 years from now, we have to wake up tomorrow morning and do what's necessary to change the course of the future. "What's necessary" includes the creation of a New Conservation Movement dedicated to gaining ground in the pursuit of ecological sustainability, and focused on the acquisition and use of political power to achieve it.

The time for involvement and action is now. Every day we wait, we pay a price. If we wait another 20 years, it may well be too late. That is the challenge and the ultimate message emanating from the IFAW Forum and from this book.

ACKNOWLEDGEMENTS

This final chapter owes a huge debt of gratitude to all the speakers at the IFAW Forum in Limerick and to all the contributors to this book. A preliminary version of this chapter was presented as the wrap-up talk in Limerick. We thank those colleagues who served as sounding boards, critics, and reality touchstones, and those who generously provided us with additional ideas and source material during the writing of this chapter. In this regard, we particularly thank Stephen Best, Ron Brooks, Ward Chesworth, Sheryl Fink, Bill Lynn, Vassili Papastavrou, Rosalind Reeve, Rob Sinclair, and Sue Wallace. We are also grateful to those who critically reviewed earlier drafts of the manuscript: Stephen Best, Ward Chesworth, Michael Earle, Sheryl Fink, Jan Hannah, Bill Lynn, Vassili Papastavrou, Rosalind Reeve and Sue Wallace. David Helton not only read a draft but edited it as well. We hasten to add that we did not always take the advice we received and therefore it should not be assumed that any of the aforementioned necessarily agrees with our interpretations, conclusions or recommendations. Any errors of fact or interpretation are of course ours.

NOTES AND SOURCES

1 This quotation comes from a speech of The Right Honourable Michaëlle Jean, Governor General of Canada, on the occasion of her installation. 27 September 2005, Ottawa, Ontario. Available at http://www.gg.ca/media/doc.asp?lang=e&DocID=4574.

2 From J.R. Whitehead. 1995. *Radar to the Future.* Epilogue. Available at http://www3.sympatico.ca/drrennie/memoirs.html. We place the emphasis on "as we know it", lest anyone assume incorrectly that either Whitehead or us

are suggesting that humans will disappear from the face of the planet within the next 100 years.

[3] North American hockey fans will recognize this exclamation from ABC broadcaster, Al Michaels, on the occasion of the unlikely victory by the US hockey team over Russia at the winter Olympics in 1980.

[4] Reviewed in Lavigne, Chapter 1, and Willison, Chapter 2, among other chapters in this book.

[5] As David Suzuki argued in his 1987 book *Metamorphosis* (Stoddard Publishing Co., Limited, Toronto, Canadà), "There is no such thing as objective reality. We select our experiences through the filters of our genes, values and belief systems".

[6] See for e.g. Lavigne, Chapter 1, Beder, Chapter 5.

[7] Rangarajan, M. 2001. *India's Wildlife History*. Permanent Black Publishers, New Delhi, India.

[8] Lavigne, Chapter 1, Table 1-4.

[9] Lavigne, Chapter 1.

[10] See for e.g., Lavigne, Chapter 1; Beder, Chapter 5; Eves, Chapter 9; Rees, Chapter 14, Brooks, Chapter 16; Oates, Chapter 18; Martin, Chapter 24.

[11] See *The Globe and Mail* (Toronto), 12 October 2004, reporting on comments made by Canada's Environment Minister, Stéphane Dion (an economist by training) on his environmental agenda.

[12] As soil scientist, Ward Chesworth points out, we continue to lose "ground" both literally (see Chesworth, Chapter 3) and metaphorically.

[13] Steve Best raised this specific issue during a coffee break at the IFAW Forum in Limerick. It reminds us of that well known quotation attributed to Santayana: "Those who do not learn from history are doomed to repeat it".

[14] Eckholm, E.P. 1976. *Losing Ground: Environmental Stress and World Food Prospects*. W.W. Norton & Company Inc., New York; Dowie, M. 1995. *Losing Ground: American Environmentalism at the Close of the Twentieth Century*. The MIT Press, Cambridge, MA.

[15] We are indebted to Ward Chesworth (see Chapter 3), for introducing us to Harrison's book (see Chesworth, W., M.R. Moss, and V.G. Thomas. Afterword: The End of History. pp. 139-140. In W. Chesworth, M.R. Moss, and V.G. Thomas (eds.). 2002. *Sustainable Development: Mandate or Mantra?* The Kenneth Hammond Lectures on Environment, Energy and Resources. 2001 Series. Faculty of Environmental Sciences, University of Guelph, Guelph, Ontario, Canada.). The original source is: Brown, H. 1954. *The Challenge of Man's Future*. Viking Press, New York. The quote is on p. 265.

[16] Leopold, A. 1966. *A Sand County Almanac with Essays on Conservation from Round River*. A Sierra Club/Ballantine Book, New York.

[17] Ibid.

[18] For readers unfamiliar with the term "back forty", it is a North American expression meaning a large, remote, often barren stretch of land, or an out-of-the-way place.

[19] This view of "shifting baselines" is from Pauly, D. 1995. Anecdotes and the shifting baseline syndrome of fisheries. Trends in Ecology and Evolution 10: 430. Also see O'Regan, Foreward; and Papastavrou and Cooke, Chapter 7.

[20] Available at http://www.millenniumassessment.org/en/Article.aspx?id=60.

[21] Wright, R. 2004. *A Short History of Progress*. House of Anansi Press, Toronto, Canada.

[22] Diamond, J. 2005. *Collapse: How Societies Choose to Fail or Succeed*. Viking, New York, NY. For a comparison of Wright and Diamond's books, see Chesworth, W. 2005. One kick at the can: *Collapse* by Jared Diamond and *A Short History of Progress* by Ronald Wright. Geoscience Canada, 32:140-144.

[23] Crossroads for Planet Earth. Scientific American, September 2005.

[24] Gibbs, W.W. 2005. How should we set priorities? Scientific American, September, pp. 108-115; but see Beder, Chapter 5; Rees, Chapter 14; Brooks, Chapter 16; and Czech, Chapter 23.

[25] The reports continued in 2006. See for e.g., Boyd, R.S. 2006. Rapidly shrinking Arctic ice could spell trouble for the rest of the world. Knight Ridder Newspapers. 10 January. Available at http://www.mercurynews.com/mld/mercurynews/news/politics/13593258.htm?template=contentModules/printstory.jsp.

[26] Pareti, S. 2005. Environment: Pacific's First Climate Change Refugees? Islands Business International. Available at http://www.islandsbusiness.com/islands_business/index_dynamic/containerNameToReplace=MiddleMiddle/focusModuleID=5548/overideSkinName=issueArticle-full.tpl Also see Doyle, A. 2005. Pacific islanders move to escape global warming. Reuters, 5 December. Available at http://www.stopglobalwarming.org/learn/read.asp?60722126 2005.

[27] Zabarenko, D. 2006. 2005 was warmest year on record: NASA. Reuters, 25 January. Available at http://www.commondreams.org/headlines06/0125-08.htm.

[28] Sydney Morning Herald.2005. Howard's baby bonus sparks jump in birth rate. 12 June. Available at http://smh.com.au/news/National/Howards-baby-bonus-sparks-jump-in-birth-rate/2005/06/12/1118347641050.html?oneclick=true.

[29] See Benedict, XVI. 2005. General Audience. Wednesday 31 August 2005. Text available at http://www.vatican.va/holy_father/benedict_xvi/audiences/2005/documents/hf_ben-xvi_aud_20050831_en.html; also see Pope urges faithful to have more children. Anon. 2005. Catholic News. September. Available at http://www.cathnews.com/news/509/3.php.

[30] Saunders, D. The growing problem of shrinking population. The Globe and Mail, 8 October 2005. p. A1, A20-A21.

[31] Mara, J. 2005. When will we tame the oceans? Nature, 436:175-176.

[32] See e.g., Holt, Chapter 4; de la Mare, Chapter 21.

[33] See Chesworth, Chapter 3.

[34] Donlan, J., H.W. Greene, J. Berger, D.E. Bock, J. Bock, D.A. Burney, J.A. Estes, D. Foreman, P.S. Martin, G.W. Roemer, F.A. Smith, and M.E. Soulé. 2005. Re-wilding North America. Nature, 436: 913-914.

[35] Robbins, J. 2005. The look of success. Conservation In Practice, 6(4):28-34. The sub-head to this article reads, "In the wake of successful wolf reintroductions, managers who once fervently defended wolves are now faced with killing them. Are we ready for modern predator management?"

[36] For background on the depletion of Atlantic cod, see Hutchings, Chapter 6.

[37] Union of Concerned Scientists. 2004. Scientific Integrity in Policy Making: Investigation of the Bush administration's abuse of science. Available at http://www.ucsusa.org/scientific_integrity/interference/reports-scientific-integrity-in-policy-making.html.

[38] Mooney, C. 2005. *The Republican War on Science.* Basic Books, New York; Wilkinson, T. 1998. *Science Under Siege: The Politician's War on Nature and Truth.* Johnson Books, Boulder, CO.

[39] See Brooks, Chapter 16.

[40] More on the 1925-Scope's Monkey Trial can be found at: http://www.law.umkc.edu/faculty/projects/FTrials/scopes/scopes.htm. This famous trial was the basis for the Jerome Lawrence and Robert E. Lee play, *Inherit the Wind* (Bantam Books, Toronto, 1960) which has become a classic in American Theater.

[41] Eldredge, N. 1982. *The Monkey Business. A Scientist Looks at Creationism.* Washington Square Press, Pocket Books, New York, NY; Ruse, M. 1982. *Darwinism Defended. A Guide to the Evolution Controversies.* Addison-Wesley Publishing Company, Reading, MA.

[42] Intelligent Design argues that life is so complex and elaborate that some greater wisdom has to be behind it. e.g., see Toland, B. 2005. 'Intelligent design' supporters to state their case in court. Pittsburg Post-Gazette, 25 September 2005, available at http://www.post-gazette.com/pg/05268/577404.stm.

[43] Interested readers can easily follow the on-going discussion on the Internet. Also see Badkhen, A. 2005. Anti-evolution teachings gain foothold in U.S. Schools. Evangelicals see flaws in Darwinism. San Francisco Chronicle, 30 November 2004, available at http://sfgate.com/cgi-bin/article.cgi?file=/c/a/2004/11/30/MNGVNA3PE11.DTL.

[44] See Easton, N.J. 2005. US judge rejects intelligent design. Broad ruling finds Pa. School board promoted religion. The Boston Globe, 21 December 2005, available at http://www.boston.com/news/nation/washington/articles/2005/12/21/us_judge_rejects_intelligent_design/ ; for Judge Jones' actual ruling, see http://www.pamd.uscourts.gov/kitzmiller/kitzmiller_342.pdf.

[45] Anonymous. 2005. Intelligent design not proper for our public schools. Reporter-Herald, Loveland Colorado. 28 December. Available at http://www.lovelandfyi.com/opinionstory.asp?ID=3408.

[46] Kenkel, B. 2006. Science is silent, not hostile, toward religion. Kentucky Kernel, Opinion. 12 January 2006.

[47] Fletcher, E. 2006. State of the Commonwealth 2006. 9 January. Available at http://governor.ky.gov/mediaroom/speeches/20060109SOC.htm.

[48] Remember, for example, James Watt's creationist view of the relationship between humans and natural resources, in Lavigne, Chapter 1.

[49] See National Center for Public Policy Research. 2005. Rep. Pombo's Endangered Species Act Reform Proposal Would Do More Harm than Good, Analysis Says. Press Release, 31 August 2005. Available at http://www.nationalcenter.org/PRTESRAEndangeredSpecies805.html.

[50] See for e.g., Defenders of Wildlife. 2005. House Guts Endangered Species Act. Release. 29 September 2005, available at http://www.saveesa.org/news/pr092905.html.

[51] Miller, J. 2006. Norton backs Endangered Species Act revamp effort. Associated Press. 6 January 2006, available at http://www.signonsandiego.com/news/nation/20060106-0017-wst-norton-esa.html.

[52] See http://www.whitehorsestar.com/auth.php?r=40697.

[53] See http://www.emagazine.com/view/?3031&printview.

[54] See for example, Mugisha and Ajarova, Chapter 10; Menon and Lavigne, Chapter 12.

[55] The devastating effects of Katrina on the city of New Orleans were chillingly predicted years before the hurricane hit. See Fishetti, M. 2001. Drowning New Orleans. Scientific American, October. Such predictions were reiterated in Bourne, J.K. Jr. 2004. Gone with the Water. National Geographic Magazine. October. Available at http://magma.nationalgeographic.com/ngm/0410/feature5/?fs=www3.nationalgeographic.com. For people who automatically dismiss "doomsday" predictions, there is a moral to this story.

[56] Lavigne, D.M. 2004. The Return of Big Brother. BBC Wildlife, May 2004. pp. 66-68. Also see O'Regan, in the Foreword to this book.

[57] Vitek, B. 2004. Abandon environmentalism for the sake of the revolution. Conservation Biology, 18:1463-1464.

[58] Shellenberger, M. and T. Nordhaus. 2004. The death of environmentalism. Available at www.thebreakthrough.org/images/Death_of_Environmentalism.pdf; Orr, D.W. 2005. Death and resurrection: the future of environmentalism. Conservation Biology, 19: 992-995. Additional discussion may be found in Grist Magazine, available at http://www.grist.org/news/maindish/2005/01/13/doe-reprint.

[59] e.g. Simon, J. 1992. Scientific establishment now agrees: Population growth is not bad for humanity. pp. 557-578. In J.H. Lehr (ed.). *Rational Readings on Environmental Concerns.* Van Norstrand Reinhold, New York, NY; Lomborg, B. 2001. *The Skeptical Environmentalist.* Cambridge University Press, Cambridge, UK; also see Giles, J. 2003. The man they love to hate. Nature, 423:216-218; Pope, C. and B. Lomborg. 2005. The State of Nature. Foreign Policy, July/August.

[60] Brown 1957.

[61] By "revolution", we mean a political or social revolution: a sudden, momentous or sweeping change in a situation, a fundamental change in the values, political institutions, social structure, leadership, and policies of a society. We tailored our definition from several dictionary sources.

[62] Cornelia Pinchot speaking at the dedication of the Gifford Pinchot Memorial Forest in Washington State, 15 October 1949, shortly after the 3rd anniversary of her husband's death; cited from Miller, C. 2001, *Gifford Pinchot and the Making of Modern Environmentalism.* Island Press/Shearwater Books, Washingon/Covelo/London. p. 11.

[63] Curtis, V. 2004. Psychology and the art of persuasion. New Scientist. 18 December 2004, p. 21.

[64] Lakoff, G. 2004. *Don't Think of an Elephant: Know your Values and Frame the Debate. The Essential Guide for Progressives.* Chelsea Green Publishing, White River Junction, Vermont. Lakoff argues that it took the Conservative right 40 years to move from relative obscurity into the position of power it enjoys today. He also argues that perhaps progressives can reinvent themselves somewhat more quickly, largely because they don't have to write the textbook as they go. The conservatives have done that for them!

[65] See Corkeron, Chapter 11.

[66] See Menon and Lavigne, Chapter 12.

[67] See Menon and Lavigne, Chapter 12, Lynn, Chapter 13, and Worcester, Chapter 17.

[68] Donovan, J.C., R.E. Morgan, and C.P. Potholm. 1981. *People, power, and politics. An introduction to Political Science.* Addison-Wesley Publishing Company, Reading, MA.

[69] O'Neill, T.A. and G. Hymel. 1994. All Politics is Local and Other Rules of the Game. Bob Adams, Inc., Holbrook, MA.

[70] Doern, B.D. 1981. *The peripheral nature of scientific and technological controversy in federal policy formation.* Science Council of Canada. Ottawa, Ontario. 108 pp.

[71] Orr, D.W. 2002. Four Challenges of Sustainability. Conservation Biology, 16:1457-1460. The quote is on p. 1459.

[72] Leopold's "ecocentrism" seems to fit the definition provided by Sterling, E. and M. Laverty. 2004. Intrinsic Value. Available at http://cnx.rice.edu/content/m12160/latest/. According to them, biocentrism or ecocentrism involves "the notion that life is the center of the universe and humans are a separate but equal part of nature".

[73] The latter definition is also from Sterling and M. Laverty 2004.

[74] See Lynn, Chapter 13; also see Lynn, W.S. 1998. Contested moralities: Animals and moral value in the Dear/Symanski debate. Ethics, Place and Environment, 1(2): 223-242.

[75] Lynn 1998, p. 231.

[76] See Worcester, Chapter 17.

[77] Policy in this context may be defined as a governing principle, plan, or course of action reflecting particular values, objectives and interpretations of knowledge, that together determine the path of administration, and provide a basis for the enactment of legislation.

[78] Henry Salt (1894) cited in Nash, R.F. 1989. *The Rights of Nature: A History of Environmental Ethics.* The University of Wisconsin Press, Madison, WI. p. 30.

[79] Patent, J. and G. Lakoff. 2004. Conceptual Levels: Bringing it Home to Values. Available at http://www.rockridgeinstitute.org/projects/strategic/conceptlevels.

[80] Lancia, R.A., T.D. Nudds and M.L. Morrison. 1993. Adaptive Resource Management: Policy as hypothesis, management by experiment/Opening comments: Slaying slippery shibboleths. Transactions of the North American Wildlife and Natural Resources Conference, 93:505-508.

[81] e.g. Lavigne, Chapter 1; Hutchings, Chapter 6; and Eves, Chapter 9.

[82] Brundtland, G. 1997. The scientific underpinning of policy. Science, 227 (5324):457.

[83] Clark, T.W. 2002. *The Policy Process. A Practical Guide for Natural Resource Professionals.* Yale University Press, New Haven, CT. p. 181.

[84] Bernabo, J.C. 1986. Perspectives on Capital Hill: Notes from a Former Fellow. Eos, 67(7), 18 February.

[85] Clark 2002, p. 9.

[86] For a recent example, see Wikelski, M. and S.J. Cooke. 2006. Conservation physiology. Trends in Ecology and Evolution, 21(2): 38-46.

[87] Donovan *et al.* 1981.

[88] For a earlier discussion of this topic, see Lavigne, D.M. 1999. The Hawaiian Monk Seal: Management of an Endangered Species. pp. 246-266. In J.R. Twiss Jr. and R.R. Reeves (eds.). *Conservation and Management of Marine Mammals.* Smithsonian Institution Press, Washington and London.

[89] e.g. Lavigne, D.M. 1995. Seals and Fisheries, Science and Politics. Invited Paper. Symposium II, The role of science in Conservation and Management. Eleventh Biennial Conference on the Biology of Marine Mammals. 14-18 December 1995. Orlando, FL (available at http://www.imma.org/orlando.html); Hutchings, J.A., C. Walters, and R.L. Haedrich. 1997. Is scientific inquiry incompatible with government information control. Canadian Journal of Fisheries and Aquatic Sciences, 54:1198-1210.

[90] Lavigne, D.M. 1985. Canada's Sealing Controversy: The Issues and the Interest Groups. Brief submitted to the Royal Commission on Seals and the Sealing Industry in Canada. La Vie Wildlife Research Associates Ltd., Rockwood, Ontario, Canada.

[91] This succinct quotation comes from Lynn, W.S. 2004. Interdisciplinary collaboration and the "moral turn" in wolf recovery. 16th Annual North American Interagency Wolf Conference. Working Collaboratively Toward Long-Term Wolf Conservation. Abstract. Also see Lynn, William S. 2004. The Quality of Ethics: Moral Causation in the Interdisciplinary Science of Geography, pp. 231-244. In R. Lee, and D. M. Smith. *Geographies and Moralities:International Perspectives on Justice, Development and Place.* Routledge, London.; Also, see Lynn, Chapter 13.

[92] See Lavigne, D.M. 1997. The role of science in fisheries management. Advisory Committee on Protection of the Sea. Conference on Oceans and Security. Panel on Oceans and Seas of the Americas. US House of Representatives. 19-21 May 1997. Washington, DC; also see Lavigne 1999; Hutchings *et al.* 1997.

[93] Hutchings *et al.* 1997; Lavigne 1997.

[94] Lavigne 1997.

[95] See Orr 2002, p. 1459.

[96] See Holt, S.J. and L.M. Talbot. 1978. New principles for the conservation of wild living resources. Wildlife Monographs, 59:1-33; Taylor, B. 1993. "Best" abundance estimates and best management: Why they are not the same. Technical Memorandum NOAA-TM-NMFS-SWFSC-188. National Oceanic and Atmospheric Administration, La Jolla, CA; Wade, P.R. 1998. Calculating limits to the allowable human-caused mortality of cetaceans and pinnipeds. Marine

Mammal Science, 14:1-37; Wade, P.R. and R.P. Angliss. 1997. Guidelines for assessing marine mammal stocks (GAMMS). Technical Memorandum NMFS-OPR-12. National Oceanic and Atmospheric Administration, Silver Springs, MD.

[97] See de la Mare, Chapter 21.

[98] Here, we follow the definition provided in Wildlife Ministers' Council of Canada. 1990. A Wildlife Policy for Canada. Canadian Wildlife Service, Ottawa, Canada: "All wild organisms and their habitats – including wild plants, invertebrates, and microorganisms, as well as fishes, amphibians, reptiles, and the birds and mammals traditionally regarded as wildlife" (p. 6).

[99] While many readers may tend to associate "intrinsic value" with an animal rights agenda, the fact is that traditional progressive conservationists have long recognized the "intrinsic value" of wildlife. (For the purposes of the present discussion, we define "intrinsic value" simply as "the inherent worth of something independent of its value to anyone or anything else"; see Sterling, E. and M. Laverty, 2004.) Recognition of the intrinsic value of animals (or wildlife) is included, for example, in the Preamble to the European Convention on the Conservation of European Wildlife and Natural Habitats (the Bern Convention, 1979); and in Wildlife Minister's Council of Canada. 1990. *A Wildlife Policy for Canada.* Minister of Environment, Canadian Wildlife Service, Ottawa. The Netherland's 1992 *Animal Health and Welfare Act* recognizes that animals were not created just for the benefit of humans and that they have intrinsic value; intrinsic value is also recognized in the Preamble of the *Convention on Biological Diversity* (1992), and in the *Earth Charter* (2000) although, in the latter, the actual words do not appear. Principle 1.1a reads, "Recognize that all beings are interdependent and every form of life has value regardless of its worth to human beings" (available at http://www.earthcharter.org/files/charter/charter.pdf). Evidence that the idea has penetrated the mainstream scientific literature may be found in May, R.M. 2001. Foreward. pp xii-xvi. In J.D. Reynolds, G.M. Mace, K.H. Redford, and J.G. Robinson (eds.). *Conservation of Exploited Species.* Conservation Biology 6. Cambridge University Press, Cambridge, UK. In this foreward, May (now Lord May) acknowledges the idea that all life forms have "inherent rights". It must be added, however, that the recognition of "intrinsic value" or "inherent rights" generally appears to have had little impact to date on the way humans have conducted their affairs. Clearly, this will have to change if the New Conservation Movement is to gain ground in the pursuit of ecological sustainability.

[100] These sentiments are captured, for example in the mission statement of Tufts University's Center For Animals and Public Policy, Tufts Cummings School of Veterinary Medicine: The mission of the Center, founded in 1983, is to support and encourage scholarly evaluation and understanding of the complex societal issues and public policy dimensions of the changing role of animals in society. Work conducted by the Center is based on the tenet that animals matter in and of themselves, that human and animal well-being are linked, and that both are improved through enhanced understanding of their interactions. Available at http://www.tufts.edu/vet/cfa/about_mission.html.

[101] The quotation is from the second World Conservation Strategy, *Caring for the Earth: A Strategy for Sustainable Living.* Published in partnership by IUCN – The World Conservation Union, UNEP – The United Nations Environment Programme, and WWF – World Wide Fund for Nature. Gland, Switzerland. p. 14, Box 2. Geist, Chapter 19, also deals with killing for frivolous purposes, and prevention of waste was one of Gifford Pinchot's three principles of Conservation, in 1910 (see Lavigne, Chapter 1, p. 3).

[102] Vassili Papastavrou (Chapter 7) has long advocated this principle and IFAW has a history of promoting protection for relatively abundant species, including harp seals, *Pagophilus groenlandicus.* But what is most encouraging is the appearance of this principle in the state of Nevada's recently announced Wildlife Action Plan. The plan is described as a proactive, progressive and cost-effective approach to conserving fish and wildlife. Available at http://www.pahrumpvalleytimes.com/2005/12/23/sports/ndow-conserve.html.

[103] Wildlife Ministers' Council of Canada 1990, p. 9.

[104] See Geist, Chapter 19.

[105] As currently enshrined in the United Nations Convention on the Law of the Sea.

[106] Ibid. Also see Geist, V. 1988. How markets in wildlife meat and parts, and the sale of hunting privileges jeopardize wildlife conservation. Conservation Biology, 2:1-12; Geist, V. 1989. Legal trafficking and paid hunting threaten conservation. Transactions of the North American Wildlife and Natural Resources Conference, 54:172-178.

[107] Ibid. Geist 1988.

[108] Wildlife Ministers' Council of Canada 1990, p. 9.

[109] We thank Vassili Papastavrou for this addition.

[110] See Earle, Chapter 15.

[111] We consider that subsistence hunting has different objectives from commercial hunting. Subsistence hunting is generally done to meet nutritional requirements of the hunting community, and not to put dead animals, their parts or derivatives into commercial trade. For further discussion of subsistence hunting in the context of modern conservation, also see Lavigne *et al.* 1999.

[112] *Caring for the Earth* 1991, p. 14, Box 2.

[113] Ibid.

[114] The creation of protected areas has always been an important part of progressive conservation. Such areas must be large enough to be of ecological significance, if they are to achieve their objectives, and the importance of corridors linking protected areas has been discussed extensively in recent years. The continuing importance of establishing protected areas to maintain biodiversity is discussed, for example, in Willison, Chapter 2, and in Oates, Chapter 18.

[115] See, for example, Willison, Chapter 2; Holt, Chapter 4; Papastavrou and Cooke, Chapter 7; Campbell and Thomas, Chapter 22.

[116] For a more detailed discussion of one such example, see Lavigne 1999.

[117] See Earle, Chapter 15.

[118] Tamara, T. and S. Ohsumi, 1999. Estimation of total food

consumption by cetacens in the world's oceans. Institute of Cetacean Research (ICR), Japan. Also see Lavigne, D.M. 2003. Marine Mammals and Fisheries: The Role of Science in the Culling Debate. Pp. 31-47. In N. Gales, M. Hindell and R. Kirkwood (eds.). *Marine Mammals:Fisheries, Tourism and Management Issues.* CSIRO Publishing, Collingwood VIC, Australia.

[119] Lavigne, Chapter 1; Beder, Chapter 5; Rees, Chapter 14; Brooks, Chapter 16; also see Beder, S. 1996. *The Nature of Sustainable Development.* Second Edition. Scribe Publications, Newham, Australia; Lavigne, D.M. 2002. Ecological footprints, doublespeak, and the evolution of the Machiavellian mind. pp. 61-91. In W. Chesworth, M.R. Moss, and V.G. Thomas (eds.). *Sustainable Development: Mandate or Mantra.* The Kenneth Hammond Lectures on Environment, Energy and Resources 2001 Series. Faculty of Environmental Sciences, University of Guelph, Guelph, Canada.

[120] See Brooks, Chapter 16.

[121] Prominent ecologist, Eugene Odum, used to say, "Energy is the common currency of ecosystems"; see Odum, E. 1971. *Fundamentals of Ecology.* Third Edition. Philadelphia: W.B. Saunders Company.

[122] See Czech, Chapter 23.

[123] For a discussion, see Dawe, N.K. and K.L. Ryan (2003) The faulty three-legged-stool model of sustainable development. Conservation Biology, 17 1458-1460.

[124] Ibid.

[125] Ehrenfeld, D.W. 1971. *Biological Conservation.* Holt, Reinhart and Winston of Canada Ltd., Toronto.

[126] Menon and Lavigne, Chapter 12, Martin, Chapter 24.

[127] Miller 2001, p. 94.

[128] Pinchot, G. 1945, cited in Miller 2001, p. 365.

[129] We thank Ward Chesworth, the author of Chapter 3, for reminding us of this important point while reviewing an earlier draft of this chapter.

[130] Meacher, M. 2004. Natural Governance. Resurgence. Issue 222:28-31. Available at http://www.resurgence.org/resurgence/issues/meacher222.htm.

[131] See D. Hazen in Lakoff 2004, p. xi and elsewhere.

[132] See for e.g., Lavigne, D.M., V.B. Scheffer, and S. R. Kellert. 1999. The evolution of North American attitudes toward marine mammals. pp. 10-47. In: J.R. Twiss and R.R. Reeves (eds.). *Conservation and Management of Marine Mammals.* Smithsonian Institution Press, Washington, DC. Also see Lavigne, Chapter 1; Beder, Chapter 5.

[133] See for e.g., Lavigne, Chapter 1, Table 1–1.

[134] See Hazen, D. in Lakoff 2004, p. xi. Also see Lavigne 2004, and Lavigne, Chapter 1.

[135] See Dowie 1995.

[136] See Lavigne, Chapter 1.

[137] Lakoff 2004. Also see Lakoff, G. 2002. *Moral Politics: How Liberals and Conservatives Think.* Second Edition. The University of Chicago Press, Chicago.

[138] See Kheiriddin, T. and A. Daifallah. 2005. *Rescuing Canada's Right: A Blueprint for a Conservative Revolution.* John Wiley & Sons, Ltd., Mississauga, Ontario.

[139] See Best, Chapter 25.

[140] Usually attributed to George Santayana, e.g. see http://en.thinkexist.com/quotation/those_who_do_not_learn_from_history_are_doomed_to/170710.html.

[141] See Willison, Chapter 2. Also see Tucker, M.E. 2004. Scanning the horizon for hope. Conservation Biology, 18: 299-300; Orr, D.W. 2004. Hope in hard times. Conservation Biology, 18:295-298.

[142] See Lakoff 2004.

[143] Ibid. p. 114.

[144] Grefe, E.A. 1981. *Fighting to Win: Business Political Power.* Law & Business, Inc/Harcourt Brace Jovanovich, Publishers.

[145] Lakoff 2004.

[146] Curtis 2004; Worcester, Chapter 17.

[147] Lakoff 2004.

[148] US Census Bureau projections, available at http://www.census.gov/ipc/www/worldpop.html.

[149] We added "restoration projects" here in response to a suggestion by one of our reviewers.

[150] United Nations Framework Convention on Climate Change; available at http://unfccc.int/essential_background/feeling_the_heat/items/2917.php.

[151] Lavigne, D.M. R.J. Brooks, D.A. Rosen, and D.A. Galbraith. 1989. Cold, Energetics, and Popultions. Pp. 403-432. In L.C.H. Wang (ed.). *Advances in Comparative & Environmental Physiology 4. Animal Adaptation to Cold.* Springer-Verlag, Berlin.

[152] Lakoff 2002, 2004.

[153] For an interesting and thought provoking discussion of the next 50 years, see Smil, V. 2004. The next 50 years: On microbes and who's on top. University of Manitoba, Winnipeg, MB. Unpublished MS.

[154] See Rising Powers: The Changing Geopolitical Landscape. In: *Mapping the Global Future: Report of the National Intelligence Council's 2020 Project.* Available at http://www.cia.gov/nic/NIC_globaltrend2020_s2.html.

[155] Collectively, Brazil, Russia, India, and China are referred to these days as the "BRIC" group. For more information see http://www.equitymaster.com/p-detail.asp?date=12/22/2005&story=1; for predictions of national "real GDPs" for 2025 and 2050, see http://www.blonnet.com/2003/10/27/stories/2003102700230900.htm.

[156] See Pauly 1995; also see O'Regan, Foreword.

[157] Kheiriddin and Daifallah 2005.

[158] Kidman Cox, R. 2004. Editorial. BBC Wildlife magazine, May, p. 5; also see Monbiot, G. 2004. Natural Aesthetes. Wildlife is wonderful. We don't need any other excuses to protect it. Guardian, 13 January 2004.

[159] Collar, N. J. 2003. Beyond value: biodiversity and the freedom of the mind. Global Ecology & Biogeography 12: 265-269; also see McCallum, I. 2005. *Ecological Intelligence: Rediscovering ourselves in Nature.* Africa Geographic, Cape Town, South Africa.

[160] Rees, W. 2004. Waking the Sleepwalkers – Globalization and Sustainability: Conflict or Convergence. pp. 1-34. In W. Chesworth, M.R. Moss, and V.G. Thomas (eds.). *The Human Ecological Footprint.* The Kenneth Hammond

Lectures on Environment, Energy and Resources. 2002 Series. Faculty of Environmental Sciences, University of Guelph, Guelph, Canada. For further discussion, also see Lavigne, D.M. The Human Ecological Footprint: A commenatary. pp. 167-190 in the same volume. Also see Holt, Chapter 4.

[161] See, for example, Brooks, R.J. 2004. Introduction. In W. Chesworth, M.R. Moss, and V.G. Thomas (eds.). *The Human Ecological Footprint.* The Kenneth Hammond Lectures on Environment, Energy and Resources. 2002 Series. Faculty of Environmental Sciences, University of Guelph, Guelph, Canada. For further discussion, also see Lavigne, D.M. 2004.

[162] Leopold, A. 1949. *A Sand County Almanac, and Sketches here and there.* Oxford University Press, Inc., New York.

[163] Krutch, J.W. 1969. Conservation is Not Enough. pp 367-384. In *The Best Nature Writing of Joseph Wood Krutch.* William Morrow & Company, Inc. It seems that this essay first appeared in an earlier book *The Voice of the Desert*, published in 1954 (see http://naturewriting.com/krutch.htm).

[164] The quotation can be found in the *Dictionary of Environmental Quotations* compiled by Barbara K. Rodes and Rice Odell, Simon and Schuster (1992).

[165] Orr, D. 1992. Environmental Literacy: Education as if the earth mattered. Twelfth Annual E.F. Schumacher Lectures. E.F. Schumacher Society, Great Barrington, MA. The Gould quote is from: Gould, S.J. 1991, Unenchanted evening, Natural History, 9/91: 4-13.

[166] The definition is available at http://www.lovemarks.com.

[167] For a discussion, see Orr, D. 2002. Four Challenges of Sustainability. Conservation Biology, 16: 1457-1460; Porritt, J. 2002. Sustainability without Spirituality: A Contradiction in Terms. Conservation Biology, 16: 1465; Christie, I. 2004. Sustainability and Spiritual Renewal: the Challenge of Creating a Politics of Relevance. Conservation Biology 16: 1466-1468; also see McCallum, 2005.

[168] Lakoff 2004.

[169] Lakoff 2004, p. xv.

[170] D. Hazen, in Lakoff 2004, p. xiii.

[171] Orwell, G. 1949. *1984.* Harcourt, Brace and Company, Inc. Reprinted 1961, Signet Classics, The New American Library of World Literature, Inc. New York.

[172] See Lavigne, Chapter 1, Table 1–1.

[173] Roberts, K. 2002. Bold as Love. View from the Top. Executive Speaker Series. Address to Stanford Graduate School of Business, 22 April. Available at http://www.saatchikevin.com/talkingit/stanford.html.

[174] As writer (and Forum participant) Barbara Kyle is quick to point out, "Human beings are feeling animals that think, not thinking animals that feel".

[175] Just as we and others have identified the need for a "revolution" in order to gain ground in pursuit of ecological sustainability, Kheiriddin and Daifallah (2005) use the same word to describe what is required to reinvent the conservative movement in Canada (p. 50). The common question thus becomes, how do you organize such a revolution?

[176] Kheiriddin and Daifallah. 2005, Chapter 4. The Path to Power: Building a Conservative Infrastructure, is a useful guide, in addition to Lakoff 2004.

[177] In its political ascendancy over the past 30 years, for example, the radical right in the United States established 43 think tanks at the cost of some 2-3 billion dollars. See the DVD: How Democrats and Progressives Can Win: Solutions from George Lakoff. Available from www.winwithlanguage.com. Also see Beder, S. 1999. Examining the role of think tanks. Engineers Australia. November, p. 66. Available at www.uow.edu.au/arts/sts/sbeder/columns/engcol19.html.

[178] Thomas, L. 1984. *Late night thoughts on listing to Mahler's Ninth Symphony.* Bantam Books, New York, NY.; also see McCallum 2005.

[179] Grefe, E.A. 1981.

[180] Dowie 1995.

[181] Britons unconvinced on evolution. BBC News, 26 January 2006. Available at http://news.bbc.co.uk/1/hi/sci/tech/4648598.stm.

[182] Also see Orr 2002, p. 1459.

[183] See Worcester, Chapter 17.

[184] e.g. Monbiot, G. An Activists Guide to Exploit the Media. Available at http://www.urban75.com/Action/media.html; also see Lakoff 2004.

[185] Kheiriddin and Daifallah. 2005.

[186] See for e.g., Gore, A. 1992. *Earth in the Balance: Ecology and the Human Spirit.* Boston: Houghton Mifflin Company; Ehrenfeld, J.R. Sustainability by Choice. Available at http://www.capitalizingonchange.org/docs/Alberta-talk.pdf ; Also see President George W. Bush's January 2006 State of the Union Address. Even he was willing to admit, that "America is addicted to oil..." Available at http://www.whitehouse.gov/news/releases/2006/01/20060131-10.html.

[187] Whiten, A. and R.W. Byrne (eds.). 1997. *Machiavellian Intelligence II. Extensions and Evaluations.* Cambridge University Press, Cambridge, U.K.

[188] Gaskin, D.E. 1982. *The Ecology of Whales and Dolphins.* Heinemann Educational Books Ltd., London, UK. p. 345.

[189] Here, we use the word "myth" to mean a traditional story that serves to explain the worldview of a people.

[190] Orr 2002.

[191] Whiten and Byrne (eds.) 1997, back cover.

[192] Donovan *et al.* 1981.

[193] Ibid, p. 5.

[194] Lavigne, D.M. 2002. Ecological footprints, doublespeak, and the evolution of the Machiavellian mind. pp. 61-91. In W. Chesworth, M.R. Moss, and V.G. Thomas (eds.). *Sustainable Development: Mandate or Mantra.* The Kenneth Hammond Lectures on Environment, Energy and Resources 2001 Series. Faculty of Environmental Sciences, University of Guelph, Guelph, Canada.

[195] Byrne, R.W. and A. Whiten. 1997. Machiavellian Intelligence. pp. 1-23. In A. Whitten and R.W. Byrne (eds.). *Machiavellian Intelligence II. Extensions and Evaluations.* Cambridge University Press, Cambridge, U.K.

[196] Bull, G. 1961. Niccolò Machiavelli. *The Prince.* Translation, with an Introduction. Reprinted 1995. London, New York: Penguin Books. p. 55.

[197] Tuchman, B.W. 1984. *March of Folly: From Troy to Vietnam.* Alfred A. Knopf, New York, NY., p. 4-5.

[198] Ibid, p. 7.

[199] Achbar, M., J. Abbott, and J. Bakan. 2003. The Corporation. Big Picture Media Corporation MMIII. For details see www.thecorporation.com; Bakan, J. 2004. *The Corporation: The Pathological Pursuit of Profit and Power.* Toronto: Viking Canada.

[200] Achbar *et al.* , 2003, also see Bakan, 2004; Anon., 2004. The lunatic you work for. The Economist. 6 May, available at http://www.economist.com.

[201] Anon., 2004.

[202] Johanson, D. and J. Shreeve. 1989. *Lucy's child. The discovery of a human ancestor.* William Morrow and Company, Inc., New York, NY.

[203] Parts of this section are drawn from a lecture given by David Lavigne, days before the IFAW Forum in Limerick. That lecture is published as Lavigne, D.M. 2005. Reducing the Agricultural Eco-Footprint: Reflections of a Neo-Darwinian Ecologist. pp. 119-166. In A. Eaglesham, A. Wildeman, and R.W.F. Hardy (eds.). *Agricultural Biotechnology: Finding Common International Goals.* National Agricultural Biotechnology Council Report 16. NABC, Ithaca, NY. Available at http://nabc.cals.cornell.edu/pubs/nabc_16/talks/lavigne_corrected.pdf.

[204] Perlas, N. 2000. *Shaping Globalization: Civil Society, Cultural Power and Threefolding.* Quezon City, Philippines: Center for Alternative Development Initiatives. For additional information, see http://www.cadi.ph.

[205] Parrish, D. 1999. Who rules the world? – the International Monetary Fund, the World Bank and the World Trade Organization. New Statesman, 2 April.

[206] Orr 2002.

[207] Laurance, W.F. 2004. The perils of payoff: corruption as a threat to global biodiversity. Trends in Ecology and Evolution, 19:399-401.

[208] Collar 2003.

[209] Chomsky, N. 2003. *Hegemony or Survival: America's Quest for Global Dominance.* New York: Metropolitan Books, Henry Holt and Company.

[210] From a column written by David Suzuki, cited in Lavigne, D.1992. Republicans vs owls. BBC Wildlife, November. p 74.

[211] For grassroots campaigns, see for e.g., Dowie, 1995.

[212] Chomsky 2003.

[213] Dowie 1995.

[214] e.g. Shah, A. 2000. Free Trade and Globalization: Multilateral Agreement on Investment. Available at http://www.globalissues.org/TradeRelated/MAI.asp; Shah, A. 2003. Free Trade and Globalization: Public Protests Around the World. Available at http://www.globalissues.org/TradeRelated/FreeTrade/Protests.asp.

[215] Chomsky 2003, p. 236,

[216] Nixon, R.M. 1969. Speech. 3 November. Available at http://www.historicaldocuments.com/RichardNixon'sVietnamizationSpeech.htm.

[217] Suroweicki, J. 2004. *The Wisdom of Crowds.* Doubleday, New York, NY.

[218] Rees 2004.

There once was a lad named Lavigne
Whose hands were never quite clean
But he closed down a forum
With excessive decorum
And concluded, "You know what I mean".

Sidney J. Holt 2004

CONTRIBUTORS

LILLY B. AJAROVA
Uganda Wildlife Authority, Plot 3 Kintu Road, Nakasero, PO Box 3530 Kampala, Uganda.
Current address: Executive Director, Chimpanzee Sanctuary & Wildlife Conservation Trust, P.O. Box 884, Entebbe, Uganda.

SHARON BEDER
Professor, School of Social Sciences, Media and Communication, University of Wollongong, New South Wales, 2522 Australia.

STEPHEN BEST
Founding Director, Environment Voters, Suite 101, 221 Broadview Ave. Toronto, Ontario, M4M 2G3, Canada.

RONALD J. BROOKS
Professor, Department of Zoology (now Department of Integrative Biology), College of Biological Science, University of Guelph, Guelph, Ontario, N1G 2W1, Canada.

MICHELLE CAMPBELL
LL.B. graduate of Osgoode Hall Law School of York University, Toronto, Canada
michellecampbell@osgoode.yorku.ca

WARD CHESWORTH
Professor Emeritus, Department of Land Resource Science, University of Guelph, Guelph, Ontario, N1G 2W1, Canada.

JUSTIN COOKE
Centre for Ecosystem Management Studies, Mooshof, 79297 Winden, Germany.

PETER J. CORKERON
Research Associate, Bioacoustics Research Program, Cornell Laboratory of Ornithology, 159 Sapsucker Woods Rd, Ithaca, NY 14850, USA.

ROSAMUND KIDMAN COX
26 Sion Hill, Clifton, Bristol, BS8 4AZ, UK.

BRIAN CZECH
President, Center for the Advancement of the Steady State Economy, 5101 South 11th St., Arlington, VA 22204, and; Visiting Assistant Professor, Virginia Polytechnic Institute and State University, National Capitol Region, 1021 Prince Street, Alexandria, Virginia 22314, USA.

WILLIAM K. DE LA MARE
Professor, School of Resource and Environmental Management, Simon Fraser University, Burnaby, British Columbia, V5A 1S6, Canada.

MICHAEL EARLE
Fisheries Advisor, Greens/EFA in the European Parliament, Rue Wiertz, 1047 Brussels, Belgium.

HEATHER E. EVES
Director, Bushmeat Crisis Task Force, 8403 Colesville Road, Suite 710, Silver Spring, Maryland 20910, USA.

VALERIUS GEIST
Professor Emeritus of Environmental Science, Adjunct Professor of Biology, Faculty of Environmental Design, University of Calgary, Calgary, Alberta, T2N 1N4, Canada.

SIDNEY HOLT
Voc. Palazzetta 68, Paciano (PG),06060, Italy.

JEFFREY A. HUTCHINGS
Professor, Department of Biology, Dalhousie University, Halifax, Nova Scotia, B3H 4J1, Canada.

ASHOK KUMAR
Vice-Chairman, Wildlife Trust of India, P.O. Box 3150, New Delhi, 110 003, India.

DAVID LAVIGNE
Science Advisor, International Fund for Animal Welfare, 40 Norwich St. East, Guelph, Ontario, N1H 2G6, Canada.

WILLIAM S. LYNN
Senior Ethics Advisor, Practical Ethics,
Current address: Assistant Professor, Center for Animals and Public Policy, Tufts University, 200 Westboro Road, North Grafton, Massachusetts, 01536, USA

ATHERTON MARTIN
Executive Director, The Development Institute (TDI), 17 Church Street, Roseau, Commonwealth of Dominica.

VIVEK MENON
Executive Director, Wildlife Trust of India, P.O. Box 3150, New Delhi, 110 003, India.

E.J. MILNER-GULLAND
Reader in Conservation Science, Imperial College London, Division of Biology, Manor House, Silwood Park Campus, Buckhurst Road, Ascot, Berkshire, SL5 7PY, UK.

ARTHUR R. MUGISHA
Executive Director, Uganda Wildlife Authority, Plot 3 Kintu Road, Nakasero, PO Box 3530 Kampala, Uganda. Current address: Fauna & Flora International, Riara Road, P.O. Box 20110 – 00200 Nairobi, Kenya.

JOHN F. OATES
Professor, Department of Anthropology, Hunter College and the Graduate School, City University of New York, New York, New York, 10021, USA.

VASSILI PAPASTAVROU
International Fund for Animal Welfare, The Old Chapel, Fairview Drive, Bristol BS6 6PW, UK.

WILLIAM E. REES
Professor, School of Community and Regional Planning, University of British Columbia, 6333 Memorial Road, Vancouver, British Columbia, V6T 1Z2, Canada.

VERNON G. THOMAS
Associate Professor, Department of Zoology (now Department of Integrative Biology), College of Biological Science, University of Guelph, Guelph, Ontario, N1G 2W1, Canada.

MICHAEL WAMITHI
Senior International Advisor for Africa, International Fund for Animal Welfare, P.O. Box 25499, Nairobi, Kenya.

MARTIN WILLISON
Professor, Department of Biology, Dalhousie University, Halifax, Nova Scotia, B3H 4J1, Canada.

SIR ROBERT WORCESTER
Founder, MORI Ltd, MORI House, 79-81 Borough Road, London, SE1 1FY, UK.

Index

A

B

C

F

I

J

K

L

M

n

o

P

Q

R

S

T

U

V

W

X, Y, & Z